UCF $99.95

Handbook of Adhesives and Sealants

Handbook of Adhesives and Sealants

Edward M. Petrie

McGraw-Hill

New York San Francisco Washington, D.C. Auckland Bogotá
Caracas Lisbon London Madrid Mexico City Milan
Montreal New Delhi San Juan Singapore
Sydney Tokyo Toronto

Library of Congress Cataloging-in-Publication Data

Petrie, Edward M.
    Handbook of adhesives and sealants / Edward M. Petrie.
       p.   cm.
    Includes index.
    ISBN 0-07-049888-1
    1. Adhesives. 2. Sealing compounds. I. Title.
IN PROCESS
668'.3—dc21   99-35410
                                                            CIP

McGraw-Hill
*A Division of The **McGraw·Hill** Companies*

Copyright @ 2000 by The McGraw-Hill Companies, Inc. All rights reserved. Printed in the United States of America. Except as permitted under the United States Copyright Act of 1976, no part of this publication may be reproduced or distributed in any form or by any means, or stored in a data base or retrieval system, without the prior written permission of the publisher.

1 2 3 4 5 6 7 8 9 0 DOC/DOC 9 0 4 3 2 1 0 9

ISBN 0-07-049888-1

The sponsoring editor for this book was Robert Esposito and the production supervisor was Pamela A. Pelton. It was set in Century Schoolbook by Pro-Image Corporation.

Printed and bound by R. R. Donnelley & Sons Company.

 This book was printed on recycled, acid-free paper containing a minimum of 50% recycled, de-inked fiber.

McGraw-Hill books are available at special quantity discounts to use as premiums and sales promotions, or for use in corporate training programs. For more information, please write to the Director of Special Sales, McGraw-Hill, Inc. 11 West 19th Street, New York, NY 10011. Or contact your local bookstore.

Information contained in this work has been obtained by The McGraw-Hill Companies, Inc. ("McGraw-Hill") from sources believed to be reliable. However, neither McGraw-Hill nor its authors guarantee the accuracy or completeness of any information published herein, and neither McGraw-Hill nor its authors shall be responsible for any errors, omissions, or damages arising out of use of this information. This work is published with the understanding that McGraw-Hill and its authors are supplying information but are not attempting to render engineering or other professional services. If such services are required, the assistance of an appropriate professional should be sought.

Contents

Preface xix

Chapter 1. An Introduction to Adhesives and Sealants — 1

- 1.1 Introduction — 1
- 1.2 Fundamentals of adhesives and sealants — 2
  - 1.2.1 Importance of adhesives and sealants — 2
  - 1.2.2 Definitions — 2
  - 1.2.3 Important factors for successfully using adhesives and sealants — 6
  - 1.2.4 Nature of the technologies related to adhesives and sealants — 8
- 1.3 Markets and applications — 10
  - 1.3.1 History — 10
  - 1.3.2 Current markets — 12
  - 1.3.3 Market trends and drivers — 14
  - 1.3.4 Adhesive and sealant industries — 17
- 1.4 Advantages and disadvantages of adhesive bonding — 18
  - 1.4.1 Competitive methods — 18
  - 1.4.2 Mechanical advantages — 23
  - 1.4.3 Design advantages — 26
  - 1.4.4 Production advantages — 27
  - 1.4.5 Other advantages — 28
  - 1.4.6 Mechanical limitations — 28
  - 1.4.7 Design limitations — 29
  - 1.4.8 Production limitations — 30
  - 1.4.9 Other limitations — 30
  - 1.4.10 Combining adhesive and mechanical fastening — 31
- 1.5 Functions of sealants — 31
  - 1.5.1 Mechanical considerations — 32
  - 1.5.2 Adhesion properties — 34
  - 1.5.3 Design considerations — 35
  - 1.5.4 Chemical effects — 36
  - 1.5.5 Production considerations — 36
- 1.6 Basic mechanisms — 36
  - 1.6.1 General requirements for all adhesives and sealants — 36
    - 1.6.1.1 Surface condition — 37
    - 1.6.1.2 Wetting the substrate — 37
    - 1.6.1.3 Solidification of the adhesives or sealant — 37

			1.6.1.4	Forming an impervious joint	38
			1.6.1.5	Joint design	39
			1.6.1.6	Selection and control of materials and manufacturing processes	39
		1.6.2	Mechanism of bond failure		39
	1.7	General materials and processes			42
		1.7.1	Materials used for adhesives and sealants		42
		1.7.2	Manufacturing processes for adhesives and sealants		44
		1.7.3	End-use processes for adhesives and sealants		46
	1.8	Sources of information			47

Chapter 2. Theories of Adhesion — 49

	2.1	Introduction			49
	2.2	Forces involved in adhesion			49
		2.2.1	Adhesive and cohesive forces		49
		2.2.2	The concept of surface energy		51
		2.2.3	Work of adhesion and cohesion		54
		2.2.4	Bond failure energy		55
		2.2.5	Surface attachment theory of joint strength		57
	2.3	Theories of adhesion			59
		2.3.1	Adsorption theory		59
		2.3.2	Mechanical theory		62
		2.3.3	Electrostatic and diffusion theories		64
		2.3.4	Weak-boundary-layer theory		64
	2.4	Stages in an adhesive's or a sealant's life			66
		2.4.1	Application and Wetting		66
		2.4.2	Setting or solidification		69
			2.4.2.1	Localized stress	70
			2.4.2.2	Setting stresses due to thermal expansion differences	71
			2.4.2.3	Setting stresses due to shrinkage of the adhesive or sealant	74
		2.4.3	Operation in service		75
			2.4.3.1	Short term effects	75
			2.4.3.2	Long term effects	77
	2.5	Special mechanisms related to sealants			79
	2.6	Polymer material interactions			81
		2.6.1	Polymeric materials		82
		2.6.2	Properties important for adhesives and sealants		82
			2.6.2.1	Properties important for adhesion	82
			2.6.2.2	Properties affecting cohesion	83
	2.7	Surface and interface interactions			87
		2.7.1	Boundary layer theory		88
		2.7.2	Interphase region		89

Chapter 3. Joint Design — 93

	3.1	Introduction		93
	3.2	Types of stress		94
		3.2.1	Tensile and compressive stress	94
		3.2.2	Shear stress	96
		3.2.3	Cleavage and peel stress	97

3.3	Maximizing joint efficiency			98
	3.3.1	Adhesive properties		99
	3.3.2	Adhesive thickness		100
	3.3.3	Effect of bond area geometry		101
	3.3.4	Adherend properties		102
3.4	General joint design rules			104
3.5	Common adhesive joint designs			105
	3.5.1	Joints for flat adherends		105
		3.5.1.1	Butt joints	105
		3.5.1.2	Lap joints	105
		3.5.1.3	Strap joints	107
	3.5.2	Stiffening joints and other methods of reducing peel stress		109
	3.5.3	Cylindrical joints		110
	3.5.4	Angle and corner joints		112
	3.5.5	Plastic and elastomer joints		114
		3.5.5.1	Flexible plastics and elastomers	114
		3.5.5.2	Reinforced plastics	115
	3.5.6	Wood joints		115
3.6	Sealant joint efficiency			115
	3.6.1	Stress distribution in butt joints		115
		3.6.1.1	Elastic sealants	118
		3.6.1.2	Deformable sealants	121
	3.6.2	Shear joints		121
3.7	Common sealant joint designs			122
	3.7.1	Butt joints		122
	3.7.2	Lap joints		122

Chapter 4. Standard Test Methods 127

4.1	Introduction			127
4.2	Reasons to test and basic principles			128
4.3	Fundamental material property tests			135
	4.3.1	Adhesives and sealants		137
		4.3.1.1	Viscosity	137
		4.3.1.2	Shelf life	139
		4.3.1.3	Working life	140
		4.3.1.4	Tack	140
		4.3.1.5	Cure rate	141
		4.3.1.6	Hardness	142
		4.3.1.7	Solids content	143
	4.3.2	Adherends and other materials used in the bonding process		143
4.4	Standard test methods for adhesive joints			145
	4.4.1	Tensile tests		146
	4.4.2	Lap-shear tests		149
	4.4.3	Peel tests		151
	4.4.4	Cleavage tests		156
	4.4.5	Fatigue tests		157
	4.4.6	Impact tests		158
	4.4.7	Creep tests		160
	4.4.8	Environmental tests		161
4.5	Standard test methods for sealant joints			163
	4.5.1	Movement capability		164

		4.5.2	Peel adhesion	165
		4.5.3	Tear strength	166
		4.5.4	Compression set resistance and creep	167
		4.5.5	Environmental tests	168
	4.6	Prototype and non-standard test methods		170
	4.7	Specifications and standards		173

Chapter 5. Quality Control and Nondestructive Tests 177

5.1	Introduction			177
5.2	Quality control			177
	5.2.1	Pre-manufacturing processes		179
	5.2.2	Quality control of the incoming materials		180
	5.2.3	Quality control of surface treatment		181
	5.2.4	Quality control of the bonding process		182
5.3	Bond inspection			183
	5.3.1	Destructive testing		183
	5.3.2	Non-destructive testing		184
		5.3.2.1	Visual inspection	184
		5.3.2.2	Tap test	186
		5.3.2.3	Proof tests	187
		5.3.2.4	Ultrasonic inspection	187
		5.3.2.5	Other NDT methods	189
5.4	Testing of failed joints for cause of failure			191

Chapter 6. Surfaces and Surface Preparation 197

6.1	Introduction			197
6.2	Nature of substrate surfaces			198
	6.2.1	Metallic surfaces		200
	6.2.2	Polymeric surfaces		203
	6.2.3	Other surfaces		205
6.3	Surface treatment			206
	6.3.1	Importance of surface treatment		207
	6.3.2	Choosing the surface treatment		209
	6.3.3	Evaluation of treated parts before and after bonding		213
	6.3.4	Substrate equilibrium		215
6.4	Passive surface preparation methods			215
	6.4.1	Passive chemical surface treatment		216
		6.4.1.1	Solvent cleaning	217
		6.4.1.2	Chemical cleaning	221
		6.4.1.3	Other cleaning methods	222
	6.4.2	Passive mechanical treatment		223
6.5	Active surface treatments			226
	6.5.1	Active chemical surface treatments		227
		6.5.1.1	Chemical treatment of metal surfaces	232
		6.5.1.2	Chemical treatment of polymeric surfaces	234
	6.5.2	Active physical surface treatments for polymeric materials		236
		6.5.2.1	Corona discharge	237
		6.5.2.2	Flame treatment	239
		6.5.2.3	Plasma treatment	240
		6.5.2.4	Other physical surface treatments for polymers	242

6.6	Specific surface treatments		246
	6.6.1	Metallic adherends	246
	6.6.2	Plastic adherends	246
	6.6.3	Polymeric composite adherends	247
	6.6.4	Elastomeric adherends	249
	6.6.5	Other adherends	250

Chapter 7. Primers and Adhesion Promoters — 253

7.1	Introduction		253
7.2	Primers		255
	7.2.1	Application and use	256
	7.2.2	Primers for metal substrates	259
	7.2.3	Primers for polymeric substrates	263
	7.2.4	Primers for unvulcanized elastomers	264
7.3	Adhesion promoters		266
	7.3.1	Silane adhesion promoters	267
	7.3.2	Titanates, zirconates, and others	273

Chapter 8. Adhesive Classification — 279

8.1	Introduction			279
8.2	Basic classification			279
	8.2.1	Function		280
	8.2.2	Chemical composition		281
		8.2.2.1	Thermosetting adhesives	281
		8.2.2.2	Thermoplastic adhesives	284
		8.2.2.3	Elastomeric adhesives	285
		8.2.2.4	Hybrid adhesives	285
	8.2.3	Method of reaction		286
		8.2.3.1	Chemical reaction	289
			8.2.3.1.1 Multiple part adhesive systems	291
			8.2.3.1.2 Single component systems that cure via catalyst or hardener	294
			8.2.3.1.3 Moisture curing adhesives	295
			8.2.3.1.4 UV/light curing adhesives	296
			8.2.3.1.5 Adhesives catalyzed by the substrate	297
			8.2.3.1.6 Adhesives in solid form (tape, film, powder)	299
		8.2.3.2	Solvent or water loss	300
			8.2.3.2.1 Contact adhesives	301
			8.2.3.2.2 Pressure sensitive adhesives	301
			8.2.3.2.3 Reactivatable adhesives	305
		8.2.3.3	Resinous solvent adhesives	306
		8.2.3.4	Hardening from the melt	306
	8.2.4	Physical form		307
		8.2.4.1	Pastes and liquids	310
			8.2.4.1.1 Two-part solventless	310
			8.2.4.1.2 One-part solventless	312
		8.2.4.2	Solvent-based adhesives	312
		8.2.4.3	Water-based adhesives	313
		8.2.4.4	Solid forms	314
			8.2.4.4.1 Tape and film	314
			8.2.4.4.2 Powdered adhesives	314
			8.2.4.4.3 Hot melt forms	315

		8.2.5	Cost	315
		8.2.6	Specific adherends and method of applications	316

Chapter 9. Adhesive Composition and Formulation — 319

9.1	Introduction			319
9.2	Adhesive composition			319
9.3	Adhesives formulation			326
	9.3.1	Controlling flow		327
	9.3.2	Extending temperature range		329
	9.3.3	Improving toughness		330
	9.3.4	Lowering the coefficient of thermal expansion		333
	9.3.5	Reducing shrinkage		334
	9.3.6	Increasing tack		335
	9.3.7	Modifying electrical and thermal conductivity		336
9.4	Commercial formulations			338

Chapter 10. Adhesive Families — 343

10.1	Introduction			343
10.2	Classification methods			343
10.3	Structural adhesives			355
	10.3.1	Epoxies		355
		10.3.1.1	Epoxy resins	360
		10.3.1.2	Curing agents	361
	10.3.2	Epoxy hybrids		366
		10.3.2.1	Toughened epoxies	367
		10.3.2.2	Epoxy-phenolic	371
		10.3.2.3	Epoxy-nylon	372
		10.3.2.4	Epoxy-polysulfide	372
		10.3.2.5	Epoxy-vinyl	373
	10.3.3	Resorcinol formaldehyde and phenol resorcinol formaldehyde		373
	10.3.4	Melamine formaldehyde and urea formaldehyde		374
	10.3.5	Phenolics		374
	10.3.6	Modified phenolics		375
		10.3.6.1	Nitrile-phenolic	375
		10.3.6.2	Vinyl-phenolic	377
		10.3.6.3	Neoprene-phenolic	377
	10.3.7	Polyaromatic high temperature resins		377
		10.3.7.1	Polyimide	378
		10.3.7.2	Bismaleimide	380
		10.3.7.3	Polybenzimidazole	381
	10.3.8	Polyesters		381
	10.3.9	Polyurethanes		382
	10.3.10	Anaerobic resins		385
	10.3.11	Cyanoacrylates		389
	10.3.12	Modified acrylics		390
10.4	Non-structural adhesives			392
	10.4.1	Elastomeric resins		393
		10.4.1.1	Natural rubber	395
		10.4.1.2	Asphalt	396
		10.4.1.3	Reclaimed rubber	396

	10.4.1.4	Butyl rubber	396
	10.4.1.5	Styrene butadiene rubber (SBR)	397
	10.4.1.6	Polychloroprene (neoprene)	397
	10.4.1.7	Acrylonitrile butadiene (nitrile)	398
	10.4.1.8	Polyisobutylene	399
	10.4.1.9	Polyvinyl methyl ether	399
	10.4.1.10	Polysulfide	399
	10.4.1.11	Silicone	400
10.4.2	Thermoplastic resins	402
	10.4.2.1	Polyvinyl acetal	403
	10.4.2.2	Polyvinyl acetate	403
	10.4.2.3	Polyvinyl alcohol	406
	10.4.2.4	Thermoplastic elastomers	406
	10.4.2.5	Ethylene vinyl acetate	406
	10.4.2.6	Cellulosic resins	407
	10.4.2.7	Polyamide	407
	10.4.2.8	Polyester	407
	10.4.2.9	Polyolefins	408
	10.4.2.10	Polysulfone	408
	10.4.2.11	Phenoxy	408
	10.4.2.12	Acrylic	409
10.4.3	Naturally occurring resins	409
	10.4.3.1	Natural organic resins	410
		10.4.3.1.1 Glues of agricultural origin	410
		10.4.3.1.2 Glues of animal origin	410
	10.4.3.2	Inorganic adhesives and cements	411
		10.4.3.2.1 Sodium silicate	411
		10.4.3.2.2 Phosphate cements	412
		10.4.3.2.3 Litharge cement	412
		10.4.3.2.4 Sulfur cement	412

Chapter 11. Selection of Adhesives 415

11.1	Introduction	415
11.2	Planning for the bonding process	416
	11.2.1 Consideration of alternative bonding methods	418
	11.2.2 Information regarding product and processing requirements	419
	11.2.2.1 Processing conditions	420
	11.2.2.2 Joint design	421
	11.2.2.3 Expected service conditions	422
11.3	Selecting the adhesive	423
11.4	Substrates	424
	11.4.1 Adhesives for metal	430
	11.4.2 Bonding plastics	432
	11.4.3 Adhesives for bonding elastomers	432
	11.4.3.1 Vulcanized elastomers	432
	11.4.3.2 Unvulcanized elastomers	434
	11.4.4 Adhesives bonding of other common substrates	435
	11.4.4.1 Wood	435
	11.4.4.2 Glass	437
11.5	Nature of the joint design	437
	11.5.1 Relation between stress and adhesive selection	437

xii Contents

	11.5.2	Relation between adhesive selection, bondline thickness, and viscosity	438
11.6	Effect of the part and production requirements on adhesive selection		440
	11.6.1	Nature of the part or assembly	441
	11.6.2	Production requirements	441
	11.6.3	Cost	445
11.7	Service conditions		445
	11.7.1	Service stress	446
	11.7.2	Service environment	446
11.8	Special considerations		448
11.9	Verification of reliability		448
11.10	Sources of information and assistance		449

Chapter 12. Sealant Classification and Composition — 451

12.1	Introduction	451
12.2	Basic sealant classifications	452
	12.2.1 Hardening and non-hardening types	452
	12.2.2 Cure type	456
	12.2.2.1 Two-part systems	456
	12.2.2.2 Single component sealants	458
	12.2.2.3 Solvent and water release sealants	459
	12.2.3 Classification by end-use	459
	12.2.4 Performance classifications	459
12.3	Sealant composition	461
	12.3.1 Primary resin	464
	12.3.2 Solvents	466
	12.3.3 Fillers	466
	12.3.4 Plasticizers	467
	12.3.5 Additives to improve adhesion	467
	12.3.6 Other additives	469
12.4	Sealant formulation	469
	12.4.1 Application properties	469
	12.4.2 Performance properties	472
12.5	Commercial products and formulations	473

Chapter 13. Sealant Families — 475

13.1	Introduction	475
13.2	Low performance sealants	476
	13.2.1 Oil- and resin-based sealants	481
	13.2.2 Asphaltic and other bituminous mastics	482
	13.2.3 Polyvinyl acetate	482
	13.2.4 Epoxy	483
	13.2.5 Polyvinyl chloride plastisol	483
13.3	Medium performance sealants	483
	13.3.1 Hydrocarbon rubber-based sealants	484
	13.3.2 Acrylic	486
	13.3.3 Chlorosulfonated polyethylene (Hypalon)	487
	13.3.4 Hot-melt sealants	487
13.4	High performance sealants	487

	13.4.1	Fluorosilicone and fluoropolymer sealants	488
	13.4.2	Polysulfides	489
	13.4.3	Polyethers	490
	13.4.4	Polyurethane	492
	13.4.5	Silicone	496
	13.4.6	Styrene butadiene copolymer sealants	497
	13.4.7	Chloroprenes	498
	13.4.8	Other specialty sealants	498

Chapter 14. Selecting and Using Sealants — 501

- 14.1 Introduction — 501
- 14.2 Nature of the joint design — 502
  - 14.2.1 Common butt and lap joints — 503
  - 14.2.2 Backup materials — 508
  - 14.2.3 Threaded joints — 508
  - 14.2.4 Gasketing — 509
  - 14.2.5 Porosity sealing — 512
- 14.3 Sealant substrates — 513
  - 14.3.1 Surface preparation — 515
  - 14.3.2 Primers — 515
  - 14.3.3 Common substrate surfaces — 515
    - 14.3.3.1 Concrete and masonry — 516
    - 14.3.3.2 Stone — 516
    - 14.3.3.3 Glass and porcelain — 517
    - 14.3.3.4 Painted surfaces — 517
    - 14.3.3.5 Unpainted metal — 517
- 14.4 Application requirements — 518
  - 14.4.1 Bulk materials — 518
  - 14.4.2 Tape sealants — 522
  - 14.4.3 Preformed gaskets — 523
  - 14.4.4 Foam sealants — 523
  - 14.4.5 Application properties — 524
    - 14.4.5.1 Rheological properties — 524
    - 14.4.5.2 Ambient temperature on application — 525
    - 14.4.5.3 Cure rate — 525
- 14.5 Performance requirements — 526
  - 14.5.1 Movement capability — 527
  - 14.5.2 Other mechanical properties — 530
  - 14.5.3 Adhesion — 531
  - 14.5.4 Durability — 532
  - 14.5.5 Appearance — 534
- 14.6 Sources of information and assistance — 534

Chapter 15. Methods of Joining Plastics Other Than With Adhesives — 537

- 15.1 Introduction — 537
- 15.2 Plastic materials — 538
- 15.3 Joining methods for plastics — 542
- 15.4 Direct heat welding — 543
  - 15.4.1 Heated tool welding — 543
  - 15.4.2 Hot gas welding — 553

	15.4.3	Resistance wire welding		557
	15.4.4	Laser welding		558
	15.4.5	Infrared welding		560
15.5	Indirect heating			560
	15.5.1	Induction welding		561
	15.5.2	Dielectric heating		564
15.6	Friction welding			565
	15.6.1	Spin welding		566
	15.6.2	Ultrasonic welding		569
	15.6.3	Vibration welding		575
15.7	Solvent cementing			577
15.8	Methods of mechanical joining			580
	15.8.1	Mechanical fasteners		581
		15.8.1.1	Machine screws and bolts	581
		15.8.1.2	Self-threading screws	583
		15.8.1.3	Rivets	585
		15.8.1.4	Spring steel fasteners	585
	15.8.2	Design for self-fastening		586
		15.8.2.1	Press-fit	586
		15.8.2.2	Snap-fit	588
15.9	More information on joining plastics			589

Chapter 16. Bonding and Sealing Specific Substrates — 593

16.1	Introduction		593
16.2	Metal bonding		594
	16.2.1	Aluminum and its alloys	595
	16.2.2	Beryllium and its alloys	602
	16.2.3	Cadmium plating	606
	16.2.4	Copper and copper alloys	607
	16.2.5	Gold, silver, platinum, and other precious metals	608
	16.2.6	Lead	608
	16.2.7	Magnesium and magnesium-based alloys	609
	16.2.8	Nickel and nickel alloys	611
	16.2.9	Plated parts (zinc, chrome, and galvanized)	611
	16.2.10	Steel and iron	612
	16.2.11	Tin	615
	16.2.12	Titanium and titanium-based alloys	615
	16.2.13	Zinc and zinc alloys	617
16.3	Plastic substrates in general		617
16.4	Thermosetting plastic substrates		620
	16.4.1	Alkyds	621
	16.4.2	Diallyl phthalate	622
	16.4.3	Epoxy	622
	16.4.4	Phenolic, melamine, and urea resins	623
	16.4.5	Polyimides	623
	16.4.6	Polyesters	624
	16.4.7	Silicones	625
	16.4.8	Thermosetting polyurethanes	626
16.5	Thermoplastic substrates		627
	16.5.1	Acrylonitrile-butadiene-styrene (ABS)	628
	16.5.2	Acetal	629
	16.5.3	Acrylics	631

	16.5.4	Cellulosics (cellulose acetate, cellulose acetate butyrate, cellulose propionate, ethyl cellulose, and cellulose nitrate)	634
	16.5.5	Chlorinated polyether	634
	16.5.6	Fluorocarbons	635
	16.5.7	Nylon (polyamide)	637
	16.5.8	Polycarbonates	640
	16.5.9	Polyolefins (polyethylene, polypropylene, polymethylpentene)	644
	16.5.10	Polyphenylene oxides	646
	16.5.11	Thermoplastic polyesters (polyethylene terephthalate, polybutylene terephthalate, and polytetramethylene terephthalate)	647
	16.5.12	Polyimide (PI), polyetherimide (PEI), polyamide-imide (PAI)	649
	16.5.13	Polyetheretherketone (PEEK), polyaryletherketone (PAEK), polyetherketone (PEK)	650
	16.5.14	Polystyrenes	650
	16.5.15	Polysulfone	652
	16.5.16	Polyethersulfone (PES)	653
	16.5.17	Polyphenylene sulfide (PPS)	654
	16.5.18	Polyvinyl chloride (PVC)	654
16.6	Composites		655
	16.6.1	Adhesives bonding of composites	657
		16.6.1.1 Types of adhesives for composites	658
		16.6.1.2 Surface preparation for adhesive bonding	661
	16.6.2	Thermoplastic composites	662
16.7	Plastic foams		662
16.8	Elastomers		666
	16.8.1	Unvulcanized bonding	669
	16.8.2	Bonding vulcanized elastomers	669
16.9	Wood and wood products		673
16.10	Glass and ceramics		677
16.11	Honeycomb and other structural sandwich panels		679

Chapter 17. Effect of the Environment   685

17.1	Introduction		685
17.2	High temperatures		689
	17.2.1	Factors affecting temperature resistance	689
	17.2.2	Adhesives and sealants for high temperature service	693
		17.2.2.1 Epoxy	693
		17.2.2.2 Modified phenolics	698
		17.2.2.3 Silicone	698
		17.2.2.4 Polyimides and other aromatic resins	699
		17.2.2.4.1 Polyimide	699
		17.2.2.4.2 Polybenzimidazole	700
		17.2.2.4.3 Bismaleimide	700
		17.2.2.5 Polysulfone thermoplastics	701
		17.2.2.6 High temperature cyanoacrylates	701
		17.2.2.7 Specialty elastomers	702
17.3	Low temperatures		702
	17.3.1	Factors affecting low temperature performance	703

		17.3.2	Low temperature adhesives and sealants	703
			17.3.2.1 Modified epoxies	704
			17.3.2.2 Modified phenolics	705
			17.3.2.3 Polyurethanes	705
			17.3.2.4 Silicones	705
			17.3.2.5 Modified acrylics	706
			17.3.2.6 Rubber-based adhesives	706
			17.3.2.7 Polyaromatic resins	706
			17.3.2.8 Sealants	706
	17.4	Humidity and water		707
		17.4.1	Effect on the bulk materials	707
		17.4.2	Effect on the interface	710
		17.4.3	Combined effects of stress, moisture, and temperature	713
	17.5	Outdoor weathering		717
		17.5.1	Non-sea coast environment	717
		17.5.2	Sea coast environment	718
	17.6	Chemicals and solvents		719
	17.7	Vacuum and outgassing		722
	17.8	Radiation		723

Chapter 18. Production Processes and Equipment   727

18.1	Introduction		727
18.2	Storage		728
18.3	Preparing the adhesive or sealant for application		732
	18.3.1	Transferring the product	732
	18.3.2	Metering	733
	18.3.3	Mixing	735
	18.3.4	Transferring to the application area	738
18.4	Application		739
	18.4.1	Liquids	741
	18.4.2	Pastes and mastics	746
	18.4.3	Powders	746
	18.4.4	Films	747
	18.4.5	Hot-melt adhesives	750
18.5	Bonding equipment		751
	18.5.1	Pressure equipment	752
	18.5.2	Heating equipment	752
18.6	Environmental and safety concerns		754

Chapter 19. Information Technology   759

19.1	Introduction		759
19.2	Information access		759
19.3	Adhesive and sealant selection		760
	19.3.1	AdhesivesMart	761
	19.3.2	Adhesives Selector	761
	19.3.3	Sealant System	762
	19.3.4	ADHESIVES	763

Appendix A. Glossary 765

A.1 Standard Definitions of Terms Relating to ADHESIVES 765
A.2 Standard Definitions of Terms Relating to SEALANTS 785

Appendix B. Other Sources of Information 793

Appendix C. Specification and Standards 803

C.1 American Society for Testing and Materials (ASTM) 803
C.2 U.S. Federal Specifications and Standards 816
C.3 Other Industry Specification and Standards 820
C.4 Other Standards Organizations 822

Appendix D. Surface Preparation Methods for Common Substrate Materials 825

D.1 Surface Preparation Methods for Metal Substrates 826
D.2 Surface Preparation Methods for Plastic Substrates 838
D.3 Surface Preparation Methods for Elastomeric Substrates 844
D.4 Surface Preparation Methods for Miscellaneous Substrates 847

Appendix E. Suppliers to the Adhesive and Sealant Industry 849

E.1 Adhesive and Sealant Suppliers 850
E.2 Production Equipment 858
E.3 Application Equipment 859
E.4 Curing and Drying Equipment 862
E.5 Plastic Welding Equipment 862
E.6 Testing and Laboratory Equipment 863
E.7 Consultants and Testing Laboratories 866

Conversion Factors 869
Index 871

Preface

Adhesives and sealants are truly remarkable materials, and I find the technology supporting these materials to be equally extraordinary and at times a bit overpowering for infrequent users of these products. I know of no other technology spread so expansively throughout our lives. It is difficult to imagine a product—in the home, in industry, in transportation, or anywhere else for that matter—that does not use adhesives or sealants in some manner.

Yet, many of us who work with adhesives and sealants, probably "back into" being "experts" almost by default. We had no early intentions of a close relationship with this technology. We went to school and trained, perhaps, to be materials specialists, engineers, designers, or manufacturers. One day our supervisor or the nature of our work demands that we discover how to assemble a product. How do we join together parts to form a functional product that will endure all possible service environments, is harmonious with the company's production processes and schedules, and has some aesthetic and environmental quality? How long will the joint survive in service? What is the repeatability of the process? How do we control the process and check for inferior or improper joints? If the joint is of poor quality, how do we get the parts apart or must they be scrapped? Will the bonding or sealing process affect the environment or the safety of the assembly workers?

In answering these questions, we may have taken a broad, initial approach and considered all forms of assembly including mechanical fasteners, welding, and adhesives. After studying the pros and cons of each process, we possibly choose an adhesive or sealant. Either by good guidance from suppliers, mentors, or just by chance, our first assembled product looks competent. It provides a decent service life, and maybe even results in a profitable product. We are now frequently consulted because we have seemingly mastered a foreboding technology that to everyone else has a somewhat sinister demeanor and magical aura about it. Then, we are truly overwhelmed when it becomes obvious that to fully understand and gain expert stature with adhesives and sealants, we must master many fundamental principles. These fundamentals are derived from diverse and alien sciences such as polymeric materials, surface chemistry, and fracture mechanics. To be successful, we know that one must master and integrate these sciences.

The above situation is, of course, magnified a bit, but I am sure that it reaches home to many who pick up this Handbook and desire an efficient method of understanding adhesives and sealants and applying them reliably. As a periodic end-user, you possibly do not have the time or the resources to thoroughly study the various technologies related to adhesives and sealants. You have little desire to wade through volumes of text and product information looking for the specific methods, processes, and possible examples that will apply to the joining application *du jour*. However, you do want such information close-by if the need arises. Above all, you hope to learn from the experiences of others. Ideally for your needs, a single reference source can provide the guidance and fundamental knowledge required for many adhesive bonding or sealing applications. I hope that I am correct in these assumptions, because this is why I have written this Handbook.

There are a number of important handbooks, journals, and papers, on adhesives and sealants already in the literature. Many of these are important works, and they serve as a foundation for this Handbook. To the authors of these remarkable vehicles, I am exceedingly grateful. Much of their work is referred to here. This Handbook may not go into the technical depth that is evident in these previous works. However, it will provide sufficient detail and a broad foundation to enable the practitioner to reliably design, select, and use adhesives or sealants.

This Handbook will be a single, comprehensive source of recent information, historical experience, and guidance for any adhesive or sealant application. The Handbook will define the universe of adhesives and sealants from the perspective of the end-user, and in that way will provide a practical and useful source of information.

My perspective is as an end-user as well. I have consulted on adhesives and sealants for over 33 years in diverse commercial industrial applications, first in Westinghouse Electric Corporation and then in Asea Brown Boveri (ABB) and other companies. Thus, I have seen many successful applications of adhesive or sealant systems and many disasters and near disasters. As an industrial consultant, one generally becomes involved with a case because there is a problem that cannot be solved with the resources at hand. Thus, horror stories as well as successful applications will litter my presentation. I hope that the readers will find these experiences educational and illuminating.

The *Handbook of Adhesives and Sealants* is a guide to the *entire* field of adhesives and sealants. Although primarily directed toward the end-user and containing important application and design data, the Handbook also provides significant information for those interested in developing, manufacturing, marketing, purchasing, or just generally becoming familiar with these important materials.

The initial chapter covers the importance of adhesives and sealants and the multi-disciplined nature of the technology. Also introduced in the first chapter are markets and applications, the functions of adhesives and sealants, and common materials and processes. Major sources of information regarding adhesives and sealants are identified for readers interested in gaining a deeper understanding or maintaining currency with this dynamic technology.

Subsequent chapters then cover important elements necessary to determine where adhesives and sealants should and should not be used and, when indicated, how to successfully use them. Chapters cover adhesion theories, joint design rules, test methods, substrate preparation methods, adhesive and sealant materials, application and processing methods, and the effects of the service environment. The final chapter describes how information technology is changing the selection and knowledge gathering process.

Throughout the Handbook, specific applications and examples illustrate the concepts being discussed. Due to the broad nature of this technology, the reader will often be directed from one section to other relevant sections within the book. References at the end of each chapter direct the interested reader to more in-depth information and understanding regarding specific subjects.

I wish to express a sincere "thank you" to all those who made my goal of putting this Handbook together a reality. Special thanks goes to Charlie Harper for his motivation and support and to Ron Sampson for his mentoring and wisdom in my early days as a young research engineer.

Edward M. Petrie

Acknowledgment

This book is dedicated to my wife, Carol—who is everything to me including proofreader, motivator, and friend; and to my son, Eddie. I also dedicate this book to all those excellent engineers and technicians that I have known in my years at Westinghouse Corporation. Thank you for the experiences, many of which are in this book, that made my early career worthwhile and never boring.

Handbook of Adhesives and Sealants

Chapter 1

An Introduction to Adhesive and Sealants

1.1 Introduction

This chapter provides an understanding of adhesives and sealants as a means for assembling and adding value to finished products. The importance and prominence that adhesives and sealants have as commercial products are highlighted. The multiple functions played by adhesives and sealants are identified as are the critical procedures required to achieve successful results. The advantages and disadvantages of using these materials are explained and compared to other methods of joining.

Basic definitions of common terms used in the adhesive and sealant industries are provided in this chapter, and a glossary of terms appears in Appendix A. The processes employed by the manufacturers of adhesives and sealants and by their end-users are described. Sources of information for further understanding and study are offered at the conclusion of this chapter and in Appendix B.

Through this chapter, the reader will gain an appreciation of the complex processes related to adhesives and sealants and the multiple sciences that form their foundation. This chapter reveals why a multi-disciplined approach is necessary for the successful application of adhesives and sealants. Most of the topics presented are again visited in detail in later chapters.

1.2 Fundamentals of Adhesives and Sealants

1.2.1 Importance of adhesives and sealants

Adhesives and sealants surround us in nature and in our daily lives. Substantial businesses exist to develop, manufacture, and market these materials, and they are used within virtually every business and industry. Applications abound from office "post-it notes®" to automotive safety glass to footwear to aerospace structures to "no-lick" postage stamps. Many products that we take for granted could never exist if it were not for adhesive bonding or sealing.

If someone could determine the total value added to our economy by the relatively small amount of adhesives and sealants that are used, the result would be staggering. Yet, with adhesives and sealants all around us, with applications extending back to at least biblical times, and with many examples of outstanding adhesion in nature (e.g., barnacles and ice on roads), why are there so many failures when we try to "engineer" the use of adhesives or sealants in practice? Why does it seem as if we must resort to trial and error, if not a bit of luck or magic? Examples of catastrophic disasters such as the 1986 Challenger space shuttle sealant problem and the 1988 Aloha Airlines 737 fuselage peeling apart in flight unfortunately also invade the history of adhesives and sealants. Perhaps no other class of materials or technology is so essential yet so ripe for potential misadventure.

The adhesives and sealants industry is bolstered by thousands of years of trial and error. This long history can be coupled with significant additions to the fundamental supporting sciences and with the development of advanced materials and processes. Consequently, society has generally progressed to a point where we actually trust not only our fortunes but also our lives to these materials. The study of adhesives and sealants and the sciences surrounding their application has never been more important.

1.2.2 Definitions

As any science that has progressed over the centuries, the science that supports adhesives and sealants has developed a jargon and language of its own. Appendix A defines terms that are commonly used in these industries. Important, basic terms necessary to develop a fundamental understanding of how and why adhesives and sealants provide value are given in this section.

Adhesives and sealants are often made of similar materials, and they are sometimes used in similar applications. These materials have comparable processing requirements and failure mechanisms, and the fundamentals of how they work are similar. Therefore, adhesives and sealants are often considered together, as they are in this Handbook. However, different specifications and test methods apply to adhesives and sealants, and most often they are designed to perform different functions. Their definitions hint at these differing functions.

Adhesive — a substance capable of holding at least two surfaces together in a strong and permanent manner.

Sealant — a substance capable of attaching to at least two surfaces, thereby, filling the space between them to provide a barrier or protective coating.

Adhesives and sealants are often considered together because they both adhere and seal; both must be resistant to their operating environments; and their properties are highly dependent on how they are applied and processed. Adhesives and sealants also share several common characteristics.

- They must behave as a liquid, at some time in the course of bond formation, in order to flow over and *wet* (make intimate contact with) the adherends.
- They form surface attachment through *adhesion* (the development of intermolecular forces).
- They must harden to carry sometimes continuous, sometimes variable load throughout their lives.
- They transfer and distribute load among the components in an assembly.
- They must fill gaps, cavities, and spaces.
- They must work with other components of the assembly to provide a durable product.

Adhesives are chosen for their holding and bonding power. They are generally materials having high shear and tensile strength. *Structural adhesive* is a term generally used to define an adhesive whose strength is critical to the success of the assembly. This term is usually reserved to describe adhesives with high shear strength (in excess of 1,000 pounds per square inch or psi) and good environmental resistance. Examples of structural adhesives are epoxy, thermosetting acrylic, and urethane systems. Structural adhesives are usually expected to last the life of the product to which they are applied.

Non-structural adhesives are adhesives with much lower strength and permanence. They are generally used for temporary fastening or

4   Chapter One

to bond weak substrates. Examples of non-structural adhesives are pressure sensitive films, wood glue, elastomers, and sealants.

Sealants are generally chosen for their ability to fill gaps, resist relative movement of the substrates, and exclude or contain another material. They are generally lower in strength than adhesives, but have better flexibility. Common sealants include urethanes, silicones, and acrylic systems.

Both adhesives and sealants function primarily by the property of adhesion. *Adhesion* is the attraction of two different substances resulting from intermolecular forces *between* the substances. This is distinctly different from *cohesion*, which involves only the intermolecular attractive forces *within* a single substance. The intermolecular forces acting in both adhesion and cohesion are primarily van der Waals forces which will be explained in the next chapter. To better understand the difference between adhesion and cohesion, consider the failed joints illustrated in Fig. 1.1. Joints fail either adhesively or cohesively or by some combination of the two.

Adhesive failure is an interfacial bond failure between the adhesive and the adherend. Cohesive failure could exist within either the adhesive material or the adherend. Cohesive failure of the adhesive occurs when stress fracture within the adhesive material allows a layer of adhesive to remain on both substrates (i.e., the attachment of the adhesive to the substrate is stronger than the internal strength of the adhesive itself, and the adhesive fails within its bulk). When the adherend fails before the adhesive and the joint area remains intact, it is known as a cohesive failure of the adherend.

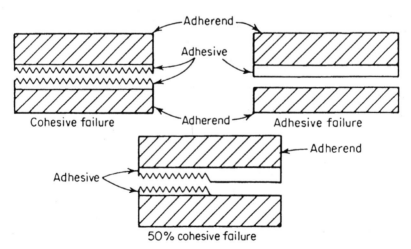

Figure 1.1 Examples of cohesive and adhesive failure.

Other important definitions may be illustrated by considering the schematic of the joint in Fig. 1.2 where two substrates are bonded together with an adhesive or sealant. The *substrate* is the material to be bonded. After bonding, the substrate is often referred to as an *adherend* (although sometimes these two terms are used synonymously).

The area between the adhesive and adherend is referred to as the *interphase region*. This interphase region is a thin region near the point of adhesive—adherend contact. The interphase region has different chemical and physical characteristics than either the bulk adhesive or the adherend. The nature of the interphase region is a critical factor in determining the properties and quality of an adhesive bond.

Different from the interphase is the interface, which is contained within the interphase. The *interface* is the plane of contact between the surface of one material and the surface of the other. The interface is often useful in describing surface energetics. The interface is also at times referred to as a *boundary layer*. Between the adhesive and adherend there can be several interfaces composed of layers of different materials. The boundary layers will be discussed in detail in the following chapters.

Sometimes a *primer* is used with adhesives or sealants. A primer is applied to a surface prior to the application of an adhesive or sealant, usually for improving the performance of the bond or protecting the surface until the adhesive or sealant can be applied. The *joint* is the

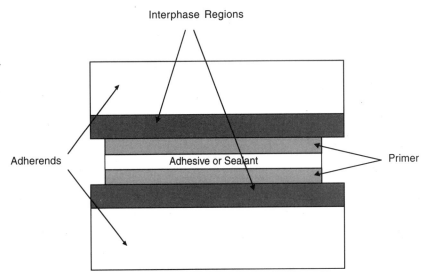

Figure 1.2 Components of a typical adhesive or sealant joint.

part of the assembly made up of adherends; adhesive or sealant; primers, if present; and all associated interphase regions as shown in Fig. 1.2.

1.2.3 Important factors for successfully using adhesives and sealants

From the complexity of the joint as described above, it should be evident that what is necessary to successfully understand and use adhesives or sealants is far broader than simply a knowledge of certain materials. The quality of the resulting application will depend on many factors, some of which are very entangled and complicated.

One of the principal factors in the success of either an adhesive or sealant is adhesion. Table 1.1 lists some of the external and internal factors that influence adhesion. An understanding of how these factors affect adhesion will determine the success of the bonding or sealing operation. Knowledge of production processes, economics, and environment and safety factors is also important.

Anyone intending to use adhesives or sealants faces the formidable tasks of selecting the correct materials and determining proper processes. The adhesive or sealant must flow onto the substrate surface and then change from a flowable liquid to a structural solid without creating harmful internal stresses in the joint. The substrate surface must have previously been cleaned and, possibly, prepared specially for maximum adhesion. The joint geometry must be correctly designed with regard to the materials selected and to the expected loads to avoid undesirable local stresses that could lead to early and premature failure. Also, the physical and chemical characteristics of the joint (adhesive/sealant, adherends, and interphase regions) must be understood and forecast in relation to the expected operating environment.

The end-user should not only be concerned with the performance of the joint immediately after bonding or sealing. The performance of the joint must also be considered throughout its practical service life. Almost all adhesive or sealant systems will undergo some change during their life. These changes could have a profound effect on the strength and permanence of the joint.

Unfortunately, substrates and adhesive/sealant materials tend to change due to external influences from the environment. These changes could occur: (a) during formation of the joint; and (b) during aging in service. Not only is the adhesive and adherend subject to change, but the interphase region could be subject to transformation as well. These simultaneously occurring, dynamic processes are one reason why it is so difficult to predict the life of a bonded joint. It may

TABLE 1.1 Factors Influencing Selection of an Adhesive or Sealant[1]

Stress

Tension	Forces acting perpendicular to the plane of the adhesive. Not commonly encountered in bonding thin plastic or metal sheets, leather, cork compositions, etc.
Shear	Forces acting in the plane of the adhesive. Pure shear is seldom encountered in adhesive assemblies; substantial tension components are usually found
Impact	Minimum force required to cause the adhesive to fail in a single blow. May be determined in tension or shear. Measures brittleness
Peel	Stripping of a flexible member fastened with adhesive to another flexible or rigid member. Stress is applied at a line; test loads are expressed in pounds per inch width. Commonly used angles of peel in tests are 90° for relatively stiff and 180° for flexible members
Cleavage	Forces applied at one end of a rigid bonded assembly which tend to split the bonded members apart. Can be considered as "peel" of two rigid members
Fatigue	Dynamic—alternate loading in shear or tension-compression. Static—maximum load sustained for long periods of time in tension or shear; tests are also used to determine creep

Chemical Factors

External	Effect of chemical agents such as water, salt water, gasoline, by hydraulic fluid, acids, alkalies, etc.
Internal	Effect of adherend on adhesive (i.e., exuded plasticizers in certain plastics and rubber); effect of adhesive on the adherend (crazing, staining, etc.)

Exposure

Weathering	Combined effect of rainfall, sunlight, temperature changes, type of atmosphere
Light	Important only with translucent adherends. Effect of artificial or natural light, or ultraviolet
Oxidation	Usually tested by exposure to ozone with the joint either unstressed or stressed, in which case deterioration is faster
Moisture	Either adhesive or adherend may be affected by high humidity or wet conditions. Cyclic testing with alternate moist and dry conditions can be valuable. May cause dimensional changes
Salt spray	Important only in coastal or marine atmospheres. Possible corrosion of adherend should also be considered

Temperature

High	Normal atmospheric variations may be encountered, or exceptional conditions. Bond strength may be affected by reactions in adhesive or adherend; decomposition or changes in physical properties of adhesive are important

TABLE 1.1 Factors Influencing Selection of an Adhesive or Sealant (*Continued*)

Low	May cause crystallization or embrittlement, detected by strength test. Cyclic testing with low or high temperatures may detect lack of durability
Biological Factors	
Bacteria or mold	Usually warm, humid tropical conditions. Can affect bond strength, and cause emission of odor or discoloration
Rodents or vermin....	Adhesives of animal or vegetable origin may be attacked by rats, cockroaches, etc.
Working Properties	
Application	Brushing, spray, trowel, or knife-spreader application characteristics are usually determined by trial and error. Consistency or viscosity may be adequate indications. Mechanical stability of emulsions and dispersions, and foaming tendency, can be important for machine application
Bonding range	Minimum drying or solvent-reactivation time before suitable bond can be obtained. Maximum allowable time before assembly. Permissible temperature range with heat-activated adhesives
Blocking.............	Tendency of surfaces coated for storage before assembly to adhere under slight pressure, or changes in humidity or temperatures
Curing rate	Minimum curing time, and effect of overcuring. May be determined as a shear or tensile-strength vs. curing-time curve at a specific curing temperature
Storage stability......	Physical and chemical changes in original unapplied state as a result of storage for extended time periods at representative storage temperatures
Coverage	Area of bond that can be formed with unit weight or volume of adhesive; expressed as pounds per 1,000 ft of bond line, or square feet per gallon. Depends on method of application; dimensions of work or of adhesive-coated area in relation to part size may affect coverage

be very difficult to know exactly the composition of the joint at any point in time. The possibility of these transformations resulting in an unacceptable material within the joint or in altering the mode of failure is great. In some applications, they could result in a catastrophic, premature joint failure.

1.2.4 Nature of the technologies related to adhesives and sealants

A multi-disciplined set of rules and a field-tested methodology are necessary to successfully negotiate the minefield of obstacles listed above. This requires consideration of fundamental concepts from a number of scientific disciplines. Figure 1.3 illustrates the various academic disciplines that are relevant. The primary sciences of physics, mechanics, and chemistry will overlap in certain areas to form the disciplines of surface science, polymeric materials, and joint design that are important to the science of adhesion. There are then further segments of these sciences such as polymer rheology and fracture mechanics, which are also highly relevant. Each of these specialized disciplines has contributed significantly to the science of adhesion and to its resulting stature in industrial products. The resulting overlap of all of these disciplines could be referred to as the "science" needed to successfully apply adhesives and sealants.

It is these various disciplines, and especially the areas where they overlap, that provide the primary subject matter of this Handbook. In today's industrial environment, usually the person responsible for integration of adhesives or sealants into an assembled product must be conversant with all of the relevant technologies. These are represented by Fig. 1.3 and by the equally important areas of product design, manufacturing, and economics. It is to this often over-burdened individual that this Handbook is focused.

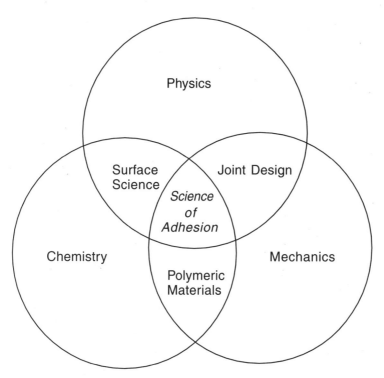

Figure 1.3 The science of adhesion requires the adaptation of multiple disciplines.

The steps necessary to achieve a practical and economic bond or seal will be developed through insight into the fundamentals of these sciences. The Handbook will attempt to identify solutions to satisfy most applications or, at least, illuminate the correct path for the end-user.

1.3 Markets and Applications

1.3.1 History

Adhesives and sealants were first used many thousands of years ago. Early hunters may have seen improvement in their aim by bonding feathers to arrows with beeswax, a primitive form of adhesive. The Tower of Babel was probably built with the aid of mortar and tar or pitch as a sealant. Carvings in Thebes (circa 1300 BC) show a glue pot and brush to bond veneer to a plank of sycamore. Until relatively recently, most adhesives and sealants evolved from vegetable, animal, or mineral substances.

In the early 1900s, synthetic polymeric adhesives began displacing many of these naturally occurring products owing to their stronger adhesion, greater formulation possibilities, and superior resistance to operating environments. However, non-polymeric materials are still widely used and represent the bulk of the total volume of adhesives and sealants employed today. Common applications for these non-polymeric materials include bonding porous substrates such as wood or paper. Casein adhesive (a dairy by-product) and soluble sodium silicate adhesives (an inorganic, ceramic material) are commonly used in the cardboard packaging industries. Naturally occurring, bitumen or asphalt materials have been accepted as sealants for many centuries.

The development of modern polymeric adhesives and sealants began about the same time as the polymer industry itself, early in the 1900s. In fact, the polymeric and elastomeric resins industry is bound very closely to the adhesive and sealant industries. Table 1.2 summarizes highlights of the historical development of adhesive and sealant products. The modern adhesives age began about 1910 with the development of phenol formaldehyde adhesives for the plywood industry. Adhesives and sealants found important markets in the construction industry, which was providing much of the growing infrastructure in the U.S. at the time.

Significant growth then again occurred in the 1940s and 1950s with the development of structural adhesives and sealants for the military aircraft industry. Because of their exceptional strength-to-weight ratio, the development of modern adhesives and sealants is closely related to the history of the aircraft and aerospace industries. Durability of adhesive joints was a problem in aircraft service until advanced

TABLE 1.2 Historical Development of Adhesives and Sealants

Approximate decade of commercial availability	Adhesive or sealant
Pre 1910	Glue from animal bones Fish glue Vegetable adhesives
1910	Phenol-formaldehyde Casein glues
1920	Cellulose ester Alkyd resin Cyclized rubber in adhesives Polychloroprene (Neoprene) Soybean adhesives
1930	Urea-formaldehyde Pressure sensitive tapes Phenolic resin adhesive films Polyvinyl acetate wood glues
1940	Nitrile-phenolic Chlorinated rubber Melamine formaldehyde Vinyl-phenolic Acrylic Polyurethanes
1950	Epoxies Cyanoacrylates Anaerobics Epoxy alloys
1960	Polyimide Polybenzimidazole Polyquinoxaline
1970	Second-generation acrylic Acrylic pressure sensitive Structural polyurethanes
1980	Tougheners for thermoset resins Waterborne epoxies Waterborne contact adhesives Formable and foamed hot melts
1990	Polyurethane modified epoxy Curable hot melts UV and light cure systems

adhesive systems were defined, introduced, and verified in the late 1970's.

With successful experiences in these industries, it was soon realized that adhesives could be used to economically replace mechanical fastening methods such as welding, brazing, or riveting. It was also realized that sealants could be used to provide additional function and value to products in industries ranging from transportation to construction.

Even the medical profession has come to use adhesives and sealants in everyday processes. A cyanoacrylate adhesive is used for closing wounds and lacerations while increasing patient comfort and reducing scarring. Light or UV cured adhesives and sealants are commonly used in dental restoration.

The science of adhesion is now well accepted, and the basic rules and methods for achieving high performance joints have been well established. The industry has a strong foundation of formulations and processes. Today, many of the new adhesive and sealant developments are focusing on production cycle time and cost; environmental enhancement; or application to new substrates, such as engineering plastics, advanced composites, and ceramics that are rapidly gaining acceptance.

Development efforts supporting adhesives and sealants are directed to optimizing the manufacturing and assembly processes. For example, automated meter, mixing, and dispensing equipment and weldbonding adhesives have been perfected to reduce production time in high-volume manufacturing operations. New adhesives and sealants are often applied with robotic equipment to further enhance productivity. UV curable adhesives have been developed to take advantage of their ease of application, elimination of mixing and heat curing, and elimination of liquid solvent. Ultrasonic and other fast thermal welding techniques have found a receptive home in the high-volume transportation and consumer product industries. Microwave assisted drying of water-based adhesives and new hot melt systems have also been developed to make bonding more agreeable to the fast-paced manufacturing world.

1.3.2 Current markets

Adhesives and sealants are used in a variety of industries: construction, packaging, furniture, automotive, appliance, textile, aircraft, and many others. A large number of manufacturers supply many different products to numerous end-users for a multitude of applications. Even though the number of adhesive and sealant companies dwindle due to industry-wide consolidation, there are over 1,500 companies in the U.S. alone manufacturing various types of adhesive products. Many of these companies are producing products for their own internal use. One study defines seven major market areas and 59 major market segments (Table 1.3). However, there are many additional markets and niche applications where adhesives surpass other methods of joining.

TABLE 1.3 Classification of Adhesive Markets[2]

Packaging
Corrugated board manufacture

Carton side-seam and closures (including glue lap and case sealing)

Composite bonding of disposable products (towel and tissue laminating, pick up and tail-tie diapers, sanitary napkins, hospital supplies)

Bags

Labels

Cups

Cigarette and filter manufacture

Envelope manufacture (excluding remoistenable gums)

Remoistenable products (stamps, envelopes, tapes, labels)

Flexible food laminates

Other flexible laminates (including labels on display cartons and other packaging lamination)

Point of sale labels

Speciality packages (cosmetics, toiletries)

Composite containers and tubes

Tapes
Packaging tapes

Industrial tapes

Surgical tapes

Masking tapes

Consumer tapes

Construction
Acoustic ceiling panels, floor tile and continuous flooring installation

Ceramic tile installation

Counter top lamination

Manufacture of prefabricated beams and trusses

Carpet layment adhesives

Flooring underlayment adhesives

Installation of prefinished panels

Joint cements (gutters, plastic pipe)

Curtain wall manufacture

Wall covering installation

Dry wall lamination adhesives

Other nonrigid bonding
Fabric combining (including stitchless sewing)

Apparel laminates

Shoe assembly — sole attachment

Other shoe manufacturing adhesives

Sports equipment

Book binding

Rug backing

Flock cements

Air and liquid filter manufacture

Consumer adhesives
Do-it-yourself products

Model and hobby supplies

School and stationery products

Decorative films

Transportation
Auto, truck, and bus interior trim attachment

Auto, truck, and bus exterior trim attachment

Vinyl roof bonding

Auto, truck, and bus assemblies (including side panels, doors, hoods, and trunk lids)

Weatherstrip and gasket bonding

Aircraft and aerospace structural assemblies

LNG tank assembly

Other rigid bonding
Shake proof fastening

Furniture manufacture

Manufacture of millwork, doors, kitchen cabinets, vanitories (excluding counter top lamination)

Appliance assembly and trim attachment

Houseware assembly and trim attachment

TV, radio, and electronics assembly

Machinery manufacture and assembly

Supported and unsupported film lamination

Manufacture of sandwich panels (road signs, etc.)

The total U.S. market for adhesive and sealant products is estimated to be near $10 billion dollars. Figure 1.4a shows the leading adhesives and sealants products in 1995. Most of the use is in relatively nonexotic areas such as general-purpose industrial assembly, hot melts, and binders. The more specialized products account for a relatively small portion of the overall market. Figure 1.4b shows the leading adhesives and sealants end-use markets. The largest markets are industrial assembly, packaging, and wood related products. The U.S. adhesive industry is difficult to define quantitatively because of its breadth and degree of fragmentation. North America dominates the world's adhesive markets with an estimated 38% share of global revenues.

The packaging and construction industries together account for 80% of adhesives demand. Construction markets are dominated by the use of phenolic and amino adhesives as binders in wood panels. Nonstructural adhesives account for the largest volume in these markets. Corrugated boxes are the single largest product for adhesives within the packaging sectors. Pressure sensitive tapes and labels are also important products within this segment. The main markets for structural adhesives are transportation, industrial assembly, and construction. However, structural adhesives occupy a relatively small segment of the total adhesives market. The household market is also sizable, particularly for polyvinyl acetate (wood glue), cyanoacrylates ("super glue"), two-part epoxies, and modified acrylic adhesives.

The main sealant market segments are the construction, consumer products, transportation, industrial, aerospace, appliance, and electronics segments. The leading market is construction, followed by transportation and industrial. The types of sealant used in each of these markets are identified in Table 1.4 and described in later chapters. Synthetic sealants account for nearly 70% of the total sealant market. Synthetic sealants are dominated by polyurethanes and silicones and benefit from a strong construction industry.

1.3.3 Market trends and drivers

The compounded annual growth rate (CAGR) of adhesives from 1996 to 2003 is expected to be 5.3%, and sealants will have an estimated CAGR of 4.5%. However, certain products will see growth at several times that average. The markets for high quality adhesives and sealants, such as epoxies, silicones, and polyurethanes, have grown faster than the markets for larger volume commodity type products. The reasons for this faster growth rate are:

- Lower level of pollutants (especially driving the growth of water borne and hot melt pressure sensitive materials)

An Introduction to Adhesives and Sealants   15

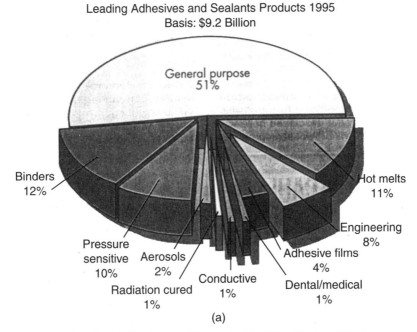

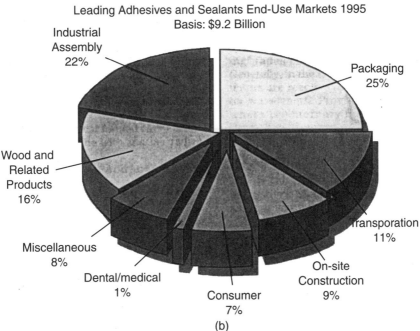

Figure 1.4 Leading adhesive and sealant (*a*) products and (*b*) end-use markets in 1995.[3]

TABLE 1.4 Sealant Usage by Generic Sealant Type[4]

Generic type	Construction	Industrial	Transportation	Appliances	Aerospace
Asphalt, bitumen	x	x			
Oleoresinous	x	x			
Butyl	x	x	x		
Hypalon	x				
EPDM	x	x	x		
Neoprene		x			
Styrene-butadiene		x			
Polyvinyl chloride		x			
Acrylic solution	x	x			
Acrylic emulsion	x	x			
Polyvinyl acetate	x	x			
Polysulfide	x	x	x	x	x
Polyurethane	x	x	x	x	x
Silicone		x	x	x	x
Fluoropolymers		x	x	x	x
Epoxies	x	x	x		
Intumescents	x				

- New users (e.g., consumer electronics, sporting goods)
- Higher standards of performance
- Newer materials (e.g., the use of nonferrous parts such as aluminum, composites, and engineering plastics on car bodies)

The demand for adhesives in the U.S. is forecast to rise to 14 billion pounds in the year 2001, with market value reaching $9 billion. There will be continuing shifts from lower cost natural products toward highly formulated synthetic adhesives. The general focus of adhesive product development will be on lowering solvent content and volatile organic compound emissions. These trends will result in many new environmentally compatible adhesive systems with higher solids contents, such as water based or hot melt products. Natural adhesives will lag total aggregated demand with almost all growth in this sector arising in starch and dextrin adhesives used in paperboard packaging.

The U.S. demand for sealants was about 2.1 billion pounds and $2.4 billion market value in 1998. Synthetic sealants dominate and will expand their share of the sealant market due to superior performance characteristics over natural sealants. Polyurethanes, silicones, and acrylics command the synthetic market, taking share from butyl rubber and polysulfide types. These products are experiencing above average growth. The overall sealant growth rate will be dictated by global economies especially in construction and transportation. The construction market increases will depend on improved nonresidential construction spending, sustained economic growth, and an aging infrastructure. The transportation equipment market has grown due to the increased production of more fuel efficient, quieter riding cars. Like the adhesives market, the industrial sealants market is fragmented among a multitude of applications and industries.

1.3.4 Adhesive and sealant industries

The industries that are most influenced by adhesives and sealants consist of four main categories:

1. Base material producers including resins, mineral fillers, extenders, etc.
2. Formulators who take the base materials and combine, process, and package them into adhesive and sealant systems that provide various levels of performance
3. End-users who take the packaged adhesives and sealants and produce assembled products
4. Associated industries such as equipment manufacturers, testing laboratories, consultants, etc.

The base material producers are usually large chemical or material companies that manufacture materials for broader markets such as petrochemicals or plastics. When demand warrants, they will produce materials specifically for the adhesive and sealant formulators.

The formulators can range from very small business with several employees, addressing small niche markets, to large international companies with several hundred products. Both small and large formulators are generally willing to modify a formulation if they believe that it will improve performance, production efficiency, or add some other value. However, a minimum volume is usually necessary before formulators will make modifications to a standard formulation or develop a new product for a specific application. Formulators have a significant knowledge base regarding adhesive or sealant systems and how they are to be applied in practice.

The end-user usually purchases its adhesive from a formulator, rather than produce it internally. It is increasingly difficult for an end-user to keep up with the continuing technological changes. Therefore, it is often best left to the specialized formulator. The end-user, however, must select the proper adhesive or sealant, substrate, joint design, and processing conditions for specific applications. Once these are selected and verified as to performance and cost, the end-user must be vigilant that none of the processes, materials, or other relevant factors change.

Several other industries are also greatly affected by adhesives and sealants. For example, equipment suppliers specialize in producing machinery for application, assembly, curing, surface preparation, etc. Equipment suppliers also specialize in developing and manufacturing the testing apparatus that can be used to measure joint strength and processing parameters. Then, there are testing laboratories and consultants who provide assistance and services to the end-user on a contractual basis.

1.4 Advantages and Disadvantages of Adhesive Bonding

This section addresses the process of choosing a method of joining. Each joining application should be considered with regard to its specific requirements. There are times when adhesives are the worst possible option for joining two substrates, and there are times when adhesives may be the best or only alternative.

Often, one must consider the time, trouble, and expense that may be necessary to use an adhesive. For example, certain plastics may require expensive surface preparation processes so that the adhesive can wet their surface. Applications requiring high temperature service conditions may call for an adhesive that requires an elevated temperature cure over a prolonged period.

On the other hand, certain applications could not exist without adhesive bonding. Examples of these are joining of ceramic or elastomeric materials, the joining of very thin substrates, the joining of surface skin to honeycomb, and numerous other applications. There are also certain applications where adhesives are chosen because of their low cost and easy, fast joining ability (e.g., packaging, consumer products, large area joints).

Sometimes conventional welding or a mechanical joining process is just not possible. Substrate materials may be incompatible for metallurgical welding due to their thermal expansion coefficients, chemistry, or heat resistance. The end product may not be able to accept the bulk or shape required by mechanical fasteners.

Usually, the choice of joining process is not all black or white. Certain processes will have distinct advantages and disadvantages in specific applications. The choice may involve trade-offs in performance, production capability, cost, and reliability. This section will provide sufficient information to make such an analysis.

1.4.1 Competitive methods

A variety of joining methods can be used to provide the assembly function. Alternative joining methods include adhesive bonding, welding, brazing, soldering, and mechanical fastening. All fastening and joining systems, including adhesives, fall into one of three general categories illustrated in Fig. 1.5.

- Periodic—the attachment of two members by occasionally placing through hole fasteners or other individual mechanisms. (This is the most widely used joining technique for structures requiring high mechanical strength and a minimum of sealing or other non-strength functions.)
- Linear—a continuous or occasional edge bead attachment, such as welding.
- Area—an attachment achieved by full-face contact and union between the two mating surfaces. (Soldering, brazing, and adhesives are examples of area attachment.)

Although adhesive bonding can be successfully employed in periodic or linear attachment applications, the main benefits and advantages are realized when adhesives are used in the "area" attachment designs.

In evaluating the appropriate joining method for a particular application, a number of factors must be considered, such as those suggested in Table 1.5. There generally is no single method of fastening that is obviously the best choice. Some fastening methods can quickly be eliminated from consideration, such as the welding of ceramic substrates or the use of an organic adhesive in an application that will see extremely high service temperatures. Adhesives are usually the proper choice when the substrates are physically dissimilar or met-

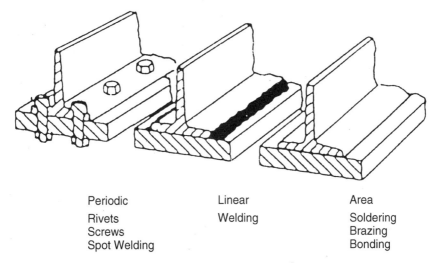

Periodic	Linear	Area
Rivets	Welding	Soldering
Screws		Brazing
Spot Welding		Bonding

Figure 1.5 Periodic, linear, and area attachment systems.[5]

TABLE 1.5 General Comparison of Joining Process Characteristics[2]

	Welding	Brazing and soldering	Mechanical fastening	Adhesive bonding
		Joint Features		
Permanence	Permanent joints	Usually permanent (soldering may be non-permanent)	Threaded fasteners permit disassembly	Permanent joints
Stress distribution	Local stress points in structure	Fairly good stress distribution	Points of high stress at fasteners	Good uniform load distribution over joint area (except in peel)
Appearance	Joint appearance usually acceptable. Some dressing necessary for smooth surfaces	Good appearance joints	Surface discontinuities sometimes unacceptable	No surface marking. Joint almost invisible
Materials joined	Generally limited to similar material groups	Some capability of joining dissimilar metals	Most forms and combinations of materials can be fastened	Ideal for joining most dissimilar materials
Temperature resistance	Very high temperature resistance	Temperature resistance limited by filler metal	High temperature resistance	Poor resistance to elevated temperatures
Mechanical resistance	Special provision often necessary to enhance fatigue resistance	Fairly good resistance to vibration	Special provision for fatigue and resistance to loosening at joints	Excellent fatigue properties. Electrical resistance reduces corrosion

Production Aspects

Joint preparation	Little or none on thin material. Edge preparation for thick plates	Prefluxing often required (except for special brazing processes)	Hole preparation and often tapping for threaded fasteners	Cleaning often necessary
Post-processing	Heat transfer sometimes necessary	Corrosive fluxes must be cleaned off	Usually no post-processing — occasionally re-tightening in service	Not often required
Equipment	Relatively expensive, bulky and often required heavy power supply	Manual equipment cheap. Special furnaces and automatic unit expensive	Relatively cheap, portable and "on-site" assembly	Only large multi-feature, multi-component dispensers are expensive
Consumables	Wire, rods, etc., fairly cheap	Some special brazing fillers expensive. Soft solders cheap	Quite expensive	Structural adhesives somewhat expensive
Production rate	Can be very fast	Automatic processes quite fast	Joint preparation and manual tightening slow. Mechanized tightening fairly rapid	Seconds to hours, according to type
Quality assurance	NDT methods applicable to most processes	Inspection difficult, particularly on soldered electrical joints	Reasonable confidence in torque control tightening	NDT methods limited

allurgically incompatible materials, thermoset plastics, ceramics, elastomers, thin materials, or very small parts. Adhesive bonding is also generally appropriate when there are large areas to join, or when adhesives can be chosen to provide improvement in manufacturing productivity.

Usually the decision of which fastening method to use involves several trade-offs. A trade-off analysis, as shown in Table 1.6, can be useful in identifying potential fastening methods. When this is performed, the possibility of using adhesives over other methods becomes apparent.

The science of adhesive bonding has advanced to a degree where adhesives must be considered an attractive and practical alternative to mechanical fastening for many applications. Adhesive bonding presents several distinct advantages over conventional mechanical methods of fastening. There are also some disadvantages which may make adhesive bonding impractical. These pros and cons are summarized in Table 1.7.

The design engineer must consider and weigh these factors before deciding on a method of fastening. However, in some applications ad-

TABLE 1.6 How Joining Methods Compare[5]

	Riveting	Welding	Brazing	Adhesive bonding
Preliminary machining	P	E	P	E
With thin metals	P	P	F	E
Limits on metal combinations	F	P	P	E
Surface preparation	E	G	F	P
Tooling	E	F	F	F
Need for access to joint	P	P	E	E
Heat requirements	E	P	P	F–G
Stress distribution	P	F–G	E	E
Sealing function	P	F	E	G
Rate of strength development	E	E	E	P
Distortion of assembly	F	P	F	E
Final machining	G–E	F	E	E
Final heat treatment	E	F	F	E
Solvent resistance	E	E	E	F
Effect of temperature	E	E	E	P
Ease of repair	G	P	P	F
Level of skill required	E	G	E	E

NOTES: E—Excellent, G—Good, F—Fair, P—Poor.

TABLE 1.7 Advantages and Disadvantages of Adhesive Bonding

Advantages	Disadvantages
1. Provides large stress-bearing area.	1. Surfaces must be carefully cleaned.
2. Provides excellent fatigue strength.	2. Long cure times may be needed.
3. Damps vibration and absorbs shock.	3. Limitation on upper continuous operating temperature (generally 350°F).
4. Minimizes or prevents galvanic corrosion between dissimilar metals.	
5. Joins all shapes and thicknesses.	4. Heat and pressure may be required.
6. Provides smooth contours.	5. Jigs and fixtures may be needed.
7. Seals joints.	6. Rigid process control usually necessary.
8. Joins any combination of similar or dissimilar materials.	7. Inspection of finished joint difficult.
9. Often less expensive and faster than mechanical fastening.	8. Useful life depends on environment.
10. Heat, if required, is too low to affect metal parts.	9. Environmental, health, and safety considerations are necessary.
11. Provides attractive strength-to-weight ratio.	10. Special training sometimes required.

hesive bonding is the only logical choice. In the aircraft industry, for example, adhesives make the use of thin metal and honeycomb structures feasible because stresses are transmitted more effectively by adhesives than by rivets or welds. Plastics, elastomers, and certain metals (e.g., aluminum and titanium) can often be more reliably joined with adhesives than with other methods. Welding usually occurs at too high a temperature, and mechanical fastening destroys the lightness and aesthetics of the final product. Certain examples of less obvious applications where adhesive bonding is a practical method of assembly are shown in Table 1.8. The following discussion on the advantages and disadvantages of adhesives should assist the user in determining the feasibility of adhesives for a specific application.

1.4.2 Mechanical advantages

The most common methods of structural fastening are shown in Fig. 1.6. Adhesive bonding does not have many of the disadvantages of other methods. Welding or brazing, useful on heavy-gauge metal, is expensive and requires great heat. Dissimilar metals usually have different coefficients of thermal expansion or thermal conductivities making them more difficult to weld. Some metals have unstable oxides that also make welding difficult. Many light metals such as aluminum, magnesium, and titanium are difficult to weld and are weakened or distorted by the heat of welding. High temperature metallurgical joining methods can cause thin sheets to distort. Beneficial properties ob-

TABLE 1.8 Examples of Applications where Bonding is a Practical Method of Assembly[6]

Application areas for adhesives	Examples
Dissimilar materials	Combinations of metals, rubbers, plastics, foamed materials, fabrics, wood, ceramics, glass, etc.
Dissimilar materials which constitute a corrosion couple	Iron to copper or brass
Heat sensitive materials	Thermoplastics, magnetic materials, glass
Laminated structures	Sandwich construction based on honeycomb materials; heat exchangers; sheet laminates, core laminates
Reinforced structures	Stiffeners for wall paneling, boxes and containers, partitions, automobile chassis parts, aircraft body parts
Structural applications	Load bearing structures in the aircraft fuselage, automotive and civil engineering industries
Bonded inserts	Plug inserts, studs, rivets, concentric shafts; tubes, frame construction; shaft-rotor joints; tools; reinforced plastics with metal inserts; paint brush bristles
Sealed joints and units	Pipe joining, encapsulation, container seams, lid seals
Fragile components	Instrumentation, thin films and foils, microelectronics components and others where precise location of parts is required
Components of particular dimensions	Where bonding areas are large or there is a need for shape conformity between bonded parts
Temporary fastening	Where the intention is to dismantle the bond later, the use of various labels, surgical and pressure sensitive tapes, adhesives for positioning and locating parts, in lieu of jigs, prior to assembly by other means

tained from metallurgical heat treating processes could be lost because of a high temperature joining process. Adhesives, on the other hand, provide a low temperature, high strength, joint with many of these substrates. They thereby avoid many of the problems commonly encountered with other methods of joining.

Many polymeric adhesives are viscoelastic and act like tough, relatively flexible materials with the ability to expand and contract. This allows the bonding of materials having greatly different coefficients of thermal expansion or elastic moduli. Toughness also provides resistance to thermal cycling and crack propagation.

Bonded structures are often mechanically equivalent to, or stronger than, structures built with more conventional assembly methods. An

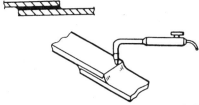

Brazing is an expensive bonding method. Requiring excessive heat, it often results in irregular, distorted parts. Adhesive bonds are always uniform.

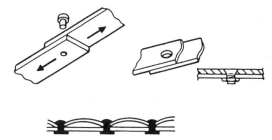

When joining a material with mechanical fasteners, holes must be drilled through the assembly. These holes weaken the material and allow concentration of stress.

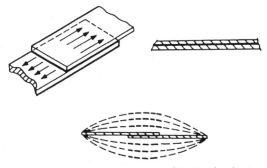

A high strength adhesive bond withstands stress more effectively than either welds or mechanical fasteners.

Figure 1.6 Common methods of structural fastening.[7]

adhesive will spread stress uniformly from one member to another, thus eliminating localized stress concentrations that can occur with other fastening systems. When using mechanical fasteners, substrates may need to be thicker or otherwise strengthened to handle the concentrated stress, thereby adding weight and cost to the final assembly. Consequently, adhesives often allow structures to be built with lower cost and less weight.

Many benefits are obtained because the adhesive joint is capable of spreading stress relatively evenly over the entire overlap region of the bond. Alternatively, mechanical fasteners and spot welding provides points of attachment in small discrete areas. Consequently, adhesive shear strengths of as high as 7000 psi can be obtained. The stress-distribution characteristics and inherent toughness of adhesives also provide bonds with superior fatigue resistance, as shown in Fig. 1.7. Generally in well-designed joints, the adherends fail in fatigue before the adhesive. Holes, needed for rivets or other fasteners, are not required for an adhesive bond, thereby avoiding possible areas of high stress concentration. The elimination of holes also maintains the integrity of the structural members.

1.4.3 Design advantages

Adhesives offer certain valuable design advantages. Unlike rivets or bolts, adhesives produce smooth contours that are aerodynamically and aesthetically beneficial. Adhesives also offer a better strength-to-weight ratio than other methods of mechanical fastening. Adhesives can join any combination of solid materials regardless of shape, thickness, or mismatch in physical properties such as coefficient of thermal expansion or elastic modulus. Certain substrates may be too thin or too small to weld reproducibly without distortion. Thus, medical prod-

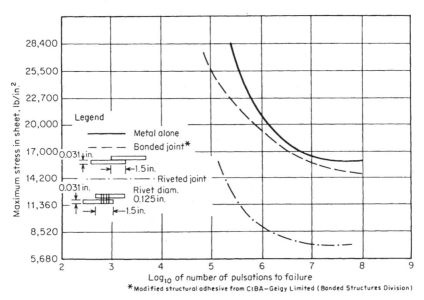

Figure 1.7 Fatigue strengths of aluminum-alloy specimens under pulsating tensile load.[8]

ucts and microelectronics are often assembled with adhesives. Nonmetallic materials, such as plastics, elastomers, ceramics, and many paper products, can be joined together and to one another more economically and efficiently with adhesive bonding than with other methods.

Adhesives may also be a good way of adding options or additions to a line of manufactured items that share a common design. This allows the elimination or reduction of extra holes for mechanical fasteners and can eliminate expensive machining or stamping steps on the common part. Versatility in product aesthetics, good mechanical reliability, and manufacturing speed are also benefits of providing design options with adhesives.

1.4.4 Production advantages

Adhesive bonding is, at times, faster and less expensive than conventional fastening methods. It is well suited for high-volume production or assemblies requiring large bonded areas. As the size of the area to be joined increases, the time and labor saved by using adhesives instead of mechanical fasteners become progressively greater because the entire joint area can be bonded in one operation. Figure 1.8 shows the economy of large area metal-to-metal bonding compared with riveting.

Some adhesives are especially well suited to applications requiring rapid assembly especially if the end-use requirements (i.e., strength, heat, and chemical resistance) are not too severe. The packaging industry and much of the decorative furniture industry uses adhesives because they are fast and consistent. In the medical products industry, use of ultraviolet curing permits rapid assembly of syringes and other articles. Certain automotive materials are chosen for their ability to

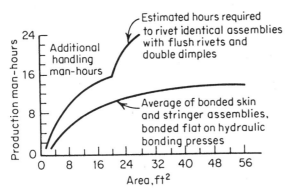

Figure 1.8 The economy of metal-to-metal bonding compared with conventional riveted structure.[9]

be ultrasonically welded, thereby, allowing efficient, fast, high volume assembly operations.

At times, adhesive bonding may be more expensive than other fastening methods. However, the overall cost of the final part may be less through reduced material requirements, weight savings, elimination of other operations such as drilling, countersinking, welding, etc., and simplified assembly. Using associated production processes such as a paint-drying oven to cure the adhesive may also save costs.

1.4.5 Other advantages

Adhesives are useful for providing secondary functions as well as the primary function of holding the substrates together. Many designers feel that one of the most valuable characteristics of adhesive bonding is their multi-functional nature. In addition to performing a mechanical fastening operation, an adhesive may also be used as a sealant, vibration damper, insulator, and gap filler—all in the same application.

Because adhesives are viscoelastic materials, they can act as vibration dampers to reduce the noise and oscillation encountered in some assemblies. Adhesives can also perform sealing functions, offering a barrier to the passage of fluids and gases. Another property of adhesives that is often advantageous is their ability to function as electrical and thermal insulators in a joint. The degree of insulation can be varied with different adhesive formulations and fillers. Adhesives can even be made electrically and thermally conductive with silver and boron nitride fillers, respectively. Since adhesives usually do not conduct electricity, they prevent galvanic corrosion when dissimilar metals are bonded.

1.4.6 Mechanical limitations

The most serious limitation on the use of polymeric adhesives is their time-dependent strength in degrading environments such as moisture, high temperatures, or chemicals. For example, organic adhesives perform well between -60 and $350°F$, but only a few adhesives can withstand operating temperatures outside that range. Chemical environments and outdoor weathering also degrade adhesives. The rate of strength degradation may be accelerated by continuous stress or elevated temperatures.

The combination of continuous stress along with high moisture conditions is of special concern. Certain adhesives will only survive in this environment if their service stress is significantly less than their ultimate strength (e.g., less than 10% of ultimate strength).

Since nearly every adhesive application is unique, the adhesive manufacturers often do not have data concerning the aging characteristics of their adhesives in specific environments. Thus, before any adhesive is established in production, a thorough evaluation should be made in either a real or a simulated operating environment.

With most structural adhesives, strength is more directional than with mechanical fasteners. Generally, adhesives perform better when stressed in shear or tension than when exposed to cleavage or peel forces. Residual stresses inside the joint can also present serious problems. Such stresses arise from shrinkage due to cure or aging, from different coefficients of thermal expansion between substrates, and from other circumstances.

The adhesive material itself should never be used as a structural substrate. Very heavy bondlines with uneven joint thickness result in undesirable concentrations of stresses. Many adhesives cure by an exothermic chemical reaction whose intensity is dependent on the mass of material. Adhesives are generally formulated to cure in thin sections. Therefore, certain epoxy adhesives, when applied in significant bulk, could over-heat due to their own crosslinking reaction and, in fact, burn or degrade when cured in thick sections.

1.4.7 Design limitations

The adhesive joint must be carefully designed for optimum performance. Design factors include the type of stress, environmental influences, and production methods that will be used. Many rigid adhesives do not work well when external stresses act to peel or cleave the substrates from one another. These stresses can often be reduced or eliminated by careful joint design. Seldom can a joint, which is designed for mechanical fastening, be used successfully for adhesive bonding without revision. Sometimes such revisions result in added expense or manufacturing steps.

There are no standards to guide the user with regard to design limits or to provide a safe design margin. These will depend on the adhesive and substrate, on the production methods, on the specific end-use environment, and on many other factors that are often not foreseen at the time of design development. Therefore, it is very difficult to predict the useful life of a bonded joint. Simple life estimation processes used in other industries (such as Arrhenius plots to predict the aging of electrical wire insulation) are not effective with adhesives because of the numerous and sometimes competing reactions that can take place within a bonded joint. The only effective method of estimating the useful life of an adhesive bond is to test prototypes under environmental conditions that will accelerate the stress on the bond.

Yet, one must be certain that these accelerated conditions do not cause reactions that are normally not experienced in the application.

1.4.8 Production limitations

Slow and critical processing requirements of some adhesives can be a major disadvantage particularly in high-volume production operations. Several production concerns must be considered when bonding operations are first projected. All adhesives require clean surfaces to obtain the best results. Depending on the type and condition of the substrate and the bond strength desired, surface preparations ranging from a simple solvent wipe to chemical etching are necessary. Adhesives should never be applied onto other coatings or over other adhesives unless the characteristics of these materials are accurately known. The resulting bond strength will be no greater than the "weakest link in the chain".

If the adhesive has multiple components, the parts must be carefully weighed and mixed. The setting operation often requires heat and pressure. Lengthy set time makes jigs and fixtures necessary for assembly. Rigid process controls are also necessary, because the adhesive properties are dependent on curing parameters and surface preparations. The inspection of finished joints for quality control is very difficult. This also necessitates strict control over the entire bonding process to ensure uniform quality. Non-destructive test techniques cannot quantitatively predict joint strength.

1.4.9 Other limitations

Since the true "general-purpose" adhesive has not yet been developed, the end-user must allow time to test candidate adhesives and bonding processes. Everyone involved in the design, selection, testing, and manufacture of adhesive bonded assemblies should be trained as to the critical requirements and processes. Adhesives are sometimes composed of material that may present personnel hazards, including flammability and dermatitis, in which case the necessary precautions must be considered. Workers must be trained how to handle these materials safely.

The following items contribute to a "hidden cost" of using adhesives, and they also could contribute to serious production difficulties:

- The storage life of the adhesive may be unrealistically short; some adhesives require refrigerated storage.
- The adhesive may begin to solidify before the worker is ready.
- The cost of surface preparation and primers, if necessary, must be considered.

- Ease of handling, waste, and reproducibility can be essential cost factors.
- Cleanup is a cost factor, especially where misapplied adhesive may ruin the appearance of a product.
- Once bonded, samples cannot easily be disassembled; if misalignment occurs and the adhesive cures, usually the part must be scrapped.

Many of these hidden costs can be minimized by the proper choice of adhesives and processes. However, it should be remembered that storage, cure, and waste disposal are seldom a concern in joining with mechanical fasteners, and with welding, the joining material is essentially free of charge.

1.4.10 Combining adhesive and mechanical fastening

There are advantages in combining adhesive bonding with mechanical fastening. The combination can provide properties that are superior to either singular method. It is also possible to reduce the number of mechanical fastening steps without sacrificing strength or reliability.

When combined with adhesives, mechanical joints can provide fixturing for the adhesive as it cures. In this way, expensive fixturing equipment and the time required to set-up such equipment are not necessary. The curing time is also eliminated as a potential bottleneck to the overall assembly process. The stress distribution characteristic of the adhesive bond also allows the designer the freedom to eliminate thicker substrates or reinforcements that may be necessary with mechanical fasteners alone.

The secondary functions of the adhesive, such as sealing, vibration damping, electrical insulation, etc., may also be used to achieve an assembly with greater value. In the automotive industry, for example, adhesives are used in combination with spot welding for joining trunk assemblies. This combination provides sound deadening and sealing in addition to a strong joint. Adhesives can also be combined with fasteners that are designed directly into the part (e.g., snap fit connectors). Here the adhesive eventually assumes the full structural load with the mechanical fastener providing the fixturing as the adhesive cures.

1.5 Functions of Sealants

Sealants are generally used as a barrier or a means of protection. In this way, sealants are used to exclude dust, dirt, moisture, and chemicals or to contain a liquid or gas. They are also often used as a coating to protect a surface or an article. They can exclude noise and vibration, improve appearance, and perform a joining function. Certain sealants, like adhesives, can be used to assemble parts, and many adhesives can be used to seal. Sealants can also be used as electrical or thermal insulators, fire barriers, and as products for smoothing, filleting or faying. No matter what the application, a sealant has three basic functions:

1. It fills a gap between two or more substrates
2. It forms a barrier by the physical properties of the sealant itself and by its adhesion to the substrate
3. It maintains its sealing property for the expected lifetime, service conditions, and environments.

Unlike adhesives, there are not many functional alternatives to the sealing process. Innovative product design can possibly accomplish the same function as a sealant. Soldering or welding can be used instead of a sealant in certain instances, depending on the substrates and the relative movement that the substrates will see in service. However, the simplicity and reliability offered by organic elastomers usually make them the apparent choice for performing these functions. Many sealants are designed for specific applications. Table 1.9 gives typical applications for various classes of sealants.

The proper application of a sealant involves more than merely choosing a material with the correct physical and chemical properties. As with adhesives, the substrates to be sealed, the joint design, performance expectations, production requirements, and economic costs must all be considered. Table 1.10 is a partial list of considerations that are often used to select sealants in the construction industry.

1.5.1 Mechanical considerations

Important mechanical properties of sealants include elongation, compressibility, tensile strength, modulus of elasticity, tear resistance, and fatigue resistance. Depending on the nature of the application, a sealant may require very little strength or great strength. The sealant must have sufficient mechanical characteristics to remain attached to the substrates during service and to provide a barrier. The substrates could move considerably, requiring that the sealant expand and contract significantly without loosing adhesion from the surface. Defining the sealant's movement capability is a complex process. Temperature, test rate, and joint configuration will influence the result.

While movement capability is very important, other consequential mechanical properties are: unprimed adhesion strength to various

TABLE 1.9 Typical Applications for Sealants[10]

Generic base	Typical uses
Oil	Small wooden window sash
Oil and resin	Metal windows
Butyl	
Noncuring	With polybutene for metal buildings, slip joints, interlocking curtain-wall joints, sound deadening, tapes
Curing	Home sealants, repair of lock-strip gaskets, tapes; with resins for hot melts on insulating glass
Polyisobutylene	Primary seal on insulating glass
Asphalts	With bitumen on gutters, driveway repair; with neoprene on gutters, waterstops, and adhesives
Acrylics	
Nonplasticized	Water-based for interior-use joints on wallboard
Plasticized	Caulks for exterior joints on low-rise housing, with good movement capabilities, excellent weathering
Solvent-based	Exterior joints on high-rise construction, around doors and windows with low movement
Block copolymer	
Solvent-based	For low-rise buildings with good movement
Hypalon	
Solvent-based	Exterior joints on high-rise construction, around doors and windows
PVC-coal-tar	As a hot melt on airfield runways and highways
Polysulfide	
One-part	High-rise building joints
Two-part	High-rise building joints, aircraft fuel tanks, boating, insulating glass sealant for remedial housing; with coal tar for airport aprons
Urethane	
One-part	High-rise buiding joints
Two-part	High-rise building joints, insulating glass sealant, with coal tar and asphalt for membrane waterproofing compounds
Silicone	
One-part	Low and medium modulus for high-rise building joints; low modulus for highways and difficult building joints; medium and high modulus for insulating glass with polyisobutylene; structural glazing; home use as bathtub caulk
Two-part	Mostly in-plant use of prefab units and insulating glass
Neoprene	Fire-resistant gaskets, lock-strip gaskets, foam gaskets
EPDM	Gaskets, lock-strip gaskets, foam gaskets
Nitrile	
Solvent-based	For small cracks and narrow joints
Epoxy	Concrete repair, complex beam construction; potting, molding, sealing transformers; high-voltage splicing, capacitor sealant; with polymers as a concrete coating on bridges
Polyester	Potting, molding, and encapsulating

TABLE 1.10 Checklist of Considerations for the Selection of Construction Sealants[11]

- Required Joint Movement
- Minimum Joint Width
- Required Strength
- Chemical Environment
- In-service Temperatures
- Temperatures at Time of Application
- Intensity of Sun and Weather In Service
- Longevity
- General Climate at Application
- Materials Cost: Initial and Lifetime
- Installation Cost
- Other
  Fungicides
  Radiation Resistance
  Insulating or Conductive Requirements
  Color
  Intrusion or Abrasion Resistance
  Cure Rate
  Below-Grade or Continuous Water Immersion
  Accessibility of Joint
  Priming
  Special Cleaning Requirements
  Dryness
  Other Restrictions

substrates, recovery from stress, and tear resistance. Significant production properties are: cure rate, low temperature flow characteristics, paint-over ability, color, self-leveling properties, non-bubbling properties, and cost. Aging properties of concern include: resistance to ultraviolet radiation; low and high temperature mechanical properties; and resistance to hydrolysis, thermal aging, and oxidation.

1.5.2 Adhesion properties

Adhesion is an important factor in determining a sealant's performance. The same rules of adhesion that apply to adhesives also apply to sealants. Adhesion is primarily affected by the physio-chemical interaction between the sealant material and the surface to which it is applied. However, in certain joints where there is great movement, strong adhesion of a sealant to a specific substrate may not be desirable. In these situations, the adhesive strength is stronger than the cohesive strength of the sealant, and the sealant may tear apart when it expands or contracts. This requires that the sealant be applied so that it does not adhere to all surfaces. To achieve this affect, a bond-breaker or release material at the bottom of the joint is generally used, as shown in Fig. 1.9.

Conditions that will influence the adhesion of sealants include water exposure, temperature extremes, movement considerations, and surface cleanliness. Often a surface conditioning process or a priming step is necessary to make a substrate compatible with a specific sealant.

1.5.3 Design considerations

When working with sealants, concerns such as crack bridging, coverage rates, color, practicality of placement, order of placement, unusual movement conditions, and aesthetics must be addressed. One consideration that is required of sealants and not generally with adhesives is appearance. A sealant material may be acceptable in all respects, but appearance problems could make it aesthetically unacceptable. Usually, sealants are easily visible whether the application is in the automotive, construction, or appliance industries. Adhesives, on the other hand, are often hidden by the substrates. The sealant material could also contain compounds that discolor surrounding areas. They

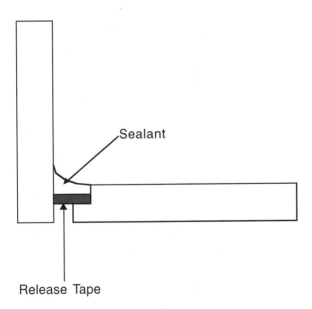

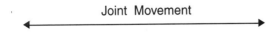

Figure 1.9 Corner sealant joint showing release tape.[11]

can also be incorrectly applied so that the flow of liquids in contact with the sealant results in a residual buildup of extraneous matter at the joint.

1.5.4 Chemical effects

Sealants can also have a chemical effect on the substrate. Chemical incompatibility could cause the sealant or substrate to soften, harden, crack, craze, inhibit cure, or cause other changes. An example of this would be the use of an acid release sealant (such as a silicone sealant that releases acidic acid on cure) on a surface like concrete, marble, or limestone. On these surfaces, an acid/base reaction can cause the formation of bond breaking salts at the bond-line.

Another example of chemical incompatibility is the bleed of plasticizers or other low molecular weight volatiles through sealants, causing them to discolor after exposure to sunlight. This happens frequently when sealants or coatings are applied over asphalt or organic rubber-based materials that are formulated with low molecular weight plasticizers.

1.5.5 Production considerations

An important consideration for any sealing operation is the relative ease of handling and applying the sealant. There are wide ranges of sealants available with varying degrees of application difficulty. There are single and two component sealants, primer and primerless sealant systems, hot melt application systems, preformed sealant tapes and sealants containing solvents. As with adhesives, the time required for the sealant to harden from a liquid state into a semi-solid with some degree of handling strength is very important.

1.6 Basic Mechanisms

1.6.1 General requirements for all adhesives and sealants

If one looks at the adhesive bonding or sealing "process" as a complete procedure, encompassing all aspects of material selection, joint design, production, etc., then the basic requirements are the same no matter what the application. These universal requirements for successful application are:

1. Cleanliness of the substrate surface
2. Wetting of the substrate surface (intimate contact of the adhesive or sealant on the substrate)
3. Solidification of the adhesive or sealant

4. Forming a "joint" structure (adhesive or sealant material, interphase regions, and adherends) that is resistant to the operating stress and environment
5. Design of the joint
6. Selection and control of materials and manufacturing processes.

1.6.1.1 Surface condition. Above all else, one must start with a clean surface. Foreign materials such as dirt, oil, moisture, and weak oxide layers must be removed from the substrate surface, or else the adhesive or sealant will bond to these weak boundary layers rather than to the substrate in question.

Various surface preparations remove or strengthen the weak boundary layer. These treatments generally involve physical or chemical processes or a combination of both. The choice of surface preparation process will depend on the adhesive or sealant, the substrate, the nature of the substrate before bonding, the required bond strength and durability, and the production processes, time, and budget available to the user. Surface preparation methods for specific substrates will be discussed in a later chapter.

1.6.1.2 Wetting the substrate. Initially, the adhesive or sealant must be either a liquid or a readily deformed solid so that it can be easily applied and formed to the required geometry within the assembled joint. It is necessary for the adhesive or sealant to flow and then conform to the surfaces of the adherends on both micro- and macroscales. Small air pockets caused by the roughness of the substrate at the interface must be easily displaced with adhesive or sealant. While it is in the liquid state, the material must "wet" the substrate surface. The term *wetting* refers to a liquid spreading over and intimately contacting a solid surface as shown in Fig. 1.10. The causes of good and poor wetting will be explained in the following chapter. One result of good wetting is greater contact area between adherend and adhesive over which the forces of adhesion can act.

1.6.1.3 Solidification of the adhesive or sealant. The liquid adhesive or sealant, once applied, must be converted into a solid. Solidification occurs in one of three ways: chemical reaction by any combination of heat, pressure, curing agent, or other activator such as UV light, radiation, etc.; cooling from a molten liquid to a solid; and drying due to solvent evaporation. The method by which solidification occurs depends on the choice of adhesive or sealant material.

When organic resins solidify, they undergo volumetric shrinkage due to the crosslinking reaction, loss of solvent, or thermal expansion coefficient (contraction on cooling from an elevated temperature cure).

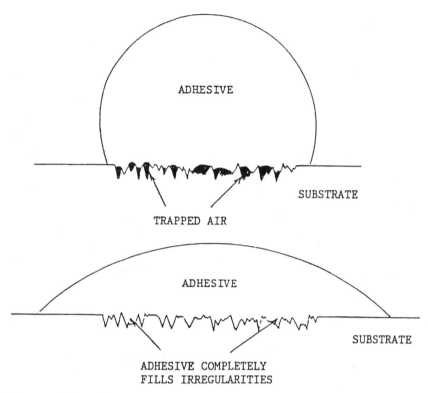

Figure 1.10 Illustration of poor (top) and good (bottom) wetting by an adhesive spreading over a surface.

In the case of adhesives or sealants, it is important that the material does not shrink excessively. Otherwise, undesirable internal stresses could develop in the joint.

1.6.1.4 Forming an impervious joint. Once solidified, the adhesive or sealant must have adequate strength and toughness to resist failure under all expected service conditions. To determine the effect of the environment on the performance of the joint, one must consider the adhesive or sealant material, the substrate, and the interphase regions that are formed before, during, and after the bonding process.

The initial performance and the durability of the joint are strongly dependent on how the substrates are prepared and on the severity of the service environment. The structure and chemistry of the surface region of the parts to be joined and their response to service environments may well govern bond performance. If these surface regions change significantly during processing of the joint or during service life, then the resulting performance characteristics of the joint may also change.

1.6.1.5 Joint design. The adhesive or sealant joint should be designed to take advantage of the high shear and tensile strength properties of most materials and to spread the resulting load over as great an area as possible. Such design considerations will be discussed in the following chapters. Although adequate adhesive-bonded assemblies have been made from joints designed for mechanical fastening, the maximum benefits of the adhesive can be obtained only in assemblies specifically designed for adhesive bonding.

1.6.1.6 Selection and control of materials and manufacturing processes. When determining which adhesives are suitable candidates for an application, a number of important considerations must be taken into account. The factors most likely to influence adhesive selection were listed in Table 1.1. With regard to these controlling factors, the many adhesives available can usually be narrowed to a few candidates that are most likely to be successful.

The appropriate manufacturing processes must then be chosen to provide consistent, high strength joints within the allotted time and cost. The exact manufacturing process will depend on many factors including the choice of adhesive or sealant and the availability of equipment. However, once chosen, the manufacturing process must be rigidly controlled with regard to the incoming adhesive or sealant materials and with regard to the incoming substrate materials. A change in processing parameters could change the degree of stress in the joint, or even the chemical nature of the interphase regions.

Should the user decide to change substrate suppliers, he or she should re-verify completely the entire bonding processes. An example is the case of a vulcanized elastomeric substrate such as neoprene. There are many formulations that an elastomer supplier can use to meet a material specification. However, the formulations may contain compounds (e.g., low molecular weight extenders, plasticizers, etc.) that drastically reduce the adhesion of any material to the surface.

1.6.2 Mechanism of bond failure

As there are general similarities regarding the development of successful adhesive and sealant joints, there are similarities regarding the nature of adhesive or sealant failure. Joints may fail in adhesion or cohesion or by some combination of the two. Adhesive failure is an interfacial bond failure between the adhesive and adherend. Cohesive failure occurs when the failure is such that a layer of adhesive or sealant remains on the adherend. When the adherend fails before the adhesive, it is known as a cohesive failure of the adherend. The various modes of possible bond failures were shown in Fig. 1.1.

Cohesive failure within the adhesive or one of the adherends is sometimes considered a preferred type of failure, because the maximum strength of the material in the joint has been reached. However, failure mode should not be used as a sole criterion for a useful joint. A cohesive type of failure does not necessarily insure a practical or economic assembly or one that will even survive the expected life. The function of the joint is a more important criterion than the mode of joint failure. However, an analysis of failure mode can be an extremely useful guide to determine if the failure was due to a weak boundary layer or improper surface preparation (see Table 1.11).

An inference to the nature of the failure can be made by examining the failure mode. For example, if the adhesive failure is interfacial, the bulk cohesive strength of the adhesive material can be assumed to be greater than the intermolecular strength of adhesion. If the overall joint strength is not sufficient, the user needs to address the "weakest link"—probably the surface condition of the substrate. Improving cohesive strength of the adhesive or sealant would not add to the quality of the joint in this case.

If on the other hand, the failure exhibits a cohesive mode (either the adhesive or the adherend), then the bond strength is stronger than the forces holding the bulk together. To improve the characteristics of this joint, one must look at the bulk adhesive or sealant material. In practice, however, one usually cannot improve the adhesive and keep the interphase area untouched. These interrelated effects make it appear as if the science of adhesion is primarily one of trial and error.

The exact cause of a premature adhesive failure is very hard to determine, because so many factors in adhesive bonding are interrelated. However, there are certain common factors at work when an adhesive bond is made that contribute to the weakening of all bonds. The influences of these factors are qualitatively summarized in Fig. 1.11. This is a very useful illustration showing why one can never

TABLE 1.11 Failure Mode as an Inference to Bond Quality

Failure mode	Inference
Adhesive failure (interfacial)	Cohesive strength > interfacial strength
Cohesive failure (bulk)	Interfacial strength > cohesive strength
Adhesives/cohesive (mixed failure mode)	Interfacial strength ≈ cohesive strength

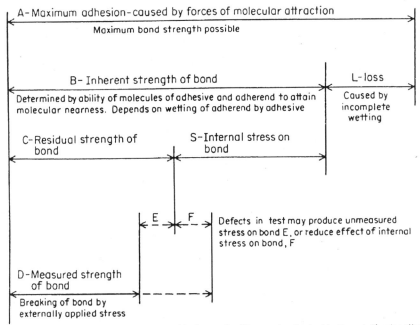

Figure 1.11 Relations between the forces involved in adhesion.[12]

achieve in practice theoretical adhesion values. It also shows clearly the factors that contribute to the reduction of adhesive strength.

If the adhesive does not wet the surface of the substrate, the maximum possible joint strength will be degraded. Internal stresses occur in the adhesive joint during production because of the different physical characteristics of the adhesive and substrate. For example, the coefficient of thermal expansion of adhesive and substrate should be as close as possible to limit stresses that develop during thermal cycling or after cooling from an elevated-temperature cure. Adhesives can be formulated with various fillers to modify their thermal-expansion characteristics and limit such internal stresses. A relatively elastic adhesive, capable of accommodating internal stress, may also be useful when thermal-expansion differences are of concern.

Once an adhesive bond is made and placed in service, other forces are at work weakening the bond. The type of stress, its orientation to the adhesive, and the rate of loading are important. The external stress could either reduce the measured bond strength further, or it could actually increase the measured bond strength by counteracting degrading internal stresses in the joint as shown in Fig. 1.11.

1.7 General Materials and Processes

This section will describe the most common materials and processes used for adhesive bonding and sealing. They are described here in the most general and functional of terms and then will be addressed again in more detail in later chapters.

1.7.1 Materials used for adhesives and sealants

Solid surfaces brought into intimate contact rarely stick to one another, but any liquid placed between them will cause some degree of adhesion. Therefore, almost anything could be used as an adhesive or sealant. However, the nature of the substrate, its surface chemistry, the processing method, and the type of the load and service environment will dictate what material is best to use. The material scientists have developed many substances of high molecular weight which give good adhesion and sealing ability to a variety of substrates.

Although adhesives and sealants are often formulated from the same types of base materials, they are usually "engineered" to have different properties. Adhesives and sealants are highly formulated materials with many components. They generally have an organic base, although there are some mineral-based adhesives that perform very well in certain applications. Table 1.12 shows the origin (natural or synthetic), basic type, and chemical families of common types of adhesive materials. The modern organic materials are all high polymers: long chains of carbon atoms that can form three-dimensional networks. The oldest adhesives and sealants, many that are still used today, are of natural origin. The most common of these are animal glues, starches, and tar or pitch. Naturally occurring adhesives and sealants also include dextrin, asphalt, vegetable proteins, natural rubber, and shellac.

For a material to be a potential adhesive or sealant, it must meet three basic criteria. First, the material has, at some stage, to be in a liquid form so that it can readily spread over and make intimate contact with the substrate. Second, the material must be capable of hardening into a solid to withstand and distribute stress. Lastly, the material must resist the environments that it will see during processing and during service.

Modern polymers are ideal materials to use as adhesives and sealants because they can be applied as low viscosity liquids and, then by various means, hardened into a strong material with relatively good resistance to stress and various environments. This hardening may occur through loss of water or solvent, cooling from the molten state,

TABLE 1.12 Origin, Basic Type, and Chemical Family of Common Adhesives[6]

Origin	Basic type	Family	Examples
Natural	Animal	Albumin Animal glue Casein Shellac Beeswax	
	Vegetable	Natural resins	Gum arabic, tragacanth, colophony, Canada balsam,
		Oils and waxes	Carnauba wax, linseed oil
		Proteins	Soybean
		Carbohydrates	Starch, dextrins
	Mineral	Inorganic minerals	Silicates, magnesia, phosphates, litharge, sulfur
		Mineral waxes	Paraffin
		Mineral resins	Amber
		Bitumen	Asphalt
Synthetic	Elastomers	Natural rubber	Natural rubber and derivatives
		Synthetic rubber	Butyl, polyisobutylene, polybutadiene blends, polyisoprenes, polychloroprene, polyurethane, silicone, polysulfide, polyolefins
		Reclaimed rubber	
	Thermoplastic	Cellulose derivatives	Acetate, acetate-butyrate, caprate, nitrate, methyl cellulose, hydroxyl ethyl cellulose, ethyl cellulose, carboxy methyl cellulose
		Vinyl polymers and copolymers	Polyvinyl acetate, polyvinyl alcohol, polyvinyl chloride, polyvinylidene chloride
		Polyesters (saturated)	Polystyrene, polyamides
		Polyacrylates	Methacrylate and acrylate polymers, cyanoacrylates
		Polyethers	Polyhydroxy ether, polyphenolic ethers
		Polysulfones	
	Thermosetting	Amino plastics	Urea and melamine formaldehydes
		Epoxies	Epoxy polyamide, epoxy bitumen, epoxy polysulfide, epoxy nylon
		Phenolic resins and modification	Phenol and resorcinol formaldehydes, phenolic-nitrile, phenolic-neoprene, phenolic-epoxy
		Polyesters (unsaturated)	
		Polyaromatics	Polyimide, polybenzimidazole, polyphenylene
		Furanes	Phenol furfural

chemical reaction by crosslinking or curing between the molecular chains, or by polymerizing from the monomer state. Table 1.13 offers examples of common adhesives and the changes that are necessary, after application, for them to solidify and become effective.

Modern synthetic organic based adhesives and sealants will be the primary topic of this Handbook. However, natural based and mineral based adhesives and sealants will also be included because they have wide use in certain applications. Adhesion occurring via metallic processes (i.e., welding, soldering, brazing, etc.) will not be included other than to discuss these as alternative joining methods. Metallic joining technologies are thoroughly described in other Handbooks.

1.7.2 Manufacturing processes for adhesives and sealants

Modern adhesives are often a complex formulation of components that perform specialty functions. The formulation of raw materials into

TABLE 1.13 Methods by Which Common Adhesives Harden[13]

Type of adhesive	Applied form	Change necessary to become effective
Carpenter's glue	Hot aqueous liquid	Loss of water into substrate (wood) and cooling
Polyvinyl acetate (white glue)	Aqueous emulsion	Loss of water into porous substrate
Hot-melt adhesive	Hot viscous liquid	Immediate cooling on contact with surfaces
Contact adhesives	Solution in organic solvent	Allowed to lose solvent until tacky, then surfaces combined
Anaerobic adhesives	Monomer of low viscosity	Polymerizes in joint when oxygen is excluded
Cyanoacrylate	Monomer with low viscosity	Polymerizes in joint with trace of moisture on surface in presence of metal ions
Urea-formaldehyde	Mixed with water immediately before use	Loss of water into substrate and setting owing to release of catalyst by water
Epoxy adhesives	Mixed with curing agent before use; applied as viscous liquid; some single component epoxies can be applied as film to set under heat and pressure	Chemical reaction either at room or elevated temperatures
Polyvinylformal with phenol formaldehyde resin	Liquid with powder or, more usually, a film with the powder set in a flexible matrix	Set with heat and pressure. The pressure is essential to prevent porosity from a small amount of water liberated on cure

serviceable adhesive bonding and sealing systems is itself a broad field of technology. Adhesives and sealants can be produced in various forms: one and two part liquids, solvent based solutions, water based emulsion, supported or unsupported film, preformed pellets or shaped extrusions, and numerous other forms. This variety of formulation possibilities and end-use forms are indicative of the advanced state of development of adhesives and sealants.

These products are generally developed and prepared for sale by formulators that range from very small operations to large international businesses. An adhesive or sealant formulation will depend on the base material that is the principal part of the formulation and on the requirements of the application. Processes and formulations are often considered proprietary to the adhesive or sealant manufacturer. However, there are formulas and processing methods that are public knowledge due to the tremendous amount of published research on adhesives and sealants. In general, an adhesive or sealant formulation consists of the following components.

The adhesive *base* or *binder* is the primary component of an adhesive that holds the substrates together. The binder is, generally, the component from which the name of the adhesive is derived. For example, an epoxy adhesive may have many components, but the primary material is epoxy resin.

A *hardener* is a substance added to an adhesive to promote the curing reaction by taking part in it. Two-part adhesive systems generally have one part, which is the base, and a second part, which is the hardener. Upon mixing, a chemical reaction ensues which causes the adhesive to solidify. A *catalyst* is sometimes incorporated into an adhesive formulation to speed-up the reaction between the base and hardener. Certain adhesive bases only need a source of energy to cure. This source may come from ultraviolet light, radiation, etc. In these cases, the adhesive may contain a catalyst but no hardener.

Solvents are sometimes needed to disperse the adhesive to a spreadable consistency. Solvents used with synthetic resins and elastomers are generally organic in nature, and often a mixture of solvents is required to achieve the desired properties. There must be some way for the solvent to escape the joint after the adhesive is applied and the assembly is made. Thus, solvents are generally only used in applications where passages are available for their escape, such as in the preparation of pressure sensitive coatings or bonding of porous substrates.

An ingredient added to an adhesive to reduce the concentration of base material is called a *diluent*. Diluents are principally used to lower the viscosity and modify the processing conditions of some adhesives and sealants. Reactive diluents chemically react with the base material during cure, become part of the product, and do not evaporate as does a solvent. Non-reactive diluents are much like solvents and could leach out of the adhesive or sealant during its life.

Fillers are relatively non-adhesive substances added to the adhesive or sealant to improve their working properties, strength, permanence, or other qualities. Fillers are also used to reduce material cost. By selective use of fillers, the properties of an adhesive or sealant can be changed tremendously. Thermal expansion, electrical and thermal conduction, shrinkage, viscosity, and thermal resistance are only a few properties that can be modified by the use of fillers.

A *carrier* or *reinforcement* is usually a thin fabric or paper used to support the semi-cured adhesive composition to provide a tape or film. The carrier may also serve as a bond-line spacer and reinforcement for the adhesive.

1.7.3 End-use processes for adhesives and sealants

It must be realized that the adhesive or sealant itself is only part of the joint and somewhat surprisingly does not play a singularly important role in determining the success of the joint. The manufacturing methods used in producing an assembly will also determine the initial degree of adhesion and service characteristics of the joint. A typical flow chart for the adhesive bonding or sealing process is shown in Fig. 1.12. It must be realized, however, that the decisions made in one process segment may affect the decision in other process segments. For example, the choice of substrate will most definitely affect the method of surface preparation used. The choice of adhesive or sealant could also affect the degree of surface preparation required. (Certain adhesives are less sensitive to contaminated or low energy surfaces than others.) The processing equipment, time required, etc. will affect the type of adhesive or sealant that can be chosen and may also affect the type of substrate that can be considered for the application. For example, certain thermoplastics are used in the auto industry because they can be thermally welded in a few seconds—a necessity in a large volume production operation. Thus, the general processes involved in adhesive bonding or sealing are interrelated and often dependent on one another. The user must find the proper processing methods for the specific application and consistently use them to ensure acceptable results.

Figure 1.12 Basic steps in the adhesive bonding or sealing process.

The joining of surfaces with an adhesive or sealant consists of a series of individual operations, each of which must be done properly to achieve the desired result. The number of operations and the method of actually making the bond will depend on several factors. The factors that must be considered are the type of materials to be bonded, the nature of the assembly design, the adhesive to be used, the facilities available, the time allotted (e.g., number of joints per minute) and the cost allocated. Details of these production processes are given in later chapters.

Adhesive bonding is somewhat unique in that there are numerous processes available. Certain methods even use the substrate itself as an adhesive. These are known as solvent welding and thermal welding. Heat or solvent can be applied in some manner to the substrate to turn part of the material into a liquid. This liquefied material then acts as the adhesive or sealant, filling in the gap at the joint and solidifying on cooling from the melt or by loss of solvent. The substrate in essence becomes the adhesive. These processes will be considered in this Handbook as a special type of adhesive bonding although they are popularly referred to as welding methods.

Many of the adhesive or sealant problems that evolve are not due to a poor choice of material or joint design but are directly related to faulty production techniques. The user must obtain the proper processing instructions from the manufacturer and follow them consistently to ensure acceptable results.

1.8 Sources of Information

Information and literature available from the manufacturers of materials can provide useful assistance regarding specific adhesive or sealant applications. These manufacturers may be classified as:

- Base raw materials suppliers to the adhesives and sealants industry
- Formulators or adhesive and sealant suppliers
- Manufacturers of raw materials that are used as substrates (i.e., aluminum, steel, plastic, composites, etc.)

These manufacturers are usually very happy to provide information and guidance related to their products. Often, they will offer to the end-user the utilization of their research and development groups and/or test laboratories to provide suggestions, develop specific formulations, or look at problems and other issues that may arise. The use of these resources is highly recommended.

Specifications and standards also provide a significant amount of information regarding adhesives and sealants. These documents describe materials, processes, and test methods. They are identified in Chapter 4 and in Appendix C.

Other helpful sources of information are identified in the reference sections at the end of each chapter. In Appendix B, the reader will find a guide to information sources categorized by: literature (books, journals, and periodicals); professional societies; specifications and standards; databases; and Internet sources.

References

1. Koehn, G. W., "Design Manual on Adhesives", *Machine Design*, April 1954.
2. Harshorn, S. R., "Introduction", Chapter 1, *Structural Adhesives: Chemistry and Technology* (New York: Plenum Press, 1986).
3. Prane, J. W., "Newly Revised Rauch Guide Remains Comprehensive Source", *Adhesives Age*, Sept. 1996; originally from "The Rauch Guide to the U.S. Adhesives and Sealants Industry, 1995–1996 Edition", Impact Marketing Consultants, Inc., Manchester Center, VT.
4. Elias, M. G., et. al., "Sealants Markets and Applications", *Adhesives and Sealants*, vol. 3 of Engineered Materials Handbook, ASM International, 1990.
5. Nielsen, P. O., "Selecting An Adhesive: Why and How", Chapter 5, *Adhesives in Manufacturing*, G. L. Schneberger, ed. (New York: Marcel Dekker, Inc., 1983).
6. Schields, J., Chapter 1, *Adhesives Handbook*, 3rd ed. (London: Butterworths, 1984).
7. Technical Bulletin, *Structural Adhesives*, 3M Company, Adhesives, Coatings, and Sealers Division.
8. Powis, C. N., "Some Applications of Structural Adhesives", in *Aspects of Adhesion*, vol. 4, D. J. Alner, ed. (London, University of London Press, 1968).
9. Cagle, C. V., *Adhesive Bonding Techniques and Applications* (New York: McGraw-Hill, 1968) at 11.
10. Panek, J. R., and Cook, J. P., Chapter 2, *Construction Adhesives and Sealants* (New York: John Wiley & Sons, Inc., 1991).
11. Dunn, D. J., "Sealants and Sealant Technology", in *Adhesives and Sealants*, vol. 3, Engineered Materials Handbook, ASM International, 1990.
12. Reinhart, F. W., "Survey of Adhesion and Types of Bonds Involved", in *Adhesion and Adhesives Fundamentals and Practices*, J. E. Rutzler and R. L. Savage, eds. (London: Soc. of Chemical Industry, 1954).
13. Wake, W. C., "Sticking Together to Last", *Industrial Research/Development*, August 1979.

Chapter 2

Theories of Adhesion

2.1 Introduction

There is no unifying single theory of adhesion on which to accurately model all interactions that take place between the adhesive and the adherend. The existing theories of adhesion presented in this chapter provide methods by which one can rationalize practical observations. They are generally useful in understanding why adhesives stick and why, at times, they fail. Adhesion theories allow us to make predictions and even obtain a qualitative realization of joint strength.

There are several theories of adhesion that have endured the test of time. Each is applicable in certain circumstances, but none are universally applicable. By being familiar with these theories, one can develop a knowledge base and an awareness of how adhesives and sealants work in practical situations.

2.2 Forces Involved in Adhesion

The forces involved in holding adhesives and sealants to their substrates or in holding adhesives and sealants together as a bulk material arise from the same origins. These same forces are all around us in nature. To understand what is happening in an adhesive or sealant joint, we must first understand the forces that bind atoms and molecules together. Although there are many kinds of forces, it is mainly those of a physical and chemical nature that are important in understanding the development of adhesive and sealant joints.

2.2.1 Adhesive and cohesive forces

Bond strength is not only the result of adhesion forces. Other forces contribute to the strength of a joint. For example, molasses may have good adhesion, but it is a poor adhesive or sealant. Its failure is usually cohesive. Cohesive strength of an adhesive or sealant is at least as important as its adhesive strength. Like a weak link in a chain, the bond will fail at the place where the intermolecular forces are the weakest.

Adhesive forces hold two materials together at their surfaces. Cohesive forces hold adjacent molecules of a single material together. Adhesive or sealant joints may fail either adhesively or cohesively. Adhesive failure is failure at the interface between adherend and the adhesive. An example would be the peeling of cellophane tape from a glass surface if the adhesive film separates cleanly from the glass. Cohesive failure is failure within the adhesive or one of the adherends. Cohesive failure would result if two metal substrates held together with grease were pulled apart. The grease would be found on the two substrates after the joint failed. The grease would have failed cohesively. Another example of cohesive failure is if two wooden panels were bonded together with an epoxy adhesive and then pulled apart. Most likely, the resulting failure would show that particles of wood fiber were left embedded in the adhesive. In this case, the wood or adherend failed cohesively.

Both adhesive and cohesive forces are the result of forces existing between atoms or molecules. These forces are the result of unlike charge attractions between molecules. The positive portion of one molecule attracts the negative portion of adjacent molecules. The more positive or negative the charged sites and the closer together the molecules, the greater will be the forces of attraction.

Adhesive or cohesive forces can be attributed to either short or long range molecular interactions. These are also referred to as primary or secondary bonds. Table 2.1 characterizes these forces. The exact types of forces that could be operating at the interface are generally thought to be the following:

- van der Waals forces (physical adsorption)
- Hydrogen bonding (strong polar attraction)
- Ionic, covalent, or co-ordination bonds (chemisorption)

Short-range molecular interactions include covalent, ionic, and metallic forces. Covalent forces result from chemical reactions such as provided by some surface treatments on glass fiber. Welding, soldering, or brazing processes form metallic bonds. However, these forces generally are not at work in the more common, everyday adhesive applications. The most important forces relative to adhesion are the sec-

TABLE 2.1 Forces at the Interface or Within the Bulk of a Material

Type of force	Source of force	Bond energy (KJ/mol)	Description
Primary or Short Range Forces	Covalent forces	60–700	Diamond or cross-linked polymers. Highly directional.
	Ionic or electrostatic	600–1000	Crystals. Less directional than covalent.
	Metallic	100–350	Forces in welded joints.
Secondary or van der Waals Forces	Dispersion	0.1–40	Arise from interactions between temporary dipoles. Accounts for 75–100% of molecular cohesion. Forces fall off as the 6th power of the distance.
	Polar	4–20	Arise from the interactions of permanent dipoles. Decrease with the 3rd power of the distance.
	Hydrogen bonding	Up to 40	Results from sharing of proton between two atoms possessing loan pairs of electrons. Longer range than most polar and dispersion bonds.

ondary or van der Waals forces. The exact nature of these forces and their influence on adhesive or cohesive strength are difficult to accurately determine. However, a general awareness of their origin and characteristics assists in understanding why strong bonds form and why they fail.

2.2.2 The concept of surface energy

The forces holding an adhesive to a substrate or maintaining the cohesive integrity of a solid can be measured as the work necessary to separate two surfaces beyond the range of the forces holding them together. In one case, the surfaces are the adhesive and substrate; in the other, they are like-molecules in the bulk of the material. This force is dependent on the intermolecular forces that exist in the material and upon the intermolecular spacing. It is sometimes referred to as the surface energy, γ (gamma).

The certainty that liquids have a surface energy is easily demonstrated by the fact that a finely divided liquid, when suspended in another medium, assumes a spherical shape. In the absence of gravitational distortion of shape (i.e., the energy associated with having a surface), the liquid tends to go to its lowest energy state — that of a sphere. The surface energy of a pure liquid is easily obtained because this is simply its surface tension, γ_{LV}.

Surface tension and surface energy are numerically identical for liquids. Surface energy is generally given in units of millijoules per meter squared (mJ/m^2), while surface tension is given in units of dynes/centimeter (dynes/cm) or Newtons per meter (N/m). The surface tensions of organic liquids and of most inorganic liquids rarely exceed the value for water (32 dynes/cm).

The surface energies of liquids are readily determined by measuring the surface tension with a duNouy ring[1] or Wilhelmy[2] plate as shown in Fig. 2.1. With the duNouy ring, a clean platinum ring is placed under the surface of the test liquid and the liquid is slowly moved downward until the ring breaks through the liquid surface. The force is recorded, and by means of appropriate conversion factors, the surface tension of the liquid is calculated. The Wilhelmy plate is a similar method which measures the force of a liquid on a plate passing through its surface.

Another method of measuring surface tension is the "drop weight/drop volume" method.[3] Here, the average volume of test liquid to cause a drop to fall from a carefully calibrated syringe is used to calculate the surface tension of the liquid.

Whereas, the surface tension of a liquid is a real surface stress, the same cannot be said of a solid surface. With a solid, work is done in stretching a surface and not in forming the surface. For a solid surface,

Wilhelmy Plate

duNouy Ring

Figure 2.1 Wilhelmy plate and du Nouy ring methods of measuring surface tension of a liquid. (Courtesy: Krüss USA).

surface energy and surface tension are not the same. Still it is often convenient to refer to γ indiscriminately as either surface energy or surface tension, but it is inaccurate because the "tension" in the surface of the solid is greater than the surface energy. It is an easy matter to measure the surface tension of a liquid in equilibrium with its vapor, γ_{LV}, but not to measure the surface energy of a solid. Measurements on high energy solids are mostly made near the melting point; whereas, it is the room temperature properties that mainly concern adhesive studies. Surface free energies of low surface energy solids (i.e., polymeric materials) have been indirectly estimated through contact angle measurement methods as explained below.

In a contact angle measurement, a drop of liquid is placed upon the surface of a solid. It is assumed that the liquid does not react with the solid and that the solid surface is perfectly smooth and rigid. The drop is allowed to flow and equilibrate with the surface. The measurement of the contact angle, θ (theta), is usually done with a goniometer that is simply a protractor mounted inside a telescope. The angle that the drop makes with the surface is measured carefully. A diagram of the contact angle measurement is shown in Fig. 2.2.

A force balance between the liquid and the solid can be written as:

$$\gamma_{LV} \cos \theta = \gamma_{SV} - \gamma_{SL}$$

where γ_{LV} is the liquid-vapor interfacial tension, θ is the contact angle, γ_{SV} is the solid-vapor interfacial tension, and γ_{SL} is the solid-liquid interfacial tension. This is known as the Young equation after the scientist who originated the analysis.[4] The γ_{SV} is the solid-vapor interfacial energy and not the true surface energy of the solid. The surface energy is related to γ_{SV} through the following relationship:

$$\gamma_{SV} = \gamma - \pi_e$$

where γ is the true surface energy of the solid and π_e is a quantity known as the equilibrium spreading pressure. It is a measure of the energy released through adsorption of the vapor onto the surface of the solid, thus lowering the surface energy.

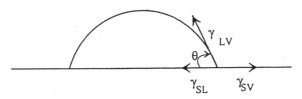

Figure 2.2 Schematic diagram of the contact angle and its surface free energy (tension) components.

A rather simple method of estimating the surface energy of solids was developed by Zisman.[5] Zisman proposed that a *critical surface tension*, γ_C, can be estimated by measuring the contact angle of a series of liquids with known surface tensions on the surface of interest. These contact angles are plotted as a function of the γ_{LV} of the test liquid. The critical surface tension is defined as the intercept of the horizontal line $\cos \theta = 1$ with the extrapolated straight line plot of $\cos \theta$ against γ_{LV} as shown in Fig. 2.3. This intersection is the point where the contact angle is 0 degrees. A hypothetical test liquid having this γ_{LV} would just spread over the substrate.

The critical surface tension value for most inorganic solids is in the hundreds or thousands of dynes/cm, and for polymers and organic liquids, is at least an order of magnitude lower than that of inorganic solids. Values of critical surface tensions for common solids and surface tensions of common liquids are shown in Table 2.2. Critical surface tension is an important concept that leads to a better understanding of wetting. This will be discussed in coming sections.

2.2.3 Work of adhesion and cohesion

If a bulk material is subjected to a sufficient tensile force, the material will break thereby creating two new surfaces. If the material is completely brittle, the work done on the sample is dissipated only in creating the new surface. Under those assumptions, if the failure is truly cohesive where both sides of the broken material are of the same composition, then

$$W_C = 2\gamma$$

where W_C is defined as the work of cohesion.

Now similarly consider separating an adhesive (material 1) from a substrate (material 2). The energy expended should be the sum of the two surface energies γ_1 and γ_2. However, because the two materials were in contact, there were intermolecular forces present before the materials were split apart. This interfacial energy can be represented as γ_{12}. W_A, the work of adhesion, may be defined by the surface energies of the adhesive and the adherend:

$$W_A = \gamma_1 + \gamma_2 - \gamma_{12}$$

This is the classical Dupre equation,[6] which was developed in 1869. This equation could also be represented as:

$$W_A = \gamma_{LV} + \gamma_{SV} - \gamma_{SL}$$

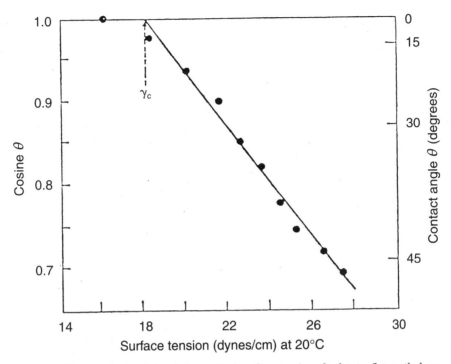

Figure 2.3 Zisman plot for determining critical surface tension of polytetrafluoroethylene. Test liquids are n-alkanes.[5]

Substitution of the Young equation into the Dupre equation results in the Young-Dupre equation that states:

$$W_A = \gamma_{LV}(1 + \cos \theta).$$

This equation relates a thermodynamic parameter, W_A, to two easily determinable quantities: the contact angle and the liquid-vapor surface tension. For conditions of perfect wetting ($\cos \theta = 1$):

$$W_A = 2\gamma_{LV} = W_C$$

2.2.4 Bond failure energy

Bond failure energy is composed of two parts: a reversible work of adhesion and an irreversible work of adhesive deformation. Thus, the strength of styrene-butadiene rubber adhesive depends on two components: a viscoelastic energy dissipation term, which is a function of test rate and temperature, and the intrinsic failure energy which agrees closely with the work of adhesion, W_A, when bond failure is apparently interfacial.

TABLE 2.2 Critical Surface Tensions for Common Solids and Surface Tensions for Common Liquids

Materials	Critical surface tension, dyne/cm
Acetal	47
Acrylonitrile-butadiene-styrene	35
Cellulose	45
Epoxy	47
Fluoroethylene propylene	16
Polyamide	46
Polycarbonate	46
Polyethylene	31
Polyethylene terephthalate	43
Polyimide	40
Polymethylmethacrylate	39
Polyphenylene sulfide	38
Polystyrene	33
Polysulfone	41
Polytetrafluoroethylene	18
Polyvinyl chloride	39
Silicone	24
Aluminum	≈500
Copper	≈1000
Material	Surface tension, dyne/cm
Epoxy resin	47
Fluorinated epoxy resin*	33
Glycerol	63
Petroleum lubricating oil	29
Silicone oils	21
Water	73

*Experimental resin; developed to wet low-energy surfaces. (Note low surface tension relative to most plastics.)

Much work in adhesion science has centered on the relationship between W_A, the calculated work of adhesion, and practical adhesion, or the real measured adhesion. Ahagon and Gent[7] indicate that practical adhesion can be related to the work of adhesion plus a function describing the energy dissipation mechanisms within an adhesive bond.

$$\text{Practical Adhesion} = W_A + f(W_A)\zeta.$$

ζ (zeta) is a factor related to the viscoelastic properties of the adhesive and, thereby, is related to the mechanical energy absorption characteristics of the joint. This is sometimes related to the amount of energy absorbed by the deformation of the joint. As shown in Fig. 2.4, the practical work of adhesion is equal to the theoretical work of adhesion as determined by interfacial effects and to the mechanical work which is absorbed within the joint. Thus, with a completely non-deformable

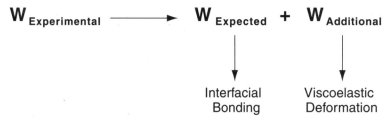

Figure 2.4 The measured work of adhesion is made up of thermodynamic and mechanical components.

adhesive, interphase, and adherend the practical work of adhesion is equal to the theoretical work of adhesion.

It should be realized that the above discussion on forces and work of adhesion is very simplistic and summarizes a great deal to a fault. There are also significant debates over the applicability and direct usefulness of these relationships. However, the following conclusions can be derived and are of significant assistance to the user of adhesives and sealants:

1. The work of adhesion is at a maximum when the contact angle, θ, equals 0 degrees, that is when the liquid spreads completely on the surface of the solid. This condition implies that there are stronger forces between the molecules of the liquid and the substrate than between the liquid molecules themselves.
2. Adhesion will tend to go to zero as the contact angle increases above 90 degrees.
3. Under conditions of perfect wetting of a surface by a liquid, $W_A = 2\gamma_{LV}$. Hence $W_A = W_C$.

These conclusions will be discussed further in the following sections.

2.2.5 Surface attachment theory of joint strength

The preceding discussion looks at adhesive failure and cohesive failure as separate modes. However, in practice they result from the same joint specimen. One can generalize on the influence of the degree of interfacial surface attachment on the adhesive joint strength and on the mode of failure. The degree of interfacial surface attachment may vary due to wetting, boundary layer effects, or other phenomena that influence the degree of adhesion at the interface. The different interfacial states of adhesion are summarized in Fig. 2.5.

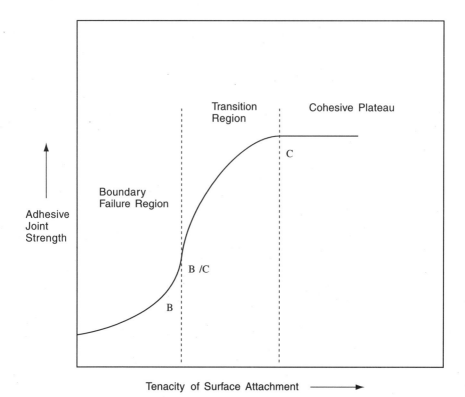

Figure 2.5 Schematic diagram of interfacial states encountered in adhesion. Here B represents boundary failure; C, cohesive failure; and B/C or C/B, mixed failure.[8]

In the boundary failure region, designated by B in Fig. 2.5, the strength of the bond is controlled by physical and mechanical forces active at the interface. In this region, the strength of interfacial contact is much less than the cohesive strength of the adhesive material. At a certain level, the degree of surface attachment begins to influence the strength of the joint. This critical degree of surface attachment may be referred to as the threshold value.

In the transition zone, the strength of the adhesive joint is very sensitive to the degree of surface attachment. In this region, mixed failure is observed. At a critical degree of surface attachment, which can be referred to as a minimal or saturation value, the adhesive joint will rupture, showing total cohesive failure, designated by region C. The attainment of this state of surface attachment would represent the ultimate strength of the joint.

Above this so-called saturation degree of surface attachment, the strength of the joint levels off into a cohesive plateau. Under these conditions further increases in the degree of surface attachment do not result in an increase in the mechanical strength of the adhesive joint.

2.3 Theories of Adhesion

The actual mechanism of adhesive attachment is not explicitly defined. Several theories attempt to describe adhesion. No single theory explains adhesion in a general, comprehensive way. Some theories are more applicable for certain substrates and applications; other theories are more appropriate for different circumstances. Each theory has been subjected to much study, question, and controversy. However, each contains certain concepts and information that are useful in understanding the basic requirements for a good bond.

The most common theories of adhesion are based on adsorption, simple mechanical interlocking, diffusion, electrostatic interaction, and weak-boundary layers.

2.3.1 Adsorption theory

The adsorption theory states that adhesion results from molecular contact between two materials and the surface forces that develop. Adhesion results from the adsorption of adhesive molecules onto the substrate and the resulting attractive forces, usually designated as secondary or van der Waals forces. For these forces to develop, the respective surfaces must not be separated more than five angstroms in distance. Therefore, the adhesive must make intimate, molecular contact with the substrate surface.

The process of establishing continuous contact between an adhesive and the adherend is known as "wetting." Figure 2.6 illustrates good and poor wetting of an adhesive spreading over a surface. Good wetting results when the adhesive flows into the valleys and crevices on the substrate surface; poor wetting results when the adhesive bridges over the valleys formed by these crevices. Obtaining intimate contact of the adhesive with the surface essentially ensures that interfacial flaws are minimized or eliminated. Poor wetting causes less actual area of contact between the adhesive and adherend, and stress regions develop at the small air pockets along the interface. This results in lower overall joint strength.

Wetting can be determined by contact angle measurements. It is governed by the Young equation which relates the equilibrium contact angle, θ, made by the wetting component on the substrate to the appropriate interfacial tensions:

$$\gamma_{LV} \cos \theta = \gamma_{SV} - \gamma_{SL}$$

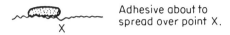
Adhesive about to spread over point X.

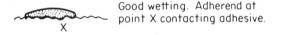
Good wetting. Adherend at point X contacting adhesive.

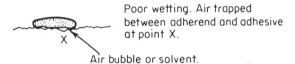
Poor wetting. Air trapped between adherend and adhesive at point X.
Air bubble or solvent.

Figure 2.6 An illustration of good and poor wetting by an adhesive spreading over a surface.[9]

The term γ_{SV} is the interfacial tension of the solid material in equilibrium with a fluid vapor; γ_{LV} is the surface tension of the fluid material in equilibrium with its vapor; and γ_{SL} is the interfacial tension between the solid and liquid materials. Complete, spontaneous wetting occurs when $\theta = 0°$, or the material spreads uniformly over a substrate to form a thin film. A contact angle of $0°$ occurs with a pure water droplet on a clean, glass slide. Therefore, complete spontaneous wetting occurs when cosine $\theta > 1.0$ or when:

$$\gamma_{SV} > \gamma_{SL} + \gamma_{LV}.$$

Wetting is favored when the substrate surface tension, γ_{SV}, or its critical surface energy, γ_C, is high, and the surface tension of the wetting liquid, γ_{LV}, is low (i.e., $\gamma_{C\ substrate} > \gamma_{adhesive}$). Low energy polymers, therefore easily wet high energy substrates such as metals. Conversely, polymeric substrates having low surface energies will not be readily wet by other materials and are useful for applications requiring nonstick, passive surfaces.

Thus, most common adhesive liquids readily wet clean metal surfaces, ceramic surfaces, and many high energy polymeric surfaces. However, common adhesives do not wet low energy surfaces. This explains why organic adhesives, such as epoxies, have excellent adhesion to metals, but offer weak adhesion on many untreated polymeric substrates, such as polyethylene, polypropylene, and the fluorocarbons.

$$\text{For good wetting:} \quad \gamma_{adhesive} \ll \gamma_{C\ substrate}$$

For poor wetting: $\gamma_{\text{adhesive}} \gg \gamma_{C\ \text{substrate}}$

A simple view of the relationship of wetting and adhesion is provided by Fig. 2.7. Here the contact angle of a drop of epoxy adhesive on a variety of surfaces is shown. The surface energy of a typical epoxy resin is about 42 mJ/m^2 (or dynes/cm). The expected bond strengths would increase as the contact angle decreases. Therefore, the bond strength of the epoxy adhesive on the epoxy substrate would be expected to be the greatest, followed by polyvinylchloride, polyethylene, and polytetrafluoroethylene in that order.

Some important concepts develop out of the premise that for good wetting to occur $\gamma_{\text{adhesive}} \ll \gamma_{C\ \text{substrate}}$. You would expect from this relationship that polyethylene and fluorocarbon, if used as adhesives, would provide excellent adhesion to a variety of surfaces including polymers and metals. In fact, they do provide excellent adhesion. However, commercial polyethylene generally has many lower molecular weight constituents that create a weak boundary layer, thus preventing practical adhesion. Fluorocarbons cannot be easily melted or put into solution. Thus, they are difficult to get into a fluid state to wet the surface and then solidify without significant internal stresses. However, polyethylene makes an excellent base for hot melt adhesives once the weak low molecular weight constituents are removed. Researchers are attempting to develop epoxy resins with fluorinated chains so that they can easily wet most surfaces.[11]

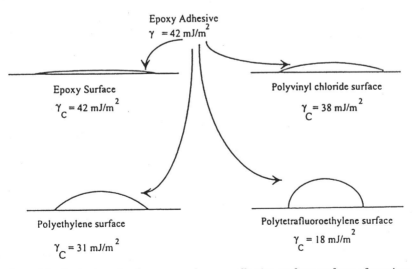

Figure 2.7 Contact angle of an uncured epoxy adhesive on four surfaces of varying critical surface tension. Note that as the critical surface tension of the surfaces decrease, the contact angle increases indicating less wetting of the surface by the epoxy.[10]

It is also easy to see why silicone and fluorocarbon surfaces provide good mold release surfaces. Most resins will not easily wet these surfaces. It is also easy to see why silicone and mineral oil provide weak boundary layers. If a very thin layer of such oil is on the substrate, the adhesive will want to spread over the oil rather than the substrate. Most adhesives would not wet a surface contaminated by these oils. It is also interesting to note that by making a coating (or adhesive) more likely to wet a substrate (by lowering its surface tension), you may be inadvertently making it more difficult for any subsequent coating or adhesive to wet this new material once it is cured. Graffiti resistant paints work in this manner.

After intimate contact is achieved between adhesive and adherend through wetting, it is believed that permanent adhesion results primarily through forces of molecular attraction. Four general types of chemical bonds are recognized as being involved in adhesion and cohesion: electrostatic, covalent, and metallic, which are referred to as *primary bonds*, and van der Waals forces that are referred to as *secondary bonds*.

2.3.2 Mechanical theory

The surface of a solid material is never truly smooth but consists of a maze of peaks and valleys. At one time, adhesion was thought to occur only by the adhesive flowing and filling micro-cavities on the substrate. When the adhesive then hardens, the substrates are held together mechanically. According to the mechanical theory of adhesion, in order to function properly, the adhesive must penetrate the cavities on the surface, displace the trapped air at the interface, and lock-on mechanically to the substrate.

One way that surface roughness aids in adhesion is by increasing the total contact area between the adhesive or sealant and the adherend. If interfacial or intermolecular attraction is the basis for adhesion, increasing the actual area of contact by a large amount will increase the total energy of surface interaction by a proportional amount. Thus, the mechanical interfacing theory generally teaches that roughening of surfaces is beneficial because it gives "teeth" to the substrate, and by virtue of roughening increases the total effective area over which the forces of adhesion can develop. Exceptions to this rule will be described in the following sections.

The mechanical theory also teaches that joint designs that have large bonding areas are better than joint designs having a smaller area. However, there is a point where increasing the joint area, for example the overlap distances of a lap joint, has diminishing returns. The next chapter will discuss this possibility.

Surface roughness generally aids in adhesive bonding by the mechanical interlocking effect. With rough surface preparation as shown schematically in Fig. 2.8a, the adhesive would have to pass through the adherend in order for separation to take place. What generally occurs is that the roughness or micro-overhangs cause the adhesive to plastically deform which in turn absorbs energy. Consequently, the strength of the adhesive joint increases.

Another benefit of mechanical interlocking is that a rough surface will provide a crack propagation barrier. Notice that in Fig. 2.8b as a wedge is driven into the edge of a sharp interface between adherends A and B, little energy dissipation is required to separate the adherends, and a clean separation of adherends is possible. The substrates will simply "unzip". However, if there is surface roughness as shown in Fig. 2.8a, then a tortuous interface between the adhering materials will act as path-breaks between the separating adherends. These excursions in crack propagation dissipate energy and increase the resulting strength of the joint.

Thus, there are many cases where the forces of adhesion and the mechanical interlocking forces are working together in the same joint. In these cases, the practical work of adhesion is equal to the work developed by adhesion mechanisms (van der Waals forces) in addition to the work developed by mechanical mechanisms (elastic deformation).

Mechanical anchoring of the adhesive appears to be a prime factor in bonding many porous substrates. Adhesives also frequently bond better to nonporous abraded surfaces than to natural surfaces. This beneficial effect of surface roughening may be due to:

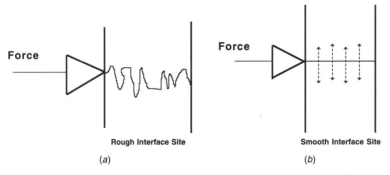

Figure 2.8 Schematics showing (a) tortuous interface between two adhering materials with rough surfaces and (b) two adherends with smooth surfaces. Note that in (a) the applied force cannot cleanly follow the path between the two adherends and as excursions are made energy is dissipated.[10]

1. Mechanical interlocking
2. Formation of a clean surface
3. Formation of a highly reactive surface
4. Formation of a larger surface

It is widely believed that although the surface becomes rougher because of abrasion, it is a change in both physical and chemical properties of the surface layer that produces an increase in adhesive strength. While some adhesive applications can be explained by mechanical interlocking, it has been shown that mechanical effects are not always of prime importance.

2.3.3 Electrostatic and diffusion theories

The electrostatic and diffusion theories of adhesion are generally not regarded as highly as the adsorption theory or mechanical theories. However, there are certain applications where each is very appropriate.

The electrostatic theory states that electrostatic forces in the form of an electrical double layer are formed at the adhesive-adherend interface. These forces account for resistance to separation. This theory gathers support from the fact that electrical discharges have been noticed when an adhesive is peeled from a substrate. Electrostatic adhesion theory is regarded as an accepted theory for biological cell adhesion. A simple form of adhesion can also arise from direct contact electrification. This has been demonstrated for thin films of metal sputtered onto polymeric surfaces.

The fundamental concept of the diffusion theory is that adhesion arises through the inter-diffusion of molecules in the adhesive and adherend. The diffusion theory is primarily applicable when both the adhesive and adherend are polymeric, having compatible long-chain molecules capable of movement. The key is that the adhesive and the adherend must be chemically compatible in terms of diffusion and miscibility. Solvent or heat welding of thermoplastic substrates is considered to be due to diffusion of molecules. Other than certain thermoplastics, situations in which the adherend and adhesive are soluble in one another are relatively rare. Therefore, the diffusion theory of adhesion can be applied in only a limited number of cases.

2.3.4 Weak-boundary-layer theory

According to the weak-boundary-layer theory, as first described by Bikerman,[12] when bond failure seems to be at the interface, usually a cohesive rupture of a weak boundary layer is the real event. This theory largely suggests that true interfacial failure seldom occurs. Failure may occur so near to the interface that it is apparently at the interface, but in most cases it is ductile plastic deformation or cohesive failure of a weak boundary layer material. Weak boundary layers can originate from the adhesive, the adherend, the environment, or a combination of any of the three.

Weak boundary layers can occur on the adhesive or adherend if an impurity concentrates near the bonding surface and forms a weak attachment to the substrate. When failure occurs, it is the weak boundary layer that fails, although failure may seem to occur at the adhesive-adherend interface.

The history of a typical adhesive joint may be divided into three time periods: application of the adhesive, setting, and the period it is in service. Weak boundary layers could develop during any one of these periods.

Before application of the adhesive, the most important task is to remove the weak boundary layers. The most common material to be removed is atmospheric air. Displacement of air by the adhesive is a process already described: wetting. Two other examples of weak boundary layers that are present at the time of adhesive application can be seen in polyethylene and metal substrates. Polyethylene substrates usually have a weak, low-molecular-weight constituent evenly distributed throughout the polymer. This weak boundary layer is present at the interface and contributes to low failing stress when untreated polyethylene is used as an adhesive or adherend. Some metallic oxides, such as aluminum oxide, are very strong and do not significantly impair joint strength. However other oxides, such as those associated with copper and copper alloys, are weak and require removal prior to application of the adhesive. Weak boundary layers such as low molecular weight polyethylene constituents and copper oxides can be removed or strengthened by various surface treatments.

During the second major time period in the life of an adhesive, solidification of the adhesive is the primary process. In many instances, new boundary layers may form during the setting stage, and if they happen to be weak, the final joint will also be weak. An example of a weak boundary layer formed during this stage may be a chemical reaction by-product of the setting reaction. Certain active metal surfaces (e.g., titanium) are known to react with some chemical hardeners used in adhesive formulations. A clean, reactive substrate surface may also become contaminated by the components in the adhesive itself. For example, the adhesive could contain water or low molecular weight constituents that preferentially adsorb on the substrate surface.

During the third period in an adhesive's life (service exposure), weak boundary layers could occur by environmental moisture diffusing through either the adhesive or the adherend and locating at the interface. Plasticizers, solvents, or other low molecular weight substances also may migrate out of the adherend or adhesive and deposit at the interface. The bonding of plasticized polyvinyl chloride, for example, can be difficult because the plasticizers will migrate to the interphase with time. Notice the way some plasticized binder covers will stick together on your bookshelf? This is due to the highly mobile nature of the plasticizer. Another weak boundary layer that can form during aging occurs by the adherend continuing to cure or chemically react with its surroundings. Corrosion of an aluminum adherend could occur at the interface and weaken the joint strength. On exposure to moderately elevated temperatures, under-cured phenolic substrates, for example, continue to liberate moisture as a by-product of the curing reaction, thereby degrading the joint strength.

2.4 Stages in an Adhesive's or a Sealant's Life

As was indicated in the previous section, the life of an adhesive bond has several important stages no matter whether it is in an adhesive or a sealant joint. These stages are:

(1) Application and wetting

(2) Setting or solidification

(3) Operating in service

The degree of interfacial adhesion is greatly determined by stages 1 and 2. Stages 2 and 3 determine the degree of cohesive strength. All stages markedly influence the overall joint performance. From a practical standpoint, the last stage is where the ultimate performance of the adhesive or sealant is measured. It includes aging of the joint materials in the service environment as well as exposure to any stresses in the application.

There are processes going on in all three stages that ultimately will affect the adhesion and the performance of the adhesive or sealant. It is important to understand these processes and the effects that they will have on the quality of the joint.

2.4.1 Application and Wetting

As explained under the adsorption theory of adhesion, an adhesive must first wet the substrate and come into intimate contact with it.

The result of good wetting is simply that there is greater contact area between adherend and adhesive over which the forces of adhesion (e.g., van der Waals forces) may act. As shown in the preceding sections, for maximum wetting, the surface energy of the liquid adhesive must be less than that of the solid adherend or $\gamma_{LV} < \gamma_C$. Table 2.2 provides surface tensions for common adhesive liquids and critical surface tension for various solids.

The wetting of surfaces by adhesives can be described by two activities: a lateral spreading of the film; and a penetration of the fluid adhesive into the surface cavities that are characteristic of the inherent surface roughness. The first activity is controlled by the relative surface energies of the adhesive and substrate. The second activity is controlled mainly by the viscosity of the adhesive and the time it is in the liquid state.

It may be useful to first consider the adhesion that occurs when only a wettable liquid is used to provide adhesion (i.e., there is no cohesive strength component). When two flat, smooth surfaces are spontaneously wet by a thin liquid, strong adhesion can result. The reason is evident from Fig. 2.9 and from an application of the classic Laplace-Kelvin equation of capillarity:[14]

$$p_1 - p_2 = \gamma_{LV} \left(\frac{1}{R_1} - \frac{1}{R_2} \right)$$

Where p_1 and p_2 are respectively the pressure within the liquid and the pressure in the vapor outside the liquid; R_1 and R_2 are respectively

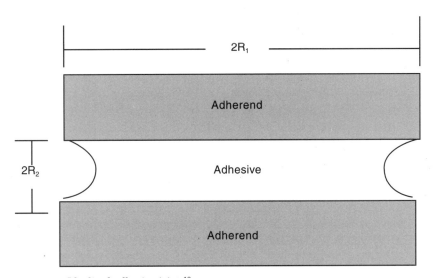

Figure 2.9 Idealized adhesive joint.[13]

the radius and half the thickness of a thin, circular layer of liquid in contact with the solid; and γ_{LV} is the surface tension of the liquid. If R_1 is greater than R_2, then p_2 will be greater than p_1. Hence, the two plates will be forced together because of the pressure difference $p_2 - p_1$. From this equation it follows that many common liquids that spontaneously wet two such solid surfaces will hold them together.

In reality, however, the assumptions stated above are never realized. Extremely close fitting, perfectly smooth, solid adherends would be very expensive to prepare. The absence of dust or other particles on the surface could be critically important in forming a strong joint. The resistance of the joint to stress is determined solely by the viscosity of the liquid film; hence, only if the viscosity is very great could the joint withstand practical loading pressures.

Since real surfaces are not smooth or perfectly flat, it is necessary to understand the effects of surface roughness on joint strength. A viscous liquid can appear to spread over a solid surface and yet have many gas pockets or voids in small surface pores and crevices. Even if the liquid does spread spontaneously over the solid, there is no certainty that it will have sufficient time to fill in all the voids and displace the air. The gap-filling mechanism is generally competing with the setting mechanism of the liquid.

Problems occur when the liquid solidifies rapidly after being applied. Two examples are fast curing epoxies and fast gelling hot melt adhesives. Very fast reacting epoxy adhesive systems generally do not have the high adhesion strength that slower curing epoxy systems have. One reason for this (there are others primarily related to the chemistry of these fast acting systems) is that the curing reaction does not provide sufficient time for the adhesive to fill the crevices on the substrate surface. Cyanoacrylate adhesives, on the other hand, are also very fast curing adhesives that provide exceptionally high bond strengths on many substrates. Although cyanoacrylate resins set rapidly, their viscosity and wetting characteristics are such that they quickly permeate the crevices and valleys on a substrate surface. When a hot melt adhesive is applied in melt form to a cold metal surface, the adhesion is much weaker than if the hot melt was applied to a preheated metal surface and then allowed to cool at a slower rate along with the substrate. When the hot melt makes contact with the cold surface the adhesive gels immediately, and there is no time for the adhesive to fill-in the cavities on the surface.

In certain cases, roughening of the surface may be undesirable— actually decreasing the resulting joint strength. There is a noticeable difference in measuring the contact angle of a liquid spreading over a clean dry surface, θ_A, and the contact angle measured when the liquid recedes from a previously wet surface, θ_R. When a difference is observed between these two angles, the most common cause is the entrapment of liquid in the valleys and pores of the surface as the liquid advances over it. When the liquid recedes, the surface uncovered usually includes wet areas. For this reason θ_R is smaller than θ_A. However, when extreme care is used in preparing clean, sufficiently smooth surfaces, no differences are found. Wenzel[15] noticed that the apparent or measured contact angle (θ') between the liquid and the surface of the solid and the true contact angle (θ) between the liquid and the surface follow the relationship:

$$r = \frac{\cos \theta'}{\cos \theta}$$

where r is a measure of roughness ($r = 1.0$ for smooth surface and generally $r = 1.5$ to 3.0 for machined and ground metal surfaces). Here θ' can be thought of as the contact angle between the liquid and its envelope to the solid surface, and θ is the contact angle between the liquid and the surface at the air-liquid-solid contact boundary. Since most organic liquids exhibit contact angles of less than 90 degrees on clean polished metals, the effect of roughening the surface is to make $\theta > \theta'$. Thus, wetting is improved, and the resulting joint strength is positively affected by the roughening. However, when the surface is wet by a liquid with a contact angle greater than 90 degrees, then the effect of roughening is to make $\theta < \theta'$ and to reduce wetting.

This effect is evident when trying to bond a low energy surface, such as polyethylene, with an epoxy adhesive. Without surface preparation, the contact angle that the liquid epoxy makes on the polyethylene is greater than 90 degrees. Therefore, epoxy adhesive joints made by roughening the surface of polyethylene will be weaker than joint strengths made with a non-roughened surface. This is sometimes disconcerting to those accustomed to bonding metal surfaces. With metal surfaces, most adhesives make contact angles less than 90 degrees, and surface roughening is usually, and correctly, looked upon as a possible way of improving joint strength. However, with untreated polyethylene, the effect is just the opposite — exactly what is not wanted.

2.4.2 Setting or solidification

When a liquid adhesive or sealant solidifies, the loss in joint strength is much greater than simply the loss of contact area due to the considerations noted above. On setting or curing, there can be possible reductions in adhesive strength due to:

1. Localized stress concentration points at the interface

2. Stress due to differences in thermal expansion coefficients of the adhesive and adherend (mainly associated with adhesives that cure at temperatures different than their normal service temperatures)
3. Stress due to shrinkage of the adhesive or sealant as it cures

2.4.2.1 Localized stress. Losses of theoretical adhesive strength arise from the action of internal stress concentrations caused by the trapped gas and voids. Griffith[16] showed that adhesive joints may fail at relatively low stress if cracks, air bubbles, voids, inclusions, or other surface defects are present. If the gas pockets or voids in the surface depressions of the adherend are all nearly in the same plane and not far apart (as is shown in Fig. 2.10 upper), cracks can rapidly propagate from one void to the next. However, a variable degree of roughness, such as shown in Fig. 2.10 lower, provides barriers to spontaneous crack propagation. Therefore, not only is surface roughening important, but the degree and type of roughness may be important as well.

It has also been shown[17] that a concentration of stress can occur at the point on the free meniscus surface of the adhesive (edge of the bond-line). This stress concentration increases in value as the contact angle, θ, increases. At the same time, the region in which the maximum stress concentration occurs will move toward the adhesive-adherend interface. The stress concentration factors for a lap joint in shear with a contact angle ranging from 0 to 90 degrees are summarized in Fig. 2.11. For contact angles less than about 30 degrees (i.e., for good wetting) the maximum stress occurs in the free surface of the adhesive away from the edges, and the stress concentration is not much greater than unity. For larger contact angles, the maximum stress occurs at the edges A, A (i.e., at the actual interface between adhesive and adherend). The stress concentration factor increases until, for a value of $\theta = 90$ degrees (non-wetting), it is greater than 2.5. Thus, poor wetting will be associated with a weak-spot at the

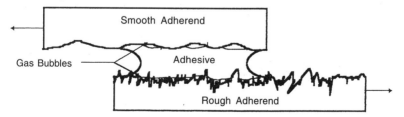

Figure 2.10 Effect of surface roughness on coplanarity of gas bubbles: upper adherend is smooth and gas bubbles are in the same plane, lower adherend has roughness and gas bubbles are in several planes.[13]

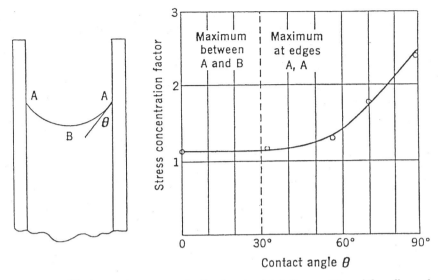

Figure 2.11 Maximum stress concentration in a lap joint. Poor wetting of the adherend produces maximum stress concentration at point of contact of adhesive, adherend, and atmosphere.[13]

adhesive–adherend interface with a consequent likelihood of premature failure at this region.

2.4.2.2 Setting stresses due to thermal expansion differences. When a liquid adhesive solidifies, the theoretical strength of the joint is reduced because of internal stress concentrations that usually develop. The most common cause of internal stress is due to the difference in the thermal expansion coefficients of the adhesive and the adherends. These stresses must be considered when the adhesive or sealant solidifies at a temperature that is different from the normal temperature to which it will be exposed in service. Figure 2.12 shows that thermal expansion coefficients for some common adhesives and substrates are more than an order of magnitude apart. This means that the bulk adhesive will move more than 10 times as far as the substrate when the temperature changes, thereby causing stress at the interface.

The stresses produced by thermal expansion differences can be significant. Take for example an annular journal bearing where a polyamide-imide insert is bonded to the internal circumference of a stainless steel housing (Fig. 2.13). Further, assume that the adhesive used is one that cures at 250°F. At the cure temperature, all substrates and the gelled adhesive are in equilibrium. However, when the temperature begins to reduce as the system approaches ambient conditions,

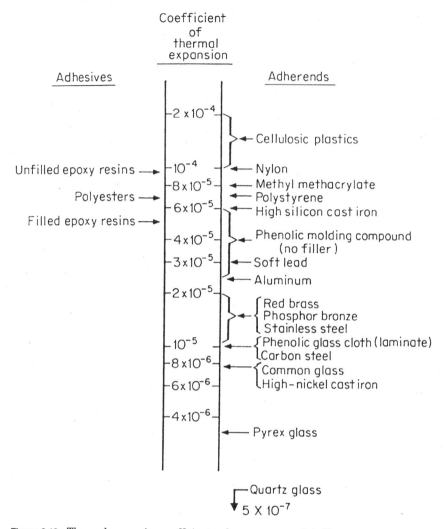

Figure 2.12 Thermal expansion coefficients of common materials.[18]

stresses in the adhesive develop because the polyamide-imide substrate wants to shrink to a greater extent than the steel. At ambient room temperature, these stresses may be significant but not high enough to cause adhesive failure. Further assume that the bonded bearing is to be placed in service with operating temperatures that will vary between 250°F and −40°F. At 250°F the internal stresses due to the miss-match in thermal expansion are reduced to zero, since we are back at the equilibrium condition (assuming that there was no

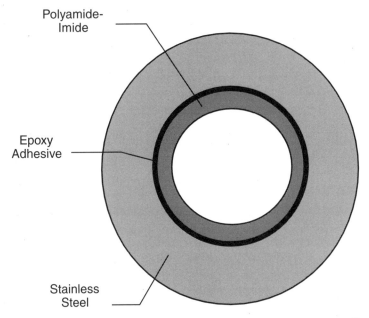

Figure 2.13 Journal bearing application. Outer cylinder (stainless steel) is bonded to inner cylinder (polyamide-imide) with an epoxy adhesive. Exposure to low temperatures causes significant stress on bond due to differences in coefficient of thermal expansion.

shrinkage in the adhesive as it cured or other stresses in the joint). However, when the service temperature reaches −40°F, the thermal expansion differences create internal stress in addition to those already there due to curing. Thus, a failure could easily occur.

A similar example is evident by a typical graph of an elevated temperature cured adhesive joint as a function of test temperature, Fig. 2.14. Notice that the bond strength actually increases with temperature to a maximum, and then falls off with increasing temperature. This is similar to the case above where the internal stresses are actually reduced by the service temperature. At some elevated temperature, the internal stresses are completely relieved and the bond strength reaches a maximum. The test temperature at which this occurs is usually very close to the curing temperature. At higher test temperatures, additional stresses develop or the effects of thermal degradation become evident, and the bond strength then decreases with increasing test temperature.

There are several possible solutions to the expansion miss-match problem. One is to use a resilient adhesive that deforms with the substrate during temperature change. The penalty here is possible creep

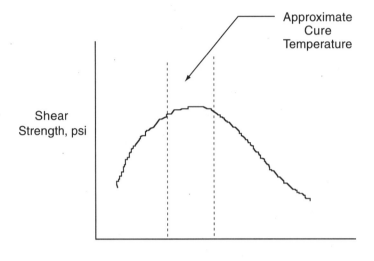

Figure 2.14 Plot of the strength of an aluminum joint (bonded with an elevated temperature curing epoxy adhesive) as a function of test temperature.

of the adhesive, and highly deformable adhesives usually have low cohesive strength. Another approach is to adjust the expansion coefficient of the adhesive to a value nearer to that of the substrate. This is generally accomplished by selection of a different adhesive or by formulating the adhesive with specific fillers to "tailor" the thermal expansion coefficient. A third possible solution is to coat one or both substrates with a primer or coupling agent. This substance can provide either resiliency or an intermediate thermal expansion coefficient that will help reduce the overall stress in the joint.

2.4.2.3 Setting stresses due to shrinkage of the adhesive or sealant. Nearly all polymeric materials (including adhesives and sealants) shrink during solidification. Sometimes they shrink because of escaping solvent, leaving less mass in the bond line. Even 100% reactive adhesives, such as epoxies and urethanes, experience some shrinkage because their solid polymerized mass occupies less volume than the liquid reactants. Table 2.3 shows typical volume shrinkage for various reactive adhesive systems during cure.

The result of such shrinkage is internal stresses and the possible formation of cracks and voids within the bond-line itself. Elastic adhesives deform when exposed to such internal stress and are less affected by shrinkage. Formulators are often able to adjust the final hardness of the adhesive or sealant to minimize stress during shrinkage.

TABLE 2.3 Shrinkage of Common Adhesives[19]

Adhesive types	% Shrinkage
Acrylics	5–10
Anaerobics	6–9
Epoxies	4–5
Urethane	3–5
Polyamide Hot Melts	1–2
Silicones (Curable)	<1

2.4.3 Operation in service

Once the solidification mechanism is complete, the joint is generally exposed to its service environment. We refer here to the "joint" rather than the adhesive or sealant because now that the bond is created, the joint is a single entity that has individual characteristics of its own.

The service environment may include cyclic exposure to temperature, stress, chemicals, radiation, or a number of other environments that are common to the application. It is important that the joint resist the environmental conditions so that a practical working strength can be maintained throughout the expected service life. The effect of service conditions on the adhesive joint occurs mainly through localized stress or environmental aging.

2.4.3.1 Short term effects.
Localized stresses are mainly due to the immediate effects of temperature and differences in thermal expansion coefficient. The effect of differing thermal expansion coefficients on internal stress generated during cure has been discussed in the preceding sections. However, thermal stresses could easily occur during the joint's service life.

If the temperature is uniform throughout the bond, the approximate stress on a thin rigid bond may be calculated from the following relationship:[20]

$$S = \frac{\Delta T (k_1 - k_2)}{(1/E_1) - (1/E_2)}$$

where S = shear stress on the adhesive due to differential thermal expansion rates of the adherends, without consideration for adhesive strain
E_1, E_2 = Young's moduli of the adherends
ΔT = temperature differential between zero stress temperature and service temperature (zero stress condition usually exists at the cure temperature)
k_1, k_2 = thermal expansion coefficients of adherends

If the adherends are assumed to be unyielding, and the adhesive is relatively flexible and thick, the greatest stress on the adhesive, occurring at the ends of the joint, may be approximated from the following relationship:

$$S = \Delta T \, (k_1 - k_2) \, (GL/2d)$$

where S = greatest shear stress in the adhesive due to differential thermal expansion of the adherends, without consideration for adherend strain
G = shear modulus of the adhesive
d = thickness of the adhesive
L = length of the joint

These theoretical expressions are approximations in that they exclude the strain capability of either the adhesive or the adherends. Such strain would tend to relieve some of the stress. The values calculated, however, are greater than the actual stress and, therefore, conservative.

Methods of reducing such stresses consist of using flexible materials or trying to better match the thermal expansion coefficients. The coefficient of thermal expansion of adhesive and adherend should be as close as possible to limit stresses that may develop during thermal cycling or after cooling from an elevated-temperature cure. As shown in Table 2.3, polymeric adhesives generally have a thermal-expansion coefficient an order of magnitude greater than metals. Adhesives can be formulated with various fillers to modify their thermal-expansion characteristics and limit internal stresses. A relatively elastic adhesive capable of accommodating internal stress may also be useful when thermal-expansion differences are of concern.

Once an adhesive bond is made and placed in service, other forces are at work weakening the bond. The type of stress involved, its orientation to the adhesive, and the rate in which the stress is applied are important. When a bond separates cohesively, it is because adjacent molecular segments have physically moved away from each other. This could occur from breaking of strong bonds within the molecular chain or from rupture of weak bonds between the chains. The bond separation is either rapid (cracking) or slow (creep).

Cracking results when a localized stress becomes great enough to physically separate adjacent molecular segments. Highly crystalline or highly crosslinked polymers are likely to crack rather than creep under stress. When bond separation appears as an adhesive failure, it is generally because a crack has followed the interface surface or because some chemical has displaced the adhesive from the adherend (see below). Cracks may result from internal or external stress.

Creep can occur when enough force is applied to a mass of linear molecules to cause them to disentangle or overcome their crystalline order. Creep is a slow process. Creep is more pronounced at temperatures above the adhesive's glass transition temperature, T_g. Polymers with low T_g cannot be used where large loads must be carried at moderate temperatures. Crosslinking reduces creep because the polymer segments are immobilized by the network structure and cannot easily slide by one another.

2.4.3.2 Long term effects. Potentially harmful external stress may be of mechanical, thermal, or chemical origin. Temperature, humidity, salt spray, fluids, gases, mechanical loads, radiation, and vacuum are the most common extreme environments. They may be sufficiently great to cause premature failure. The required environmental resistance depends on the individual application. These requirements must be established before the final joint configuration and material selections can be made. The effects of harsh environments on adhesives and sealants are detailed in Chapter 17.

Sustained loads can cause premature failure in service even though similar unloaded joints may exhibit adequate strength in the same environment. The cumulative effect of different types of stress (e.g., mechanical and chemical) may often be an overlooked factor regarding adhesive failure. Temperature variation combined with cyclic loading, for instance, can greatly reduce durability when compared with simple single parameter aging of the same sample.

There are times when a properly cured unstressed bond fails for no apparent reason. It may exhibit apparent adhesive failure under low or moderate load after extended aging. When behavior of this type is encountered, the cause is often desorption, i.e., displacement of the adhesive from the surface by a chemical from the environment or from within the substrate (Fig. 2.15). Attachment to substrate sites involves a dynamic equilibrium between the adhesive and other molecules that may be present. Water, solvent, plasticizer, and various gases and ions may all compete for the surface attachment sites. Moisture is by far the most effective desorbing substance commonly available. Bonds are able to resist desorption only if the low molecular weight substances are unable to penetrate the joint or if the substrate-adhesive bond is thermodynamically or kinetically favored over the substrate-desorber interaction. Some widely used coupling agents form thin films between the substrate and the adhesive. These have very high affinities for each material. Such substances will be difficult to displace by itinerant molecules.

An adhesive joint may contain as many as five different layers: first adherend, first boundary layer, the adhesive, second boundary layer, and second adherend. The life of a typical adhesive joint may be di-

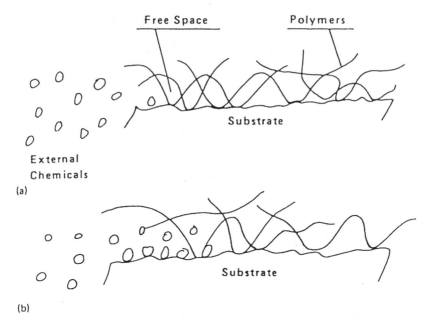

Figure 2.15 Competition between an adhesive and other chemicals for surface sites leading to displacement of the adhesive from the surface. (*a*) Adhesive adsorbed at surface sites. (*b*) Adhesive displaced from surface sites.[21]

vided into three periods: application of the adhesive, setting, and service. The behavior of each of the five layers during the three periods indicates that there may be up to 15 different mechanisms occurring during the life of a single joint. All of these could affect the performance characteristics of the joint.

The possible combinations of adhesives, adherends, stresses, and environments are so great that reliable adhesive strength and aging data are seldom available to the design engineer. Where time and funds permit, the candidate adhesive joints should always be evaluated under actual or simulated operating loads in the intended environment.

Environmental aging generally causes: additional internal stresses on the adhesive bond; degradation of the adhesive, adherend, or interphase area; or creation of new interphase regions. The service environment may produce additional stress on the adhesive bond that could contribute positively or negatively to the stresses that are already inherent in the joint.

Consider, for example, aging of an adhesive or sealant joint at moderately elevated temperatures. The adhesive or sealant, depending on the actual chemical composition, could shrink due to loss of plasticizer or increased crosslinking. It also could increase in modulus of elasticity for the same reasons. However, if the glass transition temperature was exceeded, it is likely that the effects would be different. That is, the material would expand because of increased thermal expansion coefficient, and the modulus would decrease because of increased freedom of rotation in the molecule. Many of these effects will be described in detail in later chapters. However, it is important to realize that these changes could either increase or reduce the stress on the joint. This is why the "environmental stress" effect that was shown in Fig. 1.11 can either positively or negatively affect the natural bond strength of a joint.

Aging can also increase degradation of the adhesive, adherend or interphase regions. These degradation mechanisms generally lead to a lowering of the cohesive strength of the adhesive. For example, thermal heating may first cause softening of the adhesive and, then, additional heating will cause oxidation and finally pyrolysis of the molecular chains. Exposure to moisture can cause cohesive weakening through a hydrolysis reaction mechanism.

Environmental factors may also cause new interphase areas that could provide a weak boundary layer in the joint. Environmental moisture can diffuse through the adhesive and/or adherend, accumulate at the interface and cause corrosion of metallic adherends. Plasticizer or moisture migration could also accumulate at the interface and form a weak boundary layer.

2.5 Special Mechanisms Related to Sealants

All of the modes of adhesion failure described in the preceding sections also apply to sealants. Adhesion failure, a loss of bond between the sealant material and its substrate, is a common type of sealant failure. Sealant failures most commonly occur under the following conditions.

1. The sealant is improperly formulated and, therefore, does not adhere to the substrate (i.e., the surface energetics do not allow good adhesion to occur).

2. There is a weak boundary layer either initially or formed during service, and the weak boundary layer fails in service.

3. The sealant becomes brittle and more limited in its movement capability because of aging or environmental conditions. Thus, it can no longer compensate for the required substrate movement, and bond failure results because the cohesive strength is greater than the adhesive strength.

Cohesive failure is a failure within the body of the sealant material. This failure frequently begins with a small nick or puncture of the material as shown in Fig. 2.16a. This is often because bulk sealant materials are soft. With repeated movement, usually beyond their movement capability, the sealant will rupture in the center of the joint. Spalling failure in sealant joints (Fig. 2.16b) is equivalent to a cohesive failure of the adherend. It is most common with high strength sealants that are used with concrete. Some sealants have greater cohesive strength than concrete especially when the concrete is not fully cured. Intrusion failure (Fig. 2.16c) is often seen in construction joints. The failure occurs when the sealant extends and necks down and then fills with dirt. During the compression cycle, this pocket of dirt attempts to close, and the closing action causes abrasion on the surface of the sealant. This abrasion then leads to failure during a subsequent tension cycle.

An additional common type of sealant failure is shown in Fig. 2.17a. Here the sealant is extended and held in a necked-down shape. When the joint closes, the sealant acts as a slender column and buckles. This exposes the strained sealant to the environment and causes peeling forces at the edges of the sealant. Compression failure is also common, as shown in Fig. 2.17b. The sealant is compressed and relaxes into

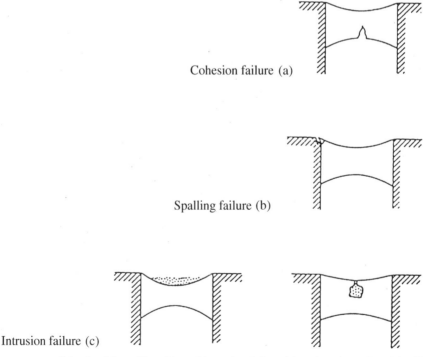

Figure 2.16 Cohesion (a), spalling (b), and intrusion failure (c) modes of a sealant joint.[22]

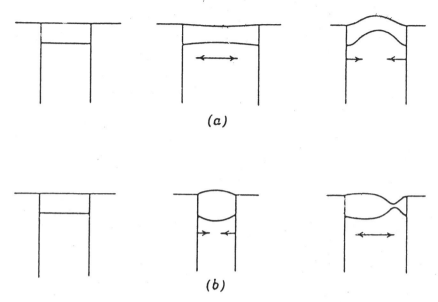

Figure 2.17 Change in sealant shape due to flow. (*a*) Viscous tension-compression effect. (*b*) Viscous compression-tension effect.[22]

equilibrium in its new shape. When the joint opens, the sealant does not return to its former rectangular shape. Instead, it yields at its minimum cross-section, which is immediately adjacent to the joint interface. This is a highly concentrated strain that will lead to failure.

Another common problem in using sealants is assuming that the important properties of the sealant are those that are measured in a fully cured state. Often because of temperature cycles such as those occurring at day and night, the movement of the substrates cannot wait for the sealant to fully cure. Excessive or premature stress on the undercured sealant will usually lead to cohesive failure. The solution is to use fast curing sealants so that a significant portion of the cure occurs before the first critical cycle of the substrate.

2.6 Polymer Material Interactions

Adhesion depends, as has been stated above, mainly on the dispersion forces possessed by all molecules, then on a means of bringing these into effect by close proximity, and lastly on the cohesive behavior of the adhesive (i.e., its strength, flexibility, and response to temperature). Some polymers will be better adhesives or sealants than others due to the enhancement of van der Waal's forces by hydrogen bonding or other donor-acceptor interactions possible at the interface. If these interfacial effects are neglected, all polymers have similar properties or characteristics whose value will determine their general usefulness as adhesives or sealants in specific applications.

2.6.1 Polymeric materials

It is natural to ask why so many polymeric materials with greatly different chemical composition and structure make good adhesives or sealants. The reason is that: all polymeric materials have fundamental properties that are important in adhesion; polymers can be made to flow and then set via several reaction mechanisms to form a cohesively strong solid, and additives can be used to modify these properties.

The performance of a polymeric adhesive or sealant is dependent upon the physical properties of the polymeric base from which it is made and the various formulating agents used to modify the base polymer. Relationships between the chemical structure of a polymer, its physical properties, and its adhesive performance have interested scientists for many years. This has resulted in many excellent technical studies on property-structure relationships and has helped to improve understanding of the factors controlling adhesive and sealant performance.

2.6.2 Properties important for adhesives and sealants

The fundamental parametric properties that are important in successful adhesives and sealants are the glass transition temperature, the solubility parameter, the surface free energy, and the viscosity. Added to these are the non-parametric qualities of microstructure or organization. Some of these properties are mainly involved in the application of adhesive to the substrate, and they play little direct part once intimate contact on a molecular scale has been achieved.

2.6.2.1 Properties important for adhesion. Surface energetics are controlled largely by the general chemical composition of the polymer molecules. Factors such as the polarity and surface functionality of the molecules affect the surface tension and often account for the great improvement in adhesion. These properties are usually determined by the base polymer in the adhesive or sealant formulation.

Addition of carboxyl functionality to polymers is well known as an adhesion promotion mechanism for polar substrates. Thus, -COOH groups improve the adhesion of polyesters, acrylics, and olefin polymers to these substrates. Other electronegative atoms, such as chlorine and nitrogen, have been noticed to improve the chances for adhesion via hydrogen bonding across the interface.

Favorable surface energetics are a necessary, but not singularly sufficient, condition for bond formation. Proper spreading kinetics and rheology are also required. The adhesive must have the necessary rheology to flow on the time scale of the bond formation. Time dependent flow of the adhesive, which is often confused with or combined with the energetic "wetting" of the substrate by an adhesive, is a major factor in application of adhesives. The major property affecting flow is viscosity and, therefore, molecular weight.

2.6.2.2 Properties affecting cohesion. In addition to good adhesion, an adhesive or sealant must have satisfactory cohesive strength. Factors that can affect cohesive strength include molecular weight, crystallinity, hydrogen bonding, crosslinking, and compatibility (in multi-phase systems).

Many properties of adhesive bonds are influenced by the mobility of the molecular chain structure as shown in Fig. 2.18. When chain segments can move easily such as when the temperature exceeds the glass transition temperature, T_g, they can deform under impact or

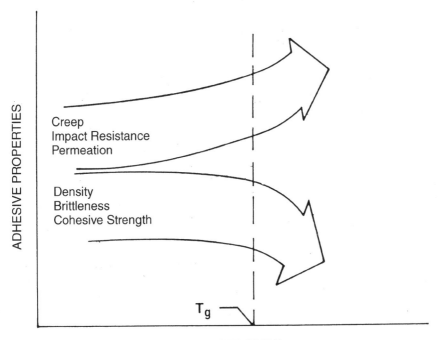

Figure 2.18 General trends of some adhesive properties related to temperature or molecular mobility. The sharp change at the T_g reflects the abrupt increase in molecular motion.[21]

assume new alignments under mechanical or thermal-expansion stress. This movement spreads the applied energy over a greater number of atoms and, thus, gives the bond a better chance to resist stress. Brittleness is, therefore, reduced and flexibility is increased. Molecular flexibility can be controlled by the following conditions.

	Effect on flexibility
Molecular weight	Negative
Crosslink density	Negative
Crystallinity	Negative
T_g	Negative
Fillers	Negative
Plasticizers	Positive
Flexibilizers	Positive

Increasing molecular weight improves cohesive strength but often weakens adhesion. Crystallinity can significantly improve the cohesive strength of a polymer up to its melting point, but too much crystallinity can also embrittle the adhesive. Heterogeneous nucleation (crystallinity occurring at the melt surface of a polymer) has been shown to improve the surface cohesive strength of FEP Teflon, Nylon-6, and polyethylene when cast against high energy surfaces.

Crosslinked adhesives have cohesive properties that are found to depend on the molecular weight between crosslinks, M_c. Shear strengths of epoxy aluminum joints decreased as M_c increased; however, flexibility and toughness are increased. Figure 2.19 shows the relationship between crosslink density and the physical state of epoxy resins.

The excellent cohesive strength of polyamides compared to other common polymers of equivalent molecular weight is due to the presence of interchain hydrogen bonding. Excellent adhesion of epoxies to aluminum, of surface treated rubber and other polymers to glass, and of polymers to cellulosics are also attributed to interfacial hydrogen bonding. Hydrogen bonding can be considered a special case of crosslinking.

Like all polymers, adhesive and sealant materials undergo constant thermally induced vibration. The amplitude of these vibrations is determined primarily by temperature, chain flexibility, and crosslinking, and to a lesser extent by fillers and physical stresses. A certain amount of chain flexibility is desirable since it imparts resiliency and toughness to the adhesive film. Too much flexibility, however, may lead to "creep" (i.e., plastic flow under load) or poor temperature resistance.

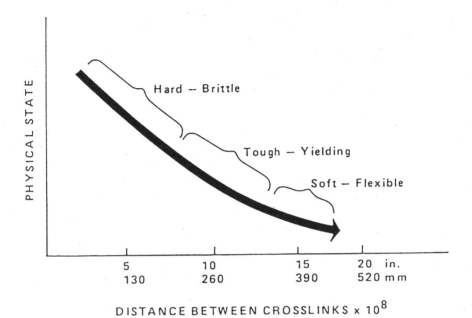

Figure 2.19 Effect of crosslink density on the physical state of epoxy resins.[21]

Adhesive formulators use the glass transition temperature, T_g, as a practical basis for compounding products with the appropriate amount of internal motion. T_g is the temperature at which the slope of a temperature-volume plot undergoes a sudden upward change shown in Fig. 2.20. It is the temperature at which there is significantly more molecular mobility than at lower temperatures (i.e., the molecules have sufficient thermal energy to be considered mobile). T_g is a prop-

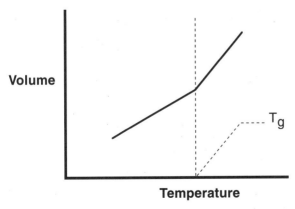

Figure 2.20 The effect of temperature on the total volume of a polymer.

erty of a polymer that depends on its chemical composition and the degree of crosslinking or molecular interaction.

For good bond strength and creep resistance, the T_g of the adhesive or sealant should be well above the maximum temperature that it will see in service. However, peel strength will be low if the T_g is high. It is very difficult to provide an adhesive that has high peel strength with good cohesive strength and environmental resistance. This problem and possible solutions for it will be discussed in detail in following chapters. A high T_g will also limit the low temperature properties of the adhesive or sealant. Bond strength at elevated temperatures can be increased by raising crystallinity, hydrogen bonding, and crosslinking. Typical glass transition temperatures for adhesive resins are shown in Table 2.4.

There is an interdependence on the rheological characteristics of the bulk adhesive and its adhesion characteristics. For example, if a cohesively weak polymeric material is being used as the adhesive, the demand on the strength of surface attachment would be much less than if the adhesive were a tough polymeric blend. Hence, the rheological strength of the adhesive polymer plays an important role in determining the magnitude of the joint strength.

Figure 2.21 presents the tensile strength of a carboxylic acid containing acrylic terpolymer as a function of temperature. The glass transition temperature of this terpolymer is 97°F. The joint strength of this adhesive is typical of noncrystalline thermoplastic adhesive systems. In the rubbery range (above T_g), the joint strengths are low because the polymer itself is cohesively weak. At lower temperatures, approaching T_g, the adhesive-joint strengths increase and are maximized near T_g. In the glassy region, it has been found that the joint strength depends on how brittle the adhesive material is. If the adhesive is brittle in the glassy state, the adhesive joint strength will decrease. With tougher plastics or resins like crosslinked epoxies, there is little or no drop in joint strength immediately below T_g. Usually a broad maximum joint strength plateau is observed.

From this data, one can now attempt to generalize the effect of the rheological state on adhesive joint strength. This generalized condition

TABLE 2.4 Glass Transition Temperatures of Common Polymers[19]

Adhesive type	Glass transition temperature, °F
Silicone	−130
Natural Rubber	−94
Polyamide	140
Epoxy	212

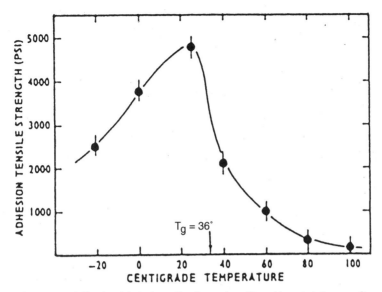

Figure 2.21 Adhesive joint strength of a carboxylic acid containing acrylic terpolymer to aluminum as a function of temperature.[8]

is presented in Fig. 2.22. It is of interest to point out the similarity between the adhesive joint strength and the modulus of the material in the various rheological states. Rubbery systems have low moduli and exhibit lower adhesive lap shear and tensile strengths; glassy polymers have higher moduli and produce generally high adhesive joint strength.

In summery, joint strengths are controlled by the fundamental properties of the polymeric material that is used to formulate the adhesive or sealant. These properties include the tenacity of surface attachment, the cohesive strength of the polymer material, and the rheological characteristics of the polymer. Rheological characteristics determine the degree of deformation that the adhesive material undergoes during rupture of the joint.

2.7 Surface and Interface Interactions

When two materials are bonded, the resultant composite has at least five elements: adherend 1/interphase 1/adhesive/interphase 2/adherend 2. Note that with application of primers or other components, such as spacers or reinforcing fibers, the number of elements can increase dramatically. The strength of the joint will be determined by the strength of the weakest member. Often the weakest member is one of the interphase regions, since this is generally where weak boundary

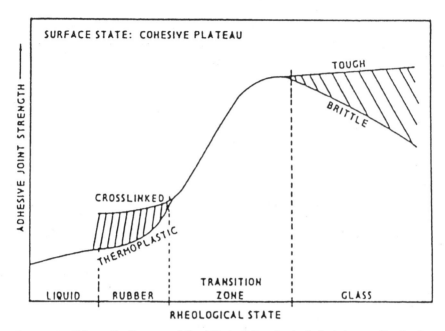

Figure 2.22 Schematic diagram of the effect of the rheological state on adhesive lap shear and tensile joint strengths.[8]

layers can occur. Examination of weak boundary layers and investigation of phenomena occurring at the interphase regions of the joint provides valuable information.

2.7.1 Boundary layer theory

One interpretation of adhesion that has been the most useful in describing adhesion phenomenon is the boundary layer theory set forth by Bikerman.[23] This theory proposes the existence at an interface of a finite boundary layer composed of adsorbed molecules that differ in nature from those in the bulk phases. A schematic diagram showing examples of weak boundary layers is presented in Fig. 2.23.

The criterion for good adhesion is merely that the boundary layer be strong enough to withstand the effects of external stress. According to Bikerman, rupture of an adhesive joint practically always proceeds through a single material phase rather than between two materials. The rupture is initiated at a point where the local stress exceeds the local strength. When failure occurs in the boundary stratum, a weak boundary layer is present.

Weak boundary layers may form due to a variety of causes. Often the formation is unpredictable, and it is difficult or impossible to determine the actual composition of the boundary layer. Examples of common weak boundary layers include:

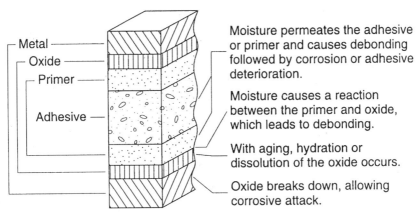

Figure 2.23 Examples of weak boundary layers.[24]

- Entrapment of air
- Impurities, or low molecular weight species that concentrate near the surface
- Cohesively weak oxide layers on the substrate surface
- Chemical deterioration of the coating, adhesive polymer, or substrate during the bonding process (e.g., catalytic air oxidation during heating)
- Chemical deterioration and/or corrosion between the adhesive and the substrate. (In some cases, for example with metals, the formation of brittle inter-metallic compounds can lead to a weak boundary layer.)

2.7.2 Interphase region

Sharpe[25] extended Bikerman's two-dimensional weak boundary layer concept into a three-dimensional interphase concept. Interphases are thin regions of the joint that have properties that are widely different from the bulk materials. These regions are thick enough to have properties, such as modulus, Poisson's ratio, tensile strength, etc., that will affect the final properties of the joint.

Interphase regions can be formed by solidification of certain polymers on certain high energy substrates. The interfacial structure will be characteristic of the composition and structure of both the polymer and the substrate, as well as the ambient conditions. Solidification preserves this organization to create an interphase with unique properties that becomes a permanent part of the joint, influencing its mechanical response. Examples include semicrystalline polymers, such as polyethylene. When this polymer is solidified from the melt while in contact with a solid substrate, a visibly different structure in the polymer near the interface—the so-called transcrystalline structure—is formed. This region has mechanical properties different from the ordinary bulk structure of the polyethylene from which it came.

Interphase regions on metals are generally complex oxides. The mechanical properties of an oxide on a particular metal substrate depend on the history of that particular piece of metal. Such interphase properties are determined by the conditions that generated them.

It is clear that interphases, which are quite thin relative to the joints in which they are present, will not have much effect on small deformation properties of the joint. However, they can have remarkable effects on the ultimate properties such as the breaking stress of the joint. This is particularly important if the interphase regions are sensitive to various environments, such as temperate and moisture.

It is also highly probable that interphases are not homogeneous in the sense that their composition, structure and, therefore, properties vary across their depth. Research needs to be directed at answering questions such as how to model the joint to include interphase regions, what are the appropriate properties of the interphase to measure, and how do you measure these properties.

References

1. DuNuoy, P. and Lecounte, J., *Gen. Physiol.*, 1 (1919) at 521.
2. Wilhelmy, L., *Ann. Phys.*, 119 (1863) at 177.
3. Harkings, W. D. and Brown, R. E., *J. Am. Chem. Soc.*, 41 (1919) at 499; also Harkings, W. D., *Physical Chemistry of Surface Films* (New York: Reinhold Publishing, 1952).
4. Young, T., *Trans. Roy. Soc.*, 95 (1805) at 65.
5. Fox, H. W. and Zisman, W. A., *J. Colloid. Science*, 5 (1950) at 514; also Zisman, W. A., "Relation of Equilibrium Contact Angle to Liquid and Solid Constitution", Chapter 1 in *Contact Angle, Wettability, and Adhesion*, R. F. Gould, ed. (Washington, D.C.: American Chemical Society, 1964).
6. Dupre, A., *Theorie Mechanique de la Chaleur* (Paris: Gauthier-Villars, 1869) at 369.
7. Ahagon, A. and Gent, A. N., *J. Polym. Sci.*, Polym. Phys. Ed., 13 (1975) at 1285.
8. Lewis, A. F. and Saxon, R., "Epoxy Resin Adhesives," Chapter 6 in *Epoxy Resins Chemistry and Technology*, C. A. May and Y. Tanaka, eds. (New York: Marcel Dekker, 1973).
9. Schneberger, G. L., "Chemical Aspects of Adhesive Bonding, Part II: Physical Properties", *Adhesives Age*, March 1970.
10. Pocius, A. V., Chapter 6, *Adhesion and Adhesives Technology* (New York: Hanser Publishers, 1997).
11. Yen Lee, Sheng, "The Use of Fluoroepoxy Compounds as Adhesives to Bond Fluoroplastics Without Surface Treatment", *SAMPE Quarterly*, vol. 19, no. 3, 1988.
12. Bikerman, J. J., *The Science of Adhesive Joints* (New York: Academic Press, 1961).
13. Baier, R. E., et. al., "Adhesion: Mechanisms That Assist or Impede It", *Science*, vol. 162, December 1968 at 1360–1368.
14. Adam, N. K., *The Physics and Chemistry of Surfaces* (London: Oxford University Press, 1941) at 9.
15. Wenzel, R. N., *Ind. Eng. Chem.*, vol. 28, 1936, at 988.
16. Griffith, A. A., *Phil. Trans. Roy. Soc. London*, Ser. A, 221, 163 (1920).
17. Mylonas, C., in *Proc. Seventh Int. Congr. Appl. Mech.*, London (1948).

18. Perry, H. A., "Room Temperature Setting Adhesives for Metals and Plastics", *Adhesion and Adhesives Fundamentals and Practice,* J. E. Rutzler and R. L. Savage, eds., (London: Society of Chemical Industry, 1954).
19. Schneberger, G. L., "Basic Bonding Concepts", *Adhesives Age,* May 1985.
20. Yurek, D. A., "Adhesive Bonded Joints", *Adhesives Age,* December, 1965.
21. Schneberger, G. L., "Polymer Structure and Adhesive Behavior", *Adhesives and Manufacturing,* Schneberger, G. L., ed. (New York: Marcel Dekker, Inc., 1983).
22. Paneck, J. R. and Cook, J. P., "Stresses and Strains in Sealants", Chapter 4 in *Construction Sealants and Adhesives,* 3rd ed. (New York: John Wiley and Sons, 1991).
23. Bikerman, J. J., "Causes of Poor Adhesion", *Industrial and Engineering Chemistry,* vol. 59, no. 9, 1967.
24. Fastening, Joining, and Assembly Reference Issue, *Machine Design,* Nov. 17, 1988.
25. Sharpe, L. H., Keynote Address, in "The Science and Technology of Adhesive Bonding", by L. H. Sharpe and S. E. Wentworth, eds., *Proceedings of the 35th Sagamore Army Materials Research Conference* (1990).

Chapter 3

Joint Design

3.1 Introduction

The strength of an adhesive or sealant joint in the absence of outside or environmental factors is determined by the mechanical properties of the materials comprising the joint, the extent of interfacial contact, and residual stresses within the joint. To design a practical joint, it is also essential to know as much as possible about the expected service requirements including anticipated stresses and environmental extremes.

The design of the adhesive or sealant joint will play a significant factor in determining how it will survive outside loads. Although it may be tempting to use joints originally intended for other methods of fastening, adhesives and sealants require joints of a special design for optimum properties. The practice of using joints designed for some other method of assembly and slightly altering them to adapt to adhesives or sealants can lead to unfavorable results.

As with most fundamental processes involving adhesives or sealants, joint design cannot be completed without consideration of numerous other factors. One must have familiarity with the adhesive or sealant materials, their physical properties, and the cure conditions that will be employed. One must also be familiar with the requirements and costs of machining and forming substrates into various joint geometries. Finally, one must be cognizant of the types of stresses anticipated in service, their magnitude and duration, and their orientation to the joint.

This chapter will define the types of stress that are common to adhesive and sealant joints. The reason why shear stress is preferred over peel or cleavage stress will be made clear. Various designs will be recommended for several common joint geometries. Stress distribution analysis will lead to methods of maximizing the efficiency of the joint.

3.2 Types of Stress

A uniform stress pattern in an adhesive or sealant joint is seldom produced by application of an external force. Rather, non-uniform stress distributions are the norm. Since fracture initiates when and where local stress exceeds local strength, stress concentrations have a large influence on the breaking strength of a joint. Residual *internal* stresses and their tendency to form stress concentration regions were discussed in the previous chapter. This chapter considers the application of *external* loads on the joint.

External loads produce local stresses that may be many times the average stress. These stress concentrations are often unexpected, and they may determine the actual force that the joint can sustain. It is the responsibility of the joint designer to compensate or to minimize these effects, but first they must be understood.

Four basic types of loading stress are common to adhesive or sealant joints: tensile, shear, cleavage, and peel. Any combination or variation of these stresses, illustrated in Fig. 3.1, may be encountered in an application.

3.2.1 Tensile and compressive stress

Tensile stress develops when forces acting perpendicular to the plane of the joint are distributed uniformly over the entire bonded area. In tension, the adhesive develops high stress regions at the outer edge (Fig. 3.2), and those edges then support a disproportionate amount of the load. The first small crack that occurs at the weakest area of one of the highly stressed edges will propagate swiftly and lead to failure of the joint. However, if the joint is properly designed, it will show good resistance to tensile loading because the loading is more easily distributed.

Proper design requires that the joint has parallel substrate surfaces and axial loads. Unfortunately in practical applications, bond thick-

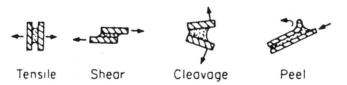

Figure 3.1 The four basic types of stress common to adhesives and sealants.

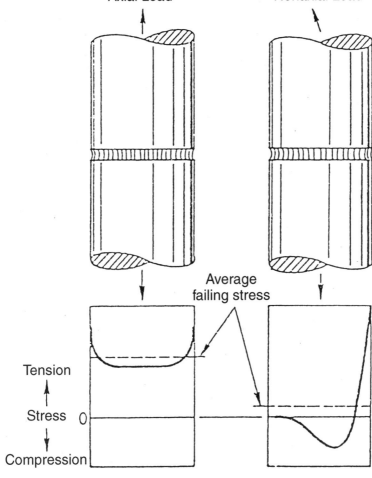

Figure 3.2 Stress distribution in a butt joint under axial and non-axial loading.[1]

ness tolerance is difficult to control, and loads are rarely axial. Undesirable cleavage or peel stresses (explained below) then tend to develop. Tensile joints should be designed with physical restraints to ensure continual axial loading. The adherends must also have sufficient rigidity so that the stress is distributed evenly over the entire bonded area.

In practice, the stress distribution in simple tensile joints, such as the butt tensile joint shown in Fig. 3.2, is far from uniform. It is uniform only if the adhesive and adherends do not deform laterally when the joint is stressed, or if they deform by the same amount.

Compression loads are the opposite of tensile. As with tensile loads, it is important to keep the loads aligned so that the adhesive will see purely compressive stress. An adhesive loaded in compression is unlikely to fail, although it may crack at weak spots due to uneven stress distribution. Actually, a joint loaded in "pure" compression hardly needs bonding of any sort. If the compression force is high enough and there is no movement of the parts, the parts will stay in position relative to one another unless the adhesive fails cohesively.

The polyamide-imide journal bearing application in Chapter 2 (Fig. 2.13) is an excellent example of this. At elevated temperatures, the internal cylinder (polyamide-imide) is compressing against the outside cylinder (steel) because of the thermal expansion coefficient differences. At elevated service temperatures an adhesive is not necessary in the classical sense, and the parts are held together by thermal fit. However, at lower operating temperatures the opposite effect occurs. The internal cylinder wants to contract more than the steel cylinder, and the adhesive is exposed to high tensile forces of uneven distribution.

Tensile or compressive stress is measured as force per bonded area and is usually given in units of pounds per square inch of bonded area (psi). For example, if two circular rods of 1 in. diameter were bonded as a butt joint, the tensile strength would be measured as the ultimate load (in pounds of force) divided by the bonded area $[\pi \times (0.5 \text{ in.})^2]$ and given in units of psi.

In SI (Système International d'Unités) units, stress is given as megapascals (MPa) One psi is equivalent to 0.006895 MPa; or IMPa is equivalent to 145 psi.

3.2.2 Shear stress

Shear stress results when forces acting in the plane of the adhesive try to separate the adherends. Joints that are dependent on the adhesive's shear strength are relatively easy to make and are commonly used in practice. Adhesives are generally strongest when stressed in shear because all of the bonded area contributes to the strength of the joint and the substrates are relatively easy to keep aligned.

The lap shear joint, shown in Fig. 3.3 left, represents the most common joint design in adhesive bonding. Shear stresses are measured similar to tensile forces, as force per bonded area, psi. By overlapping the substrates, one places the load bearing area in shear. Note that most of the stress is localized at the ends of the overlap. The center of the lap joint contributes little to joint strength. In fact, depending on the joint geometry and physical properties of the adhesive and adherends, two small bands of adhesive at each end of the overlap may

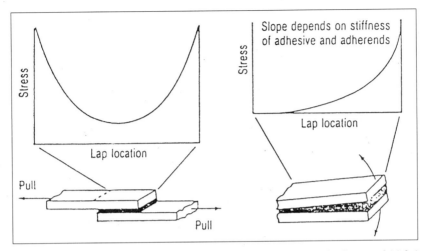

Figure 3.3 Stress distribution on an adhesive when stressed in (left) shear and (right) peel or cleavage.[2]

provide the same bond strength as when the entire overlap area is bonded with adhesive. The reason for this is provided in Section 3.3.4.

3.2.3 Cleavage and peel stress

Cleavage and peel stresses are undesirable for adhesives and sealants. Cleavage is defined as the stress occurring when forces at one end of a rigid bonded assembly act to pry the adherends apart. Peel stress is similar to cleavage, but it applies to a joint where one or both of the adherends are flexible. Thus, the angle of separation (or the angle made by the separating substrates) can be much greater for peel than for cleavage.

Joints loaded in peel or cleavage offer lower strength than joints loaded in shear because the stress is concentrated at only a very small area of the total bond. The stress distribution of an adhesive in cleavage is shown in Fig. 3.3. All of the stress is localized at the end of the bond that is bearing the load. The adhesive at the other end of the bond is providing little to the ultimate strength of the joint.

Cleavage and peel forces are measured as force per linear length of bond. Consider an application where a 1 in. wide strip of pressure sensitive adhesive tape is placed on an aluminum substrate and then peeled off the substrate to measure adhesive strength. The peel strength would be measured as pounds of force required to strip the tape off the substrate divided by the width of the bond being peeled, i.e., 1 in.

Units of peel or cleavage strength are usually expressed as pounds per inch of bond width (piw). SI units for peel are kg/m where 1 piw is equivalent to 17.858 kg/m.

Peel or cleavage stress should be avoided where possible, since the stress is confined to a very thin line at the edge of the bond (Fig. 3.3 right). Brittle adhesives are particularly weak in peel because the stress is localized at only a very thin line at the edge of the bond as shown in Fig. 3.4. The stiffness of the adhesive does not allow distribution of stress over an area much larger than the thickness of the bond-line. On the other hand tough, flexible adhesives distribute the peeling stress over a wider bond area and show greater resistance to peeling forces.

Common rigid epoxy adhesives can generally provide greater than 2000 psi shear strength, but they may only provide on the order of 2 piw of peel strength. Tough, flexible adhesives that are specifically formulated for high peel resistance could provide peel strength values in the range of 25–50 piw.

3.3 Maximizing Joint Efficiency

For maximum joint efficiency, non-uniform stress distribution should be reduced though proper joint design and selection of certain design variables that are of importance to stress distribution. The number of variables affecting the stress distribution, even in the most common joint designs, is large. The following variables are most important.

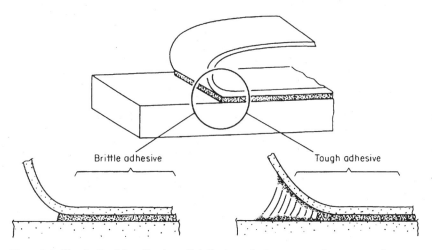

Figure 3.4 Tough, flexible adhesives distribute peel stress over a larger area.[3]

1. Adhesive material properties
2. Adhesive thickness
3. Geometry of the bond area
4. Adherend properties

The effect of non-uniform stress distribution is that the average stress (i.e., the load divided by bond area) is always lower than the maximum stress at localized areas within the joint. Only in cases where there is a near uniform stress does the average stress approach the maximum stress. Failure in the bond always begins at the maximum stress regions. Therefore, an understanding of the stress distribution in a joint is of primary importance in the design of adhesive joints.

3.3.1 Adhesive properties

The simplest example of how adhesive properties affect joint design efficiencies can be demonstrated by the stress distribution analysis of a simple lap shear or peel joint, Fig. 3.3. The maximum shear stress is dependent on the rheological characteristics of the adhesive. Tough, flexible adhesives have less of a maximum stress, but the average stress is generally higher. Since typically high elongation adhesives have lower cohesive strengths, the advantage of high elongation and peel strength is usually compromised by a corresponding decrease in the adhesive's internal shear strength.

Adhesive modulus also influences the stress distribution; however, it is not a direct effect. For two adhesives of the same strength and elongation, the higher modulus adhesive would carry more load. However, the higher elongating adhesives that have good peel and cleavage strength, tend to have lower moduli and poorer shear strength.

Crack propagation sensitivity is greater with brittle adhesives (i.e., those having low elongation and high modulus). Fatigue life is, generally, lower with brittle adhesives. If the applied fatigue stress is measured as a percentage of ultimate, then the fatigue life of joints fabricated from high elongation adhesives is considerably superior to more brittle adhesives. This is due to both uniform stress distribution and high internal energy damping with more flexible adhesives.

If high stress nonuniform distributions are expected in service either because of external loading, such as peel or cleavage, or from internal stress, such as from thermal expansion differences or shrinkage, then tough, flexible adhesives and sealants are usually better than more brittle ones. This is why flexible adhesives are often recommended for bonding plastic or elastomeric surfaces. The physical nature of these substrates often results in cleavage or peeling forces acting on the adhesive, and generally, there are significant differences in thermal expansion coefficients. Brittle adhesives would provide joints having high stress concentrations; whereas, flexible adhesives would provide more uniform stress distribution.

When an application requires both high temperature resistance and high peel strength the adhesive user faces a difficult compromise. Usually, the more heat resistant polymers are also more densely cross-linked and brittle. Flexibilized polymers have better resistance to peel, impact, and fatigue forces but reduced heat resistance. As we will see in later chapters, modern adhesive formulations, notably temperature resistant thermoset base resins with a discrete elastomeric phase, have come a long way in solving this dilemma.

In summary, usually a tough, flexible adhesive or sealant is preferred over a brittle stiff one as long as the adhesive has sufficient cohesive strength for the application. With these adhesives, it is much easier to have a joint with uniformly distributed stress. However, there are disadvantages in using tough, flexible adhesives that must be overcome. They usually have lower cohesive strength. Since the forces that hold the internal molecules together are lower, the temperature capability and environmental resistance also suffer. As a result, tough, flexible adhesives are commonly used on substrates such as plastics and elastomers where the environmental service conditions are usually not extreme and the physical properties of the adhesive closely match those of the substrates. More brittle adhesives are more densely cross-linked and usually employed in structural applications that are likely to see elevated temperatures and aggressive environmental conditioning.

3.3.2 Adhesive thickness

The most important aspects of adhesive thickness, or bond-line thickness, are its magnitude and its uniformity or homogeneity. Generally, one tries to have as thin an adhesive layer as possible without any chance of bond starvation. In practice, this translates into bond-line thicknesses from 0.002 in. to 0.008 in. Adhesive strength does not vary significantly in this range. It is usually best to attempt to have a constant bond-line thickness of about 0.005 in. With thicker adhesive bond-lines, one runs the risk of incorporating higher void concentrations into the joint. In addition, stresses at the corner of the adhesive-adherend tend to be larger because it is difficult to keep the loads axial with a very thick bond-line. It should also be remembered that adhesives are generally formulated to cure in thin sections.

Thicker sections could change the curing properties and result in increased internal stresses and non-optimal physical properties.

The substrates should be as parallel as possible, thus requiring uniformity in adhesive thickness across the bonded area. If the substrates are not parallel, the loading will not remain aligned and this condition could translate into cleavage stress on the adhesive.

There are several methods used for maintaining a constant, predetermined adhesive thickness. These methods include:

- Adjusting the viscosity of the adhesive
- Application of a precalculated amount of pressure during cure
- Using fixturing that is specifically designed for the application
- Application of a shim or insert within the bond-line so that a uniform, predetermined thickness can be maintained

These techniques will be reviewed in Chapter 18.

3.3.3 Effect of bond area geometry

For a given adhesive and adherend, the strength of a joint stressed in shear depends primarily on the width and depth of the overlap and the thickness of the adherend. Adhesive shear strength is directly proportional to the width of the joint. Strength can sometimes be increased by increasing the overlap depth, but the relationship is not linear. Since the ends of the bonded joint carry a higher proportion of load than the interior area, the most efficient way of increasing joint strength is by increasing the width of the joint.

A plot of failure load versus overlap length for brittle and ductile adhesives is shown in Fig. 3.5. Increasing overlap length increases the joint strength to a point where a further increase in bond overlap length does not result in an increase in load carrying ability. This curve is the result of the capability of ductile adhesives resisting non-uniform loads more than brittle adhesives.

Since the stress distribution across the bonded area is not uniform and depends on joint geometry, the failure load of one joint design cannot be used to predict the failure load of another design having a different geometry. The results of a particular test pertain only to joints that are exact duplicates of one another. This means that the results of laboratory tests on lap shear specimens cannot be directly converted to more complex joint geometries.

To compare different geometries, curves showing the ratios of overlap length to adherend thickness, l/t, are sometimes useful. Figure 3.6 shows an example of the effect of l/t ratio on aluminum joints bonded

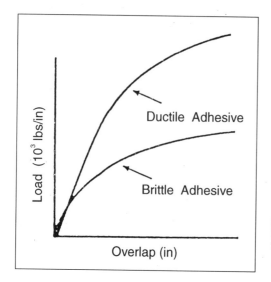

Figure 3.5 Effect of overlap length on failure load using ductile and brittle adhesives.[4]

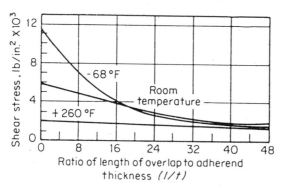

Figure 3.6 The effect of ratio of length of overlap to adherend thickness (l/t) on adhesive strength at three test temperatures for aluminum joints bonded with a nitrile rubber adhesive.[5]

with a nitrile-rubber adhesive. From such a chart, the design engineer can determine the overlap length required for a given adherend thickness to obtain a specific joint strength.

3.3.4 Adherend properties

The properties of the substrates being joined have a major influence on the stress distribution in the joint. Non-uniform stress distribution in the adhesive is caused by the relative displacement of the adherends due to the strain in the adherends. Since the adhesive must accept this displacement differential, the flexibility of the adhesive is, involved.

The stiffness of the adherend is characterized by the product of the Young's modulus, E, and the adherend thickness, t. Then Et of each adherend becomes an important factor in the shear stress distribution. As the product Et becomes large, the shear stress distribution becomes more uniform.

In a shear joint made from thin, relatively flexible adherends, there is a tendency for the bonded area to distort. This distortion, illustrated in Fig. 3.7, causes cleavage stress on the ends of the joint, and the joint strength may be considerably impaired. Thicker adherends are more rigid, and the distortion is not as much a problem as with thin-gauge adherends. Figure 3.8 shows the general interrelationship between failure load, depth of overlap, and adherend thickness for a specific metallic adhesive joint. As the adherend thickness (i.e., the relationship Et) increases, the failure load increases for identical overlap lengths. For constant adherend thicknesses or constant Et, the failure load increases with increasing overlap length up to a certain point. Beyond that overlap distance the failure load remains constant. In this region the entire load is supported by the edge region of the overlap. The central section of adhesive is not contributing to the strength of the joint.

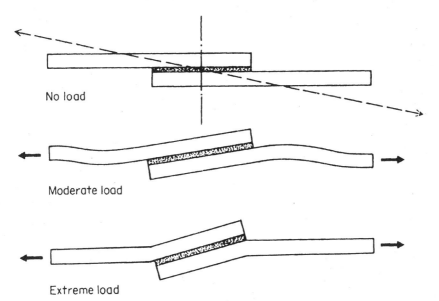

Figure 3.7 Distortion caused by loading can introduce cleavage stresses and must be considered in the joint design.[6]

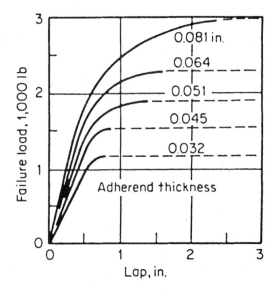

Figure 3.8 Interrelation of failure loads, depth of lap, and adherend thickness for lap joints with a specific adhesive and adherend.[7]

3.4 General Joint Design Rules

The designer should take into consideration the following rules in the design of adhesive or sealant joints. These rules are the basis for recommended joint design geometries in the next section.

1. Keep the stress on the bond-line to a minimum.
2. Whenever possible, design the joint so that the operating loads will stress the adhesive in shear.
3. Peel and cleavage stresses should be minimized.
4. Distribute the stress as uniformly as possible over the entire bonded area.
5. Adhesive strength is directly proportional to bond width. Increasing width will always increase bond strength; increasing the depth of overlap does not always increase strength.
6. Generally, rigid adhesives are better in shear, and flexible adhesives are better in peel.
7. Although typically a stronger adhesive material may produce a stronger joint, a high elongation adhesive with a lower cohesive strength could produce a stronger joint in applications where the stress is distributed nonuniformly.
8. The stiffness of the adherends and adhesive influence the strength of a joint. In general, the stiffer the adherend with respect to the

adhesive, the more uniform the stress distribution in the joint and the higher the bond strength.
9. The higher the Et (modulus x thickness) of the adherend, the less likely the deformation during load, and the stronger the joint.
10. Within reasonable limits, the adhesive bond-line thickness is not a strong influence on the strength of the joint. More important characteristics are a uniform joint thickness and void free adhesive layer.

The remainder of this chapter will provide practical applications of these rules.

3.5 Common Adhesive Joint Designs

The ideal bonded joint is one in which under all practical loading conditions the adhesive is stressed in the direction in which it most resists failure. A favorable stress can be applied to the bond by using proper joint design.

Some joint designs may be impractical, expensive to make, or hard to align. The design engineer will often have to weigh these factors against optimum adhesive performance. Common joint designs are shown in the next several sections for flat and cylindrical adherends, stiffeners, angle and corner joints, plastic and elastomer joints, and wood.

3.5.1 Joints for flat adherends

3.5.1.1 Butt joints. The simplest joint to make is the plain butt joint. However, butt joints cannot withstand bending forces because the adhesive would experience cleavage stress. If the adherends are too thick to design as simple overlap-type joints, the butt joint can be improved in a number of ways, as shown in Fig. 3.9. All the modified butt joints reduce the cleavage effect caused by side loading. Tongue-and-groove joints have an advantage in that they are self-aligning and act as a reservoir for the adhesive. The scarf joint keeps the axis of loading in-line with the joint and does not require a major machining operation.

3.5.1.2 Lap joints. Lap joints are the most commonly used adhesive joint because they are simple to fabricate, applicable to thin adherends, and stress the adhesive in shear. However, the adherends in the simple lap joint are offset, and the shear forces are not in-line, as was illustrated in Fig. 3.7. This twisting of the lap shear specimen results

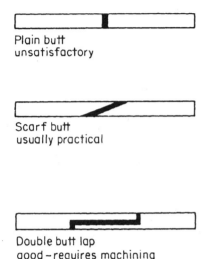

Plain butt
unsatisfactory

Scarf butt
usually practical

Double butt lap
good - requires machining

Tongue and groove
excellent - requires machining

Figure 3.9 Butt joint designs.

in cleavage stress at the ends of the joint, which seriously impairs its efficiency. Modifications of lap-joint design include:

1. Redesigning the joint to bring the load on the adherends in-line;
2. Making the adherends more rigid (thicker) near the bond area (see Fig. 3.8); and
3. Making the edges of the bonded area more flexible for better conformance, thus minimizing peel (i.e., tapering the edges of the adherend).

Modifications of lap joints are shown in Fig. 3.10. The joggle-lap joint design is the easiest method of bringing loads into alignment. The joggle lap can be made by simply bending the adherends. It also provides a surface to which it is easy to apply pressure. The double-lap joint has a balanced construction that is subjected to bending only if loads on the double side of the lap are not balanced. The beveled lap joint is also more efficient than the plain lap joint. The beveled edges, made by tapering the ends of the adherends, allow conformance during loading. This reduces cleavage stress on the ends of the joint.

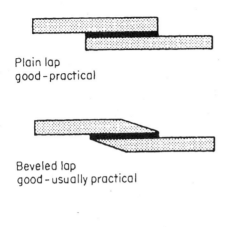

Plain lap
good-practical

Beveled lap
good-usually practical

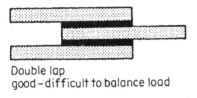

Double lap
good-difficult to balance load

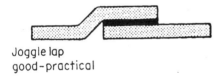

Joggle lap
good-practical

Figure 3.10 Lap joint designs.

As shown in Fig. 3.11, tapering or beveling the edges of the joint greatly improves the load bearing capacity, because it permits those regions to bend and, thus, to distribute the stress down the length of the bonded area to some degree. Comparison of curve 1 for untapered and curve 2 for tapered lap shear joints reveals how much better the load is shared along the linear axis of the adhesive bond. Figure 3.12 shows the strength advantage that can be gained by tapering the ends of the standard lap shear specimen.

3.5.1.3 Strap joints. Strap joints keep the operating loads aligned and are generally used where overlap joints are impractical because of adherend thickness. Strap joint designs are shown in Fig. 3.13. Like the lap joint, the single strap is subjected to cleavage stress under bending forces. The double strap joint is more desirable when bending stresses are encountered. The beveled double strap and recessed dou-

108　Chapter Three

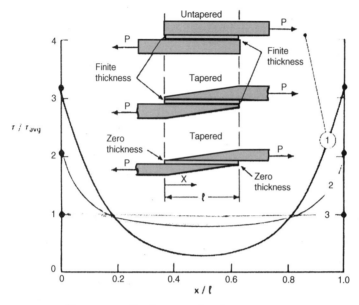

Figure 3.11 Shear stress distribution in bonded joints: untapered (1), tapered and finite thickness (2), and tapered and zero thickness (3).[2]

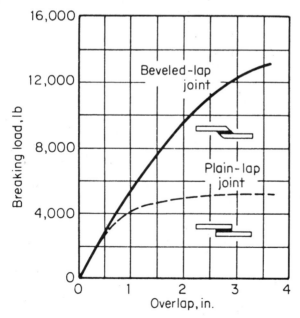

Figure 3.12 The effect of beveled and plain lap-joint design on breaking load.[8]

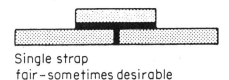

Single strap
fair - sometimes desirable

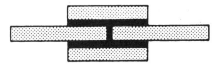

Double strap
good - sometimes desirable

Recessed double strap
good - expensive machining

Beveled double strap
very good - difficult production

Figure 3.13 Strap joint designs.

ble strap are the best joint designs to resist bending forces. Unfortunately, they both require expensive machining.

3.5.2 Stiffening joints and other methods of reducing peel stress

When thin members are bonded to thicker sheets, operating loads generally tend to peel the thin member from its base, as shown in Fig. 3.14 (top). The subsequent illustrations show what can be done to decrease peeling tendencies in simple joints.

Often thin sheets of a material are made more rigid by bonding stiffening members to the sheet. When such sheets are flexed, the

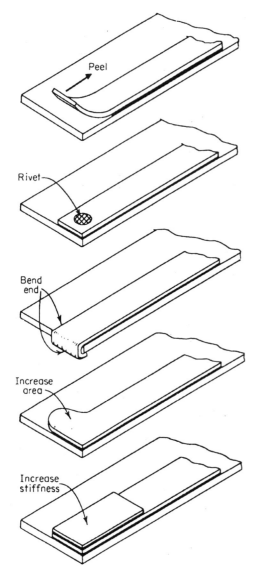

Figure 3.14 Minimizing peel in adhesive joints.[9]

bonded joints are subjected to cleavage stress. Some design methods for reducing cleavage stress on stiffening joints are illustrated in Fig. 3.15. Resistance of stiffening members to bending forces is increased by extending the bond area, providing greater flange flexibility, and increasing the stiffness of the base sheet.

3.5.3 Cylindrical joints

Several recommended designs for rod and tube joints are illustrated in Fig. 3.16. These designs should be used instead of the simpler butt

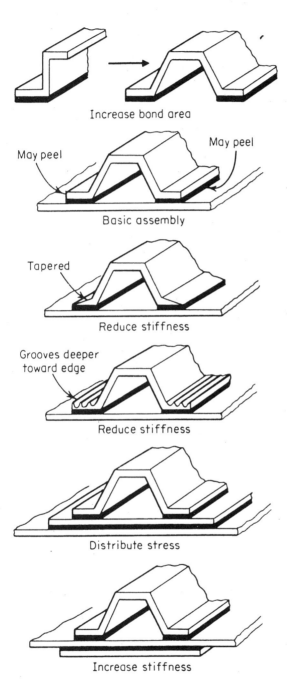

Figure 3.15 Stiffening assembly designs.[9]

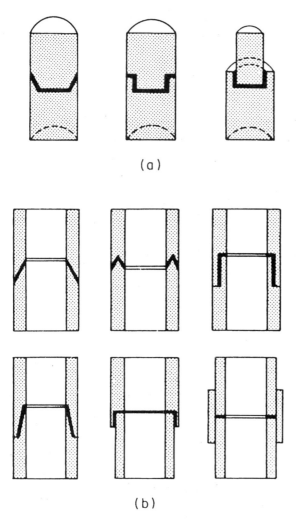

Figure 3.16 Designs for rod and tube joints. (*a*) Three designs for round bars. (*b*) Six designs for cylinders or tubes.[10]

joint. Their resistance to bending forces and subsequent cleavage is much better, and the bonded area is larger. Unfortunately, most of these joint designs require a machining operation.

3.5.4 Angle and corner joints

A butt joint is the simplest method of bonding two surfaces that meet at an odd angle. Although the butt joint has good resistance to pure tension and compression, its bending strength is very poor. Dado, L, and T angle joints, shown in Fig. 3.17, offer greatly improved properties. The T design is the preferable angle joint because of its large bonding area and good strength in all directions.

Corner joints for relatively flexible adherends such as sheet metal should be designed with reinforcements for support. Various corner-joint designs are shown in Fig. 3.18. With very thin adherends, angle joints offer low strength because of high peel concentrations. A design consisting of right-angle corner plates or slip joints offers the most satisfactory performance. Thick, rigid members such as rectangular bars and wood may be bonded with an end lap joint, but greater strength can be obtained with mortise and tenon. Hollow members, such as extrusions, fasten together best with mitered joints and inner splines.

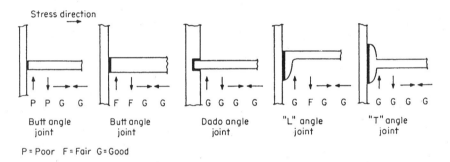

Figure 3.17 Angle joint designs. [9]

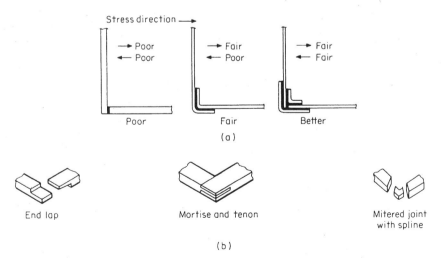

Figure 3.18 Corner joint designs. (*a*) Corner joints for relatively thin adherends. (*b*) Corner joints for thick adherends.[9,11]

3.5.5 Plastic and elastomer joints

Design of joints for plastics and elastomers generally follows the same practice as for metal. However, the designer should be aware of certain characteristics for these materials that require special consideration. Such characteristics include flexibility, low modulus, high thermal expansion coefficients, thin section availability, and anisotropy. These characteristics tend to produce significant non-uniform stress distribution in the joint. Thus, tough, flexible adhesives are usually recommended to bond plastic or elastomer substrates.

3.5.5.1 Flexible plastics and elastomers.
Thin or flexible polymeric substrates may be joined using a simple or modified lap joint. The double strap joint is best, but also the most time-consuming to fabricate. The strap material should be made out of the same material as the parts to be joined, or at least have approximately equivalent strength, flexibility, and thickness. The adhesive should have the same degree of flexibility as the adherends.

If the sections to be bonded are relatively thick, a scarf joint is acceptable. The length of the scarf should be at least four times the thickness; sometimes larger scarfs may be needed.

When bonding elastic material, forces on the elastomer during cure of the adhesive should be carefully controlled, since excess pressure will cause residual stresses at the bond interface. Stress concentrations may also be minimized in rubber-to-metal joints by elimination of sharp corners and using metal thick enough to prevent peel stresses that arise with thinner-gauge metals.

As with all joint designs, flexible plastic and elastomeric joints should avoid peel stress. Figure 3.19 illustrates methods of bonding flexible substrates so that the adhesive will be stressed in its strongest direction.

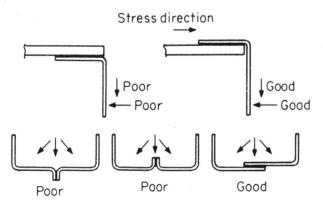

Figure 3.19 Joint designs for flexible substrates.[9]

3.5.5.2 Reinforced plastics.

Reinforced plastics are often anisotropic materials. This means their strength properties are directional. Joints made from anisotropic substrates should be designed to stress both the adhesive and adherend in the direction of greatest strength. Plastic laminates, for example, should be stressed parallel to the laminations. Stresses normal to the laminate may cause the substrate to delaminate. Single and joggle lap joints are more likely to cause delamination than scarf or beveled lap joints. The strap-joint variations are useful when bending loads occur.

3.5.6 Wood joints

Wood is an anisotropic and dimensionally unstable material. Properties differ with grain direction. Joints should be made from similar types and cuts of wood to avoid stress concentrations.

Tensile strength of wood in the line of the grain is approximately ten times greater than in a direction perpendicular to the grain. Adhesive joints should be designed to take this factor into account.

Wood changes dimensions with moisture absorption. The degree of dimensional stability is also dependent on grain direction. When wood is bonded to a much more dimensionally stable material, the adhesive must be strong enough to withstand the loads caused by dimensional distortion of the wood.

A plain scarf joint with very flat slopes is the most efficient joint design for wood, but it requires the application of lateral or transverse pressure during the time the adhesive hardens. Common wood joints that are self-aligning and require no pressure are shown in Figs. 3.20 and 3.21.

3.6 Sealant Joint Efficiency

Many factors ultimately affect the performance of a sealant joint, but the shape and dimensions of the sealant cross-section are considered of primary importance. The primary dimensions of a sealant joint are its depth and width as shown in Fig. 3.22b. As with adhesives, the design of the seal varies with the types of material being used. However, there are only two main types of sealant joint configurations: butt joints and lap joints.

3.6.1 Stress distribution in butt joints

In butt joints, as temperature rises, the sealant will go into compression due to the thermal expansion of the adherends being sealed. As the temperature declines, the adherends will contract and the joint

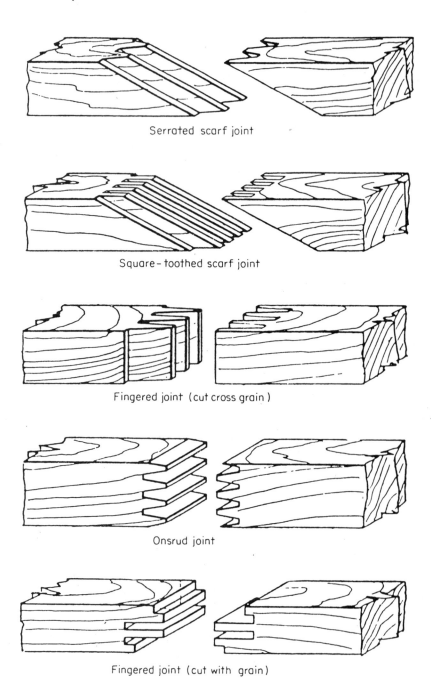

Figure 3.20 End joints for wood.[1]

Joint Design   117

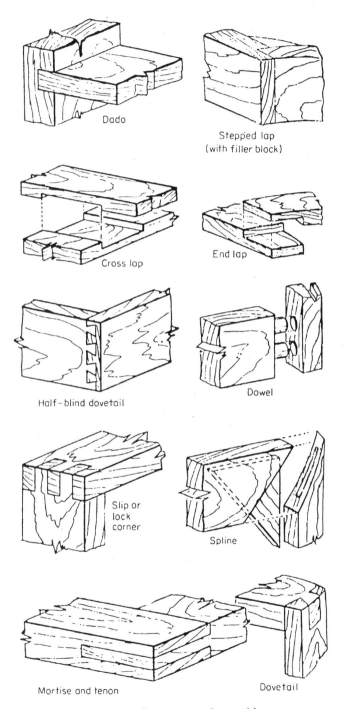

Figure 3.21 Construction adhesive joints for wood.[1]

will go into tension. The movement of a sealant in a butt joint is shown in Fig. 3.22.

Early investigators used dog-bone tensile specimens of the cured sealant itself to determine elongation capacity and tensile strength; however, these results do not compare with the behavior of the sealant in actual joints. Peterson[12] showed that this was the result of unequal stress distribution. Figure 3.23 shows the concentration of stress in a butt joint at the corners.

3.6.1.1 Elastic sealants. Elastic sealant materials are prime candidates for working joints because they return to their original shape after the removal of the imposed force. This property of recovery is desirable for sealants that must have a ±25% joint movement capability.

When large movement capabilities are required, joints with shallow depth are the best design for reducing strain. It is not possible to

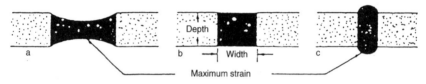

Figure 3.22 Movement of sealant in a butt joint.[13]

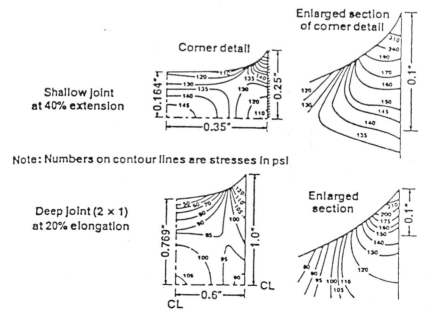

Figure 3.23 Unequal stress distributions in sealant butt joints.[13]

designate a depth relationship that holds for all sealant widths. The following rules cover joint designs and sealants commonly used in construction applications.[14]

1. The minimum size joint should be ¼ × ¼ in.
2. For widths from ¼ to ½ in., depth should equal the width.
3. For widths from ½ to 1 in., depth should be approximately ½ in.
4. For widths more than 1 in., depths may be ½ to ⅝ in., depending on the width and application area, as well as the type of sealant.

The minimum joint size is given because it is difficult to seal a joint that is any smaller unless a liquid sealant is used. A depth of at least ¼ in. is needed to ensure adhesion to the side of the joint.

Figure 3.24 shows two joints with an equal width but different depths. The sealant does not change in volume during extension. As a result, the deeper seal has to neck down farther when both joints are extended the same distance. Note also that the top and bottom surface of the sealant material in the deeper design has extended a great deal more than the corresponding surface in the shallow but more proper seal.

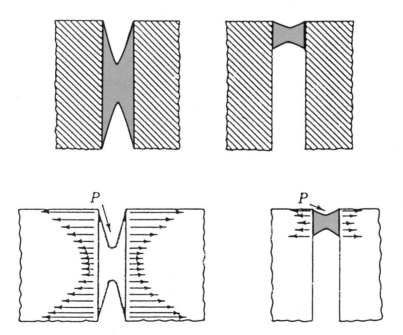

Figure 3.24 Comparisons of maximum strains (left) for a 2 inch deep seal; extension of joint 100%, sealant strain 550%; and (right) for a ¼ inch deep seal; extension of joint 100%, sealant strain 160%.[14]

A parabolic stress deformation of a sealant also occurs when the sealant is in compression. Figure 3.25 shows that a compressed sealant will extrude outside the joint area under load. This may become subject to abrasion and provide a safety hazard as well as an unsightly joint. It is advisable in horizontal joints that are subject to traffic to keep the level of the sealant a distance ⅛ in. below the substrate level.

An improvement in the rectangular cross section can be achieved by using a curved back-up material in the joint and tooling the top surface as shown in Fig. 3.26 (top). The tooling must be done with care, because if the sealant is too thin at its center it will buckle under compression as shown in Fig. 3.26 (bottom).

Elastic sealants should be bonded to only two sides of the joint in order to perform properly. The bottom surface of the sealant must be

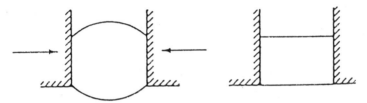

Figure 3.25 Sealant under compression.[14]

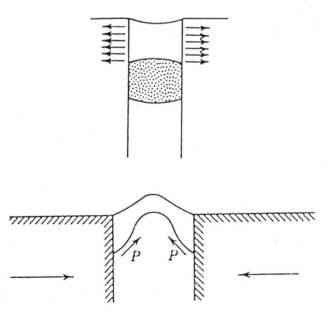

Figure 3.26 Tooled sealant showing (top) uniform substrate stress and (bottom) under compression if the sealant is too thin.[14]

free to deform. If the bottom of the joint is bonded, the sealant may rupture in order to deform.

Sealants can be installed at any temperature, and outdoor sealants can be installed at any season of the year. The ideal situation is when the sealant is installed during moderate temperatures so that it will be subjected to both compression and tension during a normal temperature cycle.

3.6.1.2 Deformable sealants. The rules described in the preceding section hold true for elastic sealants that are characterized by their recovery. Another broad class of sealants is deformable sealants that show some degree of instantaneous elasticity or recovery under short-term load, but they creep under longer-term loading. The deformable or low recovery sealants show a great deal of stress relaxation. When held in a deformed state, these materials will relax into equilibrium so that a new shape is formed in this unstressed state.

The maximum width of these sealants should never be more than ¾ in. since the sealant could begin to show some permanent change with several cycles. Another limitation for this class of sealant is that they are limited to areas of application in which the movements do not exceed ±12.5%. By limiting this group of sealants to applications requiring less than ±12.5% movement, the sealant will not become distorted with repeated temperature cycles. Another property of most sealants in this group is that with the loss of solvent, some sealants become quite tough and can get harder in time. This age hardening is not conducive to large joint movements.

3.6.2 Shear joints

The most typical types of sealant shear joints occur in glazing sealants and sealants placed between glass panels. Strains in shear joints are less than in butt joints for a given thickness. However, the movement in shear can cause catastrophic failures under certain circumstances[14].

The strain on the sealant when the shear joint moves 50% is only 25%. The proper joint shape is a square, such as ½ in. × ½ in., if a great amount of movement is expected. If a sealant with ±50% joint movement is used, the surface could be displaced 112% without extending the sealant more than 50% in either direction. With use of low modulus silicone sealant with 100% extensibility, the surfaces could be displaced up to 175% without extending the sealant more than 100%. In each case the sealant gets no compression.

Figure 3.27 illustrates the movement in lap shear joints. If greater movement is anticipated, then it is necessary only to widen the distance between the moving surfaces.

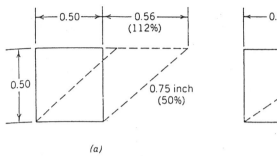 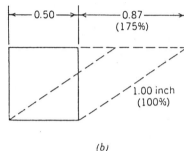

Figure 3.27 Movements in lap shear joints. Original sealant dimensions are ½ × ½ inches. (a) Displacement of 0.56 inches with sealant only extended 50%, using a ±50% capacity sealant (medium modulus). (b) Displacement of 0.87 inches with sealant only extended 100%, using sealant with +100% extension capability (low modulus).[14]

3.7 Common Sealant Joint Designs

Design of a sealant joint depends to a great deal on the type of sealant selected. Common joints are the simple butt joint, lap joint, and angle joint, as shown in Fig. 3.28.

Where a choice is available, the sealant most compatible with the part configuration should be used. For example, a free flowing liquid sealant cannot be applied to a vertical surface, such as a wall joint. In this application, a non-sagging type must be used.

3.7.1 Butt joints

There are many variations of the common sealant joint configurations. Various versions of the compound butt joint are shown in Fig. 3.29.

If sealant extension is important in a working butt joint, a release agent should be placed on the bottom inner surface so that the sealant does not bond to this face. Thus, the sealant is free to stretch when joint volume increases (Fig. 3.29b, c, and d). Figure 3.30 shows that the same basic principle applies to corner joint designs. Here a corner bead is used as a back-up material.

3.7.2 Lap joints

Lap joints can be easily sealed with tape; bead sealed, if the thickness of the joined sheets is sufficient to support a bead of sealant; and sandwich sealed. Sandwich sealing is a common method with structural adhesives. Sealant thickness is more critical when the material is applied between surfaces. In all joints, production time can be delayed considerably by excessively thick applications of the material.

Joint Design   123

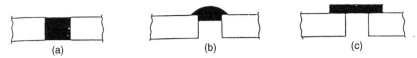

Butt joint—Use sealant if thickness of plate is sufficient (a), or bead seal if plates are thin (b). Tape can also be used. If joint moves due to dynamic loads or thermal expansion and contraction, a flexible sealant with good adhesion must be selected (c). Select flexible tape for butt joint if movement is anticipated.

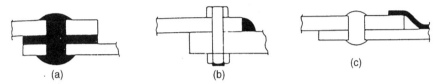

Lap joint—Sandwich sealant between mating surfaces and rivet, bolt or spot weld seam to secure joint (a). Thick plates can be sealed with a bead of sealant (b); and tape can also be used if sufficient overlap is provided as a surface to which the tape can adhere (c).

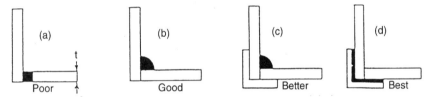

Angle joint—Simple butt joint can be sealed as shown (a) if material thickness t is sufficient. But a better choice is the bead of sealant (b), which is independent of material thickness. Supported angle joints with bead (c) or sandwich seal (d) are better choices.

Figure 3.28 Common methods for sealing butt, lap, and angle joints.[15]

If lap joints cannot be sandwich sealed, the joint should be designed with a built-in receptacle or lip to receive the sealant as shown in Fig. 3.31. Joint designs should allow for free and easy application.

If the joint (Fig. 3.31c) is the seam of a tank made from two stamped steel pieces, it can be sealed with tape on the inside, provided that the seam is accessible. The tape must be capable of holding in the fluid. Inside sealing in this fashion is augmented by fluid pressure. Liquid sealants can also be gun extruded on the inside of the seam. Putty consistency sealant can be knifed or thumbed into place. Tape can also be used on the outside, but it must be able to hold the liquid pressure and resist attack from the outside environment and mechanical wear.

Many aspects of sealant joint design need to be considered in concert with sealant selection. The reader is, therefore, referred to Chapter 14 where a more detailed discussion occurs on sealant joint design.

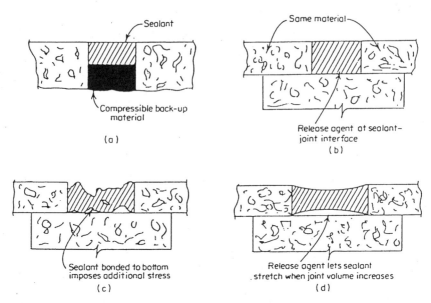

Figure 3.29 Compound butt joints: joint (a) has back-up material, which should be under compression even at maximum joint volume. Sealant must withstand compression and extension. A compound joint of similar materials (b) may cause sealant failure due to expansion and contractions, if sealant is bonded to bottom layer (c). A release agent either chemical or film, prevents sealant from bonding to bottom and allows materials to stretch normally (d).[15]

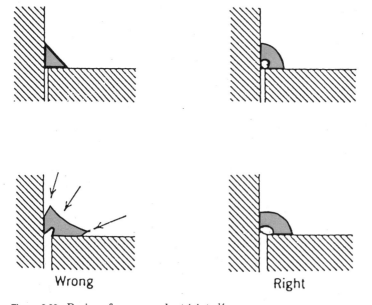

Figure 3.30 Design of corner sealant joints.[14]

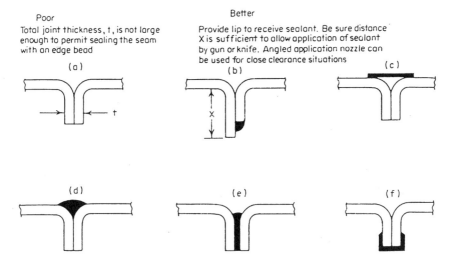

Figure 3.31 Various joint designs for a standard edge seal (*a*): lip on joint (*b*) provides handy receptacle for sealant. Seam can be inside-sealed by tape (*c*), putty (*d*), or sandwich sealed with putty or anaerobic (*e*), tape can also be used on the outside seam (*f*).[15]

References

1. Sharpe, L. H., "The Materials, Processes, and Design Methods for Assembling with Adhesives", *Machine Design*, August 18, 1966 at 178–200.
2. Chastain, C. E., "Designing Adhesive Joints", *Appliance Engineer*, vol. 8, no. 4, 1974 at 22–25.
3. Rider, D. K., "Which Adhesives for Bonded Metal Assembly", *Prod. Eng.*, May 25, 1964.
4. *International Plastics Selector.* 4th ed. (Englewood, CO: D.A.T.A. Business Publishing 1997).
5. Lunsford, L. R., "Design of Bonded Joints", in M. J. Bodnar, ed. *Symposium on Adhesives for Structural Applications* (New York: Interscience, 1962).
6. Merriam, J.C., "Adhesive Bonding?", *Mater. Des. Eng.*, September 1959.
7. Elliott, S. Y., "Techniques for Evaluation of Adhesive", *Handbook of Adhesive Bonding*, C. V. Cagel, ed. (New York: McGraw-Hill, 1973).
8. "Adhesive Bonding of Metals: What are Design Factors?," *Iron Age*, June 15, 1961.
9. Kohen, G. W., "Design Manual on Adhesives," *Machine Design*, April, 1954.
10. *Adhesive Bonding Alcoa Aluminum*, Aluminum Co. of America, 1967.
11. *Facts about Use of Adhesive and Adhesive Joints*, Hysol Division, Dexter Corporation, Bulletin Gl–700.
12. Peterson, E., *Building Seals and Sealants*, ASTM STP 606, (Philadelphia: American Society of Testing and Materials 1976) at 31.
13. Evans, R. M., Chapter 1, *Polyurethane Sealants* (Lancaster, PA: Technomic Publishing Co., Inc., 1993).
14. Paneck, J. R. and Cook, J. P., Chapter 4, *Construction Sealants and Adhesives*, 3rd ed. (New York: John Wiley & Sons 1991).
15. Amstock, J. S., "Sealants", Chapter 7 in *Handbook of Adhesive Bonding*, C. V. Cagel, ed. (New York: McGraw Hill, 1973).

Chapter 4

Standard Test Methods

4.1 Introduction

The testing of adhesives and sealants is a complex subject because there are many parameters that can affect the test result. Interactions of the adhesive and adherend are often obscure. The joint design and type of load strongly influence the final stress distribution. The parameters by which the test is conducted, such as strain rate, temperature, and environmental conditions, also have a strong impact on the test results.

Although complex, testing is an extremely important subject since one cannot reliably predict the strength of the adhesive bond based solely on characteristics of the adhesive, substrate, and the joint design. To determine the strength of the joint, one must resort to testing.

Through testing the many factors that can affect the strength of the joint can be measured. Sometimes it is difficult to separate the variables that control joint strength, but at least through testing, we have a vehicle to provide a quantitative approximation. Standard tests, such as ASTM tests, are very useful for comparing and determining the consistency of materials and processes. They are less valuable in accurately predicting the strength of specific production joints. Usually, prototype tests must be designed for this purpose.

In this chapter, standard test methods for both adhesives and sealants will be described. Similar test methods are often applied to both adhesives and sealants since adhesion is a major factor in both applications. This chapter will review fundamental tests that are used to control the consistency of materials used in these processes. The more common end-use tests and the effect of testing variables, such as adhesive thickness and strain rate, will be reviewed. Where possible, correlation of end-use test results and the fundamental properties of the adhesive or sealant will be suggested. This chapter will conclude with a discussion regarding a methodology that can be used for establishing the service life of the adhesive or sealant joint. The next chapter will address tests that are generally used for quality control and for nondestructive evaluation of the finished joint.

4.2 Reasons to Test and Basic Principles

Adhesive and sealant tests are conducted for a variety of reasons. They are used to either:

1. Choose among materials or processes, such as adhesive, adherend, or joint design;
2. Monitor the quality of production materials to make certain that they have not changed since the last time they were verified for use in the bonding process;
3. Confirm the effectiveness of a bonding process, such as surface cleaning or curing; or
4. Investigate parameters or process variables that may lead to measured differences in the performance of the bond.

There are two general categories of tests for adhesives and sealants: fundamental property tests and end-use tests. End-use tests, such as T-peel and lap shear, are those which try to simulate the type of loading and service conditions to which a joint will be subjected. These tests are relatively straight-forward, but experience is required to establish the correct specimen type and testing procedures, judge the reliability of the resulting data, and interpret the results and apply them to a practical application.

The measurement of a fundamental or bulk property, such as viscosity, hardness, or setting rate, is usually simpler and more reproducible than end-use tests. However, a correlation between fundamental properties and the results of end-use testing is difficult. Fundamental property tests are usually employed to assess the consistency of the incoming adhesive or substrate once the joint system is verified as suitable in a specific application. Often fundamental property testing is undertaken after there is a failure or unexplained occurrence to determine if a change in the incoming material may have been the possible culprit.

A number of standard tests for adhesives and sealants have been specified by the American Society for Testing and Materials (ASTM) and other professional organizations such as the U.S. Department of Defense and the Society of Automotive Engineers. By far, ASTM standards are the most commonly referenced test methods. Selected ASTM standards are presented in Table 4.1 for adhesives and in Table 4.2 for sealants. A more detailed listing of test specifications and standards may be found in Appendix C. The properties reported by suppliers of adhesives and sealants often reference ASTM standards.

A word of caution should be noted here. Standard test data are sufficient to compare strengths of various bonding systems. They can be used to compare relative effectiveness of different adhesives, surface treatments, curing schedules, and so forth. They can also be used to separate and quantitatively define the many variables that ultimately determine the performance of a joint. However, standard test results cannot be readily translated into specific strength values for an actual production joint. Actual joints generally have a complex geometry that is significantly different from the standard test specimen geometry. Under specific operating environments and depending on the kinds, frequency, and severity of the stress that the joint actually sees in service, an adhesive or sealant may perform spectacularly better or worse than what is represented by ASTM tests on a supplier's data sheet.

The most noticeable differences between standard test results and the results from an actual joint in service are due to several factors.

1. Joint design is seldom the same.
2. The mode and application of stress loading are usually different and more complex in practice.
3. Environmental aging is usually less severe in service (laboratory tests tend to "accelerate" aging so that the testing can be completed in a reasonable time). However, the effects of the actual service environment are generally more complex. For example, when in service, the joint may be simultaneously exposed to cyclic stress, cyclic temperature, and humid environments.
4. Laboratory test specimens are usually made in "controlled" environments. This control pertains to the equipment, the cleanliness (less weak boundary layer opportunity in the lab), and the personnel (training, care, and awareness).
5. The sample population is limited with actual production parts because of expense. Even with laboratory specimens, a full design-of-experiment statistical process is difficult to achieve because of the many production variables that can affect the joint strength.

Thus, the most reliable test is to measure the strength of an actual assembly under actual operating conditions. Unfortunately, such tests are often expensive or impractical. The next best method is to measure the strength of an actual assembly under simulated operating condi-

TABLE 4.1 Selected ASTM Tests for Adhesives

Aging
Resistance of Adhesives to Cyclic Aging Conditions, Test for (D 1183)
Bonding Permanency of Water- or Solvent-Soluble Liquid Adhesives for Labeling Glass Bottles, Test for (D 1581)
Bonding Permanency of Water- or Solvent-Soluble Liquid Adhesives for Automatic Machine Sealing Top Flaps of Fiber Specimens, Test for (D 1713)
Permanence of Adhesive-Bonded Joints in Plywood under Mold Conditions, Test for (D 1877)
Accelerated Aging of Adhesive Joints by the Oxygen-Pressure Method, Practice for (D 3632)

Amylaceous matter
Amylaceous Matter in Adhesives, Test for (D 1488)

Biodeterioration
Susceptibility of Dry Adhesive Film to Attack by Roaches, Test for (D 1382)
Susceptibility of Dry Adhesive Film to Attack by Laboratory Rats, Test for (D 1383)
Permanence of Adhesive-Bonded Joints in Plywood under Mold Conditions, Test for (D 1877)
Effect of Bacterial Contamination of Adhesive Preparations and Adhesive Films, Test for (D 4299)
Effect of Mold Contamination on Permanence of Adhesive Preparation and Adhesive Films, Test for (D 4300)

Blocking point
Blocking Point of Potentially Adhesive Layers, Test for (D 1146)

Bonding permanency
(See Aging)

Chemical reagents
Resistance of Adhesive Bonds to Chemical Reagents, Test for (D 896)

Cleavage
Cleavage Strength of Metal-to-Metal Adhesive Bonds, Test for (D 1062)

Cleavage/peel strength
Strength Properties of Adhesives in Cleavage Peel by Tension Loading (Engineering Plastics-to-Engineering Plastics), Test for (D 3807)
(See also Peel Strength)

Corrosivity
Determining Corrosivity of Adhesive Materials, Practice for (D 3310)

Creep
Conducting Creep Tests of Metal-to-Metal Adhesives, Practice for (D 1780)
Creep Properties of Adhesives in Shear by Compression Loading (Metal-to-Metal), Test for (D 2293)
Creep Properties of Adhesives in Shear by Tension Loading, Test for (D 2294)

TABLE 4.1 Selected ASTM Tests for Adhesives (*Continued*)

Cryogenic temperatures
Strength Properties of Adhesives in Shear by Tension Loading in the Temperature Range from −267.8 to −55°C (−450 to −67°F), Test for (D 2557)
Density
Density of Adhesives in Fluid Form, Test for (D 1875)
Durability (including weathering)
Effect of Moisture and Temperature on Adhesive Bonds, Test for (D 1151)
Atmospheric Exposure of Adhesive-Bonded Joints and Structures, Practice for (D 1828)
Determining Durability of Adhesive Joints Stressed in Peel, Practice for (D 2918)
Determining Durability of Adhesive Joints Stressed in Shear by Tension Loading, Practice for (D 2919)
(See also Wedge Test)
Electrical properties
Adhesives Relative to Their Use as Electrical Insulation, Testing (D 1304)
Electrolytic corrosion
Determining Electrolytic Corrosion of Copper by Adhesives, Practice for (D 3482)
Fatigue
Fatigue Properties of Adhesives in Shear by Tension Loading (Metal/Metal), Test for (D 3166)
Filler content
Filler Content of Phenol, Resorcinol, and Melamine Adhesives, Test for (D 1579)
Flexibility
(See Flexural Strength)
Flexural strength
Flexural Strength of Adhesive Bonded Laminated Assemblies, Test for (D 1184)
Flexibility Determination of Hot Melt Adhesives by Mandrel Bend Test Method, Practice for (D 3111)
Flow properties
Flow Properties of Adhesives, Test for (D 2183)
Fracture strength in cleavage
Fracture Strength in Cleavage of Adhesives in Bonded Joints, Practice for (D 3433)
Gap-filling adhesive bonds
Strength of Gap Filling Adhesive Bonds in Shear by Compression Loading, Practice for (D 3931)
High-temperature effects
Strength Properties of Adhesives in Shear by Tension Loading at Elevated Temperatures (Metal-to-Metal), Test for (D 2295)

TABLE 4.1 Selected ASTM Tests for Adhesives (*Continued*)

Hydrogen-ion concentration
Hydrogen Ion Concentration, Test for (D 1583)

Impact strength
Impact Strength of Adhesive Bonds, Test for (D 950)

Light exposure
(See Radiation Exposure)

Low and cryogenic temperatures
Strength Properties of Adhesives in Shear by Tension Loading in the Temperature Range from −267.8 to −55°C (−450 to 67°F), Test for (D 2557)

Nonvolatile content
Nonvolatile Content of Aqueous Adhesives, Test for (D 1489)
Nonvolatile Content of Urea-Formaldehyde Resin Solutions, Test for (D 1490)
Nonvolatile Content of Phenol, Resorcinol, and Melamine Adhesives, Test for (D 1582)

Odor
Determination of the Odor of Adhesives, Test for (D 4339)

Peel strength (stripping strength)
Peel or Stripping Strength of Adhesive Bonds, Test for (D 903)
Climbing Drum Peel Test for Adhesives, Method for (D 1781)
Peel Resistance of Adhesives (T-Peel Test), Test for (D 1876)
Evaluating Peel Strength of Shoe Sole Attaching Adhesives, Test for (D 2558)
Determining Durability of Adhesive Joints Stressed in Peel, Practice for (D 2918)
Floating Roller Peel Resistance, Test for (D 3167)

Penetration
Penetration of Adhesives, Test for (D 1916)

pH
(See Hydrogen-Ion Concentration)

Radiation exposure (including light)
Exposure of Adhesive Specimens to Artificial (Carbon-Arc Type) and Natural Light, Practice for (D 904)
Exposure of Adhesive Specimens to High-Energy Radiation, Practice for (D 1879)

Rubber cement tests
Rubber Cements, Testing of (D 816)

Salt spray (fog) testing
Salt Spray (Fog) Testing, Method of (B 117)
Modified Salt Spray (Fog) Testing, Practice for (G 85)

TABLE 4.1 Selected ASTM Tests for Adhesives (*Continued*)

Shear strength (tensile shear strength)
Shear Strength and Shear Modulus of Structural Adhesives, Test for (E 229)
Strength Properties of Adhesive Bonds in Shear by Compression Loading, Test for (D 905)
Strength Properties of Adhesives in Plywood Type Construction in Shear by Tension Loading, Test for (D 906)
Strength Properties of Adhesives in Shear by Tension Loading (Metal-to-Metal), Test for (D 1002)
Determining Strength Development of Adhesive Bonds, Practice for (D 1144)
Strength Properties of Metal-to-Metal Adhesives by Compression Loading (Disk Shear), Test for (D 2181)
Strength Properties of Adhesives in Shear by Tension Loading at Elevated Temperatures (Metal-to-Metal), Test for (D 2295)
Strength Properties of Adhesives in Two-Ply Wood Construction in Shear by Tension Loading, Test for (D 2339)
Strength Properties of Adhesives in Shear by Tension Loading in the Temperature Range from -267.8 to $-55°C$ (-450 to $-67°F$), Test for (D 2557)
Determining Durability of Adhesive Joints Stressed in Shear by Tension Loading, Practice for (D 2919)
Determining the Strength of Adhesively Bonded Rigid Plastic Lap-Shear by Tension Loading, Practice for (D 3163)
Determining the Strength of Adhesively Bonded Plastic Lap-Shear Sandwich Joints in Shear by Tension Loading, Practice for (D 3164)
Strength Properties of Adhesives in Shear by Tension Loading of Laminated Assemblies, Test for (D 3165)
Fatigue Properties of Adhesives in Shear by Tension Loading (Metal/Metal), Test for (D 3166)
Strength Properties of Double Lap Shear Adhesive Joints by Tension Loading, Test for (D 3528)
Strength of Gap-Filling Adhesive Bonds in Shear by Compression Loading, Practice for (D 3931)
Measuring Strength and Shear Modulus of Nonrigid Adhesives by the Thick Adherend Tensile Lap Specimen, Practice for (D 3983)
Measuring Shear Properties of Structural Adhesives by the Modified-Rail Test, Practice for (D 4027)
Specimen preparation
Preparation of Bar and Rod Specimens for Adhesion Tests, Practice for (D 2094)
Spot-adhesion test
Qualitative Determination of Adhesion of Adhesives to Substrates by Spot Adhesion Test Method, Practice for (D 3808)

TABLE 4.1 Selected ASTM Tests for Adhesives (*Continued*)

Spread (coverage)	
Applied Weight per Unit Area of Dried Adhesive Solids, Test for (D 898)	
Applied Weight per Unit Area of Liquid Adhesive, Test for (D 899)	
Storage life	
Storage Life of Adhesives by Consistency and Bond Strength, Test for (D 1337)	
Strength development	
Determining Strength Development of Adhesive Bonds, Practice for (D 1144)	
Stress-cracking resistance	
Evaluating the Stress Cracking of Plastics by Adhesives Using the Bent Beam Method, Practice for (D 3929)	
Stripping strength	
(See Peel Strength)	
Surface preparation	
Preparation of Surfaces of Plastics Prior to Adhesive Bonding, Practice for (D 2093)	
Preparation of Metal Surfaces for Adhesive Bonding, Practice for (D 2651)	
Analysis of Sulfochromate Etch Solution Using in Surface Preparation of Aluminum, Methods of (D 2674)	
Preparation of Aluminum Surfaces for Structural Adhesive Bonding (Phosphoric Acid Anodizing), Practice for (D 3933)	
Tack	
Pressure Sensitive Tack of Adhesives Using an Inverted Probe Machine, Test for (D 2979)	
Tack of Pressure-Sensitive Adhesives by Rolling Ball, Test for (D 3121)	
Tensile strength	
Tensile Properties of Adhesive Bonds, Test for (D 897)	
Determining Strength Development of Adhesive Bonds, Practice for (D 1144)	
Cross-Lap Specimens for Tensile Properties of Adhesives, Testing of (D 1344)	
Tensile Strength of Adhesives by Means of Bar and Rod Specimens, Method for (D 2095)	
Torque strength	
Determining the Torque Strength of Ultraviolet (UV) Light-Cured Glass/Metal Adhesive Joints, Practice for (D 3658)	
Viscosity	
Viscosity of Adhesives, Test for (D 1084)	
Apparent Viscosity of Adhesives Having Shear-Rate-Dependent Flow Properties, Test for (D 2556)	
Viscosity of Hot Melt Adhesives and Coating Materials, Test for (D 3236)	

TABLE 4.1 Selected ASTM Tests for Adhesives (*Continued*)

Volume resistivity
Volume Resistivity of Conductive Adhesives, Test for (D 2739)
Water absorptiveness (of paper labels)
Water Absorptiveness of Paper Labels, Test for (D 1584)
Weathering
(See Durability)
Wedge test
Adhesive Bonded Surface Durability of Aluminum (Wedge Test) (D 3762)
Working life
Working Life of Liquid or Paste Adhesive by Consistency and Bond Strength, Test for (D 1338)

*The latest revisions of ASTM standards can be obtained from the American Society for Testing and Materials, 100 Barr Harbor Drive, Conshohocken, PA 19428.

tions that do not stress the joint significantly different than it would normally be stressed in service.

The usefulness and limitations of standard testing methods should be understood clearly before a testing program is established. The user of adhesives and sealants must choose the test geometry, procedure, and methodology that best serves the application. To do this, one needs to understand the differences in the various test methods and the parameters that will affect the data. Once the advantages and limitations of the various standard tests are understood, the end-user may find it necessary to devise his own methods to test specific combinations of loads and environments that are anticipated.

It should be apparent by now, that there are numerous parameters that can affect the performance of a joint, and many combinations of parameters are possible. Therefore, a prime rule in any adhesive or sealant testing program is to standardize and document test variables as thoroughly as possible. The adhesive formulator, supplier, and end-user should all utilize the exact same procedures and specimen construction. Test programs that are performed at different locations in the company and by different personnel should make every effort to be identical to one another.

4.3 Fundamental Material Property Tests

Fundamental tests can be used for monitoring the consistency of incoming products that are used in the adhesive bonding or sealing operations. Such tests may be used to characterize either the adhesive, adherend, or other materials, such as primers or solvents, used in the joint or in its construction.

TABLE 4.2 Selected ASTM Tests for Sealants

Adhesion testing
T-Peel Strength of Hot Applied Sealants, Test Method for, (C 906)
Tensile Adhesion Properties of Structural Sealants, Test Method for Determining, (C 1135)
Lap Shear Strength of Hot Applied Sealants, Test Method for, (C 961)

Cracks
Melting of Hot Applied Joint and Crack Sealant and Filler for Evaluation, Practice for, (D 5167)

Applications
Seal Durability of Sealed Insulating Glass Units, Test Methods for, (E 773)
Evaluating Pipeline Coating and Patching Materials, Method for, (G 55)
Seal Strength of Flexible Barrier Materials, Test Method for, (F 88)
Preparing Concrete Blocks for Testing Sealants for Joints and Cracks, Practice for, (D 1985)
Use in Selection of Liquid Applied Sealants, Guide for, (C 1299)
Performance Tests of Clear Floor Sealers, Method for, (D 1546)
Sealing Seams of Resilient Sheet Flooring Products by Use of Liquid Seam Sealers, Practice for, (F 693)

Glossaries
Building Seals and Sealants, Terminology of, (C 717)
Formed in Place Sealants for Joints and Cracks in Pavements, Terminology Relating to, (D 5535)

High temperatures
Testing Polymeric Seal Materials for Geothermal and/or High Temperature Service Under Sealing Stress, Test Method for, (E 1069)
Effects of Heat Aging on Weight Loss, Cracking, and Chalking of Elastomeric Sealants, Test Method for, (C 792)
Weight Loss After Heat Aging of Preformed Sealing Tape, Test Method for, (C 771)

Laboratories
Laboratories Engaged in the Testing of Building Sealants, Practice for, (C 1021)

Sealing
Fluid Tightness Ability of Adhesives Used on Threaded Fasteners, Test Method for, (D 5657)
Seal Strength of Flexible Barrier Materials, Test Method for, (F 88)
Sealability of Gasket Materials, Test Methods for, (F 37)

TABLE 4.2 Selected ASTM Tests for Sealants (*Continued*)

Sealant backer material
Determination of Water Absorption of Sealant Backup (Joint Filler) Material, Test Method for, (C 1016)
Determining the Outgassing Potential of Sealant Backing, Test Method for, (C1253)
Solvent release
Use of Solvent Release Type Sealants, Standard Practices for, (C 804)
Immersion Testing Nonmetallic Sealant Materials by Immersion in a Simulated Geothermal Test Fluid, Test Method for, (E 1068)

Fundamental tests are also referred to as bulk property tests because they define the properties of the material in the bulk and not in the adhesive or sealant joint. These tests include ultimate tensile strength and elongation, modulus of elasticity, hardness, tear resistance, abrasion resistance, toxicity, electrical properties, and color retention. It is easy to test for these properties using simple sample sections.

Once a combination of materials has been verified to provide acceptable joints, the materials are tested and a "fingerprint" is generally made of the resulting fundamental properties. Samples taken from subsequent lots or suspicious materials can then be subjected to test and the results compared to the original fingerprint. A significant difference in a measured fundamental property could be evidence supporting further investigation into the reason for the change and for determining how the change affects adhesion.

4.3.1 Adhesives and sealants

The most commonly used fundamental property tests measure the viscosity, shelf life, working life (setting characteristic in bulk), tack, cure rate (setting characteristic in the joint), hardness, and percent solids. Unless otherwise specified, the conditions surrounding the specimens at least 24 hours prior to and during the test are controlled to 73.5 ± 2°F and 50 ± 4% relative humidity.

4.3.1.1 Viscosity. Viscosity is defined as the resistance of a liquid material to flow. It is usually measured in fundamental units of poise or centipoise. The unit of centipoise (cps) is sometimes confusing unless one is familiar with these particular units. The following comparisons of common liquids may be of assistance in understanding centipoise values.

1 cps	Water
400 cps	#10 Motor oil
1,000 cps	Castor oil
3,500 cps	Karo syrup
4,500 cps	#40 Motor oil
25,000 cps	Hershey Chocolate Syrup

There are also a number of specialty viscosity tests that employ their own relative units. These tests have been developed for specific industries such as paints and coatings. Generally, their viscosity units can be directly converted to poise.

Adhesive viscosity is an indication of how easily the product can be pumped or spread onto a surface. It reveals information, together with the liquid's setting rate and surface tension, that is pertinent to the wetting characteristics of the adhesive. The viscosity also reveals information regarding the age and compounding of the adhesive. Through the relatively easy measurement of viscosity, changes in density, stability, solvent content, and molecular weight can be noticed. Viscosity measurements for free flowing adhesives or sealants are usually based on one of the following methods described in ASTM D 1084.

The most popular test for products ranging in viscosity from 50 to 200,000 cps is by a rotating spindle instrument such as the Brookfield viscometer. The simple equipment used for this measurement is shown in Fig. 4.1. The instrument measures the resistance of the fluid on a spindle of certain size that is rotating at a predetermined rate. The method is relatively simple and quick. It can be adapted to either the laboratory or production floor.

Another test for determining the viscosity of liquid adhesives measures the time it takes the test liquid to flow by gravity completely out of a cup with a certain size hole in the bottom. These consistency cups are designed to expel 50 ml of sample in 30–100 secs under controlled temperature and relative humidity conditions. The number of seconds for complete flow-out of the sample is determined. There are different cup volumes and hole sizes that can be used and conversions exist for relating the viscosity measured in one cup to another. This test is commonly used in the paint industry for adjusting the solvent content in paint systems.

The viscosity of thixotropic materials that exhibit a shear rate dependency is usually determined by the procedure described in ASTM D-2556. The viscosity is determined at different shear rates with a viscometer. From this plot, apparent viscosity associated with a particular rotation speed and spindle shape can be obtained. Materials with thixotropic characteristics include Vaseline jelly and toothpaste.

Standard Test Methods   139

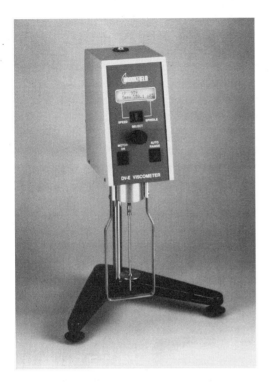

Figure 4.1 Digital Brookfield viscometer allows laboratory technician to measure viscosity and torque. (Photo courtesy of Brookfield Engineering Laboratories)

They are materials that tend to have high viscosity characteristics and exhibit no flow at low shear rates. However, when pressure is applied (higher shear rates), the material flows easily, exhibiting a characteristic of lower viscosity. Such materials are very common in the adhesive and sealant industries. Thixotropic materials can be pumped through a nozzle, mixed, or applied to a surface with little resistance. When applied to a vertical surface, they will not flow under their own weight. Yet, they can be easily spread or tooled before hardening with only slight pressure.

The viscosity of non-flowable products is determined by an extrusion test. A Semco 440 nozzle[1] or its equivalent is attached to a standard adhesive/sealant cartridge filled with the material to be tested. The cartridge is then placed in an air operated sealant gun set for a pressure of 90-95 psi. The weight of material that is extruded in 10 sec is measured, and the extrusion rate is reported in gms/min. Care must be taken to clear trapped air from the cartridge and nozzle.

4.3.1.2 Shelf life. When a polymeric resin is stored for a considerable length of time, physical and chemical changes may occur within the material that will affect its performance as an adhesive or sealant.

The shelf life is generally defined as the time that an adhesive or sealant can be stored, preferably under controlled conditions, and remain unchanged at least from a fundamental property point of view. ASTM D 1337 provides a method for determining the shelf life of an adhesive or sealant. The changes in consistency (viscosity) or bond strength are measured after various storage periods at a specified temperature.

All of the materials tested for shelf life are stored in unopened containers. Storage in containers that are once opened exposes the resin to oxygen and humidity that, depending on the type of resin, could drastically reduce the shelf life and affect final properties. Most polymeric adhesives and sealants have a shelf life greater than 6 months at room temperature. However, some one-component adhesives need to be stored at refrigerated conditions to have a practical shelf life.

Certain adhesives or sealants may also have limited life when stored at cold temperatures or when exposed to repeated freeze/thaw cycles. Generally, these are water based products, but freeze/thaw stability problems are not restricted to them. The resistance to freeze/thaw cycles or to low temperature storage is measured using procedures similar to the shelf life measurement discussed above. Even though an adhesive or sealant may have good low temperature storage properties, it should be brought to room temperature before mixing or application.

4.3.1.3 Working life. The working life of an adhesive or sealant is the time from when the product is ready for use (i.e., mixed and ready to apply to a substrate) and the time when it is no longer usable because the setting mechanism has progressed to such an extent that the product is no longer workable. This characteristic is also known as the pot life of the adhesive. ASTM D 1338 establishes two procedures for determining working life. One method uses viscosity change, and the other uses shear strength as the criteria for determining when the effective working life has expired.

Working life is usually determined on a volume of adhesive or sealant material that is practical and normally used in production. The volume of the tested material must be defined in the test report because many adhesives and sealants have a working life that is dependent on sample mass.

4.3.1.4 Tack. Tack is the property of an adhesive to adhere to another surface on immediate contact. It is the "stickiness" of the adhesive while in the fluid state. Sometimes tack is also referred to as "green strength". Tack is an important property for many pressure sensitive adhesives and preformed sealants.

Tackiness is generally determined with a tensile testing machine and test blocks. The blocks are pulled apart immediately after the adhesive is applied and the blocks are joined. The result is reported in force required per square inch of bonded area (psi) to separate the blocks. Various instruments have also been developed to measure tack for specific applications. Table 4.3 lists and compares various test methods that have been used.

Other important characteristics related to tack and commonly measured are:

- Dry tack—a property of certain adhesives to stick to one another even though they seem dry to the touch
- Tack range—the time that an adhesive will remain in a tacky condition

4.3.1.5 Cure rate. Structural adhesives usually require curing by either the application of heat, the addition of a catalyst, the addition of pressure, or a combination of the three. The strength developed in the adhesive joint at various times during the curing process may be measured by lap shear tensile specimens. This test is commonly used to determine when an adhesive or sealant is fully cured or when the system reaches a "handling" strength so that the assembled product can be moved with moderate care.

TABLE 4.3 Comparison of Test Methods to Determine Tack of an Adhesive[2]

Organization	Method	Common name	Notes
ASTM	D 2979	Probe Tack	
ASTM	D 3121	Rolling Ball Tack	1
TLMI	LIB 1	Loop Tack	2
TLMI	LIB 2	Loop Tack	3
PSTC	5	Quick Stick	4
PSTC	6	Rolling Ball Tack	1
FINAT	FTM 9	Loop Tack	5
AFERA	4015	Quick Stick	4

NOTES:
1-The methods described in ASTM D 3121 and PSTC 6 are virtually identical.
2-This method describes Loop Tack measurement using a specially designed piece of test equipment.
3-This method describes Loop Tack measurement using an adapted tensile tester.
4-PSTC 5 and AFERA 4015 describe nearly the same testing technique. The AFERA method describes the use of a lightweight (25 grams) roller to apply the test specimen to the panel, whereas the PSTC method uses only the weight of the test strip to accomplish lamination.
5-The FINAT method is similar to the TLMI method for Loop Tack, except that the FINAT method uses glass rather than stainless steel as the test surface.
6-ASTM (American Society for Testing and Materials)
  TLMI (Tag and Label Manufacturers Institute)
  PSTC (Pressure Sensitive Tape Council)
  FINAT (European Assn. for the Self Adhesive, Labeling Industry)
  AFERA (Association des Fabricants Europeens de Rubans, Auto-Adhesifs)

Cure rate is an important factor when the expense of jigs and fixturing equipment is high or fast production rates are critical. It is also used as a quality control test to determine if the adhesive's or sealant's curing mechanism has changed from lot to lot or if it may have been spoiled by storage, moisture contamination, etc. ASTM D 1144 provides a recommended practice for determining the rate of bond strength development for either tensile or lap shear specimens.

There are also several methods of determining cure rate on the bulk material. These are generally analytical procedures that are common in most polymeric material laboratories. With these methods, fundamental properties, such as dielectric loss, mechanical damping, and exotherm, are measured as a function of time and temperature. Several of these test methods are described in the next chapter.

Cure time is very important for sealants as well as adhesives. Often the sealant will be required to function as a barrier or resist the movement of substrates very soon after it is applied. With construction sealants, for example, it may not be possible to delay the environmental conditions until after the adhesive cures. Thus, curing time becomes a critical parameter in selection of the sealant. ASTM C 679 is a method for determining the time that a mechanic can work the sealant into the joint before the sealant starts to skin or solidify.

4.3.1.6 Hardness. Hardness of the adhesive or sealant may be used as an indication of cure. It may also be used as a quality control check on certain substrates. Hardness may be determined in several ways: resistance to indentation, rebound efficiency, and resistance to scratching or abrasion. The indentation method is the most commonly used technique.

There are several ways of measuring indentation, but they only differ in the type of equipment used. Basically, they all measure the size of indentation produced by a hardened steel or diamond tool under a defined pressure. A durometer is an instrument for measuring hardness by pressing a needle-like instrument into the specimen. Durometers are available in several scales for measuring hard materials to soft elastomers. The two types appropriate for most cured adhesives and sealants are the Shore Type A and Shore Type D. ASTM C 661 offers a method for measuring indentation hardness of elastomeric type sealants.

Lower hardness readings than expected may be an indication of under-cure or of a formulation change in the product. It may also be an indication of entrapped air in the adhesive or sealant or an unwanted chemical reaction with the environment. Higher hardness readings than expected may be an indication of over-cure.

A simple, but not very quantitative, hardness test has been used for hundreds of years—the fingernail indentation test. The indentation that a fingernail makes in the edge of an adhesive bond or in the body of a sealant can often be used as an approximate indication of hardness of the material.

4.3.1.7 Solids content. The solids content of an adhesive or sealant should be checked to assure that formulation or dilution errors have not been made. "Solids" can refer either to the non-volatile component of the adhesive or the inorganic component of the adhesive.

The nonvolatile solids content is usually determined by weighing a small amount of material in a clean container, heating or curing until a constant weight is obtained, and then weighing the container again. The percentage of solids may be determined as the ratio of the sample weights before and after curing × 100. The percent solids is an indirect measurement of the amount of volatile material in the sample that was driven off during the heating cycle. The volatile material may be solvent, water, or other additives. ASTM D 1489 offers a test method for determining the non-volatile content of aqueous adhesives.

The solids content provides a measurement of the non-volatile materials that are contained within the formulation. Addition of solvent can be used to "correct" the viscosity of adhesives to account for variabilities in the manufacturing process or inconsistent base materials in the resin formulation. Measurement of solids content will generally identify such practices.

The ash test is a measure of the total inorganic content in a sample. In this test, a weighed sample is placed in a muffle furnace at approximately 1000°F for 30 min or at a temperature and time long enough to completely pyrolize any organic matter. The remaining material is inorganic filler, reinforcement, etc. ASTM D 5040 provides a method of measuring the ash content of adhesives. Ash content will determine the amount of inorganic fillers that are in an adhesive or sealant sample. The manufacturer can adjust the filler content, like the solids content, to correct for errors or inconsistency in formulation. Higher concentrations of solvent or filler are also often used to lower the cost of the adhesive or sealant system.

4.3.2 Adherends and other materials used in the bonding process

Fundamental property tests are also commonly used to characterize the substrate and provide a "signature" for lot to lot comparisons. In all too many cases, the substrate is considered to be a constant throughout the life of a production part. In practice, there are a number of common reasons for substrates to change.

Metallic substrates change primarily with regard to their surfaces. Different oxide layers can be formed that are depending on the chemistry of the surface and the way it was treated during fabrication of the part. The degree of surface cleaning or surface roughness may also change throughout a production run due to differences in processing chemicals, or procedures, or to contamination of the materials that are used.

For organic substrates, the possible types of change are numerous and potentially more significant. Polymeric substrates may change in modulus due to aging, loss of plasticizer, or continued chemical reaction. They may change because of slight formulation variations that take place from lot to lot. These changes may not be sufficient to cause a change in the bulk physical properties. Therefore, the change may go unnoticed to the quality control department because they are looking at only bulk properties and not at subtle formulation variations that could affect adhesion properties. Such substrate modifications could result in drastic changes in the surface properties of the material.

A case in point is an elastomeric substrate, such as compounded nitrile rubber. Generally, the user will specify the rubber part by compressive strength, extension, hardness, or other standard bulk property tests. The supplier can formulate virtually thousands of formulations that will meet these specified properties. In doing so he may use an incompatible plasticizer, or a low molecular weight extender, or a new filler that is more hydroscopic. These small changes could have a very large effect on the resulting joint strength. Therefore, any change in substrate formulation or processing must be re-verified with regard to its adhesion characteristics.

Test methods used to determine the uniformity of substrates are numerous and vary with the type of material. They are generally the same tests used to characterize the material or to determine its fundamental physical properties. Tests that are commonly employed are hardness, tensile strength, modulus, and surface characteristics such as roughness or contact angle measured with a standard liquid. Often a test similar to the non-volatile test mentioned above is used to determine if there are any compounds in the substrate that are capable of outgasing on exposure to elevated temperatures. Internal moisture content of certain polymers, such as nylon and polycarbonate, is also known to affect adhesion.

4.4 Standard Test Methods for Adhesive Joints

The physical testing of standard adhesive joints provides a method of comparison for materials and processes that are being evaluated. Standard tests also provide a means to control the adequacy of the bonding process, once it is established, and of assessing its conformance to specification.

Standard test methods are only useful if they can be reproduced. It is important that the same results can be measured by both the adhesive developer and the end-user. It is also important that the results are reproducible with time and with different testing personnel. The accuracy and reproducibility of test results depend on the conditions under which the bonding process is performed. The following variables must be strictly controlled.

1. Procedures for cleaning, etching, and drying the surface of the substrates prior to application of adhesive.

2. The time between surface preparation and application of adhesive and the environmental conditions present during this period. This includes the temperature and percent relative humidity. Usually standard atmospheric conditions are specified (73.5° ± 2°F and 50 ± 4% relative humidity).

3. Complete procedures for mixing the adhesive components.

4. Conditions and methods for application of the adhesive to the substrate surface.

5. Curing conditions, including the pressure, temperature, and time of the curing cycle. It should be specified whether or not the temperature is measured within the glue line, or on some other point on the substrate, or at some location within the curing oven. The temperatures could vary significantly at these different locations depending on the weight and size of the assembly. When an adhesive producer specifies a temperature and time for cure, it refers to the conditions of the actual adhesive within the bond-line.

6. Conditioning procedures for specimens after curing and prior to testing.

7. The rate at which the sample is loaded during test. Peel and impact tests especially are dependent on the speed at which the sample is tested.

A standard test report usually documents the resulting measurements such as tensile shear strength, peel strength, etc. It should also indicate all of the pertinent conditions that are required to assure reproducibility in subsequent testing. It is also very useful to describe the failure mode of the tested specimens. An analysis of the type (or mode) of failure is a valuable tool to determine the cause of adhesive or sealant failure. The failed joint should be visually examined to determine where and to what extent failure occurred. The percent of the failure in the adhesion and cohesion mode should be provided. A description of the failure mode itself (location, percent coverage, uniformity, etc.) is also useful. The purpose of this exercise is to help identify the weak link in the joint to better understand the mechanism of failure.

Numerous standard test methods have been developed by various government, industrial, and university investigators. Many of these have been prepared or adopted under the auspices of the ASTM Committee D 14 on Adhesives, D 24 on Sealants, or other professional societies. Reference to the appropriate standards will adequately equip one with the background necessary to conduct the test or a version of it. Several of the more common standard tests will be described in this section. Numerous variations exist for specific applications or materials. In these descriptions, the emphasis will be on understanding the reasons for the test, its relationship to a specific adhesive property, advantages and limitations of the test, and possible variations or extrapolations of the test method. The detailed description of the test mechanics will be kept to a minimum, since they are adequately covered in the existing standards and specifications.

4.4.1 Tensile tests

The tensile strength of an adhesive joint is seldom reported in the adhesive supplier's literature because pure tensile stress is not often encountered in actual production. An exception to this is the tensile test of the bonds between the skin and core of a honeycomb or composite sandwich. However, the tensile test is not only useful as a quality control test for metal and sandwich adhesives; it can also be employed to yield fundamental and uncomplicated tensile strain, modulus, and strength data for the adhesive.

The ASTM D 897 tensile button test is widely used to measure tensile strength of a butt joint made with cylindrical specimens (Fig. 4.2). The tensile strength of this bond is defined as the maximum tensile load per unit area required to break the bond (psi). The cross-sectional bond area is specified to be equal to one square inch. The specimen is loaded by means of two grips that are designed to keep the loads axially in-line. The tensile test specimen requires considerable machining to ensure parallel surfaces.

A similar specimen design uses a sandwich construction with a dissimilar material bonded between the two cylindrical halves of the but-

Standard Test Methods   147

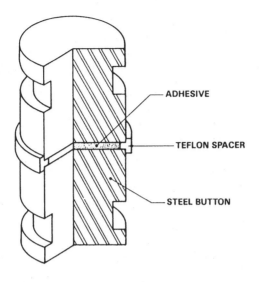

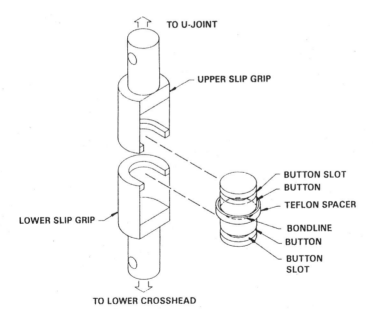

Figure 4.2 Standard button tensile test specimen and tensile test grips.[3]

ton specimen. This design is commonly used to measure the tensile strength of adhesives between dissimilar materials. This method is also useful if the adherend cannot be machined into the shape of the button specimen. With some modifications in the dimensions, the button tensile test has also been adapted for testing adherence of honeycomb-cover sheets to the core (ASTM C 297).

Although developed and used primarily for testing wood joints, the cross type tension test, shown in Fig. 4.3, can be used to test other substrates or honeycomb specimens (Fig. 4.3c). This simple cross lap specimen is described in ASTM D 1344. This test method is attractive in that it does not involve significant machining or high specimen cost, as the button tensile test specimens do. It is very important that the specimens are thick and rigid enough to resist bending. With only moderate bending, the loads will quickly go into peel or cleavage stress. Because of a high degree of variation arising out of possible bending modes, a sample population of at least ten is recommended for this test method.

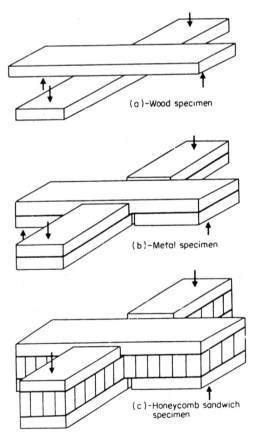

Figure 4.3 Cross-lap tension test specimen.[4]

Generally, high modulus adhesives possess the highest tensile strengths. There appears to be an inverse relationship between thickness and tensile bond strength for particular adhesive-adherend combinations. Thin bonds are best as long as the bond-line is not starved of adhesive. Thicker bonds give lower strengths because cleavage forces more readily occur due to non-axial loading, or internal stresses in the adhesive.

4.4.2 Lap-shear tests

The lap-shear or tensile-shear test measures the strength of the adhesive in shear. It is the most common adhesive test because the specimens are inexpensive, easy to fabricate, and simple to test. However, at times it is difficult to minimize or eliminate bending stresses in common shear joint specimens. Because the standard lap shear tests introduce some degree of peel into the adhesive joint, values obtained for the lap shear strength of epoxy adhesives may average 4000–5000 psi; whereas, values for bulk tensile strengths have been reported up to 12000 psi.[5]

The common lap-shear test method is described in ASTM D 1002, and the standard test specimen is shown in Fig. 4.4. This is the most commonly used shear test for structural adhesives. However, due to the nonuniform stress distribution in the adhesive arising from the joint configuration, the failure strength values are of little use for engineering design purposes.

Testing is carried out by pulling the two ends of the overlap in tension causing the adhesive to be stressed in shear. Hence, this test is frequently called the tensile-shear test. Since the test calls for a sample population of five, specimens can be made and cut from larger test panels illustrated in Fig. 4.4b.

The width of the lap shear specimen is generally one inch. The recommended length of overlap, for metal substrates of 0.064 in. thickness, is 0.5 ± 0.05 in. The length should be chosen so that the yield point of the substrate is not exceeded. In lap shear specimens, an optimum adhesive thickness exists. For maximum bond strengths, the optimum thickness varies with adhesives of different moduli (from about 2 mils for high modulus adhesives to about 6 mils for low modulus adhesives).[4]

The lap shear specimen can be used for determining shear strength of dissimilar materials in a manner similar to that which was described for the laminated button tension specimen. Thin or relatively weak materials such as plastics, rubber, or fabrics are sandwiched between stronger adherends and tested.

Two variations are used to avoid the bending forces that occur with simple ASTM 1002 specimens: the laminated lap shear specimen

150   Chapter Four

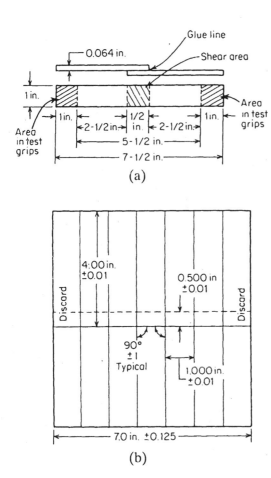

Figure 4.4 Standard lap shear test specimen design: (a) form and dimensions of lap shear specimen; (b) standard test panel of five lap shear specimens. From ASTM D 1002

(ASTM D 3165) shown in Fig. 4.5a and the double lap specimen (ASTM D 3528) shown in Fig. 4.5b. These specimens minimize joint eccentricity and provide higher strength values than does the single overlap specimen. For the specimen in Fig. 4.5a, the overlap joint can be made from saw cuts in the top and bottom substrates of a bonded laminate. This process negates the effects of extruded adhesive at the edges of the lap and the sheared edge of the standard type of lap shear specimen. As a result, the chances of deformation and uneven surface preparation are lessened.

Compression shear tests are also commonly used. ASTM D 2182 describes a simple compression specimen geometry and the compres-

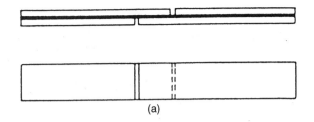

(a)

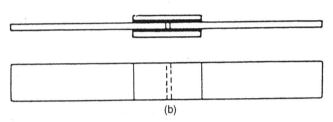

(b)

Figure 4.5 Modified lap shear specimens used to maintain axial loading: (a) single saw cut specimen, (b) double lap specimen.[4]

sion shear test apparatus. The compression shear design also reduces bending and, therefore, peeling at the edges of the laps. Higher and more realistic strength values are obtained with the compression shear specimen over the standard lap shear specimen.

4.4.3 Peel tests

A well-designed joint will minimize peel stress, but not all peel forces can be eliminated. Because adhesives are notoriously weak in peel, tests to measure peel resistance are very important. Peel tests involve stripping away a flexible adherend from another adherend that may be flexible or rigid. The specimen is usually peeled at an angle of 90 or 180 degrees. The most common types of peel test are the T-peel, the floating roller peel, and the climbing-drum methods. Representative test specimens are shown in Fig. 4.6. The values resulting from each test method can be substantially different; hence it is important to specify the test method employed.

Peel values are recorded in pounds per inch of width (piw) of the bonded specimen. They tend to fluctuate more than any other adhesive test result because of the extremely small area at which the stress is localized during loading. Even during the test, the peel strength values tend to fluctuate depending on the type of adhesive, adherend,

152   **Chapter Four**

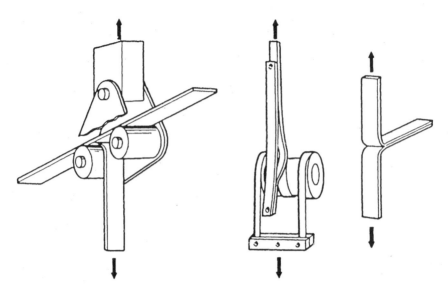

Figure 4.6   Common types of adhesive peel tests: (left) floating roller peel, (center) climbing drum peel, (right) t-peel.[6]

and condition of the test. In preparing the samples, care must be taken to produce void-free laminated bond-lines. A typical load curve for the T-peel test is shown in Fig. 4.7. Peel strength is taken as the average value of the center portion of the curve, usually over at least a 5 in. length of crosshead travel.

The rate of peel loading is more important than in lap-shear loading, and should be known and controlled as closely as possible. The rate at which the load is applied is usually specified in the ASTM test procedure. Adhesive thickness has a significant effect on peel-strength values as does the angle of peeling. The relative effects of these parameters are also dependent on the elasticity of the adhesives.

Figure 4.8 shows the effect of peel speed and angle on the strength of an epoxy adhesive. With elastomeric adhesives, thicker bond-lines

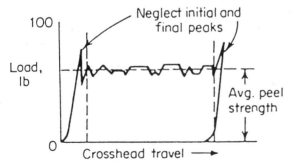

Figure 4.7   Peel test record.[4]

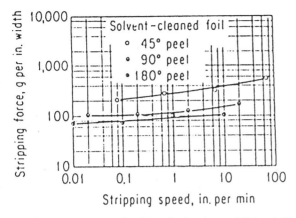

Figure 4.8 Peel strength of 3 mil aluminum foil bonded with DETA cured DGEBA epoxy adhesive.[9]

will generally result in higher peel strengths. The elongation characteristics of these adhesives permit a greater area of the bond to absorb the applied load. With more rigid adhesives, the thicker bond-line lowers the peel strength (as shown in Fig. 4.9 for an epoxy adhesive) because of stresses at the edge of the bond. Peel strength is also a function of the stiffness or modulus of the substrates and of the adhesive, and it is also a function of the thickness of the two substrates being peeled.[7,8]

The T-peel test is described in ASTM D 1876 and is the most popular of all peel tests. The T-peel specimen is shown in Fig. 4.10. Generally, this test method is used when both adherends are flexible. Because the angle of peel is uncontrolled and dependent on the properties of the adherends and the adhesive, the test is less reproducible than other peel tests. A sample population of at least ten is required, whereas most other ASTM tests require a minimum of five.

The floating roller peel test is used when one adherend is flexible and the other is rigid. The flexible member is peeled through a spool arrangement to maintain a constant angle of peel. Thus, the values obtained are generally more reproducible than the T-peel test method. The floating roller peel resistance test is designated ASTM D 3167.

The climbing-drum peel specimen is described in ASTM D 1781. This test method is intended primarily for determining peel strength of thin metal facings on honeycomb cores, although it can be used for joints where at least one member is flexible. The fixtures of the floating roller peel and drum peel tests help control the angle of peel so that they generally provide more reproducible peel values for a given adhesive than the T-peel method.[10,11]

A variation of the T-peel test is a 180 deg stripping test illustrated in Fig. 4.11 and described in ASTM D 903. This method is commonly

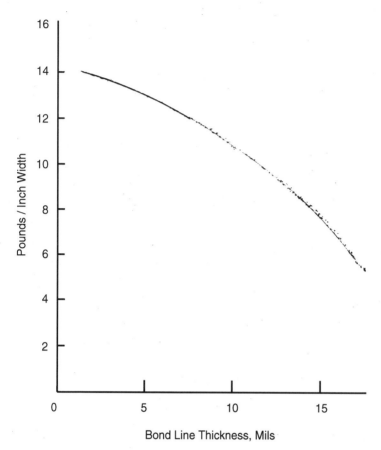

Figure 4.9 Effect of adhesive thickness on peel resistance of Resiweld 7007 epoxy adhesive.[7]

used when one adherend is flexible enough to permit a 180 deg turn near the point of loading. This test offers more reproducible results than the T-peel test because the angle of peel is maintained constant although it is dependent on the nature of the adherend.

Considerable work has been done, notably by D. H. Kaeble[12,13] and J. J. Bikerman,[14,15] to develop formulae for the force required to peel a flexible tape from a rigid substrate. Figure 4.12 shows those equations for a peeling force. An examination of the Kaeble formula would indicate that the peel force is directly proportional to the adhesive thickness. Work has shown, however, that the relationship is logarithmic as seen in Fig. 4.13. Considerable study has been made of the effect of backing on peel force.[8] The peel angle and the joint strength depend on the type of backing and its thickness.

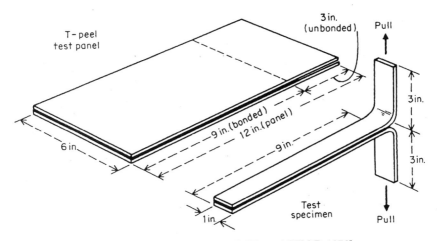

Figure 4.10 Test panel and specimen for T-peel. [From ASTM D 1876]

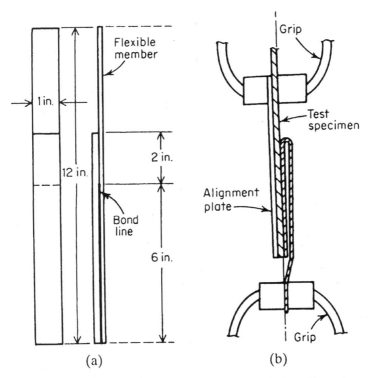

Figure 4.11 180° peel test specimens: (a) specimen design, (b) specimen under test. [From ASTM D 903]

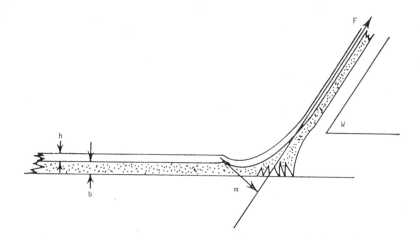

Bikerman—Peel force at 90° from substrate

$$F = 0.3799a\,(E)^{1/4}b^{1/4}h \over (E_1)$$

Kaelble—Peel force at any angle

$$F = \frac{K^2 a b \sigma_0^2}{2E_1(1-\cos^w)}$$

where $K = \dfrac{Bm_c}{Bm_c + \sin^w}$ $B = \dfrac{E_1 a}{(4\,E I a)^{1/4}}$

At 180°, $K = 1$, when the Peel Force relation reduces to

$$F = \frac{\sigma_0^2\,ab}{4\,E_1}$$

where—

F = Peel Force; a = Bond Width; b = Thickness of Adhesive; h = Thickness of Flexible member; w = Peel angle; σ_0 = Adhesive Stress at failure; E = Young's Modulus of Flexible Member; E_1 = Young's Modulus of Adhesive; I = Moment of Inertia of Flexible Member; m = Moment Arm; m_c = $-m\cdot h\cos^w$; K = Complex dimensionless parameter depending on geometry and moduli.

Figure 4.12 Equation for peel force required to strip a flexible adherend from a rigid substrate.[8]

4.4.4 Cleavage test

Cleavage tests are conducted by prying apart one end of a rigid bonded joint and measuring the load necessary to cause rupture. Cleavage tests are used in place of peel tests when both adherends are rigid. The test is also a qualitative measure of the fracture toughness of the adhesive. Data obtained are adaptable to engineering design.

The cleavage test utilizes a specimen similar to a compressive shear specimen except that the load is intentionally placed on one edge of the bonded area. The test method is described in ASTM D 1062. The specimen is usually loaded at a rate of 0.05 in/min until failure occurs. The failing load is reported as breaking load per unit area in psi units. A standard test specimen is illustrated in Fig. 4.14. Because cleavage test specimens involve considerable machining, peel tests are usually preferred where possible.

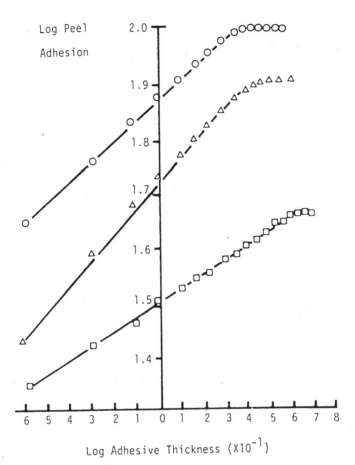

Figure 4.13 Comparison of peel force with adhesive thickness (0.5 mil polyester film used as backing in all tests).[8]

4.4.5 Fatigue tests

Fatigue testing places a given load repeatedly on a bonded joint. Lap-shear or other specimens are tested on a fatiguing machine capable of inducing cyclic loading (usually in tension or a combination of tension and compression but also in bending) on the joint. ASTM D 3166 provides procedures for testing and measurement of the fatigue strength of lap specimens.

The fatigue strength of an adhesive is reported as the number of cycles of a known load necessary to cause failure. Cycles to failure and the corresponding loads are plotted on coordinates of stress versus the logarithm of the number of cycles. The point at which the smooth curve connecting the points of minimum stress crosses the 10 million

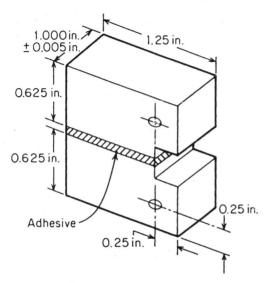

Figure 4.14 Cleavage test specimen. [From ASTM D 1062]

cycle line is usually reported as the fatigue strength. Fatigue strength is dependent on adhesive, curing conditions, joint geometry, mode of stressing, magnitude of stress, and frequency and amplitude of load cycling.

Lap shear fatigue data are limited in engineering design because of the stress distribution patterns of the lap shear specimen configuration relative to practical joint designs. However, fatigue testing of the actual part or of the joint itself provides useful engineering design values.

4.4.6 Impact tests

Impact testing is of importance because adhesives, like most polymeric materials, are sensitive to high rates of applied force. The resistance of an adhesive to impact can be determined by ASTM D 950. This test is analogous to the Izod impact test method used for impact studies on materials.

The specimen is mounted in a grip shown in Fig. 4.15 and placed in a standard impact machine. One adherend is struck with a pendulum hammer traveling at 11 ft/s, and the energy of impact is reported in pounds per square inch of bonded area. It is often difficult to achieve reproducible results with impact testing, and as a result, the test is not widely used in production situations. Impact data indicate that as the thickness of the adhesive film increases, its appar-

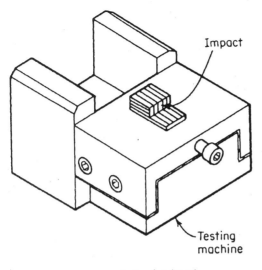

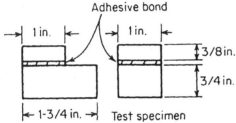

Figure 4.15 Impact test specimen. [From ASTM D 950]

ent strength also increases.[16] This suggests that a portion of the energy required to rupture the bond is absorbed by the adhesive layer and is independent of adhesion.

The impact strength of a viscoelastic adhesive is affected by the rate of which the impact occurs. Often it is very difficult to achieve very high rates of impact with conventional laboratory testing. One example of this is an adhesive system used to bond shock mounting pads to electrical equipment aboard submarines. Impact testing of specimens done in a laboratory using pendulum or drop-weight impact fixtures showed that the adhesive would not fail under the loads expected. However, the proof test was to place the electrical equipment, with shock mounting attached, aboard a barge and then set-off explosive charges at various depths under the barge. The proof test showed that the viscoelastic adhesive failed when explosive-induced high rates of impact were experienced. The adhesive acts like a brittle polymer at high rates of loading because the molecular chains within the ma-

160 Chapter Four

terial do not have time to slip by one another and absorb a great deal of the energy. Special, high speed impact tests have been developed for certain applications. These tests generally use chemical explosive force or electromagnetic energy to establish the high speeds required.

4.4.7 Creep tests

In practical joints, adhesives are not always loaded for short periods of times. Often the application requires that the adhesive joint survive continuous loading or stress. The dimensional change occurring in a stressed adhesive over a long time period is called *creep*. Creep data are seldom reported in the adhesive supplier's literature because the tests are time-consuming and expensive. This is very unfortunate, since sustained loading is a common occurrence in adhesive applications. All adhesives tend to creep, some much more than others. With weak adhesives, creep may be so extensive that bond failure occurs prematurely. Adhesives have also been found to degrade much more rapidly when environmentally aged in a stressed rather than an unstressed condition. This phenomenon will be investigated more closely in Chapter 17.

Creep-test data are accumulated by loading a specimen with a predetermined stress and measuring the total deformation as a function of time or measuring the time necessary for complete failure of the specimen. Depending on the adhesive, loads, and testing conditions, the time required for a measurable deformation may be extremely long.

ASTM D 2294 defines a test for creep properties of adhesives utilizing a spring-loaded apparatus to maintain constant stress. With this apparatus (Fig. 4.16) once loaded, the elongation of the lap shear specimen is measured by observing the separation of fine razor scratches across its polished edges through a microscope. A typical creep curve of an adhesive bonded lap shear specimen is shown in Fig. 4.17. There

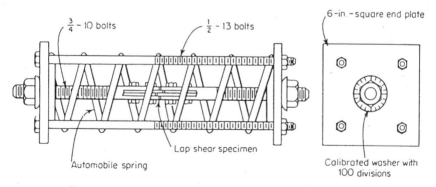

Figure 4.16 Compressive spring with tension creep-shear specimen.[4]

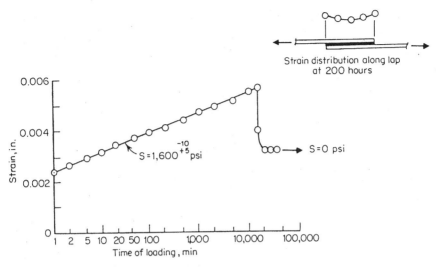

Figure 4.17 Typical creep-time curve for a lap shear adhesive joint bonded with a viscoelastic adhesive.[4]

is a substantial increase within the first time period followed by a gradual increase with time. Once the load is removed at the conclusion of the test, there will be a gradual recovery of strain with time. Since the capacity of adhesive to recover will vary, the basis on which to judge and compare creep behavior should not be the total strain at the conclusion of the test, but the irrecoverable strain or permanent set. The strain of an adhesive increases with thickness of film, and hence the creep per unit thickness is the recommended measure of the creep of an adhesive.

4.4.8 Environmental tests

It is desirable to know the rate at which an adhesive bond will lose strength due to environmental factors in service. Strength values determined by short-term tests do not always give an adequate indication of an adhesive's performance during continuous environmental exposure. Laboratory-controlled aging tests seldom last longer than a few thousand hours. To predict the permanence of an adhesive over a 20-year product life requires accelerated test procedures and extrapolation of data. Such extrapolations are extremely risky because the causes of adhesive-bond deterioration are complex. Unfortunately no universal method has yet been established to estimate bond life accurately from short-term aging data.

Adhesives and sealants may experience many different and exotic environments. Laboratory environmental aging is accomplished by exposing a stressed or unstressed joint to simulated operating conditions. Exposure is typically to elevated temperature, water, salt spray, or various chemical solutions that are representative of the service conditions. A number of standard chemicals that are used to soak bonded specimens for seven days at room temperate are described in ASTM D 896 (see Table 4.4). The method merely outlines the manner in which the test specimens are conditioned before testing. After suitable exposure, bonds may be tested in whatever manner seems appropriate.

The type of stress and environment should be selected to be a close approximation of real-life conditions. Two simultaneous environments, such as heat and moisture, may cause the adhesive to degrade much faster than when exposed in any single environment because one condition could accelerate the effect of the other.

Stressed aging tests are important because they more accurately simulate service conditions than do simple tests where the specimens are merely hung or placed in a test environment. The Alcoa Stress

TABLE 4.4 Standard Test Exposure of Adhesives[17]

Test Exposure Number	Temperature, F(a)	Moisture conditions
1	−70	As conditioned
2	−30	As conditioned
3	−30	Presoaked (b)
4	32	As conditioned
5	73.4	50% relative humidity
6	73.4	Immersed in water
7	100	88% relative humidity
8	145	Oven, uncontrolled humidity
9	145	Over water (c)
10	145	Immersed in water
11	158	Oven, uncontrolled humidity
12	158	Over water (c)
13	180	Oven, uncontrolled humidity
14	180	Over water (c)
15	212	Oven, uncontrolled humidity
16	212	Immersed in water
17	221	Oven, uncontrolled humidity
18	300	Oven, uncontrolled humidity
19	400	Oven, uncontrolled humidity
20	500	Oven, uncontrolled humidity
21	600	Oven, uncontrolled humidity

NOTES:
a-The tolerance for test temperature shall be ±1.8F up to 180F and ±1% for temperatures above 180F.
b-Presoaking shall consist of submerging specimens in water and applying vacuum at 20 in. of mercury until weight equilibrium is reached.
c-The relative humidity will ordinarily be 95 to 100%.

Test Fixture (Fig. 4.18) is a device for applying stresses of various magnitudes to lap shear specimens and then aging these stressed specimens in different environments. A plot of stress to time of failure can give important information about the loss of bond strength under stressed conditions. This aging phenomenon and a typical stress-to-failure plot are shown in Chapter 17.

There are several analytical tools that provide methods of extrapolating rate and temperature test data. One of these tools is the Williams, Landel, Ferry (WLF) transformation.[18] This method uses the principle that the work expended in deforming a flexible adhesive is a major component of the overall practical work of adhesion. The materials used as flexible adhesives are usually viscoelastic polymers. As such, the force of separation is highly dependent on their viscoelastic properties and is, therefore, rate and temperature dependent. Test data, taken as a function of rate and temperature, then can be expressed in the form of master curves obtained by WLF transformation. This offers the possibility of studying adhesive behavior over a sufficient range of temperature and rate for most practical applications. High rates of strain may be simulated by testing at lower rates of strain and lower temperatures.

4.5 Standard Test Methods for Sealant Joints

Many of the standard adhesive tests described in the previous section also pertain to sealants, especially if the need is to determine the adhesion characteristics of the sealant. The fundamental property tests used to measure consistency or working characteristics of the adhesive before it is placed in a joint are equally relevant to sealants. However, high performance sealants generally require test methods that are different from those used with adhesives.

Many test methods have been developed specifically by the industries that utilize sealants, such as automotive and construction. There are various industrial and professional organizations that are attempting to standardize the large number of sealant tests that have been developed over the years. The International Standards Organization (ISO) is writing test methods and specifications on a global basis. As with adhesive joints, many of the most useful and popular tests for sealant joints are defined in terms of ASTM test specifications. The more important properties and methods of testing sealants are described below.

There are several important comprehensive specifications that are used for sealant materials. These specifications describe test methods

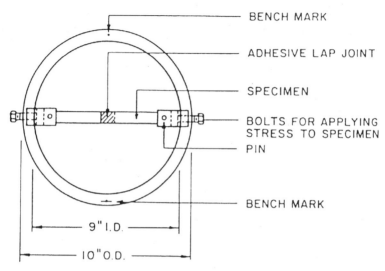

Figure 4.18 Alcoa stressing fixture for measuring simultaneous effect of stress and environmental conditions.[18]

and also provide minimum values for several different properties of sealants. Unlike test methods for adhesives, which are separate documents, these specifications for sealants combine several test methods. They also provide minimum values for categories of sealants. ASTM C 920 is such a comprehensive document. It describes the minimum acceptable properties required to meet the specification, as well as movement capabilities of the sealant.

4.5.1 Movement capability

The movement capability of a sealant joint is perhaps one of its most important characteristics. At a minimum, the sealant joint will move with response to daily and seasonal temperature changes. This movement capability is determined by the specimen geometry and the coefficient of thermal expansion of the substrates.

The movement capability is always stated as a ± percentage value that indicates the amount of movement the sealant can take in extension and compression. Movement capability must consider the environment and length of exposure time (i.e., prolonged elongation at low temperatures, prolonged compression at high temperatures, and combinations of the two).

ASTM C 719 is a sealant test procedure that defines several conditions for which the sealant must operate in both extension and compression. The test involves the cycling of sealant joints at a movement rate of 0.125 in/hr after various exposures to water and hot and cold temperatures. These conditions are shown in Table 4.5. Typical specimen geometries are shown in Fig. 4.19. The ISO Committee on Construction has also developed a similar test method that differs somewhat from ASTM C 719. It requires extended periods of sustained elongation (Table 4.6). There are a number of generally recognized categories of movement capability in which these test methods attempt to place sealants, for example ±5%, ±10%, ±12.5%, ±25%, ±50%, and +100%/−50%.

ASTM C 920 describes a bond cohesion test using specimens similar to ASTM C 719. The test measures initial adhesive strength and cycling after compression and water immersion. Another test includes cycling after UV exposure. Failure is defined as an adhesive or cohesive separation exceeding 1/10 in. in depth. Values can be obtained for various extensions. Measured strengths of 5–10 piw are common for low modulus sealants with extensions of up to 100%; medium modulus sealants, 15–30 psi; and high modulus sealants, 25–50 psi.

4.5.2 Peel adhesion

The peel adhesion test in combination with the tensile adhesion test above gives a good indication of how a sealant might perform in service. Peel specimens are formed from a substrate of interest and the sealant which is reinforced with a strip of cotton fabric or fine stainless wire mesh. Once cured, the cloth or wire is folded back 180 degrees and peeled at a rate of separation of 2 in/min as shown in Fig. 4.20. The load is reported in lbs per inch of width. The test is usually run after the test specimens are immersed in water for three weeks. This

TABLE 4.5 ASTM Movement Capability Test Method C 719[19]

Cure cycle	Conditions
7 days	Standard conditions.
7 days	37.8F, 95% relative humidity.
7 days	Standard conditions.
7 days	Distilled water.
Interim test	Bend 60 deg, examine for failure, if OK, continue.
7 days	Compress specimens to required compression (e.g., 25% for ±25%). Hold at 70C. After 7 days cool to ambient for 1 hr.
10 cycles	Extension/compression at 0.125 in/hr.
10 cycles	Place in −26C compartment. Compress while cooling. Remove, allow to reach room temperature.

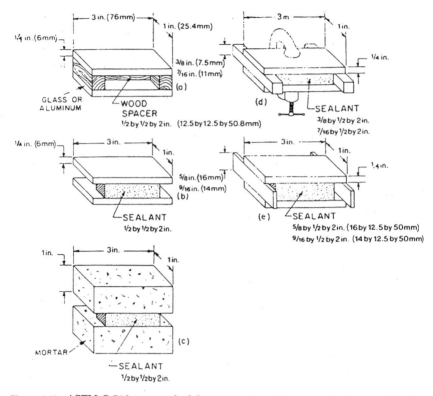

Figure 4.19 ASTM C 719 test method for measuring adhesion and cohesion of elastomeric joint sealants under cyclic movement.[19]

is a quick and easy way to determine adhesion either in the laboratory or at a job site. Minimum values of 5 piw are considered satisfactory in ASTM C 920.

4.5.3 Tear strength

The tear strength of a sealant is important because often the sealant must resist abrasion. Abrasion can come from outside sources, such as automobile tires continually running over a concrete sealant. Abrasion can also come from internal sources, such as when dirt particles become trapped in the sealant and then cause abrasion each time the sealant goes through an expansion-contraction cycle.

Tear strength is measured by ASTM D 624. This is a general measure of the toughness and abrasion resistance of the sealant. However, one can also get a general idea of the tear strength of a sealant by simply scraping at it with a fingernail.

TABLE 4.6 ISO Movement Capability Test Method[19]

Cycle 1	Cycle 2
1 hr cool to −20C	1 hr heat to 70C
2 hrs cool to −20C extend	2 hrs compression at 70C
1 hr heat to 70C	1 hr cool to −20C
2 hrs compress at 70C	2 hrs extend at −20C
1 hr cool to −20C	1 hr heat to 70C
17 hrs extend to −20C	17 hrs compress at 80C

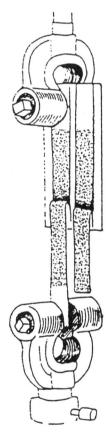

Figure 4.20 Peel adhesion test for sealants.[19]

4.5.4 Compression set resistance and creep

To test compression set resistance or stress relaxation, two parallel plates hold a section of cured sealant compressed under a known set of conditions. Good recovery is characteristic of high performance adhesives.

Compression set characteristics are determined by first placing a test specimen under load for a period of time at temperature and then subjecting the specimen to cyclic testing. Such a test method has been adopted in ASTM C 719 and specification C 920. With certain sealants, e.g., polysulfides, it has been determined that compression set often leads to failure in tension.[21,22]

The stress relaxation test is conducted by extending the sealant 25–50% in a tensile testing machine. The specimen is then locked into position at a constant deformation. The stress relaxation may be reported as a curve (Fig. 4.21) of stress versus time at constant deformation or as a stress relaxation time (the time required for the stress in the specimen to decay to 36.8% of its initial value). ASTM C 920 includes a compression cycle in part of the test exposure.

Similarly, creep properties can also be measured. The creep of a sealant is generally reported as a curve of deformation versus time under constant load (Fig. 4.22).

4.5.5 Environmental tests

Many of the environmental tests used with sealants have their foundations in the construction industry because of the significant amount of sealants used there. Water immersion of the specimens before testing is a part of most specifications because water or moisture is generally encountered in outdoor sealant applications. Three weeks is the time period recommended for most immersion tests.

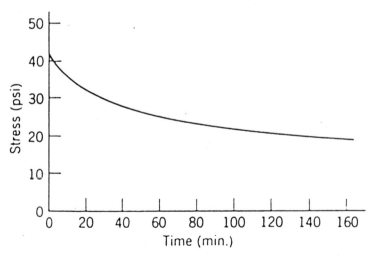

Figure 4.21 Typical stress relaxation curve for sealant.[20]

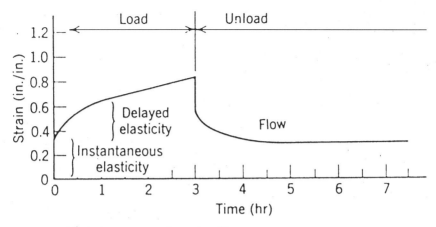

Figure 4.22 Typical creep curve for sealant.[20]

Ozone and ultraviolet (UV) radiation are factors in environmental exposure of many sealants. Major cities now exhibit 50 pphm of ozone in their atmospheres, which is the level set in standard ozone environmental tests. Ozone at this level will cause surface cracking after reasonable exposure. UV will similarly cause some hardening of certain sealants and will cause some to lose adhesion. UV is of special concern for sealants that are used in glass applications. Resistance to surface degradation after several hours in a standard weatherometer is indicative of good performance for a sealant.

Resistance to solvents or chemicals is not a standard requirement for sealants. However, the tests are easy to run, and most sealant manufacturers have such data. The time for immersion varies considerably, and it should be recognized that the absorption of chemical into the sealant is generally not very fast.

The effect of heat aging at moderate temperatures on sealants is generally through the loss of volatile plasticizers. Higher temperatures result in increased crosslinking, oxidation, and pyrolysis of the sealant. This, in turn, results in increased hardness and reduction of movement capability. ASTM C 920 specification requires a maximum percentage of weight loss of 10–12% after heat aging for two weeks at 158°F. The conditioning generally specified is the application of accumulated time at temperature expected in service. For example, the total hours of service expected at 200°F is estimated over the sealant's life and incorporated into the testing methodology. It is unwise to attempt to accelerate testing by increasing temperature above the actual service temperature without knowledge of the material characteristics of the sealant. At higher temperatures, additional reaction mechanisms may take place related to the thermal movement of molecules within the sealant.

4.6 Prototype and Non-standard Test Methods

One must keep in mind that the final product will never be a T-peel specimen or a lap shear specimen. ASTM and other standard test methods are excellent tests to offer relative comparisons of adhesives and/or bonding conditions. They also offer a valuable starting point for adhesive evaluation. However, there may be better ways to obtain information for predicting performance of the actual joint in service. The standard test methods are well defined and, therefore, can be well controlled. However, their relationship, if any, to performance in the product must be established by trial and error or advanced analytical means.

The methodology of the test and evaluation program requires careful thought. A critical feature is that the tests must have a known relationship to the final product. Often this requires either developing creative, non-standard tests that stress the part in a mode that is more indicative of its service load, or producing actual prototype specimens with the adhesives and bonding conditions that are intended to be used in production. These prototype specimens then would be subjected to simulated service environments. The environmental exposure can be accelerated to reduce testing time. Caution needs to be exerted so that the acceleration does not cause reactions or mechanisms within the materials or bond-line that would not actually be present in the intended real environment. These "non-standard" test methods should be controlled so that the tests are repeatable and the variability is low. Among the obvious variables that need to be controlled are surface cleaning, joint geometries, method and extent of material mixing, method of application, fixtures utilized, and cure conditions.

To develop a joint with an adequate service life and with a realistic design margin, the use of a "Mathes" ladder[23] is suggested to establish a testing hierarchy. In this process, shown in Fig. 4.23, testing proceeds from simple, standard tests of basic materials where well-defined test specifications are available to increasingly complex tests. Depending on the application and the type of information available from the lower rungs of the ladder, the need for more complex testing may be reduced or even eliminated. However, the need to completely understand the simpler tests is mandatory. Unexpected failure in service is often associated with a lack of understanding of the effects of the service environment on the basic materials or on a lack of understanding of the test variables (e.g., rate of loading).

The most difficult failure situations to predict are those that result from interactive effects. Thus, it is important to consider and evaluate the adhesive or sealant joint as a "system". There is a thought pro-

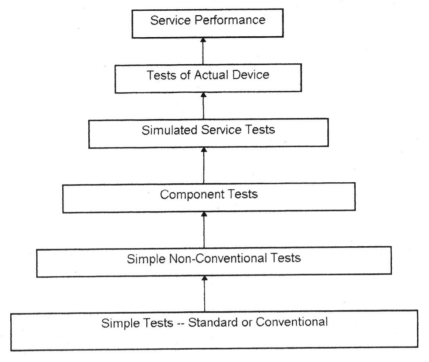

Figure 4.23 Testing hierarchy as illustrated by the "Mathes ladder".[23]

cesses which will help to guide the engineer toward systematically getting the information required.

1. Define extent of service condition variables (upper limits of conditions to be encountered).
2. Identify specific failure mode(s).
3. Determine rate of damage for each failure mode at the extreme service conditions.
4. Define critical failure mode.
5. Establish endurance limit of the system.
6. Determine reliability of endurance limit value(s).
7. Plan margin of safety for engineering design from established endurance limit.

The testing and characterization of adhesives or sealants is a key to their successful application. However, testing must be combined with understanding of the principles of adhesion and the many processes that are involved.

TABLE 4.7 Examples of Government Adhesive Specifications and Standards

	Military specifications
MIL-A-928............	Adhesive, Metal to Wood, Structural
MIL-A-1154............	Adhesive, Bonding, Vulcanized Synthetic Rubber to Steel
MIL-C-1219............	Cement, Iron and Steel
MIL-C-3316............	Adhesive, Fire Resistant, Thermal Insulation
MIL-C-4003............	Cement; General Purpose, Synthetic Base
MIL-A-5092............	Adhesive, Rubber (Synthetic and Reclaimed Rubber Base)
MIL-A-5534............	Adhesive, High Temperature Setting Resin (Phenol, Melamine and Resorcinol Base)
MIL-C-5339............	Cement, Natural Rubber
MIL-A-5540............	Adhesive, Polychloroprene
MIL-A-8576............	Adhesive, Acrylic Monomer Base, for Acrylic Plastic
MIL-A-8623............	Adhesive, Epoxy Resin, Metal-to-Metal Structural Bonding
MIL-A-9117............	Adhesive, Sealing, for Aromatic Fuel Cells and General Repair
MIL-C-10523	Cement, Gasket, for Automobile Applications
MIL-S-11030...........	Sealing Compound, Noncuring Polysulfide Base
MIL-S-11031...........	Sealing Compound, Adhesive; Curing, Polysulfide Base
MIL-A-11238	Adhesive, Cellulose Nitrate
MIL-C-12850	Cement, Rubber
MIL-A-13554	Adhesive for Cellulose Nitrate Film on Metals
MIL-C-13792	Cement, Vinyl Acetate Base Solvent Type
MIL-A-13883	Adhesive, Synthetic Rubber (Hot or Cold Bonding)
MIL-A-14042	Adhesive, Epoxy
MIL-C-14064	Cement, Grinding Disk
MIL-P-14536..........	Polyisobutylene Binder
MIL-I-15126	Insulation Tape, Electrical, Pressure-Sensitive Adhesive and Pressure-Sensitive Thermosetting Adhesive
MIL-C-18726	Cement, Vinyl Alcohol-Acetate
MIL-A-22010	Adhesive, Solvent Type, Polyvinyl Chloride
MIL-A-22397	Adhesive, Phenol and Resorcinol Resin Base
MIL-A-22434	Adhesive, Polyester, Thixotropic
MIL-C-22608	Compound Insulating, High Temperature
MIL-A-22895	Adhesive, Metal Identification Plate
MIL-C-23092	Cement, Natural Rubber
MIL-A-25055	Adhesive; Acrylic Monomer Base, for Acrylic Plastics
MIL-A-25457	Adhesive, Air-drying Silicone Rubber
MIL-A-25463	Adhesive, Metallic Structural Honeycomb Construction
MIL-A-46050	Adhesive, Special; Rapid Room Temperature Curing, Solventless
MIL-A-46051	Adhesive, Room-Temperature and Intermediate-Temperature Setting Resin (Phenol, Resorcinol, and Melamine Base)
MIL-A-52194	Adhesive, Epoxy (for Bonding Glass Reinforced Polyester)
MIL-A-9067C	Adhesive Bonding, Process and Inspection Requirements for

TABLE 4.7 Examples of Government Adhesive Specifications and Standards (*Continued*)

	Military specifications (*Continued*)
MIL-STD-401	Sandwich Construction and Core Materials, General Test Methods

	Federal specifications
MMM-A-121	Adhesive, Room-Temperature and Intermediate-Temperature Setting Resin (Phenol, Resorcinol and Melamine Base)
MMM-A-00185	Adhesive, Rubber
MMM-A-00187	Adhesive, Synthetic, Epoxy Resin Base, Paste Form, General Purpose
MMM-A-132	Adhesives, Heat Resistant, Airframe Structural, Metal-to-Metal
MMM-A-134	Adhesive, Epoxy Resin, Metal to Metal Structural Bonding
A-A-1556	Sealing Compound (Elastomeric Joint Sealants)
TT-S-277B	Sealing Compound, Rubber Base, Two Component for Caulking, Sealing, and Glazing in Building Construction
Federal Test . . . . Method 175	Adhesives, Methods of Testing

4.7 Specifications and Standards

An adhesive or sealant specification, like all material specifications, is a document that specifies values for all the important properties together with limits of variability and methods for determining these values. There are many adhesives and sealant specifications, of which the most prominent are the industrial and government specifications, which describe and establish the technical and physical characteristics or performance requirements of these materials. The most common sources of standards and specifications for the adhesives and sealants industry are the following:

- American Society of Testing and Materials (ASTM)
- International Organization for Standardization (ISO)
- U.S. Department of Defense (military and federal specifications and standards)
- National Aeronautics and Space Administration (NASA)
- Society of Automotive Engineers (SAE)
- Technical Association of the Pulp and Paper Industry (TAPPI)

Relevant ASTM specifications and standards were presented in Tables 4.1 and 4.2. A variety of federal and military specification describing adhesives and test methods have been prepared. Selected gov-

TABLE 4.8 Sources of Common Specifications and Standards for Adhesives and Sealants

1. Military
   Naval Publication and Forms Center
   5801 Tabor Ave.
   Philadelphia, PA 19120

2. Federal Standards
   Specification Sales
   3FRSBS Bldg. 197
   Washington Navy Yard
   General Services Administration
   Washington, DC 20407

3. Society of Automotive Engineers
   SAE, Inc.
   400 Commonwealth Drive
   Warrendale, PA 15096

4. American Society for Testing and Materials (ASTM)
   ASTM
   100 Barr Harbor Dr.
   Conshohocken, PA 19428

5. American Society for Nondestructive Testing
   1711 Arlingate Ln.
   P.O. Box 28518
   Columbus, OH 43228

6. Technical Association for the Pulp and Paper Industries (TAPPI)
   Technology Park/Atlanta
   P. O. Box 105113
   Atlanta, GA 30348

7. Pressure Sensitive Tape Council
   401 N. Michigan Ave.
   Chicago, IL 60611

ernment specifications are described in Table 4.7. Table 4.8 identifies several sources of specifications for adhesives, sealants and related equipment. A more complete list of specifications and standards are provided in Appendix C.

Certain specifications and standards provide excellent tutorials on adhesives and sealants. MIL-HDBK-691 offers a handbook on adhesive bonding, and MIL-HDBK-725 provides a guide to the properties and uses of adhesives. ASTM C 962 provides an excellent source of information regarding sealant joint design and the types of sealants that are appropriate for various substrates. Although this specification is primarily for construction sealants, much of the information that it contains is generally useful for other sealant applications. A useful guide to proper sealant application procedures is also available from the Sealant, Water-Proofers, and Restoration Institute (SWRI) as an application manual and a videotape.[24]

References

1. Semco Packaging and Applications Systems, Courtaulds Aerospace, Glendale, CA.
2. Muny, R. P., "Getting the Right Results", *Adhesives Age*, August, 1996 at 20–23.
3. Anderson, G. P., et. al., "Effect of Removing Eccentricity from Button Tensile Adhesion Tests", *Adhesively Bonded Joints: Testing, Analysis, and Design*, ASTM STP 981, W. S. Johnson, ed. (Philadelphia: American Society of Testing and Materials 1988).
4. Elliot, S. Y., "Techniques for Evaluation of Adhesives", in Chapter 31, *Handbook of Adhesives Bonding*, C. V. Cagle, ed. (New York: McGraw Hill, 1973).
5. Mathews and Silver, Methods for Testing the Strength Properties of Adhesives and Test Data, NEVOID Report 3923, 1962.
6. DelLollis, N. J., *Adhesives for Metals Theory and Technology*, (New York: Industrial Press, 1970).
7. Vazirani, H. N., "How to Monitor Joint Quality with a Peel Resistance Test", *Adhesives Age*, January, 1970 at 40–43.
8. Johnston, J., "Peel Adhesion Testing of Pressure Sensitive Tapes", *Adhesives Age*, April, 1968 at 20–26.
9. Snoddon, Metal to Resin Adhesion as Determined by a Stripping Test, ASTM STP 201 (Philadelphia: American Society of Testing and Materials, 1957).
10. DelLollis, N. J., *Adhesives, Adherends, Adhesion*, (Huntington, N.Y.: Robert E. Krieger Publishing Co., 1980).
11. Gent, A. N., "More on the Peel Test Riddle", *Adhesives Age,* February, 1997 at 59.
12. Kaelble, D. H., "Theory and Analysis of Peel Adhesion: Mechanisms and Mechanics," *Transactions of the Society of Rhelogy*, vol. III, at 161–182, (1959).
13. Kaelble, D. H., *Transactions of the Society of Rhelogy*, vol. IV, at 45–73, (1960).
14. Bikerman, J. J., "Theory of Peeling Through a Hookean Solid", *Journal of Applied Physics*, vol. 28, at 1484–1485, 1957.
15. Bikerman, J. J., "Experiments in Peeling", *Journal of Applied Polymer Science*, vol. II, no. 5, at 216–224, 1959.
16. Silver, I., *Modern Plastics*, vol. 26, no. 9, 1949, at 95.
17. "How to Test Adhesive Properties", *Materials Engineering*, March 1972 at 60–64.
18. Miniford, J. D., "Permanence of Adhesive–Bonded Aluminum Joints", Chapter 23 in *Adhesives in Manufacturing*, G. L. Schneberger, ed. (New York: Marcel Dekker, 1983).
19. Evans, R. M., *Polyurethane Sealants Technology and Applications* (Lancaster, PA: Technomic Publishing Co., 1993).
20. Paneck, J. R. and Cook, J. P., Chapter 1, *Construction Sealant and Adhesives* (New York: John Wiley and Sons, 1991).
21. Burstrom, P. G., "Aging and Deformation Properties of Building Joint Sealants", Report TVBM-1002 (University of Lund, Sweden, 1979).
22. Karpati, K., *Journal of Coatings Technology*, 1984 at 719.
23. Mathes, K. N., "Functional Evaluation of Insulating Material", *Trans. AIEE*, 67: 1236 (1948).
24. Sealant, Waterproofing and Restoration Institute, 2841 Main, Kansas City, MO 64108.

Chapter 5

Quality Control and Nondestructive Tests

5.1 Introduction

In this chapter, various methods of controlling the consistency of the sealing or bonding operation will be introduced. These processes usually fall into the categories of quality control or nondestructive testing. Although some are sophisticated, many of these processes do not require advanced equipment or knowledge. Simple equipment, visual examination, and common sense are the main tools in most quality control departments. They are supported by strong specifications and proper training.

The control processes that are practiced by the end-user of the adhesive or sealant will be the main subject of this chapter. Important processes, such as training and plant cleanliness, that need to be considered even before the manufacturing operation are included in this discussion. Quality control methods that are commonly used for checking incoming materials and controlling the various processes are reviewed. Bond inspection techniques will include both destructive and nondestructive tests. Analytical methods will also be described that can be used to examine failed joints for cause of failure.

5.2 Quality Control

Quality control encompasses all of the processes and activities that ensure adequate quality in the final product. Quality control is very important when using adhesives or sealants because once fully bonded, joints are difficult to take apart or correct. By the time it takes to notice that one step in the bonding process is out of control, significant costs could occur. For adhesives and sealants, quality control must be defined in its broadest sense. This consists of defining the means to prevent problems whether they be training, controlling manufacturing procedures, incoming inspection, or visual and physical examination of the finished product.

A flow chart for a quality control system is illustrated in Fig. 5.1. This system is designed to ensure reproducible bonds and, if a substandard bond is detected, to make suitable corrections. However, good quality control will begin even before the receipt of materials. This usually begins with proper training of personnel and conditioning of the manufacturing area.

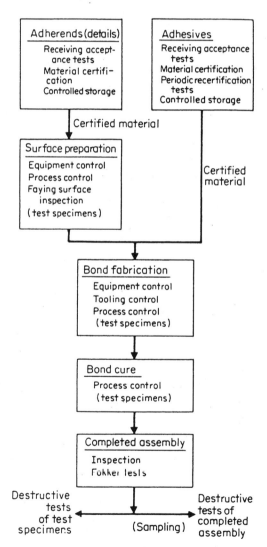

Figure 5.1 Flow chart of a quality control system for adhesive bonding.[1]

5.2.1 Pre-manufacturing processes

The human element enters the adhesive-bonding process probably more than in any other fabrication technique. An extremely high percentage of defects can be traced to poor workmanship or lack of understanding regarding adhesion. This generally appears first in the surface-preparation steps, but may also arise in any of the other steps necessary to achieve a bonded assembly. This problem can be largely overcome by proper motivation and education. All employees from design engineer to laborer to quality-control inspector should be familiar with adhesive bonding technology and be aware of the circumstances that can lead to poor joints.

A great many defects can also be traced to poor design engineering. Table 5.1 lists the polled replies of experts to a question: "Where do design engineers most often err in designing bonded joints?" The most common errors relate to training, design considerations, and production line problems. The probability of many of these problems occuring can be minimized or eliminated through proper training and education.

Specifications are a necessary part of a quality control program. A specification simply is a statement of the requirements that the adhesive, sealant, or process must meet in order to be accepted for use. A product specification is an agreement between supplier and user.

TABLE 5.1 Where Design Engineers Commonly Make Errors[2]

Adhesive technology	Design consideration	Production-line problems
Low peel strength	Using a butt joint when lap joint would be stronger	Lack of careful surface preparation
Overlooking such factors as pot life, curing time, operating temperatures	Loads causing unsuspected cleavage forces	Expecting prototype performance from bonds made on assembly line
Failure to get technical help from supplier in selecting an adhesive	Overlooking effect of increased service temp in decreasing resistance to chemicals	Failure to keep surfaces clean until adhesive is applied
Assuming that strongest adhesive is always the best without considering cost or processing	Failure to check coefficients of expansion when unlike materials are bonded	Failure to consider the application method and equipment when designing joint
Lack of care in test procedures	Calling for heat-curing adhesive on a part that will not stand the heat	
	Overdesigning by asking for more strength or heat resistance than is needed	

Conformance to a specification does not mean that the adhesive or sealant will perform perfectly in service. It only means that the product conforms to the specification. Bonding or sealing specifications should not only account for the adhesive or sealant, but they should also define the adherends and the ancillary processes for preparing the adherends and the joint assembly.

The product specification writer must try to put into the specification the requirements that, if met, will provide the greatest likelihood of success. These requirements should be standard tests that are agreed upon by both the supplier and the user. The tests should be indicative of how the adhesive is used in production and how the finished joint is to be used in service. Tests that are not directly related to the specific application should not be included. Tests should not be used simply because they are standard test methods or have been used in the past. Specifications from other sources (e.g., ASTM, military, etc.) may be used if they are applicable.

Specifications may require different categories of testing. For example, there may be extensive series of tests that are required for initial verification or qualification of the adhesive and supplier. These tests would be used to approve a certain product at the onset. Other receiving tests may be used to verify the consistency of the product from lot to lot. A typical specification has the following format: title; scope; general requirements; performance requirements; test methods; controls; reference documents; and approved source list.

In addition to the staff, the operational facilities must be well prepared before the use of adhesives or sealants. The plant's bonding area should be as clean as possible prior to receipt of materials. The basic approach to keeping the assembly area clean is to segregate it from the other manufacturing operations either in a corner of the plant or in isolated rooms. The air should be dry and filtered to prevent moisture or other contaminants from gathering at a possible interface. The cleaning and bonding operations should be separated from each other. If mold release is used to prevent adhesive flash from sticking to bonding equipment, it is advisable that great care be taken to assure that the release does not contaminate either the adhesive or the adherends. Spray mold releases, especially silicone release agents, have a tendency to migrate to undesirable areas.

5.2.2 Quality control of the incoming materials

Acceptance tests on adhesives or sealants as well as adherends should be directed toward assurance that incoming materials are identical from lot to lot. The tests should be those which can quickly and accurately detect deficiencies in the product's physical or chemical properties. ASTM lists various test methods that are commonly used for adhesive or sealant acceptance. Many of the more common test methods were described in the previous chapter.

A receiving inspection program consists of comparison of the purchase order with what is received. If the vendor's test reports are required with the shipment, it is verified that they were received and that the test values are acceptable as defined by the original material specification.

It may also be necessary to test the incoming bulk product in-house for fundamental properties. These inspections usually consist of an evaluation of physical and chemical properties such as: color, viscosity, percent solids, weight per gallon, pot life, open time, and flow.

Test specimens may also be made to verify strength of the adhesive. These specimens should be stressed in directions that are representative of the forces which the bond will see in service, i.e., shear, peel, tension, or cleavage. If possible, the specimens should be prepared and cured in the same manner as actual production assemblies. If time permits, specimens should also be tested in simulated service environments, e.g., high temperature and humidity.

Because of cost and time required for extensive in-house testing, the trend today is to have most of the quality control tests done by the supplier. The supplier then provides a certified test report with shipment of the product. The test program used by the supplier, his internal controls, etc. are usually verified and approved by the user on a periodic basis, such as once per year.

Once identified and approved for receipt, all incoming material should be labeled with a receipt date. This date will be prominently shown on the container or material while it is in inventory. Precautions must be taken to assure that the oldest material is used first and that the shelf life of the product does not expire before it is used. The date stamp on the product is the most reliable method of doing this.

5.2.3 Quality control of surface treatment

Generally, some sort of surface preparation is required for reliable adhesion. The extent of the actual surface preparation process will vary depending on the performance characteristics desired, the nature of the adherend, and time and cost considerations.

Surface preparation processes must be carefully controlled for reliable production of bonded parts. If a chemical surface treatment is required, the process must be monitored for proper sequence, bath temperature, solution concentration, and contaminants. If sand or grit blasting is employed, the abrasive must be changed regularly. An adequate supply of clean wiping cloths for solvent cleaning is also mandatory. Checks should be made to determine if cloths or solvent containers may have become contaminated.

The surface preparation process can be checked for effectiveness by the water-break free test. After the final treating step, the substrate surface is checked for a continuous film of water that should form when deionized water droplets are placed on the surface. A similar test for treatment of polymeric fabric is shown in Fig. 5.2. A surface which is uniformly wet by distilled water will likely also be wet by the adhesive since the specific surface energy of water is 72 dynes/cm and most organic adhesives is 30–50 dynes/cm. However, this test tells little about weak boundary layers or other contaminants that may be present on the substrate's surface but still be capable of wetting with water.

After the adequacy of the surface treatment has been determined, precautions must be taken to assure that the substrates are kept clean and dry until bonding. The adhesive or primer should be applied to the treated surface as quickly as possible.

5.2.4 Quality control of the bonding process

All parts should be fitted together first without adhesive or sealant to minimize production problems due to fit. The suitability of fit is either established by visual inspection or direct measurement with a gauge or shim. It is desirable that the extremes in mechanical tolerances also be noted and that test specimens be made with the worst possible fit to assure that the bonding process will always provide reliable joints.

The adhesive metering and mixing operation should be monitored by periodically sampling the mixed adhesive and testing it for adhesive properties. Simple viscosity measurements, flow tests, or visual inspection of consistency are the best methods of monitoring conformance. A visual inspection can also be made for air entrapment. The quality-control engineer should be sure that the oldest adhesive is used first and that the specified shelf life has not been exceeded.

During the actual assembly operation, the cleanliness of the shop and tools should be verified. The shop atmosphere should be controlled as closely as possible. Temperature in the range of 65 to 90°F and relative humidity from 20 to 65% is best for almost all bonding operations.

The amount of applied adhesive and the final bond-line thickness must also be monitored because they may have an effect on joint strength. Curing conditions should be monitored for pressure, heat-up rate, maximum and minimum temperature during cure, time at the required temperature, and cool-down rate.

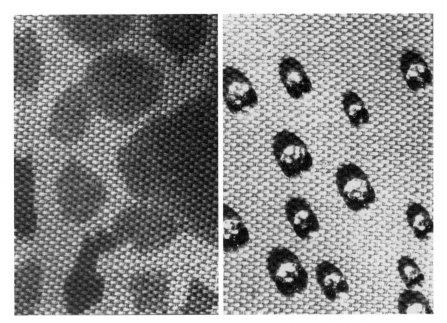

Figure 5.2 Example of water break free test for treated polymeric fabric. (Photo courtesy of Krüss USA).

5.3 Bond Inspection

After the adhesive or sealant is cured, the joint area can be inspected to detect gross flaws or defects. This inspection procedure can either be destructive or nondestructive in nature. The nondestructive type of tests can be either visual or use advanced analytical equipment. These types of bond inspections are described below.

5.3.1 Destructive testing

Destructive testing generally involves placing samples of the production run in simulated or accelerated service and determining if it has similar properties to a specimen that is known to have a good bond and adequate service performance. It is desirable to fabricate a standard test specimen in the same cycle as the part actually being bonded. The test specimen could be either test coupons (standard ASTM test specimens), extensions of actual parts (i.e., tabs that can be removed from the part and tested), or special test specimens that are close to the actual part design but amenable to mechanical testing. If a special test specimen is used, the specimen should be designed for a test method that is easy to perform and indicative of the way the part will be loaded in service, but it should also be designed so that the geometry and mass does not depart too severely from the actual production part. The test specimen is then either tested immediately after bonding or after a simulated environmental cycle.

After testing, the joint area and mode of failure should be examined closely. This generally leads to clues that are indicative of problems. The causes and remedies for faults revealed by such mechanical tests and subsequent visual inspection are described in Table 5.2. Similar visual techniques have been developed to determine the causes of contact adhesive failures.[3]

Test specimens such as those mentioned above are often used to verify the quality of the first article through the production line and then to periodically test articles for conformance. This type of testing will detect discrepancies affecting the entire lot, but it cannot evaluate factors that affect individual joints or specific areas of a particular joint.

5.3.2 Non-destructive testing

Non-destructive testing (NDT) is usually far more economical than destructive test methods, and every assembly can be tested if desired. However, nondestructive testing primarily offers qualitative information reguarding the strength of the joint. Several non-destructive test methods are used to check appearance and quality of structures made with adhesives or sealants. The main methods are simple ones like visual inspection, tap, and proof testing. More advanced nondestructive monitoring such as ultrasonic or radiographic inspection is also used in critical applications. The most difficult defect to find are those related to improper curing and surface treatments. Therefore, great care and control must be given to these processes.

5.3.2.1 Visual inspection.
A trained eye can detect a surprising number of faulty joints by close inspection of the adhesive around the bond area even if the substrate is not transparent or translucent. Close examination of the visually apparent adhesive or sealant (generally around the edges of the joint) can lead to useful conclusions. Unfilled areas and voids can sometimes be detected by noting lack of adhesive or sealant material. Misalignment of parts are readily visible. The texture of the adhesive around the edges of the joint can also be a clue to the effectiveness of the curing process and whether air was entrapped in the adhesive. The adhesion of the flash to the substrates can be qualitatively measured by attempting to pry the flash away from the substrate. This can indirectly be a measure of surface cleanliness. Table 5.3 lists the characteristics of faulty joints that can be detected visually, their cause, and possible remedies.

TABLE 5.2 Faults Revealed by Mechanical Tests

Fault	Cause	Remedy
Thick, uneven glue line	Clamping pressure too low	Increase pressure. Check that clamps are seating properly
	No follow-up pressure	Modify clamps or check for freedom of moving parts
	Curing temperature too low	Use higher curing temperature. Check that temperature is above the minimum specified throughout the curing cycle
	Adhesive exceeded its shelf life, resulting in increased viscosity	Use fresh adhesive
Adhesive residue has spongy appearance or contains bubbles	Excess air stirred into adhesive	Vacuum-degas adhesive before application
	Solvents not completely dried out before bonding	Increase drying time or temperature. Make sure drying area is properly ventilated
	Adhesive material contains volatile constituent	Seek advice from manufacturers
	A low-boiling constituent boiled away	Curing temperature is too high
Voids in bond (i.e., areas that are not bonded), clean bare metal exposed, adhesive failure at interface	Joint surfaces not properly treated	Check treating procedure; use clean solvents and wiping rags. Wiping rags must not be made from synthetic fiber. Make sure cleaned parts are not touched before bonding. Cover stored parts to prevent dust from settling on them
	Resin may be contaminated	Replace resin. Check solids content. Clean resin tank
	Substrates distorted	Check for distortion; correct or discard distorted components. If distorted components must be used, try adhesive with better gap-filling ability
Adhesive can be softened by heating or wiping with solvent	Adhesive not properly cured	Use higher curing temperature or extend curing time. Temperature and time must be above the minimum specified throughout the curing cycle. Check mixing ratios and thoroughness of mixing. Large parts act as a heat sink, necessitating larger cure times

TABLE 5.3 Visual Inspection of Faulty Bonds

Fault	Cause	Remedy
No appearance of adhesive around edges of joint or adhesive bond line too thick	Clamping pressure too low	Increase pressure. Check that clamps are seating properly
	Starved joint	Apply more adhesive
	Curing temperature too low	Use higher curing temperature. Check that temperature is above the minimum specified
Adhesive bond line too thin	Clamping pressure too high	Lessen pressure
	Curing temperature too high	Use lower curing temperature
	Starved joint	Apply more adhesive
Adhesive flash breaks easily away from substrate	Improper surface treatment	Check treating procedure; use clean solvents and wiping rags Make sure cleaned parts are not touched before bonding
Adhesive flash is excessively porous	Excess air stirred into adhesive	Vacuum-degas adhesive before application
	Solvent not completely dried out before bonding	Increase drying time or temperature
	Adhesive material contains volatile constituent	Seek advice from manufacturers
Adhesive flash can be softened by heating or wiping with solvent	Adhesive not properly cured	Use higher curing temperature or extend curing time Temperature and time must be above minimum specified. Check mixing

5.3.2.2 Tap test. One of the first non-destructive methods used to evaluate the quality of an adhesive joint was by tapping the bonded joint and assessing the resulting tone. Tone differences indicate inconsistencies in the bonded joint. This could be due to insufficient cure, voids, or other problems. Simple tapping of a bonded joint with a coin or light hammer can indicate an unbonded area. Sharp clear tones indicate that adhesive is present and adhering to the substrate in some degree; dull hollow tones indicate a void or unattached area.

The success of the tap test depends on the skill and experience of the operator, the background noise level, and the type of structure.

Some improvement in the tap test can be achieved by using a solenoid operated hammer and a microphone pickup. The resulting electrical signals can be analyzed on the basis of amplitude and frequency. However, the tap test, in its most successful mode, will only measure the qualitative characteristics of the joint. It will tell whether adhesive is in the joint or not, providing an acoustical path from substrate to substrate, or it will tell if the adhesive is undercured or filled with air, thereby, causing a damped path for the acoustical signal. The tap test provides no quantitative information and little information about the presence and/or nature of a weak boundary layer.

5.3.2.3 Proof tests. If a high degree of reliability is required, it is necessary to proof test the production unit. The proof test should simulate actual service conditions in the manner in which the joint or structure is loaded and the stress level should be higher than that expected in service. The duration of the proof test should reflect the expected life of the joint, but usually this is not possible. The proof test should be designed so that it is normally a non-destructive test, unless the bond is unexpectedly weak. Care must be taken to design the proof test so that it does not overstress the part and cause damage that will result in a reduced service life.

A common example of a proof test is to apply a cleavage load to a bonded honeycomb sandwich by placing an instrument between the face and core and applying a predetermined force perpendicular to the core. If there is no bond disruption due to this test, it is supposed that the product will meet all its service requirements. A common proof evaluation used with sealants is leak testing with a mobile and easily detected gas such as helium or application of hydrostatic pressures.

5.3.2.4 Ultrasonic inspection. The success of the tap test, although limited, led to the use of ultrasonics to determine bond quality. Ultrasonic methods are at present the most popular NDT technique for use on adhesive joints. Ultrasonic testing measures the response of the bonded joint to loading by low-power ultrasonic energy. Short pulses of ultrasonic energy can be introduced on one side of the structure and detected on the other side. This is called through-transmission testing. An unbonded area, void, or high damped adhesive (undercured or filled with air) prevents the ultrasonic energy from passing efficiently through the structure.

A number of different types of ultrasonic inspection techniques using pulsed ultrasound waves from 2.25 to 10 MHz can be applied to bonded structures.[4] The most common methods are:

- Contact pulse echo—the ultrasonic signal is transmitted and received by a single unit;

- Contact through transmission—the transmitting search unit is on one side of the bonded structure and the receiving unit is on the other; and

- Immersion method—the assembly is immersed in a tank of water; the water acts as a coupling mechanism for the ultrasonic signal.

Figure 5.3 illustrates these various NDT ultrasonic methods.

Pulse echo techniques are perhaps the easiest to use in production. The sound is transmitted through the part, and reflections are obtained from voids at the bond interface. The result is generally considered only qualitative because a poorly bonded joint will show as a good joint as long as it is acoustically coupled. A thin layer of oil or water at the interface may act as a coupling and disguise an unbonded

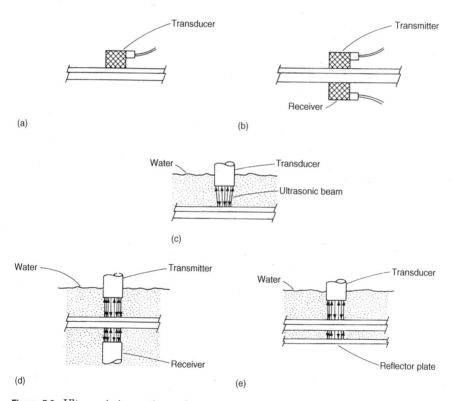

Figure 5.3 Ultrasonic inspection techniques. (a) Contact pulse echo with a search unit combining a transmitter and receiver. (b) Contact through transmission. Transmitting search unit on top and receiving search unit on bottom. (c) Immersion pulse echo with search unit (transmitter/receiver) and part inspected under water. (d) Immersion through transmission with both search units (transmitter and receiver) and part under water. (e) Immersion reflector plate. Same as (c) but each unit requires a reflector plate below the part being inspected.[4]

area. Shear waves can also be introduced into the structure with a wedge shaped transducer.[5] This technique is effective in analyzing sandwich structures.

Bonded structures that are ultrasonically tested by the immersion method often use a C-scan recorder to record the test. This recorder is an electrical device that accepts signals from the pulser/receiver and prints out a plan view of the part. The ultrasonic search unit is automatically scanned over the part. The ultrasonic signals for bond or unbond are detected from built-in reference standards. C-scan NDT techniques are used extensively by aircraft manufacturers to inspect bonded parts.

One of the oldest and best known ultrasonic testing system is the Fokker Bond Tester originally manufactured by the Fokker Aircraft Company in Germany. This method uses a sweep frequency resonance method of ultrasonic inspection. Some degree of quantitative analysis is claimed with the Fokker Bond Tester in the aircraft industry. Table 5.4 indicates the degree of correlation between predicted and actual bond strength of the Fokker Bond Tester compared with two other NDT instruments.

5.3.2.5 Other NDT methods. Radiography (x-ray) inspection can be used to detect voids or discontinuities in the adhesive bond. This method is more expensive and requires more skilled experience than ultrasonic methods. The adhesive must contain some metal powder or other suitable filler to create enough contrast to make defects visible.

TABLE 5.4 Correlation between Predicted and Actual Bond Strengths Made by Three Ultrasonic NDT Instruments[1]

Type of bond	Adhesive material	Nondestructive test instrument		
		Fokker, lb/in.^{2}	Coinda scope, lb/in.^{2}	Stubmeter, lb/in.^{2}
Overlap	FM-47	±950	±1,300	±1,400
Overlap	HT-424	±360	±1,100	±950
Overlap	FM-58	±740	±1,175	±1,050
Overlap	Metlbond 4021	±730	±600	±1,150
Honeycomb	FM-47	±250		
Honeycomb	HT-424	±200		
Overlap	FM-47	±750		
Honeycomb	Epoxy-phenolic	±20 to ±150		
Overlap	Redux 775	±570		

95% confidence limits.

This method is applicable to honeycomb sandwich structures as well as metal and non-metal joints.

Thermal transmission methods are relatively new techniques for adhesive inspection. Heat flow is determined by monitoring the surface temperature of a test piece immediately after external heating or cooling has been applied. Subsurface anomalies will alter the heat flow pattern and, thereby, affect the surface temperature. The surface temperature difference can be detected by thermometers, thermocouples, or heat sensitive coatings. Liquid crystals applied to the joint can make voids visible if the substrate is heated.

Thermal transmission testing is an excellent way for detecting various types of anomalies such as surface corrosion under paint before the corrosion becomes visually evident. Thin single layer structures, such as aircraft skin panels, can be inspected for surface and subsurface discontinuities. This test, as shown in Fig. 5.4, is simple and inexpensive, although materials with poor heat transfer properties are difficult to test, and the joint must be accessible from both sides. For nonmetallic materials, the defect diameter must be on the order of four times its depth below the surface to obtain a reliable thermal indication. For metals, the defect diameter must be approximately eight times its depth. Some bright surfaces such as bare copper and aluminum do not emit sufficient infrared radiation and may require application of a dark coating on their surface.

Thermal wave inspection is also a relatively new technique for studying adhesive disbonding.[7] With this method, heat is injected into the test object's surface from a hot gas pulse. The resulting surface temperature transient is analyzed to determine the bond quality in nearly real time. The surface temperature transients are sensed using a noncontacting, emissivity independent infrared sensor or video camera. This method is not adversely affected by surface blemishes or roughness.

Table 5.5 shows the types of defects that can be detected with various NDT techniques. Table 5.5a shows the correlation of NDT results for built-in defects for laminate panels, and Table 5.5b shows a similar correlation for honeycomb structures. A universal NDT method for evaluating all bonded structures is not currently available. Generally, the selection of a test method is based on:

- Part configuration and materials of construction
- Types and sizes of flaws to be detected
- Accessibility to the inspection area
- Availability and qualifications of equipment and personnel

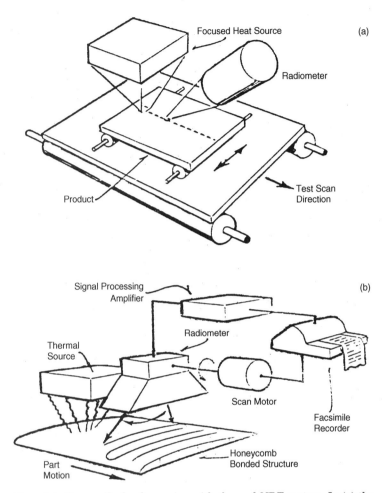

Figure 5.4 Two methods of scanning with thermal NDT system. In (a) the focused thermal source and radiometer sensor are stationary while the product is moved in two directions to produce the scanning pattern. In (b) the product moves in one direction as the radiometer detector oscillates to provide the scanning motion.[6]

- Through-put rate required of the NDT process
- Required documentation of the process and test results.

5.4 Testing of Failed Joints for Cause of Failure

Close examination of failed joints can sometimes lead to an explanation of why the specimen failed. With visual or microscopic examina-

TABLE 5.5 Correlation of NDT Results for Built-in Defects in Laminate Panels and Correlation of NDT Results for Built-in Defects in Honeycomb Structures[4]

Laminate defects	Radiography		Ultrasonic bond testers				NDE method(a) Ultrasonic inspection(b)					Coin tap test	Remarks
	Low-kilovolt x-ray	Neutron radiography	Fokker bond tester	Sondicator	Harmonic bond tester	210 sonic bond tester	Contact pulse echo	Contact through transmission	Immersion C-scan pulse echo	Immersion C-scan through transmission	Immersion C-scan reflector plate		
1. Void	(c)	D	D	D	D	D	D	D	D(d)	D	D	PD(e)	c,d,e,f
2. Void (C-14 repair)	(c)	D	D	D	D	D	D	D	D(d)	D	D	PD(e)	d,e
3. Void (9309 repair)	(c)	D	D	D	D	D	D	D	D(d)	D	D	PD(e)	d,e
4. Lack of bond (skin to adhesive)	(c)	ND	D	D	D	D	D	D	D(d)	D	D	PD(e)	d,e
5. Manufacturer sheet (FM123-41)	(c)	PD(g)	D	D	ND	PD	PD	PD	D(d)	D(g)	D(g)	ND	d,g
6. Thick adhesive (1, 2, 3 ply)	(c)	ND	D	ND	ND	D	ND	ND	ND(e)	PD(d)	D(d)	ND	d,e
7. Porous adhesive	(c)	D	D	ND	ND	D	D	D	D	D	D	ND	c
8. Burned adhesive	(c)	ND	D	ND	ND	D	D	D	PE	D	D	ND	h
9. Corroded joint	D	D	D	PD	PD	D	D	D	PD	D	D	PD	i

(a) ND, not detected; PD, partial detection; D, detected. (b) Contact surface wave was tried but did not detect any built-in defects. (c) Panels were made using FM 73, which is not x-ray opaque. With x-ray opaque adhesive, defects 1, 2, 3, 5, 6, 7 are detected. (d) Method suffers from ultrasonic wave interference effects caused by tapered metal doubles or variations in adhesive thickness. (e) MIL-C-88286 (white) external topcoat and PR1432G (green) plus MIL-C-83019 (clear) bilge topcoat dampened the pulse-echo response. (f) Minimum detectable size approximately equal to size of probe being used. (g) Manufacturer separator sheet not detectable but developed porosity and an edge unbond during cure cycle, which was detectable. (h) Caused by drilling holes or band sawing bonded joints. (i) Moisture in bond joint (Armco 252 adhesive, Forest Products Laboratory etch)

	NDE method(a)											
	Radiography		Ultrasonic bond testers				Ultrasonic inspection(b)					
Honeycomb defects	Low-kilovolt x-ray	Neutron radiography	Fokker bond tester	Sondicator	Harmonic bond tester	210 sonic bond tester	Contact pulse echo	Contact through transmission	Contact shear wave	Immersion C-scan pulse pulse echo	Immersion C-scan through transmission	Coin tap test	Remarks
1. Void (adhesive to skin)	ND(b)	D	D	D	D	D	D	D	ND	D	D	D	Replacement(b) standard
2. Void (adhesive to skin) repair with C-14	...	...	...	...	...	...	...	...	...	...	...	...	No void improperly made
3. Void (adhesive to skin) repair with 9309	...	...	...	...	...	...	...	...	...	...	...	...	No void improperly made
4. Void (adhesive to core)	ND	D	D	...	D	D	ND	D	ND	D	D	D	Replacement standard
5. Water intrusion	D	D	ND	ND	ND	ND	ND	D	ND	ND	D	D	...
6. Crushed core (after bonding)	D(c)	PD	PD(d)	ND	PD(d)	ND	ND	PD(d)	ND	PD(d)	PD(d)	D	(c, d)
7. Manufacturer separator sheet (skin to adhesive)	ND(e)	D	ND	ND	PD	ND	PD	D	ND	D	D	PD	(e)
8. Manufacturer separator sheet (adhesive to core)	ND(e)	D	ND	ND	ND	D	ND	ND	D	PD	D	(e)	
9. Void (foam to closure)	D	D	D	D	D	D	D	D	ND	D	D	D	...
10. Inadequate tie-in of foam to core	D	D	D	D	D	D	ND	D	ND	D	ND	D	...
11. Inadequate depth of foam at closure	D	D	D	D	ND	PD(f)	ND	ND	ND	ND	ND	PD	...
12. Chemically milled step void	ND(e)	D	ND	ND	ND	ND	ND	ND	ND	ND	ND	ND	(e)

(a) ND, not detected; PD, partial detection; D, detected. (b) Panels were made using FM 73, which is not x-ray opaque. (c) The 0.13 and 0.25 mm (0.005 and 0.010 in.) crushed core detected by straight and better by angle shot. (d) Detects 0.25 mm (0.010 in.) crush core. (e) Has been detected by x-ray when adhesive was x-ray opaque (FM 400). (f) Discloses defect at a very high sensitivity.

tion it is sometimes evident why adhesive failure occurred. Evidence may suggest improper wetting at the original interface or some new interface.

In comparing surface features after bond failure with the original adherend surface, the maximum resolution of about 1×10^{-8} m (100 Å) for scanning electron microscopes may not always be sufficient to detect a thin film of adhesive closely reproducing the original surface profile. Optical and staining methods described by Brett[8] to determine the presence of such films are mainly applicable to fairly thick films since optical techniques use interference phenomenon. Films a few angstroms thick are still largely undetected.

The sciences of microphotography and holography have also been used for NDT of adhesive bonds. Magnification and photography of the failed substrate will often lead to useful clues to the cause of failure. Holography is a method of producing photographic images of flaws and voids using coherent light such as that produced by a laser. The major advantage of holography is that it photographs successive "slices" through the scene volume. A true three dimensional image of a defect or void can then be reconstructed.

The use of highly specialized surface characterization tools has greatly improved the opportunity for deducing the surface chemical composition. These tools have been developed for the purpose of analyzing both the adherend and the adhesive. For adherends, analytical examination generally centers on either:

(1) The surface chemistry by elemental analysis, chemical species, or analysis of contaminates and boundary layers, or

(2) On analysis of failed surface for evidence of interfacial failure, failure within the adherend (e.g., metal oxide, composite matrix or fiber, etc.) or within a primer or other boundary layer. Analysis of the adhesive generally consists of characterizing the cured film, curing agents, and failed specimen surface chemistry.

Over the last 20 years, analytical tools have become available that allow for the characterization of the elemental and chemical composition of solid surfaces. The application of these analytical tools has increased our understanding of surface properties and successfully characterized surface layers. Table 5.6 shows a comparison of analytical capabilities of some surface sensitive analytical tools. The most popular of these are secondary ion mass spectroscopy (SIMS), ion scattering spectroscopy (ISS), and Auger electron spectroscopy (AES).[10–14] These tools have proven practical even when the surface films are only on the order of atomic dimensions or when the failure occurred near the original interface and included parts of both the adhesive and the adherend.

TABLE 5.6 Techniques for Studying Surface Structure and Composition[9]

	Technique	Probe Species	Description
ESCA	Electron Spectroscopy for Chemical Analysis	X-rays	Chemical information through line shape analysis. Useful for insulators and conductors. Widely used on polymers. Sample may be damaged due to X-rays.
SIMS	Secondary Ion Mass Spectrometry	Ions	Semi-quantitative elemental analysis. Useful only for conductors or Ionic insulators. Sample is severely damaged.
AES	Auger Electron Spectroscopy	Electrons	Quantitative elemental analysis tool for determining surface composition of semiconductors and conductors. Limited to non-insulators.
ATR-FTIR	Attenuated Total Reflectance Fourier Transform Infrared Spectroscopy	Infrared light	Detailed chemical information. Non-vacuum method. Specimens must be flat and capable of intimate contact with the necessary crystal.
ISS	Ion Scattering Spectroscopy	Ions	Information gained on surface composition. Elemental analysis with single layer atomic resolution.
XES	X-Ray Emission Spectroscopy	X-rays	Information on energy levels and chemical state of adsorbed molecules; surface composition.

By itself, SIMS has been shown to be a powerful tool for elemental surface characterization by Benninghoven[15] and Schubert and Tracy[16]. However, uncertain or rapidly changing secondary ion yield due to changes in chemical bonding make quantitative analysis virtually impossible using SIMS alone.[17,18] SIMS is most helpful when combined with other techniques, such as ISS and AES.

The greatest strength of ESCA (Electron Spectroscopy for Chemical Analysis) lies in its ability to provide information on the surface chemistry of polymers or organics.[19] ESCA is also known as x-ray photoelectron spectroscopy (XPS). The surface excitation produced by ESCA analysis is relatively non-destructive. AES is a useful high spatial resolution technique for the analysis of metals, alloys, and inorganic materials. Polymer and organic surfaces pose problems because of beam damage and sample charging. ISS is also an elemental analysis technique with single atomic layer resolution and is often used in conjunction with other surface techniques.

References

1. Smith, D. F., and Cagle, C. V., "A Quality Control System for Adhesive Bonding Utilizing Ultrasonic Testing", in *Structural Adhesive Bonding*, M. J. Bodanr, ed. (New York: Interscience, 1966).
2. Hiler, M. J., "Adhesive Failures: Reasons and Preventions", *Adhesives Age*, January 1967.
3. Hauser, R. L., et al., "Physical Analysis of Contact Adhesives", *Adhesives Age*, December 1994.
4. Hagemaier, D. J., "End Product Nondestructive Evaluation of Adhesive Bonded Metal Joints", *Adhesives and Sealants*, vol. 3, Engineered Materials Handbook, ASM International, 1990.
5. Botsco, R. T. and Anderson, R. T., "Nondestructive Testing: Assuring Reliability in Critically Bonded Structures", *Adhesives Age,* May 1984 at 19–21.
6. Hitt, W. C., "Practical Aspects of Nondestructive Testing", *Automation,* June 1971 at 67–71.
7. Thomas, R. L. and Favro, L. D., "Thermal Wave Inspection of Adhesive Disbonding", *41st International Sample Symposium,* March 1996 at 579.
8. Brett, C. L., *J. Appl. Sci.*, 18:315 (1974).
9. Dillard, J. G., "Microscopic/Spectroscopic Studies in Adhesion Related to Durability of Adhesive Bonded Metals and Composites", *Adhesion Principles and Practice Course,* Kent State University, May 1998.
10. Baun, W. L., et al., "Chemistry of Metal and Alloy Adherends by Secondary Ion Mass Spectroscopy, Ion Scattering Spectroscopy, and Auger Electron Spectroscopy", *ASTM STP 596,* American Society of Testing and Materials, Philadelphia, March 1975.
11. McDevitt, N. T., and Baun, W. L., "Some Observations of the Relation Between Chemical Surface Treatments and the Growth of Anodic Barrier Layer Films", AFML-TR-76-74, June 1976.
12. McDevitt, et al., "Surface Studies of Anodic Aluminum Oxide Layers Formed in Phosphoric Acid Solutions", AFML-TR-77-55, May 1977.
13. McDevitt, et al., "Accelerated Corrosion of Adhesively Bonded 7075 Aluminum Using Wedge Crack Specimens", AFML-TR-77-184, October 1977.
14. Baun, W. L., et al. "Pitting Corrosion and Surface Chemical Properties of a Thin Oxide Layer on Anodized Aluminum", AFML-TR-78-128, September 1978.
15. Benninghoven, A., *Surface Sci.*, 28:541 (1971). In German.
16. Schubert, R. and Tracy, J. C., "A Simple and Inexpensive SIMS Apparatus", *Rev. of Sci. Inst.,* 44:487 (1973).
17. Schubert, R., "The Analysis of 301 Stainless Steel By SIMS", *J. Vacuum Sci. Tech.,* 12(1):505 (1975).
18. Werner, H. W., "Investigation of Solids by Means of an Ion Beam Bombardment Mass Spectrometer", *Dev. Appl. Spectroscopy,* 7A:219 (1969).
19. Zhuang, H. and Gardella, J. A., "Spectroscopic Characterization of Polymer Surfaces", *MRS Bulletin,* 1996.

Chapter 6

Surfaces and Surface Preparation

6.1 Introduction

Since adhesives and sealants must function by surface attachment, the nature and condition of the substrate surface are critical to the success of any bonding or sealing operation. Four common criteria are generally recognized for an ideal bonding surface: cleanliness, continuity, stability, and wetting of the surface by the adhesive or sealant.

Cleanliness does not necessarily mean the absence of all surface films, since some surface films are very strongly attached to the bulk substrate and offer a suitable surface for adhesion. However, cleanliness does require the removal of unwanted or weak boundary layers such as oil, dirt, or corrosion. The purpose of cleaning the surface is to remove any weakly attached materials and to provide a surface that is relatively consistent from part to part.

Discontinuities on the adherend surface, whether chemical or physical, may adversely affect the apparent strength of the joint by creating localized regions of poor bonding and stress concentration within the joint. Discontinuities may also make surface cleaning or treating processes non-homogeneous. These discontinuities could be due to inconsistent manufacturing processes or chemical inhomogeneity within the substrate.

Stability of the substrate surface is important before bonding as well as after bonding. Unwanted boundary layers could form during the time between surface preparation and application of the adhesive, depending on the shop environment and the reactivity of the surface. Boundary layers could also form during the time period after the adhesive is applied and before it sets, depending on the reactivity of the surface with the components in the adhesive or sealant. Certain boundary layers can also form after the adhesive or sealant is cured, depending on the nature of the bond and the type of aggressive environment to which the joint is exposed. The boundary layers that form after the assembled joint is in service may be the most perplexing because they are often unexpected and may lead to catastrophically early bond failure.

As discussed in Chapter 2, wetting of the adherend surface is a required and important process in establishing adhesion. There will be various degrees of wetting, dependent on the chemistry of the surface that comes into contact with the adhesive. Along with the wettability of the surface, surface roughness and topology also influence the strength of bonded joints. The suitability of the bonding surface will also depend on the type and degree of cleaning or surface treatment that was performed before application of the adhesive or sealant.

This chapter introduces the critical surface factors that affect the formation, strength, and durability of adhesive and sealant joints. It is important to have a general knowledge of substrate surfaces and the factors that cause them to change. The surface conditions will dictate the success of most bonding operations. This chapter will also describe various commercial methods that have been developed to control the surface and to provide a suitable foundation for attachment of the adhesive or sealant. Different cleaning and surface treatments are described as well as their common applications. Appendices C-1 through C-4 offer recipes and processing procedures that have been reported to be effective surface treatments for metals, plastics, elastomers, and other substrates. These surface treatments have proved successful in many adhesive and sealant applications. Whereas this chapter looks at surfaces in a general way, the surface characteristics and recommended surface treatments for specific substrates are covered in detail in Chapter 16.

6.2 Nature of Substrate Surfaces

The term "surface" in adhesive science is usually defined as that portion of the adherend with which the adhesive interacts. The surface is defined by both area and depth of interaction. For a freshly cleaved single crystal, this interaction region might be only one or two atomic layers in depth. For anodized aluminum, a low viscosity adhesive might reach a depth of several hundred nanometers or more. For a very porous surface, such as wood, the interaction region may be several millimeters in depth.

When a supposedly smooth solid surface is examined closely under a microscope, it is found to contain irregularities. It is not flat and smooth but contains many surface asperities, such as peaks and valleys, with a certain degree of roughness. A rough surface provides more bonding area than a smooth one of the same gross dimensions. The greater effective surface area offers a larger area for the forces of adhesion to operate, thereby providing a stronger joint. However, a greater degree of surface roughness could also contribute to stress concentrations in the adhesive joint, which reduces its strength, similar to a notch effect in metals. This effect depends on how well the adhesive wets the surface and penetrates into the surface roughness.

Surfaces are full of surprises, and they are seldom what they seem. Often they contain constituents that are very different from the bulk material. For metals and alloys, these surfaces may consist of oxides and adsorbed gases. For many nonmetals, they may be moisture, migrating additives, or adsorbed films, such as shop contaminants. These outer layers can either be loosely bound or tightly adhered to the base material, and they may have high or low cohesive strength. Two surface characteristics that can hurt adhesion are when: (1) the chemical nature provides a low surface energy; and (2) the surface is either cohesively weak or weakly attached to the base substrate. When either one of these conditions are present, the substrate surface must be treated in some manner to either increase the surface energy or strengthen the surface layer.

It does not take much contamination to affect adhesion. A single molecular layer of contaminant can prevent proper wetting of the substrate by the adhesive or sealant. The adhesive or sealant will try to wet the contaminant surface layer rather than the substrate itself. Since most contaminants (oils, greases, fingerprints, mold release, etc.) have a low surface energy, the adhesive will not wet the surface nor will it form a continuous film.

Certain surfaces also may have weakly attached surface or boundary layers. Examples of these are contaminant films, oxide layers, rust, corrosion, scale, and loose surface particles. A weak substrate boundary layer can provide the "weak-link" for reduced bond strength or premature failure as shown in Fig. 6.1.

In an ideal bonded assembly, the substrate should be the weakest link. In most assemblies that are properly bonded, the adhesive is the weak link because the forces of adhesion are greater than the forces holding the adhesive material together. Usually, the internal strength of the substrate and adhesive or sealant system is well understood and can be controlled. However, when the surface region becomes the weakest link, it may result in low failure strength and an inconsis-

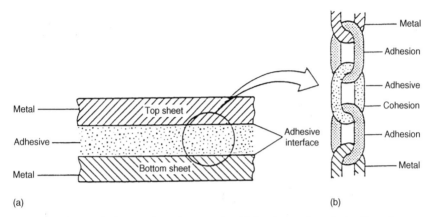

Figure 6.1 In an ideal joint the substrate should be the weakest link. The adhesive joint (a) can be divided into at least five regions that are similar to the links (b) in a chain.[1]

tency in failure values. Thus, it is imperative that these surface characteristics be understood and controlled in some manner. The characteristics of generic substrate surfaces will be described in the following sections.

The chemical constituents that are present on the surface may be different in structure and chemistry from those that are in the bulk material. This difference can be due to adsorption of contaminants from the environment or from segregation of bulk constituents at the surface during the material's processing. Contaminants and segregated bulk components are not always homogeneously distributed in the surface region, and inhomogeneities can lead to bond discontinuities that can concentrate stress. For polymeric materials, there may be greater crystallinity or chemical orientation at the surface due to the way in which the material was processed. There may also be a greater concentration of absorbed species such as water, polar substances, or oxidized polymer at the polymer's surface as opposed to the bulk material.

6.2.1 Metallic surfaces

The "surface" of metals, such as steel or aluminum alloys, might consist of several regions having no clearly defined boundaries between them as shown in Fig. 6.2a. Moving outward from the bulk metal, there will likely be a region that is still metallic but which has a chemical composition different from the bulk region due to segregation of alloying elements or impurities. Next, there will be a mixed oxide of the metals followed by a hydroxide layer and probably an absorbed

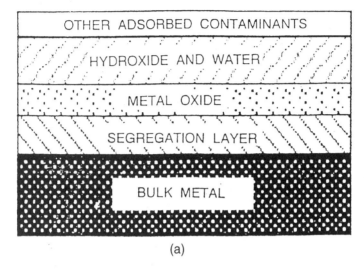

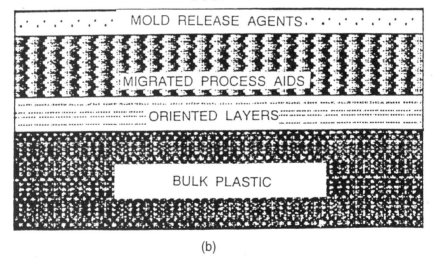

Figure 6.2 Schematic representation of metal (*a*) and polymer (*b*) substrate surface regions.

water layer. In addition, there will be contaminants absorbed from the atmosphere that might include sulfur, nitrogen, halogen, or other compounds depending on the reactivity of the metal and the pollutants in the substrate's manufacturing environment. The contamination layer will also depend on how the metal had been stored and handled. There could also be processing aids on the surface such as rolling oils, lubricants, drawing compounds, and corrosion inhibitors. Finally, the mechanical working of the metal will probably have, to some degree, mixed all these regions together into a non-homogeneous mixture.

Virtually all common metal surfaces exist as hydrated oxides as shown in Fig. 6.3. Even materials such as stainless steels, nickel, and chromium are coated with transparent metal oxides that tenaciously bind at least one layer of water. Thus, the adhesive or sealant used for these materials must be compatible with the firmly bound layer of water attached to the surface metal oxide layer. The nature of the oxide layer will depend on the metal beneath the surface and the conditions that caused the oxide surface to grow. Certain adhesives or sealants will interact more effectively with certain oxide layers. For

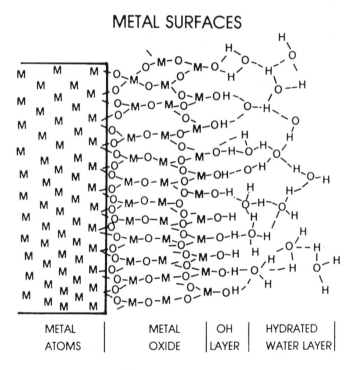

Figure 6.3 Metal surfaces are actually hydrated metal oxides.[2]

example, low viscosity adhesives are more likely to penetrate a porous oxide layer than high viscosity adhesives.

When working with metal adherends, one must recognize that the nature of the surface can be significantly different for the same type of metal. The surface characteristics will depend on how the metal was processed and heat treated. They may also be determined by the secondary finishing processes (machining, forming, etc.) and how the material is handled and stored. The surface characteristics will also depend on the type and conditions of any prebond surface preparation process. For example with certain surface prebond treatments for aluminum, impurities in the rinsing water or etch solution and variations in the treating temperature can critically change the nature of the resulting surface.

The kind and degree of surface treatment required for optimal adhesion will depend on many factors. It is sometimes desired, although not always practical or wise, to have the pure, bulk adherend material exposed directly to the adhesive, with no intervening layer of oxide film, anodizing coating, or contaminant. However, with some substrates, such as aluminum, the oxide layer is thin, dense, and strong and will retard diffusion and further oxide growth. Certain oxide layers actually protect the aluminum surface from corrosion and transformation during aging in service. Therefore, it may not be necessary to remove a clean aluminum oxide layer. However, other oxide layers, such as those formed on copper alloy surfaces, are cohesively weak and should be removed before application of the adhesive or sealant. Depending on the nature of the application and the substrate, at times it may be best to completely remove the original (and unknown) oxide layer that was delivered with the substrate, and "manufacture" a known, protective oxide layer before bonding. Various chemical conversions and other prebond treatments have been developed to perform this task.

6.2.2 Polymeric surfaces

The situation with organic substrates, such as plastics or elastomers, is even more complex than with metals. These materials have lower surface energies and lower tensile strength than metals, and most importantly, polymeric surfaces are more dynamic and likely to change than metals. Thus, there is a greater probability of variation in the surface. As shown in Fig. 6.2b, polymeric surfaces have the potential for low molecular weight fragments, oxidation products, plasticizers, processing aids, lubricants and slip aids, adsorbed water, and organic contaminants along with various other surprises for the end-user. These could all be present in the surface region. They will affect the resulting bond strength without having a significant influence on the bulk properties of the material.

Components within the polymeric bulk material can also migrate to the surface. It is common to find low molecular weight polymers or oligomers, plasticizers, pigments, mold release agents, shrink control agents, and other processing aids as well as adsorbed contaminants in the surface region. More so than with metals, the surface regions of plastics are dynamic regions, continuously establishing new equilibrium internally with the bulk material and externally with the surroundings. In flexible amorphous plastics above the glass transition temperature, low molecular weight components are able to diffuse out of the bulk and to the surface region while elements of the surroundings can diffuse into the adherend.

A well-known example of this effect is the migration of plasticizer from flexible polyvinyl chloride. The plasticizer can migrate from the bulk adherend to the joint region and then to the interface. If the adhesive is an effective barrier to plasticizer migration, this will create a weak boundary layer at the interface. If the adhesive is not an effective barrier, then the plasticizer will migrate through the interface and into the adhesive and thereby possibly change the physical properties of the adhesive.

The nature of the polymeric surface can change rapidly in response to its surroundings. Even when the bulk material is in the glassy state (below its glass transition temperature), the surface region can be quite mobile owing to the presence of low molecular weight polymer constituents and contaminants. Polymers, having both polar and non-polar regions in their molecular chain, can present different chain segments at the surface depending on whether the surroundings are polar or not. Wiping a surface with an ionic solution will cause the polar groups to orient toward the surface. While the same treatment with a non-polar solvent, such as hexane, can bring the non-polar components to the surface. Exposure to heat after surface treatment could cause fresh, untreated molecular species to appear on the surface, thereby losing the beneficial characteristics of the surface treatment.

As a result of these dynamic reactions, it is difficult to be confident about the surface of any polymeric material. The actual surface to which we are bonding is not always the surface that we anticipate. It is also possible that the surface could change once the bond is made and the assembled joint is placed into service. Thus, a weak boundary layer that is not present during the bonding process may form during the joint's operating life and contribute to a weakening of the interface. Although these dynamic processes are not always damaging to the bond strength or to the integrity of the assembled joint, they need to be considered early in the assembly design process. If such surface interactions are considered possible, then they should be fully tested with prototype joints made in the production process.

6.2.3 Other surfaces

Many natural and man-made surfaces vary significantly in characteristics important for bonding or sealing. Wood and cellulose based products, for example, will differ widely in surface roughness, pH, porosity and moisture content even within a single sample. The nature of these surfaces will also change with aging and oxidation. The presence of sap, pitch, resins, or preservatives will also affect bond strength. Generally, freshly cut wood substrates are ideal for bonding because of their porosity. However, care must be taken to remove loose sawdust.

Ceramic materials have smooth, glass-like surfaces, usually with very high surface energy. Since ceramics have high surface energies, they are usually easy to bond under normal conditions. However, many commercially important ceramics have glazed (glass-like) surfaces. This glazed surface could provide another interface in the joint that must be addressed. The polar nature of the bonds between atoms in a ceramic material means that there will likely be an adsorbed layer of water and hydroxide ions. This layer is tightly held to the ceramic surface. Adhesives used with ceramics, as those used with metals, must be compatible with the surface moisture layer.

Concrete is a substrate whose surface characteristics are likely to be affected by the environment in which it cures. Once cured, concrete has several surface characteristics that are hazardous for bonding or sealing. The concrete surface is extremely alkaline and will destroy any hydrolysis sensitive materials that are at the interface. It often has a weak, powdery surface layer that must be penetrated or removed. Thus, sealers are commonly used to moisture proof and strengthen the concrete surface prior to bonding.

Bonding to painted or plated parts presents a problem not encountered with other adherends. It is not recommended to bond to painted surfaces because the resulting bond is only as strong as the adhesion of the paint to the base material. Generally, the paint must be removed or abraded sufficiently so that any weakly attached areas are removed. Plated surfaces should also be tested before bonding to determine how strongly the plating is attached to the base substrate. Plated surfaces are often porous and usually exhibit poor resistance to moisture, especially if the bonding agent does not seal the joint.

Modern polymeric composites are being considered for light weight vehicles because of their high strength and low weight. Surfaces of these materials are usually liberally coated with mold-release agents such as silicone or fluorocarbon to aid release from the mold. Mold-release films, such as cellophane, have also been used. It is essential that these surface layers are removed before bonding. Just as the mold release agents provide release of the composite from its mold, if not removed from the part before bonding, they will also provide release from the adhesive or sealant. Polymeric composites may also be fabricated to have a resin rich surface for a glassy appearance called a gel coat. This will provide a weaker surface layer than the material in the bulk of the composite.

Elastomeric surfaces are very similar to plastic surfaces. However, the more fluid nature of the elastomer's molecules allow easy diffusion of contaminants and low molecular weight fragments to the surface region. These could find their way to the interface and cause a weak boundary layer. Elastomeric substrates are especially susceptible to this problem, because the formulations are generally more complex, and they have more low molecular weight constituents than plastics. Like plastics, many elastomeric surfaces are low energy surfaces and require treatment to raise the surface energy prior to bonding.

6.3 Surface Treatment

The main purpose of surface preparation is to ensure that adhesion develops to the extent that the weakest link in the joint is either in the adhesive or sealant or in the adherend. With optimum surface treatment, failure should not occur at the interface because of a weak boundary layer or insufficient wetting. As a general rule, all substrates must be treated in some manner prior to bonding if not to remove or prevent the formation of weak boundary layers then to provide a consistent surface. Certain low energy surfaces must be modified chemically or physically prior to bonding so that the adhesive or sealant can adequately wet the surface and provide the attractive forces necessary for good adhesion. Surface preparation can range from simple solvent wiping to a combination of mechanical abrading, chemical cleaning, and acid etching.

Surface preparation can provide several principal functions:

- Remove weak boundary layers that impede wetting of the substrate and create "weak links" at the interface. Common weak boundary layers are greases, oils, scale, rust, tarnish, and other oxides.
- Protect the substrate surface so that weak boundary layers do not develop during processing of the joint or during aging in service.
- Influence the surface energy of the substrate so as to reduce the contact angle between the adhesive and substrate.

To make an economical and practical joint, the surface preparation methods must also meet several other requirements. They must be safe to handle and should not be flammable or toxic. They should be inexpensive and provide fast processing times. The processes should be easy to monitor and control in a production situation. In addition, the surface treating process should not in itself leave a weak boundary layer. If chemical solutions are used, they should rinse off easily and not continue to react with the surface. The surface treating process should allow for practical time between preparation and application of the adhesive or sealant. Finally, the surface provided by the treatment should not change once the assembled joint is made and placed into service.

6.3.1 Importance of the surface treatment

Surface preparation of substrates prior to bonding is one of the most important factors in the adhesive bonding process. Prebond treatments are intended to provide cohesively strong and easily wettable surfaces. The strength of an adhesive joint is significantly increased when loose deposits such as rust, scales, flaking paint, and organic contaminants are removed from the surface so that the adhesive can more easily wet the substrate. Table 6.1 shows the effect of surface preparations on adhesive-joint strength of several metallic adherends. The bond strength provided by the surface preparation is dependent on the type of adherend as well as the type of adhesive or sealant used.

Surface preparations enhance the quality of a bonded metal joint by performing one or more of the following functions: remove contaminants; control adsorbed water; control oxide formation; poison surface atoms which catalyze polymer breakdown; protect the adhesive from the adherend and vice versa; match the adherend crystal structure to the adhesive molecular structure; and control surface roughness.[8] Surface preparations enhance the quality of polymeric joints in a similar manner. However, polymeric surface preparations may also chemically alter the surface to raise the surface energy of the substrate.

Surface treatments control and protect the substrate surface before bonding, and they protect the surface from changing after the assembly is placed in service. Thus, surface preparations affect the permanence of the joint as well as its initial strength. Figure 6.4 illustrates the effect of surface treatment on the performance of aluminum alloy epoxy adhesive joints after various times of exposure to water at 122°F.

Plastic and elastomeric adherends are even more dependent than metals on surface preparation. Most of these materials have complex

TABLE 6.1 Effect of Substrate Pretreatment on Strength of Adhesive Bonded Joints

Adherend	Treatment	Adhesive	Shear strength, lb/in.^{2}	Ref.
Aluminum	As received	Epoxy	444	3
	Vapor degreased		837	
	Grit blast		1,751	
	Acid etch		2,756	
Aluminum	As received	Vinyl-phenolic	2,442	4
	Degreased		2,741	
	Acid etch		5,173	
Stainless steel	As received	Vinyl-phenolic	5,215	4
	Degreased		6,306	
	Acid etch		7,056	
Cold-rolled steel	As received	Epoxy	2,900	5
	Vapor degreased		2,910	
	Grit blast		4,260	
	Acid etch		4,470	
Copper	Vapor degreased	Epoxy	1,790	6
	Acid etch		2,330	
Titanium	As received	Vinyl-phenolic	1,356	4
	Degreased		3,180	
	Acid etch		6,743	
Titanium	Acid etch	Epoxy	3,183	7
	Liquid pickle		3,317	
	Liquid hone		3,900	
	Hydrofluorosilicic acid etch		4,005	

Note: See references at end of chapter.

formulations, and their surfaces are often contaminated with mold-release agents or other additives. These contaminants must be removed before bonding. Many plastics and plastic composites can be treated prior to bonding by simple mechanical abrasion or alkaline cleaning to remove such contaminants.

In some cases, however, it is necessary that the polymeric surface be physically or chemically modified to encourage wetting and achieve acceptable bonding. Usually the critical surface energy of the substrate must be raised to a level where it is equivalent to or greater than the surface energy of the adhesive. This applies particularly to crystalline thermoplastics such as polyolefins, linear polyesters, and fluorocarbons. These and certain other polymeric substrates are generally unsuitable for adhesive bonding in their natural state. Methods used to improve the bonding characteristics of these polymeric surfaces include:

1. Oxidation via chemical or flame treatment
2. Electrical (corona) discharge to leave a more reactive surface

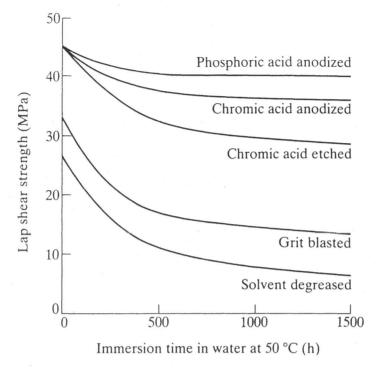

Figure 6.4 Effect of surface pretreatments on the performance of aluminum joints bonded with a toughened epoxy adhesive and subjected to aging in water at 50°C.[9]

3. Ionized inert gas treatment which strengthens the surface by a chemical change (e.g., crosslinking) or physical change and leaves it more reactive
4. Metal-ion treatment that removes fluorine from the surfaces of fluorocarbons
5. Application of primers, adhesion promoters, and other wettable chemical species

These processes have been developed over time and are conventionally used in production applications. Other, newer surface preparation processes are now being developed specifically for the increased usage of engineering plastics and composites for light weight, energy saving vehicles in the automotive and aerospace industries. Safety and environmental regulations are also driving the development of new pre-bond treatments for metal and polymeric surfaces.

6.3.2 Choosing the surface treatment

The degree to which adherends must be prepared is related to the service environment and the ultimate joint strength required. Surface preparations can range from simple solvent wiping to a combination of mechanical abrading, chemical cleaning, and acid etching. In many low- to medium-strength applications, extensive surface preparation may be unnecessary. However, where maximum bond strength, permanence, and reliability are necessary, carefully controlled surface-treating processes are required. As shown in Fig. 6.5, high strength durable bonds generally require aggressive and expensive surface treatment processes. These optimized processes also require prolonged production time and provide safety and environmental concerns. Thus, one should be careful not to over-specify the surface treatment required. Only the minimal process necessary to accomplish the functional objectives of the application is required. The following factors should be considered in the selection of a surface preparation:

1. The ultimate initial bond strength required
2. The degree of permanence necessary and the service environment
3. The amount of and type of contamination initially on the adherend
4. The type of adherend and nature of its surface
5. Production factors such as cost, cycle time, safety and environmental compliance, training, monitoring and control

Any surface treatment used for bonding or sealing requires the completion of one or more of the following operations: cleaning, mechan-

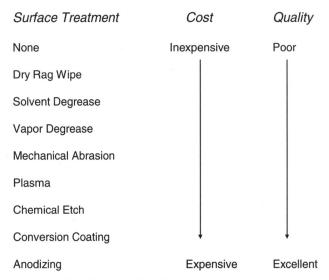

Figure 6.5 The objective of surface preparation is to provide a consistent, reproducible adherend surface that when bonded meets the strength and durability required for the application.

ical abrasion, or active surface modification. *Passive* surface treatment processes do not alter the chemistry of the surface but only clean and remove weakly attached surface layers (i.e., solvent washing, mechanical abrasion). *Active* surface treatment processes cause a chemical change to the surface (i.e., anodizing, etching, plasma treatment). Tables 6.2 and 6.3 characterize surface treatments for metallic and polymeric substrates respectively.

More than one surface treatment may be required for optimum joint properties. A four-step process that is often recommended for achieving high strength joints with many substrates consists of degreasing, mechanical abrasion, repeated degreasing, and chemical treatment or etching. Table 6.4 shows the relative bond strengths that can be realized when bonding aluminum after various surface treatment processes. Note that when low or medium strength is sufficient for the application, only minimal surface preparation is required.

TABLE 6.2 Characterization of Common Surface Treatments for Metals[10]

Pretreatment type	Possible effects of pretreatment
Solvent	Removal of most of organic contamination
Mechanical	Removal of most of organic contamination. Removal of weak or loosely adhering inorganic layers, e.g. mill scale. Change to topography (increase in surface roughness). Change to surface chemistry
Conversion coating	Change to topography (increase in surface roughness). Change to surface chemistry, e.g. the incorporation of a phosphate into the surface layers
Chemical (etching, anodizing)	Removal of organic contamination. Change to topography (increase in surface roughness). Change to surface chemistry. Change in the thickness and morphology of metal oxide

TABLE 6.3 Characterization of Common Surface Treatments for Polymers[11]

Pretreatment type	Possible effects of pretreatment
Solvent	Removal of contaminants and additives. Roughening (e.g. trichloroethylene vapor/polypropylene). Weakening of surface regions if excessive attack by the solvent
Mechanical	Removal of contaminants and additives. Roughening
Oxidative	Removal of contaminants and additives. Introduction of functional groups. Change in topography (e.g. roughening with chromic acid treatment of polyolefins)
Plasma	Removal of contaminants and cross-linking (if inert gas used). Introduction of functional groups if active gases such as oxygen are used. Grafting of monomers to polymer surface after activation, e.g. by argon plasma

TABLE 6.4 Surface Treatments for the Adhesive Bonding of Aluminum[12]

Surface treatment	Type of bond
Solvent wipe (MEK, MIBK, trichloroethylene)............	Low to medium strength
Abrasion of surface, plus solvent wipe (sandblasting, coarse sandpaper, etc.)	Medium to high strength
Hot-vapor degrease (trichloroethylene).................	Medium strength
Abrasion of surface, plus vapor degrease................	Medium to high strength
Alodine treatment	Low strength
Anodize..	Medium strength
Caustic etch*..	High strength
Chromic acid etch (sodium dichromate–sulfuric acid)†	Maximum strength

*A good caustic etch is Oakite 164 (Oakite Products, Inc., 19 Rector Street, New York, N.Y.).
†Recommended pretreatment for aluminum to achieve maximum bond strength and weatherability:
1. Degrease in hot trichloroethylene vapor (160°F).
2. Dip in the following chromic acid solution for 10 min at 160°F:
   Sodium dichromate ($Na_2Cr_2O_7 \cdot 2H_2O$............................. 1 part/wt.
   Conc. sulfuric acid (sp. gr. 1.86)....................................10 parts/wt.
   Distilled water ....30 parts/wt.
3. Rinse thoroughly in cold, running, distilled, or deionized water.
4. Air-dry for 30 min, followed by 10 min at 150°F.

The sequence of these surface treating steps is important. The substrate is initially degreased to remove gross organic contamination. It is then subjected to mechanical abrasion to remove strongly attached weak boundary layers. A second degreasing step is then performed to clean the substrate of residue and possible contaminants leftover from the abrasion processes. Note that the first degreasing step is necessary; otherwise, the contaminants would be driven further into the substrate by the mechanical abrasion process. The abrasive medium could itself become contaminated and spread a weak boundary layer from article to article. Once the substrate is clean, the final step, if necessary, is usually one that acts on the pure substrate surface. This process is intended to alter the physical or chemical nature of the substrate. Its goal is to provide better wetting, to passivate the surface so that weak boundary layers cannot develop, or to simply give the substrate more "teeth" for mechanical interlocking with an adhesive.

Table 6.5 shows the effect of various combinations of aluminum surface preparations on lap shear strength. With this particular combination of adhesive, adherend, and processing conditions, optimum bond strength (3,000 psi) on aluminum occurs when a treatment consisting of vapor degreasing, abrading, alkaline cleaning and acid etching is used. However, simple solvent wiping and abrasion results in moderate and relatively consistent bond strength (1,500–2,000 psi).

TABLE 6.5 Surface Preparaton of Aluminum Substrates vs. Lap Shear Strength[3]

Group treatment	$\bar{X}$, lb/in.^{2}	s, lb/in.^{2}	C_v, %
1. Vapor degrease, grit blast 90-mesh grit, alkaline clean, $Na_2Cr_2O_7$–H_2SO_4, distilled water	3.091	103	3.5
2. Vapor degrease, grit blast 90-mesh grit, alkaline clean, $Na_2Cr_2O_7$–H_2SO_4, tap water	2,929	215	7.3
3. Vapor degrease, alkaline clean, $Na_2Cr_2O_7$–H_2SO_4, distilled water	2,800	307	10.96
4. Vapor degrease, alkaline clean, $Na_2Cr_2O_7$–H_2SO_4, tap water	2,826	115	4.1
5. Vapor degrease, alkaline clean, chromic–H_2SO_4, deionized water	2,874	163	5.6
6. Vapor degrease, $Na_2Cr_2O_7$–H_2SO_4, tap water	2,756	363	1.3
7. Unsealed anodized	1.935	209	10.8
8. Vapor degrease, grit blast 90-mesh grit	1,751	138	7.9
9. Vapor degrease, wet and dry sand, 100 + 240 mesh grit, N_2 blown	1,758	160	9.1
10. Vapor degrease, wet and dry sand, wipe off with sandpaper	1,726	60	3.4
11. Solvent wipe, wet and dry sand, wipe off with sandpaper (done rapidly)	1,540	68	4
12. Solvent wipe, sand (not wet and dry), 120 grit	1,329	135	1.0
13. Solvent wipe, wet and dry sand, 240 grit only	1,345	205	15.2
14. Vapor degrease, aluminum wool	1,478		
15. Vapor degrease, 15% NaOH	1,671		
16. Vapor degrease	837	72	8.5
17. Solvent wipe (benzene)	353		
18. As received	444	232	52.2

$\bar{X}$ = average value. s = standard deviation. C_v = coefficient of variation. Resin employed is EA 934 Hysol Division, Dexter Corp.; cured 16 h at 75°F plus 1 h at 180°F.

6.3.3 Evaluation of treated parts before and after bonding

The common goals of surface treatment are to produce a clean and wettable surface. There is, unfortunately, no standard procedure or equipment available to tell when a surface is clean. Thus, the term "clean" is difficult to define. One can try to define clean as no visible soil or foreign matter on the surface when inspected with the naked eye. Since this is very subjective, the quality of the surface treatment process will ultimately be dependent upon rigid process controls and well-trained operators. These process controls include the monitoring of critical parameters such as solvent purity, temperature, and time; equipment maintenance; the accumulated number of parts or bond area being treated with the same solution; and the handling and storage of the clean parts prior to application of adhesive or sealant. Periodically through a production run, the bond strengths of prototype parts should be tested to determine if the surface treatment process is still working as expected.

On many nonporous surfaces a useful and quick method for testing the effectiveness of the surface preparation is the "water-break test." If distilled water beads when sprayed on the surface and does not wet the substrate, the surface-preparation steps should be repeated. A break in the water film may signal a soiled or contaminated area. If the water wets the surface in a uniform film, an effective surface operation may be assumed. However, the water-break test only provides a rough approximation of the surface condition. For quantitative interpretation of the cleanliness of a substrate, one could also measure contact angle directly on the treated substrate with a drop of reference liquid. The reference liquid could be distilled water or a liquid having a surface tension similar to the adhesive or sealant that will be used.

Another test to determine cleanliness of the substrate involves wiping a clean white towel across the prepared surface to see if any gross contaminants are present. This technique is often used to check flat surfaces and surfaces that are not easily cleaned because of the part geometry. Similarly, a small strip of common office tape could be applied to the surface, peeled off and then examined on a white background for evidence of contamination. Of course, any residue left by the tape will then need to be cleaned from the substrate. Certain forms of contamination, notably oils, can also be more easily seen under ultraviolet (UV) light. An ultraviolet detection process has been suggested which requires soiling the substrate with a fluorescent oil, normal cleaning, and then inspecting the surface under ultraviolet light.[13] The degree of cleanliness can be quantified through photoelectron emission or reflectance measurements. The higher the reflectance, the cleaner the surface is. Other advanced analytical methods for detecting surface contamination are described in Chapter 5.

The objective of treating an adherend prior to bonding is to obtain a joint where the weakest link is the adhesive layer and not the interface. Thus, destructively tested joints should be examined for the mode of failure. If failure is in the cohesive-mode (within the adhesive layer or adherend), the surface treatment may be considered to be optimum for that particular combination of adherend, adhesive, and testing conditions. If an adhesion-mode of failure appears to be at the interface, one may assume that additional surface optimization is necessary if higher bond strengths are required. It must also be realized that specimens may exhibit cohesive failure initially and then interfacial failure after aging for a period of time. In these cases, a developing weak boundary layer can be considered a possible culprit. Both adhesive and surface preparations need to be tested with respect to initial bond strength and permanence in the intended service environment.

6.3.4 Substrate equilibrium

After the surface-preparation process has been completed, the substrates may have to be stored before bonding. Typical storage lives for various metals subjected to different treatments are shown in Table 6.6.

The maximum allowable time between surface treatment and application of the adhesive will be dependent on how soon the substrate surface can change in the shop environment and how strong the subsequent reformed surface is attached to the base material. For example, after its natural surface is exposed, copper alloys can form oxide layers relatively quickly. In the case of copper, this boundary layer is weakly attached to the base metal. On the other hand, aluminum oxide layers are formed very fast: however, they are generally tightly attached to the base metal and do not provide a significant problem for adhesion. Certain polymeric surface treatments lose their effectiveness very quickly because of the reactive and mobile nature of the polymer molecules. Because of the relatively short storage life of many treated materials, the bonding operation should be conducted as soon as possible after the surface preparation process.

If prolonged storage is necessary, either (1) the parts should be carefully protected and stored in a controlled, clean environment, or (2) a compatible organic primer may be used to coat the treated substrates immediately after surface preparation. The primer will protect the treated surface during storage and interact with the adhesive during bonding. Many primer systems are sold together with adhesives for this purpose. Certain primers have also been formulated specifically for corrosion resistance and, thereby, continue to protect the surface after the joint is placed in service. Such primers are described in the next chapter.

6.4 Passive Surface Preparation Methods

The surface preparation processes described in this section are classified as passive processes. They do not actively alter the chemical nature of the surface. Passive processes only clean the substrate and remove weak boundary layers in the form of contamination. Solvent washing, chemical cleaning, and mechanical abrasion are considered passive processes. Depending on the degree of adhesion and permanence required, passive processes may be used as either the only surface preparation or as the initial step in a more detailed surface treatment process.

TABLE 6.6 Maximum Allowable Time between Surface Preparation and Bonding or Priming of Metal Substrates[14]

Metal	Surface	Time
Aluminum	Wet-abrasive-blasted	72 h
Aluminum	Sulfuric–chromic acid etched	6 days
Aluminum	Anodized	30 days
Stainless steel	Sulfuric acid etched	30 days
Steel	Sandblasted	4 h
Brass	Wet-abrasive-blasted	8 h

Passive *mechanical* processes use physical or mechanical means to remove some of the surface material and its contaminants, thereby exposing a fresh, clean, and chemically active surface. Mechanical surface treatments are usually not sufficient by themselves. Some form of chemical or solvent cleaning is also necessary to remove organic contaminants from the surface. Chemical or solvent cleaning is performed before mechanical abrasion and again immediately afterward to remove dust and other remnants from the abrasion process. Sanding, abrasive scrubbing, wire brushing, grit blasting, grinding, and machining are common examples of mechanical processes. Surface abrasion is important because it increases the substrate surface area that is in contact with the adhesive in addition to removing weak boundary layers. Although mechanical abrasion processes are fast and quickly expose the bulk material, they often have a high material cost and labor content. Care must be observed regarding the contamination of the abrasive media and possible recontamination of cleaned substrates. Therefore, the abrasive media must be checked and changed often.

Passive *chemical* cleaning processes merely remove obvious surface contamination, including soil, grease, oil, fingerprints, etc. by chemical means without altering the parent material. Vapor degreasing, solvent washing, alkaline and detergent cleaning, and ultrasonic cleaning are typical examples. As with any process utilizing cleaning solutions, sudden or gradual contamination is always a possibility and must be considered in a quality control plan. There are significant new developments occurring with these processes due to safety and health issues and new environmental regulations.[15,16] New equipment and processes have been designed to eliminate or reduce harmful emissions, and new "safety solvents" have been developed to replace harsh cleaning solvents and chemicals.

6.4.1 Passive chemical surface treatment

Passive chemical surface treatments remove soil and organic contaminants from the surface. They include such common processes as solvent wiping, vapor degreasing, and chemical cleaning. Contaminants removed by passive cleaning include dirt, oil, mold release, moisture, grease, fingerprints and other foreign substances on the surface. All cleaning methods are improved by additional agitation that can take the form of scrubbing, aggressive stirring, or ultrasonic agitation. Table 6.7 provides an indication of the wide variation in surface cleanliness as a function of cleaning medium and level of agitation.

Most cleaning methods require the use of solvents or chemicals; thus, safety and consideration of the environment are of prime importance. Toxicity, flammability, materials incompatibility, and hazardous equipment are all important safety factors that must be considered in choosing the proper cleaning or surface treating system. Environmental factors that must be considered are volatile emissions and waste handling, storage, and disposal.

6.4.1.1 Solvent cleaning. Solvent cleaning is the process of removing soil and organic contaminants from a substrate surface with an organic solvent. Where loosely held dirt, grease, and oil are the only contaminants, simple solvent wiping alone will provide surfaces for weak- to medium-strength bonds. Solvent cleaning is widely used and should precede any chemical or abrasive surface preparation. However, it is the least effective substrate treatment in that it only cleans the surface of organic contaminants and loosely held particles.

Volatile solvents such as toluene, acetone, methyl ethyl ketone, methyl alcohol, and trichloroethylene are acceptable. However, the local and most recent safety and environmental restrictions must be consulted before selecting any solvent. For many years 1,1,1-trichloroethane was the work-horse for most cleaning processes because of its excellent solvency, low toxicity, nonflammability, and high permissible exposure levels. However, because it is a depleter of stratospheric ozone, it has been phased out.[18] In its place are substitutes like aqueous cleaners and chlorinated solvents such as methylene chloride, perchloroethylene, and trichloroethylene. Today, trichloroethylene and mineral spirits are perhaps the most commonly used sol-

TABLE 6.7 Efficiency of Degreasing as a Function of Process[17]

Degreasing method	Cleaning efficiency (%)
1. Pressure washing with detergent solution	14
2. Mechanical agitation in petroleum solvent	30
3. Vapour degreasing in trichloroethylene	35
4. Wire brushing in detergent solution	92
5. Ultrasonic agitation in detergent solution	100

vents for cleaning substrates prior to bonding. They will not attack steel, copper, zinc, or other metals, and they are economical. They have many of the same advantages as trichloroethane without the problem of ozone depletion.

The solvent industry has also made significant strides in developing newer grades and blends of solvents for a variety of applications that are either biodegradable and/or EPA compatible. New low volatile solvents are taking the place of the older, less environmentally safe solvents in the adhesive and sealant industries. Substitute solvents for methyl ethyl ketone; 1,1,1-trichloroethane, and freon 113, have been found acceptable for surface cleaning in many industrial applications.[19]

Since the surface cleanliness is difficult to measure, special precautions are necessary to prevent the solvent from becoming contaminated. For example, the wiping cloth should never touch the solvent container, and new wiping cloths must be used often. A clean cloth should be saturated with the solvent and wiped across the area to be bonded until no signs of residue are evident on the cloth or substrate. With solvent wiping, the cleanliness of the surfaces tends to be dependent on the training and attention given by the operator. Automated spray or immersion processes are less dependent on the operator's skill. After cleaning, the parts should be air-dried in a clean, dry environment before being bonded. Usually they are placed in a drying oven with circulating warm air.

The solvent immersion method is more suited for production volumes, and it is often sufficient to remove light contamination and soil. In this method, the part is immersed in a container of solvent and mildly agitated by tumbling, solvent mixing, brushing, or wiping. After being soaked and scrubbed, the parts must be rinsed by a clean flowing liquid or spray. A number of different solvents may be used in this process. It is important to note that the parts will be no cleaner than the final rinse solvent.

The multiple bath method of solvent immersion, Fig. 6.6, is the most common immersion method. The first tank is the wash tank in which scrubbing may be performed. The second and third tanks are rinse tanks. With this method, one must prevent contamination of the cleaner solvents by continually changing the scrub and rinse solvents.

The spray method of solvent cleaning is very efficient due to the scrubbing effect produced by the impingement of high speed particles on the surfaces being cleaned. The spray causes flow and drainage on the surface of the substrate which washes away loosened soil. Trichloroethylene and perchloroethylene are generally used for spraying.

Vapor degreasing is a form of solvent cleaning that is attractive when many parts must be prepared. This method is also more repro-

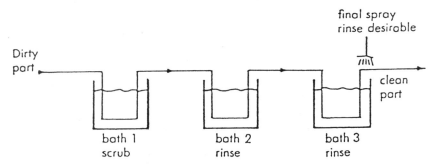

Figure 6.6 Three bath method of solvent immersion cleaning.[20]

ducible than solvent wiping. It will remove soluble soils and contaminants from a variety of metallic and non-metallic parts. Vapor degreasing consists of suspending the adherends in a container of hot chlorinated solvent, such as trichloroethylene (boiling point 250°F), for about 30 secs. When the hot vapors come into contact with the relatively cool substrate, solvent condensation occurs on the substrate which dissolves the organic contaminants from the surface. Vapor degreasing is preferred to solvent wiping because the surfaces are continuously being washed in distilled uncontaminated solvent. It is recommended that the vapor degreaser be cleaned, and a fresh supply of solvent used when the contaminants dissolved in the old solvent lower the boiling point significantly. Extremely soiled parts may not be suitable for vapor degreasing unless initially cleaned by other methods. Plastic and elastomeric materials should not be treated by the vapor-degreasing process without testing because the hot solvent may detrimentally affect the part. For these materials, carefully selected solvents or detergent solutions are acceptable cleaning liquids.

A simple vapor degreaser consists of a large beaker on a hot plate under a fume hood with a cooling coil suspended several inches from the top of the beaker. Cold tap water is circulated through the coil. The solvent is added to the beaker and heated until boiling. The parts are suspended in a wire basket or equivalent and held in the solvent vapors until clean. Usually commercial vapor degreasing equipment is used. Vapor degreasing can be combined with immersion and spray cleaning or even with ultrasonic cleaning. Figure 6.7 illustrates representative processes.

Vapor degreasing requires both the proper types of solvent and degreasing equipment. The solvents used must have certain properties, including the following[20]:

- High solvency of oils, greases, and other soils
- Non-flammable, non-explosive, and non-reactive

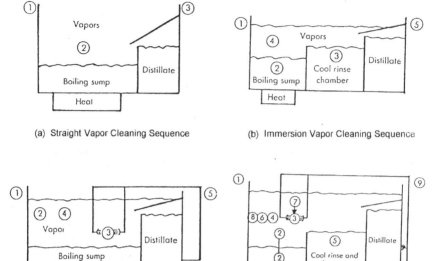

Figure 6.7 Representative methods of vapor degreasing: (*a*) straight vapor cleaning, (*b*) immersion vapor cleaning, (*c*) spray vapor cleaning, and (*d*) ultrasonic vapor cleaning. Numbers represent the sequence in the processing steps.[20]

- High vapor density compared to air
- Low heat of vaporization and specific heat in order to maximize condensation and heat consumption
- Chemical stability
- Safe to operate
- Low boiling point for easy distillation but high enough for easy condensation
- Conformance to air pollution control legislation

Perchloroethylene and trichloroethylene are the most commonly used of the vapor degreasing solvents. Although non-flammable, these solvents are still toxic in both their liquid and vapor forms. Proper ventilation and safety precautions must be taken. These solvents are considered to have high boiling temperatures and are generally used for metal parts contaminated with greases, oils, or processing lubricants. Methylene chloride, Freon TF, Freon TA, and Freon TE are generally used for temperature sensitive parts, particularly electronic and delicate mechanical components and assemblies. These solvents are considered to have low boiling temperatures. The temperature at which they condense on the substrate surface is generally too low to affect the part.

6.4.1.2 Chemical cleaning. Strong detergent solutions are often used to emulsify surface contaminants on both metallic and nonmetallic substrates. These methods are popular on polymeric surfaces where solvent cleaning may degrade the part or on parts where the contamination is more easily removed by an aqueous cleaner (e.g., salt films, dirt). Chemical cleaning is generally used in combination with other surface treatments. Chemical cleaning by itself will not remove heavy or strongly attached contaminants such as rust or scale. All of the chemical cleaning processes should be preceded and followed by solvent cleaning (when possible) or water rinsing.

Detergents, soaps, and other cleaning chemicals are the least expensive and easiest cleaning agents to handle. Two basic environmentally friendly cleaning products are available: aqueous cleaners and non-chlorinated solvents.

Aqueous cleaners are generally manufactured in concentrated forms and are diluted with water before use. They come in three major classes: acidic, neutral, and alkaline. The alkaline cleaners are generally used for cleaning metal surfaces prior to bonding. The following types of aqueous cleaners are available.[21]

- Caustics (sodium or potassium hydroxide): Caustics work well on steel, removing scale, smut, light rust and heavy oils, but cannot be used on aluminum. They are also dangerous if not handled carefully.
- Silicates (sodium meta silicate): Silicated cleaners work well on aluminum, but often are too aggressive on brass or copper.
- Amines (triethanolamine, monoethanolamine): Amines are good on all metals, but pose odor concerns.
- Phosphates (trisodium phosphate, tetra potassium pyro phosphate): Phosphates are very safe for the user (9.5 to 10.5 pH), but often are not aggressive enough for heavy cleaning unless they are combined with large amounts of wetting agents and/or solvents.
- Acids (phosphoric, hydrofluoric, citric, etc.): Acids are very effective on metals that require oxide removal, but they often require strict safety precautions.
- Chelates (EDTA): In small doses, chelates can help extend the life of cleaning solutions. However, they often prompt wastewater concerns because of the dissolved metals in the spent waste.

There are many detergents that are capable of cleaning substrates prior to bonding. Generally, one to five ounces of liquid detergent per gallon of tap water is used. Many commercial detergents such as Oakite (Oakite Products, Inc., 19 Rector Street, New York) and Sprex (DuBois Co., 1120 West Front, Cincinnati, OH) are also available.

Formulation and processing methods for a mild alkaline cleaning process and a heavy duty alkaline cleaning process are shown in Table 6.8. The parts are immersed for 8–12 minutes in the hot, agitated solution. As dirt and other contaminants collect in the bath, more alkalinity must be added to restore the pH factor to a suitable level. The parts are finally rinsed with tap water followed by distilled water.

A wet abrasive method of cleaning consists of scouring the surface of the adherend with detergent. This process consists of thoroughly scrubbing the substrate with a clean cloth or non-metallic bristle brush and a detergent solution maintained at 110–140°F. It is a common surface cleaning method used for many polymeric substrates.

6.4.1.3 Other cleaning methods. Vapor-honing and ultrasonic cleaning are efficient treating methods for small, delicate parts. When the substrate is so delicate that abrasive treatment is too rough, contaminants can be removed by vapor honing. This method is similar to grit blasting except that very fine abrasive particles are suspended in a high-velocity water or steam spray. Sometimes solvents are used as the liquid medium in vapor-honing operations. Thorough rinsing after vapor honing is usually not required.

Ultrasonic cleaning employs a bath of cleaning liquid or solvent ultrasonically activated by a high-frequency transducer. The part to be cleaned is immersed in the liquid, which carries the sonic waves to the surface of the part. High frequency vibrations then dislodge the

TABLE 6.8 Mild and Heavy Duty Alkaline Cleaning Processes for All Substrates

	Mild alkaline cleaning method	Heavy duty akaline cleaning method
Cleaning solution	■ tetrasodium pyrophosphate 15 pbw ■ sodium metasilicate 80 pbw ■ Nacconol 40F (a surfactant) 5 pbw	■ sodium metalsilicate 46 pbw ■ trisodium phosphate 23 pbw ■ sodium hydroxide 23 pbw ■ Nacconol 40F 8 pbw
Mixing	Mix 6–8 oz of above into one gallon of water	Mix 6–8 oz of above into one gallon of water
Immersion temperature	160–180°F	160–180°F
Immersion time	8–12 min	8–12 min

contaminants. Ultrasonic cleaning is ideal for lightly soiled parts with intricate shapes, surfaces, and cavities that are not easily cleaned by spray or immersion techniques. Commercial ultrasonic cleaning units are available from a number of manufacturers.

Electrolytic cleaning is a modification of alkaline cleaning in which an electric current is imposed on the part to produce vigorous gassing on the surface to produce release of solids and contamination. With anodic cleaning the gas bubbles are oxygen, and with cathodic cleaning they are hydrogen. The effect of this cleaning process on metal surfaces and its influence on long term strength on metal-adhesive bonding is superior to normal industrial cleaning processes.[22]

6.4.2 Passive mechanical treatment

Mechanical methods for surface preparation include abrasive blasting, wire brushing, and abrasion with sandpaper, emery cloth, or metal wool. These methods are most effective for removing heavy, loose particles such as dirt, scale, tarnish, and oxide layers. Cleaning is generally required both before and after mechanical surface preparation. The parts should be degreased before abrasive treatment to prevent contaminants from being rubbed into the surface. Solid particles left on the surfaces after abrading can be removed by blasts of clean, dry air and solvent wiping.

Dry abrasion consists of lightly and uniformly sanding the surface of the substrate material with medium (180–325 grit) abrasive paper. Composite abrasive materials, such as "Scotch-Brite®" (3M Company), have proved popular for mechanical surface preparation.[23] These abrasive materials are available in pad and sheet form, and they can conform easily to the shape of a surface. When combined with water flushing, they can provide clean almost oxide free surfaces. Hand sanding, wire brushing, and other methods that are highly related to the operator's skill and patience must be carefully controlled. These methods provide inconsistency and should be used only when no other method is possible.

Abrasive blasting is generally the preferred method for removing contamination from most metal surfaces. It is particularly appropriate for removal of rust, oxide layers, old coatings, and other heavy contamination. Blasting is a fast, efficient, and easily controlled process. It requires containment of the blast media and the resulting dust. The abrasive medium must be regularly renewed and/or cleaned to maintain efficiency and avoid contamination. Blasting is impractical for thin or delicate parts because of warping and possible physical damage to the part. Blasting is also a line-of-sight process, and certain part geometries may be inappropriate.

Dry abrasive blast consists of a uniform blasting of the adherend surface with a clean, fine, non-metallic grit such as flintstones, silica, silicon carbide, aluminum oxide, or glass beads. The particle size will vary with the surface and the material. The degree of blasting for metal substrates should be a "white metal blast"—complete removal of all visible rust, mill scale, paint, and foreign matter. Written definitions and visible standards as to the degree of surface abrasion are available. A white metal blast is defined in SSPC-SP 5 (Steel Structures Painting Council) or NACE No. 1 (National Association of Corrosion Engineers).[24] The degree of blasting for polymeric substrates should be sufficient to remove the surface glaze from the part.

Each substrate reacts favorably with a specific range of abrasive sizes. In many applications joint strength generally increases with the degree of surface roughness.[3,25] Often the joint performance is more dependent on the type of mechanical abrasion rather than the depth of the abrasion. Table 6.9 shows that for stainless steel and aluminum, sandblasted surfaces provide greater adhesion than when the surface is only machined. However, excessively rough surfaces also increase the probability that voids will be left at the interface, causing stress risers that may be detrimental to the joint in service. Table 6.10 pre-

TABLE 6.9 Effect of Surface Roughness on Butt Tensile Strength of Joints Bonded an Epoxy Adhesive[25]

Adherend		Adherend surface†	Butt tensile strength, lb/in.^{2}
6061	Al	Polished	4,720 ± 1,000
6061	Al	0.005-in. grooves	6,420 ± 500
6061	Al	0.005-in. grooves, sandblasted	7,020 ± 1,120
6061	Al	Sandblasted (40–50 grit)	7,920 ± 530
6061	Al	Sandblasted (10–20 grit)	7,680 ± 360
304	SS	Polished	4,030 ± 840
304	SS	0.010-in. grooves	5,110 ± 1,020
304	SS	0.010-in. grooves, sandblasted	5,510 ± 770
304	SS	Sandblasted (40–50 grit)	7,750 ± 840
304	SS	Sandblasted (10–20 grit)	9,120 ± 470

*74°C/16 h cure.
†Adherend surfaces were chromate-etched.

TABLE 6.10 Recommended Abrasive Methods and Size for Various Metal Substrates[14]

Material	Method	Size
Steel	Dry blast	80–100 grit abrasive
Aluminum	Wet blast	140–325 grit abrasive
Brass	Dry blast	80–100 grit abrasive
Brass	Wet blast	140–325 grit abrasive
Stainless steel	Wet blast	140–325 grit abrasive

sents a range of abrasive sizes and methods that have been found favorable for abrasive cleaning of several common metal substrates. Recommended abrasive sizes for metals are in the 80–325 grit range; sizes recommended for polymeric substrates are in the 120–220 grit range.

Certain low energy surfaces, such as polyolefins and fluorocarbons, should generally not be abraded at all prior to application of an adhesive or sealant. Abrasion and the resulting roughness on a low energy surface will only increase the probability of air pockets being trapped in the crevices and valleys at the interface. These air pockets contribute to stress concentration points and a generally weaker joint. A general rule of thumb is that unless the adhesive makes a contact angle of 90 degrees or less with the substrate, mechanical abrasion and surface roughening should be avoided.

A wet-abrasive blasting process can be employed which may be more adaptable than conventional dry blasting. With this process a 20–325 grit aluminum oxide or glass bead abrasive slurry is used as the blasting medium. Generally, three parts of water by volume to one part by volume of the abrasive is used. Wet-blasting units can handle a wider range of abrasive materials, and a spray rinse automatically removes the blasting residue.

Several specialty mechanical blasting systems have been developed over the years for applications where standard grit blasting is not appropriate. These methods include cryogenic, hydrodynamic, and polymeric blasting.

Cryogenic and hydrodynamic blasting have been used as an abrasive-free surface treatment prior to adhesive bonding. Both processes are ideal when conventional abrasive media provides a contaminant or when the collection and reprocessing of the abrasive are prohibitive. These processes are often used in the field for surface preparation of structures in the need of repair.

The cryogenic process uses peletized carbon dioxide at $-100°F$ as a fluidized abrasive cleaning agent for surface preparation and removal of corrosion and old coatings.[26] Not only does this provide an abrasion mechanism, but certain inorganic salts and organic contaminants can be dissolved with supercritical carbon dioxide.

High pressure water blast has been used for prebond surface treatments to eliminate hazardous materials. The combination of high pressure water abrasion with subsequent application of an adhesive promoter/primer has been found to provide high strength and durable aluminum bonds.[27,28]

Polymeric blasting media has also been used for removal of paint, coatings, and other contaminants for a variety of different applications.[29] The abrasive medium consists of hard plastic material. Various types of grits are available with a relatively wide hardness range. The grit hardness is usually sufficient to remove paint coatings and other organic contaminants from the surface, but it is not hard enough to cause abrasion of the base metal or damage to relatively delicate parts. Although polymer blasting is generally thought of as a coating removal process rather than a surface preparation process, it has found use as a prebond process in electrical and electronic applications where a conductive blast medium is not acceptable.

6.5 Active Surface Treatments

Active surface treatments are chemical or physical processes that not merely clean the surface or remove weak boundary layers, but they also transform the inherent surface chemistry. They either improve wetting or modify the boundary layer to be more receptive to bonding. Acid etching, oxidation, anodizing, and pickling processes are examples of active surface treatments.

Active surface treatments of metal substrates are usually chemical treatments that cause the formation of a predetermined type of oxide layer or surface structure that is strong, stable, and receptive to adhesives or sealants. Active surface treatments for polymeric surfaces are usually chemical or physical treatments that alter the chemistry of the surface to make it more wettable. Tables 6.11 and 6.12 summarize the active surface preparations that are commonly used for these substrates.

Active surface treatments are usually the last step in the sequence of surface preparation processes and are only used when maximum strength and permanence are required in a joint. It is always preceded and followed by surface cleaning via one of the passive processes defined in the last section.

Active surface treatments provide improved bond strength and durability through several processes.

1. Remove the weak boundary layer or alternately modify the boundary layer to provide a cohesively strong layer that is well bonded to the bulk, stable, and receptive to common adhesives.
2. (Primarily for polymeric surfaces) Increase the surface energy of the natural surface so that it is greater than the surface energy of the adhesive used.
3. Improve the surface topography to enable the capillary action of the adhesive to maximize joint strength.

TABLE 6.11 Common Active Surface Treatments for Metallic Substrates

Treatment	Metal	Result
Acid etch: FPL (Forest Product Lab) chromic sulfuric acid etch P2 etch: chromate free etch Phosphoric acid Nitric–phosphoric acid Hydrofluoric Sulfuric acid	Aluminum Aluminum Steel Steel Stainless steel Stainless steel	Micro-rough oxide morphology that is appropriate for adhesion
Anodization Phosphoric acid Chromic acid Sodium hydroxide	 Aluminum Aluminum and titanium Titanium	Development of a protective oxide layer that is resistant to corrosion
Alkaline Etch Alkaline peroxide solution	Titanium	Protective oxide layer
Specialty copper etch solutions Ebonol C Alkaline chlorite	Copper and its alloys	Thick matte black oxide with significant microfiberous roughness for interlocking
Phosphate conversion coatings: Zinc Iron	Steel	Precipitates crystallites onto the surface which provide good bonding morphology

4. Protect the surface or provide a new surface that is more resistant to environmental influences once the joint is in service.

6.5.1 Active chemical surface treatments

Chemical treatments change the physical and chemical properties of the surface to produce one that is highly receptive to adhesion. Specific chemical treatments have been developed for various metallic and nonmetallic surfaces. The chemicals used are acidic or alkaline in nature. Common chemical treatments include the use of sulfuric acid–sodium dichromate, phenol, sodium naphthalene, ferric chloride–nitric acid, and nitric–hydrofluoric acid solutions.

In all cases, extreme care and good laboratory practice should be used in handling these materials. Personnel need to be trained in the handling and use of acid and alkaline solutions and must wear the proper clothing as the chemicals could be very harmful if they come into contact with the skin. Ventilation and spill containment are particularly important safety considerations. Individuals must be aware

TABLE 6.12 Common Active Surface Treatments for Polymeric Substrates

(1) Common Active Chemical Treatments

Treatment	Polymer	Result

Oxidizing acids and acid mixtures:
  Chromic
  Nitric
  Sulfuric
  Formic
	Polyolefins, ABS, polycarbonate, nylon, polyphenylene oxide, and acetal	Oxidation of the surface
		
		Reactive groups (hydroxyl, carbonyl, carboxylic acid and hydrogen sulfite) are introduced
		
		Cavities formed to provide interlocking sites

"Satinizing"—mildly acidic solution of perchloroethylene, p-toluenesulfonic acid and colloidal silica	Delrin acetal	Produces uniformly distributed "anchor" points on the part's surface
Sodium naphthalane solution	Fluorinated plastics	Dissolves amorphous regions on the surface and removes fluorine atoms Increases mechanical interlocking by microroughening Unsaturated bonds and carbonyl bonds introduced
Iodine	Nylon	Alters surface crystallinity from alpha to beta form
"Cyclizing"—concentrated sulfuric acid	Natural rubber, styrene butadiene rubber, and acrylonitrile butadiene rubber elastomers	Hairline fractures on the surface increases mechanical interlocking

of disposal regulations and the cost of disposing these solutions properly. Like cleaning solutions, active surface preparation solutions become contaminated and lose their efficiency with time. Continuous monitoring and quality control are required to yield reproducible results.

Figure 6.8 shows a common flow chart for surface preparation for a metallic substrate that is already clean of loose boundary layers such as scale and rust. The first degreasing operation is to remove gross organic contaminants from the surface prior to chemical treatment.

TABLE 6.12 Common Active Surface Treatments for Polymeric Substrates (*Continued*)

(2) Common Active Physical Treatments

Treatment	Polymer	Result
Flame	Polyolefins, nylon, other low surface energy plastics	Oxidizes the surface introducing polar groups (carbonyl, carboxyl, amide, and hydroperoxide)
Corona or electrical discharge	Polyolefins, polyethylene terephthalate, PVC, polystyrene, cellulose, fluorocarbons	Oxidation and introduction of active groups such as carbonyls, hydroxyls, hydroperoxides, aldehydes, ethers, esters, and carbonoxylic acids Increased surface roughness
Plasma discharge	Nearly all low energy surfaces including most thermoplastics Silicone rubber and other low surface energy elastomers	Crosslinking of the surface Surface oxidation with the formation of polar groups Grafting of active chemical species to the surface Halogenation of the surface
Ultraviolet radiation	Polyolefins, polyethylene terephthalate, EPDM rubber, other low energy polymers	Chain scission of surface molecules followed by crosslinking Surface oxidation
Laser treatment	Polyolefins, engineering plastics, sheet molding compounds	Removal of surface contamination and weak boundary layers Roughening of filled surfaces Soften the surface of thermoplastics
Ion beam etching	Fluorocarbons	Creation of needles or spires on the surface for improved interlocking

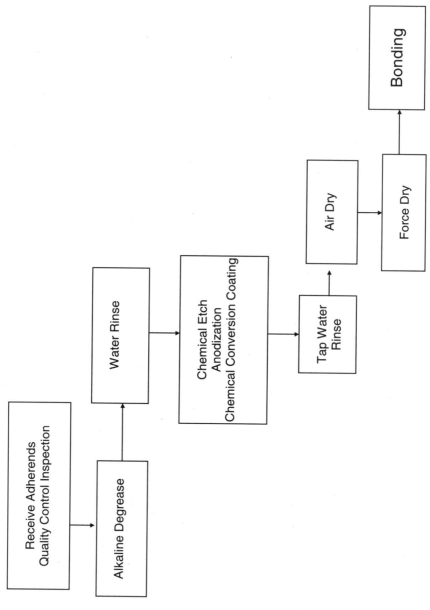

Figure 6.8 Flow chart for the surface preparation of metallic parts by active chemical means.

During the surface treatment, the part or area to be bonded is usually immersed in a chemical solution at room or elevated temperature for a matter of minutes. Tank temperature and agitation must also be controlled. Chemical solutions must be changed regularly to prevent contamination and ensure repeatable concentration. After chemical treatment, the parts are immediately rinsed with water and dried. The rinse operations are to remove the residue of the surface treatment steps. Chemical treatments are usually specified by the following parameters:

- *Solution used.* Such as caustic etch, sulfuric–dichromate, sodium etch, ferric chloride–nitric acid, etc.
- *Solution temperature.* Usually room temperature to 100°C.
- *Immersion time.* Usually seconds to minutes.
- *Type of rinse.* Usually running tap water followed by distilled water.
- *Type of drying.* Usually in air at room temperature. However, elevated temperatures or other conditions are not uncommon.

Chemical treating solutions should be prepared in containers of glass, ceramic, or chemical resistant plastic. Stirring rods should be made of the same material. Metals others than those to be treated should not touch the solution. For solutions containing chemicals that attack glass and polyolefin containers (e.g., hydrofluoric acid, fluorides), TFE fluorocarbon containers should be used. Solutions in plastic trays can be heated by immersion in hot water baths. Hot plates or infrared heaters can be used for glass and porcelain trays.

Many chemical treating solutions have been developed over the years for both metallic and non-metallic substrates. These are described in detail in the final sections of this chapter and in Appendices C-1 through C-4. Many proprietary metal surface treatment solutions have also been developed by companies that specialize in this field. These usually contain mixtures of acids and specialty chemicals to control over-etching. Since these solutions come already mixed, their hazardous nature is less than if one had to formulate a solution from raw materials. However, the commercial solutions are still hazardous and provide environmental problems particularly in regard to the disposal of spent liquid. Some paste-type etching products are also available that simultaneously clean and chemically treat surfaces. They react at room temperature and need only be applied to the specific area to be bonded. Thus, these treatments are very useful for complex part geometries and for parts that cannot be immersed in a chemical solution. However, these paste etchants generally require much longer treatment time than acid-bath processes.

6.5.1.1 Chemical treatment of metal surfaces. The purpose of chemically treating a metal surface is generally either to remove an unwanted oxide or other weak boundary layer, or to protect the surface from corrosion. Chemical treatments also provide surface roughening on a microscale which improves mechanical adhesion. Chemical treatment processes generally remove the complex elements that are on the substrate surface and replace them with a more uniform, more tightly held structure.

Metal surfaces are usually some combination of oxides, sulfides, chlorides, acid salts, absorbed moisture, oil, and atmospheric gases. These weak boundary layers are formed by the process used to fabricate the metal, and their surface characteristics are dependent on the parameters of these processes. The resulting surface structure is usually weakly bonded to the base metal and prone to crack or flake off. With many metal substrates, it is not sufficient to only remove these weak layers. The pure, bare metal surface may be very reactive, and unwanted oxide layers and corrosion products could quickly form. Thus, the surface preparation must not only remove the original surface, but replace it with a surface coating that will protect the interface during further processing and during the joint's service life.

The best method of removing surface material is by mechanical processes described previously. However, chemical pickling or acid descaling are chemical methods to remove mass surface material. With these processes, the metal oxide surface is effectively and rapidly dissolved leaving bare bulk metal. Pickling may employ a dip or spray system. Hydrochloric, sulfuric, and nitric acids are often used. The particular acid will depend upon the metal and the type of oxide being treated. The rinsing and drying of the substrate, once pickling is complete, is a very important step so that all acid and acid by-products are completely removed from the surface. It is important to use inhibited acid cleaners to avoid corrosion of cleaned surfaces. A limitation of acid descaling is that some dimensional change will occur in addition to oxide removal.

Chemical etching processes are similar to pickling except that a more complex solution is generally used. This surface preparation process not only removes surface layers, but it also transforms the surface, making it more chemically active and more receptive to bonding with certain adhesives. Common etching processes for aluminum alloys use chromic–sulfuric or phosphoric acids. Sulfuric acid based etching processes are commonly used for steel. These multi-stage processes require careful control, and safety and environmental issues are often difficult.

Corrosion of many metal surfaces begins immediately after removal of the protecting oils or contamination. Corrosion can quickly affect the bond by providing a weak boundary layer before the adhesive is applied. Corrosion can also occur after the joint is made and, thereby, affect its permanence. Mechanical abrasion or solvent cleaning can provide adhesive joints that are strong in dry conditions. However, this is not the case when joints are exposed to water or water vapor. This fact is illustrated in Fig. 6.9 for aluminum. Resistance to water is much improved if metal surfaces can be treated with a protective coating before being bonded.

A number of techniques have been developed to convert corrosion prone, clean surfaces to less reactive ones. Three common conversion processes are phosphating, anodizing, and chromating. These processes remove the inconsistent, weak surface on metal substrates and replace it with one that is strong, permanent, and reproducible. The type of conversion processes will depend on the substrate, the nature of the oxide layer on its surface, and the type of adhesive or sealant used. The formation of a nonconductive coating on a steel surface will minimize the effect of galvanic corrosion.

The crystalline nature of a phosphate coating will normally increase the bonding properties of a surface. The two most common types of phosphate coatings are iron and zinc. They are produced by treating the surface with acid solutions of iron or zinc phosphate.[31] Iron phosphate coatings are easier to apply and more environmentally acceptable. Zinc phosphate coatings provide better corrosion protection. An important consideration when using these conversion coatings is es-

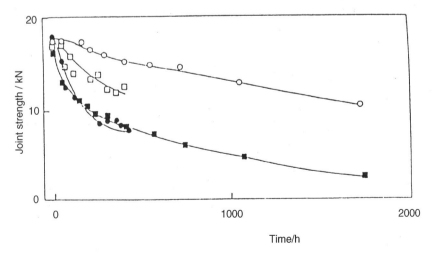

Figure 6.9 Effect of high humidity (97% relative humidity at 43°C) on the strength of aluminum joints bonded with an epoxy-polyamide adhesive. Surface treatments are (○) chromic–sulfuric acid etch, (□) alkaline etch, (●) solvent degrease, and (■) phosphoric acid anodize.[30]

tablishing how much coating weight is required. This is best done by the actual testing of finished parts.

For aluminum, anodizing provides the most water durable adhesive joints. It is used by many automotive and aerospace suppliers. The corrosion protection is provided by anodizing the clean deoxidized aluminum surface in either chromic or phosphoric acid electrolytic baths. In the U.S., phosphoric acid anodizing is often used because of its lower toxicity and easier disposal. Anodizing creates an oxide under controlled voltage and temperature conditions, thereby creating a more protective surface.[31] As with phosphating, care must be taken to ensure the optimum coating thickness.

Chromate conversion coatings are used with some metals, such as aluminum, zinc, magnesium, and copper alloys, to enhance adhesion and to protect the surface. They are tough, hydrated gel structures that are usually applied by immersion in heated solutions of proprietary, chromium containing compounds. Chromium compounds have also been elecrolytically applied to steel.

6.5.1.2 Chemical treatment of polymeric surfaces.
The chemical modification of low energy polymer surfaces may be carried out by treatment with chromic acid, metallic sodium complex dispersions, bleach/detergents, potassium iodate/sulfuric acid and other mixtures. The chemical used and the treating conditions are dependent on the type of polymer being treated and the degree of adhesion required.

Chemical treatment of polymeric surfaces is generally more difficult than metallic surfaces and requires special considerations. Polymeric products often contain pigments, antioxidants, slip agents, mold release agents, etc. that can migrate to the surface and interfere or alter a surface treatment process. Slight changes in the polymer formulation or its fabrication process may result in changes in the surface condition and the effectiveness of treating operations. Depending on the exact formulation and method of manufacture, the surfaces can be considerably different on parts manufactured from the same generic plastic and having the same bulk physical properties. Polymers that are molded against a hot metallic surface may have very different surface characteristics, for example, than the same polymer that is formed without contact to a metal surface. Melt temperature and cooling rate can also affect the surface properties of the polymer. The control of surface crystallization during plastic processing is important. Molding conditions can alter the surface crystalline content when compared to the bulk polymer. Normally crystalline regions are much less susceptible to etching than amorphous regions.

Most important, however, is the fact that surface treatments on polymeric materials are subject to the degree of handling after treatment. Certain treatments can be removed by rubbing or scuffing or by exposure of the part to elevated temperatures before the adhesive is applied. It is best to bond polymeric parts as soon as possible after they are treated. The actual "shelf life" of treated polymeric parts will depend on the nature of the part, the handling and storage conditions, and the type of surface treatment that was administered.

Liquid etchants can be used for chemical modification or dissolving surface contamination. Etchants effectively treat irregularly shaped objects that are difficult to treat by corona or flame treatments. A number of etching solutions and procedures have been developed for specific polymeric surfaces. The choice of the liquid etchant depends on the polymer. Polyolefins are usually treated by oxidizing acids such as chromic, sulfuric, nitric, or mixtures of these. Fluorocarbons are usually treated by sodium-napthalene etching solution.

To facilitate the etching process, removal of organic surface contaminants must first be accomplished by passive chemical cleaning. Sometimes solvent degreasing is used prior to etching to gently soften and swell the polymer surface. Chemical etching solutions for polymeric substrates can be recipes that are mixed from raw materials or they can be proprietary, commercial solutions. The temperature of these chemical etching treatments is generally above 122–140°C to ensure fast processing; however, effective treatments can also be carried out at room temperature. These acids chemically alter the polymer surface as well as produce roughened or even porous-like surfaces. The nature of the surface modification is dependent on the polymer and the etchant used. Chromic acid, for example, will oxidize polyolefin surfaces. Whereas, sodium naphthalene etching processes will remove the surface fluorine atoms from a fluorocarbon surface. Extensive etching of the surface can lead to a cohesively weakened surface structure that is undesirable for bonding purposes. As a result, surface treating processes must be followed explicitly. Complete removal of the etching chemicals by a rinse step is essential. Highly reactive etchants can continue reacting with the surface after application of the adhesive and, thereby, degrade the chemical and physical stability of the surface.

Several novel active chemical treating processes have been recently developed for polymeric substrates. Much of this development is due to the increased usage of engineering plastics in certain industries and concerns over environmental and safety issues with conventional chemical processes. The newer processes include fluorine surface treatment, grafting, and adhesive abrasion.

The fluorine treatment of plastics for improved adhesion is a relatively new process. Through this treatment a new surface is applied to the polymer. The new surface that will eventually be in contact with the adhesive is a polar surface having higher surface energy. Such "engineered" surface treatments of polypropylene and polyethylene claim to give a six fold increase in peel strength when bonded with an epoxy adhesive.[32] Evidence indicates that covalent bond formation occurs between the fluorinated surface and the amine component of the epoxy hardener. It is claimed that nearly all plastic surfaces can be effectively improved by fluorine treatment.[33,34]

Another approach to active surface treatment is to graft polar monomers onto surfaces of low energy polymers to alter the surface by making them more polar. Very thin coatings can be applied by plasma technology. Significant improvements in bond strength have been made with polyolefin substrates and epoxy adhesives.[35–39] However, on a commercial scale these procedures would require considerable capital investment and the use of hazardous chemicals. The use of conventionally applied surface primers and adhesion promoters to treat polymeric surfaces has proved to be a more desirable alternative on the production line. These primers are discussed in Chapter 7.

Adhesive abrasion is a process developed for polymeric substrates such as the fluorocarbons.[40] It is primarily used in the microelectronics industry. With this process, the plastic surface is abraded while immersed in a liquid adhesive. The abraded substrates are then immediately mated and allowed to cure. Abrasion in the presence of the liquid adhesive produces free radicals that react directly with the adhesive before they can be scavenged by atmospheric oxygen.

Although bond strength can be improved significantly by these chemical methods, most are time consuming batch processes, often taking hours to treat, wash, and dry parts. There is also the problem of disposing of hazardous waste from these processes and the operator's safety. Thus, the application has been mostly for small volume, high value parts.

There are many industries performing fast, high volume assembling or sealing operations on substrates with low surface energy. Many of the high volume assembly applications selectively use thermoplastic substrate materials so that thermal welding or solvent cementing assembly processes can be used which do not require active surface treatment. These processes are described in Chapter 15.

6.5.2 Active physical surface treatments for polymeric materials

Because of the main disadvantages of chemical treatments (hazardous nature and a slow, batch type process), a number of other active surface treatments have been developed for polymeric materials. These processes utilize the reactivity of the polymeric surface to gain change that is favorable for adhesion. Rather than chemical solutions, these surface modifications are usually made by physical means such as flame, electric discharge, UV light, or laser.

6.5.2.1 Corona discharge. Corona discharge treatment is a popular method of dry surface preparation of polymer films. The purpose of the treatment is to make the polymer surface more receptive to inks or coatings; however, it has also been used effectively as a pretreatment for adhesives. The treatment is believed to oxidize the surface of the polymer so that the ink, coating, or adhesive can permeate the roughness of the thin oxidized layer. The most common methods of treating surfaces by oxidizing are corona treatment and flame treatment (see next section). For film, corona treatment is the preferred method of the two.

The schematic in Fig. 6.10 illustrates two basic arrangements for treating film and coated paper with corona treating equipment. The electric equipment consists basically of a high frequency generator (10–20 kHz), a stationary electrode electrically connected to the generator, and a dielectrically covered treater roll that serves as the grounded electrode. The material to be treated is carried over this roller for continuous processing. A suitable voltage, typically 20 kV, is developed between the electrodes. This produces a spark or corona discharge by ionizing the air in the gap between the electrodes. The ionized particles in the air gap bombard and penetrate into the molecular structure of the substrate. Free electrons and ions impact the substrate with energies sufficient to break the molecular bonds on the surface of most polymeric substrates. This creates free radicals that react rapidly with oxygen to form polar chemical groups on the substrate surface and increase the surface energy to a point where many adhesives, paints, and coatings can wet the substrate.

Corona treating equipment is inexpensive, clean, and easily adapted to in-line operations. Parameters that positively affect the efficiency of the treating process include power input and temperature. Whereas, line speed and humidity negatively affect the efficiency. Stored film is generally considered more difficult to treat than film that has just been processed (e.g., exiting an extruder). The opportunity of low molecular weight additives and contaminants to migrate to the surface is greater for stored film. These surface contaminants impede the treating efficiency of the corona.

Corona treatment is mainly suitable for films although thin containers have been treated by nesting them on a shaped electrode and rotating the part adjacent to a high voltage electrode. Approximately 25 mils (0.025 in) is the maximum thickness of sheet that can be treated by the corona discharge method. However, high frequency arc

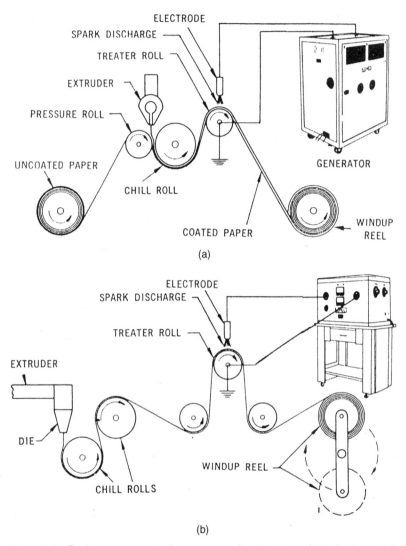

Figure 6.10 Basic arrangements for corona surface treating of (a) plastic coated paper and (b) plastic film.[41]

treatment has been applied to automotive trim parts of up to 1/8 inch thick. Significant bond strength improvements were noticed by treating talc filled polypropylene parts before bonding with hot melt adhesives.[42]

Corona discharge treatments have been commonly employed to treat substrates such as films or polymer coated paper for printing. The polymers commonly treated are polyethylene, polypropylene, and polyethylene terephthalate, or Mylar®. Fluorocarbon film surfaces have also been treated with corona in environments other than air. Metal foil surfaces have been treated with corona, but the effect is only moderate, probably due to removal of organic contaminants by the oxidizing exposure.

6.5.2.2 Flame treatment. Flame treatment consists of exposing a surface to a gas flame for less than several seconds. Flame treatment burns-off contaminants and oxidizes the surface of the polymer similar to corona treatment. Flame treatment is used as a surface treatment for many low energy polymeric parts prior to bonding, printing, or painting. Figure 6.11 shows the flame treatment of a plastic bottle.

Flame treatment is believed to provide a polar surface that is conducive to adhesion. A brief exposure to the flame oxidizes the surface through a free radical mechanism, introducing higher surface energy groups (hydroxyl, carbonyl, carboxyl, and amide groups) to the part's

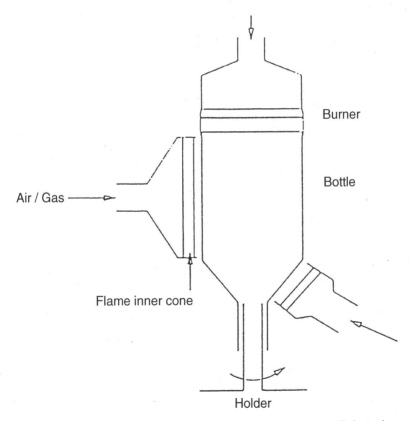

Figure 6.11 Flame treatment of a plastic bottle. The treatment will depend on the gas mixture and flow rate, position of part relative to flame, and exposure time.

surface. Molecular chain scission and crosslinking also occur depending on the polymer and on the nature of the flame treatment. This process is widely used to prepare polyolefin surfaces for painting, printing, or adhesive bonding. Flame treatment can be used for both film and shaped products and can be used for both continuous and bulk processing.

In the flame treating process gas burners are fed from the facility's mains (chief component methane) or bottled gas (propane or butane). Of importance in the operation of a flame treater is the gas/air mix ratio. Depending on the level of gas in the mix, the flame can have substantially different characteristics. A slight excess of oxygen over that required for complete combustion is recommended. This treating method increases the resulting adhesive strength of joints made with polyolefins and other low energy plastics. The flame oxidizes the surface, resulting in an increase in critical surface energy. The time that the flame is applied and its nearness to the surface are also important quality control factors. A surface is typically exposed to the flame region just above the blue cone until it becomes glossy. It is important not to overexpose the plastic because warping or other damage of the part may result.

Automated flame processing equipment are available, similar to the corona treaters. Hand-held equipment such as a torch or Bunsen burner can also be used, although uniform treatment is more difficult. The effect of polymer additives on the treatment efficiency are not as great as with corona treatment.

6.5.2.3 Plasma treatment.

A gas plasma treating process has been developed for surface treatment of many polymeric materials. It is a dry process that is becoming a common method of treating many different engineering plastics when maximum joint strength is required. Low energy materials, such as polyolefins, polytetrafluoroethylene, polyethylene terephthalate, nylon, silicone rubber, etc. are readily treated with gas plasma. Relative bond strength improvements of ten to several hundred times are possible depending on the substrate and gas plasma. Plasma treatment has become a very popular treatment for small to medium sized parts that can fit into a vacuum chamber and for production volumes that are amenable to a batch type process.

Operationally, a plasma differs from corona and flame treatment in that the process is completed at less than atmospheric pressure and with gases other than air. Because of the necessity for partial vacuum, plasma treatment is essentially a batch process. The type of plasma gas can be selected to initiate a wide assortment of chemical reactions, including:

1. Surface crosslinking
2. Surface oxidation or reduction
3. Grafting of active radicals to material surfaces
4. Halogenation of surfaces
5. Deposition of inorganic and organic films

Gases or mixtures of gases used for plasma treatment of polymers include nitrogen, argon, oxygen, nitrous oxide, helium, tetrafluoromethane, water, and ammonia. Each gas produces a unique surface treatment process. It should be noted that surface chemistry modification by plasma treatment can make polymer surfaces totally wettable or non-wettable. Non-wettable plasma treatments generally involve the deposition of fluorine containing chemical groups to produce medical products.

Gas plasma is an extremely reactive ionized gas. The main difference between plasma surface treatment and corona surface treatments is the nature of the plasma (specialty gas vs. air) and the operating pressure of the plasma (0.1 to 10 torr vs. 760 torr). With the plasma treatment technique, a low-pressure inert gas is activated by an electrodeless radio-frequency discharge or microwave excitation to produce metastable excited species that react with the polymeric surface. The plasma treatment produces changes only to the depth of several molecular layers. Generally, only very short treating times (secs to mins) are necessary. Commercial instruments are available from several manufacturers to plasma treat parts prior to bonding. Because of the necessity for very low pressures, a glass or ceramic vacuum container is generally used. Thus, plasma treatment is generally thought of as a batch type process for parts of up to moderate size. Continuous plasma treaters have been developed for processing film and fiber and large volume chambers have been built for treating large parts. However, capital expense has limited these applications to specialty markets.

It is generally believed that the plasma treating process provides surfaces with greater stability than chemical etch, corona, flame, or other common polymeric treatment processes. Plasma treated parts can be stored for weeks or longer in a clean, dry storage area. Exposure to temperatures near the polymer's glass transition temperature will deteriorate the surface treatment. Once well bonded, the surfaces of treated polymers are relatively stable excluding the effects of outside environmental influences.

With plasma treatment, surface wettability can be readily induced on a variety of normally non-wettable materials as shown in Table

6.13. Certain polymeric surfaces, such as the polyolefins, become crosslinked during plasma treatment. The surface skin of polyethylene, for example, will become crosslinked so that if the part were placed on a hot plate of sufficient heat, the interior would turn to a molten liquid while the crosslinked outer skin held a solid shape. Other polymers are affected in different ways. Plasma treated polymers usually form adhesive bonds that are 2 to 4 times the strength of untreated polymers. Table 6.14 presents bond strength of various plastic substrates that were pretreated with gas plasma and bonded with an epoxy or urethane adhesive.

6.5.2.4 Other physical surface treatments for polymers. Other surface treatments have been reported to enhance adhesion of low energy polymers. Bond strengths of some materials can be improved if the mating surfaces are etched or otherwise textured prior to joining. Ion beam etching and excimer laser radiation are two ways of doing this.

Ion beam etching has been used on stainless steel, graphite, and fluorocarbon surfaces.[44] The resulting surface is composed of needles or spires that allow improved mechanical bonding. Bonds are stronger both in tension and in shear than those made on chemically etched plastics. An additional benefit of ion beam treatment is that its effect does not diminish with time as in the case of chemical etching.

Excimer laser surface treatment has been used for preparing polyester sheet molding compounds (SMC) for adhesive bonding in the automotive industry. The excimer laser preparation of SMC surfaces occurs through the following stages: ablation of surface contaminates, selective ablation of calcium carbonate filler from the SMC, and removal of polyester resin from the SMC.[45,46]

UV irradiation has also been applied as a prebond surface treatment to a variety of plastics.[47] Basically, this process involves applying a 5% solution of benzophenone (a light sensitizer) to the surface of the part. The part is then briefly exposed to an ultraviolet light source. This causes chemical changes in the polymer surface, including an increase in wettability and a certain amount of crosslinking that reportedly strengthens the surface and improves the physical interaction necessary for adhesion. It is likely that this process promotes controlled degradation of the surface of the polymer, offering a better surface on which to bond. The process was originally developed for application to EPDM rubber, but it may also be applicable for polypropylene, polyethylene, and other thermoplastics.

Primers are widely recognized for preparing metals for adhesive bonding. It has also been reported that certain polymeric surfaces, notably polyolefins, can be modified by applications of primer solutions such as chlorinated polyethylene. The probable mechanism by which

TABLE 6.13 Effect of Plasma Exposure on Contact Angle and Wetting[43]

		Trademark or generic name	Initial surface energy (dynes/cm)	Final surface energy (dynes/cm)	Initial water contact angle (degrees)	Final water contact angle (degrees)
Hydrocarbons:						
PP	Polypropylene	Marlex, Profax	29	>73	87	22
PE	Polyethylene	Alathon, Dylan	31	>73	87	42
PS	Polystyrene	Styron, Lustrex	38	>73	72.5	15
ABS	Acrylonitrile/butadiene/styrene copolymer	Cycolac	35	>73	82	26
—	Polyamide (Nylon)	Zytel	<36	>73	63	17
PMMA	Polymethyl methacrylate		<36	>73	—	—
PVA/PE	Polyvinyl acetate/polyethylene copolymer	Elvax	38	>73	—	—
—	Epoxy	Araldite, Epon	<36	>73	59.0	12.5
—	Polyester		41	>73	71	18
PVC	Rigid polyvinylchloride	Geon	39	>73	90	35
PF	Phenolic		—	>73	59	36.5
Fluorocarbons:						
ETFE	ETFE/PE Copolymer	Tefzel	37	>73	92	53
FEP	Fluorinated ethylene propylene	Teflon, Halon, Fluon	22	72	96	68
PVDF	Polyvinylidene fluoride	Kynar	25	>73	78.5	36

TABLE 6.13 Effect of Plasma Exposure on Contact Angle and Wetting (Continued)

		Trademark or generic name	Initial surface energy (dynes/cm)	Final surface energy (dynes/cm)	Initial water contact angle (degrees)	Final water contact angle (degrees)
Engineering Thermoplastics:						
PET	Polyethylene terephthalate	Rynite, Mylar, Melinex	41	>73	76.5	17.5
PC	Polycarbonate	Lexan, Merlon	46	>73	75	33
PI	Polyimide	Kapton	40	>73	79	30
—	Polyaramid	Kevlar		>73	—	—
—	Polyaryl etherketone	PEEK	<36	>73	92.5	3.5
—	Polyacetal	Delrin, Celcon	<36	>73	—	—
PPO	Polyphenylene oxide	Noryl	47	>73	75	38
PBT	Polybutylene terephthalate	Valox, Tenite, Celanex	32	>73	—	—
—	Polysulfone	Udel	41	>73	76.6	16.5
PES	Polyethersulfone	Victrex	50	>73	92	9.
—	Polyarylsulfone	Radel	41	>73	70	21
PPS	Polyphenylene sulfide	Ryton	38	>73	84.5	28.5
Elastomers:						
SR	Silicone	Silastic, SR, Rhodorsil	24	>73	96	53
—	Natural rubber		25	>73	—	—
—	Latex		—	>73	—	—
PUR	Polyurethane	Cyanaprene, Adiprene	—	>73	—	—
SBR	Styrene Butadiene Rubber	Sanoprene	48	>73	—	—
Fluoroelastomers:						
FPM	Fluorocarbon copolymer elastomer	Vilton, Fluorel	<36	>73	87	51.5

TABLE 6.14 Lap Shear Strength for Several Plasma Treatmented Polymers

Adherends	Adhesive	Bond strength, psi		Source
		Control	After plasma treatment	
High density polyethylene–aluminum	Epoxy–polyamide	315	3500	1
Low density polyethylene–aluminum	Epoxy–polyamide	372	1466	1
Nylon 6–aluminum	Epoxy–polyamide	846	3956	1
Polystyrene–aluminum	Epoxy–polyamide	566	4015	1
Mylar–aluminum	Epoxy–polyamide	530	1660	1
Polymethylmethacrylate–aluminum	Epoxy–polyamide	410	928	1
Polypropylene–aluminum	Epoxy–polyamide	370	3080	1
Tedlar, PVF–aluminum	Epoxy–polyamide	278	1370	1
Celcon acetal–aluminum	Epoxy–polyamide	118	258	1
Cellulose acetate butyrate–aluminum	Epoxy–polyamide	1090	2516	1
Thermoplastic polyester PBT	Epoxy	520	1640	2
Thermoplastic polyester	Polyurethane	190	960	2
Polyetherimide	Epoxy	190	2060	2
Polycarbonate	Epoxy	1700	2240	2
Polycarbonate	Polyurethane	540	1140	2
Delrin acetal	Epoxy	160	650	2
Polyester PET	Epoxy	683	6067	3
Fluorocarbon, ETFE	Epoxy	10	293	4

SOURCES:
1. Hall J. R., et al., "Activated Gas: Plasma Surface Treatment of Polymers for Adhesive Bonding", *Journal of Applied Polymer Science*, 13, 2085–2096, 1969.
2. Kaplan, S. L., and Rose, P. W., "Plasma Treatment Upgrades Adhesion of Plastic Parts", *Plastics Engineering*, May 1988.
3. Sangiuolo, S., and Hansen, W. L., International Coil Winding Association Technical Conference, Rosemont, IL, 1990.
4. Hansen, G. P., Rushing, R. A., et al., "Achieving Optimum Bond Strength with Plasma Treatment", Technical Paper AD89-537, Society of Manufacturing Engineers, Dearborn, Mich., 1989.

these primers operate is diffusion into the plastic and the creation of partially chlorinated surface, which can interact more strongly with polar adhesives. A resorcinol formaldehyde primer substantially improves the adhesion to nylon. Isocyanate primers give large improvements in the adhesion between urethane adhesives and styrene butadiene elastomers. Several diverse materials such as transition metal complexes, triphenylphosphine, and cobalt acetylacetonate have been found to be excellent primers to polyolefins bonded with a cyanoacrylate adhesive. In these cases, the strength of the joint is often greater than the strength of the parent plastic. Primers and adhesion promoters are discussed in detail in the next chapter.

6.6 Specific Surface Treatments

Appendix C lists recommended surface-treating procedures for most common substrates. There are several text books that provide excellent reviews regarding specific surface treatments for a variety of substrates. These are listed here.

Shields, J., *Adhesives Handbook*, 3rd edition, (London: Butterworths 1984)
Skeist, I. (ed.), *Handbook of Adhesives*, 3rd edition (New York: Van Nostrand Reinhold 1990)
Wegman, R. F., *Surface Preparation Techniques for Adhesive Bonding*, (Park Ridge, NJ: Noyes Publications 1989)
Snogren, R. C., *The Handbook of Surface Preparation*, (New York: Palmerton Publishing 1974)
Landrock, A. H., *Adhesives Technology Handbook*, (Park Ridge, NJ: Noyes Publications 1985)
Cagle, C. V. (ed.), *Handbook of Adhesive Bonding*, (McGraw Hill: New York 1973)
MIL-HDBK-691B, *Adhesive Bonding*, Department of Defense, March 1987.

In Chapter 16, common metallic, plastic, elastomeric and other substrates are covered in depth with respect of their unique surface characteristics and recommended bonding processes. These processes have been found to provide high bond strength and durability.

6.6.1 Metallic adherends

Appendix C-1 lists common recommended surface-treating procedures for metallic adherends. These methods have been specifically found to provide reproducible structural bonds and fit easily into the bonding operation.

ASTM D 2651 describes practices that have proved satisfactory for preparing various metal surfaces for adhesive bonding. Surface preparations are included for aluminum alloy, titanium alloy, copper and copper alloys. Formulation and procedures are described in detail that are based on commercial practice of numerous agencies and organizations.

6.6.2 Plastic adherends

There are many surface treatments available for plastic parts. These have been developed by the resin manufacturers, assembly manufacturers, and adhesive developers. ASTM D 2093 describes recommended surface preparations for plastic adherends. Appendix C-2 lists common recommended surface treatments for plastic adherends. Table 6.15 provides a comparison of general surface treatment techniques.

Solvent and heat welding are methods of fastening plastics that do not require chemical alteration of the surface although cleaning or degreasing is recommended. These welding procedures will be discussed in Chapter 15 of this Handbook. The plastic materials commonly used in bonded structures, their unique characteristics, and successful surface treatments and bonding process are more fully described in Chapter 16.

6.6.3 Polymeric composite adherends

All of the surface preparations described in Appendix C-2 for polymeric substrates are also applicable for when they are reinforced and made to form a composite material. However, care must be taken so that the liquid surface preparation chemicals do not wick into the composite along the interface between the fibers and the resin matrix. As a result, immersion treatments are usually not used. Surface wipe with a solvent or cleaning agent and abrasion followed by another surface wipe is usually all that is necessary for treating the high energy composite substrates prior to bonding. For low energy composite substrates, chemical etching (protecting the cut edges of the composite) or plasma treatment are sometimes used.

Plasma treatment has been found to give significantly improved adhesion properties to thermoplastic based carbon composites (polyetheretherketone and polyphenylene suflide). Whereas, thermosetting composites (e.g., epoxy) provide sufficient joint strength with only light abrasion and solvent cleaning.[48]

The nature of the composite surface treatment will depend on the resin matrix, the permanence characteristics required, the nature of the cut surfaces where reinforcement may be exposed to the environment, and the production facilities that are available.

Many surface roughening approaches have been tried for composites, and all have some merit. One method that has gained wide acceptance is the use of a peel ply. In this technique, a densely woven nylon or polyester cloth is used as the outer layer of the composite during its manufacture (Fig. 6.12). This ply is then torn or peeled away just before bonding. The tearing or peeling process fractures the resin matrix coating and exposes a clean, virgin, roughened surface for the bonding process.

In the cases where the peel ply is not used, some sort of light abrasion is required to break the glazed resin finish on the composite

TABLE 6.15 Comparison of Surface Treatment Techniques for Polymeric Parts

Technology	Bond strength	Advantages	Disadvantages	Consistency	Versatility	Capital cost	Labor cost	Environment impact
None	Low	Fastest, no training	Poor and inconsistent bond strength	Poor	—	None	None	None
Mechanical Abrasion	Good	Fast, no chemicals, moderate training	Dust and particles must be controlled, surfaces must first be clean	Poor	Very Good	Little	Medium	Dust
Solvent Wipe	Medium–Good	Fast, moderate training	Chemicals must be used and kept uncontaminated	Fair	Good	Low	Medium	Organic vapor
Vapor Degreasing	Good	Fast, moderate training	Chemicals must be used and kept uncontaminated; not useful on certain plastics	Good	Good	Medium	Medium	Organic vapor
Flame, thermal	Good	Fast, no chemicals required	Only good for simple shapes, bonds must be made rapidly; only useful on certain plastics	Fair	Poor	Low	Low	Open flame
Acid etch	Very good	High strength and durability	Slow, chemicals required and processing steps are critical; training required	Good	Fair	High	High	Fumes, chemicals, disposal costs
Corona	Good	Fast	Suitable mostly for plastic film and sheet, training required	Good	Poor	High	Medium	Ozone
Plasma	Very good	Fast, high bond strength on low energy plastics	Batch operation, equipment is expensive especially for large parts	Very good	Poor	High	Medium	Low, bottled gases
Surface primers	Good	Fast, protects parts until they are bonded; silanes provide good strength and durability	Parts often must be surface treated before primer is applied, primer step requires time	Very good	Fair	Low	Medium	Solvent vapor

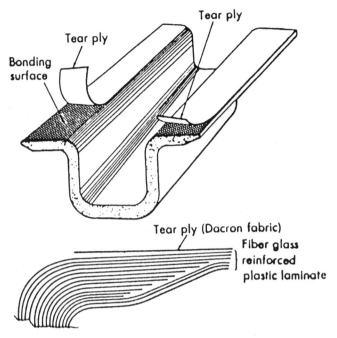

Figure 6.12 Structural reinforced plastic laminate with tear ply to produce fresh bonding surface.[20]

surface. The surface glaze should be roughened without damaging the reinforcing fibers or forming subsurface cracks in the matrix.

6.6.4 Elastomeric adherends

As shown in Appendix C-3, solvent washing and abrasion are common treatments for most elastomers, but chemical treatment may be required for maximum properties. Vulcanized rubber parts are often contaminated with mold release and plasticizers or extenders that can migrate to the surface. Solvents for cleaning must be carefully chosen to avoid possible swelling of the elastomer and entrapment of solvent in the bulk material. Certain synthetic and natural rubbers require "cyclizing" with concentrated sulfuric acid until hairline fractures are evident on the surface. Fluorosilicone and silicone rubbers must be primed before bonding. The primer acts as an intermediate interface, providing good adhesion to the rubber and a more wettable surface for the adhesive.

The elastomers commonly used in bonded structures, their unique characteristics, and successful surface treatments and bonding process are more fully described in Chapter 16.

6.6.5 Other adherends

Appendix C-4 provides surface treatments for a variety of materials not covered in the preceding tables. Other adherends commonly used in bonded structures, their unique characteristics, and successful surface treatments and bonding process are more fully described in Chapter 16.

References

1. Hagemaier, D. J., "End Product Nondestructive Evaluation of Adhesive Bonded Metal Joints", *Adhesives and Sealants,* vol. 3, Engineered Materials Handbook, ASM International, 1990.
2. Schneberger, G. L., "Adhesives for Specific Substrates", Chapter 21, in *Adhesives in Manufacturing,* G. L. Schneberger, ed. (New York: Marcel Dekker, Inc., 1983).
3. Chessin, N. and Curran, V., "Preparation of Aluminum Surface for Bonding", in *Structural Adhesive Bonding,* M. J. Bodnar, ed. (New York: Interscience, 1966).
4. Muchnick, S. N., "Adhesive Bonding of Metals", *Mech. Eng.,* January 1956.
5. Vazirani, H. N., "Surface Preparation of Steel and Its Alloys for Adhesive Bonding and Organic Coatings", *J. Adhesion,* July 1969.
6. Vazirani, H. N., "Surface Preparation of Copper and Its Alloys for Adhesive Bonding and Organic Coatings", *J. Adhesion,* July 1969.
7. Water, R. E., Voss, D. L., and Hochberg, M. S., "Structural Bonding of Titanium for Advanced Aircraft", in Proc. Nat. SAMPE Tech. Conf., vol. 2, *Aerospace Adhesives and Elastomers,* 1970.
8. Thelen, E., "Adherend Surface Preparation", in *Symposium on Adhesives for Structural Applications,* M. J. Bodnar, ed. (New York: Interscience, 1966).
9. Kinloch, A. J., *Adhesion and Adhesives,* (London: Chapman and Hall, 1987), at 339.
10. Brewis, D. M., "Pretreatment of Metals Prior to Bonding", in *Handbook of Adhesion,* D. E. Backham, ed. (Essex, England: Longman Scientific and Technical, 1992).
11. Brewis, D. M., "Pretreatment of Polymers", in *Handbook of Adhesion,* D. E. Backham, ed. (Essex, England: Longman Scientific and Technical, 1992).
12. Reynolds Metals Co., *Adhesive Bonding Aluminum,* Richmond, VA, 1966.
13. Gruss, B., "Cleaning and Surface Preparation", *Metal Finishing,* Organic Finishing Guidebook and Directory Issue, May 1996.
14. Rogers, N. L., "Surface Preparation of Metals", in *Structural Adhesive Bonding,* M. J. Bodnar, ed. (New York: Interscience, 1966).
15. Rosty, R., Martinelli, D., et al., *SAMPE Journal,* July/August, 1987.
16. Tira, J., *SAMPE Journal,* July/August 1987.
17. Bulat, Metal Progress, 68, 94–5 (1955) or Watts, J. F., "Degreasing", in *Handbook of Adhesion,* D. E. Backham, ed. (Essex, England: Longman Scientific and Technical, 1992).
18. "Solvents Can be Environmentally Friendly", *Manufacturing Engineering,* May 1997.
19. Mackay, C. D., "Good Adhesive Bonding Starts with Surface Preparation", *Adhesives Age,* June 1998.
20. Snogren, R. C., *Handbook of Surface Preparation* (New York: Palmerton Publishing Company, Inc., 1974).
21. Knipe, R., "Green Cleaning Technologies", *Advanced Materials and Processes,* August 1997.
22. Ismail, A., et al., "Metal Surface Treatment by Electrochemical Cleaning and Its Influence on Long Term Strength of Metal Adhesive Bonds", *Surface Chemical Cleaning and Passivation for Semiconductor Processing,* Materials Research Society Symposium Proceedings, 1993, vol. 315, at 131–136.
23. Pocius, A. V. and Clus, J. J., *Proc. 13th National SAMPE Tech. Conf.* (1981), SAMPE, Covina, CA, at 629–639.

24. Hansel, D., "Abrasive Blasting Systems", *Metal Finishing,* Organic Finishing Guidebook and Directory Issue, May 1996.
25. Jenning, C. W., "Surface Roughness and Bond Strength of Adhesives", *J. Adhesion,* May 1972.
26. Attia, F. G. and Hartsell, D. R., "Cryogenic Surface Preparation for Environmental Compliance on Offshore Structures", Proceedings of the Energy Sources Technology Conference, New Orleans, LA, 1994.
27. Howlett, J. J. and Dupuy, R., "Ultrahigh Pressure Water Jetting for Deposit Removal and Surface Preparation", *Materials Performance*, vol. 32, no. 1, January 1993.
28. Keohan, F. L. and Phillips, P.G., "New Prebond Surface Treatments for Aluminum Using Abrasive Free High Pressure Water Abrasion", 41st International SAMPE Symposium, March 24–28, 1996, at 318–329.
29. Drust, B. E., "Dry Blast for UV and Coatings Removal", 35th Annual Technical Conference Proceedings, Society of Vacuum Coaters, Baltimore, MD, 1992, at 211–217.
30. Comyn, J., Chapter 2, "Surface Treatments for Adhesion and Adhesion", in *Adhesion Science* (Cambridge, England: Royal Society of Chemistry, 1997).
31. Schneberger, G. L., "Chemical and Electrochemical Conversion Treatments", *Encyclopedia of Chemical Technology*, 3rd ed., vol. 15, (New York: Wiley, 1981).
32. Florotec, Inc., Menomonee Falls, WI.
33. Bliefert, C., "Fluoridation of Plastic Surfaces", *Metalloberflaeche,* vol. 41, no. 8, August 1987.
34. Kranz, G., et al., *Int. J. Adhesion and Adhesives,* 14 (4), 1994, at 243.
35. Yamada, K. et al, *J. Appl. Polym. Sci.,* 44 (6), 1993, at 993.
36. Novak, I. and Poliak, V., *Plasty a Kaucuk,* 30 (3), 1993, at 80.
37. Mercx, F. P. M., *Polymer,* 34 (9), 1993, at 1981.
38. Lin, C., *J. Mater. Sci. - Letters,* "Modification of Polypropylene by Peroxide Catalyst Grafting of Maleic Anhydride for Adhesive Bonding: Surface and Interface", 12 (8), 1993, at 612.
39. Novak I., et al., *Fibers and Textiles in Eastern Europe,* 3 (1), 1995, at 41.
40. *The Loctite Design Guide for Bonding Plastics,* LT-2197, Loctite Corporation.
41. Lepel High Frequency Laboratories, Inc.
42. Blitshteyn, M., McCarthy, B. C., et al., "Electrical Surface Treatment Improves Adhesive Bonding", *Adhesives Age,* December 1994.
43. Rose, P. W. and Liston, E., "Gas Plasma Technology and Surface Treatment of Polymers Prior to Adhesive Bonding", SPE Annual Technical Conference, 1985.
44. "Ion-Beam Etching for Strong Adhesive Bonds", *Mech. Eng.,* November 1980; adapted from "Adhesive Bonding of Ion Beam Textured Metals and Fluoropolymers", NASA TM-79004, and "Ion Beam Sputter Etching and Deposition of Fluorocarbons", NASA TM-78888, by B. A. Banks, J. J. Mirtich, and J. S. Sovey, NASA Lewis Research Center.
45. Uddin, M. N., et al., "Laser Surface Engineering of Automotive Components", *Plastics Engineering,* February 1997.
46. Doyle, D. J., *Structural Adhesives Engineering III – Conference Proceedings,* PRI Adhesives Group, 1992.
47. Novak, I. and Poliak, V., "Modification of Adhesive Properties of Isotactic Polypropylene", *Int. Polym. Sci. Tchnol.,* 20(5).
48. Blackman, B. R. K., Kinloch, A. J., et al., "The Plasma Treatment of Thermoplastic Fiber Composites for Adhesive Bonding", *Composites,* May 1994.

Chapter 7

Primers and Adhesion Promoters

7.1 Introduction

Some adhesives and sealants may provide only marginal adhesion to certain substrates. This could be due to the low surface energy of the substrate relative to the adhesive (e.g., epoxy bonding polyethylene) or to a boundary layer that is cohesively weak (e.g., powdery surface on concrete). The substrate may also be pervious, allowing moisture and environmental chemicals to easily pass through the substrate to the adhesive interface, thereby degrading the bond's permanence. Generally, attempts are made to overcome these problems through adhesive formulation and by substrate surface treatments. When these approaches do not work, additional bond strength and permanence may possibly be provided by primers or adhesion promoters.

Primers and adhesion promoters work in a similar fashion to improve adhesion. They add a new, usually organic, layer at the interface such as shown in Fig. 7.1. The new layer is usually bifunctional and bonds well to both the substrate and the adhesive or sealant. The new layer is very thin so that it provides improved interfacial bonding characteristics, yet it is not thick enough so that its bulk properties significantly affect the overall properties of the bond.

Both primers and adhesion promoters are strongly adsorbed onto the surface of the substrate. The adsorption may be so strong that instead of merely being physical adsorption, it has the nature of a chemical bond. Such adsorption is referred to as *chemisorption* to distinguish it from reversible physical adsorption.

The main difference between primers and adhesion promoters is that primers are liquids that are applied to the substrate as a relatively heavy surface coating prior to the application of the adhesive. Adhesion promoters are liquids that form a very thin (usually mono-

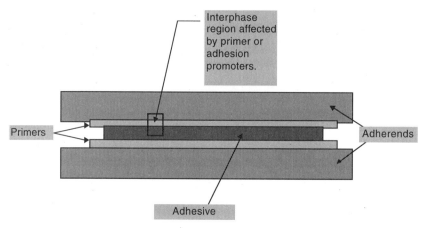

Figure 7.1 Primers and adhesion promoters provide a stronger interphase region having improved adhesion and permanence.

molecular) layer between the substrate and the adhesive. Usually chemical bonds are formed between the adhesion promoter and the adhesive, and between the adhesion promoter and the substrate surface. These bonds are stronger than the internal chemical bonds within the adhesive. These new bonds also provide an interface region that is more resistant to chemical attack from the environment. Adhesion promoters are also often referred to as coupling agents.

Adhesion promoters can be applied by either incorporating them directly into the adhesive formulation or by applying them to a substrate, similar to a primer. When applied "in-situ", through the adhesive formulation, the adhesion promoter migrates to the interface region and attaches itself between the adhesive molecule and the substrate before the adhesive cures.

Adhesion promoters or coupling agents are also used in applications other than conventional adhesives or sealants. They give plastic compounders a way of effectively improving properties and reducing the overall cost of the compound. Adhesion promoters can be applied to particulate fillers for reinforcing polymeric molding compounds and to fibers for reinforcing composites. Thus, when discussing adhesion promoters, the term "substrate" takes on the added possibilities of fillers, reinforcements, etc. as well as conventional adherends.

Specific adhesion promoters have been developed for bonding glass fibers to polyester resin, carbon fibers to epoxy resins, talc particles to nylon, and reinforcing tire cord to rubber tire compounds. In these applications, adhesion promoters not only improve the bond strength and permanence of the interface but also increase the physical properties of the resulting bulk material. Bulk properties, such as tensile strength and modulus can be significantly improved. Virtually all glass fibers used in glass fiber reinforced composites are silane treated, and the resistance to deterioration by moist environments is greatly enhanced in this way.

Classifying primers and adhesion promoters is as difficult as classifying adhesives. Usually, these materials are grouped by their chemical composition, such as silane, titanate, epoxy/amine, phenolic, and chlorinated polyethylene. The selection of a primer or adhesion promoter is normally less of a problem than selecting the adhesive since the adhesive supplier can usually offer a complete package or make recommendations. Often a primer is chemically similar to the adhesive material; only the primer has a much lower viscosity so that it can be painted or brushed on the substrate in thin coatings. Adhesion promoters are generally already included in the formulation or they are supplied as an additional "primer" material along with the adhesive or sealant.

In this chapter, we will first consider primers and how they can be used to improve the performance of bonded joints. Examples will be provided as to their application and use on various substrate surfaces. The discussion of adhesion promoters will be relative to their function and use. Various chemical classifications will be described that are commonly used for both surface application and "in-situ" application within formulated adhesive and sealant products. Several common uses of these systems in adhesive and sealant application and in polymer material reinforcement will be provided.

7.2 Primers

Primers are liquids that may be applied to a substrate prior to application of an adhesive or a sealant. The reasons for their use are varied and may include, either singularly or in combination, the following:

- Protection of surfaces after treatment (primers can be used to extend the time between preparing the adherend surface and bonding)
- Adjusting the free surface energy by providing a surface that is more easily wettable than the substrate
- Dissolving low levels of organic contamination that otherwise would remain at the interface as a weak boundary layer
- Promoting chemical reaction between adhesive and adherend
- Inhibiting corrosion of the substrate during service
- Serving as an intermediate layer to enhance the physical properties of the joint and improve bond strength (e.g., adjustment of the rhe-

ological properties at the interface or strengthening weak substrate regions)

Being lower in viscosity than the adhesive or sealant, primers can be used to penetrate porous or rough surfaces to provide better mechanical interlocking and for sealing such surfaces from the environment. Primers are often applied and appear as protective surface coatings. Application processes and equipment for applying primers are similar to those used in applying a paint coating to a substrate.

The application of a primer is an additional step in the bonding process, and it comes with associated costs and quality control requirements. Therefore, primers should only be used when justified. The most likely occasions when a primer is needed are: when the adhesive or sealant cannot be applied immediately after surface preparation, when the substrate surface is weak or porous, or when the adhesive-adherend interface requires additional protection from environments such as moisture.

7.2.1 Application and use

Unlike substrate surface treatments described in Chapter 6, primers always add a new organic layer to the surface and two new interfaces to the joint structure. Most primers are developed for specific adhesives, and many are developed for specific adhesive/substrate combinations.

Primers are applied quickly after surface preparation and result in a dry or slightly tacky film. It is generally recommended that they have a dried coating thickness range from tenths of a mil to approximately two mils. It is necessary to control the primer thickness, since if the primer layer becomes too thick its bulk properties may predominate, and the primer could become the weakest part of the joint.

Primers usually require solvent evaporation and several curing steps before the adhesive or sealant can be applied. Adhesive primers are usually not fully cured during their initial application. They are dried at room temperature and some are forced-air dried for 30-60 min at 150°F. This provides a dry, nontacky surface that can be protected from contamination and physical damage by good housekeeping practices until the substrate is ready to be bonded with an adhesive. Full primer cure is generally achieved during the cure of the adhesive.

Primers developed to protect treated surfaces prior to bonding are generally proprietary formulations manufactured by the adhesive producer to match the adhesive. These usually consist of a diluted solution (approximately 10% by weight) of the base adhesive in an organic solvent. Like the adhesive formulation, the primer may also contain wetting agents, flow control agents, and toughening compounds. If the primer is for a metal surface, corrosion inhibitors such as zinc and strontium chromate and other inorganic chromate salts may also be added to the primer formulation.

The application of corrosion resistant primers has become standard practice for the structural bonding of aluminum in the automotive and aerospace industries. The adhesive/primer combinations are chosen to provide maximum durability in severe environments in addition to providing higher initial joint strength. Improved service life is typically achieved by establishing strong and moisture resistant interfacial bonds and protecting the substrates' surface region from hydration and corrosion.

Primers can also be used to protect both treated metal and nonmetal substrates after surface treatment. The use of a primer as a shop protectant may increase production costs, but it may also provide enhanced and more consistent adhesive strength. The use of a primer also greatly increases production flexibility in bonding operations. Usually primer application can be incorporated as the final step in the surface preparation process. The primer is applied as soon as possible after surface preparation and usually no more than a few hours later. The actual application of the adhesive may then be delayed for up to several months.

With steel, for example, the maximum safe surface exposure time (SET) interval between mechanical surface preparation and bonding is 8 hours. Many other substrates have maximum SETs that are less than this. By utilizing an adhesive primer, the SET may be extended to days and even months depending on the particular adhesive/primer system used and the storage conditions prior to bonding. This process allows a shop to prepare the surface of a large number of parts, prime them, and store them for relatively long periods prior to bonding. It also enables an assembly shop to outsource the more hazardous surface preparation processes. The primer provides protection of the treated joints during transportation between the treating shop and the bonding shop. With primers, scheduling of the entire assembly operation is not dependent on the type of surface preparation.

As with metallic substrates, primers may be used to protect treated non-metallic substrates. After surface treatment, a high energy substrate has an active surface that will readily adsorb atmospheric contamination. The primer protects the treated surface until the time when the adhesive or sealant is applied. Primers are especially useful for the protection of polymeric parts that are treated by flame or corona discharge. Primers also find benefit on polymeric substrates in that their solvents will soften the surface, and some of the primer resin will diffuse into the bulk of the substrate, thereby increasing adhesive strength by molecular diffusion.

Primers also have production advantages with bonded assemblies having many sub-sections. The nature of the assembly may not be suitable for immersion in pretreating solution as a single structure. With primers, individual sub-sections can be treated, primed, and then fit into place before the bonding step without regard to time. This allows the entire assembly to be bonded at one time.

Sometimes primers can take the place of surface treatments. Two examples of this are with porous substrates and with certain plastic substrates. With weak porous substrates, such as wood, cement, or porous stone, the primer can be formulated to penetrate and bind weakly adhering material to provide a new, tightly anchored surface for the adhesive. Chlorinated polyolefin primers will increase the adhesion of coatings and adhesives to polypropylene and to thermoplastic olefins. The chlorine atoms on the outer surface of the primer increase surface energy and enhance adhesion. Examples of these are discussed in the next sections.

Low viscosity primers can also easily fill the irregularities on the substrate surface and displace air and fill hollows. This can improve the wetting of the adhesive "system". For example, if the adhesive is a hot melt applied to a bare, roughened metallic surface, the adhesive will gel before it gets a chance to efficiently wet the surface and mechanically interact with any surface roughness. However, if a dilute primer is first applied to the substrate and dried, the hot melt adhesive could bond directly to the primer which in turn has bonded to the interstices of the substrate, thus providing excellent adhesion.

There are several possible reasons for a primer to fail when a joint is placed in service. The most common causes of primer failure relate to production issues as shown in Table 7.1. Fortunately many of these

TABLE 7.1 Possible Reasons for Primer Failure

- Single layer application which is too thick
- Too long an open time (time between application and curing) allowing hydrolysis and contamination
- Frothy coating and porous layer caused by too fast a heat-up curing rate
- Low crosslink density allowing plasticizer or low molecular weight agents to migrate from the primer to the interface
- Too high a crosslink density causing a brittle primer that cannot take flexing or thermal changes
- Incompatibilty with the substrate surface or the adhesive
- Undercured or overcured primer
- Attack by solvents in the adhesive
- Attack by temperature of the adhesive curing process

problems can be detected soon after the primer is applied and before great expense is incurred in the bonding operation. Ideally the primer is a bridge to transfer stress between the adhesive and the adherend. With properly applied primers joint failures should be cohesive within the adhesive or sealant material.

Examples of some commercial primers for structural adhesives are shown in Table 7.2. It should be noted that these are formulated for specific adhesives and applications (e.g., aluminum bonding), and they have recommended curing processes and coating thickness that must be followed for optimum benefit. The substrate surface treatment, which provides the base for the primer, is generally one of the processes commonly used for adhesives or sealants as described in Chapter 6.

7.2.2 Primers for metal substrates

When a corrosive medium contacts the edge of a bonded joint, and finds an extremely active surface such as that produced by a fresh acid treatment to improve adhesion, corrosion at the metal-adhesive interface can occur. This initial corrosion and its subsequent penetration can take several forms. These are discussed more fully in Chapters 16 (bonding aluminum) and 17 (effects of the environment).

Some primers will inhibit the corrosion of metal adherends during service. By protecting the substrate's surface area from hydration and corrosion, these primers suppress the formation of weak boundary layers that could develop during exposure to wet environments. Primers that contain film forming resins are sometimes considered interfacial water barriers. They keep water out of the joint interface area and prevent corrosion of the metal surfaces. By establishing strong, moisture resistant bonds, the primer protects the adhesive-adherend interface and lengthens the service life of the bonded joint. However, moisture can diffuse through any polymeric primer, and eventually it will reach the interface area of the joint. Therefore, the onset of corrosion and other degradation reactions can only be *delayed* by the application of a primer unless the primer contains corrosion inhibitors or it chemically reacts with the substrate to provide a completely new surface layer that provides additional protection.

Representative data are shown in Fig. 7.2 for aluminum joints bonded with an epoxy film adhesive and a standard chromate containing primer. Until recently standard corrosion resistant primers contained high levels of solvent, contributing to high levels of volatile organic compounds (VOCs), and chromium compounds, which are considered to be carcinogens. As a result, development programs have been conducted on water borne adhesive primers that contain low

TABLE 7.2 Examples of Commercial Primers for Structural Adhesives[1]

Primer	Adhesive	Thickness	Application process	Properties
BR 127 epoxy phenolic (a)	FM 73 toughened epoxy	2.5–5.0 micrometers	Air drying 30 min; curing at 120C for 30 min	Corrosion inhibiting; protects metal oxide from hydrolysis
EA 2989 (b)	EA 9689 epoxy	5–10 micrometers	Air drying 60 min; curing at 175C for 60 min	Water based primer that is corrosion inhibiting
EC-2320 (c)	AF-111, AF-126 and AF-125-2 film adhesives	1.3–5.1 micrometers	Air drying up to 120C	Improves shear and peel strengths as well as environmental resistance
Nitrilotrismethylene phosphonic acid (NTMP)	Epoxy adhesives	Single molecular layer	Application in aqueous solutions of 1-1000 ppm	Improve durability of aluminium joints in wet conditions
Chlorinated polypropylene	Paint, adhesives, sealants	Thin, uniform coating	Application in organic solvent	Chlorine atoms in the primer increase polarity and enhance paint adhesion to molded polypropylene

Manufacturers:
(a) American Cyanamid
(b) Hysol Division, Dexter Corp.
(c) 3M Company

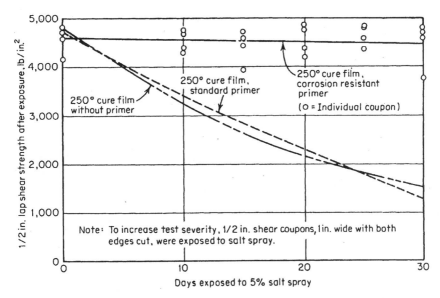

Figure 7.2 Effect of primer on lap-shear strength of aluminum joints exposed to 5 percent salt spray.[2]

VOC levels and low or no levels chrome. Data are presented on several of these newer primers in Table 7.3.

Recent advances in primer development have included water-based primer systems, primers that can be applied by the electrodeposition process, and primers with more effective corrosion resistance properties. The water-based systems have been the result of recent developments in water soluble polymers such as epoxy and phenol/resorcinol formaldehyde novolac polymers. Low and zero level VOC primers have been developed which meet the requirements of the aerospace industry.[3–5] An electrodeposited primer system has been developed primarily by the automobile industry. This is pollution-free and offers uniform film distribution, controlled film thickness, and rapid application.[6] The U.S. military has developed improved primers for structural bonding of aluminum and titanium.[7] These include environmentally safe primers.

Plasma applied coatings have also been evaluated as surface treatments for aluminum substrates being prepared for adhesive bonding. Plasma sprayed aluminum-silicon/polyester primers gave results superior to those of etched or anodized specimens.[8] Plasma spraying has also shown excellent high temperature bond performance with titanium.[9] The plasma spray process involves the rapid heating of powdered material to the molten or semi-molten state and then propelling it against the substrate at high velocities. These treatments eliminate

TABLE 7.3 Tensile Lap Shear Strengths Using Environmentally Acceptable Adhesive Primers for 250°F Curing Adhesives[3]

Primer (source)	Adhesive	Tensile lap shear strength, psi				
		73F	180F (1)	180F wet (2)	30 day salt fog (3)	60 day salt fog (3)
BR 127 (Cyanamid)	FM 73	6005	4312	2622	6105	5846
BR 250 (Cyanamid)	FM 73	5685	3498	2162	5479	—
BR 250-2 (Cyanamid)	FM 73	7980	4145	2614	5853	5722
BR X250-NC (Cyanamid)	FM 73	6163	4186	2722	6062	5927
EC 3982 (3M)	FM 73	6182	4395	2891	6427	6051
XEA 9290 (Hysol)	FM 73	5255	—	—	4791	—
BR 127 (Cyanamid)	AF163-2K	6421	4617	3499	6505	6400
BR 250 (Cyanamid	AF163-2K	6420	4636	3234	6393	—
BR250-2 (Cyanamid)	AF163-2K	6564	4358	3304	6465	6366
BRX250-NC (Cyanamid)	AF163-2K	6589	3528	3299	6634	6397
EC 3982 (3M)	AF163-2K	6733	4271	3633	6711	6573
XEA 9290 (Hysol)	AF163-2K	6217	—	—	6079	—

NOTES: (1)–Heat soaked at 180F for 10 min
(2)–Wet specimens conditioned at 140F and 95-100% RH for 60 days. Heat soaked at 180F for 4 min
(3) 50% salt fog at 95F

liquid and gaseous wastes and provide bond strength and durability comparable to that provided by conventional chemical treatments.

In addition to providing corrosion protection, primers may also be used as mediating layers between the adherend and adhesive under conditions where the adherend and adhesive are incompatible. The adhesive or sealant may not be able to chemically react with the substrate surface; however, the primer may be able to react with both the adhesive and the adherend. The primer may also be able to modify the physical characteristics of the joint. For example, elastomeric primers may be used with rigid adhesives to provide greater peel or impact resistance. A flexible interface may also provide for less internal stress due to thermal expansion differences.

Epoxy based primers are commonly used in the aerospace and automotive industries. These primers have good chemical resistance and provide corrosion resistance to aluminum and other common metals. Polysulfide based primers have been developed for applications where a high degree of elongation is necessary. These systems are used where the joint is expected to encounter a high degree of flexing or thermal movement. Resins, curing agents, and additives used in primer formulation are much like adhesive or sealant formulations except for the addition of solvents or low viscosity resins to provide a high degree of flow.

7.2.3 Primers for polymeric substrates

Primers can also be an effective way of protecting the surface of treated polymeric substrates until an adhesive can be applied and a bond formed. With polymeric substrates, the permanence of the surface treatment before adhesive application is of critical importance. Because of the dynamic and mobile nature of the polymer molecule, the treated surface molecules can turn inward into the bulk polymer and become ineffective. Conventional plastic surface treatments, such as flame treating and corona treatment, have especially short exposure times during which the treatments are effective. A primer can be used to protect the treated surface and lengthen the production window between surface treatment and application of the adhesive or sealant.

Primers generally cannot be used as a substitute for surface preparation. However, there are several instances with polymeric surfaces that primers have provided excellent adhesion without having to go through the process of surface preparation. This is a distinct advantage because surface treatment methods may be hazardous, inconvenient, time consuming, and often expensive. The use of a surface primer, although an extra step in the bonding process, is a desirable alternative for use on the production line.

Cyanoacrylate adhesives (or super-glues) do not wet or adhere well to polyolefins. The surface tension of the adhesive is much higher than that of the substrate. However, polyolefins can be primed for adhesion to cyanoacrylates by certain chemical compounds normally considered to be activators for cyanoacrylate polymerization. Materials such as long chain amines, quaternary ammonium salts, and phosphine can be applied in either pure form or in solution to the surface of the polyolefin. These primers are simply sprayed or brushed onto the substrate. After drying of the primer, the cyanoacrylate adhesive is conventionally applied and bonds extremely well to the substrate.[10,11] Several companies have discovered primers that interact with cyanoacrylates.[10,12,13] Triphenylphosphine or cobalt acetylacetonate primers used with cyanoacrylate adhesives produce adhesive bonds with polypropylene and low density polyethylene that are sufficiently strong to exceed the bulk shear strength of the substrate. They are also sufficiently durable to withstand immersion in boiling water for long periods of time.[14]

It appears that one of the main reasons for improved adhesion by primers with cyanoacrylate adhesives is that the solvents in the primer wet-out and swell the polyolefin. This facilitates interpenetration of the low viscosity cyanoacrylate resin.

A similar effect seems to work with free radical cured acrylic adhesive systems. In this system, the primer consists of a solution of a copper salt, and the adhesive is based on methyl methacrylate monomer. The bonds formed on low density polyethylene result in substrate failure, and an interphase of mixed adhesive and polyethylene is formed up to 1.5 mm thick.[15]

Chlorinated polyolefins are used for priming low energy surfaces such as polyolefins. These primers are based on either chlorinated polyethylene or polypropylene and are usually used as a solvent based solution. The application of chlorinated polyolefin as a primer on the surface of polypropylene provides a significant improvement in the adhesion of a latex topcoat to the polypropylene.[16,17] A primer based on a solution of chlorinated polypropylene has been used to adhere paint to polypropylene automobile bumpers with some interdiffusion of primer into the part. Chlorine atoms in the outer surface of the primer are believed to increase polarity and enhance paint adhesion in this case.[1] Solvent-free, water borne chlorinated polyolefin primers based on emulsions and dispersions of chlorinated polyolefin have also been developed.[18,19] They provide increase in bond strength and water resistance for polypropylene and other thermoplastic polyolefin joints.

7.2.4 Primers for unvulcanized elastomers

Adhesion primers have also found significant application in treating the surface of inserts that are to be embedded in elastomeric parts. There are many instances when unvulcanized rubber is required to bond to itself or to another substrate such as a metal insert. Primers are generally used in these applications to provide a strong bond between the elastomer and the adherend. An example is the manufacture of roller wheels. Here the metal insert (bearing assembly) needs to be well bonded to the surrounding elastomer that serves as the elastomeric wheel.

The primer is coated on the part and then air or heat dried. In the case of the roller wheel, the metal hub must be surface treated, primed and dried, and then carefully placed in the mold. The molten elastomeric resin is forced into the mold, and it contacts and wets the surface of the hub. As the elastomer cures in the mold, a strong adhesive bond forms between the metal and elastomer by virtue of the primer.

Primers are applied as intermediate layers between the insert adherend and elastomeric resin under conditions where the adherend and resin are incompatible or where a chemical reaction is required for strong adhesion. The primer is able, by its chemical formulation, to react with both adherend surface and the elastomer. A very strong bond is formed between the curing elastomer compound and the primer system during cure of the rubber. In these cases, the primer may be considered an adhesive. Although, more accurately, it is a surface primer, and the unvulcanized rubber provides the adhesion. The primer provides a protected surface that is easy to wet; but more importantly, it provides a reactive surface that is chemically compatible with the elastomer.

Often solutions of unvulcanized rubber in polymeric resin are used for the primer. For example, a typical primer formulation may include chlorinated rubber, phenolic resin, and solvent. However, several water borne elastomer bonding agents have been developed and are commercially available. The resin component provides cohesive strength and enhances adhesion to the metal. The rubber component provides toughness to the system and assists in reducing residual bond-line stress. Proprietary primers for elastomer-to-insert bonding include Chemlok, Ty Ply, Thixon, and Cilbond brand names. Table 7.4 gives details of several proprietary products.

The most common elastomers to be bonded in this way include nitrile, neoprene, urethane, natural rubber, SBR, and butyl rubber. Less

TABLE 7.4 Proprietary Primers for Bonding Unvulcanized Elastomers to Metal Inserts[20]

Product	Manufacturer	Suitable to bond	Type product
Chemlok 205	Hughson Chemicals	NBR	Metal primer
Ty Ply UP		CR	
Ty Ply T		NBR, ACM, OT	
Thixon P5	Dayton Coatings and Chemicals	NBR, CR, CO	Metal primer
Thixon P6		NBR	
Thixon P7		NBR	
Thixon P9		NBR, CR	
Cilbond 10	Compounding Ingredients Limited (UK)	NBR	Metal primer
Cilbond 17		NBR ACM	

NBR-Nitrile rubber
CR-Neoprene
ACM-Polyacrylate rubber
OT-Polysulfide rubber
CO-Epichlorohydrin rubber

common unvulcanized elastomers such as the silicone, fluorocarbon, chlorosulfonated polyethylene, and polyacrylate are more difficult to bond. However, recently developed adhesive primers also improve the bond of these elastomers to metal. Proprietary primers are available from the manufacturers listed in Table 7.4. Suggestions for specific primers to be used for particular elastomers and for certain applications can usually be obtained from the adhesive/primer manufacturer.

Surface treatment of the adherend before priming should, of course, be according to good standards. Care must be taken so that the primer/adhesive system is not wiped off the substrate during the flow of the unvulcanized elastomer during molding. Equal care must be exerted to not allow the elastomeric fluid to wash away mold release from protected surfaces onto critical surfaces where adhesion will be necessary.

Plastic inserts can similarly be bonded to unvulcanized rubber with selected primers. The plastic part will need to be treated by one of the processes discussed in the last chapter. The plastic must also be able to withstand the rubber vulcanization temperatures. Primers are also used for elastomer to fiber bonding. Applications include reinforced hose, belts, and tires. Resorcinol formaldehyde resins are commonly used as primers in these applications.

7.3 Adhesion Promoters

Adhesion promoters or coupling agents are a group of specialty bifunctional compounds that can react chemically with both the substrate and the adhesive. The adhesion promoter forms covalent bonds across the interface that are both strong and durable. Adhesion promoters can be applied directly to the substrate, similar to primers, or they can be mixed with the adhesive itself. When mixed with the adhesive, the coupling agent is capable of migrating to the interface and reacting with the substrate surface as the adhesive cures. When applied directly to the substrate, adhesion promoters are applied in a very thin coating that ideally is only one molecular layer thick.

Adhesion promoters usually consist of molecules with short organic chains having different chemical composition on either end of the chain. On one end is an organofunctional group that is particularly compatible with the given adhesive material. At the other end of the chain is an inorganic functionality that is especially compatible with a given substrate. The adhesion promoter, therefore, acts as a chemical bridge between the adhesive and the substrate.

Adhesion promoters were first used to treat glass fibers and other fillers before they are incorporated into liquid resin to make composite materials. In the fiber industry, adhesion promoters are also known as finishes. Certain finishes have been specially developed to match a particular fiber with a resin matrix. Without adhesion promoters, the interfacial resin-to-glass fiber adhesion is relatively weak, and water can diffuse along the interface with catastrophic results to the end-properties of the composite.

7.3.1 Silane adhesion promoters

Silanes are the most common commercial adhesion promoter. They are commonly used to enhance adhesion between polymeric and inorganic materials.[21,22] They usually have the form $X_3Si\text{-}R$, where X is typically a chlorine or alkoxy group and R is the organofunctionality. The organofunctional portion bonds with the resin in the adhesive or the organic medium, and the silane portion bonds to the inorganic or substrate surface. Silane coupling agents are commonly used between the adhesive and the adherend, between resin matrix and reinforcing fibers in composites, and between resin matrix and mineral fillers in plastic compounds. The resulting interface provides:

- A chemical bridge between the surface and organic polymer or between organic polymers

- A barrier to prevent moisture penetration to the interface

- Transfer of stress from the resin to the substrate or inorganic filler component thereby improving joint strength or bulk properties

- Effective dispersion of fillers and reduction in the apparent viscosity of the system

These chemicals are usually applied to fibrous reinforcements or to the substrate surface as an aqueous solution. The solutions usually are very dilute, only 0.01 to 2% by weight of silane to keep the highly reactive hydrolyzed molecules from reacting with one another. The bond strength enhancement increases with silane concentration up to a maximum of about 2%, and then the enhancement falls-off with additional concentration. Silane coupling agents react with water in aqueous solutions to form hydrolyzed silanes, which react with the surface of the inorganic substrate. The bound silanes then polymerize, building up layers outward from the substrate with the organic functionality oriented toward the adhesive. This process is shown in Fig. 7.3.

Silanes form strongly adsorbed polysiloxane films on ceramic and metal surfaces. The chemical and mechanical integrity of these films are highly dependent on application parameters such as solution concentration, solution pH, and drying time and temperature. The char-

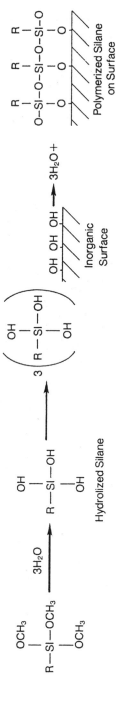

Figure 7.3 Organosilane coupling agents react with water in aqueous solutions to form hydrolyzed silanes, which react with the surface of the inorganic reinforcements or adherend. The bound silanes polymerizes, building-up layers outward from the adherend with the organic functionality oriented into the matrix.[23]

acter of the substrate may also influence the polysiloxane film structure.[24] Silanes are applied as primers to the surface of the substrate by wiping, spraying, brushing, or dipping. The film thickness is generally less than 0.1 mil. The solvent in the silane system is removed by drying at 122–140°F for 10 min. The advantage of the primer application method is that it efficiently utilizes the silane material, and there are minimal stability problems. The disadvantages are that it is a two step process, and it is difficult to see the clear silane coating unless it is pigmented. Therefore, complete surface coverage may be in doubt.

Tests have shown that silanes arrange themselves in layers with a high degree of order, influenced to a great extent by the surface of the substrate. The molecules order themselves virtually perpendicular to the surface to which they attach, and subsequent layers arrange themselves step-like in a head-to-head fashion. A rough surface can break-up the first ordered layer, preventing the formation of the second. The thickness of the silane interphase has important effects on mechanical performance. Thin, but continuous, layers seem to provide stronger and more durable adhesive bonds.

Three primary mechanisms have been suggested for enhanced adhesion via silane coupling agents.[25] The classical explanation is that the functional group on the silane molecule reacts with the adhesive resin. Another possibility is that the polysiloxane surface layer has an open porous structure and the liquid adhesive penetrates this and then hardens to form an interpenetrating interphase region. The third mechanism applies only to polymeric adherends. It is possible that the solvent used to dilute and apply the silane adhesion promoter opens the molecular structure on the substrate surface, allowing the silane to penetrate and diffuse into the adherend.

The coupling agent interphase may be hard or soft and could affect mechanical properties of the interphase region. A soft interphase, for example, can significantly improve fatigue and other properties. Soft interlayers reduce stress concentrations in the adhesive or in the resin matrix if the system is a composite. In composites, a rigid interlayer improves stress transfer of resin to the fiber and improves interfacial shear strength. Coupling agents generally increase adhesion between the resin matrix and substrate, thus raising the fracture energy required to initiate a crack. However, the same bond prevents the resin matrix from stretching. Because of this reduction in tearing capacity, the interface is less able to resist crack propagation once a crack is initiated.

There are a number of silane adhesion promoters available, and they differ from each other in the degree of their reactivity. Silanes may be produced with amine, epoxy, mercaptan, and other functionalities. Some examples are given in Table 7.5.

TABLE 7.5 Recommended Silane Coupling Agents for Various Resins. (Union Carbide, Dow Corning Corp. and General Electric Co. Supply Silane Coupling Agents)

Silane functionality	Applications
Vinyl	Free radical cure systems: crosslinked polyethylene, peroxide cured elastomers, polyesters. Polyethylene. Polypropylene
Epoxy	Epoxy, acrylics, urethanes, polysulfide
Methacryl	Unsaturated polyester, acrylic
Amino	Epoxy, phenolic, melamine, urethane, butyl rubber
Mercapto	Epoxy, sulfur cure rubbers, urethane, polysulfide
Ureido	Phenolic, urethane

Virtually all glass fibers used in fiber glass composites are treated with silane "finishes". The resulting physical properties and the resistance of the composite to deterioration by water immersion are greatly enhanced by the addition of the coupling agent. Table 7.6 shows the effect of silane treatment to reinforcing glass fibers on the dry and wet flexural strength of several composite materials. The moisture resistance properties of filled molding compounds are also enhanced by the treatment of the fillers with silane adhesion promoters prior to compounding. Silane promoters on wollastonite fillers in thermoplastic polyester molding compounds (50% filled) will improve the flexural strength after 16 hrs in 122°F water by as much as 40%.

Silanes are also generally effective in improving adhesion to metals, including aluminum, steel, cadmium, copper, and zinc. Table 7.7 shows the relative influence of the type of substrate on the effectiveness of the silane coupling agent in improving adhesion. It should be noted

TABLE 7.6 Effect of Various Silanes on Glass Reinforced Thermoset Resins[26]

Resin system	Silane	Flexural strength	
		Dry	Wet
Polyester	control	60,000	35,000
	A-174	87,000	79,000
Epoxy	control	78,000	29,000
	A-186	101,000	66,000
Malamine	control	42,000	17,000
	A-187	91,000	86,000
		Dry	High temperature[a]
Phenolic	control	69,000	14,000
	A-1100	85,000	50,000

[a] Aged 100 hr at 500°F and tested at 500°F.

TABLE 7.7 Effect of Substrate on Silane Adhesion Effectiveness[27]

Silane Effectiveness		Substrate
↑	Excellent	Silica
		Quartz
		Glass
		Aluminum
		Copper
		Alumina
		Inorganics
	Good	Alumino-Silicates (Clays)
		Mica
		Talc
		Inorganic Oxides
		Steel, Iron
		Asbestos
	Slight	Nickel
		Zinc
		Lead
		Chalk (Calcium Carbonate)
		Gypsum (Calcium Sulfate)
		Barytes (Barium Sulfate)
	None	Graphite
		Carbon Black

that smooth, high energy substrates are excellent substrates for silane attachment. Rough, discontinuous substrates show very little benefit.

Silane adhesion promoters increase the initial bond strength and also stabilize the surface to increase the permanence of the joint in moist aging environments.[21,24] The effect of a silane adhesion promoter on the durability of mild steel joint bonded with an epoxy adhesive is shown in Fig. 7.4. Silane based coupling agents are also capable of increasing the environmental resistance of aluminum,[29] titanium,[30] and stainless steel[31] joints.

Although the best results can be obtained in using silanes as substrate primers, they can also be added to the adhesive with some effect. The lap-shear values in Table 7.8 show the improvement in bond strength when silane coupling agents are incorporated into the adhesive formulation, specifically nitrile phenolics. The integral blend method of applying the silane involves adding 0.1 to 2.0% by weight of the silane to the polymer matrix prior to application. The advantage of this method is that it does not require a separate substrate coating

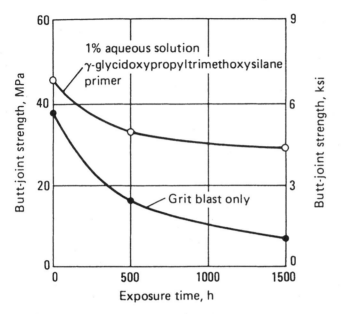

Figure 7.4 Effect of silane adhesion promoter on the durability of a mild steel joint bonded with epoxy adhesives.[32]

TABLE 7.8 Lap-shear Strengths of Several Structural Adhesives Incorporating Silane Adhesion Promoters[28]

Adhesive type	Adhesion promoter	Percentage of promoter	Lap shear, psi
Nitrile phenolic	None	—	1450
Nitrile phenolic	Gamma-aminopropyltriethoxysilane	1	2350
Nitrile phenolic	Gamma-mercaptopropyltrimethoxysilane	1	3120
Epoxy	None	—	1410
Epoxy	Gamma-aminopropyltriethoxysilane	1	1675
Epoxy	Gamma-mercaptopropyltrimethoxysilane	1	1580

2024T3 aluminum, 0.063 inch thickness, 1/2 inch overlap
Tests run at 73F; appropriate cure schedule used

step. The main disadvantage is that there could be stability problems in storage due to hydrolysis and polymerization. Figure 7.5 shows the improvement in T-peel strength to aluminum caused by the introduction of silanes into a urethane adhesive. Silanes are also commonly used in caulks and sealants to improve adhesion. The addition of 0.25% by weight silane (A-187) is reported to improve the wet and dry peel strength of an acrylic caulk to both glass and wood substrates.[27] Water-stable epoxy silane promoters have recently been developed to

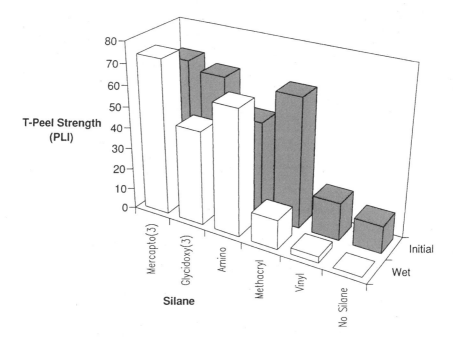

Isocyanate Terminated Urethane, Adhesion to Aluminum

1. urethane is Adiprene L-100 (DuPont)
2. substrates degreased and etched
3. most commonly recommended for pot life

Figure 7.5 Effect of silane addition to an isocyanate terminated urethane adhesive.[27]

enhance the wet strength of water borne adhesives and sealants.[33] Conventional silanes are limited to high solids and solvent adhesive and sealant applications in which moisture is not encountered until use.

7.3.2 Titanates, zirconates, and others

While silanes predominate as coupling agents, the use of titanate, zirconate, and other agents are growing. These coupling agents also allow increased impact strength and other physical properties in composites and highly filled plastic compounds. Like the silanes, they react with surface hydroxyl groups, but there is no condensation polymerization to produce a polymer network at the interface. Adhesion promoters based on titanium and zirconium have been used to improve filler-polymer adhesion and to pretreat aluminum alloy for adhesive bonding.[37]

Titanate adhesion promoters can provide a dual function of improving processability and bonding. Titanates have been used predominantly to modify the viscosity of filled systems.[17] It has been shown that a small percentage of titanate in a heavily filled resin system can reduce the viscosity significantly. Thus, titanate adhesion promoters allow higher fillings of particulate matter to either improve properties or lower the cost of the system without having a negative effect on the viscosity of the system. Improved bond strength to halocarbon substrates and improved hydrolytic stability are also claimed.[34]

The titanate structure may be tailored to provide desired properties through the six functionalities on the basic $(RO)_m\text{-}Ti\text{-}(O\text{-}X\text{-}R_2\text{-}Y)_n$ structure. The $(RO)_m$ is the hydrolyzable portion that attaches to the inorganic substrate (Fig. 7.6). It also controls dispersion, adhesion, viscosity, and hydrophobicity. The X group enhances corrosion protection and acid resistance and may provide antioxidant effects, depending on chemistry. The R_2 provides entanglements with long hydrocarbon chains and bonding via van der Waals forces. The Y group provides thermoset reactivity, chemically bonding the filler to polymer.

Examples of titanate coupling agents are shown in Table 7.9. New neoalkoxy technology allows liquid, powder and pellet titanate coupling agents to be processed like color concentrates without pretreat-

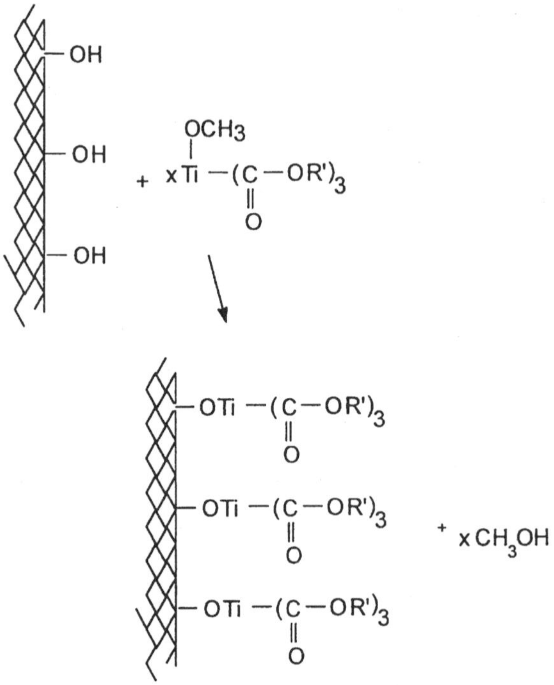

Figure 7.6 Titanate condensation on a hydroxyl containing surface.[36]

TABLE 7.9 Properties of Titanate/Zirconate Coupling Agents[23]

Titanate or Zirconate Type	Applications/ Advantages	Chemical Structure
Monoalkoxy Titanate	Stearic Acid Functionality; Aids Dispersion of Mineral Fillers in Polyolefins	$CH_3-CH(CH_3)-O-Ti(O-P(=O)(OH)-O-P(=O)(OC_8H_{17})_2)_3$
Chelate Titanate	Greater Stability in Wet Environments	$\begin{array}{c} O \\ \parallel \\ C-O-Ti(O-P(=O)(OH)-O-P(=O)(OC_8H_{17})_2)_3 \\ \mid \\ CH_2-O \end{array}$
Quat Titanate	Water Soluble	$\begin{array}{c} O \\ \parallel \\ C-O-Ti(O-P(=O)(OH)-O-P(=O)(OC_8H_{17})_2)_3 \\ \mid \\ CH_2-OH \ O^{\ominus} \ \overset{R}{\underset{R-N-R}{\oplus}} \end{array}$
Coordinate Zirconate	Phosphite Functionality; Reduces Epoxy Viscosity w/o Accelerating Cure	$(RO)_4Ti \cdot (HP(=O)(OC_{13}H_{27})_2)_2$

TABLE 7.9 Properties of Titanate/Zirconate Coupling Agents[23] (Continued)

Type	Properties	Structure
Neoalkoxy Titanate	Eliminates Pretreatment Associated with High Temperature Thermoplastics and Polyurethanes	R*—O—Ti(O—P(=O)(OH)—O—P(=O)(OC$_8$H$_{17}$)$_2$)$_3$
Neoalkoxy Zirconate	Accelerates Peroxide- and Air-based Cures; e.g., Polyester SMC/BMC	R*—O—Zr(O—P(=O)(OH)—O—P(=O)(OC$_8$H$_{17}$)$_2$)$_3$
Cycloheteroatom Titanate	Ultra-high Thermal Properties for Specialty Applications	R–O–Ti(–O–)(–O–)–R (cyclic)

ment steps.[35] Typical dosage is 0.2% by weight of the polymer. They can also be used to treat aluminum alloys for adhesive bonding.

Zirconate coupling agents have a structure very similar to titanates. Zirconium propionate is used as an adhesion promoter in printing ink formulations for polyolefins that have been treated by corona-discharge. The coupling agent is believed to form hydrogen bonds with the nitrocellulose in the ink. Surface COOH groups seem to be the most likely attachment site to the polyolefin surface. Like the titanates, zirconate coupling agents are useful in improving the dispersion characteristics of fillers in polymer systems. Other examples of zirconate coupling agents and their applications are shown in Table 7.9.

Chrome complexes have been formed as adhesion promoters by the reaction of chromium chloride with methacrylic acid. The chromium oxide portion of the adhesion promoter reacts with a substrate while the methacrylic portion reacts with a free radical curing outer layer. Chrome based adhesion promoters are commonly used as a primer for aluminum foil to increase the strength and durability of aluminum/polyethylene interfaces.[25] The polyethylene is typically extrusion coated as a hot melt onto the treated aluminum for applications as diverse as packaging and communications cable shielding. Other types of coupling agents include 1,2-diketones for steel,[38] nitrogen heterocyclic compounds such as benzotriazole for copper,[39,40] and some cobalt compounds for the adhesion of brass plated tire cords to rubber.[41]

References

1. Comyn, J., *Adhesion Science,* (Cambridge: Royal Society of Chemistry, 1997) at 27–28.
2. Krieger, R. B., "Advances in Corrosion Resistance of Bonded Structures", in Proc. Nat. SAMPE Tech. Conf., vol. 2, *Aerospace Adhesives and Elastomers,* 1970.
3. Kuhbander, R. J. and Mazza, J. J., "Evaluation of Low VOC Chromated and Non-chromated Primers for Adhesive Bonding", 38th Annual SAMPE Symposium, May 1993 at 785–795.
4. Sweet, D. E., "Metlbond X6747 100% Water Based Primer for Aerospace Bonding Applications", 38th International SAMPE Symposium, May 1993 at 1241.
5. Lindner-Meyler, K. and Brescia, J. A., "Effectiveness of Water Borne Primers for Structural Adhesive Bonding of Aluminum and Aluminum Lithium Surfaces", *J. of Adhesion Science and Technology,* vol. 9, no. 1, 1995 at 81–95.
6. Foister, R. T., Gray, R. K., et al., "Structural Adhesive Bonds to Primers Electrodeposited on Steel", *J. Adhesion,* vol. 24, 1989 at 17.
7. Reinhart, T. J., "Novel Concepts for priming Metallic Adherends for Structural Adhesive Bonding", in *Adhesion 2,* K. W. Allen, ed. (London: Elsevier Science Publishers, 1978).
8. Davis, G. D., Whisnant, P. L., et al., "Plasma Sprayed Coatings as Surface Treatments of Aluminum Adherends", 41st International SAMPE Symposium, March 1996 at 291–301.
9. Clearfield, H. M., et al., "Surface Preparation of Ti-6Al-4V for High Temperature Adhesive Bonding", *J. Adhesion,* 29, 81 (1989).
10. Okamoto, Y., and Klemarczyk, P. T., "Primers for Bonding Polyolefin" *J. Adhesion,* 40 (1993) at 81.
11. Yang, J. and Garton, A. J., *J. of Applied Polymer Science,* 48 (1993) at 359.

12. Nicolaisen, H., Rehlling, A., et al., WO 9118956.
13. Nicolaisen, H., and Rehling, A., DE 4017802.
14. Yang, J. and Garton, A., "Primers for Adhesive Bonding to Polyolefins", *J. of Applied Polymer Science,* vol. 48 (1993) at 359–370.
15. Fields, T. J., et al., *J. Adhesion Science and Technology,* vol. 9, no. 5 at 627.
16. Lawniczak, J., et al., "Chlorinated Polyolefins as Adhesion Promoters for Plastics", Proceedings of the ACS Division of Polymeric Materials Science and Engineering, vol. 68, 1993 at 28–29.
17. Waddington, S. and Briggs, D., Plym. Comm. 32, 1991 at 506.
18. Dechent, W. L., and Stoffer, J. O., "Waterbone Chlorinated Polyolefin Adhesion Promoters", Polymeric Materials Science and Engineering, Proceedings of the ACS, Division of Polymeric Materials Science and Engineering, vol. 69, 1993 at 380–381.
19. Lawniczak, F. J., Greene, P. J., et al., "Water Reducible Adhesion Promoters for Coatings on Polypropylene Based Substrates", *J. of Coatings Technology,* Dec. 1993 at 21–26.
20. Symes, T., and Oldfield, D., "Technology of Bonding Elastomers", *Treatise on Adhesion and Adhesives,* vol. 7, J. D. Minford, ed. (New York: Marcel Dekker, 1991).
21. Plueddemann, E. P., *Silane Coupling Agents,* (New York: Plenum Press, 1991).
22. Mittal, K. L., *Silane and other Coupling Agents,* (Utrecht, VSP, 1992).
23. English, L. K., "Fabricating the Future with Composite Materials, Part IV: The Interface", *Materials Engineering,* March 1987.
24. Bascom, W. D., "Primers and Coupling Agents", *Adhesives and Sealants,* vol. 3, ASM Engineered Materials Handbook, ASM International, 1990.
25. Kinloch, A. J., *Adhesion and Adhesives,* (New York: Chapman and Hall, 1987), pp. 156–157.
26. Marsden, J. G., and Sterman, S., "Organofunctional Silane Coupling Agents", in *Handbook of Adhesives,* I. Skiest, ed., 3rd ed. (New York: Van Nostrand Reinhold, 1990).
27. Waldman, B. A., "Organofunctional Silanes as Adhesion Promoters and Crosslinkers", Kent State University, *Adhesion Principles and Practice Course,* May 1997.
28. Kothari, V. M., "Specialty Adhesives Can Be Compounded for Application in Many Market Niches", *Elastomerics,* October 1989.
29. Patrick, R. L., et al., *Appl. Polymer Symp.* 16:87, 1981.
30. Boerio, F. J. and Dillingham, R. G., in *Adhesive Joints,* K. Mittal, ed. (New York: Plenum Press, 1984) at 541.
31. Chovelon, J. M., et al., "Silanization of Stainless Steel Surfaces: Influence of Application Parameters", *J. Adhesion,* vol. 50, 1995 at 43–58.
32. Kinloch, A. J., "Predicting and Increasing the Durability of Structural Adhesive Joints", in *Adhesion 3,* K. W. Allen, ed. (London: Elsevier Science Publishers, 1978) at 1.
33. Huang, M. W., and Waldman, B. A., "Water Stable Epoxysilanes Enhance Wet Strength of Sealants and Adhesives", *Adhesives and Sealants Industry,* Oct./Nov. 1998.
34. Monte, S. J., "Titanates", in *Modern Plastics Encyclopedia,* (New York: McGraw Hill, 1989) at 177.
35. Monte, S. J., and Sugerman, G., "New Neoalkoxy Titanate Coupling Agents Designed to Eliminate Particulate Pretreatment", 125th Meeting of the Rubber Division, Indianapolis, IN, May 8–11, 1984.
36. Pocius, A. V., *Adhesion and Adhesives Technology,* (New York: Hanser Publisher, 1997) at 139.
37. Calvert, P. D., et al., in *Adhesion Aspects of Polymeric Coatings,* K. L. Mittal, ed. (New York: Plenum Press, 1983) at 457.
38. Nicola, A. J., and Bell, J. P., in *Adhesion Aspects of Polymeric Coatings,* K. L. Mittal, ed. (New York: Plenum Press, 1983) at 443.
39. Yoshida, S., and Ishida, H., "A FT-IR Reflection Absorption Spectroscopic Study of Epoxy Coatings on Imidazols Treated Copper", *J. of Adhesion,* 16, 217, 1984.
40. Cotton, J. B., et al., U.S. Patent 3,837,964, 1974.
41. van Ooij, W. J., Biemond, M. E. F., *Rubber Chem. Tech.*, 57, 688, 1984.

Chapter 8

Adhesive Classifications

8.1 Introduction

There are many types and variations of commercial adhesive materials to choose from for any specific application. There are also a seemingly unlimited number of adhesive composition possibilities available to the formulator for the "engineering" of a custom product. These numerous alternatives can make the selection of the proper adhesive system a difficult process.

This chapter is the first of four chapters that will guide the reader to the proper choice of adhesive. In this chapter, the methodology that is commonly used to characterize adhesives will be identified. These classifications are commonly used throughout the industry. Chapter 9 will describe how adhesive systems are formulated to meet specific functional characteristics. Chapter 10 will describe the common chemical families that are used as adhesives and how the processing and performance of the ultimate adhesive are related to the base resin. Chapter 11 will then highlight the processes involved in selecting the correct adhesive for a given application. A similar analysis for sealant materials is presented in subsequent chapters.

8.2 Basic Classification

Adhesives are classified by many methods, and there can be many hierarchical levels to these classification schemes. Perhaps, the broadest classification scheme is to categorize an adhesive as being manufactured from materials that are either synthetic or naturally occurring. Synthetic adhesives are manufactured from man-made materials such as polymers. Natural adhesives are manufactured from naturally occurring materials such as animal or agricultural by-products.

Many adhesives are made from organic polymers. There are also adhesive systems with inorganic origin. The oldest polymers used for adhesives were of natural origin. Often naturally occurring adhesives are thought to be inferior to synthetic polymers because of their lower strength and limited freedom in processing. However, in many applications, such as bonding of paper and wood where the emphasis may be on the adhesive being biodegradable or repulpable, naturally occurring adhesives find a strong market. Modern epoxies, urethanes, acrylics, and other adhesive systems that are used in demanding structural applications are made from synthetic polymers.

The classification of adhesives into synthetic and naturally occurring categories is usually far too broad for many practical purposes. The industry has settled on several common methods of classifying adhesives that satisfy most purposes. These classifications are by:

- Function
- Chemical composition
- Mode of application or reaction
- Physical form
- Cost
- End-use

All of these classifications and distinctions overlap to some degree.

8.2.1 Function

The functional classification defines adhesives as being either structural or nonstructural. Structural adhesives are materials of high strength and permanence. Generally, structural adhesives are defined as those having shear strengths in excess of 1000 psi and resistance to most common operating environments. Their primary function is to hold structures together and be capable of resisting high loads without deformation. Structural adhesives are generally presumed to survive the life of the application.

Conversely, nonstructural adhesives are not required to support substantial loads, but they merely hold lightweight materials in place. Nonstructural adhesives creep under moderate load and are often degraded by long term environmental exposures. They are often used for temporary or short term fastening. Nonstructural adhesives are sometimes referred to as holding adhesives. Certain pressure sensitive adhesives, hot melt, and water emulsion adhesives are examples of nonstructural adhesives because they have moderately low shear strength, high creep, and poor resistance to temperature and chemicals. However, at times these materials could possibly be used in long term applications, depending on the severity of the application.

Nonstructural adhesives are sometimes used with other types of fasteners including mechanical fasteners. In these applications, the adhesive bond is considered a secondary fastener. The use of a nonstructural adhesive in concert with mechanical fasteners may allow one to reduce the number of mechanical fasteners that would normally be used and also provide additional value in the assembly such as vibration damping, sealing, or insulation.

8.2.2 Chemical composition

The classification of adhesives by chemical composition describes adhesives in the broadest sense as being either thermosetting, thermoplastic, elastomeric, or alloys (hybrids) of these. These classifications are described in Table 8.1. Usually, the chemical composition is divided further into major chemical types or families within each group, such as epoxy, urethane, neoprene, and cyanoacrylate. The polymeric families of resins that are useful in developing adhesive formulations are described in detail in the Chapter 10.

8.2.2.1 Thermosetting adhesives. Thermosetting adhesives are materials that cannot be heated and softened repeatedly after their initial cure. Once cured and crosslinked, the bond can be softened somewhat by heat, but it cannot be remelted or restored to the flowable state that existed before curing. Thermoset materials form infusible and insoluble materials. These adhesives generally degrade and weaken upon heating at high enough temperatures because of oxidation or molecular chain session.

Thermosetting adhesive systems cure by an irreversible chemical reaction at room or elevated temperatures, depending on the type of adhesive. This chemical reaction is often referred to as crosslinking. The crosslinking that occurs in the curing reaction is brought about by the linking of two linear polymers, resulting in a three dimensional rigidized chemical structure. A model crosslinked structure is shown in Fig. 8.1. Crosslinking usually occurs by chemical reaction. Some reactions require heat for initiation and completion; others can be completed at room temperature. With certain "room temperature" curing adhesives, it is the internal heat of reaction generated by the curing mechanism, called the exotherm, which actually provides the energy required to completely cure the polymeric material.

Substantial pressure may also be required with some thermosetting adhesives, yet others are capable of providing strong bonds with only

TABLE 8.1 Adhesives Classified by Chemical Composition

Classification	Thermoplastic	Thermosetting	Elastomeric	Alloys
Types within group	Cellulose acetate, cellulose acetate butyrate, cellulose nitrate, polyvinyl acetate, vinyl vinylidene, polyvinyl acetals, polyvinyl alcohol, polyamide, acrylic phenoxy	Cyanoacrylate, polyester, urea formaldehyde, melamine formaldehyde, resorcinol and phenol-resorcinol formaldehyde, epoxy, polyimide, polybenzimidazole, acrylic, acrylate acid diester	Natural rubber, reclaimed rubber, butyl, polyisobutylene, nitrile, styrene-butadiene, polyurethane, polysulfide, silicone, neoprene	Epoxy-phenolic, epoxy-polysulfide, epoxy-nylon, nitrilephenolic, neoprene-phenolic, vinyl-phenolic
Most used form	Liquid, some dry film	Liquid, but all forms common	Liquid, some film	Liquid, paste, film
Common further classifications	By vehicle (most are solvent dispersions or water emulsions)	By cure requirements (heat and/or pressure most common but some are catalyst types)	By cure requirements (all are common); also by vehicle (most are solvent dispersions or water emulsions)	By cure requirements (usually heat and pressure except some epoxy types); by vehicle (most are solvent dispersions or 100% solids); and by type of adherends or endservice conditions
Bond characteristics	Good to 150–200°F; poor creep strength; fair peel strength	Good to 200–500°F; good creep strength; fair peel strength	Good to 150–400°F; never melt completely; low strength; high flexibility	Balanced combination of properties of other chemical groups depending on formulation; generally higher strength over wider temp range
Major type of use	Unstressed joints; designs with caps, overlaps, stiffeners	Stressed joints at slightly elevated temp	Unstressed joints on lightweight materials; joints in flexure	Where highest and strictest end-service conditions must be met; sometimes regardless of cost, as military uses
Materials most commonly bonded	Formulation range covers all materials, but emphasis on nonmetallics — esp. wood, leather, cork, paper, etc.	For structural uses of most materials	Few used "straight" for rubber, fabric, foil, paper, leather, plastics films; also as tapes. Most modified with synthetic resins	Metals, ceramics, glass, thermosetting plastics; nature of adherends often not as vital as design or end-service conditions (i.e., high strength, temp)

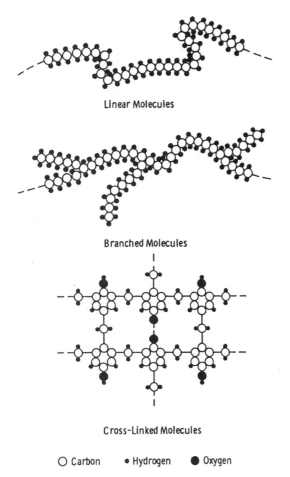

Figure 8.1 Some possible molecular structures in polymeric resins used as adhesives.[1]

moderate contact pressure. Thermosetting adhesives are sometimes provided in a solvent medium to facilitate application by brush or spray. However, they are also commonly available as solventless liquids, pastes, and solid shapes. Epoxy and urethane adhesives are examples of common adhesives that are in the thermoset chemical family.

Thermosetting adhesives may be sold as multiple or single part systems. Multiple part systems have their reactive components separated. They are weighted-out and mixed together at the time of application. Multiple part adhesives generally have longer shelf lives, and they are usually cured at room temperature or more rapidly at elevated temperatures. Once the adhesive components are mixed, the working life (sometimes defined as pot life or gel time) is limited. Single part systems have all the reactive components in a single premixed product. Generally, the single part adhesives require elevated temperature cure. They have a limited shelf life, often requiring refrigeration. Single part thermosetting systems are also available that cure at room temperature by chemical reaction with the moisture in the air, by exposure to radiation (visible, UV, electron beam, etc.), or by catalytic reaction with a substrate surface.

Because molecules of thermosetting resins are densely crosslinked, their resistance to heat and solvents is good, and they show little elastic deformation under load at elevated temperatures. As a result, most structural adhesives tend to be formulated with polymeric resins having a thermosetting molecular structure.

8.2.2.2 Thermoplastic adhesives. Thermoplastic adhesives differ from thermosets in that they do not cure or set under heat. Thermoplastics are originally solid polymers merely soften or melt when heated. Their molecular structure is either linear or branched as shown in Fig. 8.1.

Since thermoplastic molecules do not cure into a crosslinked structure, they can be melted with application of heat and then applied to a substrate. Thermal aging, brought about by repeated exposure to the high temperatures required for melting, causes eventual degradation of the material through oxidation, and this limits the number of reheat cycles. Once applied to the substrate, the parts are joined and the adhesive hardens by cooling. Hot-melt adhesives, commonly used in packaging, are examples of a solid thermoplastic material that is applied in a molten state. Adhesion develops as the melt solidifies during cooling.

Thermoplastics can also be dissolved in solvent to produce a flowable solution and then rehardened on evaporation of the solvent. Thermoplastic resins can also be dispersed in water as latex or emulsions. These products harden on evaporation of the water. Wood glues, a common household item, are thermoplastic resins that are dispersed in water as either a latex or emulsion. They harden by evaporation of the water and coalescence of the resin into a film form.

Thermoplastic adhesives are also preapplied to a substrate so that they can be "activated" at an appropriate time. Some thermoplastic adhesives make good use of this characteristic and are marketed as heat, solvent, or moisture activated adhesives. The best known example of a moisture activated adhesive is the mailing envelope adhesive that is activated by moisture to become tacky and somewhat flowable. When the moisture evaporates, the bond sets.

Thermoplastic adhesives have a more limited operating temperature range than thermosetting types. Although certain thermoplastics may provide excellent tensile shear strength at relatively moderate temperatures, these materials are not crosslinked and will tend to creep under load at lower temperatures. This creep or long term deformation under load can occur at room temperature or even at lower temperatures depending on the adhesive. Long term creep is often the characteristic that prevents these adhesives from being used in structural applications. Thermoplastic adhesives also do not have general resistance to solvents or chemicals as do the thermosetting adhesives.

8.2.2.3 Elastomeric adhesives. Because elastomeric adhesives have unique rheological characteristics, they are given their own classification. Elastomeric adhesives are based on synthetic or naturally occurring elastomeric polymers having great toughness and elongation. These adhesives are made from polymeric resins that are capable of high degrees of extension and compression. They return rapidly to their initial dimensions and shape after the load is removed. As a result, elastomeric adhesives have great energy absorbing characteristics and offer high strength in joint designs having nonuniform loading. Elastomeric adhesives may be either thermosetting or thermoplastic. The thermosetting types can be used in certain structural applications.

Elastomeric adhesives may be supplied as solvent solutions, water dispersions, pressure-sensitive tapes, and single or multiple part solventless liquids or pastes. The form and curing requirements vary with the type of elastomeric resin used in the adhesive formulation.

Elastomeric adhesives can be formulated for a wide variety of applications. Because elastomers are highly viscoelastic materials, they are characterized by a high degree of elongation, low modulus, and high toughness. This provides adhesives with high peel strength and a high degree of flexibility to bond to substrates with different expansion coefficients. Elastomers are also commonly used in adhesive formulation for sealants, vibration dampers, and sound enclosures.

8.2.2.4 Hybrid adhesives. Adhesive hybrids are made by combining thermosetting, thermoplastic, or elastomeric resins into a single adhesive formulation. Hybrids have been developed to capitalize on the most useful properties of each component. Generally high temperature, rigid resins are combined with flexible, tough elastomers or thermoplastics to provide improved peel strength and energy absorption. However, early attempts at these combinations usually resulted in an adhesive that was never better than its weakest constituent. The good high temperature properties of the base resin were always sacrificed by the addition of the flexibilizing additive.

The earliest approach to combat brittle failure was to develop adhesive formulations by blending a flexibilizing resin into the body of another resin. Thus, nitrile-phenolic, epoxy-polysulfide, and other resin blends provided resilience and toughness due to the elastomeric ingredients in the formulation. A limitation to these systems was that the elastomeric component usually lowered the glass transition temperature and degraded the elevated temperature and chemical resistance that was characteristic of the more rigid resin. These alloy blends are still in wide use today especially in the aerospace and transportation adhesive markets. They are commonly available in solvent solutions and as supported or unsupported film. Some single and two part liquid systems are also available.

More recently, advanced hybrid adhesive systems have been developed in an attempt to improve peel strength and toughness of thermosetting resins without reducing high temperature properties. These systems consist of:

1. Reactive hybrids, where two liquid components are reacted together rather than being merely blended together
2. Dispersed phase hybrids, where the flexibilizing medium is incorporated as discrete particles in a resin matrix

In reactive hybrid systems, the flexibilizing resin is reacted with the base resin to provide flexibility and toughness without significant reduction of other properties. A typical example of this system is the epoxy-urethane adhesive. Another method of providing toughness is by introducing a specific microstructure within the adhesive. This microstructure consists of a physically discrete but chemically interlinked elastomeric phase as shown in Fig. 8.2. Vinyl and carboxy terminated acrylonitrile butadiene (CTBN) polymers can be dispersed in either epoxy or acrylic resin phases.

These hybrid adhesive systems possess high peel, impact, and shear strengths without sacrificing elevated temperature or chemical resistance properties. They also have a tendency to bond well to oily substrates. It is believed that the oil on the substrate is adsorbed into the formulation and acts as another flexibilizer in the adhesive system.

8.2.3 Method of reaction

Another distinction that can be made between adhesives is the manner in which they react or solidify. There are several methods by which adhesives can solidify:

- By chemical reaction (including reaction with a hardener or reaction with an outside energy source such as heat, radiation, surface catalyst, etc.)

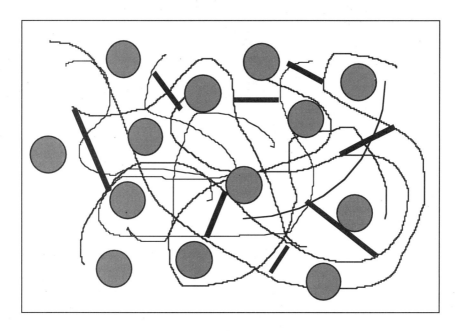

Figure 8.2 Toughened epoxy and acrylic adhesives have discrete particles of nitrile elastomer dispersed throughout the resin matrix.

- By loss of solvent
- By loss of water
- By cooling from a melt

As shown in Table 8.2, some adhesives solidify simply by losing solvent while others harden as a result of heat activation or chemical reaction. Some systems may require several mechanisms to harden. For example, thermosetting solvent based adhesives require the solvent to first evaporate and then chemical cross-linking to occur before a bond is achieved. Pressure-sensitive or contact adhesives are usually applied to the substrates as a solvent solution. Once the solvent is evaporated, the parts are mated and pressure causes the adhesive to flow. The bond then becomes stable when the pressure is absent.

The method by which adhesives cure may be an important factor in the selection process. The method of cure could significantly limit the

TABLE 8.2 Adhesives Classified by Activation and Curing Requirements

Requirement	Types available	Forms used	Remarks
Heat	Room temp to 450°F types available; 250 to 350°F types most common for structural adhesives	Formulated in all forms; liquid most common	Applying heat will usually increase bond strength of any adhesive, even room temp types
Pressure	Contact to 500 lb/in.^{2} types available; 25 to 200 lb/in.^{2} types most common for structural adhesives	Formulated in all forms; liquid or powder most common	Pressure types usually have greater strength (not true of modified epoxies)
Time	Types requiring a few seconds to a week available; ½ to 24-h types most common for structural adhesives	Formulated in all forms; solvent solutions common	Time required varies with pressure and temp applied and immediate strength
Catalyst	Extremely varied in terms of chemical catalyst required; may also contain thinners, etc.	Two components—paste (or liquid) + liquid	Sometimes catalyst types may require elevated temp (<212°F) and/or pressure instead of, or in addition to, a chemical agent

characteristics of the adhesive in the particular application. For example, if the processing requirements dictate that only a room temperature curing adhesive system can be used, then high temperature resistance will generally be sacrificed. There are very few room temperature curing adhesives that exhibit good elevated temperature resistance because of the lack of extensive crosslinking and the nature of the polymers formed with room temperature curing formulations. The assembly engineer is often constrained in selection of an adhesive by available plant facilities, production speed requirements, cost, and other important business parameters. Cure conditions required for major adhesives and sealants are shown in Table 8.3.

If grouped by method of reaction, adhesives can be further classified by type or how they are used. For example, within the group that cures by chemical reaction there are systems that cure by reacting with a hardener, by reacting with moisture in the air, by reacting with a source of radiation, or by reacting with the surface ions on the substrate. This hierarchy can be summarized as follows:

1. Adhesives that harden by chemical reaction:
   a. Two part systems
   b. Single part, cured via catalyst or hardener
   c. Moisture curing adhesives
   d. Radiation (light, UV, electron beam, etc.) curing adhesives
   e. Adhesives catalyzed by the substrate
   f. Adhesives in solid form (tape, film, powder, etc.)
2. Adhesives that harden by solvent or water loss:
   a. Contact adhesives
   b. Pressure sensitive adhesives
   c. Reactivatable adhesives
   d. Resinous adhesives
3. Adhesives that harden by cooling from the melt:
   a. Hot melt adhesives
   b. Hot melt applied pressure sensitive and thermosetting adhesives

The following sections will describe the adhesive systems that reside within each of these types.

8.2.3.1 Chemical reaction. Most thermosetting adhesives crosslink and cure by two primary chemical reactions.[3]

1. Those formed by condensation reaction, usually with water as a by-product. This group includes the phenolic and amino resins, which

TABLE 8.3 Cure Conditions for Major Adhesives and Sealants[2]

Material	Cure condition
Acrylic	Heat, catalyst or ultraviolet light
Acrylic, modified	RT, handling str in minutes, full str in hrs
Acrylic esters (anaerobics)	300 F (422 K) for 45 sec, 3 hr at RT
Amino resins	
Melamine formaldehyde	Up to 300 F (422 K)
Urea formaldehyde	RT, must use catalyst
Butadiene-acrylonitrile	2 to 10 psi (0.013 to 0.069 MPa) for 24 hr
Butadiene-styrene	20 to 100 psi (0.138 to 0.69 MPa) pressure
Butyl, sealant	Long set time to achieve reg str
Cellulose ester	Air dry
Chloroprene, sealant	7 to 14 days at 75 F (297 K)
Cyanoacrylates	0.5 to 5 min at RT
Epoxy	1 min to days; RT to 350 F (450 K) depending on formulation
Ethylene-vinyl acetate	Hot melt, cures on cooling
Ionomer	Hot melt, cures on cooling
Phenolic	Preheat 15 min at 325 F (436 K), 250 psi (1.72) MPa) and 325 F (436 K) for 15 min
Polyamide	Hot melt, cures on cooling
Polyester	Minutes to hours at RT
Polyimides	90 min at 550 to 700 F (561 to 644 K) and 500 psi (3.44 MPa)
Polysulfide	Moderate temp, low pressure
Polyurethane	Up to 300 F (422 K), low pressure
Polyvinyl acetals	Clamping for 30 min
Polyvinyl acetate & copolymers	Hot melt, cures on cooling; emulsion air dry
Polyvinyl alcohol	Water activated, RT
Silicone	Low pressure, RT

NOTE: Actual cure conditions may vary considerably, depending on preparation of the adherend, temperature, humidity and other variables.

are the oldest of the synthetic adhesives. It is usually necessary to also apply pressure to overcome the deleterious effects of the volatile by-products.

2. Those formed by addition polymerization without by-product formation. Among this group are polyesters, epoxies, urethanes, cyanoacrylates, anaerobics, and radiation curable polymers. Most elastomers and acrylics are also in this category. Only contact pressure is necessary for these adhesives, and so they are well suited for bonding large articles.

In order to achieve cure with these adhesive systems, a curing agent or catalyst is necessary. The curing reaction then proceeds at room temperature or at elevated temperatures depending on the nature of the reaction. In some cases, these adhesives will cure on activation with radiation (visible light, UV, or other form of energy). Anaerobic adhesives are catalyzed by metal ions on the surface of the substrate and cure rapidly when air is excluded from the bondline. Some polyurethane and silicone adhesives are cured by reacting with the moisture in the air.

The main advantage of adhesives that cure by chemical reaction is that they can be used for relatively large areas. With room temperature curing formulations, an oven or press is not needed. Being thermosetting in nature, these adhesives also provide a high degree of cohesive strength, temperature resistance, and resistance to common environments.

There are many adhesives that cure by chemical reaction. Chemically reactive adhesives can be further subdivided into two groups:

- Single component systems—moisture, heat activated, and free radical cure
- Multiple component systems—mix-in and no-mix[4]

These adhesives are used in both structural and non-structural applications. The most widely used chemically reactive structural adhesives are epoxy, polyurethane, modified acrylic, cyanoacrylate, and anaerobic systems. Table 8.4 describes the advantages and limitations of these five adhesives. Typical application and end-use properties are also presented.

8.2.3.1.1 Multiple part adhesive systems. Multiple component systems consist of two or more components that must be kept separate until just before the bonding operation. These components must be metered in the proper ratio, mixed, and then dispersed. Once mixed according to the manufacturer's instructions, the adhesive will have a finite

TABLE 8.4a Advantages and Limitations of the Five Most Widely Used Chemically Reactive Adhesives[5]

	Epoxy	Polyurethane	Modified acrylic	Cyanoacrylate	Anerobic
Advantages	High strength Good solvent resistance Good gap filling properties Good elevated temperature resistance Wide range of formulations Relatively low cost	Varying cure times Tough Excellent flexibility and low temperatures One or two component, room temperature or elevated temperature cure Moderate costs	Good flexibility Good peel and shear strengths No mixing required Will bond dirty (oily) surfaces Room temperature cure Moderate cost	Rapid room temperature cure One component High tensile strengths Long pot life Good adhesion to metal Dispense easily from package	Rapid room temperature cure Good solvent resistance Good elevated temperature resistance No mixing Indefinite pot life Nontoxic High strength on some substrates Moderate cost
Disadvantages	Exothermic reaction Exact proportions needed for optimum properties Two component formulations require exact measuring and mixing One component formulations often require refrigerated storage and an elevated temperature cure Short pot life	Both uncured and cured are moisture sensitive Poor elevated temperature resistance May revert with heat and moisture Short pot life Special mixing and dispensing equipment required	Low hot temperature strength Slower cure than with anerobics or cyanoacrylates Toxic Flammable Odor Limited open time Dispensing equipment required	High cost Poor durability on some surfaces Limited solvent resistance Limited elevated temperature resistance Bonds to the skin	Not recommended for permeable surfaces will not cure in air as a wet filet Limited gap cure

TABLE 8.4b Properties of the Five Most Widely Used Chemically Reactive Adhesives[5]

	Epoxy	Polyurethane	Modified acrylic	Cyanoacrylate	Anerobic
Substrates bonded	Most	Most smooth, nonferrous	Most smooth nonporous	Most nonporous metals or plastics	Metals, glass, thermosets
Service Temperature C	−55 to 121	−157 to 79	−73 to 121	−55 to 79	−55 to 149
Impact Resistance	Poor	Excellent	Good	Poor	Fair
Tensile shear, psi	2200	2200	3700	2700	2500
T-peel, piw	3	80	30	3	10
Heat cure or mixing required	Yes	Yes	No	No	No
Solvent resistance	Excellent	Good	Good	Good	Excellent
Moisture resistance	Excellent	Fair	Good	Poor	Good
Gap limitation, in. maximum	None	None	0.030	0.010	0.025
Odor	Mild	Mild	Strong	Moderate	Mild
Toxicity	Moderate	Moderate	Moderate	Low	Low
Flammability	Low	Low	High	Low	Low

working time that will depend on the type of adhesive, the chemical reaction involved, and the conditions present in the bonding environment, mainly temperature. The working life may vary from several minutes to several hours. Thus, working life could be a significant parameter in the scheduling of production, and it could be a significant cost item if significant quantities of adhesive have to be scrapped because its working life has expired.

Epoxies, polyurethanes, acrylics, phenolics, and silicones are examples of common adhesive systems that are available in multiple parts. Most of these adhesives will cure at room temperature and heat can be used to accelerate the curing process. Certain multiple part adhesive formulations must be cured at elevated temperatures.

Certain thermosetting acrylic adhesives are unique in that they are multiple part systems that do not require mixing. Multiple part "no-mix" adhesives are systems that do not require careful metering and no mixing is involved. Part "A" (e.g., resin portion) of the adhesive can be applied to one substrate and Part "B" (e.g., catalyst portion) applied to the other substrate. Cure then occurs at room temperature when the coated substrates are mated. Cure can also be achieved by coating the substrates with an accelerator/primer. Once coated, the substrates can be stored for a significant period of time. When bonding is required, the substrates are removed from storage and only the single resin component needs to be applied. Once the resin is applied to the primed substrates, free radical polymerization begins. So in essence, we have a single component adhesive that requires a primer to cure. Figure 8.3 shows the "accelerator cure" concept and the "honeymoon" cure concept where Parts A and B are applied to different substrates.

The primary advantages of these thermosetting acrylic adhesives is their ease of application and fast setting characteristics relative to epoxy or urethane. The cure occurs at room temperature and depending on the nature of the specific adhesive formulation, full cure can be completed in minutes to several hours. Cure can also be accelerated by heat.

8.2.3.1.2 Single component systems that cure via catalyst or hardener. Most single component systems consist of two or more premixed components. The curing agent or hardener is incorporated with the resin base and all the other fillers and additives. This eliminates end-user metering and mixing processes that are required of two part systems. Consistency is also usually better with the single component systems. However, refrigerated storage conditions may be required, and these adhesives have a limited shelf life.

Single component heat curing adhesives incorporate a latent curing agent or hardener in the formulation. This hardener is activated with

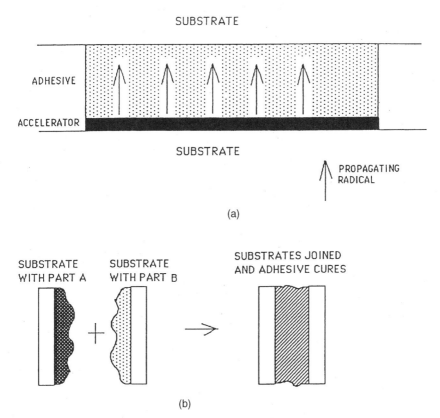

Figure 8.3 Reactive acrylic adhesives can cure via an "accelerator cure" mechanism (*a*) or by a "honeymoon cure" mechanism (*b*).[6]

heat, and then on continued heating it reacts with the base resin to produce a thermosetting molecular structure. Conventional single component epoxies that cure in about 60 min at 250°F are examples of this type of adhesive. The hardener is often dicyandiamide which is a powdered solid and required only in concentrations of a few parts per hundred. Other single component adhesive families that cure by heating include epoxy, epoxy hybrids, polyurethane, polyimide, polybenzimidazole, phenolic, and phenolic hybrids.

8.2.3.1.3 Moisture curing adhesives. Moisture curing adhesive systems use the humidity in the ambient air to react with the base resin in the adhesive formulation. The moisture vapor diffuses into the adhesive and reacts with the molecules to form a solid structure. Thus, it is important that the adhesive have access to the air. With non-porous substrates, these adhesives may cure only at the joint edges that are exposed to air. The center area of the joint may never see sufficient moisture to completely cure. As the adhesive cures on the outer edges it provides a barrier for the moisture to diffuse to the inner regions of the joint.

Moisture curing adhesives generally can cure completely at room temperature. The cure rate depends on the base polymer and its reaction chemistry and on the humidity of the curing environment. Generally, moisture cured adhesives cure more slowly than their two component reactive counterparts. Moisture cured adhesives can be accelerated by curing in a high humidity enclosure, and some types can be accelerated by heating.

Polyurethanes and silicones are common single component adhesives that cure by reacting with environmental moisture. These are available as pastes and liquids. For certain applications requiring unusually high bond strength and setup time, reactive polyurethane hot melt adhesives have been developed.[7,8] These systems are applied like a hot melt adhesive, and the parts can be rapidly assembled. As a hot melt, this adhesive has a high degree of early handling strength. The unique feature, however, is that these adhesives will continue to cure and crosslink by reacting with the moisture in the air so that when completely cured they have the strength and durability typical of most structural adhesives.

8.2.3.1.4 UV/light curing adhesives.
Adhesives that cure by exposure to radiation such as a UV or light have become essential to many assembly operations where high throughput or delicate components are prevalent. The features and benefits of UV curing are described in Table 8.5. These adhesives require the presence of the correct wavelengths of light at sufficient intensity to form durable structural bonds. UV and light cure adhesives use photoinitiators to activate the free radical curing mechanism. The free radicals formed by the pho-

TABLE 8.5 Features and Benefits of UV Curing[9]

Features	Benefits
■ cures in seconds	■ increase productivity
■ cures only upon exposure to light	■ increased design and process control
■ structural strength	■ improves integrity of the total assembly
■ 100% solvent free	■ reduced regulatory cost
■ USP Class VI certification	■ medical grades available
■ "worker friendly" longwave UV light	■ easy to introduce
■ range of viscosities for easy dispensing	■ meet most production requirements
■ range of properties—rigid to flexible	■ meet many design requirements
■ deep section curing for potting	■ ⅛–¼ in. typical for clear grades
■ optically clear	■ for easy inspection
■ grades available with secondary cure mechanism	■ cures where not exposed to light

toinitiators are able to crosslink with other components in the adhesive formulation. The adhesive that is not directly exposed to light or UV will not cure. Therefore, the substrates are generally limited to those that are transparent or to where there is only one substrate (e.g., wirebonding). Figure 8.4 illustrates bonding concepts where UV cured adhesives have been successful.

Aerobic acrylic adhesives are generally used in UV curing applications. These adhesives can cure by either direct exposure to UV or by using a preapplied activator to the substrate. Figure 8.5 shows that complete curing using the preapplied activator requires 24–48 hrs at room temperature. UV cure, however, reaches the same degree of bond strength in only seconds.

The promise of light curing adhesive has not been fully realized because some clear substrates do not transmit sufficient spectral light. Special resins are available that cure synergistically by combining UV and visible curing mechanisms. These products also have faster and deeper cures. They have the ability to cure through to deep (1/4 in.) layers to form structural strength bonds (3,000 psi) on steel.[9] They also can cure completely in shadowed areas.

8.2.3.1.5 Adhesives catalyzed by the substrate. Certain single component adhesives cure via a chemical reaction mechanism that uses the active metal ions (e.g., iron and copper) on the substrate as a catalyst. These adhesive systems are called anaerobic adhesives because they cure only in the absence of oxygen. Anaerobic is a term borrowed from biology that refers to life forms that live in the absence of oxygen.

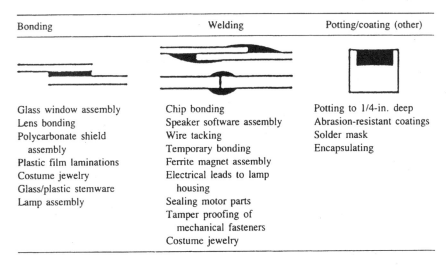

Bonding	Welding	Potting/coating (other)
Glass window assembly	Chip bonding	Potting to 1/4-in. deep
Lens bonding	Speaker software assembly	Abrasion-resistant coatings
Polycarbonate shield assembly	Wire tacking	Solder mask
Plastic film laminations	Temporary bonding	Encapsulating
Costume jewelry	Ferrite magnet assembly	
Glass/plastic stemware	Electrical leads to lamp housing	
Lamp assembly	Sealing motor parts	
	Tamper proofing of mechanical fasteners	
	Costume jewelry	

Figure 8.4 Bonding categories using UV aerobic adhesives.[10]

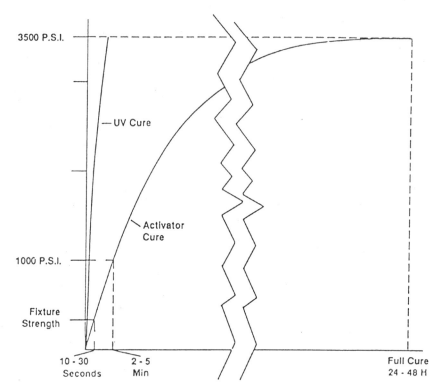

Figure 8.5 Typical aerobic acrylic adhesive tensile strength on cold rolled steel. Cure is accomplished by UV radiation and a preapplied activator.[10]

Oxygen inhibits the cure of these adhesives; so they are stored in containers with copious supply of air. Once placed between substrates, the air is eliminated and the adhesive reaction is catalyzed by the metal ions on the substrate surface. When cured, these adhesives have high strength and good resistance to moisture, solvents, and elevated temperatures.

These adhesives were developed in the 1950s and 1960s by Loctite Corporation, which continues to be a leading manufacturer and supplier of anaerobic adhesives and sealants. Anaerobic adhesives are formulated mostly with methacrylate monomers. They are available with viscosities from water thin to thixotropic pastes. They are widely used for threadlocking, retaining, and gasketing. Figure 8.6 illustrates several applications for anaerobic adhesives. Very low viscosity anaerobics are used to seal porosity in metal castings.

Anaerobic adhesives cure very fast depending on the substrate. Cure speeds range from several minutes to 24 hrs. Cure rate can be accelerated by primers or the application of heat. Clean metal and ther-

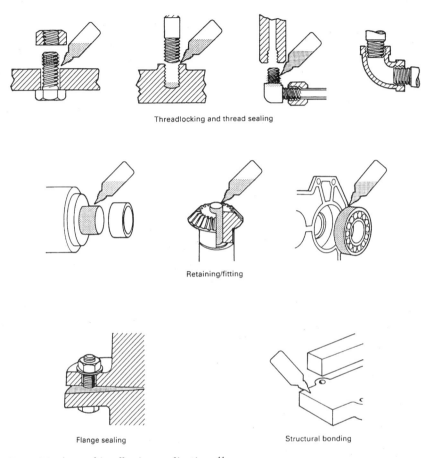

Figure 8.6 Anaerobic adhesive applications.[11]

moset surfaces produce the fastest cure. Primers or heat application may be required to achieve cure on inactive surfaces such as thermoplastics and certain metals. Photocuring has also been built into anaerobic adhesives to allow light induced hardening outside adhesive joints.

8.2.3.1.6 Adhesives in solid form (tape, film, powder). Single component solid adhesive systems are made in several ways. One method is by completely formulating the adhesive system, including resins, fillers, curing agents, etc., in the liquid state and then converting it into a solid state that is not completely cured. This conversion can be through cooling of a melt form, or by the removal of solvent from a solution, or by semicuring or "B-staging". B-staged adhesives are partially cured to a solid state that is not completely crosslinked and will still flow and bond with the addition of heat and pressure.

For tape or film adhesives, this hardening process is usually accompanied by extruding, calendering, or casting the adhesive formulation into thin films that are typically 5–10 mils thick. The product is B-staged to a condition where it is either tack free or slightly tacky to help application to vertical and contoured substrates. These films may be in the form of either unsupported sheet or they may be reinforced with glass fabric, paper, or another reinforcing media. These latter systems are called "supported" tape and film adhesives. B-stage supported film is also called prepreg. Once the tape or film is in place between substrates, the joint is heated under pressure so that the B-stage adhesive becomes slightly fluid, wets the substrate, and flows into the micro-roughness on the substrate. With additional heat and time, the adhesive completely cures to a thermosetting structure.

B-staged solid adhesive formulations may also be ground to a powder form and applied to the substrate by sifting onto the surface, or the powder might be electrodeposited or coated on the surface in another manner. The powder can also be formed into shapes or preforms by the application of pressure and a die, much like how pharmaceutical tablets are made. In this way shaped preforms can be made that will conform to a specific joint geometry.

8.2.3.2 Solvent or water loss. Solvent solutions and water based latexes and dispersions harden by the evaporation of their carrier material—either solvent or water. The carrier material's function is simply to lower the viscosity of the adhesive so that it can be easily applied to the substrate. Once applied, the water or solvent must be removed either by evaporation into the air or by diffusion into a porous substrate. Thus, solvent and water based systems often find use in applications with porous substrates such as wood, paper, leather, and fabric.

Once applied and dried, the adhesive can then provide a bond in a number of ways. It could simply harden into a cohesive resin mass, such as polyvinyl acetate or wood glues. It could form a film that is then reactivated by solvent or water or by heat as in the case of a laminating adhesive. It could also form a film that when dry has a high degree of tack so as to be a pressure sensitive adhesive. There are primarily four types of adhesives that harden by loss of solvent or water:

- Contact adhesives—The adhesive is applied to both substrates, solvent is removed, and the substrates are mated under pressure so that the adhesive coatings knit together.
- Pressure sensitive adhesives—The adhesive is applied to one or both substrates or to a carrier (film, cloth, etc.). Once the solvent is

removed, the adhesive has aggressive and permanent tackiness. The substrates can be mated with very little pressure.

- Reactivatable adhesives—The adhesive is applied to the substrate and the solvent is evaporated. At this stage the substrate may be stored or transported with a dry adhesive coating. At the time of joining, the coating is moistened with the solvent and the adhesive becomes tacky with pressure sensitive characteristics.
- Resinous solvent adhesives—The adhesive is a resinous material dissolved in the solvent. It is applied to porous substrates that are then joined. The solvent evaporates leaving the resin to mechanically lock the substrates together.

8.2.3.2.1 Contact adhesives. Contact adhesives are applied to both substrates by spray or roll coating. Usually the solvent is allowed to evaporate under ambient conditions, but sometimes heat is applied to accelerate drying. After some portion of the solvent evaporates, the adherends are joined, and the adhesive rapidly bonds or knits to itself with the application of contact pressure. There is a window of time in which there is just enough tack exhibited by the adhesive to bond.

Contact adhesives generally have relatively high shear and peel strengths. Strength and durability approach that of structural thermosetting adhesives. A typical contact adhesive application is the bonding of decorative surface materials to wood for kitchen counter tops. The contact adhesives generally have less "green strength" or immediate tack than pressure sensitive adhesives and cannot be applied to only one substrate.

8.2.3.2.2 Pressure sensitive adhesives. Pressure sensitive adhesives are applied much like contact adhesives and instantly provide a degree of tackiness. Unlike contact adhesives, their tackiness is permanent and there is no optimal time range when the substrates must be joined. Pressure sensitive adhesives are usually based on elastomer or thermoplastic solvent solutions. They are coated on the substrate or on a film backing that is used as a carrier. Pressure sensitive tape (e.g., office tape) is made in this manner. Once the solvent evaporates, the tape is ready to be applied or it is packaged so that it can be dispensed at the point of assembly.

Most pressure sensitive adhesives are made from formulations based on elastomer (natural, butyl, nitrile and styrene butadiene thermoplastic), acrylate, or silicone resins. The advantages and limitations of each of these pressure sensitive types are shown in Table 8.6. Pressure sensitive adhesives are specifically formulated for good flexibility, tack, and peel strength.

Pressure sensitive adhesives can be applied from solvent solutions, water dispersions, or hot melts. Table 8.7 shows the properties and

TABLE 8.6 Advantages and Limitations of Rubber, Acrylate and Silicone Pressure Sensitive Adhesives[5]

Chemical family	Advantages	Limitations
Rubbers	Good flexibility	Low tack and adhesion (without additives)
	High initial adhesion (better than acrylics)	Poor aging, subject to yellowing
	Ease of tackification (with additives)	Limited upper service temperature use
	Lowest cost	Moderate service life
	Good shear strength	
	Good adherence to low- and high-energy surfaces	
	Suitable for temporary or permanent holding	
Acrylates	Good UV resistance	Poor creep resistance (compared to rubbers)
	Good hydrolysis resistance (better than rubbers)	Fair initial adhesion
	Excellent adhesion buildup	Moderate cost (compared to rubbers, silicones)
	Good solvent resistance	
	Good temperature use range (-45 to $121°C$, or -50 to $250°F$)	
	Easier to apply (than rubbers)	
	Good shear strength	
	Good service life	
Silicones	Excellent chemical and solvent resistance	Highest cost (compared to rubbers, acrylates)
	Wide temperature use range (-73 to $260°C$, or -100 to $500°F$)	Lack of aggressive behavior
	Good oxidation resistance	
	Good adherence to low- and high-energy surfaces	

characteristics of these various methods of applications for elastomeric pressure sensitive adhesives. Thermoplastic elastomers are usually employed in hot melt pressure sensitive adhesives.

Most pressure sensitive adhesives are applied to plastic, paper, foil or fibrous material as suitable backings. Pressure sensitive tapes are made with adhesive on one side or on both sides (double faced tape). Double faced foam tapes are available to suit applications where substrates have surface irregularities, contours, and significant gaps in the joint area.

Several very high bond strength adhesive tapes have been developed for semi-structural applications.[13,14] Use of foam carrier provides not

TABLE 8.7 Properties and Characteristics of Elastomeric Pressure Sensitive Adhesives[12]

Property	Solvent-borne adhesives	Water-borne adhesives	Hot-melt adhesives
Solids, content, %	Usually 10–30%, with some in 40–50% range	Typically 40–60%	100%
Bonding methods	Contact, pressure sensitive, heat and solvent reactivated, wet bond	Same as solvent borne	Wet bond, heat activated, pressure sensitive
Drying rate	Usually fast; drying time adjustable by choice of solvents	Slow to dry; however, high solids content means less to dry	No volatiles to evaporate
Open time	Wide range for contact adhesives; indefinite for pressure sensitive	Short range for contact types, unless compounded for open time; indefinite for pressure-sensitive types	Short range for thin film due to fast cool-down; indefinite for pressure sensitive
Surface wet-out	Excellent on most surfaces, even with minor contamination	Good on porous substrates, poor to fair on nonpolar surfaces like plastics	Good on most surfaces; may require heating the surface before application
Water resistance	Usually good to excellent	Fair to good	Excellent
Green strength	Excellent; allows early assembly handling	Fair; slow strength development	Fast bond development as bond line loses heat
Ultimate strength	Excellent, especially for curing types; highest film strength	Good to excellent; curing compounds provide high strength	Poor to fair (mostly noncuring types)
Application procedures	Wide range, including use of brush, spray, roll, extrusion, trowel	Wide range; may get misting with some spray techniques	Usually with brush, trowel, extrusion, or roll coat; may require preheating before applying
Temperature range	−55 to 150°C (−65 to 300°F) for end-use, freeze-thaw stable in liquid form	−40 to 120°C (−40 to 250°F) for end-use; requires temperature control to keep from freezing	−45 to 120°C (−50 to 250°F) for end-use; no storage or shipping problems; upper temperature range dependent on hot-melt type

TABLE 8.7 Properties and Characteristics of Elastomeric Pressure Sensitive Adhesives[12] (Continued)

Property	Solvent-borne adhesives	Water-borne adhesives	Hot-melt adhesives
Surface attack	May cause degradation of some plastics and paint	Causes shrinkage and wrinkling of fabric and paper; corrosive to metal	Usually no problem; heat may deform thin plastics
Flexibility	Excellent	Excellent	Excellent
Hazards	Requires vapor control to OSHA limits; many are flammable and may require emissions control	Usually no problem; some have trace levels of chemicals that require control of vapors	Few problems other than working with hot dispensing equipment
Cleanup	Solvent	Water when wet, solvent when dry	Solvent
Cost	Low-to-moderate cost per gallon, high cost per dry pound	Moderate cost per gallon, low cost per dry pound	Moderate cost per dry pound

only a gap filler but also helps to distribute the stress within the joint area, thus, improving the ultimate strength of the joint. These tapes offer bond strengths up to 10 times those of conventional pressure sensitive adhesive while maintaining enough cohesive strength to support modest loads for long periods of time.[14]

Pressure sensitive adhesives provide a relatively low strength bond upon the brief application of slight pressure (usually by hand). They may be applied to any clean, dry surface. Since they are capable of sustaining only light loads because of creep, they are not considered structural adhesives. Surface cleanliness is very important in utilizing pressure sensitive adhesives.

Pressure sensitive adhesives build a stronger bond over time. One should not test pressure sensitive adhesive immediately after they are applied and the joint is made. Full strength may take several days to develop as shown in Fig. 8.7. Pressure sensitive tapes require relatively close mating substrates. The pressure sensitive adhesive film

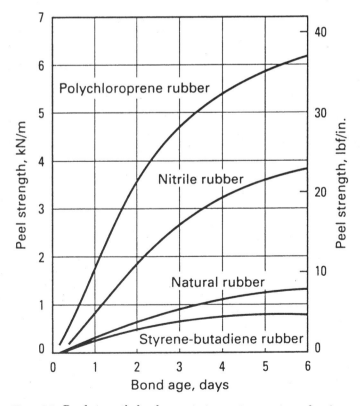

Figure 8.7 Bond strength development at room temperature of various pressure sensitive adhesives.[12]

thickness may be only several mils thick. If there are gaps larger than that between the substrates, the adhesive will not bond because there is no applied pressure in the area of greatest gap. Pressure sensitive tapes using foam (polyethylene or vinyl) carriers are used where gap tolerances may be great. These adhesives are easily recognized in automobile protective side strips, mounting tape for wall dispensers, etc.

8.2.3.2.3 Reactivatable adhesives. Certain adhesives are known as solvent activated adhesives. These systems can be applied to a substrate via solvent coating and then dried to a non-tacky state. The coated substrate is then usually stored or it proceeds to the next step in the manufacturing operation. When it is time to bond the coated substrate to another substrate, the coated piece is exposed to solvent either by wiping it across a cloth saturated with solvent or by spraying a small amount of solvent onto the coated surface. This partially liquefies the adhesive, and it becomes tacky so that the substrates can be mated under contact pressure.

Labels, nameplates, stamps, etc. are common applications for reactivatable adhesives. Care must be exerted not to apply too much or too little solvent to the dry adhesive film. The reactivating solvent is generally applied by brush, spray, or roller coating. Too much solvent could wash away most of the resinous component resulting in a starved joint. Too little solvent could result in lack of sufficient tack to bond to the other substrate. These adhesives are normally highly formulated thermoplastic resins.

8.2.3.3 Resinous solvent adhesives. Many solvent adhesives are used for their mechanical interlocking ability. They are formulated with a relatively rigid, tough thermoplastic resin. The substrate is generally porous. Once applied, the solvent evaporates into the air and through the substrate so that the adhesive resin interlocks the pores of the substrate together. Typical applications are the bonding of wood, cardboard, fabric, leather, and other porous substrates.

More dilute solvent based systems can be used to bond certain thermoplastics. In this application the solvent diffuses into the substrates leaving the resinous portion of the adhesive in the joint. More information is given on solvent cementing of thermoplastics in Chapter 15.

8.2.3.4 Hardening from the melt. Adhesives can also harden by cooling from a melt condition. Hot melt adhesives are the most common example of this type of adhesive. These are generally thermoplastic adhesives that soften and melt when heated, and they harden on subsequent cooling. The hot melt system must achieve a relatively low viscosity when in the molten state to achieve wetting, and it must not cool too rapidly or it will not have time to completely wet the roughness of the substrate surface. Once the hot melt adhesive is applied and in the molten condition, the substrates must be joined immediately. The adhesive can be applied and then the adhesive coated substrate is placed into storage for later activation. At a later date, the coated substrates could then be removed from storage, reheated to soften the adhesive, and joined under slight pressure.

When hardened, the hot melt adhesive can have various degrees of tackiness depending on the formulation. A completely pressure sensitive adhesive could result from a hot melt formulation. Some pressure sensitive tapes and films are manufactured in this manner. This method eliminates the expense and environmental hazards of using solvent simply to reduce the viscosity of a pressure sensitive resin for application to a carrier.

Materials that are primarily used as hot melt adhesives include ethylene and vinyl acetate copolymers, polyvinyl acetates, polyethylene, amorphous polypropylene, block copolymers (styrene butadiene rubber), polyamides, and polyester. In general, hot melt adhesive formulations are solids at temperatures below 175°F. Typical application temperatures are 300–400°F. Typical properties of hot melt adhesives are shown in Table 8.8.

Special dispensing equipment must be used to apply hot melts. At one time hot melts were mainly applied from a molten bath that must be kept under an inert gas to prevent oxidation of the adhesive. New application methods and forms of hot melt adhesives now avoid the need for such equipment. Cylinders or cords of hot melt can be applied to the substrate though hand guns that have a heated nozzle. Hot melt adhesive formulations must be chosen carefully based on the application equipment available and the speed at which the joint can be made. Open time of the hot melt is especially critical with substrates that are thermally conductive such as metals.

Foamable hot melt adhesives have been available since the 1980s. These adhesives use nitrogen or carbon dioxide gas to increase the volume of the adhesive by 20–70% as it is applied to the substrate. The foaming process increases open time and provides for good gap filling properties. The elastic foam also tends to relieve stresses that might develop internally within the joint.

8.2.4 Physical form

A widely used method of distinguishing between adhesives is by their physical form. Adhesive systems are available in a number of forms. The most common forms are:

- Multiple part solventless (liquid or paste)
- One part solventless (liquid or paste)

TABLE 8.8 Typical Properties of Hot Melt Adhesives[5]

Property	EVA/polyolefin homopolymers and copolymers	Polyvinyl acetate	Polyurethane	Polyamides	Polyamide copolymer	Aromatic polyamide
Brookfield viscosity, Pa·s	1–30	1.6–10	2	0.5–7.5	11	2.2
Viscosity test temperature, °C (°F)	204 (400)	121 (250)	104 (220)	204 (400)	230 (446)	204 (400)
Softening temperature, °C (°F)	99–139 (211–282)	...	...	93–154 (200–310)	...	129–140 (265–285)
Application temperature, °C (°F)	...	121–177 (250–350)	...	...	...	...
Service temperature range, °C (°F)	−34 to 80 (−30 to 176)	−1 to 120 (30 to 248)	...	−40 to 185 (−40 to 365)	...	...
Relative cost(a)	Lowest	Low to medium	Medium to high	High	High	High
Bonding substrates	Paper, wood, selected thermoplastics, selected metals, selected glasses	Paper, wood, leather, glass, selected plastics, selected metals	Plastics	Wood, leather, selected plastics, selected metals	Selected metals, selected plastics	Selected metals, selected plastics
Applications	Bookbinding, packaging, toys, automotive, furniture, electronics	Tray forming, packaging, binding, sealing cases and cartons, bottle labels, cans, jars	Laminates	Packaging, electronics, furniture, footwear	Packaging, electronics, binding	Electronics, packaging, binding

(a) Relative to other hot-melt adhesives

- One part solution (liquid)
- Solid (powder, tape, film, etc.)

Some specific types of adhesives, such as epoxy, can be available in several forms through slight variation in the composition. The characteristics and advantages of the various adhesive forms are summarized in Table 8.9.

Often the desired form of the adhesive is dictated by the complexity of the parts to be bonded and the availability of application equipment. Two-part solvent systems would, for example, require metering and mixing equipment, spray application equipment, and fixturing equipment to hold the parts together until the adhesive dries and cures. On the other extreme, a solid tape adhesive may have the catalysts already introduced into the formulation. Thus, there is no need to meter or mix components, and application does not require any special equipment. Furthermore, there is little waste with a tape adhesive, and contact with liquid chemicals, a source for dermatitis and other health problems, is eliminated. Solvent adhesives are generally used where the adhesive can be uniformly sprayed on large areas. However, with solvent based adhesives, provision must be made for evaporation of the solvent with resulting environmental concerns.

Adhesives can be simply classified as being a liquid (solvent containing or 100% solids) or a solid. Within each class there are important types of adhesive systems. Table 8.10 outlines the various liquid and solid adhesives that are in common use today.

TABLE 8.9 Adhesives Classified by Form

Type	Remarks	Advantages
Liquid......	Most common form; practically every formulation available. Principally solvent-dispersed	Easy to apply. Viscosity often under control of user. Major form for hand application
Paste.......	Wide range of consistencies. Limited formulations; principally 100% solids modified epoxies	Lends itself to high-production setups because of less time wait. High shear and creep strengths
Powder.....	Requires mixing or heating to activate curing	Longer shelf life; mixed in quantities needed
Mastic......	Applied with trowel	Void-filling, nonflowing
Film, tape ..	Limited to flat surfaces, wide range of curing ease	Quick and easy application. No waste or run-over; uniform thickness
Other.......	Rods, supported tapes, precoated copper for printed circuits, etc.	Ease of application and cure for particular use

TABLE 8.10 Various Types of Adhesives by Form

1. Solid adhesive
    a. Film adhesives (unsupported)
    b. Tape adhesives (supported)
    c. Solid powders and preforms
    d. Solvent based adhesives and primers
    e. Hot melt adhesives
2. Liquid adhesives (100% solids paste and liquid)
    a. One-component: long shelf life
        i. heat cured
        ii. cured by surface or anaerobic catalysts
        iii. cured by exposure to ultraviolet (UV) light, radiation, or some other energy source
    b. Two-component: short pot life
        i. room temperature cured
        ii. heat cured
3. Liquid solvent containing adhesives (similar to 100% solids systems above but with solvent for viscosity reduction)
    a. Solvent based contact adhesives
    b. Water based adhesives
        i. contact or pressure sensitive adhesives
        ii. solutions or emulsions
    c. Pressure sensitive adhesives

8.2.4.1 Pastes and liquids.

The difference between paste and liquid adhesive lies mainly in viscosity and method of application. Liquid adhesives are free flowing fluids that can be applied in thin films. They tend to flow, spread, and sag during cure, especially at elevated temperatures when their viscosity is even further reduced. Paste adhesives are heavily bodied, often thixotropic systems that must be applied with spreading equipment such as a trowel or caulking gun. Paste adhesives are formulated for gap filling and other bonding applications where sag resistance is important during the cure of the adhesive. The following sections will describe these various adhesive forms in more detail.

8.2.4.1.1 Two-part solventless.

Two-part solventless adhesive systems usually require proportioning and mixing immediately prior to use. Most can cure at low temperatures (e.g., room temperature) or in shorter times at elevated temperatures. Once mixed, two-part adhesives have a limited working life. As the mixture cures, the viscosity increases to the extent that it can no longer be applied. If the adhesive system has a large exotherm, the working life is also dependent on the mass of adhesive that is mixed.

Once applied and the substrates are joined, there is a certain time period for which the adhesive reaches a handling strength (i.e., a strength sufficient to release holding pressure and move the product). The time for the adhesive to reach maximum strength is usually significantly longer than the time to reach handling strength. The working time, time to reach handling strength, and time to reach maximum strength are dependent on the type of adhesive and the method by which it is cured. A typical hardware store variety two component epoxy (epoxy cured with a polyamide-amine) will have a working life of perhaps 30 min, will reach handling strength in from several hours to 24 hrs, and will develop full strength in 3–7 days. Figure 8.8 shows a curing curve of a typical room temperature curing epoxy adhesive.

Two-part systems may introduce problems in production scheduling, and some waste is unavoidable. Solventless liquid adhesives can be employed in all bonding operations without regard to size, shape, orientation, or fit of substrates. The viscosity of the adhesive generally determines how it is to be applied. Liquid adhesives lend themselves to easy handling via mechanical spreaders or spray and brush. Paste adhesives have high viscosity to allow application on vertical surfaces with little danger of sag or drip. These bodied adhesives also serve as gap fillers between two poorly mated substrates.

Two-part systems can also be used to form non-solid tape adhesives by knifing or brushing the mixed adhesive onto a supporting web (e.g., glass fabric) laid either on the substrate or on a separator sheet. When spread on a separator sheet, the supporting web can be used as a carrier to transfer the adhesive to the substrate surface.

8.2.4.1.2 One-part solventless. One-part solventless liquid adhesive systems are supplied in a ready-to-apply condition. Most of these systems are heat curing or cure by reaction of moisture in the air. Some

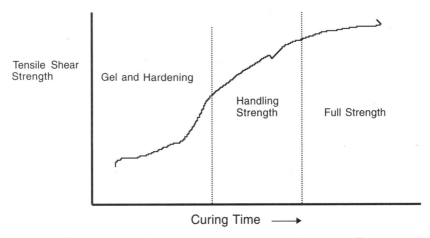

Figure 8.8 Typical curing curve for a room temperature curing epoxy adhesive.

systems cure by ultraviolet, anaerobic, or some other catalytic reaction mechanism. One-part solventless adhesives are popular in many production operations because there is no need for mixing, and waste from mixing more adhesive than required is eliminated.

Most one-part thermosetting systems contain a latent hardener that activates and cures the adhesive at elevated temperatures. The time required for cure can be a few seconds to several hours for heat curing systems. Usually, one-part systems have a limited shelf life, and refrigerated storage may be required. When stored under refrigerated conditions, it is important to warm the adhesive to room temperature before it is applied to the substrate. If a cold adhesive is applied to a warmer substrate, such as aluminum, water from humid air may condense on the substrate causing a weak boundary layer.

Certain one part adhesive systems cure by reacting with the moisture in the air. Polyurethanes and silicones react in this manner. When using these systems on non-porous substrates, it is important that the bond area not be too large or else the moisture in the air cannot get to the center of the bondline.

Anaerobic adhesives are single component adhesives that are unique in that they are shelf-stable, solventless, ready-to-use adhesives that cure at room temperatures. Their cure is inhibited by the presence of air (oxygen) in the package and during application. Once the joint is assembled and air is excluded from the liquid adhesive, curing begins. Certain anaerobic adhesives are catalyzed by the metal substrate on which they cure.

Some adhesive systems can cure by exposure to radiation. They can be cured by ultraviolet light, visible light, or electron beam radiation. Radiation cured adhesives are limited to applications where the bondline is relatively thin and the substrates are transparent to the energy source. Thus, coating applications are more common than adhesive applications. Radiation cured adhesives commonly use acrylate or unsaturated polyester base resins.

8.2.4.2 Solvent-based adhesives. Solvent-based adhesives use organic solvent to reduce viscosity for easier application. The solvent is simply a carrier for the "solids" part of the adhesive. Because of concern for health, safety, and the environment, solvent adhesives must be carefully considered before their introduction into an assembly operation. Many water-based adhesives and solventless adhesive systems have recently been developed to replace conventional solvent systems.

Solvent adhesives are applied by spraying, dipping, or brushing, but the solvent must be driven off before the joint is assembled. Solvent containing adhesive systems must have a way of letting the solvent escape from the joint. Either the solvent is allowed to evaporate before the articles are mated, or if one of the substrates is porous, the solvent can evaporate through the pores.

Solvent-based adhesives may be chemically crosslinked, usually through the application of a curing agent (either incorporated directly into single part systems or added as a second part to the base resin) and heat. The application of a solid thermosetting epoxy via solution coating is an example. Such an adhesive may be used to coat and then bond laminations of electrical steel for motor applications. The adhesive is applied and then dried through solvent evaporation. Once the parts are ready to be bonded, they are stacked under pressure. Heat is then applied to achieve flow and final chemical crosslinking.

Solvent solution adhesives may also be uncrosslinked as in the case of many pressure sensitive adhesives. In this case, a solid polymeric resin, such as polyisobutylene, may be applied to a backing material via solvent solution. Once the solvent is driven away or evaporated, the resin increases in viscosity to a consistency in which it can operate as a pressure sensitive adhesive.

8.2.4.3 Water-based adhesives. Water-based adhesives are usually emulsions of thermoplastic resins. The properties of the emulsion are derived from the polymer employed as well as the system used to emulsify the polymer in water. There are several additives necessary to stabilize and protect the emulsion. Other additives are used to adjust tack, drying time, viscosity, storage stability, etc.

Like the solvent-based adhesives, the water carrier is evaporated through the air or diffused into the porous substrate. When dry the resulting adhesive can be either a brittle, hard resin or a flexible, tacky film depending on the adhesive formulation. Water-based adhesives are like solvent-based adhesives in that they are formulated as contact, pressure sensitive, reactivatable, and resinous adhesive systems. They are used much the same way as their solvent based counterparts.

Perhaps the most widely used emulsion-based adhesive is the polyvinyl acetate-polyvinyl alcohol copolymer that is known as "white" glue or wood glue. This adhesive hardens to a relatively rigid solid when the water diffuses through the substrate.

Water-based adhesives are considered as replacements for solvent-based adhesives for the purpose of reducing volatile organic emissions in a manufacturing operation. Although there are several water-based contact and pressure sensitive adhesives on the market, they are slower drying than solvent-based adhesives, requiring about three times more heat to dry. Forced drying of water-based adhesives costs energy and causes corrosion problems in the ovens that are used. Water-based contact adhesives also have a lack of immediate bonding capability. When cured, water-based adhesives do not have the moisture resistance that solvent-based contact adhesives have.

8.2.4.4 Solid forms. Common forms of solid adhesives are tape or film, powder, or solid shape forms. The main advantages of these single component adhesives are that metering and mixing is not required, and they can be applied uniformly to a substrate with little or no waste. Thus, solid adhesives are popular in many production applications. Both thermoplastic and thermosetting adhesives can be formulated as solid systems.

8.2.4.4.1 Tape and film. Tape and film are terms used loosely and interchangeably for adhesives in sheet form. More correctly, tapes are supported on a web of paper, nonwoven fabric, or on open weave scrim of glass cotton or nylon. Supporting fibers are useful in that they provide a positive stop under bonding pressure. This can be used to control bond-line thickness and often to help distribute stresses. Films, on the other hand, are free of reinforcing fibers and consist only of the adhesive in thin, sheet form.

Tape and film adhesives are poor gap fillers especially if the gap between parts varies in depth across the bonding surface. However, they offer a uniformly thick bond-line and easy dispensing. Being a single component system there is no need for metering. Tape and film adhesives can be used in bonding large areas such as in the aerospace industry. For example, the joining of aluminum honeycomb structure to flat metal sheets is best accomplished with thermosetting film adhesives.

8.2.4.4.2 Powdered adhesives. Another form of solid adhesive is powder or granules, which must be first heated or solvent-activated to be made liquid and flowable. Being a single component system, powdered adhesives also eliminate proportioning and mixing errors, but uniform distribution over large assemblies is difficult.

Solid powder adhesives may also be fused or compacted into various preforms such as sticks, gaskets and beads. They can then be easily applied to a specifically shaped part. Certain types of thermosetting epoxy powder can be fused into small rings that are then used in rod and tube type joints. The ring is slipped over one of the substrates before assembly. When exposed to high curing temperatures, the epoxy in the ring melts, flows, and wets the substrate. Additional heating is used to fully cure the adhesive into a thermosetting part. In this type of application, a well or trough must be designed into the joint to hold the liquid adhesive in place while it cures.

8.2.4.4.3 Hot melts forms. Solid hot melt adhesives are formulated with thermoplastic resins. They can be applied in several ways. The adhesive can be applied to the surface in molten form using a heated tank and heated hoses. In this case, the substrate is generally preheated so that the adhesive will not gel immediately on contact with the cooler substrate. The surfaces are then quickly mated together and held under contact pressure until the adhesive solidifies. The substrate can also be precoated with the adhesive prior to bonding.

The hot melt adhesive can also be applied to the joint as a preformed thermoplastic film or mat. The joint is assembled, and heat and pressure are applied. The joint is then cooled to solidify the adhesive and achieve a bond.

Hot melt products are available in several shapes and forms. Sticks and cord that can be fed through heated extrusion guns avoid the necessity of having a heated bath of molten hot melt and heated hoses to pump the hot melt from the bath to the joint area.

8.2.5 Cost

Cost is usually not used as a method of classifying adhesives; however, it is an important factor in the selection of a specific adhesive and a factor in determining whether adhesives should be used at all. Thus, cost becomes a means of classification and selection, if not directly, at least indirectly. The raw materials cost or "first cost" of an adhesive can vary significantly.

Adhesive price is dependent on development costs and volume requirements. Adhesives that have been specifically developed to be resistant to adverse environments are usually more expensive than general-purpose adhesives. Adhesive prices range from pennies a pound for inorganic and animal-based systems to several hundred dollars per pound for some heat-resistant synthetic types. Adhesives in film or powder form require more processing than liquid or paste types and are usually more expensive.

When estimating the cost of using adhesives, one must not only consider the price of the adhesive but also the cost of everything required to obtain a reliable, complete joint. Therefore, the cost of fastening with adhesives must also include the cost of labor, the cost of equipment, the time required to cure the adhesive, and the economical loss due to rejects of defective joints. The following parameters may be important in analyzing the real cost of an adhesive system:

- Efficiency of coverage in relation to bonding area or number of components

- Ease of application and processing equipment needed (jigs, ovens, presses, applicators, etc.)
- Total processing time (for assembly, for preparation of adherends, for drying, for curing, etc.)
- Cost of labor for assembly and inspection of the bonded parts
- Waste of adhesive contributes to material costs and environmental costs for disposal
- Amount of rejected material as compared with other methods of joining

Adhesive-materials cost should be calculated on a cost per bonded area basis. Since many adhesives are sold as dilute solutions, a cost per unit weight or volume basis may lead to erroneous comparisons.

8.2.6 Specific adherends and method of applications

Adhesives may also be classified according to their end-use. Thus, metal adhesives, wood adhesives, and vinyl adhesives refer to the substrates to which they will bond. Similarly, acid-resistant adhesives, heat-resistant adhesives, and weatherable adhesives indicate the environments for which each is best suited. In most of this Handbook, adhesives are classified according to either their end-use or chemical composition.

Adhesives are also often classified by the method in which they are applied. Depending on viscosity, liquid adhesives can be considered to be sprayable, brushable, or trowelable. Heavily bodied adhesive pastes and mastics are considered to be extrudable; they are applied by syringe, caulking gun, or pneumatic pumping equipment.

References

1. Meier, J. F., "Fundamentals of Plastics and Elastomers", *Handbook of Plastics, Elastomers, and Composites,* 3rd ed., C. A. Harper, ed. (New York: McGraw Hill, 1996).
2. "Adhesives and Sealants", Manual 260, *Materials Engineering,* May 1976.
3. Skiest, I., and Miron, J., "Introduction to Adhesives", *Handbook of Adhesives,* 3rd ed., I. Skiest, ed. (New York: Van Nostrand Reinhold Publishing, 1990).
4. Gauthier, M. M., "Sorting Out Structural Adhesives", *Advanced Materials and Processes,* July, 1990.
5. Gauthier, M. M., "Types of Adhesives", *Adhesives and Sealants,* vol. 3, Engineered Materials Handbook, ASM International, 1990.
6. Damico, D. J., "Reactive Acrylic Adhesives", in *Handbook of Adhesive Technology,* A. Pizzi and K. L. Mittal, eds. (New York: Marcel Dekker, 1994).
7. DeGray, D., "PUR Adhesives Offer Solutions for Assembly Challenges", *Adhesives Age,* May 1998.

8. David, F. E., and Fromwiller, J. K., "Reactive Urethane Hot Melts for Textiles", *Adhesives and Sealants Industry,* March 1998.
9. Bachmann, C., "Expanding Capabilities with UV/Visible Light Curing Adhesives", *Adhesives Age,* April, 1995.
10. Bachmann, A. G., "Anaerobic Acrylics: Increasing Quality and Productivity with Customization and Adhesive/Process Integration", *Handbook of Adhesives Technology,* A. Pizzi and K. L. Mittal, eds. (New York: Marcel Dekker, 1994).
11. Melody, D. P., "Anaerobics", in *Adhesives and Sealants,* vol. 3, Engineered Materials Handbook, ASM International, 1990.
12. Harrington, W. F., "Elastomeric Adhesives", in *Adhesives and Sealants,* vol. 3, Engineered Materials Handbook, ASM International, 1990.
13. Smith, J., "Very High Bond Adhesives Tapes", *Assembly Engineering,* December 1987.
14. Bennett, G., et. al., "Structural Bonding Tape: An Innovation in Adhesive Bonding", *Adhesives Age,* September 1996.

Chapter 9

Adhesive Composition and Formulation

9.1 Introduction

Generally, it is not necessary to completely understand the ingredients in an adhesive formulation or the manufacturer's reasoning behind the specific formulation. One can often select a commercial adhesive based on functional properties, curing procedures, and other relevant information that are supplied by the adhesive manufacturer. However, a knowledge of the common materials used in formulating adhesives, and compositional techniques used to solve common application problems are especially important to anyone engaged in developing adhesive systems. A general knowledge of these factors is also important to the casual adhesive user so that he or she can more easily converse with the adhesive manufacturer regarding modifications that may be required or regarding certain characteristics of the adhesive system.

This chapter is intended to provide a very basic understanding of the components that are commonly used in adhesive formulation. The relationship between these materials and the properties exhibited by the adhesive system are explained. The customary techniques used to overcome several everyday problems are also discussed. These techniques include formulation processes and materials to control flow, extend temperature range, improve toughness, match thermal expansion coefficients, reduce shrinkage, increase tack, and modify electrical or thermal conductivity.

9.2 Adhesive Composition

Modern-day adhesives are often fairly complex formulations of components that perform specialty functions. Very few polymers are used without the addition of some modifying substance such as plasticizer, tackifier, or inert filler. The selection of the actual ingredients will depend on the end-properties required, the application and processing requirements, and the overall cost target. The various components that constitute an adhesive formulation include the following:

- Base or binder
- Catalyst or hardener
- Accelerators, inhibitors, and retarders
- Solvents
- Diluents
- Extenders
- Fillers
- Carriers and reinforcements
- Plasticizers
- Tackifiers
- Thickeners and thixotropic agents
- Film formers
- Antioxidants, antihydrolysis and antifungal agents, and stabilizers
- Soaps, surfactants, and wetting agents

The adhesive *base* or *binder* is the principal component of an adhesive. The binder provides many of the main characteristics of the adhesive such as wettability, curing properties, strength and environmental resistance. The binder is often by weight the largest component in the adhesive formulation, but this is not always the case especially with highly filled adhesive or sealant systems. The binder is generally the component from which the name of the adhesive is derived. For example, an epoxy adhesive may have many components, but the primary material or base is an epoxy resin. Once the binder is chosen, the other necessary ingredients then can be determined. The next chapter describes in detail the various polymeric resins that are commonly used as bases or binders in adhesive formulations.

A *hardener* or *curing agent* is a substance added to an adhesive to promote the curing reaction by taking part in it. They cause curing by chemically combining with the base resin. Hardeners or curing agents are specifically chosen to react with a certain resin. They will have a significant effect on the curing characteristics and on the ultimate properties of the adhesive system.

Reactive polyamide is an example of a common hardener used in two part epoxy systems. The criticality of the amount of hardener used in the adhesive formulation is dependent on the chemistry of the specific reaction involved. By over or under using a polyamide hardener, the resulting formulation will have more or less of the characteristics of the hardener, but a usable adhesive system will generally result. For example, 10–25% by weight additional polyamide hardener in a two part epoxy formulation will result in a system with greater flexibility and peel strength but lower tensile strength and environmental resistance due to the flexible nature of the polyamide molecule. Less hardener will provide higher shear strength and environmental resistance but poorer peel strength. It must be noted, however, that certain hardeners (e.g., amines, acids, and anhydrides) do have critical mixing requirements, and slight deviations from the manufacturers' mixing instructions could significantly affect the adhesive system.

Catalysts remain unchanged in the curing reaction, causing the primary resin to crosslink and solidify. Acids, bases, salts, sulfur compounds and peroxides are commonly used. Only small quantities are usually required to influence curing. Unlike hardeners, the amount of catalyst used is critical, and poor bond strengths can result when resins are over or under catalyzed.

An *accelerator, inhibitor,* or *retarder* is sometimes incorporated into an adhesive formulation to accelerate or de-accelerate the curing rate. These are critical components that control the curing rate, storage life, and working life of the adhesive formulation.

Solvents are sometimes needed to disperse the adhesive to a consistency that is more easily applied such as by brush or spray. Solvents are also used to aid in formulating the adhesive by reducing the viscosity of the base resin so that additions of other components and uniform mixing may be more easily achieved. Solvents used with synthetic resins and elastomers are generally organic in nature, and often a mixture of different solvents is required to achieve the desired properties. Polar solvents are required with polar resins; nonpolar solvents with nonpolar resins. When solvents are used in the adhesive formulation, they must be completely evaporated prior to cure. Otherwise bubbles could form in the bond-line causing a weak joint. The substrate must also be tested so that the chosen solvents do not attack or degrade it.

Water is sometimes used as a solvent for water soluble resins. Certain adhesives are also available as a water based emulsion or latex formulation. In the early 1970s, during the time of the petroleum crisis, water based adhesives were thought of as a possible replacement for solvent based adhesives systems. However, water based adhesives never met these lofty expectations primarily because of the time and energy required to remove water from the bond-line, the corrosion that the water causes in drying ovens, and the poor moisture resistance of cured water based adhesives.

An ingredient added to an adhesive to reduce the concentration of base resin or binder is called a *diluent*. Diluents are principally used to lower the viscosity and modify the processing conditions of some adhesives. The degree of viscosity reduction caused by various diluent additions to a conventional epoxy adhesive is shown in Fig. 9.1. Diluents do not evaporate as do solvents, but they become part of the final adhesive. Reactive diluents react with the resin base during cure, so that the final adhesive characteristics are determined by the reaction product of the binder and diluent. Nonreactive diluents do not react with the resin or curing agent and, therefore, generally dilute the final physical properties. Coal and pine tar are common nonreactive diluents.

Extenders are substances which usually have some adhesive properties and are added to reduce the concentration of other adhesive components and the cost of the overall formulation. Extenders also have positive value in modifying the physical properties of the adhesive. Common extenders are flours, soluble lignin, and pulverized partly cured synthetic resins.

Fillers are relatively non-adhesive substances added to the adhesive formulation to improve its working properties, strength, permanence, or other qualities. The improvements resulting from the use of common fillers are listed in Table 9.1. Fillers are also used to reduce material cost. By selective use of fillers, the properties of an adhesive can

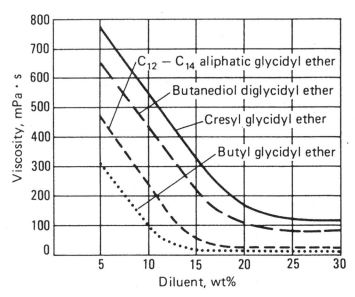

Figure 9.1 Reduction of diglycidyl ether of bisphenol A (DGEPA) epoxy resin by reactive diluents.[1]

TABLE 9.1 Fillers for Common Epoxy Adhesive Formulations[2]

Filler	Improvement in epoxy adhesive
Aluminum	Machinability, impact resistance
Alumina	Abrasion resistance, electrical
Aluminum silicate	Extender
Aluminum trioxide	Flame retardant
Asbestos fibers	Reinforcement
Barium sulfate	Extender
Calcium carbonate	Extender
Calcium sulfate	Extender
Carbon black	Color, reinforcement
Copper	Machinability, elec conductivity
Glass fibers	Reinforcement
Graphite	Lubricity
Iron	Abrasion resistance
Kaolin clay	Extender
Lead	Radiation shielding
Mica	Electrical resistance
Phenolic microspheres	Decrease density
Silica sand	Abrasion, electrical properties
Silicon carbide	Abrasion resistance
Silver	Electrical conductivity
Titanium dioxide	Color
Zinc	Adhesion, corrosion resistance
Zirconium silicate	Arc resistance

be changed significantly. Thermal expansion, electrical and thermal conduction, shrinkage, viscosity, and thermal resistance are only a few properties that can be modified by use of fillers. Common fillers are wood flour, silica, alumina, titanium oxide, metal powders, china clay and earth, slate dust, and glass fibers. Some fillers may act as extenders.

A *carrier* or *reinforcement* is usually a thin fabric, cloth, or paper used to support the semicured adhesive composition to provide a tape or film. In pressure sensitive tapes, the carrier is the backing on which the adhesive is applied. The backing may be used for functional or decorative purposes. In films or structural tape, the carrier is usually porous and the adhesive saturates the carrier. Glass, polyester, and nylon fabric are common carriers for supported B-staged adhesive films. In these cases, the carrier provides a method of applying the adhesive and also acts as reinforcement and an internal "shim" to control the final thickness of the adhesive.

Plasticizers are incorporated into an adhesive formulation to provide flexibility and/or elongation. Plasticizers may also reduce the melt viscosity of hot melt adhesives. Similar to diluents, plasticizers are nonvolatile solvents for the base resin. By being incorporated into the formulation, plasticizers separate the polymer chains and enable their deformation to be more easily accomplished. Plasticizers generally affect the viscoelastic properties of the base resin, whereas diluents simply reduce the viscosity of the system. Whereas diluents result in brittle, hard adhesive systems, plasticizers result in increased flexibility and lower modulus. The temperature at which polymers exhibit rubbery properties, the glass transition temperature, can be reduced with plasticizers.

Certain resinous materials which act as plasticizers are also well noted for increasing the tack of the formulation. Traditional *tackifiers* were based on naturally occurring resins such as pine tar. Tackifiers used in modern adhesive formulations include aliphatic and aromatic hydrocarbons, terpenes, and rosin esters. Tackifying resins that are commonly used in adhesive formulations are identified in Table 9.2. Tackifiers are useful in pressure sensitive adhesives or adhesives which require aggressive tack or "green strength" to assist in assembly of the product. In addition to increasing tack, increased peel strength and decreased shear strength also result from the addition of tackifiers in the adhesive formulation. Figure 9.2 shows the dependence of tack, peel, and shear on an adhesive formulation's resin/tackifier ratio.

Various solvents, soluble agents, resinous powders, or inert fillers can be used to control the viscosity of the adhesive. Viscosity of adhesive systems may need to be increased or decreased depending on the application. Viscosity control is one method commonly used to maintain a consistent product and bondline thickness. *Thickeners* and *thixotropic agents* are used to maintain reasonable thickness of the glue line through viscosity adjustment. Thixotropic fillers are mate-

TABLE 9.2 Incomplete Listing of Commercially Available Tackifying Resins and Illustrative Properties[3]

Commercial resin	Type	Glass transition temperature (°C)	Softening temperature °C (°F)
Piccolyte HM-85 (Hercules)	Styrenated terpene	−35	181 (358)
Nirez K-105 (Reichhold)	Polyterpene	54	192 (378)
Regalrez 1094 (Hercules)	Hydrogenated aromatic	37	186 (367)
Piccovar AP-25 (Hercules)	Aromatic	−25	148 (298)
Wingtack 10 (Goodyear)	C-5	−28	139 (283)
Escorez 1310 (Exxon)	C-5	40	186 (367)
Foral 85 (Hercules)	Glycerine rosin ester	40	181 (358)
Foral 105 (Hercules)	Hydrogenated pentaerythritol rosin ester	57	192 (378)
Zonarez A-100 (Arizona)	Terpene	55	193 (379)
Escorez 2101 (Exxon)	Mixed aliphatic/aromatic	36	185 (365)

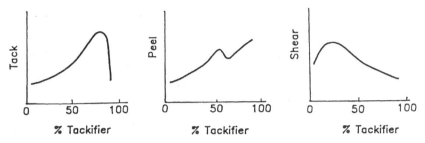

Figure 9.2 Dependence of tack, peel, and shear on resin/tackifier ratio.[4]

rials which when added to the adhesive increase the viscosity when it is under rest. Therefore, thixotropic agents provide sag resistance and the ability for an adhesive to remain in place on a vertical substrate. However, when a slight force is applied, such as in the act of stirring or extruding the adhesive, the system then acts as if it had a lower viscosity and flows with relative ease. Thixotropy will be discussed further in the next section. Scrims, carriers and woven reinforcements are other methods commonly used to control bondline thickness.

The removal of solvent from a true solution of a polymer leaves a film of the polymer. If the same polymer is available as an emulsion, removal of the liquid by evaporation does not necessarily leave a coherent film. The individual globulars of polymer will only merge to form a film if the polymer is well above its glass transition temperature. Drying of some emulsions will produce a powder unless its glass transition has first been lowered by the addition of a plasticizer or *film former* to the emulsion. Film formers are carefully chosen to lower the glass transition temperature without appreciably lowering the strength of the film.

Many polymers have a limited life and are subject to aging processes even before they are used in production. To delay these aging processes, *antioxidants, antihydrolysis agents,* and *stabilizers* are added to adhesive formulations. *Antifungal agents* or *biocides* are used in many water-based adhesive systems.

Aqueous suspensions comprise a wide range of adhesives. These contain as additives the various *soaps, surfactants,* and *wetting agents* necessary to stabilize the emulsion or latex. Additives are also incorporated into aqueous formulations to provide system stability under repeated freeze-thaw cycles during storage.

This description of various additives used in the formulation of adhesives is generic and brief. There are literally thousands of additives that can be used in adhesive systems. The choice depends on the composition of the adhesive system, how it is to be used, system cost, and the properties that need to be obtained. The following section offers a brief description of how certain additives and formulation parameters are used to control the characteristics of adhesive systems.

9.3 Adhesive Formulation

The amount of published material to guide the prospective adhesive formulator is fairly sparse. A few books[5–8] plus scattered information in other texts devoted to adhesives and sealants offer information on the topic. Much of the available information is by way of adhesive formulation recipes with a description of the resulting properties. There is little basic information to train the adhesive formulators or to guide their choice of materials relative to the desired end-properties.

Valuable information can be gained from literature provided by raw material suppliers, who often develop nonproprietary formulations employing their materials. In addition, similar information can be obtained by attending conferences and reading journals and periodicals.

The following description of adhesive formulation technology is only intended to give the reader a general understanding of the opportunities that are available to the willing adhesive formulator. The number of possibilities and the occasions for innovation seem to be endless. However, formulation skills are honed through education and experience (with major emphasis on experience). The foundation of this art comes from knowing which materials to choose based on their compatibility and the end-requirements desired. Knowledge of how to incorporate those materials together into a practical, workable formulation is also required.

Modern adhesives are very complex formulations that rely on their material composition and curing characteristics to provide a practical bond. Raw, unformulated polymeric resins, used with the appropriate hardening system, may provide a satisfactory adhesive without modification, but this is very unusual. The application of pure, unmodified polymeric resins as adhesives is complicated by the tendency of the resin to flow excessively, particularly during cure. This results in irregularities in the glue line and starved areas on the bonding surface. The temperature range of pure resins is also limited; the thermal expansion coefficient is high; the viscosity may not be sufficiently low to provide the required wetting action; shrinkage rates may be too high for many applications; and peel strengths are poor. Through formulation and modification with other materials, these characteristics may be improved selectively, although often only at the sake of other properties. Addressing such tradeoffs is the day-to-day task of the adhesive formulator.

9.3.1 Controlling flow

Controlling flow is an important part of the adhesive formulation process. If the adhesive has a propensity to flow easily before and during cure, then one risks the possibility of a final joint that is starved of adhesive material. If the adhesive flows only with the application of great amount of external pressure, then one risks the possibility of entrapping air at the interface or too thick of a bondline. These factors could result in localized high stress areas within the joint and reduction of the ultimate joint strength.

Flow characteristics can be regulated by the incorporation of fillers, by the use of scrims or woven tapes as "internal shims" within the adhesive itself, or by the careful regulation of the cure cycle. All of these options along with a few more are available to the adhesive formulator.

Generally, fillers are incorporated to control the viscosity of the adhesives as well as other properties. The type and amount of fillers are chosen so that a practical bondline thickness will result after application of the necessary pressure (usually only contact pressure, approximately 5 psi) during cure. Ordinarily, the objective is a bondline thickness of 2–10 mils. Consideration, of course, must be given to the curing temperature. Viscosity of the formulation could drastically be reduced at elevated temperatures. Unless there is a furrow designed into the joint to contain the adhesive, much of the adhesive could flow out of the joint area before the adhesive is completely cured.

Thixotropic adhesive pastes can be formulated which will not flow during cure even at elevated temperatures. Thixotropic adhesives are useful for bonding loose fitting joints. The addition of asbestos at one time provided excellent thixotropic adhesive formulations, but health and environmental regulations have severely limited the use of this material. Today, fumed silica, precipitated calcium carbonate, certain clays, and cellulose and other fibers offer thixotropic properties at relatively low levels of loading. Table 9.3 shows thixotropic epoxy adhesive formulations and resulting properties using fumed silica and reinforcing thixotropic (RT) cellulose fiber additives.

Glass, nylon, polyester, and cotton fabric or mat are often used as a carrier in tape or film adhesive systems. In addition to reinforcement and a way of distributing stress within the joint, the strands of the fabric offer an "internal shim" so that the bondline cannot be thinner than the thickness of these strands. Sufficient pressure need only be

TABLE 9.3 Thixotropic Fillers (Fumed Silica and Reinforcing Thixotropic Fibers) in Epoxy Adhesive Formulations[9]

	Control	Experimental
Formulation		
Epoxy resin 1	47	40
Epoxy resin 2	22	18.7
Dicyandiamide	3.7	3.1
Cure accelerator	1.3	1.1
Ground limestone	20	28.4
Fumed silica, surface treated	6	3
Cellulose reinforcing fibers	—	5.6
Appearance–Relative		
Surface	8.5	8
Gloss	9	9
Uniformity	9	8
Viscosity–Relative		
Initial	86	91
Aged	105	120
Slump resistance (½ × ½ × 4 in.)	No movement	No movement
Shear overlap strength		
Initial	1770 (100% cohesive failure)	1801 (100% cohesive failure)
25 h humidity	1579 (62% cohesive failure)	1727 (95% cohesive failure)
250 h salt spray	1288 (62% cohesive failure)	1592 (97% cohesive failure)

applied to cause the adhesive to flow so that the "shims" meet the substrate surface to provide a positive stop. Paper, mat, and other carrier materials may also be used for this purpose. Glass or polymeric microballoons, incorporated directly into the adhesive formulation, can also provide the shimming function. Here the diameter of the microballoons are used to determine the thickness of the bondline. Another option is to design mechanical shims into the joint itself. The parts to be assembled are designed with lips or stops so that the adhesive cannot flow out of the joint area or that a certain predetermined adhesive thickness is always maintained.

In certain cases it is necessary for the adhesive formulator to reduce the viscosity of the adhesive system to achieve better wetting characteristics. Wetting, as measured by the contact angle that the adhesive makes on the surface, is not governed by the viscosity of the adhesive. However, the rate and ease which the adhesive wets the surface of the substrate and fills in the peaks and valleys on the surface is a function of viscosity. Highly viscous adhesives could require an impractical amount of time to adequately wet the surface of a substrate.

Wetting speed can be improved by incorporating reactive or nonreactive diluents into the system. Diluents are liquid resins that usually lower the viscosity of the adhesive system to make a thin uniform coating possible. The diluent must be compatible with the base resin and not degrade the inherent wetting characteristics of the adhesive. Diluents reduce viscosity and reduce the time it takes for the adhesives to effectively wet the substrate. The reduced viscosity also aids the removal of entrapped air and aids the capillary action of the adhesive in filling pores and cavities that may be on the substrate surface. However, the addition of diluents, especially unreacted liquid resins, usually results in lowered crosslink density which in turn results in reduced elevated temperature strength and environmental resistance.

Solvents can also be used to lower the viscosity of the adhesive to achieve faster wetting of the substrate. However, solvents must be completely evaporated before the adhesive hardens, or else voids could develop in the joint area due to the solvent trying to escape at a later time.

9.3.2 Extending temperature range

Often the adhesive formulator must provide a formulation that must resist either high or low temperatures during service (and sometimes both high *and* low temperatures). In these cases, the trade-offs that must be faced are very formidable. It is difficult to achieve high temperature resistance without a system that cures at elevated temperatures. So certain processing advantages provided by room temperature curing adhesives must be sacrificed if a high degree of temperature resistance is needed. High temperature resistant adhesives are generally brittle and lack high peel strength. The development of an adhesive formulation that has high temperature resistance and high peel strength has long been a "holy grail" to adhesive formulators. With the arrival of certain elastomer modified hybrid systems, this goal has been achieved for the most part, as will be described in the next section.

Sometimes it is necessary to provide a formulation that endures both high and low temperatures. The thermal tiles on the NASA space shuttle are an example. In this application, silicone elastomer based adhesives are used. Shear strength is not as good as high temperature epoxies, for example, but the application does not require excessively high shear strength and a willing trade-off was made to get higher peel strength and resistance to both high and low temperatures.

Flexibilizers can be used to lower the glass transition temperature and improve the low temperature bond strengths of many adhesives. Good properties can be obtained down to $-40°F$; however, adhesive properties in the cryogenic range (<150°F) dictate specially selected primary resins (see Chapter 17). Flexibilizers are also recommended for bonding materials having differing rates of thermal expansion. They provide a degree of elongation in the joint, which allows the adhesive to move with the substrate. This is particularly important when the rate of heating or cooling of the joint is fast.

For high temperature resistance, the selection of the base resin and curing agent are crucial. The cured system must provide a high glass transition temperature. Unfortunately, this usually means that the formulation requires an elevated temperature cure. Also, peel strength is generally low because the high temperature resins tend to have little elongation. Resins and curing agents that provide high crosslinking density and a high degree of molecular aromaticity are usually chosen. Ideally, elevated temperature resistant adhesives should be processed in such a way as to minimize oxygen entrapment during application; they should be as dense as possible; and they should contain antioxidants to improve thermal stability.

Sometimes it is not the high or low temperature itself that causes degradation of the bond, but it is the excursion or process of getting to these temperatures that causes the degradation. Because of differences in thermal expansion coefficient and thermal conductivity, the adhesive does not expand or contract evenly with the substrate as the environmental temperature changes. Thus, it is important that the formulator either modify the thermal expansion coefficient of the adhesive to better match the substrates or provide sufficient flexibility within the adhesive to survive the temperature excursions and the resulting stresses caused within the joint. For adjusting the thermal expansion coefficient, fillers are normally used as discussed in a later section. To provide greater flexibility, the adhesive formulator uses flexibilizers or blends the base resin with an elastomeric resin.

It also should be noted that high temperature adhesives provide somewhat lower room temperature strengths than adhesives formulated specifically for lower temperature service. This is primarily due to the greater toughness of adhesives that have been formulated to cure at room temperature and their resulting higher resistance to deformation in a tensile shear joint (see Chapter 3).

9.3.3 Improving toughness

The ability of an adhesive to absorb energy without catastrophic failure can be increased through toughening. This results in enhanced resistance to fracture, impact, and thermal stress with minimal change in the gross properties of the matrix resin. Improved toughness also results in higher peel strength. The adhesive formulator can improve toughness of the adhesive formulation by various means.

A common method of toughening adhesives is blending the primary resin with other polymers, including thermoplastics or elastomers. However, this technique usually combines the good and bad characteristics of each resin system. The epoxy-nylon adhesives provide a major improvement in toughness over a pure epoxy formulation, but they have sensitivity to moisture because of the nylon (polyamide) constituent. Nitrile-phenolic adhesives have good high temperature properties due to the phenolic constituent, and they also have good peel strength due to the nitrile constituent. However, neither the peel strength nor the temperature resistance is as good as it could have been if the adhesives were formulated from the pure nitrile or phenolic base resins respectively. In these conventional hybrid systems, the added toughening resin reduces the overall glass transition temperature of the system, thereby, reducing elevated temperature performance and environmental resistance of the more brittle constituent.

However, newer adhesives systems have been developed with improved toughness without sacrificing other properties. When cured, these structural adhesives have discrete elastomeric particles embedded in a resin matrix. This can be accomplished in a variety of ways. One approach with epoxy and acrylic adhesives is by incorporating small rubber inclusions into the adhesive formulation. The rubbery polymer is dissolved in the resin binder, and when the adhesive cures, the rubber precipitates as droplets of about one micron in diameter. These rubber inclusions absorb energy and stop a crack from propagating throughout the bond-line. This allows enhanced resistance to fracture, impact, and thermal stress with minimal change in the gross properties of the matrix resin.

The most common toughened hybrids using discrete elastomeric particles are acrylic and epoxy adhesive systems. One such toughened acrylic system is based on a crosslinked network of polymethyl methacrylate with vinyl terminated acrylonitrile butadiene copolymer rubber particles existing as the dispersed phase. Similarly, adhesive systems containing epoxy resins with carboxyl-terminated butadiene acrylonitrile (CTBN) liquid polymers are available. These systems provide a balance of shear, peel, and environmental performance properties. The CTBN epoxy resin modifications are used to toughen both adhesives[10] and composites.[11]

Figure 9.3 shows comparison of a control epoxy adhesive formulation and one containing 15 phr CTBN. With the CTBN formulations, lap shear strength is higher from −50 to 75°C (−60 to 167°F), and peel strength is four times higher at room temperature. The glass transition temperature for these systems are 135°C (275°F) for the control and 129°C (264°F) for the CTBN modified adhesive.

Another class of reactive hybrid adhesive has been developed combining resin mixtures of epoxy and urethane chemistry. These are not

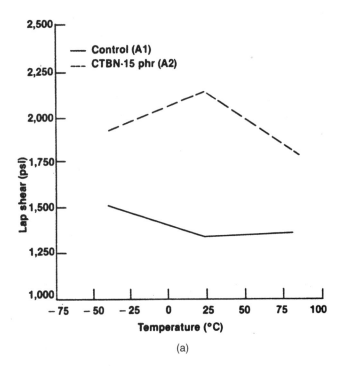
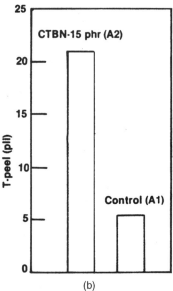

Figure 9.3 Lap shear (a) and peel strength (b) comparison of control epoxy adhesive formulation and one containing 15 phr CTBN elastomer.[11]

as strong at high temperatures as the rubberized epoxy systems described above. However, they offer low-cost formulations that can be applied as hot melts. They crosslink by reacting with moisture in the air to provide quick handling strength and latent thermosetting capability. They have a very high degree of elongation and resist hydrolysis unlike other resins with high elongation. Tensile elongation of the films can vary from 15% to more than 100% compared to 2% for most unmodified epoxy adhesives.[12]

Reactive hybrid adhesives have also been found to bond well to oily cold rolled steel as well as many other substrates. Apparently the elastomeric phase adsorbs much of the oil that is on the surface of the substrate without deterioration to the adhesive or cohesive strength of the system.

9.3.4 Lowering the coefficient of thermal expansion

Depending on the substrate, the curing temperature, and the service temperatures that are expected, the adhesive formulator may want to adjust the coefficient of thermal expansion of the adhesive system. This will lessen internal stresses that occur due to differences in thermal expansion between the substrate and the adhesive. These stresses degrade the joint strength. There are several common occasions when the difference in coefficient of thermal expansion between the substrate and adhesive will result in internal stresses in the joint. One such occurrence is when the cured joint is taken to a temperature that is different from its curing temperature.

Ideally, the coefficient of thermal expansion should be lowered (or raised) to match that of the material being bonded. With two different substrate materials, the adhesive's coefficient of thermal expansion should be adjusted to a value between those of the two substrates. This is usually done by using fillers as shown in Fig. 9.4. It is usually not possible to employ sufficiently large filler loadings to accomplish the degree of thermal expansion modification required. High loading volumes increase viscosity to the point where the adhesive cannot be easily applied or wet the substrate. For some applications and some fillers, loading volumes up to 200 phr may be employed, but optimum cohesive strength values are usually obtained with lesser amounts.

When temperature limits permit, it is useful to compensate for differences in the coefficient of thermal expansion by the use of flexibilizers to absorb the internal stresses during thermal cycling. When bonding materials with greatly differing rates of thermal expansion, severe warping may be encountered at temperature extremes in the absence of holding fixtures. To survive this type of stress, a very flexible adhesive may be required rather than one that is highly filled.

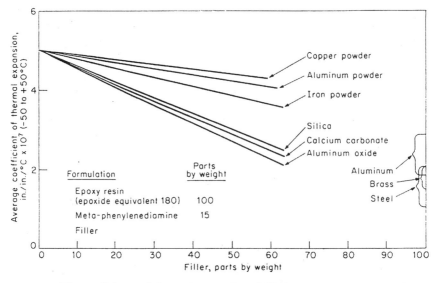

Figure 9.4 The coefficients of thermal expansion of filled epoxy resins compared with those of common metals.[13]

9.3.5 Reducing shrinkage

All resins will shrink on hardening or curing. This shrinking process could cause internal stresses within the joint that degrade its strength. Depending on the primary base resin, the adhesive formulator may need to reduce the amount of shrinkage when the adhesive hardens. This can be accomplished in several ways.

Fillers reduce the rate of shrinkage by bulk displacement of the resin in the adhesive formulation. Fillers may improve operational bond strength by 50 to 100%. The effect of various fillers and loading ratios on the strength properties of epoxy adhesive formulations is indicated in Fig. 9.5. The effect of different fillers loaded at a constant 100 phr is indicated in Table 9.4 for shear strength on phenolic laminates and aluminum.

Another way of reducing shrinkage is by blending the base resin with one that does not shrink as much on cure. In fact, there are certain monomers that have been developed (although expensive and not commercially available) that will expand on curing.[16] These monomers when blended with more common resins can provide a resulting adhesive system with either zero net shrinkage or actual volume expansion during cure. It is believed that when formulated into adhesive systems, these expanding monomers will provide exceptional bond strength and mechanical interlocking capability that previously was not realized with conventional adhesive technology.

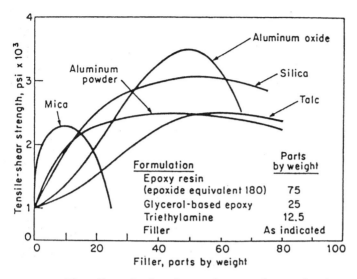

Figure 9.5 The effect of mineral extenders on the tensile shear strength of epoxy adhesives (the substrate is aluminum).[14]

TABLE 9.4 Effect of Fillers on Block Shear Strength of Adhesive Formulations[15]

(In pounds per square inch)

Fillers at 100 phr	Phenolic linen laminates tested at				Aluminum blocks tested at	
	23°C	75°C	90°C	105°C	23°C	105°C
Aluminum powder.....	2,790	....	1,470	1,390		
Ignited Al$_2$O$_3$..........	4,600	1,360	1,195	530	1,650	1,150
Short-fiber asbestos, 24 phr..............	1,740	1,270	580	510	2,180	3,910
Carbon black..........	2,000	555	980	910	2,000	1,170
Silica.................	2,840	1,600	1,250	830	1,530	600
Zinc dust	2,510	600	300	225	4,080	3,865

9.3.6 Increasing tack

In adhesive formulations, it is sometimes necessary to achieve some degree of initial tack in order for the substrates to remain in position until the adhesive hardens. Tack can be achieved by proper selection of adhesive composition ingredients, usually by the use of higher molecular weight resins.

A number of liquid and solid tackifiers are possible for use in adhesive formulations. The role of tackifiers in uncured adhesives is well understood.[17,18] Pressure sensitive adhesive performance, for example, depends heavily on tackifiers and the proper balance of viscoelastic properties. The tackifier provides the viscous nature to the adhesive whereas the base polymer provides the elastic nature. The effect of tackifiers on cure characteristics and end-properties must be understood since the effects can be significant.[19]

9.3.7 Modifying electrical and thermal conductivity

The electrical and thermal conductivity properties of adhesive systems can also be modified by formulation composition. Unmodified polymeric resins are natural insulators and do not exhibit electrical conductivity. There are certain applications, notably in the electrical and electronics industry, where electrically conductive adhesives provide a significant value. One application involves the use of conductive adhesives as an alternative for wire or circuit board soldering.

Similarly, most unmodified polymeric resins have very low thermal conductivity. There are certain applications where high levels of thermal conductivity are important in adhesives. One example is the attachment of heat sinks that must be tightly bonded to power electronic components.

Appropriate fillers have been used to produce adhesive with high thermal or electrical conductivity for specialized applications. It should be noted that, regardless of the adhesive system itself, electrical or thermal conductivity is improved by minimizing the adhesive bondline and by minimizing the organic or non-conductive part of the adhesive.

Electrically conductive adhesives owe their conductivity as well as their high cost to the incorporation of high loadings of metal powders or other special fillers of the types shown in Table 9.5. Virtually all high performance conductive products today are based on flake or powdered silver. Silver offers an advantage in conductivity stability that cannot be matched by copper or other lower cost metal powders.[14] Where elastomeric properties are required, silver filled epoxy and silver filled silicone rubber systems are commercially available. Conductive carbon (amorphous carbon or fine graphite) can also be used in conductive adhesive formulations if the degree of conductivity can be sacrificed for lower cost.

Figure 9.6 shows the mechanism for current flow through a metal filled polymer. If enough metal particles are added to form a network within the polymer matrix, electrons can flow across the particle contact points making the mixture electrically conductive.

The metal powder filled adhesives conduct both heat and electricity. Some applications, however, must conduct heat but not electricity. For example, power electronic devices and other heat generating components are bonded to heat sinks and other metal surfaces. Here the

TABLE 9.5 Volume Resistivity of Metals, Conductive Plastics and Various Insulation Materials at 25°C[20]

	Specific gravity (gms/cm^3)	ρ = volume resistivity (ohm cm)
Silver	10.5	1.6×10^{-6}
Copper	8.9	1.8×10^{-6}
Gold	19.3	2.3×10^{-6}
Aluminum	2.7	2.9×10^{-6}
Nickel	8.9	10×10^{-6}
Platinum	21.5	21.5×10^{-6}
Eutectic solders	—	$20\text{–}30 \times 10^{-6}$
Best silver-filled inks and coatings	—	1×10^{-4}
Best silver-filled epoxy adhesives	—	1×10^{-3}
Graphite	—	1.3×10^{-3}
Low cost silver-filled epoxy adhesives	—	1×10^{-2}
Graphite or carbon-filled coatings	—	10^2 to 10
Oxide-filled epoxy adhesives	1.5–2.5	$10^{14}\text{–}10^{15}$
Unfilled epoxy adhesives	1.1	$10^{14}\text{–}10^{15}$
Mica, polystyrene & other best dielectrics	—	10^{16}

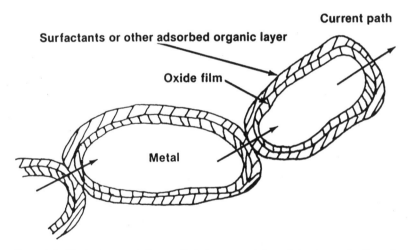

Figure 9.6 Typical current flow path between metal particles in conductive adhesives. Adsorbed organic molecules and oxide films prevent or impede passage of electrons across contact points.[20]

adhesive must permit high transfer of heat plus electric isolation. Fillers used for achieving thermal conductivity alone include aluminum oxide, beryllium oxide, boron nitride, and silica. Theoretically boron nitride is an optimum filler for thermally conductive adhesives. However, it is difficult to fill systems to greater than 40% by weight with boron nitride. Beryllium oxide is high in cost and its thermal conductivity drops drastically when mixed with organic resins. Aluminum oxide is commonly used. Table 9.6 lists thermal conductivity values for several metals as well as for beryllium oxide, aluminum oxide, and for several filled and unfilled resins. Figure 9.7 shows the thermal conductivity values for an epoxy resin as a function of volume fraction of several heat conductive fillers.

9.4 Commercial Formulations

There are literally hundreds of thousands of commercial adhesive products of various types available to all market sectors. Some of these commercial adhesive products are described in the next chapter. Adhesive products are made available from many thousands of adhesive suppliers that range from very small business organizations of several

TABLE 9.6 Thermal Conductivity of Metals, Oxides, and Conductive Adhesives[20]

	Thermal conductivity at 25°C (Btu/hr°F ft^2/ft)
Silver	240
Copper	220
Beryllium oxide	130
Aluminum	110
Steel	40
Eutectic solders	20–30
Aluminum oxide	20
Best silver-filled epoxy adhesives	1 to 4
Aluminum-filled (50%) epoxy	1 to 2
Epoxy filled with 75% by wt. Al_2O_3	0.8 to 1
Epoxy filled with 50% by wt. Al_2O_3	0.3 to 0.4
Epoxy filled with 25% by wt. Al_2O_3	0.2 to 0.3
Unfilled epoxies	0.1 to 0.15
Foamed plastics	0.01 to 0.03
Air	0.015

Table for Conversion of Thermal Conductivity Units

g cal/cm^{2}sec °C/cm	w/cm^2 °C/cm	Btu/ft^{2}hr°F/ft	Btu/ft^{2}hr°F/in
1.0	4.19	242	2900
0.23	1.0	58	690
4.13×10^{-3}	0.0173	1.0	12.0
3.44×10^{-4}	1.44×10^{-3}	0.083	1.0

Heat transfer formula:
$$q = \frac{k\Delta T}{x}$$

k = thermal conductivity
ΔT = temperature drop cross material
q = heat flow/unit area
x = material thickness

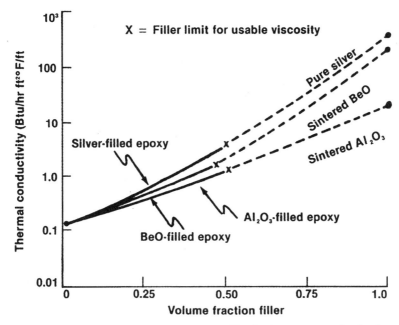

Figure 9.7 Thermal conductivity increases as filler loading increases, but loading is limited by mixture viscosity to a maximum of 40–50% by volume.[20]

people to large, multi-national companies. Their commercial formulations are proprietary and consist of many major and minor components that have been selected, tested, and field applied to meet specific applications or problems.

The intelligence behind a formulation is based on the adhesive formulator's knowledge of many of the principles that are discussed in this Handbook, knowledge of polymeric materials, and skill at applying that knowledge to meet a specific need. However, it must be remembered that these successful formulations are the results of possibly hundreds of formulations that didn't work. Certain formulations have been developed to work with specific applications; they may not work as well in another application. Although we like to think that adhesive technology is a science, it is very much trial and error with continual adjustment and fine tuning. The prime value of the formulator is the amount of time that he can reduce in the trial and error period so that one can zero-in on solutions with minimal resources.

Adhesive suppliers are eager to help with an application. Usually, the more information that they have the easier it will be to choose a current formulation, make a refinement, or custom formulate a new adhesive system. Information that the adhesive supplier must know before he can make an informed recommendation is summarized in Fig. 9.8.

Often it is tempting for the novice or adhesive user to try to formulate custom adhesives to meet a specific application. However, this is usually not a very efficient approach unless one is very skilled in both the science and the art. Formulating skills cannot generally be developed through reading or academic work. They can only be developed by working at the bench, selecting ingredients, compounding them, and then characterizing the end result with close cooperation with the end-user. This multilevel experience is the value that the many adhesive suppliers have to offer.

Information regarding adhesive formulations is available from several sources. There are several books that offer limited information on the topic plus scattered information in periodicals, journals, confer-

WHAT **MATERIALS** ARE BEING JOINED? **1**

MY APPLICATION IS TO: ☐bond ☐coat ☐seal ☐pot

____

TO ______________________________

Please be specific—say "aluminum to foamed styrene" NOT "metal to plastic." Submit samples if possible.

WHAT ARE THE **END-USE** SERVICE REQUIREMENTS? **2**

MY ASSEMBLY WILL BE EXPECTED TO WITHSTAND:

A. Temperature extremes of ______________ °F ☐continuously ☐intermittently

B. These chemicals or solvents ______________

☐immersion ☐vapor ☐continuously ☐intermittently

C. These conditions: ☐moisture ☐sunlight ☐outdoor weathering

D. Color requirements: ______________

E. These loads: (Describe amount and type of strength required. Sketch, if possible) ______________

WHICH **PRODUCTION METHOD** WILL BE USED? **3**

A. I prefer the following form of adhesive: ☐liquid ☐paste ☐film ☐powder

B. This will be ☐hand application. ☐machine application.

C. My equipment makes it desirable to use: ☐brush ☐extruder ☐knife or trowel ☐caulking gun ☐spray ☐roll coater ☐other ______________

D. These are my application requirements:

Viscosity desired: ______________

Drying time required: ______________

Curing temp. available: ______________

Curing time available: ______________

Pressure available: ______________

E. If necessary, we ☐can ☐cannot use a 2-part adhesive.

Estimated monthly volume: ______________ Seasonal: ☐no ☐yes

Price limitations: ______________

Present products used: (Samples if possible) ______________

Why unsatisfactory: ______________

Remarks: ______________

Figure 9.8 Problem analysis form for selection of adhesives.[21]

ence proceedings, and other texts, which describe adhesive formulations for specific applications. Often the user can try to employ an adhesive that was successfully developed for an application that has many of the same requirements as the current application. Valuable information can also be gained from the raw materials suppliers. They often develop adhesive formulations using their materials and freely offer these to both the adhesive suppliers and to the general public. Often these formulations become the basis of commercially available systems.

References

1. Behm, D. T., and Gannon, J., "Epoxies", in *Adhesives and Sealants,* vol. 3, Engineered Materials Handbook, ASM International, 1990.
2. *Materials Engineering,* May 1976, at 73.
3. Pocius, A. V., *Adhesion and Adhesives Technology,* Chapter 9, "The Chemistry and Physical Properties of Elastomer-Based Adhesives" (New York: Hanser Publishers, 1997).
4. Goulding, T. M., "Pressure Sensitive Adhesives", in *Handbook of Adhesive Technology,* Pizzi, A. and Mittal, K. L., eds. (New York: Marcel Dekker, 1994).
5. Skiest, I., ed. *The Handbook of Adhesives,* 3rd edition, (New York: Van Nostrand Reinhold Publishing, 1990).
6. Satas, D., ed., *The Handbook of Pressure Sensitive Adhesives,* 2nd edition, (New York: Van Nostrand Reinhold Publishing, 1989).
7. Flick, E., ed., *Adhesives and Sealant Compound Formulation,* 2nd edition, (Park Ridge, NJ: Noyes Publications, 1984).
8. Ash, M., and Ash, I., eds., *Formulary of Adhesives and Sealants,* (New York: Chemical Publishing Co., 1987).
9. Gerace, M. J., and Sullivan, C., "Cellulose Fibers as a Reinforcing Thixotrope in Adhesives and Sealants", *Adhesives Age,* July 1995.
10. Bolger, J. C., "Structural Adhesives: Today's State of the Art", in *Adhesives in Manufacturing,* G. L. Schneberger, ed. (New York: Marcel Dekker, 1983).
11. Siebert, A. R., et al., "CTBN Modified Epoxies Work in Poor Bonding Conditions", *Adhesives Age,* July 1986.
12. Guthrie, J. L., and Lin, S. C., "One Part Modified Epoxies for Unprimed Metal, Plastics", *Adhesives Age,* July 1985.
13. Lee, H., and Nevel, K., *Epoxy Resins, Their Applications and Technology* (New York: McGraw-Hill, 1957) at 151.
14. Burgman, H. A., "Selecting Structural Adhesive Materials", *Electro-Technology,* June 1965.
15. Lee, H., and Nevel, K., *Handbook of Epoxy Resins,* Chapter 21, "Epoxy Resin Adhesives" (New York: McGraw Hill, 1982).
16. Sadhir, R. K., and Luck, R. M., *Expanding Monomers: Synthesis, Characterization, and Applications* (Boca Raton, FL.: CRC Press, 1992).
17. Schlademan, J. A., *Handbook of Pressure Sensitive Adhesive Technology,* 2nd ed., D. Satas, ed. (New York: Van Nostrand Reinhold, 1989) at 527–544.
18. Aubrey, D. W., *Rubber Chem. Tech.,* 61:448, 1988.
19. Schladerman, J. A., "Tackifiers and Their Effect on Adhesive Curing", *Adhesives Age,* September 1999.
20. Bolger, J. C., and Morano, S. L., "Conductive Adhesives: How and Why They Work", *Adhesives Age,* June 1984.
21. Problem Analysis Form, National Starch Corp. Bulletin.

Chapter 10

Adhesives Families

10.1 Introduction

This chapter will review the common chemical families of materials that are employed in adhesive formulations. They consist of synthetic polymeric resins, naturally occurring resins, and inorganic products. These materials form the base resin portion of the adhesive formulation. Along with additives and modifiers, they govern the chemical and physical characteristics of the adhesive system. Families of materials that are normally used in sealant formulations will be described in Chapter 13.

10.2 Classification Methods

There are many ways of classifying adhesives. In this chapter they will primarily be grouped as either structural or non-structural. This grouping is admittedly subjective and broad, but it represents the type of application where the adhesive is commonly used. We can also classify adhesives by whether or not they form crosslinked molecules (i.e., thermosetting or thermoplastic), by their flexibility (i.e., rigid or elastomeric), and by their form and curing conditions. Table 10.1 identifies and classifies common adhesive families that are covered in this chapter.

Structural applications are those requiring high strength or permanence. Structural adhesive joints are expected to provide a useful service life that is equivalent to that of the product in which the adhesive joint is a part. Non-structural applications are those that generally require only low strength or temporary fastening. Examples of non-structural adhesives are pressure sensitive tapes, hot melts, or packaging adhesives. Some define structural adhesives as systems

TABLE 10.1 General Classification of Common Polymeric Resin Families Used in Adhesive Formulations

Chemical family	Thermoset or thermoplastic		General use			Flexibility		Form	Cure
	TS	TP	Structural Yes	Structural No	Sealant	Rigid (T = Tough)	Elastomeric		
Thermosetting resins									
Epoxy	x		x			x		F, S, 1C, 2CC	RT, ET
Modified epoxies									
Toughened epoxy hybrids	x		x			x		2CC	RT, ET
Epoxy phenolic	x		x			x		F, 1C	ET
Epoxy nylon	x		x			T		F, 1C	ET
Epoxy polysulfide	x		x		x	T		2CC	RT
Resorcinol and phenol formaldehyde	x		x			x		2CC	RT
Melamine and urea formaldehyde	x		x			x		1WR	RT, ET
Modified phenolics									
Nitrile phenolic	x		x			T		F, 1S	ET
Vinyl phenolic	x		x			T		F, 1S	ET
Neoprene phenolic	x		x			T		F, 1S	ET
Aromatic high temperature systems									
Polyimide	x		x			x		F, 1S	ET, P, V
Bismaleimide	x		x			x		F, 2CC	ET
Polybenzimidizole	x		x			x		F	ET, P, V
Polyesters	x		x			x		2CC	RT
Polyurethane	x		x				x	2CC	RT
Anaerobic resins	x		x		x	x		1C	RT
Cyanoacrylates	x		x			x		1C	RT
Modified acrylic	x		x			x		2CC	RT

Material							Form	Cure
Elastomeric resins								
Natural rubber		x	x			x	PS, CT, 1S	RT
Asphalt		x	x			x	1S, CT	RT
Butyl rubber		x	x			x	1S, CT	RT
Chloroprene (neoprene)	x	x	x			x	1S, CT	RT
Acrylonitrile butadiene (nitrile)	x	x	x			x	1S, CT	RT
Polyisobutylyene		x	x			x	PS, 1S	RT
Polyvinyl methyl ether		x	x			x	1S	RT
Polysulfide	x	x	x			x	2CC	RT
Silicone	x	x	x			x	1C, 2CC	RT
Thermoplastic resins								
Polyvinyl acetal		x	x				HM, 1S	RT
Polyvinyl acetate		x	x	x			HM, 1E	RT
Polyvinyl alcohol		x	x				HM, 1E	RT
Thermoplastic elastomers		x			x		HM, PS	RT
Ethylene vinyl acetate		x	x		T		HM, 1S	RT
Cellulosic resins		x	x		x		1S	RT
Polyamide		x	x		T		HM, 1S	RT
Polyester (saturated)		x	x		x	T	HM, 1S	RT
Polysulfone		x	x		T		HM, 1S	RT
Phenoxy		x	x		T		HM, 1S	RT
Acrylic		x	x	x	x		1S, PS	RT
Other materials used as adhesives								
Agricultural glues	x		x		x		1WR	RT
Animal glues	x		x		x		1WR	RT
Sodium silicate	x		x		x		1WR	RT
Phosphate cements	x		x		x		1WR	RT
Litharge cement	x		x		x		1WR	RT
Sulfur cement	x		x		x		1WR	RT

NOTES: Form: CT—Contact, PS—Pressure sensitive, F—Film (supported or unsupported), S—Solid (powder or preform); HM—Hot melt, 1C—One part (curing), 1E—One part (emulsion), 1S—One part (solvent), 1WR—One part (water required), 2CC—Two part liquid or paste (chemical cure); ET—Elevated temperature cure.

Cure: RT—Room temperature cure, P—Pressure required, V—Vacuum required depending on system

TABLE 10.2 Thermosetting Adhesives

Adhesive	Description	Curing method	Special characteristics	Usual adherends	Price range
Cyanoacrylate	One-part liquid	Rapidly at RT in absence of air	Fast setting; good bond strength; low viscosity; high cost; poor heat and shock resistance; will not bond to acidic surfaces	Metals, plastics, glass	Very high
Polyester	Two-part liquid or paste	RT or higher	Resistant to chemicals, moisture, heat, weathering. Good electrical properties; wide range of strengths; some resins do not fully cure in presence of air; isocyanate-cured system bonds well to many plastic films	Metals, foils, plastics, plastic laminates, glass	Low-med
Urea formaldehyde	Usually supplied as two-part resin and hardening agent. Extenders and fillers used	Under pressure	Not as durable as others but suitable for fair range of service conditions. Generally low cost and ease of application and cure. Pot life limited to 1 to 24 h	Plywood	Low
Melamine formaldehyde	Powder to be mixed wth hardening agent	Heat and pressure	Equivalent in durability and water resistance (including boiling water) to phenolics and resorcinols. Often combined with ureas to lower cost. Higher service temp than ureas	Plywood, other wood products	Medium
Resorcinol and phenolresorcinol formaldehyde	Usually alcohol-water solutions to which formaldehyde must be added	RT or higher with moderate pressure	Suitable for exterior use; unaffected by boiling water, mold, fungus, grease, oil, most solvents. Bond strength equals or betters strength of wood; do not bond directly to metal	Wood, plastics, paper, textiles, fiberboard, plywood	Medium

Adhesive	Form	Cure	Properties	Substrates	Cost
Epoxy	Two-part liquid or paste; one-part liquid, paste, or solid; solutions	RT or higher	Most versatile adhesive available; excellent tensile-shear strength; poor peel strength; excellent resistance to moisture and solvents; low cure shrinkage; variety of curing agents/hardeners results in many variations	Metals, plastic, glass, rubber, wood, ceramics	Medium
Polyimide	Supported film, solvent solution	High temp	Excellent thermal and oxidation resistance; suitable for continuous use at 550°F and short-term use to 900°F; expensive	Metals, metal foil, honeycomb core	Very high
Polybenzimidazole	Supported film	Long, high-temp cure	Good strength at high temperatures; suitable for continuous use at 450°F and short-term use at 1000°F; volatiles released during cure; deteriorate at high temperatures on exposure to air; expensive	Metals, metal foil, honeycomb core	Very high
Bismaleimide	One part paste of film	High temperature	Suitable for long term exposure to 450°F and short term exposure to 400°F; low peel strength	Metals, circuit boards, composites	High
Acrylic	Two-part liquid or paste	RT	Excellent bond to many plastics, good weather resistance, fast cure, catalyst can be used as a substrate primer; poor peel and impact strength	Metals, many plastics, wood	Medium
Anaerobic	One-part liquid or paste	RT or higher in absence of air	Chemically blocked, anaerobic type; excellent wetting ability; useful temperature range −65 to 300°F; withstands rapid thermal cycling; high-tensile-strength grade requires cure at 250°F, cures in minutes at 280°F	Metals, plastics, glass, wood	Very high

TABLE 10.3 Thermoplastic Adhesives

Adhesive	Description	Curing method	Special characteristics	Usual adherends	Price range
Cellulose acetate, cellulose acetate butyrate	Solvent solutions	Solvent evaporation	Water-clear, more heat resistant but less water resistant than cellulose nitrate; cellulose acetate butyrate has better heat and water resistance than cellulose acetate and is compatible with a wider range of plasticizers	Plastics, leather, paper, wood, glass, fabrics	Low
Cellulose nitrate	Solvent solutions	Evaporation of solvent	Tough, develops strength rapidly, water-resistant; bonds to many surfaces; discolors in sunlight; dried adhesive is flammable	Glass, metal, cloth, plastics	Low
Polyvinyl acetate	Solvent solutions and water emulsions, plasticized or unplasticized, often containing fillers and pigments. Also dried film which is light-stable, waterwhite, transparent	On evaporation of solvent or water; film by heat and pressure	Bond strength of several thousand lb/in.^{2} but not under continuous loading. The most versatile in terms of formulations and uses. Tasteless, odorless; good resistance to oil, grease, acid; fair water resistance	Emulsions particularly useful with porous materials like wood and paper. Solutions used with plastic films, mica, glass, metal, ceramics	Low
Vinyl vinylidene	Solutions in solvents like methyl ethyl ketone	Evaporation of solvent	Tough, strong, transparent and colorless. Resistant to hydrocarbon solvents, greases, oils	Particularly useful with textiles; also porous materials, plastics	Medium

Polyvinyl acetals	Solvent solutions, film, and solids	Evaporation of solvent; film and solid by heat and pressure	Flexible bond; modified with phenolics for structural use; good resistance to chemicals and oils; includes polyvinyl formal and polyvinyl butaryl types	Metals, mica, glass, rubber, wood, paper	Medium
Polyvinyl alcohol	Water solutions, often extended with starch or clay	Evaporation of water	Odorless, tasteless and fungus-resistant (if desired). Excellent resistance to grease and oils; water soluble	Porous materials such as fiberboard, paper, cloth	Low
Polyamide	Solid hot-melt, film, solvent solutions	Heat and pressure	Good film flexibility; resistant to oil and water; used for heat-sealing compounds	Metals, paper, plastic films	Medium
Acrylic	Solvent solutions, emulsions and mixtures requiring added catalysts	Evaporation of solvent; RT or elevated temp (two-part)	Good low-temperature bonds; poor heat resistance; excellent resistance to ultraviolet; clear; colorless	Glass, metals, paper, textiles, metallic foils, plastics	Medium
Phenoxy	Solvent solutions, film, solid hot-melt	Heat and pressure	Retain high strength from 40 to 180°F; resist creep up to 180°F; suitable for structural use	Metals, wood, paper, plastic film	Medium

TABLE 10.4 Elastomeric Adhesives

Adhesive	Description	Curing method	Special characteristics	Usual adherends	Price range
Natural rubber	Solvent solutions, latexes and vulcanizing type	Solvent evaporation, vulcanizing type by heat or RT (two-part)	Excellent tack, good strength. Shear strength 3–180 lb/in^2; peel strength 0.56 lb/in. width. Surface can be tack-free to touch and yet bond to similarly coated surface	Natural rubber, masonite, wood, felt, fabric, paper, metal	Medium
Reclaimed rubber	Solvent solutions, some water dispersions. Most are black, some gray and red	Evaporation of solvent	Low cost, widely used. Peel strength higher than natural rubber; failure occurs under relatively low constant loads	Rubber, sponge rubber, fabric, leather, wood, metal, painted metal, building materials	Low
Butyl	Solvent system, latex	Solvent evaporation, chemical cross linking with curing agents and heat	Low permeability to gases, good resistance to water and chemicals, poor resistance to oils, low strength	Rubber, metals	Medium
Polyisobutylene	Solvent solution	Evaporation of solvent	Sticky, low-strength bonds; copolymers can be cured to improve adhesion, environmental resistance, and elasticity; good aging; poor thermal resistance; attacked by solvents	Plastic film, rubber, metal foil, paper	Low
Nitrile	Latexes and solvent solutions compounded wth resins, metallic oxides, fillers, etc.	Evaporation of solvent and/or heat and pressure	Most versatile rubber adhesive. Superior resistance to oil and hydrocarbon solvents. Inferior in tack range; but most dry tack-free, and advantage in precoated assemblies. Shear strength of 150–2,000 lb/in.^{2}, higher than neoprene, if cured	Rubber (particularly nitrile) metal, vinyl plastics	Medium

Type	Form/Composition	Cure	Properties	Typical Uses	Cost
Styrene butadiene	Solvent solutions and latexes. Because tack is low, rubber resin is compounded with tackifiers and plasticizing oils	Evaporation of solvent	Usually better aging properties than natural or reclaimed. Low dead load strength; bond strength similar to reclaimed. Useful temp range from −40 to 160°F	Fabrics, foils, plastics film laminates, rubber and sponge rubber, wood	Low
Polyurethane	Two-part liquid or paste	RT or higher	Excellent tensile-shear strength from −400 to 200°F; poor resistance to moisture before and after cure; good adhesion to plastics	Plastics, metals, rubber	Medium
Polysulfide	Two-part liquid or paste	RT or higher	Resistant to wide range of solvents, oils, and greases; good gas impermeability; resistant to weather, sunlight, ozone; retains flexibility over wide temperature range; not suitable for permanent load-bearing applications	Metals, wood, plastics	High
Silicone	Solvent solution: heat or RT curing and pressure-sensitive; and RT vulcanizing solventless pastes	Solvent evaporation, RT or elevated temp	Of primary interest is pressure-sensitive type used for tape. High strengths for other forms are reported from −100 to 500°F; limited service to 700°F. Excellent dielectric properties	Metals; glass; paper; plastics and rubber, including silicone and butyl rubber and fluorocarbons	High–very high
Neoprene	Latexes and solvent solutions, often compounded with resins, metallic oxides, fillers, etc.	Evaporation of solvent	Superior to other rubber adhesives in most respects—quickness; strength; max temp (to 200°F, sometimes 350°F); aging; resistant to light, weathering, mild acids, and oils	Metals, leather, fabric, plastics, rubber (particularly neoprene), wood, building materials	Medium

TABLE 10.5 Hybrid Modified Adhesives

Adhesive	Description	Curing method	Special characteristics	Usual adherends	Price range
Epoxy-phenolic	Two-part paste, supported film	Heat and pressure	Good properties at moderate cures; volatiles released during cure; retains 50% of bond strength at 500°F; limited shelf life; low peel strength and shock resistance	Metals, honeycomb core, plastic laminates, ceramics	Medium
Epoxy-polysulfide	Two-part liquid or paste	RT or higher	Useful temperature range −70 to 200°F, greater resistance to impact, higher elongation, and less brittleness than epoxies	Metals, plastic, wood, concrete	Medium
Epoxy-nylon	Solvent solutions, supported and unsupported film	Heat and pressure	Excellent tensile-shear strength at cryogenic temperature; useful temperature range −423 to 180°F; limited shelf life	Metals, honeycomb core, plastics	Medium
Nitrile-phenolic	Solvent solutions, unsupported and supported film	Heat and pressure	Excellent shear strength; good peel strength; superor to vinyl and neoprene-phenolics; good adhesion	Metals, plastics, glass, rubber	Medium
Neoprene-phenolic	Solvent solutions, supported and unsupported film	Heat and pressure	Good bonds to a variety of substrates; useful temp range −70 to 200°F, excellent fatigue and impact strength	Metals, glass, plastics	Medium

Vinyl-phenolic	Solvent solutions and emulsions, tape, liquid and coreacting powder	Heat and pressure	Good shear and peel strength; good heat resistance; good resistance to weathering, humidity, oil, water, and solvents, vinyl formal and vinyl butyral forms available, vinyl formal-phenolic is strongest	Metals, paper, honeycomb core	Low-medium
Rubber modified epoxy	One part liquid or paste	Heat and pressure	Good peel strength without sacrifice of elevated temperature properties; good impact strength and low temperature properties	Metals, thermosets, thermoplastics and oily surfaces	Medium
Rubber modified acrylic	One or two part liquid or pastes	RT or higher	Good shear and impact resistance at temperatures of −150 to +250°F	Metals, thermosets, thermoplastics and oily surfaces	Medium
Epoxy urethane	One or two part paste	RT or higher	Flexible bonds with service temperature of −20 to +200°F; High impact and peel strength	Metals, thermosets, thermoplastics and oily surfaces	Medium

having tensile shear strength greater than 1000 psi and properties that do not change significantly with moderate aging. Non-structural adhesives are those with less than 1000 psi shear strength and durability that requires a sheltered or relatively mild environment.

Several adhesives are difficult to classify in this manner. For example, is a hot melt textile adhesive used in clothing manufacture a structural or non-structural adhesive? It has moderate strength and sufficient heat and chemical resistance to withstand washings and perhaps dry-cleaning. The clothing manufacturer may consider a hot melt adhesive to be as "structural" a fastener as thread or stitching. Hot melts are perfectly adequate for the application. However, hot melts are not generally considered structural adhesives in that they are not used as a primary fastener in a structural application requiring long lasting exposure to a variety of environments. On the other hand, cyanoacrylate adhesives, more commonly known as superglues, are usually classified as structural adhesives. However, their resistance to heat and moisture is relatively weak. Thus, classification of adhesives as either structural or non-structural should only be used to broadly define adhesive families and not to judge functional differences.

Polymeric adhesive families can also be classified by their chemical/physical nature as being either thermoplastic, thermosetting, elastomeric, or a hybrid modification depending on the polymer chemistry of the main constituent resins. Thermoplastic adhesives are not crosslinked and revert to a fluid state when heated. Thermosetting adhesives result in crosslinked molecular structures and do not soften with heat or solvents. Elastomeric adhesives are formulated from natural or synthetic rubbers; they are characterized by high elongation and usually low tensile strength. Elastomeric adhesives can be either thermosetting or thermoplastic. Because of their relative molecular mobility, thermosetting elastomeric adhesives soften with heat but do not become fluid as do the thermoplastics. Therefore, they are not considered in the same classification as thermosetting adhesives. Hybrid adhesives are a class of adhesives compounded from a mixture of polymers for specific property adjustment. They generally consist of two or more resin constituents each of which has characteristics that are valued in the application but are difficult to obtain with a single resin (e.g., high temperature resistance and toughness). Usually, hybrids are thermosetting in character and are considered to be structural adhesives. The adhesive families represented by these four broad classifications are described in Tables 10.2 through 10.5.

The types of polymeric resins that are listed under thermosetting and hybrid adhesives are noted for high strength, creep resistance, and resistance to environments such as heat, moisture, solvents, and oils. Their physical properties are well suited for structural adhesive applications.

Elastomeric and thermoplastic adhesives are generally not used in joints requiring continuous loading because of their tendency to creep under stress. They are also degraded by many common service environments. These adhesives find greatest use in low-strength applications. Elastomeric adhesives represent a separate classification because they exhibit unique rheological properties. Elastomeric adhesives are generally thermosetting, although there are also thermoplastic elastomers that are used in adhesive formulations. Thermoplastic adhesives are generally used as hot melts or in solvent solutions, emulsions, or dispersions. They are generally noted by their convenience and quick setting characteristics.

10.3 Structural Adhesives

Structural adhesives consist of adhesive compositions that are normally used in high strength, permanent applications. Often they provide the primary means of attachment in structural applications. They are generally formulated from thermosetting resins that require chemical crosslinking either with the addition of a curing agent and/or heat. The types of polymeric resin families that are formulated into structural adhesives are usually classified as thermosets or modified hybrids. Some high strength thermosetting elastomers, such as polyurethane, can also be classified as a structural adhesive. Common polymeric resin families that are used to formulate structural adhesives are described in the following sections.

10.3.1 Epoxies

Epoxy adhesives were introduced commercially in 1946 and have a wide application of use in the automotive, industrial, and aerospace markets. Epoxies are probably the most versatile family of adhesives because they bond well to many substrates and can be easily modified to achieve widely varying properties. This modification usually takes the form of:

1. Simple additions of organic or inorganic fillers and components
2. Selection of the appropriate epoxy resin or combination of resins of which many are available
3. Modification of the resulting resin through the choice of curing agent and reaction mechanism

Therefore, it is incorrect to define "epoxy" adhesives in a generic manner as if all these adhesives had similar properties. Depending on the type of resin and curing agent used and on the specific formulation, "epoxy" adhesives can offer the user an almost infinite assortment of end-properties as well as a wide diversity of application and curing characteristics.

Because of their good wetting characteristics, epoxy adhesives offer a high degree of adhesion to all substrates except some low surface energy, untreated plastics and elastomers. Cured epoxies have thermosetting molecular structures. They exhibit excellent tensile-shear strength but poor peel strength unless modified with a more resilient polymer. Epoxy adhesives offer excellent resistance to oil, moisture, and many solvents. Low shrinkage on curing and high resistance to creep under prolonged stress are characteristics of many high quality epoxy adhesives. Epoxy resins have no evolution of volatiles during cure and are useful in gap-filling applications.

Commercial epoxy adhesives are composed primarily of an epoxy resin and a curing agent. The curing agent may be incorporated into the resin to provide a single component adhesive, or else it may be provided in a separate container to be mixed into the resin prior to application. Epoxy adhesives are commercially available as liquids, pastes, films, and solids. Epoxy adhesives are generally supplied as 100% solids, but some are available as sprayable solvent systems. A list of selected epoxy adhesives and their properties are presented in Table 10.6.

Most commercially available epoxy adhesives are either single component, heat curing adhesives or multiple component adhesives that cure at either room or elevated temperature. Generally, epoxy systems that cure at elevated temperatures have a higher crosslinking density and glass transition temperature than systems that are formulated to cure at room temperature. This provides the elevated temperature curing epoxies with better shear strength, especially at elevated temperatures, and better environmental resistance. However, they usually have poor toughness and peel strength.

Depending on the epoxy resin and curing agent used, room-temperature-curing adhesive formulations can harden in as little as several minutes at room temperature, but most systems require from 18 to 72 hrs to reach full strength. The room temperature curing epoxy adhesives can also be cured at elevated temperatures, if faster curing times are required. The curing time is greatly temperature-dependent, as shown in Fig. 10.1.

Once the curing agent is added to the epoxy resin the adhesive must be used in a time period that is dependent on the type of resin and curing agent and on the ambient conditions. With most room temperature systems, the time period is short. The maximum time from mixing to application is considered the pot-life of the system. For most single component chemistries, the allowable time from mixing the

TABLE 10.6 Properties of Selected Epoxy Based Adhesives

Epoxy adhesives					
Adhesive tradename	Adhesive supplier	Form	Curing conditions	Properties	Applications
Araldite AV138 + HV998	Ciba-Geigy	2 part; paste	8 hrs @ 77F (1 hr @ 77F)	Shore D89	General purpose
EA 934	Hysol	2 part; paste	6 days @ 77F (40 min @ 77F)	Service temp: −423 to +350 1000 psi Resists most fluids	General purpose
Ecobond 45	Emerson and Cuming	2 part; paste	24 hrs @ 77F (3 hrs @ 77F)	Controllable flexibility and hardness through catalyst ratio Good shock and peel 3000 psi	General purpose
104 A/B	Furane	2 part; paste	3 days @ 77F (20 min @ 77F)	Good gap filling Shore D79 1500-2500 psi	Concrete patch, sealant
Scotchweld 2216 B/A	3M	2 part; paste	7 days @ 75F (1 hr @ 77F)	Service temp: −67 to +300 Shore D65 400-4930 psi High flexibility and peel	General purpose
Permabond E01	Permabond International	2 part; paste	24 hrs @ 77F (3 min @ 77F)	Fast set Cures at low temperature Color change on cure	General purpose

TABLE 10.6 Properties of Selected Epoxy Based Adhesives *(Continued)*

Epoxy adhesives

Adhesive tradename	Adhesive supplier	Form	Curing conditions	Properties	Applications
Tra Bond 2101	Tra Con	2 part; heavy liquid	24 hrs @ 77F (30 min @ 77F)	Shore D80 Service temp: −76 to +248F	General purpose Automotive and aerospace
7251	Furane	1 part; paste	5 min @ 320F	Thermally conductive and electrical insulating	Electronics industry
Magnobond 6297	Magnolia	1 part; paste	60 min @ 250F	High service temperature 1280-2790 psi	Bonding fiberglass and plastics
Scotchweld 2214 Regular	3M	1 part; paste	40 min @ 250F	Aluminum filled High strength Mil-A-132 Ty1 Cl3	General purpose
ESP308	Permabond International	1 part; paste	45 min @ 300	Service temp: −65 to +350F 3800-6000 psi Free flowing	General purpose
Meltbond 1137-1	BASF/ Narmco	Supported film	50 min @ 250F @ 45 psi	Good tack High peel	Commercial aircraft
Plastilok 660	B. F. Goodrich	Supported film	1 hr @ 250F @ 50 psi	950-1470 psi Low exotherm	Aircraft
Cybond FM 1000	American Cyanamid	Unsupported film	1 hr @ 350F @ 40 psi	2900-6600 psi High peel 130% elongation	General purpose
AF-191	3M	Supported film	1 hr @ 350F @ 45 psi	High shear and peel High strength @ 350F Mil-A-132A Ty2	Aerospace

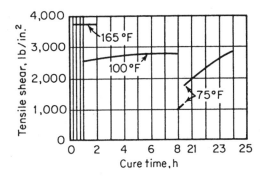

Figure 10.1 Characteristics of a particular epoxy adhesive under different cure time and temperature relationships.[1]

hardener into the resin (done by the adhesive manufacturer) to application is relatively long and dependent on the storage conditions. This time period is defined as the adhesive's shelf life.

An exothermic reaction occurs on curing of epoxy adhesive systems. Exotherm is the heat that is self-generated by the curing reaction. If the exotherm is not controlled, it can lead to drastically reduced pot life and even to a dangerous situation where the exotherm can be great enough to cause combustion. The exotherm can be minimized by lowering the temperature of the mixed components and by limiting the mixing batch size.

Epoxy adhesives are often given a post-cure. Room temperature curing and elevated temperature curing epoxy adhesives are cured to a state in which the resin is hard and the joint is capable of movement (i.e., handling strength). This provides the opportunity of releasing the fixturing used to hold the parts in place and moving the bonded component to the next manufacturing stage. Once handling strength has been reached, the joint is post-cured without fixturing by placing it in an oven or other heat source until it achieves full strength. Often post curing can be combined with some other post-bonding process such as paint curing or component drying. The epoxy adhesive manufacturer should be consulted to determine if post-curing is a viable option.

Many epoxy adhesives are capable of being B-staged. A B-stage resin is one in which a limited reaction between the resin and hardener has taken place so that the product is in a semi-cured but solid state. In the B-staged state, the polymeric adhesive is still fusible and soluble. On additional heating, the adhesive will progress from the B-stage to a completely cured state. This will usually be accompanied by moderate flow. The advantage of B-stage resins is that they permit the formulation of one component adhesives such as films, powders, and preforms. The user can then purchase an adhesive product that does not require metering or mixing. There is very little waste associated with B-staged products, and they generally provide better consistency. B-staged epoxy resins are formulated using curing agents such as aliphatic amines and aromatic polyamines along with solid forms of epoxy resins.

Secondary ingredients in epoxy adhesives include reactive diluents to adjust viscosity; mineral fillers to lower cost, adjust viscosity, or modify coefficient of thermal expansion; and fibrous fillers to improve thixotropy and cohesive strength. Epoxy resins are often modified with other resins to enhance certain properties that are necessary for the application. Often these modifications take the form of additions of elastomeric resins to improve toughness or peel strength.

10.3.1.1 Epoxy resins. The "signature" of an epoxy molecule is the three membered ring containing oxygen.

$$R - CH \overset{O}{-} CH_2$$

Most epoxy formulations are centered around the use of the diglycidyl ether of bisphenol A (DGEBA) as the base epoxy resin (Fig. 10.2). The DGEBA type of epoxy resin used in most adhesive formulations is derived from the reaction of bisphenol A and epichlorohydrin. When the repeating unit, n, equals 0.1 to 1.0, the epoxy resins are normally viscous liquids. When n exceeds 2.0, the resins are solids and need to be melted or dissolved in solvent prior to processing. Another epoxy resin that is often used is diglycidyl ether of Bisphenol F (DGEBF). This resin has a lower viscosity that allows greater possibilities in formulation. There are many other epoxy resins available to the adhesive formulator. Some of these resins have multiple functionality. Generally, a higher functionality improves heat and environmental resistance, but it may result in a more brittle product.

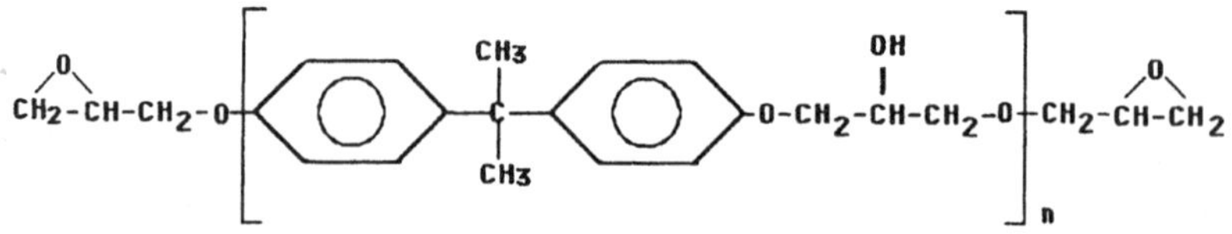

DGEBA

Where n= 0.1 - 1.0, the epoxy resins are viscous liquids. When n exceeds 2.0, the resins are normally solids and need to be melted prior to processing.

Figure 10.2 General structure of DGEBA (diglycidyl ether of bisphenol A) epoxy resin molecule.

Cycloaliphatic resins, epoxidized cresol novolacs, and other resins derived from the epoxy backbone are used in specialty applications. These are often used to take advantage of specific properties such as low viscosity, increased flexibility, clarity, improved color stability, or increased reactivity. New high temperature fluorine epoxy resins[2] may extend the useful temperature range of these adhesives. Glass transition temperatures of up to 554°F have been reported for some polymers made from these resins.

10.3.1.2 Curing agents. Epoxy resins are the most versatile of structural adhesives because they can be cured and co-reacted with many different resins to provide widely varying properties. Table 10.7 describes the influence of curing agents on the bond strength of epoxy adhesives to various adherends. Curing agents affect cohesive strength, hardness, and durability more than adhesion (controlled primarily by surface energies). Thus, differences in shear strength, peel strength, and environmental resistance may be attributable to the choice of curing agent. DGEBA resin can be cured with amines or polyamides for room-temperature setting systems; anhydrides for elevated temperature cure; or latent curing agents, such as boron trifluoride complexes, for use in single component, heat-curing adhesives. The characteristics of curing agents used with epoxy resins in adhesive formulations are summarized in Table 10.8. Epoxies can undergo various reactions. Some of the more important reactions are shown in Fig. 10.3.

Polyamide curing agents are used in most "general-purpose" epoxy adhesives. They provide a room-temperature cure and bond well to many substrates including plastics, glass, and elastomers. The polyamide-cured epoxy also offers a relatively flexible adhesive with fair peel strength, moisture resistance, and thermal-cycling properties. Mixing ratios are generally not critical. Within limits the greater the amount of polyamide in an epoxy formulation the greater the flexibility, impact strength and peel strength. However, glass transition temperature is lowered as are the shear strength and temperature resistance. There are several polyamides with varying viscosity. The reaction with conventional DGEBA epoxy resins produces a relatively low degree of exotherm.

Aliphatic primary amines provide a faster room temperature cure but have poor elevated temperature resistance. They produce adhesive systems that have a high exotherm. Primary amines are generally considered to be a strong irritant and proper ventilation and personal safety equipment must be employed. The amine/epoxy reaction occurs at room temperature and it can be accelerated with an elevated tem-

TABLE 10.7 The Influence of Epoxy Curing Agents on the Bond Strength Obtained with Various Base Materials[3]

			Tensile-shear strength, lb/in.^{2}				
Curing agent	Amount*	Cure cycle, h at °F	Polyester glass-mat laminate	Polyester glass-cloth laminate	Cold-rolled steel	Aluminum	Brass	Copper
Triethylamine	6	24 at 75, 4 at 150	1,850	2,100	2,456	1,810	1,765	655
Trimethylamine	6	24 at 75, 4 at 150	1,054	1,453	1,385	1,543	1,524	1,745
Triethylenetetramine	12	24 at 75, 4 at 150	1,150	1,632	1,423	1,675	1,625	1,325
Pyrrolidine	5	24 at 75, 4 at 150	1,250	1,694	1,295	1,733	1,632	1,420
Polyamid amine equivalent 210–230	35–65	24 at 75, 4 at 150	1,200	1,450	2,340	3,120	2,005	1,876
Metaphenylenidiamine	12.5	4 at 350	780	640	2,150	2,258	2,150	1,650
Diethylenetriamine	11	24 at 75, 4 at 150	1,010	1,126	1,350	1,420	1,135	1,236
Boron trifluoride monoethylamine	3	3 at 375	...	...	1,732	1,876	1,525	1,635
Dicyandiamide		4 at 350	530	432	2,680	2,785	2,635	2,550
Methyl nadic anhydride	85	6 at 350	600	756	2,280	2,165	1,955	1,835

Epoxy resin used was derived from bisphenol A and epichlorohydrin and had an epoxide equivalent of 180 to 195; the adhesives contained no filler.
*Per 100 parts by weight of resin.

TABLE 10.8 Characteristics of Curing Agents Used with Epoxy Resins in Adhesive Formulations[3]

Curing agent	Physical form	Amount required*	Cure temp, °F	Pot life at 75°F†	Complete cure conditions	Max use temp, °F
Triethylenetetramine	Liquid	11–13	70–275	30 min	7 days (75°F)	160
Diethylenetriamine	Liquid	10–12	70–200	30 min	7 days (75°F)	160
Diethylaminopropylamine	Liquid	6–8	83–300	5 h	30 min (75°F)	185
Metaphenylenediamine	Solid	12–14	150–400	8 h	1 h (185°F) 2 h (325°F)	300
Methylene dianiline	Solid	26–30	150–400	8 h	1 h (185°F) 2 h (325°F)	300
Boron trifluoride monoethylamine	Solid	1–4	275–400	6 months	3 h (325°F)	325
Methyl nadic anhydride	Liquid	80–100	250–100	5 days	3 h (320°F)	325
Triethylamine	Liquid	11–13	70–200	30 min	7 days (75°F)	180
Polyamides:						
Amine value 80–90	Semisolid	30–70	70–300	5 h	5 days (75°F)	‡
Amine value 210–230	Liquid	30–70	70–300	5 h	5 days (75°F)	‡
Amine value 290–320	Liquid	30–70	70–300	5 h	5 days (75°F)	‡

*Per 100 parts by weight; for an epoxy resin with an epoxide equivalent of 180 to 190.
†Five hundred grams per batch; with a bisphenol A-epichlorohydrin derived epoxy resin with an epoxide equivalent of 180 to 190.
‡Highly dependent on concentration.

Epoxy / Amine Reaction

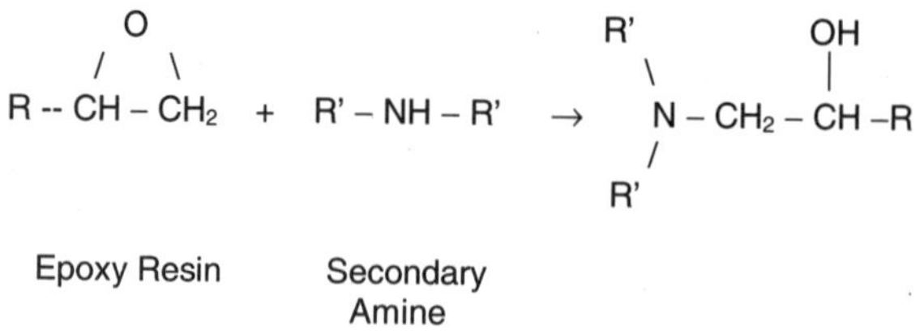

$$R\text{--}CH\underset{\underset{O}{\diagdown\diagup}}{-}CH_2 + R'-NH-R' \rightarrow \underset{\underset{R'}{\diagup}}{\overset{\overset{R'}{\diagdown}}{N}}-CH_2-\overset{\overset{OH}{|}}{CH}-R$$

Epoxy Resin   Secondary Amine

Epoxy / Anhydride Reaction

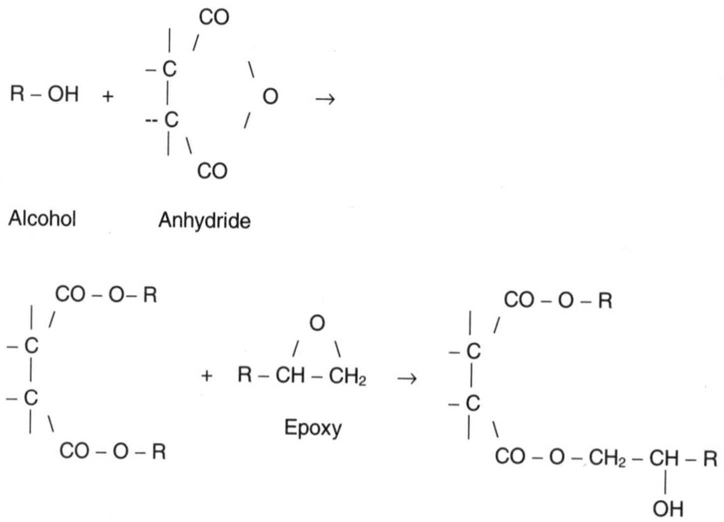

$$R-OH + \text{anhydride} \rightarrow$$

Alcohol   Anhydride

$$\begin{array}{c} \quad CO-O-R \\ -C \\ -C \\ \quad CO-O-R \end{array} + R-CH\underset{\underset{O}{\diagdown\diagup}}{-}CH_2 \rightarrow \begin{array}{c} \quad CO-O-R \\ -C \\ -C \\ \quad CO-O-CH_2-\overset{\overset{}{}}{CH}-R \\ | \\ OH \end{array}$$

Epoxy

Epoxy / Mercaptan Reaction

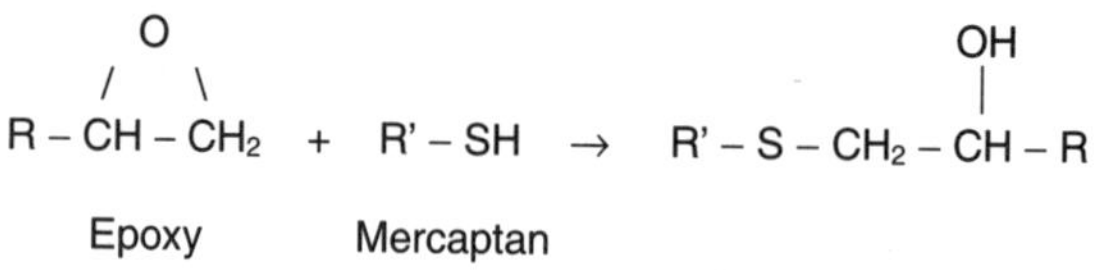

$$R-CH\underset{\underset{O}{\diagdown\diagup}}{-}CH_2 + R'-SH \rightarrow R'-S-CH_2-\overset{\overset{OH}{|}}{CH}-R$$

Epoxy   Mercaptan

Figure 10.3 Common epoxy reaction mechanisms.[4]

perature cure. The mixing ratio is more critical than with the polyamides. Greater or less curing agent than called for by the manufacturer's instruction could lead to a significantly weakened system with poor adhesion. Common primary polyamines are diethylenetriamine (DETA) and triethylenetetramine (TETA).

Aromatic amines offer improved heat and chemical resistance. For room temperature cures, they must be used with catalysts. Tertiary

Epoxy Ring - Opening Polymerization

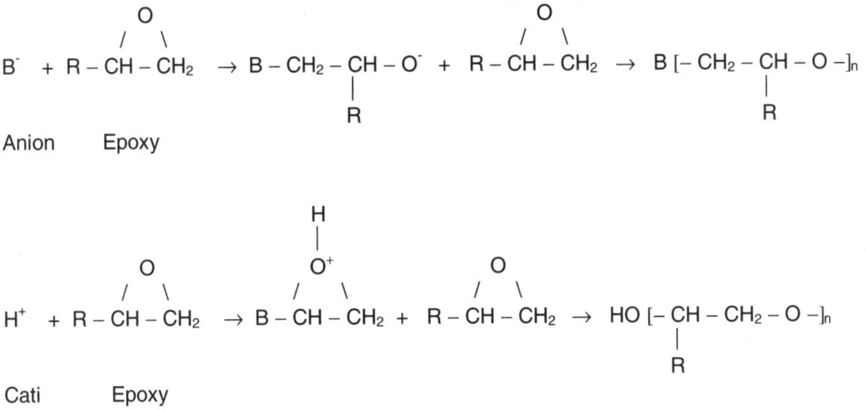

$$B^- + R-CH\underset{\underset{O}{\diagdown\diagup}}{-}CH_2 \rightarrow B-CH_2-\underset{R}{CH}-O^- + R-CH\underset{\underset{O}{\diagdown\diagup}}{-}CH_2 \rightarrow B[-CH_2-\underset{R}{CH}-O-]_n$$

Anion Epoxy

$$H^+ + R-CH\underset{\underset{O}{\diagdown\diagup}}{-}CH_2 \rightarrow B-CH\underset{\overset{O^+}{|}}{-}CH_2 + R-CH\underset{\underset{O}{\diagdown\diagup}}{-}CH_2 \rightarrow HO[-\underset{R}{CH}-CH_2-O-]_n$$
$$\overset{|}{H}$$

Cati Epoxy

Figure 10.3 (*Continued*)

amines are primarily used as catalysts but can be used as a room temperature or elevated temperature curing agent on their own. Aromatic amines are generally solid curing agents and can be used to provide a B-staged cure with most epoxy resins. Typical aromatic amines include m-phenylenediamine (MDA) and diaminodiphenyl sulfone (DDS).

The amido-amine curatives are polyamides with amine functionality and lower viscosity. They cure relatively quickly at room temperature, and they develop tougher bonds than do the amine or mercaptan cured systems. Their properties are very similar to the polyamide cured epoxy systems. The amido-amines are generally less corrosive to skin than standard amine curatives. These curing agents are used in the standard two component epoxy adhesive that have a one-to-one mixing ratio (resin and hardener by weight), and they will gel in about 30–60 mins with 12–24 hrs required for full cure. Such systems can readily be found commercially in viscosities ranging from motor oil to clay-like.

The epoxy/mercaptan reaction is normally catalyzed with a tertiary amine and is used for the standard two component, five minute curing epoxies that can be found in hardware stores. These fast curing products have a tendency to be somewhat brittle and may perform quite poorly in the peel mode. Mercaptan cured epoxy adhesives have a strong sulfur-like odor that can be an irritant.

Anhydride cured epoxy systems exhibit low viscosity, long pot life, and low exotherm. The epoxy/anhydride reaction normally must use heat to complete the cure. Anhydride cures can be accelerated with tertiary amines such as tris (dimethylaminomethyl) phenol (DMP30). They form good high temperature resistant products, but have a tendency toward lower peel strengths. Anhydride cured epoxy adhesives when cured at elevated temperatures have lower cure shrinkage than most other epoxy systems. The anhydride component must be kept covered when not in use because it will react with the moisture in the air and result in poor physical properties. Common anhydrides are phthalic anhydride, nadic methyl anhydride, and chloroendic anhydride.

Epoxy can also homopolymerize using strong acids or strong bases. Certain Lewis acids, such as boron trifluoride monoethanolamine (BF3), can polymerize epoxy resins, resulting in high temperature resistance and gel times of less than 60 sec. However, low peel properties and poor impact resistance result.[5] The cured products have high temperature and chemical resistance, but are brittle without modification to the formulation. The fast rate of cure is often considered a liability because of high exotherm, inconsistency in mixing, and application problems.

Single component epoxy systems usually work in one of two ways. In one method, a curing agent is used which is mostly insoluble in the resin and, therefore, inactive at room temperature. The curing agent liquefies and diffuses into the resin at the cure temperature, thereby activating the cure mechanism. An example of this type of curing agent is dicyandiamide. It can cure in 40 mins at 250°F and has a 6 month shelf life at room temperature. Dicyandiamide reacts with the epoxy only on heating, and the reaction stops on removal of the heat. There are several popular single component epoxy adhesives that use dicyandiamide as the curing agent. Another mechanism for achieving a one part system is to have the curing agent chemically blocked. An example of this is cyanourea compounds such as toluene bis-dimethylurea that cures epoxy formulation in 60 mins at 250°F. These single component epoxies normally have better environmental resistance than do two component, elevated temperature curing adhesives.

Imidazoles are also curing agents that are appropriate for single component epoxy systems. Like dicyandiamide, the imidazoles react with epoxide groups and provide catalytic activity resulting in high heat resistance.

10.3.2 Epoxy hybrids

A variety of polymers can be blended and co-reacted with epoxy resins to provide certain desired properties. The most common of these are nitrile, phenolic, nylon, and polysulfide resins. These blends can take a number of different forms. The added resin may be reacted with the epoxy resin, or it may be included as an unreacted modifier. The modifier may be blended into a continuous phase with the epoxy resin or precipitated-out as a discrete phase within the epoxy resin matrix.

Properties and curing characteristics of commercially available epoxy hybrid adhesives are presented in Table 10.9.

The most common types of epoxy hybrid adhesives are epoxy resins that are toughened with elastomeric resins and alloyed blends consisting of epoxy-phenolic, epoxy-nylon, and epoxy-polysulfide adhesives. These are described in the following sections.

10.3.2.1 Toughened epoxies. Because of the normally brittle nature of the epoxy resin, epoxy adhesives have been toughened with many different resins during their history. These include thermoplastic particles, nylon, and various elastomers. The most successful of these are epoxy adhesives that embody discrete nitrile rubber regions as the elastomeric toughening component. Small rubber inclusions impart toughness and high peel strength to structural epoxy adhesives. Both DGEBA and DGEBF epoxy resins have been optimized for toughness with additions of nitrile rubber. Dicyandiamide curing agents provide one component toughened epoxy formulations. Toughened epoxy adhesives are available as liquids, pastes, and films.

Many one component epoxies contain rubber toughening agents such as carboxylic acid terminated butadiene nitrile rubber (CTBN).[6-8] These elastomers are pre-reacted at the 5-15% level with the epoxy resin at high temperature. During the final cure, the rubber must precipitate out of solution into discrete particles within the epoxy resin matrix. Ideally, a distinct phase will toughen the epoxy without significantly reducing the high temperature characteristics or glass transition temperature (Tg) of the original epoxy resin. The small distinct regions of elastomer within the epoxy matrix provide crack inhibitors and internal stress relief mechanisms. These result in a general increase in toughness and impact strength and moderate improvements in peel strength without significant reduction of the glass transition temperature of the adhesive. The high impact strength and elasticity is retained at low temperatures.

Toughened epoxy adhesives are generally available as single component, elevated temperature curing systems. Typical cure requirements are one hour at 250°F. CTBN modified epoxy adhesives tend to be slower reacting and higher in viscosity than conventional epoxy systems. Two component toughened epoxy adhesives have also been developed to cure at ambient or mildly elevated temperatures. In these systems, CTBN or amine terminated acrylonitrile butadiene (ATBN) are used as the elastomeric phase and amine hardeners are used to achieve cure.[9,10] Table 10.10 shows typical single and multiple component epoxy adhesive formulations using CTBN toughening agents.

Table 10.11 also shows the bulk properties and peel strength of an adhesive formulation with and without CTBN modification. Tensile shear strength and temperature resistance of these adhesives are rel-

TABLE 10.9 Suppliers of Selected Modified Epoxy Adhesives

Adhesive tradename	Adhesive supplier	Form	Curing conditions	Properties	Applications
Epoxy phenolic					
Y-390PG	Sterling	1 part; solvent	3 min @ 325F	Chemical resistance Pressure and vacuum resistance UL 1446	Insulating varnish and adhesive
Araldite XUGY281 + Polyamine	Ciba-Geigy	2 part; paste	18 hrs @ 68F	Flexibility Chemical resistance	Industrial storage tanks Sealant
Epoxy nylon					
EA 951	Dexter	Supported and unsupported film	60 min @ 350F	3300–6500 psi on Al High peel and shear strength Mil-A-132 Ty1 Cl1	Aerospace industry
Epoxy urethane					
109-43	Creative Materials	1 part; thick paste	4 min @ 266F	Silver filled for electrical conductivity Service temp: −67 to 392F	Attaching SMDs to flex and rigid printed circuit boards
Stycast 2760	Emerson and Cuming	2 part; paste	24 hr @ 77F (1 hr @ 77F)	Shore D82 Low viscosity	Bonding to vinyl and other plastics
Dolphin 6099	Dolphin	1 part; sealant	2 days @ 77F	Air curing and one part Service temp: −20 to 200F	Anti-corrosion sealant for metals and plastics
XMH 8516	Ciba-Geigy	2 part; paste	16 hrs @ 77F (45 min @ 77F)	Shore D67 Flexibility and chemical resistance	
EP72	Master Bond	2 part; paste	18 hrs @ 75F (37 min @ 77F)	Service temp: −67 to 300F Flexible Will withstand temperature cycling High impact	

TABLE 10.10 Single Part and Two Part Epoxy Adhesive Formulations with (Formula B) and without (Formula A) Additions of CTBN.[11]

	Formula A	Formula B
Single Component		
DGEBA (eposy reisn, EEW = 190)	100	75
DGEBA/CTBN (1300X13) adduct[a]	—	25
Tubular alumina (filler)	40	40
Cab-O-Sil (thixotroping agent)	5	5
DICY (curing agent)	6	6
Melamine (accelerator)	2	2
Lap shear at ambient temperature (MPa)	18.5	20.5
T-peel at ambient temperature (kN/m)	1.1	5.5
Two Component		
DGEBA (epoxy resin)	100	77.5
Atomite whiting ($CaCO_3$ filler)	30	30
DGEBA/CTBN (Hycar 1300X8) adduct[b]	—	37.5
Ancamine AD (amine hardener)	60	59.3
Lap shear strength at −40°C (MPa)	6.9	22.7
Lap shear strength at 25°C (MPa)	8.6	17.9
Lap shear strength at 75°C (MPa)	4.4	5.5
180° T-peel strength (kN/m)	0.5	5.0

[a]Adduct contains 40% CTBN (Hycar (1300X13).
[b]Adduct contains 40% rubber.

TABLE 10.11 Bulk Properties and Peel Strength of a Control Adhesive and a CTBN Modified Epoxy Adhesive[6]

	Formulation A1 (control)	Formulation A2
Formulation, pbw		
Liquid DGEBA (EEW = 190)	100	77.5
CTBN 1300 × 13/Epoxy adduct (EEW = 340)	—	37.5
CaCO3 filler	30	30
Fumed silica thixotrope	3	3
Dicyandiamide	5	6
3-phenyl-1, 1-dimethyl urea	2	2
Bulk Properties		
Tensile strength, MPa	68.9	57.8
Elongation, %	2.5	3.2
Modulus, MPa	3848	2845
Tg, C (Determined via DSC)	135	129

atively high. Peel strength is also high and does not decrease rapidly with lower temperatures. The durability of CTBN modified epoxies are satisfactory as measured by most long term moisture tests, but it does not match the durability of the vinyl-phenolic or nitrile-phenolic types.[12] Nitrile-epoxy adhesive systems should not be used in marine environments or under continuous immersion in water.[13]

Another toughened epoxy formulation comes from mixtures of epoxies and urethane oligomers with pendant epoxy groups.[14] Curing of the pendant epoxy groups unites the urethane and non-urethane components through conventional epoxy reactions to give tough, durable films. These adhesives are generally referred to as "reactive" epoxy urethanes. This approach, however, does not provide high temperature properties as does the CTBN particle approach. These reactive hybrid epoxy-urethane adhesives were developed initially for bonding to oily cold rolled steel, but they have given good results on other substrates as well. Table 10.12 shows lap shear values obtained after curing 20 mins at 350°F on various untreated substrates.

A third approach is to use surface activated pre-cured rubber particles directly in the adhesive formulation. This process does not involve separating out a rubbery phase during cure. Elastomers that have typically been used in this approach have been urethanes and reclaimed tire products.[15] These materials are specially treated so that they provide good adhesion to the epoxy resin matrix. The main benefits of these formulations are low cost and moderate improvement in toughness and flexibility.

Although the urethane oligomer and pre-cured rubber particle formulations provide good peel strength and toughness, they do not maintain the high thermal properties inherent in the epoxy matrix as does the discrete CTBN or ATBN phase approach to toughening.

It also should be noted that soft, flexible epoxy resins have been developed through modification of the epoxy molecule. These systems do not have high temperature properties, but they are marketed for their cost effectiveness and extremely low durometer. These flexibilized epoxies are found in applications having moderate temperature requirements. They are primarily used in flexible adhesives, sealants,

TABLE 10.12 Cured Lap Shear Strength of Uncleaned Substrates Bonded with an Epoxy Urethane Reactive Hybrid System[14]

Substrate	Substrate thickness (mils)	Lap shear (psi)
Oily steel	58	3,400
Galvanized steel	75	2,770
Polyester-painted steel	18	1,440
Painted steel/galvanized steel	18/75	1,760
Electrogalvanized steel	28	1,930
Organic-coated steel	33	2,160
SMC (polyester-glass)	100	600*
R 303 aluminum	65	3,340

*Substrate breaks

caulking compounds, encapsulants, potting compounds, conformal coatings and molding compounds for rubbery parts.[16]

Several other approaches have been developed for improving the toughness and flexibility of epoxy adhesives. Modification of epoxy resin with alkyl resin is reported as a means of flexibilizing epoxies for adhesive applications.[17] Modification of epoxy resins through the direct addition of engineering thermoplastic, such as polysulfone, has also been proposed to enhance the toughness of epoxy adhesives.[18] It is likely that the search for toughened epoxy adhesives will continue in the future.

10.3.2.2 Epoxy-phenolic. Epoxy-phenolic adhesives are made by blending epoxy resins with phenolic resins to improve the high temperature capabilities of the standard epoxy resins. Adhesives based on epoxy-phenolic blends are good for continuous high-temperature service in the 350°F range or intermittent service as high as 500°F. They retain their properties over a very high temperature range, as shown in Fig. 10.4. Shear strengths of up to 3,000 psi at room temperature and 1,000 to 2,000 psi at 500°F are available. Resistance to weathering, oil, solvents, and moisture is very good. Because of the rigid nature of the constituent materials, epoxy-phenolic adhesives have relatively low peel and impact strength and limited thermal-shock resistance.

These adhesives are available as pastes, solvent solutions, and film supported on glass fabric. The films generally give better strengths than do liquid systems. Cure requires a temperature of 350°F for 1 hr under 15–50 psi pressure. Most adhesives of this type have limited storage life at room temperature.

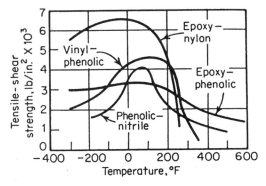

Figure 10.4 The effect of temperature on the tensile shear strength of modified epoxy adhesives (substrate is aluminum).[3]

Epoxy-phenolic adhesives were developed primarily for bonding metal joints in high-temperature applications. They were one of the first adhesives developed for high temperature aircraft applications in the early 1950s. They are also commonly used for bonding glass, ceramics, and phenolic composites. Because of their relatively good flow properties, epoxy phenolics are also used in bonding honeycomb sandwich composites.

10.3.2.3 Epoxy-nylon. Epoxy-nylon adhesives, introduced in the 1960s, were one of the first structural adhesives designed specifically to have high shear strength and extremely high peel strength. These characteristics were achieved by blending epoxy resins with co-polymers from the polyamide (or nylon) resin family that is noted for toughness and tensile properties. The nylons used as modifiers for epoxy adhesives are soluble co-polymers made from conventional nylon monomers. Conventional crystalline nylon polymers, such as nylon 6 or nylon 66, would be incompatible with epoxy resins.

Epoxy-nylon resins are one of the best materials to use in film and tape adhesives for applications where the service environment is not severe. Their main advantages are better flexibility and a large increase in peel strength compared with unmodified epoxy adhesives. Epoxy-nylon adhesives offer both excellent tensile lap shear and peel strength. In addition, epoxy-nylon adhesives have good fatigue and impact resistance. They maintain their tensile lap shear properties at cryogenic temperatures, but the peel strength is poor at low temperatures. They are limited to a maximum service temperature of 180°F, and they exhibit poor creep resistance. Possibly their most serious limitation is poor moisture resistance because of the hydrophilic nylon (polyamide) constituent. The degradation by exposure to moisture occurs with both the cured and uncured adhesives.

Epoxy-nylon adhesives are generally available as film or as solvent solutions. A pressure of 25 psi and temperature of 350°F are required for 1 hr to cure the adhesive. Because of their excellent filleting properties and high peel strength, epoxy-nylon adhesives are often used to bond aluminum skins to honeycomb core in aircraft structures.

10.3.2.4 Epoxy-polysulfide. Polysulfide resins combine with epoxy resins to provide adhesives and sealants with excellent flexibility and chemical resistance. These adhesives bond well to many different substrates. Shear strength and elevated temperature properties are low. However, resistance to peel forces and low temperatures are very good. Epoxy-polysulfides have good adhesive properties down to −150°F, and they stay flexible to −85°F. The maximum service temperature is about 122°F–180°F depending on the epoxy concentration in the formulation. Temperature resistance increases with the epoxy content of the system. Resistance to solvents, oil and grease, and exterior weathering and aging are superior to that of most thermoplastic elastomers.

The epoxy-polysulfide adhesive is usually supplied as a two-part, flowable paste that cures to a rubbery solid at room temperature. These systems can be heavily filled without adversely affecting their properties. The polysulfide elastomer will cure epoxy resins by itself, but the reaction is extremely slow. A tertiary amine is usually used as a catalyst for room temperature curing. A sulfur odor is noticeable during processing, making ventilation important.

Epoxy-polysulfides are often used in applications requiring a high degree of elongation. They are generally used to bond concrete in floors, roadways, and airport runways. Other principal uses include sealing applications, bonding of glass, potting, and bonding of rubber to metal.

10.3.2.5 Epoxy-vinyl. Epoxy resins may be blended with certain vinyl polymers to improve the impact strength and peel strength of the adhesives. Polyvinyl acetals, such as polyvinyl butyral and polyvinyl formal, and polyvinyl esters are commonly used to modify DGEBA resins at a 10–20% addition. Elevated temperature resistance is sacrificed by the addition of the low Tg vinyl resins. These adhesives are generally available as films or solvent solutions. They are commonly used as laminating adhesives for film or metallic foil.

10.3.3 Resorcinol formaldehyde and phenol resorcinol formaldehyde

Resorcinol resins come from the phenolic resin family. They are more expensive than their phenolic counterparts, but they are useful when a room-temperature cure is necessary. They find application in bonding wood, cellulose acetate, molded urea plastics, nylon, and various plastic laminates to wood core. They also bond to porous materials such as paper, textiles, leather, and fiberboard. They do not usually bond to metal, although they will bond metal to wood if the metal is first primed.

Resorcinol adhesives are primarily used for bonding wood structures. Adhesive bonds are usually as strong as the wood. Resorcinol adhesives are suitable for exterior use, and they are resistant to boiling water, oil, many solvents, and mold growth. Their service temperature ranges from −300 to +350°F. Because of high cost, resorcinol formaldehyde resins are often modified by the addition of phenolic resins to form phenol resorcinol. Resorcinol formaldehyde resins are also commonly used as a primer for bonding nylon with epoxy or polyurethane adhesives. These primers are also often used to improve the adhesion of nylon tire cord to rubber.

Resorcinol and phenol-resorcinol adhesives cure on the addition of formaldehyde. Commercially they are available as two part systems. The resorcinol part is in liquid form (generally resin in a mixture of alcohol and water), and it is mixed with a powder hardener (paraformaldehyde with fillers) before application. Once mixed, the pot life is limited, and the mixed resin generates considerable heat of exotherm. Generally, these adhesives are cold-setting, but they can also be applied to a substrate, dried, and then hot-pressed. Curing at room temperature normally takes 8–12 hrs. For maximum durability, a cure at moderately elevated temperature is recommended.

10.3.4 Melamine formaldehyde and urea formaldehyde

Melamine-formaldehyde resins are used as colorless adhesives for wood. These thermosetting resins are condensation products of unsubstituted melamine and formaldehyde. Because of their high cost, they are sometimes blended with urea formaldehyde. Melamine formaldehyde is usually supplied in powder form and reconstituted with water; a hardener is added at the time of use. Temperature of about 200°F is necessary for cure. Adhesive strength is greater than the strength of wood.

Urea-formaldehyde adhesives are not as strong or as moisture-resistant as the resorcinols. However, they are inexpensive, and both hot-and-cold setting types are available. Maximum service temperature of a urea formaldehyde adhesive is approximately 140°F. Cold-water resistance is good, but boiling water resistance may be improved by the addition of melamine-formaldehyde or phenol-resorcinol resins. Urea-based adhesives are used mainly in plywood manufacture.

10.3.5 Phenolics

Phenolic resins are the condensation product of phenol and formaldehyde. Phenolic or phenol formaldehyde is primarily used as an adhesive for bonding wood. Because of their good heat resistance and dimensional stability, they have also been used in brake linings, abrasive wheels, sandpaper, and foundry molds. Solvent solutions of phenolic resins filled with clay have been used as glass to metal adhesive for attaching metal bases to light bulbs. Phenolics are relatively inexpensive.

They are an important class of adhesive used in the manufacture of outdoor grade plywood. In most of these applications the adhesive is applied as a solution in alcohol, acetone, or water. It is coated on the substrate, dried, and then cured under heat and pressure. However, several forms of phenolic adhesives are available including spray dried powders that are dissolved in water for application and films. Curing is accomplished at temperature of approximately 280°F for several minutes.

Adhesives for plywood are essentially solutions of low molecular weight condensation products of phenol and formaldehyde in aqueous sodium hydroxide. They provide bond strengths that are generally greater than the strength of the wood substrate. Durability is good, and the bond is essentially unaffected by boiling water, mold, or fungus. These adhesives are suitable for exterior use, and their properties are not appreciably affected by low temperatures. In the presence of alkaline catalysts, crosslinking is brought about at elevated temperatures. Acid catalysts give room temperature cures, but also cause degradation of wood and paper. Phenol formaldehyde resin is available in the form of glue film, carried on tissue paper. Although expensive, this form of adhesive is useful for faying of very thin or highly porous veneers.

Phenolic resins have also been used to bond metal to glass. The bond is somewhat brittle, however, and tends to shatter under impact or vibration. As a result, phenolic resins are often modified by the addition of elastomeric resins to improve toughness and peel strength.

10.3.6 Modified phenolics

By modifying phenolic resin with various synthetic rubbers and thermoplastic materials, flexibility is greatly improved. The modified adhesive is well suited for structural bonding of many materials, notably metals. The most common type of modified phenolic adhesives are vinyl-phenolic, nitrile-phenolic, and neoprene-phenolic. Some commercially available phenolic-alloy adhesives are described in Table 10.13.

10.3.6.1 Nitrile-phenolic. Certain blends of phenolic resins modified with nitrile rubber produce adhesives useful up to 300°F continuously. Typical formulations may contain up to 50% by weight nitrile rubber. The major advantage of incorporating nitrile into phenolic resin is to improve the peel strength of the phenolic without significant reduction in high temperature strength.

On metals, nitrile-phenolics offer shear strength of up to 5,000 psi along with excellent peel and fatigue properties. Good bond strengths can also be achieved on rubber, plastics, and glass. These adhesives have high impact strength and elevated temperature capability during service. Because of their good peel strength and elevated-temperature

TABLE 10.13 Suppliers of Selected Modified Phenolic Adhesives

Adhesive tradename	Adhesive supplier	Form	Curing conditions	Properties	Applications
Vinyl phenolic					
Raybond 84048	Raybestos	1 part; solvent		Heat resistant	Automobile Friction parts
Nitrile phenolic					
Plastilok 601	B. F. Goodrich	Film	60 min @ 350F	Chemical resistance and high temperature capability	Porous and nonporous applications
Plastilok 606	B. F. Goodrich	1 part; solvent	20 min @ 180F followed by 20 min @ 300F @ 100 psi	Resistant to high temperature, chemicals, and water Flexibility	
R-81005	Raybestos	1 part; solvent	15 min @ 350F	1750 psi Heat shock resistant	Automotive; friction parts
Neoprene phenolic					
Raybond R-82007	Raybestos	1 part; solvent	1 hr @ 77F @ 250 psi	Quick set at RT	Lamination

properties, nitrile-phenolic adhesives are used commonly for bonding linings to brake shoes. They are also used in the aircraft industry for bonding of aluminum facings to honeycomb cores. Other applications include electronics, footwear, and furniture assembly.

Nitrile-phenolic adhesives are generally available as solvent solutions as well as supported and unsupported firm. They require heat curing at 300–500°F under pressure of up to 200 psi. The nitrile-phenolic systems with the highest curing temperature have the greatest resistance to elevated temperatures during service.

10.3.6.2 Vinyl-phenolic. Vinyl-phenolic adhesives are based on a combination of phenolic resin with polyvinyl formal or polyvinyl butyral resins. Because of their excellent shear and peel strength, vinyl-phenolic adhesives are one of the most successful structural adhesives for metal. Room-temperature shear strength as high as 5,000 psi is available. Maximum operating temperature, however, is only 200°F because the thermoplastic constituent softens at elevated temperatures. Chemical resistance and impact strength are excellent.

Vinyl-phenolic adhesives are supplied in solvent solutions and as supported and unsupported film. The adhesive cures rapidly at elevated temperatures under pressure. They are generally used to bond metals, rubbers, and plastics to themselves or each other. A major application of vinyl-phenolic adhesive is the bonding of copper sheet to plastic laminate in printed-circuit-board manufacture.

10.3.6.3 Neoprene-phenolic. Neoprene-phenolic alloys are used to bond a variety of substrates. Normal service temperature is from −70 to +200°F. Because of high resistance to creep and most service environments, neoprene-phenolic joints can withstand prolonged stress. Fatigue and impact strengths are also excellent. Shear strength, however, is lower than that of other modified phenolic adhesives.

Temperatures over 300°F and pressure greater than 50 psi are needed for cure. Neoprene-phenolic adhesives are available as solvent solutions and film. During cure these adhesives are quite sensitive to surface contamination from atmospheric moisture and other processing variables.

10.3.7 Polyaromatic high temperature resins

Polyimide, bismaleimide, polybenzimidazole, and other high temperature resins belong to the aromatic heterocycle polymer family that is noted for its outstanding thermal resistance. These resins are known as aromatic polymers because of their closed ring structure which when polymerized leads to a "ladder" type of polymeric structure. The polyimide molecule shown in Fig. 10.5 is a good example of a polymer with an aromatic ring or ladder structure. High temperatures must cause at least two molecular chains to break before the molecule undergoes chain scission and resulting degradation of physical properties.

In addition to their high temperature resistance, polyaromatic resins are unfortunately also noted for their difficult processing conditions. Many of these resins must be made with aggressive, high temperature solvents that are difficult to remove from the final product. Many aromatic resins harden via a condensation mechanism, producing a water byproduct as the resin crosslinks. Voids or bubbles in the bond line can be commonly encountered either from the solvent or from water liberated due to the condensation reaction.[19]

Polyaromatic types of adhesives were developed for high temperature aerospace applications, but they have limited use due to their high cost and difficult processing parameters. These adhesives are generally provided as a supported film, although polyimide resins are also available in solvent solution. During cure, temperatures of 550–650°F and high pressures are required. Often a multi-step cure process is recommended where slight pressure is initially applied until the adhesive begins to crosslink and then higher pressures are applied throughout the remainder of the cure cycle. Volatiles are released during cure which contribute to a porous, brittle bond line with relatively low peel strength. Often a vacuum is required during cure to eliminate the volatiles that are formed. Table 10.14 lists physical properties and curing characteristics of these high-temperature adhesive systems.

The toughening of these high temperature adhesives can provide a difficult challenge, since the service temperatures usually exceed the degradation point of most elastomeric toughening additives. Improvements in the toughening of high temperature epoxies and other high temperature thermosets have been accomplished through the incorporation of high temperature thermoplastics such as polyarylene ether ketone (PEK) and polyaryl ether sulfone (PES).

10.3.7.1 Polyimide. Polyimides are well known resins that have been developed into aerospace adhesive formulations mainly by NASA. They have a glass transition temperature of at least 200°F greater

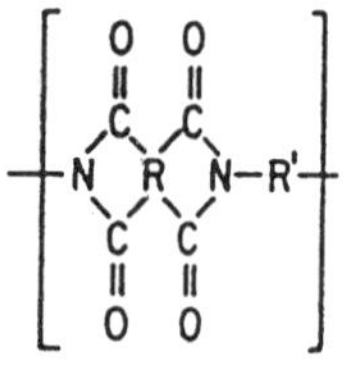

Figure 10.5 Aromatic polymeric structure (polyimide) provides unique high temperature resistance.

TABLE 10.14 Suppliers of Selected High Temperature Adhesives

Adhesive tradename	Adhesive supplier	Form	Curing conditions	Properties	Applications
Bismaleimide					
Redux 326	Ciba-Giegy	Paste	2 hr @ 350F	Good properties at 400F	Metal and composite bonding
Metlbond 6604-1	Cytec	Foaming, unsupported film	4hr @ 350F	Low density Radar transparent	Aerospace core splicing, edge sealing of honeycomb
EA 9369	Dexter	1 pt paste	60 min @ 350F	Structural properties to 450F No outgassing	Potting and edge filing
EA 9673	Dexter	Supported film	60 min @ 350F	Continuous service at 400F Moisture resistant	Honeycomb construction
Polyimide					
FM 57	Cytec	Supported film	90 min @ 350F	Condensation cured Radar transparent	Aerospace metals and composites, repairs
FM 55	Cytec	Supported film	2 hr @ 350F	Continuous service at 500F Addition cure Glass transition temperature over 500F	Aerospace metals and composites
PLD-700	BLH Electronics	Solution	2 hr @ 500F	Thin coating material	Strain gauge attachment
Epotek 390	Epoxy Technology	Solution	60 min @ 400F	Thin coating material	Electronic attachment
7501	Furane	Solution	5 min @ 320F	Thin coating material	Electronic attachment

than epoxy resins. As a result, polyimide adhesives can operate at higher temperatures than epoxies or phenolics. However, they are expensive and require high curing temperatures. Two types of polyimides (PI) are currently in use: (1) those that cure by condensation reaction and (2) those that cure by addition reaction mechanisms.

Condensation PI resins were the first type commercially introduced. Polymeric reactions that cure via condensation produce moisture as a by-product of the reaction. This moisture can result in voids in the bond-line and general weakening of the bond strength. Condensation PIs are based on the reaction product of aromatic diamines and aromatic dianhydrides and are capable of withstanding prolonged exposure to temperatures of 500–600°F and short term exposure to temperatures up to 1000°F. Because of their condensation reaction and necessary solvents to make the adhesive pliable and capable of flow, these polymers are often processed under vacuum conditions. Condensation PIs can be cured under pressure at 350°F. After about two hours at 350°F they develop sufficient strength to permit post curing at 500–600°F without pressure. The resulting polyimide is insoluble and infusible.

Addition reaction PIs are also available. These are low molecular weight polyimides with acetylene end-groups, which cure on heating by addition polymerization. Polymers that cure by addition reactions do not liberate moisture during cure. However, they still require high temperature solvents to make the adhesive film pliable. Typical processing conditions involve a stepped heating from room temperature to 400°F under light contact pressure, and then a final cure at 550°F. The oxidative stability of addition reaction PIs is not quite equal to the condensation type; however they are somewhat easier to apply and cure. For prolonged exposure, the temperature limit is about 500°F, and it is 600°F for short term exposure.

Thermoplastic polyimide adhesives have also been developed.[20] These adhesives have less stringent high temperature cure requirements, but they are not as resistant to high temperatures as the conventional polyimides. They have been used in microelectronics for attachment of components, lid sealing and wire bonding.

10.3.7.2 Bismaleimide. Bismaleimide (BMI) resins are used in electronic circuit board manufacture and in applications in which higher temperature resistance is required than can be provided by standard epoxy systems, but where the extreme high temperature resistance of a polyimide are not required. BMI adhesives are suitable for long term exposure to temperatures up to 400°F and short term exposure up to 450°F. Systems currently available are quite rigid and for this reason have low peel properties.

BMI adhesives offer significant improvement in processability over polyimides. They are more like epoxy systems than aromatic polyimide systems with respect to their curing requirements. BMI resins cure by addition reactions, and so no volatiles are given-off during cure. These systems are usually cured for several hours at 350°F under pressure and then post cured at or near the maximum service temperature without pressure.

10.3.7.3 Polybenzimidazole. The polybenzimidazole (PBI) adhesive has shear strength on steel of 3,000 psi at room temperature and 2,500 psi at 700°F. In comparison polyimide adhesives offer a shear strength of approximately 3,000 psi at room temperature, but polyimide adhesives do not have the excellent strength at 700–1000°F that is characteristic of PBI. Generally for high temperature service, polybenzimidazoles are chosen when the maximum short term resistance to the temperatures are required. Polyimides are chosen when long term resistance to high temperatures are required. Polybenzimidazole adhesives retain good mechanical properties at temperatures as low as −300°F. Therefore, they represent one of only a few adhesives that exhibit good adhesion characteristics at high temperatures as well as in the cryogenic temperature range

Volatiles are released during cure of polybenzimidazole adhesives that contribute to a porous, brittle bond line with relatively low peel strength. Venting or curing under partial vacuum helps alleviate such problems. Curing requirements consist of high temperatures (approximately 550–600°F) and pressures (200 psi). Post cure is generally required in an inert environment. As a result, large area bonding is very difficult.

Polybenzimidazole adhesive was originally developed in the early 1960s for the aerospace industry as a supported film. It was thought to be an ideal adhesive for spacecraft application where temperatures could range from very high to very low depending on the craft's solar orientation. As a result of their adverse processing requirements, polybenzimidazole adhesives are only used in special applications and their commercial use is limited. Today, polyimides and BMI adhesives have replaced polybenzimidazole adhesives in most high temperature applications.

10.3.8 Polyesters

Polyesters are a large class of synthetic resins having widely varying properties. Polyester resin-based adhesives are relatively low cost formulations that have found niche applications. They may be divided into two distinctive groups: saturated (thermoplastic) and unsaturated (thermosetting).

Unsaturated polyester resins are commonly used for casting, glass fiber laminates, and adhesive systems. Unsaturated polyesters are fast-curing, two-part systems that harden by the addition of catalysts, usually peroxides. Styrene monomer is generally used as a reactive diluent for polyester resins. Cure can occur at room or elevated temperatures depending on the type of catalyst. Accelerators, such as cobalt naphthalene, are sometimes incorporated into the resin to speed cure.

Unsaturated polyester adhesives exhibit greater shrinkage during cure and poorer chemical resistance than epoxy adhesives. Certain types of polyesters are inhibited from curing by the presence of air leaving a tacky resin surface. However, they cure fully when protected from air by covering or enclosing between two substrates. Depending on the type of polyester resin used, the adhesives can be quite flexible or very rigid. Tensile shear strengths from several hundred to thousands of psi can be realized. Polyester adhesives bond to metals, ceramics, and glass reinforced laminates. Applications include patching kits for the repair of automobile bodies and repair cement for concrete flooring. Polyester adhesives also have strong bond strength to glass-reinforced polyester laminates.

Saturated polyester resins exhibit high peel strength and are used to laminate plastic films such as polyethylene terephthalate (Mylar). They also offer excellent clarity and color stability and have been used for optical equipment. This type of polyester adhesive, available in solution or solid film form, can be used as a high performance hot melt system. They can also be chemically crosslinked with curing agents such as the isocyanates for improved thermal and chemical stability. Typical applications for saturated polyester adhesive films are the lamination of flexible printed circuitry material where the metallic conductor foil is bonded to the polymeric dielectric film in a continuous lamination process.

10.3.9 Polyurethanes

Like epoxies, polyurethane reactive adhesives include systems that are available as 100% solids or solvent based and as one or two part formulations. These adhesives can cure at room or elevated temperatures. Unlike epoxies, the polyurethane adhesives are flexible and have both relatively high shear and peel strength.

Polyurethane adhesives are made with isocyanate resins as building blocks. Most adhesives are made with either toluene diisocyanate

(TDI) or methylene diphenyl isocyanate (MDI). Most newer urethane adhesive formulations use MDI, as a result of toxicological concerns associated with TDI. The aliphatic isocyanates, being more expensive, are less commonly used than MDI. Isocyanates can react with active hydrogen compounds. Some of the major reactions of single component urethane adhesives are shown in Fig. 10.6. Reaction mechanism of a two component urethane adhesive is shown in Fig. 10.7.

One-component systems can be formulated to cure when exposed to moisture at room temperature, or by heating. Most one component polyurethanes are based on an isocyanate terminated prepolymer that is catalyzed with a tin salt or with a tertiary amine. The moisture in the air is sufficient to cure the adhesive, but the substrates must have some permeability to allow the moisture to get to the adhesive. A second type of single component polyurethane adhesive is based on "blocked isocyanates". Diisocyanates or isocyanate prepolymers can be reacted with certain active hydrogen containing compounds called blocking agents. Common blocking agents are nylon phenol, caprolactam, and methylethyl ketoxime.[21] After the blocking reaction is complete, the blocked isocyanate is stable and can be formulated with polyols and certain chain extenders without additional reaction. Once heated during cure, the blocked isocyanate thermally

One Component Moisture Curing Urethane Reaction

$$2\ O=C=N\sim\sim\sim N=C=O\ +\ H_2O\ \xrightarrow{\text{Catalyst}}\ \sim\sim\sim[-NH-CO-OH]\ \rightarrow$$

Prepolymer   Carbanic Acid

$$\sim\sim\sim NH_2\ +\ CO_2\ \rightarrow\ \sim\sim\sim NCO\ +\ \sim\sim\sim NH-CO-NH\sim\sim\sim$$

Polyurea

One Component Blocked Urethane Reaction

$$B-CO-NH-R-NH-CO-B\ +\ HO\sim\sim\sim OH\ \xrightarrow{\text{Heat}}$$

Blocked Isocyanate

$$\sim\sim NH-CO-O\sim\sim\sim O-CO-NH-R\sim\sim$$

Polyurethane

Figure 10.6 Reaction mechanisms of single component polyurethane adhesives.[4]

Component A: $\quad\quad\quad\quad$ O = C – N – R – N = O

$\quad\quad\quad\quad\quad\quad$ Diisocyanate or Isocyanate Prepolymer

+

Component B: HO~~~~OH $\quad$ + $\quad$ H$_2$N – R' – NH$_2$ $\quad$ + $\quad$ Catalyst

$\quad\quad\quad\quad\quad$ Polyol $\quad\quad\quad\quad$ Amine (Hydroxyl)
$\quad\quad\quad\quad\quad\quad\quad\quad\quad\quad$ Chain Extender

↓

Cured Adhesive:

~~~O – CO – NH – R – NH – CO – NH – R' – NH = CO – NH – R – NH – CO – O~~~

$\quad\quad\quad\quad\quad\quad\quad\quad$ Polyurea $\quad$ Polyurethane

Figure 10.7 Typical two component polyurethane adhesives reaction.[4]

reverts back to its original components. The isocyanate is regenerated and can react with the polyol and/or chain extender to form a polymer network.

The chemistry of a typical two component polyurethane is summarized in Fig. 10.7. The pot lives of these two component urethanes can vary from as little as 15 secs to as long as 16 hrs, depending on the type of reactant and catalysts. Many structural urethane adhesive formulations set quickly and require no primer or other surface preparation.

Polyurethane-based adhesives form tough bonds with high peel strength. They have better low temperature strength than other adhesives. They have exceptionally high strength at cryogenic temperatures. Only silicone adhesives have better performance at lower temperature, but their tensile strengths are much lower than the polyurethanes. Polyurethane adhesives also have good flexibility, abrasion resistance, and toughness. They have good chemical resistance, although not generally as good as epoxies or acrylics. Urethanes do not have high temperature resistance. Maximum upper use limits are about 300°F, and most formulations are used below 250°F.

Some polyurethane adhesives degrade substantially when exposed to high-humidity environments. This moisture sensitivity occurs with both the cured adhesive and the uncured components. Once a polyurethane component is open to the air it should be used immediately. To prevent moisture contamination, many large volume polyurethane users resort to keeping dry nitrogen blankets on the adhesive components before they are metered and mixed.

Because of their wetting ability and good flexibility, polyurethane adhesives bond well to many substrates including hard-to-bond plastics. Since they are very flexible, polyurethane adhesives are often used to bond films, foils, and elastomers. Urethane structural adhesives are used extensively in the automotive industry because of good adhesion properties to sheet molding component (SMC) and urethane reaction injection molded (RIM) plastics. Structural polyurethane adhesives are widely used in SMC assembly and other plastics bonding in the automotive market. Polyurethanes also bond exceptionally well to wooden substrates. The properties of several commercial polyurethane adhesives are presented on Table 10.15.

Recently reactive polyurethane hot-melt adhesives and sealants have been introduced.[22,23] These hot melts develop 2,500–4,000 psi shear strength and elongation of 500–700%. They are produced from polyurethane polymers with terminal isocyanate groups. Initially, polyurethane hot melts function similarly to conventional thermoplastic hot melts and eliminate the need for extended fixturing of assemblies. They are generally applied in molten form at temperatures as low as 215°F. However, once the polyurethane material has been applied to the substrate, the terminal isocyanate groups react with moisture in the substrate and surrounding air to form a thermoset material. During this cure cycle, the product evolves from a thermoplastic to a thermoset adhesive.

Once the hot melt polyurethane is applied, the parts should be joined rapidly. Green strength develops in seconds, and full cure will occur in one to three days. The bond will continually strengthen to far surpass the strength and environmental durability of conventional hot melts. In addition to the automotive industry, these adhesives are widely used in the textile industry for laminating fabric, seaming and stitch replacement, and producing foam and fabric laminates.[24]

10.3.10 Anaerobic resins

Acrylate monomer-based adhesives are called "anaerobic" adhesives because they cure when air is excluded from the resin. They are essentially monomeric, thin liquids that polymerize to form a tough plastic bond when confined between closely fitting metal joints. The term anaerobic is derived from the science of biology and refers to life forms that live in the absence of air. Polymerization is essentially a free radical type addition polymerization as shown in Fig. 10.8.

These materials will bond all common metals, glass, ceramics, and thermoset plastics to each other. Anaerobics can be applied to three types of surfaces: active, inactive, and inhibiting. Active surfaces, such as clean metals and thermoset plastics, will produce the fastest cures. Metal surfaces accelerate the polymerization in the absence of air. Primers or heat must be used to achieve curing on inactive surfaces

TABLE 10.15 Suppliers of Selected Anaerobic, Modified Acrylic and Polyurethane Adhesive

Adhesive tradename	Adhesive supplier	Form	Curing conditions	Properties	Applications
Anaerobic					
Retaining Compound 620	Loctite	1 part	24 hrs @ 77F or 30 min @ 250F	Solvent resistant RT or heat cure Service temp: −65 to 450F 3000-3500 psi	
Speed bonder 325	Loctite	1 part	15 min @ 250F	Solvent resistant Tough Accelerator cure available 2000-6000 psi	
Permalok HH120	Permabond International	1 part	8 hrs @ 77F	Service temp: −65 to 300F Medium viscosity	Bold and stud locking
Acrylic					
Versilok 204	Lord	2 part or 1 part plus accelerator	2 hr @ 77F (13 min @ 77F)	Service temp: −40 to 250F Semi-flexible Impact resistant	General purpose
M890	Bostik	1 part plus accelerator	4 min @ 77F	Service temp: −65 to 250F	General purpose
Depend 330	Loctite	1 part	1 min @ 77F	Service temp: −65 to 250F Tough, flexible High peel, impact strength	
EA9366	Dexter	2 part paste or 1 part plus accelerator	10 hrs @ 77F	High peel and shear 1000-3500 psi	General purpose

Polyurethane

22005	Swift	1 part; moisture cure	16 hrs @ 70F		Lamination of clear plastic or film
Pliogrip 6600	Ashland	2 part paste	Differing grades for different cures	High strength Resistant to elevated temperature and most fluids	RV, marine, industrial applications
7070	Bostik	2 part; paste	8 hrs @ 77F	Flexible Chemical resistance	General purpose, textiles, upholstery
Conathane EN-4	Conap	2 part; paste	7 days @ 77F (30 min @ 77F)	Hydrolytic stability Thermal shock resistance Low temperature elasticity	Electrical industry, underwater
Scotchweld 3532 B/A	3M	2 part; paste	2 days @ 75F (9 min @ 65F)	Very fast cure Tough; impact resistant Shore D65 250-2390 psi	

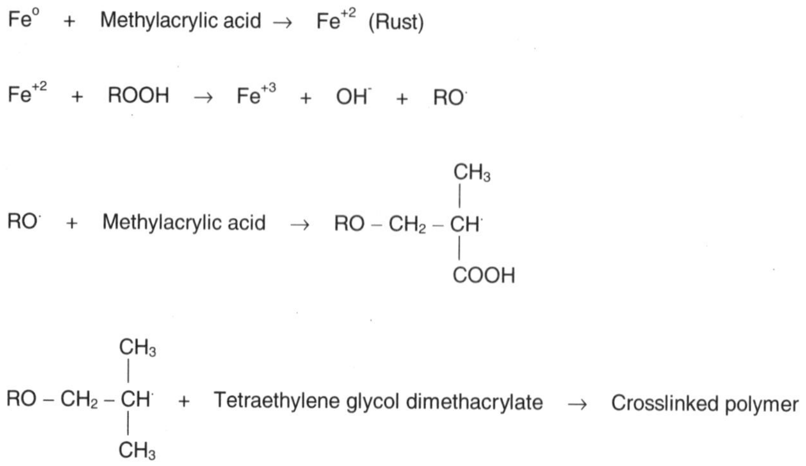

$$Fe^0 + \text{Methylacrylic acid} \rightarrow Fe^{+2} \text{ (Rust)}$$

$$Fe^{+2} + ROOH \rightarrow Fe^{+3} + OH^- + RO^\cdot$$

$$RO^\cdot + \text{Methylacrylic acid} \rightarrow RO-CH_2-\underset{\underset{COOH}{|}}{\overset{\overset{CH_3}{|}}{C}}H^\cdot$$

$$RO-CH_2-\underset{\underset{CH_3}{|}}{\overset{\overset{CH_3}{|}}{C}}H^\cdot + \text{Tetraethylene glycol dimethacrylate} \rightarrow \text{Crosslinked polymer}$$

Figure 10.8 Proposed mechanism for cure of anaerobic adhesives.[4]

(some metals and plastics) and on inhibiting surfaces, such as bright platings, chromates, oxides, and certain anodizes. Phenolic plastics and some plated metals, such as cadmium and zinc, require a primer such as ferric chloride.

Table 10.15 lists properties and curing characteristics of these adhesives. Anaerobic resins are noted for being simple to use, one-part adhesives, having fast cure at room temperature and very high cost on a dollars/pound basis. However, the cost is equivalent to only fractions of a cent per square inch of bonded area because only a small amount of adhesive is required. Anaerobic adhesives do not cure when gaps between adherend surfaces are greater than approximately 30 mils. These adhesives are available in various viscosities. Cure speed is largely dependent on the chemical nature of the parts being joined. The anaerobic adhesive formulations can cure in minutes to hours without primer at room temperature. Application of primer and/or heat will speed up the cure. The primers are generally catalysts dissolved in degreasing solvent.

Anaerobic adhesives are available in machinery and structural grades. Machinery grades provide high strength in threaded assemblies and other closed machinery joints. Structural grades provide high shear strength characteristics in flat joints. The most important use of anaerobic adhesives is as liquid lock washers for screws and bolts, flange sealing, and retaining. Because of their strong penetrating ability, they can be applied either before or after assembly of the mechanical fasteners. These resins have also been used as sealants to seal porosity in metal castings.

Certain anaerobic adhesives can withstand continuous aging at 450°F. However, generally their service temperature limit is 300°F.

They have good resistance to moisture and solvents, and salt spray resistance is excellent. Tough anaerobic adhesives have been prepared from resins having hard but flexible properties.[25]

10.3.11 Cyanoacrylates

Cyanoacrylate adhesives became widely known after their introduction in the early 1970s in the consumer market as "superglue". Cyanoacrylate adhesives are generally methyl or ethyl cyanoacrylate-base, single component liquids that have fast cure, excellent tensile shear strength, and good shelf life. Although very similar in curing characteristics to the anaerobic adhesives, cyanoacrylate resins are more rigid and less resistant to moisture.

Cyanoacrylate adhesives are available only as low-viscosity liquids that cure in seconds at room temperature without the need of a primer. Cyanoacrylates normally polymerize by means of an anionic mechanism that requires moisture in the air to initiate polymerization. The cure mechanism does not work well on acidic substrates that can inhibit or terminate the polymerization or on wet substrates which result in formation of many low molecular weight chains. Factors that influence curing include the amount of moisture on the surface to be bonded, relative humidity, pH, and bond-line thickness. The curing mechanism is water initiated. Lack of complete moisture penetration in thick bond-lines can prevent curing of the center section of the bond.

Cyanoacrylate adhesives bond well to many substrates as shown in Table 10.16. When bonding metals and other rigid surfaces, methyl cyanoacrylate bonds are stronger and more impact resistant than ethyl cyanoacrylate bonds. However, ethyl cyanoacrylate is preferred on rubber or plastic surfaces.

Cyanoacrylates normally have poor heat and moisture resistance; their peel and impact strengths are also low. Although cyanoacrylate adhesives cure, they have many of the characteristics of a thermoplastic resin. They find use as structural adhesives where there is minimal environmental stress and great value is placed on fast setting times. Cyanoacrylate adhesives also have relatively high cost.

A new generation of cyanoacrylates claims to have improved toughness and high shear strength (3,500 psi).[27] New cyanoacrylate resin monomers have been introduced to provide faster cures, higher strengths with some plastics, and greater thermal and impact resistance. Products have been developed which have 5–10 times higher peel strength than conventional compositions. Rubber toughening has been accomplished using various tougheners.[28,29] The use of phthalic anhydride has been reported to improve both the moisture resistance and heat resistance of the cyanoacrylates.[30] Cyanoacrylates are now available in gel form in addition to the conventional low viscosity liq-

TABLE 10.16 Performance of Cyanoacrylate Adhesives on Various Substrates[26]

Substrate	Age of bond	Shear strength, lb/in.^{2} of adhesive bonds
Steel–steel	10 min	1,920
	48 h	3,300
Aluminum–aluminum	10 min	1,480
	48 h	2,270
Butyl rubber–butyl rubber	10 min	150*
SBR rubber–SBR rubber	10 min	130
Neoprene rubber–neoprene rubber	10 min	100*
SBR rubber–phenolic	10 min	110*
Phenolic–phenolic	10 min	930*
	48 h	940*
Phenolic–aluminum	10 min	650
	48 h	920*
Aluminum–nylon	10 min	500
	48 h	950
Nylon–nylon	10 min	330
	48 h	600
Neoprene rubber–polyester glass	10 min	110*
Polyester glass–polyester glass	10 min	680
Acrylic–acrylic	10 min	810*
	48 h	790*
ABS–ABS	10 min	640*
	48 h	710*
Polystyrene–polystyrene	10 min	330*
Polycarbonate–polycarbonate	10 min	790
	48 h	950*

*Substrate failure.

uids. These systems provide the ability to bond to substrates such as wood, leather, and fabrics which have been notoriously difficult substrates for bonding with cyanoacrylate adhesives.

10.3.12 Modified acrylics

Modified acrylic adhesives are thermosetting systems that are also called reactive acrylics to separate them from other acrylic resins that are normally used in thermoplastic, pressure sensitive applications.

Thermosetting acrylic adhesives are two-part systems that provide high shear strength to many metals and plastics as shown in Table 10.17. Acrylic structural adhesives use the same types of chemical components in their formulation as the original anaerobic adhesive. Modern formulations are based on crosslinked polymethyl methacrylate grafted to vinyl terminated nitrile rubber. Carboxy terminated rubbers, similar to those used in toughened epoxy adhesives, and neoprene have also been used. Chlorosulfonated polyethylene has also

TABLE 10.17 Tensile Shear Strength of Various Substrates Bonded with Thermosetting Acrylic Adhesives[31]

Substract*	Avg lap shear, lb/in.^{2} at 77°F		
	Adhesive A	Adhesive B	Adhesive C
Alclad aluminum, etched	4,430	4,235	5,420
Bare aluminum, etched	4,305	3,985	5,015
Bare aluminum, blasted	3,375	3,695	4,375
Brass, blasted	4,015	3,150	4,075
302 stainless steel, blasted	4,645	4,700	5,170
302 stainless steel, etched	2,840	4,275	2,650
Cold-rolled steel, blasted	2,050	3,385	2,135
Copper, blasted	2,915	2,740	3,255
Polyvinyl chloride, solvent wiped	1,375†	1,250†	1,250†
Polymethyl methacrylate, solvent wiped	1,550†	1,160†	865†
Polycarbonate, solvent wiped	2,570†	960	2,570†
ABS, solvent wiped	1,610†	1,635†	1,280†
Alclad aluminum-PVC	1,180†		
Plywood, ⅝-in. exterior glued (lb/in.)	802†	978†	
AFG-01 gap fill (¹⁄₁₆-in.) (lb/in.)	....	1,083†	

*Metals solvent cleaned and degreased before etching or blasting.
†Substrate failure.

been used to improve flexibility and impact resistance in thermosetting acrylic adhesives. Even with the addition of elastomeric modifiers, these thermosetting acrylics are relatively rigid adhesives with low-to-moderate peel strength and poor strength properties at low temperatures. Unmodified acrylic adhesives are particularly noted for their weather and moisture resistance as well as their fast cure at room temperature. Selected thermosetting acrylic adhesives are listed in Table 10.15.

One manufacturer has developed an acrylic adhesive system where the hardener is applied to the substrate as a primer.[32] The substrate can then be stored for up to 6 months. When the parts are to be bonded, only the acrylic resin needs be applied between the already primed substrates. Cure can occur in minutes at room temperature depending on the type of acrylic resin used. In effect, this system offers a fast-reacting, one-part adhesive with indefinite shelf life. Another method of achieving cure with the acrylic systems is to apply the hardener to one surface and the resin to the other surface. When the two surfaces are bonded together, a fast cure occurs, developing handling strength in 60 secs.[33] Of course, acrylic adhesives can also be metered, mixed, and applied in the more conventional way. The fast cure and variety of application alternatives make this adhesive valuable in many production processes.

Because of their rapid cure and high strengths (3,000–5,000 psi), acrylic adhesives are used in high speed automated assembly operations. Modified acrylics have good peel, impact, and tensile lap shear strengths between temperatures of -160 and 250°F. High bond strengths are obtained on metals and plastics and on oily surfaces. Bonds have good resistance to high humidity. Many acrylic resins have a pungent, monomeric odor making them objectionable to certain users. Recent advances have eliminated some of these problems.

Certain acrylic formulations are easily converted to UV or light curing formulations with the addition of photoinitiators to activate the free radical crosslinking mechanism. These single component adhesives are usually termed "aerobic" adhesives. They have diminished sensitivity to air inhibition. Aerobic adhesives cure rapidly at room temperature to form tough, durable bonds with structural strength. In order for the UV or visible light to get through to the adhesive, the substrates must be transparent to light or else the adhesive is used as a coating/sealant such as in the application of wire bonding on circuit boards. Generally, the adhesive must be in the direct path of the light source. Several recent formulations have been developed where the adhesive will cure in deep (1/4 in) sections. With these systems indirect UV light is sufficient to cure the adhesive.

An acrylic-polyurethane adhesive has been introduced recently that has a urethane backbone crosslinked with acrylic.[34] This type of adhesive develops flexible bonds in plastic and non-plastic assemblies such as office furniture and architectural laminate. It has good weathering properties and low odor because of the incorporation of urethane. Recently, a new thermosetting acrylic adhesive was also introduced with good cold temperature resistance and improved peel adhesion to metals.[35] With conventional toughened acrylic adhesives, cold temperature resistance is limited to approximately 0°F.

10.4 Nonstructural Adhesives

Non-structural adhesives are characterized by relatively low strength and poor creep resistance at slightly elevated temperatures. The most common non-structural adhesives are based on elastomers and thermoplastics. Although these systems have low strength, they usually are easy to use and fast setting. Because of these characteristics, most non-structural adhesives are used in assembly-line fastening operations where the ultimate joint strength is not required and the environmental service conditions are not severe. The common types of non-structural adhesives are pressure-sensitive adhesives, contact adhesives, mastics, hot melts, and thermoplastic emulsions.

Although they may be classified as "non-structural" for the purpose of this discussion, many of these adhesives can be used in structural applications where the substrate, environmental conditions, and strength characteristics are suitable. For example, water emulsions, such as polyvinyl acetate, are used to bond wood in the furniture industry. They are used in "structural" applications but the porous substrate and generally mild service conditions do not put a severe burden on the adhesive. Non-structural adhesives may receive little notice because of their pedestrian applications. However, they enjoy by far the major portion of the overall adhesive market.

10.4.1 Elastomeric resins

Natural- or synthetic-rubber-based adhesives usually have excellent peel strength but low shear strength. Their resiliency provides good fatigue and impact properties. Temperature resistance is generally limited to 150–200°F, and creep under load occurs at room temperature. The basic types of rubber-based adhesives used for nonstructural applications are shown in Table 10.18.

These systems are generally supplied as solvent solutions, latex cements, and pressure-sensitive tapes. The first two forms require driving the solvent or water vehicle from the adhesive before bonding. This is accomplished by either simple evaporation or forced heating. Some of the stronger and more environmental resistant rubber-based adhesives require an elevated-temperature cure. Once the solvent is removed, these adhesives are used as either contact or pressure sensitive adhesives. The ease of application and the many different properties that can be obtained from elastomeric adhesives account for their wide use.

A contact adhesive is one that is apparently dry (after solvent evaporation) to the touch. Many formulations, however, will adhere to itself instantaneously upon contact with very slight pressure. Other formulations require the application of pressure or heat and pressure to create a bond. Contact adhesives form relatively strong bonds with high shear and peel strengths. They are often used for metal to metal bonding applications and to bond decorative sheet to wood cores. Although they will creep at elevated temperatures, depending on the nature of the base resin, contact adhesives are used in many structural applications.

Pressure-sensitive adhesives are permanently tacky and flow under pressure to provide intimate contact with the adherend surface. Pressure-sensitive tapes are made by placing these adhesives on a backing material such as rubber, film, canvas, or cotton cloth. After pressure is applied, the adhesive tightly grips the part being mounted as well as the surface to which it is affixed. Pressure-sensitive adhesives are

TABLE 10.18 Properties of Elastomeric Adhesives Used in Nonstructural Applications[3]

Adhesive	Application	Advantages	Limitations
Reclaimed rubber	Bonding paper, rubber, plastic and ceramic tile, plastic films, fibrous sound insulation and weather-stripping; also used for the adhesive on surgical and electrical tape	Low cost, applied very easily with roller coating, spraying, dipping, or brushing, gains strength very rapidly after joining, excellent moisture and water resistance	Becomes quite brittle with age, poor resistance to organic solvents
Natural rubber	Same as reclaimed rubber; also used for bonding leather and rubber sides to shoes	Excellent resilience, moisture and water resistance	Becomes quite brittle with age; poor resistance to organic solvents; does not bond well to metals
Neoprene rubber	Bonding weather stripping and fibrous soundproofing materials to metal; used extensively in industry; bonding synthetic fibers, i.e., Dacron	Good strength to 150°F, fair resistance to creep	Poor storage life, high cost; small amounts of hydrochloric acid evolved during aging that may cause corrosion in closed systems; poor resistance to sunlight
Nitrile rubber	Bonding plastic films to metals, and fibrous materials such as wood and fabrics to aluminum, brass, and steel; also, bonding nylon to nylon and other materials	Most stable synthetic-rubber adhesive, excellent oil resistance, easily modified by addition of thermosetting resins	Does not bond well to natural rubber or butyl rubber
Polyisobutylene	Bonding rubber to itself and plastic materials; also, bonding polyethylene terephthalate film to itself, aluminum foil and other plastic films	Good aging characteristics	Attacked by hydrocarbons; poor thermal resistance
Butyl	Bonding rubber to itself and metals; forms good bonding with most plastic films such as polyethylene terephthalate and polyvinylidene chloride	Excellent aging characteristics; chemically cross-linked materials have good thermal properties	Metals should be treated with an appropriate primer before bonding; attacked by hydrocarbons

truly non-structural systems and will creep extensively under load even at low temperatures. They are generally used in low strength or temporary applications.

In addition to pressure-sensitive adhesives, elastomers go into mastic compounds that find wide use in the construction industry. Neoprene and reclaimed-rubber mastics are used to bond gypsum board and plywood flooring to wood-framing members. Often the adhesive bond is much stronger than the substrate. These mastic systems cure by evaporation of solvent through the porous substrates.

Elastomer-adhesive formulation is more complex than that of other adhesives because of the need for antioxidants, tackifiers, and other additives to the base resin. Each formulation is generally designed to fulfill a specific set of properties. As a result, commercial elastomeric adhesives have widely differing properties.

10.4.1.1 Natural rubber. Natural rubber is one of the earliest materials used in formulating adhesives. The long tack retention properties of natural rubber have made it ideal for pressure sensitive adhesive formulations. Chemical catalysts or accelerators may be used at ambient temperature, or heat curing can be used to vulcanize the natural rubber to improve the strength and temperature resistance. The relative strength of vulcanized natural rubber is low relative to other adhesives. Natural rubber adhesive formulations are heavily modified with synthetic resins and additives to obtain tailored properties.

Natural rubber adhesives are commonly used in bonding nonmetallic materials such as leather, fabrics, paper, and other rubber products. Natural rubber has good resistance to water, but poor resistance to oils, organic solvents, and chemical oxidizing agents. The natural rubber molecule crystallizes at about $-30°F$, and water latex emulsions must be protected from freezing.

Often chlorinated natural rubber is used in adhesive formulations. The outstanding property of chlorinated rubber is its resistance to water, some oils, salt spray, and acids and alkalis. Resistance to organic solvents in general is poor, and the practical temperature limit is 280°F. In adhesive formulations, chlorinated rubber has been used to bond natural and synthetic rubbers to metals. Such adhesives have been used to prime metal surfaces prior to bonding with neoprene or nitrile based adhesives.

Other methods of processing natural rubber for use in adhesive formulations are: by cyclizing (treating with acids or metal salts) to form hard, tough adhesives for metal surfaces; and by reacting with hydrogen chloride (rubber hydrochloride adhesives) to form thin, transparent adhesives used in the packaging industry.

10.4.1.2 Asphalt. Asphalt adhesives are low cost thermoplastic formulations that generally consist of asphaltic resin compounded with thermoplastic rubber such as butyl or polyisobutylene. These adhesives are inherently tacky. Adhesion is very good to a wide variety of materials, such as concrete, glass, metal, and paper products.

Asphalt adhesives should be used only where there is little or no stress. Bond strengths are low and fall rapidly as temperature is raised. The high temperature properties of asphalt adhesives are dependent on the particular asphalt used and on the compounding of the adhesive. Asphalt adhesives become brittle at about 20–30°F. Water and alkali resistance of asphalt adhesives are very good. Thick films are impermeable to moisture vapor. Oils, greases, and common solvents will soften asphalt products.

Water emulsion, solvent solutions, and hot melts are widely used forms of asphalt adhesives. Such mixtures are useful as sealants, but asphalt emulsions have also been used for laminating paper and other packing materials where a water barrier is needed. They are used in many applications where strength is secondary to resistance to water, alkali, or acid. Hot melts have been used for bonding insulation and for lamination of paper and foil. Emulsions have been used as binders in roads, for installation of asphalt tile, and for bonding of waterproof packages. Another important application is in low cost roofing and flooring.

10.4.1.3 Reclaimed rubber. Adhesives based on reclaimed rubber are manufactured primarily from devulcanized rubber. They are generally made from scrap rubber products, such as automobile tires, and various additives to provide characteristics required for the application. Often new elastomeric resins will be blended with the devulcanized rubber to achieve better properties. The primary characteristic of reclaimed rubber adhesives is their low cost.

Formulations mostly consist of solvent solutions or water dispersions. They have properties similar to natural rubber. However, their elasticity is not as great as natural rubber. Reclaimed rubber adhesives are generally not considered for structural applications because of their relatively weak shear strength and poor resistance to sunlight and to temperatures greater than 158°F. As a result, they find applications in low strength, cost sensitive applications. Reclaimed rubber adhesives are used in the automobile industry for bonding carpets to floors, body sealers and sound deadeners, and bonding flexible trim to metal. They generally bond well to most metals, wood, paper, fabric, leather, and natural rubber.

10.4.1.4 Butyl rubber. Butyl rubber is an elastomer used in adhesives and sealants. It is a copolymer of a diene such as butadiene and an olefin such as isobutylene. It is used both as a primary resin and as an additive for tack. Butyl rubber is generally used in solvent solution. It is often formulated as a pressure sensitive adhesive. Butyl adhesives are normally low strength and tend to creep under load. They are useful in packaging applications because of their low permeability to gases, vapors, and moisture.

Butyl rubbers contain a small amount of unsaturation (double bonds in the molecular chain). Polyisobutylene is a completely saturated form of butyl rubber. (See polyisobutylene adhesives below.) Butyl rubber is often formulated with polyisobutylene to provide the required adjustments relative to the degree of cohesive strength and creep resistance (favorable to butyl rubber) vs. the degree of flexibility, tack, and peel strength (favorable to polyisobutylene).

10.4.1.5 Styrene butadiene rubber (SBR).
Styrene butadiene rubber is often used as the primary component in elastomeric adhesives, and it is also used as a major additive to adjust properties of other elastomeric resin based adhesives. SBR does not have the good adhesive properties of natural, nitrile, or neoprene rubber. However, it is lower in cost than these resins. Compared to natural rubber, SBR adhesives have better heat aging properties and lower moisture absorption. They have a useful temperature range of −40F to 160°F. However, they do not have the oil, solvent, or oxidation resistance of neoprene or nitrile based products. SBR resins do not naturally have a high degree of tack. Often tackifiers and plasticizers are added to SBR adhesive formulation when a high degree of tack is needed in the application.

SBR adhesives are available in a variety of formulations. Water dispersions are very common as are solvent solutions and hot melt formulations. SBR adhesives are used as pressure sensitive adhesives and laminating adhesives for paper, textile, leather, plastic films, and metal foils.

10.4.1.6 Polychloroprene (neoprene).
Neoprene has become the popular term for polychloroprene elastomer. Neoprene may be designed for curing or noncuring properties. Curing cements are formulated with crosslinking agents. Metal oxides and antioxidants are commonly found in neoprene adhesive formulations. They are responsible for the excellent resistance of neoprene films to ozone and oxidation.

Neoprene is generally used in organic solvents either as a contact adhesive (sprayed or brushed) or as an extrudable mastic. The rate of strength development is rapid for neoprene adhesives, and they can sustain low loads at high temperatures soon after bonding. However, tack retention and solvent or heat reactivating is generally inferior to other rubber adhesives. As a contact adhesive, neoprene cements are applied to both substrates, the solvent is allowed to evaporate, and the bonding surfaces are carefully assembled and forced together under pressure. Because of the exceptionally high early strength of the neoprenes, it is difficult to correct a misalignment once the prepared surfaces are in contact.

Neoprene is similar in physical properties to natural rubber, but it is stronger and has better aging and high temperature properties. Neoprene bonds are useful over a temperature range of -70 to $+180°F$. Mechanically, neoprene bonds absorb vibration and display good shear and peel strengths. Shear strengths on metal substrates can be as high as 3,000 to 5,000 psi. Cured neoprene films are more rigid than thermoplastic adhesives, but not as rigid as epoxy or phenolic adhesives. Neoprenes deform slightly under stress and, therefore, exhibit excellent fatigue and impact properties.

Neoprenes have good resistance to water, salt spray, commonly encountered chemicals, and biodeterioration. They are, therefore, often used in building construction for decorative plastic laminates, wood working, plywood and hardboard paneling to walls, etc. However, neoprenes should be used in applications where the constant shear stress is limited to 300 psi, because they are liable to cold flow. For structural applications, neoprene elastomers are blended with synthetic resin to promote mechanical strength and heat stability. Like nitrile elastomers, neoprene is often blended with phenolic resins.

Certain neoprene adhesives have been designed specifically for bonding metal to metal for structural purposes. Good bonds are formed with a variety of metals such as aluminum, magnesium, steel, and stainless steel. Some compounds are relatively poor on copper, zinc, and chromium. Neoprene is the basic adhesive used in installing laminated plastic counter tops in kitchens. The shoe industry is also a large outlet for the neoprene cements.

10.4.1.7 Acrylonitrile butadiene (nitrile). Acrylonitrile butadiene or nitrile adhesives are the most versatile of the general purpose rubber types, but they are less popular than neoprene based adhesives. Nitrile rubber adhesives are a copolymer of butadiene and acrylonitrile. The adhesive properties can be increased by increasing the nitrile content. Nitrile resins are very compatible with a wide variety of fillers and modifiers, thus giving the formulator a wide variety of properties.

Nitrile adhesives are generally available in solvent solutions of varying viscosity. Once dried on a substrate, the adhesive can be reactivated with heat or solvent before joining. Although an adequate cure can be obtained at 325°F in about 20 mins, better strengths at elevated temperatures are obtained if the adhesive is cured at higher temperature. Nitrile adhesives are also available in tape form which can be activated under heat and pressure.

Nitrile adhesive bond strength and creep resistance vary considerably with the compounding agents used. High bond strengths are obtained in the range of 3,000–4,000 psi. They may be used in continuous service at 250°F. Nitrile adhesives are not suitable for structural applications under continuous loading exceeding 100 psi. They have good resistance to oil and grease, water, and many organic solvents.

The main applications for nitrile adhesives involve high temperatures, such as the bonding of brake linings to brake shoes, clutch facings, and the attachment of grinding stones to metal holders. Nitrile rubber is blended with phenolic resins to form important structural adhesive for metals. Nitrile adhesives are also used to bond vinyl, other elastomers, and fabrics where good wear and oil and water resistance are important. Nitrile adhesives are also used for bonding vulcanized and unvulcanized rubbers to various surfaces.

10.4.1.8 Polyisobutylene. Polyisobutylene resins are commonly used in pressure sensitive adhesive formulations. Polyisobutylene is an amorphous polymer that relies mainly on molecular chain entanglement for cohesive strength. Thus, polyisobutylene adhesives are characterized by their high degree of flexibility and tack and by their low cohesive strength. They have better aging characteristics than natural rubber and have moderately good environmental resistance and elasticity. However, polyisobutylene adhesives have poor thermal resistance, and they are attacked by many solvents. Butyl rubber is simply polyisobutylene with a very low degree of crosslinking. Thus, butyl additions can be used to bring about improved cohesive strength and creep resistance.

Polyisobutylene adhesives are generally available as solvent solutions or as a pressure sensitive film on a carrier. Many pressure sensitive tape products are formulated with polyisobutylene resins. These products remain permanently tacky.

10.4.1.9 Polyvinyl methyl ether. Polyvinyl methyl ether resins provide a unique class of colorless, water soluble adhesives. They are generally used in pressure sensitive formulations. Their most common application is moisture activated pressure sensitive labels. They have good peel characteristics and do not embrittle on exposure to UV light or on excursions to low temperatures. As a result, paper labels that are manufactured with polyvinyl methyl ether adhesive will not curl and can adhere to glossy surfaces.

10.4.1.10 Polysulfide. Polysulfide elastomers are often considered to be sealers rather than adhesives because of their low strength and high degree of elongation. Polysulfide elastomers are available as liquids that can be easily formulated for a variety of applications.

Polysulfide adhesives are generally two part high viscosity solvent free systems that may require a catalyst to accelerate cure time. These systems cure at room temperature, but do not reach full strength for 3–7 days. Since little or no solvents are used in the formulations, shrinkage during setting is negligible. Hence, polysulfides are used in a variety of sealing applications.

Because of their relatively low strength, creep tendency under low loads, and low resistance to elevated temperatures, these materials are primarily used as sealants that are applied with an extrusion gun. Resistance to water, organic solvents, grease, oils, and salt water is better than that of the other thermoplastic rubber adhesives. Low temperature properties are excellent with retention of flexibility to −80°F. However, high temperature resistance is poor, and the material usually softens at 160–200°F with very little strength over 250°F.

Some polysulfide elastomers have been used to reduce the rigidity of two part thermosetting epoxies. These modified adhesives have been used to bond aluminum and have very good durability although low tensile strength relative to the unmodified epoxy adhesive.

10.4.1.11 Silicone. Silicone resins have low shear properties but excellent peel strength and heat resistance. Silicone adhesives are available as either solvent solutions for pressure sensitive adhesives or as one or two part liquids.

Silicone adhesives are generally supplied as solvent solutions for pressure-sensitive application. These systems cure via a condensation and radical polymerization process. The resulting adhesives are very tacky and exhibit only moderate peel strengths as a result of their very poor cohesive strength. The adhesive reaches maximum physical properties after being cured at elevated temperature with an organic peroxide catalyst. A lesser degree of adhesion can also be developed at room temperature.

Silicone adhesives retain their adhesive qualities over a wide temperature range, and after extended exposure to elevated temperature. They are very tacky materials that bond to a wide variety of substrates. Because silicones have a relatively low surface energy, they bond well to many low surface energy plastics such as polyethylene and fluorocarbons. Table 10.19 shows typical adhesive-strength properties of pressure-sensitive tape prepared with aluminum-foil backing.

Pressure sensitive silicone adhesives are often used in the form of pressure sensitive tapes. A large application is pressure sensitive tape used in the electronics industry for various applications on printed circuitry. They are also often used as adhesives on silicone rubber backing for gasketing on ovens and other high temperature apparatus.

Room-temperature-vulcanizing (RTV) silicone-rubber adhesives and sealants form flexible bonds with high peel strength to many sub-

TABLE 10.19 Effect of Temperature and Aging on Silicone Pressure Sensitive Tape (aluminum foil backing)[36]

Testing temperature		Adhesive strength, oz/in.	
°C	°F	Uncatalyzed adhesive	Catalyzed adhesive (1% benzoyl peroxide)
−70	−94	Over 100*	Over 100*
−20	−4	Over 100*	Over 100*
100	212	60	48
150	302	60	45
200	392	Cohesive failure	40
250	482	Cohesive failure	35
Heat-aging cycle prior to testing (tested at 25°C, 77°F):			
No aging................		60	52
1 h at 150°C (302°F)		55	50
24 h at 150°C (302°F)		60	50
7 days at 150°C (302°F) ..		70	50
1 h at 250°C (482°F)		60	50
24 h at 250°C (82°F)		65	65
7 days at 250°C (482°F) ..		Cohesive failure	65

*Maximum limit of testing equipment.

strates. These adhesives are one-component pastes that cure by reacting with moisture in the air. Because of this unique curing mechanism, adhesive bond lines should not overlap by more than 1 in. RTV silicone materials cure at room temperature in about 24 hrs. With most RTV silicone formulations, acetic acid is released during cure. Consequently corrosion of metals, such as copper and brass, in the bonding area may be a problem. For corrosive substrates, certain RTV silicone formulations are available that cure by liberating methanol rather than acetic acid.

Two-part RTV systems that do not liberate byproducts during cure are also available. These adhesive systems are generally cured by a stannous octoate or dibutyltindilaurate catalyst at room or elevated temperatures. These two-part systems normally contain water in the formulation for more effective cure. Two-part RTV adhesives can cure in thick sections, where the one-part system cannot because of a cured skin that forms quickly, inhibiting atmospheric moisture from gaining access to the center section of the bond-line.

Fully cured RTV adhesives can be used for extended periods up to 450°F and for shorter periods up to 550°F. Silicone adhesives are soft and compliant and have good chemical and environmental resistance. The low temperature limit for these adhesives is generally −80°F because of crystallization that can occur at that temperature. Figure 10.9 illustrates the peel strength of an RTV adhesive on aluminum as a function of heat aging.

RTV silicone rubber bonds best to clean metal, glass, wood, silicone resin, vulcanized silicone rubber, ceramic, and many plastic surfaces. The low shrinkage and excellent gap filling capabilities make these materials ideal sealants for many applications. They are best known as sealants for use in kitchens and bathrooms.

10.4.2 Thermoplastic resins

Thermoplastic adhesives represent a class of adhesives that can set without chemical reaction. They can be repeatedly melted and solidified without greatly hindering their properties. They also can be dissolved in solvent or suspended in a water-based emulsion. Thus, these adhesives are generally formulated so that they harden by either evaporation of solvent or on cooling from a hot melt.

Table 10.3 describes the most common types of thermoplastic adhesives. These adhesives are useful in the −20 to 150°F temperature range. Their physical properties vary with chemical type. Some resins, like the polyamides, are quite flexible, while others are very rigid. Being uncrosslinked, all thermoplastic adhesives will creep at moderate loads and temperatures.

Thermoplastic adhesives are generally available as solvent solutions, water-based emulsions, or hot melts. Solvent- and water-based systems are useful in bonding porous materials such as wood, plastic foam, and paper. Contact adhesives are probably the best known solvent-based adhesives. Water-based systems are especially useful for bonding foams that could be degraded by solvent-based adhesives. Once hardened, thermoplastic adhesives will soften when exposed to the solvent in which they were originally supplied.

Hot-melt systems are usually flexible and tough. They are used extensively for sealing applications involving paper, plastic films, and

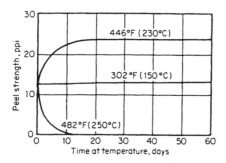

Figure 10.9 Peel strength of RTV silicone rubber bonded to aluminum as a function of heat aging.[37]

metal foil. Table 10.20 offers a general comparison of hot-melt adhesives. Hot melts can be supplied as tapes or ribbons, films, granules, pellets, blocks, or cards which are melted and pressed between the substrate. The rate at which the adhesive cools and sets is dependent on the type of substrate and whether it is preheated or not. Table 10.21 lists the advantages and disadvantages associated with the use of water-based, solvent-based, and hot-melt thermoplastic adhesives.

10.4.2.1 Polyvinyl acetal. Polyvinyl acetal adhesives generally are formulated from either polyvinyl butyral or polyvinyl formal resins. The polyvinyl butyral has a lower melt viscosity and is a more flexible polymer providing better peel strength. Polyvinyl butyral adhesives are made as solvent solutions of high viscosity and low solids content. They are water white and transparent, light stable, and have excellent adhesion to glass. Water resistance is relatively poor, but humidity resistance is adequate for most outdoor applications. Polyvinyl butyral is commonly used in laminating safety glass.

Polyvinyl formal is used in flexibilizing thermosetting resins. Hybrid adhesives such as vinyl-phenolics and vinyl-epoxies use polyvinyl formal as the flexibilizing agent.

10.4.2.2 Polyvinyl acetate. Polyvinyl acetate is the most versatile of the vinyl resin containing adhesives. Polyvinyl acetate adhesives are generally available as solvent solutions or emulsions. They are either plasticized or unplasticized. Their formulations may contain fillers and pigments, depending on the application. This polymer is perhaps best noted as an emulsion stabilized with a few percent of polyvinyl alcohol. Such a polyvinyl acetate latex is the basis for the common household "white glue".

For bonding wood a set time of 10 mins to 3 hrs at 68°F and contact to moderate pressure (145 psi) is recommended. One to 7 days conditioning is recommended before aggressive handling of bonded assemblies. In the furniture industry, microwave heating is sometimes employed to speed the drying. Bonded substrates can be disassembled by heat and then reactivated by solvent. The adhesive should be applied at temperatures of 60 to 90°F. The latex can be ruined by repeated freeze-thaw cycles, and shelf life is generally limited.

The polyvinyl acetate adhesive is relatively flexible especially at low temperatures. Resistance to most solvents and moisture is poor although these adhesives will withstand contact with grease, oils, and petroleum fluids. The cured films tend to soften at temperatures approaching 113°F. With proper formulation, these can be low cost adhesives with high initial tack and good gap filling properties. Polyvinyl acetates tend to creep under substantial load.

TABLE 10.20 General Comparison of Common Hot-Melt Adhesives[38]

Property	Ethylene vinyl acetate	Ethylene ethyl acetate	Ethylene acrylic acid	Ionomer	Phenoxy	Polyamide	Polyester	Polyethylene	Polyvinyl acetate	Polyvinyl butyral
Softening point, °C	40	60	70	75	100	100	....	....	65–195	....
Melting point, °C	95	90	....	....	....	....	267	137	....	130
Crystallinity	L	L	M	L	L	L	H	H or L	L	L
Melt index	6 (2–20)	3 (80–100)	3 (0.5–400)	2	2.5	2	5	5 (0.5–20)	....	....
Tensile strength, lb/in.^{2}	2,750	2,000	3,300	4,000	9,500	2,000	4,500	2,000	5,000	6,500
% elongation	800	700	600	450	75	300	500	150	10	100
Lap shear on metal, lb/in.^{2}	....	....	1,700	....	3,500	1,050	....	....	....	....
Film peel, lb/in.	1	4.5	12	12	....	....	5	2	....	....
Cost	M	M	M	M	H	M	H	L	M–L	M
Usage	H	M	L	L	L	H	H	L	M	H

H = high. M = medium. L = low.

TABLE 10.21 Advantages and Disadvantages of Thermoplastic Adhesive Forms[39]

Form	Advantages	Disadvantages
Water base	Lower cost	Poorer water resistance
	Nonflammable	Subject to freezing
	Nontoxic solvent	Shrinks fabrics
	Wide range of solids content	Wrinkles or curls paper
	Wide range of viscosity	Can be contaminated by some metals used for storage and application
	High concentration of high-molecular-weight material can be used	Corrosive to some metals
	Penetration and wetting can be varied	Slow drying
		Poorer electrical properties
Solvent base	Water-resistant	Fire and explosive hazard
	Wide range of drying rates and open times	Health hazard
		Special explosion proof and ventilating equipment needed
	Develop high early bond strength	
	Easily wets some difficult surfaces	
Hot melt	Lower package and shipping cost per pound of solid material	Special applicating equipment needed
	Will not freeze	Limited strength because of viscosity and temperature limitations
	Drying and equipment for drying are unnecessary	
	Impervious surfaces easily bonded	Degrades on continuous heating
		Poorer coating weight control
	Fast bond-strength development	Preheating of adherends may be necessary
	Good storage stability	
	Provides continuous water-vapor impermeable and water-resistant adhesive film	

Products of this type are good adhesives for applications where at least one substrate is porous. Paper, plastics, metal foil, leather, and cloth can be bonded with polyvinyl acetate emulsions. The major use is in packaging. Polyvinyl acetates are also used in hot-melt adhesive formulations. In this form they are commonly used for bookbinding and lamination of foils. Organic solvent solutions are also used, although not as much as the latex.

10.4.2.3 Polyvinyl alcohol. Polyvinyl alcohol is a water soluble thermoplastic synthetic resin that is supplied generally as a formulated water solution for the bonding of porous materials such as paper and textiles. With wood and paper, polyvinyl alcohol adhesives lose water rapidly and consequently set quickly, allowing pressure release in minutes after assembly. Wet tack properties, combined with quick set, make them quite useful in automatic labeling operations, packaging, and continuous laminating of foil and paper. Vinyl alcohol adhesives can be heat sealed.

Water resistance is only fair but sufficient for most interior applications. Resistance to grease and oil is excellent. The adhesives are also odorless and tasteless. They consequently find application in the packaging of food.

The addition of polyvinyl alcohol to polyvinyl acetate emulsions is often used to slow down the speed of set and to improve flexibility and tack. However, the heat resistance and resistance to long term creep of the pure polyvinyl acetate resin are generally lowered.

10.4.2.4 Thermoplastic elastomers. Thermoplastic elastomers are block copolymers. The best known materials are styrene-butadiene. They do not have the good adhesion characteristics of nitrile, neoprene, or natural rubber. However, they are lower in cost and have better heat aging resistance than natural rubber. Styrene-butadiene rubber (SBR) is the largest volume synthetic rubber manufactured in the U.S. These materials are not useful as adhesives in their natural state and are often heavily compounded with plasticizers, fillers, and other resins.

SBR adhesives have very good toughness and low water absorption. Formulated compositions have very good tensile strength characteristics. They are used commonly as pressure sensitive adhesives, hot melt adhesives, sealants, and binders. They are used for the splicing of tire threads and as adhesives to bond fabrics to themselves as well as other materials.

10.4.2.5 Ethylene vinyl acetate. This random copolymer is often formulated as a hot-melt adhesive containing up to 30% vinyl acetate in polyethylene. As with all hot melt formulations, melt viscosity is very dependent on molecular weight of the base resin. Tackifiers are added to reduce viscosity and improve wetting and adhesion. Waxes are added to lower cost and reduce viscosity. Inorganic fillers are used to lower cost and increase viscosity. Antioxidants are required to protect the adhesive during application and during its service life.

Use includes cardboard, book binding, and iron-on clothing articles. The adhesives will creep at relatively low stress. This can be lessened by crosslinking through the addition of a peroxide or by irradiation. Curable hot melt adhesive systems are under development.

10.4.2.6 Cellulosic resins. Cellulose derivatives are one of the oldest and most versatile of adhesives to be originally introduced for household applications. Cellulose nitrate adhesives are solvent solutions that bond well to many thermoplastics. They are water resistant and form a tough bond to a variety of materials such as metal, glass, fabric and plastics. Cellulose nitrate adhesive formulation was first introduced for bonding plastic models and toys.

Cellulose acetate adhesives are more heat resistant and exhibit better aging. However, they are not as moisture resistant as the cellulose nitrate counterparts. The principal use is bonding cellulose acetate plastics.

Ethyl cellulose is available as solvent solutions or as hot melts. The properties can vary considerably with the type of resin used and the adhesive formulation. Good bonds are formed to porous materials, but adhesion on nonporous substrates is not outstanding. Water resistance is only fair.

10.4.2.7 Polyamide. Polyamide hot melt adhesives have lower melting points than the polyamides used for engineering plastics (nylon). They have greater temperature resistance than ethylene vinyl acetate adhesives and require fewer additives. However, they are generally more expensive. Polyamide terpolymers are used for bonding textile fabrics and have good resistance to dry cleaning fluids. There are a great many resins in the polyamide family. They can be varied to provide hot melts of almost any desired temperature over a span of several hundred degrees.

Polyamide resins are also commonly employed as modifiers for both structural and nonstructural adhesives to improve flexibility and resistance to impact.

10.4.2.8 Polyester. Polyesters, chemically similar to those used in the synthetic fiber industry, are also used in hot-melt adhesive formulations. One of the earliest applications for hot-melt polyester adhesives was in the manufacture of shoes. Polyester hot-melt formulations are commonly found in preformed sticks or coils that are used with hot-melt extrusion guns. The adhesive is pushed through the heated nozzle of the gun, and molten product is applied to the substrate. Hot-melt polyester adhesives are commonly used for bonding fabrics such as decorative trim, draperies, etc. Polyester adhesives can be used in their natural state, but often they are compounded heavily to provide for increased toughness, peel strength, and open time.

Polyester hot-melt adhesives are based on terephthalic acid, but other diacids are also used. Polyester hot-melt resins have a relatively high melting point of approximately 500°F. Thus, they are generally used as a heat activated film. They can be formulated to have high tensile strength and are often employed where high strength and high temperature resistance are required.

10.4.2.9 Polyolefins. Polyolefins are generally not thought of as adhesives because they are so hard to bond as a substrate. However, polyolefin resins make excellent hot melt adhesives because they do have a low surface energy and wet most polymeric and metallic substrates. However, commercial grade standard polyolefin resins cannot be used in these applications. Standard polyolefin products (film for example) have low molecular weight constituents that will contribute to a weak boundary layer. Polyolefins that are used for adhesives are relatively pure compounds with high molecular weight.

Adhesives that are based on polyolefin resins are used mainly as hot melt adhesives in either block or film form. In film form, they provide excellent peel strength as a hot laminating adhesive for plastic films and metallic foils. Polyethylene adhesive films are also used in extrusion coating. In this application, a ply of polyethylene is extruded onto a structure to provide a heat sealing medium for later bonding.

10.4.2.10 Polysulfone. Polysulfone resins, like polyesters, are temperature resistant thermoplastic adhesives that require fairly high temperature for heat activation after all solvents have been removed. These very tough, high strength resins have excellent resistance to creep. They maintain their structural strength (more than 60% of their room temperature strength) up to 375°F. They also have creep resistance at 300°F. Tensile shear strengths of over 4,000 psi have been obtained with stainless steel after pressing at 700°F. Polysulfone adhesives are resistant to strong acids and alkalis. They are attacked by polar organic solvents and aromatic hydrocarbons.

Polysulfone adhesives have good gap filling properties. They are often chosen for applications where the high temperature structural properties of thermosetting adhesives are required, but where the processing speed and toughness of a thermoplastic are also necessary.

10.4.2.11 Phenoxy. Phenoxy adhesives are thermoplastics similar to the epoxy family. They are a one component material supplied in powder, pellet, or film form. They can be used as a hot melt or applied as a solvent solution.

Phenoxy adhesives withstand weathering well and have excellent resistance to water, alcohols, and aliphatic hydrocarbons. Service temperature ranges from −52 to +180°F. Resistance to creep is high, even at 176°F. These adhesives provide rigid, tough glue lines with high adhesive strength. In hot melt form, plasticizers are generally used to reduce the melt temperature. Good adhesion has been obtained with copper, brass, steel, aluminum, wood, and many other substrates.

Phenoxy resins are used as structural adhesives for rapid assembly of metals and rigid materials, for continuous lamination of metal to metal or wood, and for flexible substrates. Other applications are in pipe forming, assembly of automotive components, and bonding of polymeric materials.

10.4.2.12 Acrylic. Thermoplastic acrylics have excellent adhesive qualities. Acrylic adhesives are available as solutions, as emulsions, or as polymer-monomer mixtures, which may be cured by ultraviolet light, heat, or chemical catalysts.

Pressure sensitive adhesives are based on solvent systems. Bonding consists of drying to the point of tackiness and then joining the two tacky surfaces. The acrylics have medium water resistance, good resistance to oil and thermal shock, and poor resistance to organic solvents. Temperature resistance varies with the nature of the base resin. Acrylic adhesives have excellent outdoor weathering and aging characteristics. One of their main applications is on automobile "bumper stickers" or for outdoor name tags, labels, signs, etc.

Polyacrylate elastomers are used to make high performance pressure sensitive adhesives. These are used for holding a wide variety of substrates including, automobile identification tags, mounting tape, protective moldings, etc. These adhesives have excellent strength and outdoor aging resistance.

Acrylic hot-melt pressure-sensitive adhesives are relatively new. These adhesives can be applied as a hot melt, but once on the substrate they act as a pressure-sensitive adhesive. In this method of application, the use of solvent simply as a carrier to apply resin to the substrate is eliminated. Acrylic hot-melt adhesives provide higher thermal, oxidative, and UV stability than other hot-melt or pressure-sensitive adhesives.

10.4.3 Naturally occurring resins

Naturally occurring adhesives are still used today in many applications. These are usually based on natural organic products or inorganics. They are used primarily when a low cost adhesive is required. However, some non-polymeric adhesives such as ceramic cements have very specific properties (e.g., high temperature resistance) that find use in niche applications.

10.4.3.1 Natural organic resins. Natural organic adhesives are high polymers like their synthetic counterparts. Many of these adhesives are used because of their early history and the number of industries that have adopted and continued to use these systems. They are noted mainly by their lower cost. In certain applications, they dominate over synthetic adhesives.

10.4.3.1.1 Glues of agricultural origin. These adhesives are soluble or dispersible in water and are produced or extracted from natural sources. There are several common adhesives that are derived from agricultural sources. These are starch, dextrin, soybean, and oleoresinous adhesives.

Starch adhesives are derived primarily from the cassava plant, although other sources may be used. The starch is usually heated in alkaline solutions then cooled to room temperature to prepare the dispersion. After cooling, they are applied as cold press adhesives. They develop their strength by loss of water into porous substrates. Tack develops rapidly. Normal setting on wood substrates requires 1–2 days at room temperature with contact pressure. Starch adhesives are used for paper cartons, bottle labeling, and stationery applications. Joint strengths are low compared to the vegetable types. However, the adhesive is resistant to water and biodeterioration.

Dextrins are degradation products of starch. They are produced by heating starch in the presence or absence of hydrolytic agents, usually basic or acid producing substances. They are often mixed with catalytic agents. These adhesives can bond to many different substrates. Paper and paperboard are the most common substrates. Laminating adhesives are usually made from highly soluble white dextrins and contain fillers such as clay.

Soybean glue is derived from protein. These adhesives are cheap and can be used for making semi-water resistant plywood and for coating some types of paper. They set at room temperatures.

The most common form of oleoresinous adhesive is made from the oleoresin of the pine tree. This materials is used either in solvent solution or as a hot melt mastic. It has poor resistance to water and oxidation. Bond strengths are moderate and develop rapidly. This adhesive is often used for temporary fastening.

10.4.3.1.2 Glues of animal origin. Many of the most important natural adhesives occur from animal origin. Almost half of the animal glues produced go into remoistenable gummed tapes. Common animal-based glues are derived from casein, blood, bone, and hide byproducts and fish.

Casein glue is the protein of skim milk. Dry mix casein glues are simply mixed with water before use. They are used at room temperature and set by loss of water through a porous substrate and by a certain degree of chemical conversion. Thus, mixed casein adhesives have a finite pot life. They are used for paper labels, woodworking, and other interior applications. They cannot be used outdoors although they are resistant to dry heat up to 158°F. Resistance to organic solvents is generally good. Certain casein glues are used to bond high voltage paper insulation. Casein has a unique property in that it can be impregnated with insulating oil and thus provide excellent insulation characteristics.

Blood albumen adhesives (blood glues) are used in much the same way as casein glues. The blood proteins can be hardened by hot pressing as well as by loss of water. Porous materials require only several minutes at 176°F to set. These adhesives are used to a limited degree to make softwood plywood. The major applications are bonding porous materials such as paper, cork, leather, and textiles.

Animal glues are made from bone and hide byproducts. These glues are supplied as liquids, jellies, or solids in the form of flakes, cubes, granules, etc. for reconstitution with water. They are used primarily for furniture woodworking, but are also used for leather, paper and textiles. These glues set at temperatures in the range of 176–194°F or they may set in longer times at room temperature. The bond strength usually exceeds the strength of wood and fibrous substrates.

Fish glues are by-products of fish skins, usually cod. They have properties similar to animal glues. Fish glues were the forerunners of all household glues. Fish glue is usually used in cold setting liquid form. These glues may be exposed to repeated freezing and thawing cycles without adverse effects. Initial tack is excellent on remoistening. Fish glues bond well to glass, ceramics, metals, wood, cork, paper, and leather. The main uses are in the preparation of gummed tapes.

10.4.3.2 Inorganic adhesives and cements.

Adhesives based on inorganic compounds generally have low cost and high temperature properties. They are usually applied as water based adhesives. These materials form strong bonds in applications where at least one substrate is porous. The advent of synthetic organic polymer adhesives has led to a decline in the use of many of these adhesives except for specialty applications. The most common inorganic adhesives are sodium silicate, phosphate cement, litharge cement, and sulfur cement.

10.4.3.2.1 Sodium silicate.

Commonly know as water-glass, this colorless low cost inorganic material is usually supplied as a viscous water solution. These adhesives display little tack, and holding pressure must be applied until the bond is sufficiently dry. The dry adhesive is brittle and water sensitive. In fact the adhesive may be re-dissolved by water. Water resistance can be improved by applying suitable aluminum salts to substrates such as paper prior to bonding. Addition of sugar, glycerin, and other materials promotes retention of moisture in the film and increases its flexibility, tackiness, and toughness. Kaolin clay is frequently added to raise the viscosity and prevent excessive penetration into porous substrates. Sodium silicate adhesives are very resistant to high temperatures. Some can resist up to 2,000°F. Their main use is to bond porous substrates such as paper and cardboard where an inexpensive, fast processing adhesive is required. Other applications for silicates include wood bonding, adhering glass to porous substrates, glass fiber insulation bonding, fabrication of foundry molds, and abrasive wheel cements.

10.4.3.2.2 Phosphate cements. These cements are based on the reaction products of phosphoric acid with other materials such as sodium silicate, metal oxides and hydroxides, and the salts of basic elements. Zinc phosphate is one of the most important metal phosphate cements and is used as a dental adhesive. Metallic phosphates of aluminum, magnesium, chromium, and zirconium develop excellent thermal stability when heated to 570°F. They are virtually insoluble in hot water once fully set. Other applications are the bonding of refractory materials.

10.4.3.2.3 Litharge cement. Mixtures of glycerin and litharge (lead oxide) are used as adhesives in the repair of tubs and sinks, pipe valves, glass, stoneware, and common gas conduits. A mixture of one part slightly diluted glycerin with two to three parts of lead oxide requires approximately one day to form a crystalline compound. The resulting cement resists weak acids and nitric acid, but reacts with sulfuric acid. These materials have been used as ceramic seals in potting electronic equipment. They are not used much today because of the toxic effects of lead.

10.4.3.2.4 Sulfur cement. Liquid sulfur (melting point 730°F) can really be considered as an inorganic hot melt adhesive. However, this material should not be exposed at temperatures greater than 200°F because of a marked change in the coefficient of expansion at 205°F as a result of phase change. The addition of carbon black and polysulfides improves the physical properties of this adhesive. Adhesion to metals, particularly copper is good. Tensile strengths of about 580 psi are possible. The principal use of sulfur cement is for acid tank construction, where resistance to high temperature oxidizing acids are required.

References

1. Austin, J. E., and Jackson, L. C., "Management: Teach Your Engineers to Design Better with Adhesives", *SAE J.*, October 1961.

2. Hull, C. G., "Future Trends in Epoxy Technology - 1989", joint presentation by Shell Chemical and Pacific Anchor Chemical Companies.
3. Burgman, H. A., "Selecting Structural Adhesive Materials", *Electrotechnology,* June 1965.
4. Frisch, K. C., "Bonding and Chemistry of Structural Adhesives", from *Adhesion Principles and Practice Course,* Kent State University, May 1997.
5. Wright, C. D., and Mugge, J. M., "Epoxy Structural Adhesives", Chapter 3 in *Structural Adhesives – Chemistry and Technology,* S. R. Harshorn, ed. (New York: Plenum Press, 1986) at 128.
6. Sieber, A. R., et al., "CTBN-Modified Epoxies Work in Poor Bonding Conditions", *Adhesives Age,* July 1986 at 19–23.
7. Barlet, P., et al., "Relationships Between Structure and Mechanical Properties of Rubber Modified Epoxy Networks Cured with Dicyandiamide Hardener", *Journal of Polymer Science,* 30: 2955–2966, 1985.
8. Drake, R. and Siebert, A., "Elastomer Modified Epoxy Resins for Structural Applications", *SAMPE Quarterly,* vol. 6, no. 4, July 1975.
9. Murphy, W. T., "Elastomer Modified Epoxies in Adhesive Applications", SPI-ERF Spring Meeting, Willamsburg, VA, May 18–20, 1983.
10. Murphy, W. T., and Siebert, A. R., "Development of Resins for Damage Tolerant Composites–A Systematic Approach", Natl. SAMPE Symp. (Proc.), 29, *Technology Vectors,* 422–36, 1984.
11. B. F. Goodrich Company, "Hycar CTBN-Modified Epoxy Adhesives, AB-8", Cleveland, OH.
12. Bolger, J. C., "Structural Adhesives for Metal Bonding", Chapter 1 in *Treatise on Adhesion and Adhesives,* vol. 3, R. L. Patrick, ed., (New York: Marcel Dekker, 1973).
13. DeLollis, N. J., "Durability of Adhesive Bonds (A Review)", Natl. SAMPE Symposium (Proc.), 22, *Diversity Technology Explosion,* 673–698, 1982.
14. Guthrie, J. L., and Lin, S. C. "One Part Modified Epoxies for Unprimed Metal, Plastics", *Adhesives Age,* July 1985 at 23–25.
15. Gerace, M. J., et al., "Surface Activated Rubber Particles Improve Structural Adhesives", *Adhesives Age,* December 1995.
16. Hartenstein, R. R., and Kimball, P. H., "Soft Epoxies Offer New Opportunities for Adhesives", *Adhesives and Sealants Industry,* February 1997 at 24–35.
17. Ward, R. J., and Bobalek, E. G., "Some Correlations Among Thermal, Electrical, and Mechanical Properties of Alkyl-Epoxy Copolymer Adhesives", *Ind. Eng. Chem. Prod. Res. Develop.,* 2, 85, 1963.
18. Hayes, B. J. and Mitra, P., South African Patent 68.00870, to Ciba, Ltd.
19. Subrahmanian, K. P., "High Temperature Adhesive", in *Structural Adhesives – Chemistry and Technology,* S. R. Harshorn, ed., (New York: Plenum Press, 1986).
20. Ying, L., "A Reworkable High Reliability Thermoplastic Die Attached Adhesive", *Proc. International Symposium on Microelectronics,* 1986 at 621.
21. Wicks Jr., Z. W., *Prog. Org. Coatings,* 9, 3, 1981.
22. Chamber, J., et al., "Full Reactive PU Hot Melts Offer Performance Advantages", *Adhesives Age,* August 1998.
23. De Gray, D., "PUR Adhesives Offer Solutions for Assembly Challenges", *Adhesives Age,* May 1998.
24. David, F. E., and Fromwiller, J. K., "Reactive Urethane Hot Melts for Textiles", *Adhesives and Sealants Industry,* March 1998.
25. Baccei, L. J., U.S. Patent 4,309,526, 1982.
26. Eastman Chemical Company, Leaflet R-206A.
27. Prism Adhesive, *Loctite Worldwide Design Handbook 1996 / 97,* Loctite Corporation, 1995.
28. Kato, K., et al., Japanese Patent, 47-51087, 1972.
29. Gleave, E. R., U.S. Patent 4,102,945, 1978.
30. Harris, S. J., U.S. Patent 4,450,465, 1984.
31. B.F. Goodrich General Products Bulletin, GPC-72-AD-3.
32. Lord Corp., Chemical Product Div., Cary, NC.
33. Loctite Corp., Rock Hill, CT.
34. Permabond International, Div. of National Starch and Chemical Co., Englewood, NJ.

35. Morganelli, R., and Cheng, H., "New Methacrylates Match Two Part Urethane and Epoxy Performance", *Adhesives Age,* June 1998.
36. Pressure Sensitive Adhesives, Dow Corning Bulletin, 02-032.
37. Silastic RTV Silicone Rubber, Dow Corning Tech. Bull. 61-015a.
38. Bruno, E. J., ed., *Adhesives in Modern Manufacturing,* Society of Manufacturing Engineers, 1970 at 29.
39. Lichman, J., "Water-based and Solvent-based Adhesives", *Handbook of Adhesives,* 2nd ed. I. Skeist, ed. (New York: Reinhold, 1977).

Chapter 11

Selection of Adhesives

11.1 Introduction

This chapter will describe the processes involved in selecting an adhesive for a particular application. This is not as easy an endeavor as it might originally appear. To achieve optimum performance when bonding two materials, one must carefully plan every stage of the bonding process. The selection of an adhesive is a critical factor that will influence each step. The adhesive selection will be dependent primarily on:

- The type and nature of substrates to be bonded
- The methods of curing that are available and practical
- The expected environments and stresses that the joint will see in service

The adhesive selection process is difficult because many factors must be considered, and there is no universal adhesive that will fulfill every application. It is usually necessary to compromise when selecting a practical adhesive system. Some properties and characteristics that are desired will be more important than others, and a thoughtful prioritization of these criteria will be necessary in selecting an adhesive. One must first go about finding an adhesive that will satisfy the high priority requirements of the application. The lower priority "requirements" may then need to be compromised to find the best fit.

One needs to optimize the entire bonding process and not merely a single part of the process. Considerations need to be given at the same time to the substrates, joint design, surface pretreatment, quality control, application and curing methods, and other processes. Seldom is everything except the adhesive fixed. Very often, nothing is rigidly established, and it is up to the design engineer to not only choose the adhesive but also to decide on all of the peripheral processes. Alternative substrates, processes, etc. should always be considered. For example, a slight change in the application methods could open the door for consideration of an entirely new family of adhesives.

This chapter will identify some of the critical deliberations that are necessary in making such compromises. It will be shown how choices regarding

- Substrate
- Joint design
- Production processes
- Service requirements

affect the choice of the adhesive. A planning process is suggested to find the optimal adhesive system that results in a reliable joint at the least cost.

11.2 Planning for the Bonding Process

In choosing the correct adhesive and developing an optimized bonding process, there are several critical decisions that need to be made as shown in Fig. 11.1. Each of the steps, such as substrate selection and joint design, needs to be analyzed and optimized with respect to their influence on the final, desired result. A significant problem is that each

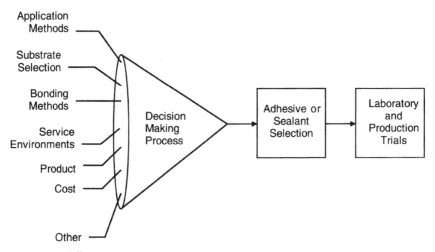

Figure 11.1 Adhesive and sealant selection considerations.

of these decisions cannot be made independently. The substrate selection, for example, may influence the processing conditions and joint design that is necessary. Therefore, the adhesive selection must be made considering all of the parameters involved in the bonding process.

What should be considered first? If you choose an adhesive based on the maximum adhesion to the substrates selected, you may be severely limiting yourself with respect to curing processes or environmental resistance. Likewise, if you select an adhesive based only on its resistance to service conditions, you may have to redesign the joint and find that the selected substrates are now not possible (or cost effective) to manufacture in the required configuration. This provides the individual selecting the adhesive system with a formidable problem. What characteristics are most important and how should one go through the process of selecting an appropriate adhesive?

To achieve optimum results, every stage of the bonding process must be well planned and thought out. After this analysis, if the estimates of strength, durability, and cost do not appear to provide an adequate safety margin, then one must go back through the entire process, making refinements where they can be made to end with a truly optimized result. The following approach is often recommended when using engineering adhesives.[1]

1. Consider alternative assembly and bonding processes. There are several joining processes available to the design engineer. These need to be considered before time and resources are committed to any singular joining method.

2. Gather all information possible regarding the resulting product and its requirements. This will include identification of possible alternative substrates, joint designs, and available processing facilities. Determine the environmental conditions and types of loading that the joint must resist. Plan all phases of the joining process, from selecting the substrates to deciding on quality control methods.

3. Design the joint especially for the method of bonding that is chosen. Consider the service conditions and stresses that the joint is expected to encounter in service.

4. Choose the appropriate adhesive in cooperation with an adhesive supplier. Estimate the required adhesive strength and durability.

5. Ensure the reliability of the adhesive bond by engaging in a testing program to verify the durability of the joint by simulating actual assembly practices and aging environments. This testing program should also determine the optimum bond dimensions and thickness.

Steps 1 and 2 are considered planning processes and they are described below. Steps 3 through 5 are production decisions that utilize the information acquired in the planning process. These steps are influenced by the selection of the adhesive. They are described in the next section of this Chapter.

11.2.1 Consideration of alternative bonding methods

As early as possible, one should determine if there are any other methods that could be used to manufacture the assembly. Alternative methods should be examined and compared with respect to cost, performance, and reliability. Often adhesive bonding provides advantages that do not exist with mechanical fastening, soldering, or welding. Advantages and disadvantages of adhesive bonding relative to other fastening methods were described in Chapter 1. In addition, several disadvantages could arise in choosing adhesives. Sometimes the disadvantages of adhesive bonding outweigh the advantages, and an alternative process such as welding, riveting, or mechanical fasteners should be chosen. The earlier the designer realizes this, the better it will be, for it can be a long, torturous, and expensive road to a successful adhesive bonding process.

After considering other methods of fastening and if bonding is chosen as the most appropriate assembly method, then several generic bonding methods may be considered. These methods of bonding are standard adhesive bonding (assembly and insert bonding) and diffusion methods such as solvent and thermal welding. Solvent and thermal welding methods are used primarily for the assembly of thermoplastics. They are described in detail in Chapter 15.

Assembly bonding is what one normally thinks of when considering adhesive bonding. In assembly bonding, the substrates are already formed, possibly even in their final state. The adhesive is then applied directly to the substrate and a bond is made. An example of assembly bonding is the adhesion of an automobile trunk lid to a brace for the purpose of mechanical strength and noise damping. Each of the substrates are already in their near final state (perhaps painting is all that is required after assembly), and an adhesive is used to fasten the two sections together. Another example, may be the lamination of a copper foil to a polymeric film for flexible printed circuitry. The two parts are in there near final state before being bonded together.

Insert bonding is another bonding method. Insert bonding is generally a possibility when at least one of the substrates is polymeric in nature. Insert bonding takes place by molding one substrate directly to another substrate. With insert bonding, one or both of the substrates are being shaped or formed at the same time that it is being bonded. Sometimes the material being used to fabricate the substrate can also be used as the adhesive. An example is the manufacture of a urethane roller skating wheel with a metallic hub. The liquid urethane casting resin is cast into a mold that already contains the hub. The urethane fills the mold and cures, thereby forming the roller wheel and bonding directly to the hub. The urethane that is used to fabricate the wheel is also the adhesive. A primer may be needed on the insert to obtain maximum adhesion. With insert bonding there often is no reason to select a "third party" adhesive; however, an adhesive or primer must sometimes be applied to the shaped substrate.

There are other types of bonding where a "third party" adhesive is not necessary. These are *solvent cementing* and various forms of *thermal welding*. These methods are commonly used on certain thermoplastics. In such cases, the polymeric material in one of the substrates becomes the adhesive. The thermoplastic substrate material is made fluid by either dissolving the surface with a solvent or by using elevated temperatures to create a melt zone. The resulting fluid polymer is used as the "adhesive" to make the bond. Once the solvent evaporates or the melt zone cools and solidifies, a bonded joint is formed.

Whether bonding conventionally with adhesives or using insert bonding, solvent cementing, or heat sealing, the principles of adhesives bonding hold true. Close attention must be given to surface preparation, to entrapped air in the bond line, to stresses within the bond, and to weak boundary layers.

11.2.2 Information regarding product and processing requirements

One should first carefully gather all of the information possible about the product being bonded and the processing capabilities that are available. This is a fact-finding step to accumulate critical information on the application and the restrictions that may be placed on the bonding process. The question of joint design, service environment, or specific adhesive is not considered directly at this time. This first step in the assembly process asks the following:

1. What materials are to be joined? Are there alternative materials available?
2. What surface preparation is required? This may depend on the choice of adhesive and the service environment that is expected.
3. What are the overriding cost factors that must be considered?
4. What methods of adhesive application are suitable and workable?
5. What methods are available to harden or cure the adhesive? Will holding fixtures be available? Ovens? Other special equipment?

6. What is the maximum cycle-time permitted for the assembly operation? How critical is cycle-time?
7. What safety and environmental issues need to be considered?
8. What quality control processes are practical and affordable? How important are consistency, joint strength, and durability to the application?

The parameters that are gathered during this fact-finding may need to be reconsidered and compromised as the adhesive selection process proceeds and various adhesive systems are matched to the list of requirements. Once these questions are answered, the process then centers around designing the joint and choosing the adhesive.

11.2.2.1 Processing conditions. The conditions under which the adhesive is to be bonded are important parameters in optimizing the process. Certain factory circumstances restrict the type of adhesives that can be considered. Production resources may just not be available to consider certain types of adhesive systems without major capital investments. Cycle time restrictions, very common in fast, high volume operations, may further limit the number of adhesives that can be considered. Production speed and budget will also likely limit the type of surface treatment processes that are possible. Often the processing characteristics of the adhesive (i.e., how it is applied and cured) are factors that override all others in selecting an adhesive system. This is especially true for applications where ultimate strength and durability are not necessary.

Typical production considerations that are involved in the assembly process include:

- The form of the adhesive
- Method of preparation and use
- Shelf or storage life
- Working life
- Method or machinery necessary for bonding
- Permitted time between coating and bonding
- Drying time and temperature (for solvent or water based adhesives)
- Temperature required for application and curing
- Rate of strength development for the purpose of moving the article to the next step in the process
- Ability for hardening or cure to take place in an associated process (i.e., painting or drying steps)

- Inflammability and toxicity of the adhesive and accessory materials

Of these factors probably the most important are the method chosen for application of the adhesive and curing parameters such as temperature and pressure.

The method chosen for the application of an adhesive to the workpiece is determined by the size and shape of the parts being bonded and the number of components to be bonded. This consideration will often lead to the form of adhesive that is most appropriate for the application. For example, films and spray coatings are very common for large parts. Pastes are commonly used for joints requiring narrow beads. UV cured adhesives may be best for high volume parts where strength is not critical and the adhesive can be directly exposed to a light source.

Curing temperature and pressure is often a critical parameter in the bonding process. Sub-components may not be resistant to the temperatures and pressures that are required for certain adhesives. High curing temperatures may cause internal stress that will crack or distort delicate parts. Low temperature cures may require too long for practical assembly operations. The main problem is that high strength and temperature resistance usually coincide with high temperature cures.

Another critical processing parameter is the surface preparation that may be required. The degree of preparation will depend on the type of substrate, its condition, and the adhesive chosen in addition to the joint strength and durability required for the application.

11.2.2.2 Joint design. Once the method of bonding and the substrates are chosen, the specific joint design can be addressed. The most important consideration in the joint design may be the ease in which the substrates can be fabricated. For some substrates, it may be easy and inexpensive to design a joggle overlap joint. With other substrates (composites, for example) it may be very difficult. Joint designs should consider:

- The type of substrate
- The cost associated with manufacturing the joint design
- Ease at which the joint can be fabricated and assembled
- The ease in which the joint can be inspected after bonding is complete
- Most importantly, the type of stress that the joint will see in service

One must consider the type of stress that the joint is expected to encounter in service. Make certain that peel and cleavage stresses will be minimized through the proper design of the joint. If peel or cleavage forces will be present because of the nature of the application or the joint design, this could severely limit the candidate adhesive systems. Use joint design practices that are described in Chapter 3 of this Handbook. Design the joint to alleviate joint fabrication or assembly problems. The joint design should be chosen to provide production efficiency with the selected bonding methods. Joint designs that are subject to easy visual inspection after bonding are best.

11.2.2.3 Expected service conditions.
Estimate the duration and nature of the stress (peel, shear, etc.), the required bond strength, and the environmental conditions that the assembly must endure. Several factors to be considered in this step are listed below.

1. *Type of stress loading (tension, compression, flexural, torsion, shear, peel, cleavage, or oscillating).* Ideally, if there is a failure, it should be away from the bondline and in the substrates and at some stress level much higher than the expected load requirements. If failure does not occur in the substrate, then it should occur cohesively within the adhesive material. Possibly, the worst mode of failure is an adhesion failure at the interface.

2. *Degree of toughness required to resist impact, peel, or cleavage forces.* If toughness is required, relatively flexible adhesives with low glass transition temperatures may be chosen. Unfortunately, these adhesives do not possess the high degree of temperature and chemical resistance that more rigid formulations possess.

3. *Operating temperature range (maximum, minimum, total amount of time at the maximum temperature, rates of temperature change).* The adhesive should have a glass transition temperature above the normal operating temperatures. Most adhesives can tolerate excursions on the order of 10s of degrees above their glass transition temperature, depending on the loads and duration of time at temperature. Usually adhesives that are more rigid have the higher glass transition temperatures. The total amount of time at elevated temperatures is an important criterion. If the joint will only make several excursions of limited time to its maximum operating temperature, then a less temperature resistant adhesive may suffice. If the joint will spend most of its time at the temperature maximum, the elevated temperature resistance and resistance to high temperature oxidation will be important criteria in the choice of an adhesive.

4. *Chemical resistance (resistance to acids, bases, solvents, or other chemicals).* Not only does the type of chemical medium need to be considered but also the temperature of the joint at the time of contact

with the chemical environment. Certain adhesive systems will have very good resistance to certain chemicals and poor resistance to others. Guidance is often required from the adhesive manufacturer or through a test program. The most resistant adhesives are generally those most densely crosslinked, resulting in relatively rigid adhesives that must be cured at elevated temperatures. A straightforward method of improving chemical resistance is to protect the joint from the chemical medium.

5. *Weather and climatic resistance; resistance to humidity.* For metallic joints the most common mode of failure in outdoor environments is corrosion of the substrate and resulting weak boundary layers at the interface due to moisture penetrating the adhesive and adsorbing on the substrate surface. Adhesives usually withstand moderate outdoor weathering relatively well. However, high humidity may be disastrous to certain adhesive systems, especially if the humidity is combined with moderate temperatures and loads. These issues will be discussed thoroughly in Chapter 17.

6. *Differences in thermal expansion rates.* The thermal expansion and thermal conductivity of the adhesive will generally be different from that of the substrates. Consequently, there will be stresses induced during thermal excursions. This could occur immediately after curing, when the joint is removed from the curing oven and returned to ambient temperatures. Tough, flexible adhesives best resist such internal stress. Another approach is to try to match the adhesive's thermal expansion coefficient with that of the substrate.

In addition to understanding the effects of the service environment on the adhesive and the adhesive joint, one should also try to determine the consequences of catastrophic failure, and understand the most likely conditions that could lead to such a scenario.

11.3 Selecting the Adhesive

As one can also see from the previous chapters, the number of available adhesives to choose from can be very great. However, with a knowledge of the functional product requirements and possible production capabilities, one can start the process of selecting the adhesive system.

The various adhesive families overlap somewhat with regard to their characteristics. Several adhesive families can provide similar properties. There is also a significant variation available within certain adhesive families. Specifying a "thermosetting epoxy" adhesive for example is not sufficient. Virtually thousands of formulations are available within this family with different characteristics regarding end-properties, application requirements, curing processes, and so forth.

There is also no general-purpose adhesive. There is no perfect solution for any particular adhesive problem. However, with the great many types of adhesives available and with the many available variations that can be provided, a "good" solution can usually be found. However, these solutions generally result in a compromise where some lesser requirements may have to be sacrificed for other more critical requirements.

The best adhesive for a particular application will depend on the materials to be bonded, the service and assembly requirements, and the economics of the bonding operation. By using these factors as criteria for selection, the many commercially available adhesives can be narrowed down to a few possible candidates.

Choose the appropriate adhesive in cooperation with an adhesive supplier. Once the general product requirements, joint design, and service requirements are known, adhesive systems with good durability can be selected. The desired form of the adhesive and method of application can then be chosen based on the formulation possibilities available within the class of adhesive chosen and on the availability of production equipment and scheduling requirements.

The tendencies to over specify the adhesive requirements must be avoided. Requirements for higher strength or greater heat resistance than is actually necessary for a specific application will exclude from consideration many formulations that may be equally acceptable for the job, at lower cost, and with simpler production capabilities. Usually, the objective in specifying an adhesive is to select a material that will cause the joint to fail in cohesion rather than adhesion. Ideally, the substrate fails before the adhesive. Low strength materials such as paper, fabric, etc. may be weaker than most structural adhesives, and the use of high strength adhesives could be an expensive overspecification. Lower cost and more easily applied adhesives may result in the same failure mode—cohesive failure of the substrate.

Re-evaluate the preceding stages in the bonding process to determine that the selection of adhesive does not violate any of the critical requirements. Make appropriate changes in the plan to accommodate the adhesive of choice, or select another adhesive.

11.4 Substrates

Generally, the first consideration in making an adhesive selection is the type of substrates that must be held together. Often a two-dimensional array, such as shown in Table 11.1, is used to indicate the adhesive family that best matches a particular substrate or a combination of substrate materials. A number in a given square indicates the particular adhesive that will bond to a cross-referenced substrate. This information is intended only as a guideline to show common adhesives that have been used successfully in various applications.

Although a good starting point, such a simple tool is subject to several limitations. It generally suggests adhesive systems that are chemically compatible with the substrate or have a modulus similar to the modulus of the substrate being bonded. These are important criteria, especially for chemically active, low modulus substrates such as many plastics and elastomers. However, there are important attributes regarding the substrate that must be understood in order to make an effective adhesive selection. The adhesive selections in Table 11.1 are listed without regard to strength or service requirements. Lack of a suggested adhesive for a particular substrate does not necessarily mean that a poor bond will result, only that information is not commonly available concerning that particular combination.

If two different materials are to be bonded, the recommended adhesives in Table 11.1 are those showing identical numbers under both substrates. Although adhesives may bond well to different substrates, joints may be inferior because of mismatch in thermal-expansion coefficients. In these cases, a resilient adhesive should be carefully chosen to compensate for stress.

The type and nature of the materials to be bonded and the surface preparation that they may require are prime factors in determining which adhesive to use. Some adherends such as aluminum or wood can be successfully bonded with many adhesive types; other adherends such as nylon can be bonded with only a few. The substrate type, condition, porosity, finish, acidity, alkalinity, etc. will all influence the adhesive selection. The surface material, its processing history and the condition at the time of bonding may be more important than the bulk substrate material.

The adhesive selection process begins with a general knowledge of the materials being bonded. This sometimes offers an indication of the nature of the surface to which the adhesive will bond. A general rule of selection is to choose an adhesive that will wet the substrate. The adhesive's surface energy should be less than the surface energy of the substrate. When cured, the adhesive should have a modulus and thermal expansion coefficient similar to the substrate or else have toughness to accommodate stresses caused by thermal movements. Differences in flexibility or thermal expansion between the adherend and the adhesive can introduce internal stresses into the bondline.

TABLE 11.1 Selection of Adhesives for Use with Various Substrates (Based on Reference 2)

Adherend	Adhesives (number key given in Table 11.5)

Adherend																		
Plastics																		
Phenoxy	1	2				11												
Polyamide (nylon)	1	2	3	7 8		11	15		23			30	32					40
Polycarbonate	1	2	3		9	11	14 15						32					
Polychlorotrifluoroethylene					9 10	11								34 35				
Polyester (fiber composite)	2	3			9	11	14		24		27 28	29 30	32		36			40
Polyether (chlorinated)												30			36			40
Polyethylene					9	11			23			30	32		36			40
Polyethylene (film)	2								23			29 30	32		35 36			40
Polyethylene (chlorinated)	2	5										30			35 36			
Polyethylene terephthalate (Mylar)	2	5			10			19	23			29 30			35			
Polyformaldehyde	2				9	11						30						
Polyimide					9 10	11 12			24			30		34 35	36 37		39	40
Polymethyl methacrylate	1	5				11	14 15					30		34 35	36			40
Polyphenylene						11		19	24									
Polyphenylene oxide	2	3			9	11	14					29 30	32		36			40
Polypropylene						11						30						
Polypropylene (film)	2					11			23			30			35			
Polystyrene	1	3					14 15		23 24									
Polystyrene (film)	1						15	19										
Polysulfone							14											
Polytetrafluoroethylene					9 10	11						29 30	32	34 35	36			40
Polyurethane	1	2				11						30	32		36			40
Polyvinyl chloride	1	2			9	11	14 15	19	24			29 30	32		35			40
Polyvinyl chloride (film)	2	3						19	24			30	32					40
Polyvinyl fluoride	2					11						30	32 33					
Polyvinylidene chloride	2								23			30	32					
Silicone	2													34 35				
Styrene acrylonitrile						11	14					30						
Urea formaldehyde			4															
Foams																		
Epoxy					9	11				26 27		30						
Latex	1	2			9	11						30	32					

426

Material	Applicable Code Numbers
Phenol formaldehyde	2, 5, 9, 11, 19, 27, 28, 32, 36, 41, 43
Polyethylene	9, 11
Polyethylene–cellulose acetate	
Polyphenylene oxide	2, 9, 11, 14, 21, 28, 30
Polystyrene	2, 9, 11, 14, 19, 30, 31, 36
Polyurethane	2, 9, 11, 28, 30, 32, 35, 36
Polyvinylchloride	2, 3, 9, 11, 15, 19, 30, 32, 33, 36
Silicone	24, 34
Urea formaldehyde	4, 7

Rubbers

Material	Applicable Code Numbers
Butyl	1, 15, 28, 32, 35, 40
Butadiene nitrile	1, 2, 7, 9, 11, 15, 30, 32, 35
Butadiene styrene	1, 2, 9, 10, 15, 30, 31, 32, 35
Chlorosulfonated polyethylene	
Ethylene propylene	11
Fluorocarbon	1, 9, 10, 11, 15, 34, 35
Fluorosilicone	34
Polyacrylic	24
Polybutadiene	2, 8, 10, 11, 28, 31, 35, 36
Polychloroprene (neoprene)	1, 2, 9, 10, 11, 15, 30, 31, 32, 33, 34, 35
Polyisoprene (natural)	1, 2, 7, 9, 10, 11, 15, 26, 27, 30, 31, 32, 33
Polysulfide	15, 28, 29, 30, 32, 35, 36
Polyurethane	1, 2, 11, 15

Wood, Allied Materials

Material	Applicable Code Numbers
Cork	2, 3, 4, 5, 6, 7, 8, 9, 11, 16, 17, 18, 19, 21, 23, 26, 27, 28, 30, 31, 32, 33, 34, 36, 40, 41
Hardboard, chipboard	4, 7
Wood	2, 3, 4, 5, 6, 7, 8, 9, 11, 14, 15, 16, 17, 18, 19, 21, 23, 25, 26, 28, 29, 30, 31, 32, 33, 34, 36, 40
Wood (laminates)	2, 3, 4, 5, 6, 7, 8, 9, 11, 16, 17, 18, 19, 21, 23, 25, 26, 28, 29, 30, 31, 32, 33, 34, 36, 40

Fiber Products

Material	Applicable Code Numbers
Cardboard	2, 3, 4, 5, 6, 7, 8, 9, 11, 16, 17, 18, 19, 21, 22, 23, 24, 26, 27, 28, 29, 30, 31, 32, 33, 34, 36, 37, 40, 41
Cotton	2, 3, 7, 9, 11, 16, 17, 18, 19, 23, 24, 26, 27, 30, 32, 33, 34, 35, 36, 40, 41
Felt	3, 4, 5, 7, 9, 11, 16, 17, 18, 19, 23, 24, 26, 27, 30, 32, 33, 34, 35, 36, 40, 41
Jute	2, 3, 4, 5, 7, 9, 11, 16, 17, 18, 19, 23, 24, 26, 27, 30, 31, 32, 33, 34, 36, 40, 41
Leather	2, 3, 7, 9, 11, 16, 17, 18, 19, 21, 23, 24, 26, 27, 28, 29, 30, 31, 32, 33, 34, 36, 40, 41
Paper (bookbinding)	2, 3, 5, 9, 11, 16, 17, 18, 19, 21, 22, 23, 26, 27, 28, 29, 31, 32, 33, 36

TABLE 11.1 Selection of Adhesives for Use with Various Substrates (Based on Reference 2) (Continued)

Adherend	Adhesives (number key given in Table 11.5)
Fiber Products (Continued)	
Paper (labels)	2 16 17 18 19 22 23 24 26 27 29 30 31 32 33 34 36
Paper (packaging)	2 3 16 17 18 19 21 22 23 24 26 27 28 29 30 31 32 33 36
Rayon	2 3 4 5 7 9 11 16 17 18 19 22 24 26 27 28 30 31 32 33 34 35 36 40
Silk	2 3 4 5 7 9 11 16 17 18 19 23 24 26 27 28 30 31 32 33 34 35 36 40 41 43
Wool	2 3 4 5 7 9 11 16 17 18 19 23 24 26 27 28 30 31 32 33 34 35 36 40 41
Inorganic Materials	
Asbestos	2 3 5 7 8 9 10 11 14 16 18 19 21 27 28 29 30 31 32 33 34 35 37 40 43
Carbon	9
	11
Carborundum	9
	11
Ceramics (porcelain, vitreous)	1 2 3 4 7 8 9 10 11 14 15 16 17 18 19 21 27 28 29 30 31 32 33 34 35 36 37 40 41
Concrete, stone, granite	2 3 4 7 8
	9
	11 14 35
Ferrite	9
	11
Glass	1 2 3 5 9 10 11 14 15 16 17 18 21 28 30 31 32 34 35 37 40
Magnesium fluoride	
Mica	11 14 35
Quartz	9 11 40 43
Sodium chloride	11
Tungsten carbide	9 11

Such stresses can lead to premature failure of a bond. Thus, rigid heavily filled adhesives are often chosen for bonding metals. Flexible adhesives are often chosen for bonding plastics and elastomers.

For adherend materials of different composition (e.g., steel and plastics), it has been suggested[2,3] that an adhesive of modulus close to $\frac{1}{2}(E_1 + E_2)$ and a total relative elongation close to $\frac{1}{2}(L_1 + L_2)$, would satisfy minimum stress requirements for adherends with elastic moduli E_1 and E_2 and total relative elongations L_1 and L_2. (L = change in length per unit length on application of a stretching force.)

Adhesives and substrates should also have similar chemical resistance properties. The swelling of one component of a joint in a service environment results in stress adjacent to the adhesive–adherend interface. These stresses could result in premature failure of the bonded components. Incompatible adhesive and adherend can also lead to damage of an assembly. Examples of this include:

- The corrosion of metallic parts by acidic adhesives
- The migration of plasticizers into the adhesive from flexible plastic substrates with consequent loss of adhesion at the interface
- The action of adhesive solvents and volatiles on plastic adherends (particularly thin plastic films)[2]

Chemical incompatibility is often enhanced by poor control over the adhesive's mixing and curing conditions. It is often advisable to test the compatibility of adhesive and adherends.

The porosity of the substrate to be bonded can place additional demands on an adhesive system. Excessive penetration can make low viscosity adhesives unsuitable. Adhesion to coated or treated surfaces such as painted or plated steel must take into account the surface coating, not only the base substrate. It is essential to completely understand the substrate surface and to know how this surface could change with time. The reader is referred to Chapters 2 and 6 for discussion of weak boundary layers and surface preparation processes for improved bonding.

Substrate characteristics will also place requirements on the selection of the proper bonding equipment. Elastomeric substrates, for example, cannot be subjected to high holding pressure during an adhesive's cure, or else degrading internal stresses could become frozen into the bondline. Wood or cellulosic substrates cannot be exposed to high cure temperatures without dimensional changes of the substrate that will depend on the grain orientation and initial moisture content of the substrate.

11.4.1 Adhesives for metal

The chemical types of structural adhesives for metal bonding were described in the preceding chapter. Adhesives for bonding metals may be divided into two main groups:

- Structural—for use in continuously stressed structures
- Non-structural—for use in low strength applications

The heat and environmental resistance are often of great importance, especially in the case of structural adhesives.

Since organic adhesives readily wet most metallic surfaces, the adhesive selection does not depend as much on the type of metal substrate as on other requirements. However, virtually all commonly bonded metal surfaces exist as hydrated oxides. This situation is discussed fully in Chapter 6. Adhesives used for bonding metal are really bonding metal oxide surfaces. They, therefore, must be compatible with the firmly bonded layer of water attached to the surface of most metal oxide crystals. The base metal greatly influences the properties of the hydrated oxide surface. Thus, certain metals will possess surfaces that interact more effectively with one type of adhesive than another.

Thermal and oxidative stability as well as corrosion and water resistance, depend on the adherend surface as well as on the adhesive itself. Epoxy and phenolic based adhesives degrade less rapidly at elevated temperatures when in contact with aluminum or glass than when in contact with copper, nickel, magnesium or zinc surfaces. The divalent metals (Zn, Cu, Fe^{+2}, Ni, Mg, etc.) have a more basic oxide surface than the higher valence metal oxides (Al, Fe^{+3}, Si, etc.) and hence promote dehydrogenation reactions which lead to faster degradation.

Selecting a specific adhesive from a table of general properties is difficult because formulations within one class of adhesive may vary widely in physical properties. General physical data for several common structural metal adhesives are presented in Table 11.2. This table may prove useful in making preliminary selections or eliminating obviously unsuitable adhesives. Once the candidate adhesives are restricted to a few types, the designer can search more efficiently for the best bonding system.

Nonstructural adhesives for metals include elastomeric and thermoplastic resins. These are generally used as pressure-sensitive, contact, or hot-melt adhesives. They are noted for fast production, low cost, and low to medium strength. Typical adhesives for nonstructural bonding applications were described previously. Most nonstructural

TABLE 11.2 Properties of Structural Adhesives Used to Bond Metals

Adhesive	Service temp, °F Max	Service temp, °F Min	Shear strength, lb/in.²	Peel strength	Impact strength	Creep resistance	Solvent resistance	Moisture resistance	Type of bond
Epoxy-amine	150	−50	3,000–5,000	Poor	Poor	Good	Good	Good	Rigid
Epoxy-polyamide	150	−60	2,000–4,000	Medium	Good	Good	Good	Medium	Tough and moderately flexible
Epoxy-anhydride	300	−60	3,000–5,000	Poor	Medium	Good	Good	Good	Rigid
Epoxy-phenolic	350	−423	3,200	Poor	Poor	Good	Good	Good	Rigid
Epoxy-nylon	180	−423	6,500	Very good	Good	Medium	Good	Poor	Tough
Epoxy-polysulfide	150	−100	3,000	Good	Medium	Medium	Good	Good	Flexible
Nitrile-phenolic	300	−100	3,000	Good	Good	Good	Good	Good	Tough and moderately flexible
Vinyl-phenolic	225	−60	2,000–5,000	Very good	Good	Medium	Medium	Good	Tough and moderately flexible
Neoprene-phenolic	200	−70	3,000	Good	Good	Good	Good	Good	Tough and moderately flexible
Polyimide	600	−423	3,000	Poor	Poor	Good	Good	Medium	Rigid
Polybenzimidazole	500	−423	2,000–3,000	Poor	Poor	Good	Good	Good	Rigid
Polyurethane	150	−423	5,000	Good	Good	Good	Medium	Poor	Flexible
Acrylate acid diester	200	−60	2,000–4,000	Poor	Medium	Good	Poor	Poor	Rigid
Cyanoacrylate	150	−60	2,000	Poor	Poor	Good	Poor	Poor	Rigid
Phenoxy	180	−70	2,500	Medium	Good	Good	Poor	Good	Tough and moderately flexible
Thermosetting acrylic	250	−60	3,000–4,000	Poor	Poor	Good	Good	Good	Rigid

adhesives can be used on any clean metal surface and on many plastics and elastomers.

11.4.2 Bonding plastics

Bonding of plastics is generally more difficult than metal bonding because of the lower surface energy of polymeric substrates. To have acceptable joint strength, many low energy plastic materials require surface treatments prior to bonding. Appendix C-2 offers a guideline for which treatments to use. Fortunately, there are other methods of bonding plastics than by using adhesives and many of these methods do not require lengthy or costly surface treatment processes. Many thermoplastic materials may be satisfactorily joined by various forms of solvent and thermal welding. These alternatives to adhesive bonding will be described in detail in Chapter 15. The plastic resin manufacturer is generally the leading source of information on the proper methods of joining a particular plastic.

For high surface energy thermosetting plastics, such as epoxies, polyesters, and phenolics, adhesive bonding is the easiest and usually the best method of fastening one type of plastic to another type or to a different substrate. Thermoplastic and rubber-based contact adhesives are often used to bond plastics when the application is nonstructural. These adhesives are often less sensitive to heat and solvents than the parent plastic. Tough thermosetting adhesives, such as epoxies, polyurethanes, or acrylics, are commonly used for structural applications. When bonding plastics to almost any other type of material, the mismatch in thermal expansion coefficients can impose significant destructive stress on the bond during thermal cycling. Thus, adhesives that are recommended for plastics are generally resilient, tough materials to allow distribution of stress at temperature extremes.

Recommended adhesive selections for various plastic substrates are listed in Table 11.3. Suitable adhesives for various common thermoplastic materials are listed, and an indication is given when solvent or thermal welding techniques are also applicable. With thermosetting plastics, heat and solvent welding techniques are not possible, and this type of plastic is generally bonded with thermosetting adhesives.

11.4.3 Adhesives for bonding elastomers

11.4.3.1 Vulcanized elastomers. Bonding of cured or vulcanized elastomers to themselves and other materials is often completed by using a pressure sensitive contact adhesive derived from an elastomer resin similar to the one being bonded. Flexible thermosetting adhesives such as epoxy-polyamide or polyurethane also offer excellent adhesive

TABLE 11.3 Adhesives for Bonding Common Thermosets and Thermoplastics

Thermoplastic

Acrylic plastics (polymethyl methacrylate) — May also be heat or solvent welded; with solvent welding crazing may occur.
  Thermoplastics acrylic (in solvent solution)
  Thermosetting acrylic
  Epoxy
  Neoprene or nitrile synthetic rubber
  Resorcinol formaldehyde
  Polyurethane

Cellulose acetate plastic — May also be heat or solvent welded
  Cellulose acetate (in solvent solution)
  Cellulose nitrate
  Epoxy
  Polyvinyl acetate
  Resorcinol formaldehyde
  Polyurethane

Cellulose nitrate plastics — May also be heat or solvent welded
  Epoxy
  Resorcinol formaldehyde

Polyamide plastics (Nylon) — May also be heat welded; not generally solvent welded
  Epoxy
  Phenolic blends
  Resorcinol formaldehyde
  Synthetic rubber-based
  Polyurethane

Polystyrene plastics — May be solvent or heat welded; low boiling solvents may cause crazing
  Polystyrene in solution
  Epoxy
  Polyurethane

Polyethylene plastics — May be heat welded; not solvent welded
  Phenolic blends
  Synthetic rubber-based
  Epoxy (when treated)

Polytetrafluoroethylene (PTFE, Teflon) — May be heat welded with difficulty under pressure
  Phenolic blends
  Synthetic rubber-based
  Epoxy (when treated)
  Polyurethane (when treated)

Polyvinyl chloride plastics (PVC) — May be heat welded; not generally solvent welded
  Thermosetting acrylics
  Synthetic rubber (neoprene or nitrile) — nitrile for heavily plasticized vinyl
  Phenolic blends

Thermosets

All thermosets including composites and most engineering thermoplastics
  Epoxy
  Furane
  Phenolic blends
  Phenol and resorcinol formaldehyde
  Polyester
  Synthetic rubber based
  Urea formaldehyde
  Polyurethane
  Thermosetting acrylic

strength to most elastomers. Surface treatment consists of washing with a solvent, abrading, or acid cyclizing as described in Appendix C-3.

Elastomers vary greatly in their formulation from one manufacturer to another. Fillers, plasticizers, antioxidants, and other components may affect the adhesive bond. Adhesives should be thoroughly tested with a specific elastomer and then reevaluated if the elastomer manufacturer or formulation is changed even though the physical property specification may remain unchanged. Table 11.4 shows common types of adhesives used for bonding elastomers.

11.4.3.2 Unvulcanized elastomers. Uncured or unvulcanized elastomers may be bonded to metals and other rigid adherends as part of their curing process by first priming the adherend with a suitable air- or heat-drying adhesive. The liquid elastomer is then injected or poured around the substrate surface, and then it is vulcanized or cured. Bonding of rubber to metal hubs or inserts is commonly accomplished this way. The elastomers that can be bonded in this manner include nitrile, neoprene, urethane, natural rubber, SBR, and butyl rubber. Less common unvulcanized elastomers, such as the silicones, fluorocarbons, chlorosulfonated polyethylene, and polyacrylate, are

TABLE 11.4 Common Adhesives for Bonding Rubber

Chlorinated rubber—used to some extent for bonding natural rubber

Epoxy—used to bond some rubbers

Epoxy-polyamide—used to bond some rubbers

Natural rubber solution—for bonding natural rubber

Natural rubber latex—for bonding sponge rubber

Neoprene synthetic rubber—for bonding natural and synthetic rubbers, especially neoprene

Nitrile synthetic rubber—for bonding natural and synthetic rubbers

Phenolic-neoprene—for most types of rubber

Phenolic-nitrile—for most types of rubber

Phenolic-vinyl—cycled rubbers, i.e., those surface treated with sulfuric acid

Polyurethanes—used to some extent for rubbers especially urethane

Reclaimed rubbers—for most types of rubber

Resin-rubber blends—numerous formulations are available for bonding most types of rubber

Resorcinol formaldehyde—has been used to some extent for bonding rubbers

Silicones—for bonding silicone rubbers

more difficult to bond. However, recently developed adhesive primers improve the bond of these elastomers to metal substrates. Surface treatment of the substrate before priming should be according to good standards. Figure 11.2 illustrates a bondability index of common elastomers and plastics.

11.4.4 Adhesive bonding of other common substrates

11.4.4.1 Wood. Wood adhesives are probably the oldest adhesives known to man. Adhesives used for wood include animal or hide glues, starch, casein, soybean, blood glues, and fish glues as well as modern synthetic resins.

Resorcinol and resorcinol-formaldehyde resins are cold setting adhesives for wood structures. They have been shown to be at least as durable as wood itself under all conditions. Urea-formaldehyde adhesives, commonly modified with melamine formaldehyde, are less durable. They are used in the production of plywood and in wood veneering for interior applications. Phenol-formaldehyde and resorcinol-formaldehyde adhesive systems have the best heat and weather resistance. Polyvinyl acetates are quick drying, water based adhesives commonly used for the assembly of furniture.[4] This adhesive

	1	2	3	4	5	6	7	8	9	10
Plastics										
Thermosetting plastics (epoxy, phonemic, polyester)										x
PVC								x		
Engineering plastics (polyimide, PPO, PPS, acetal)								x		
Rubber Modified Plastics (polycarbonate, polybutylene terephthalate)							x			
Nylon						x				
Polyolefin plastics (polypropylene, polyethylene)		x								
Fluoroplastics (polytetrafluoroethylene)	x									
Elastomers										
Nitrile Rubber										x
Neoprene										x
Urethane									x	
Styrene butadiene rubber						x				
Hypalon						x				
Natural Rubber						x				
Butyl					x					
Ethylene propylene rubber	x									
Fluorelastomers	x									
Silicone	x									

Figure 11.2 Bondability index for plastics and elastomers; 10 = best; 1 = worst.

produces bonds stronger than the wood itself, but they are generally not resistant to moisture or high temperature. Common adhesives used for bonding wood are identified in Chapter 16 (Table 16.23).

An accelerated aging test, ASTM D-3434, is convenient for evaluating adhesive performance and predicting durability of bonded wood joints. Samples are subjected to 800 cycles of exposure to boiling water. This is equivalent to approximately 16 years of outdoor exposure for a good wood adhesive. Figure 11.3 compares the performance of various wood adhesives in this test as well as solid, unbonded wood.

Epoxies have been used for certain specialized applications such as metal to wood bonds. Newer adhesives have also been introduced in the industry with very strong and reliable bonds in addition to processing ease. These include hot melt polyurethane adhesives, moisture cure urethanes, and silicone adhesives.

Rubber based contact adhesives are also used for wood, especially for laminating decorative panels. Mastic adhesives have been used in construction industry and are usually applied by caulking guns. These are based on elastomers, including reclaimed rubber, neoprene, butadiene-styrene, polyurethane, and butyl rubber.

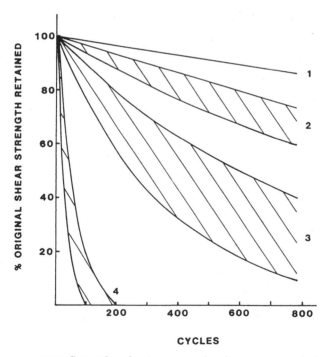

Figure 11.3 Strength reduction curves for the automation boil test: (1) solid wood; (2) fast cure epoxy, slow cure epoxy, phenolic-resorcinol, urethane, phenolic; (3) casein, melamine-urea, melamine; (4) urethane, urea.[5]

11.4.4.2 Glass. Adhesives for bonding glass and related substrates are selected based on polarity, available functional groups, and compatibility. They include polyvinyl butyral, phenolic butyral, phenolic nitrile, neoprene, polysulfide, silicone, vinyl acetate, epoxy, and acrylic adhesives.[6] Commercial adhesives used to bond glass and their physical characteristics are identified in Chapter 16 (Table 16.24).

Adhesives used to bond glass are generally transparent heat-setting resins that are water-resistant to meet the requirements of outdoor applications. They must have either a degree of flexibility and toughness or thermal expansion coefficient that is close to that of glass. Flexibilized room temperature curing adhesives are often used. Many glass substrates have been broken by using a rigid adhesive and then exposing the bond to temperature excursions. The adhesive bonds so well to the glass that the glass fails cohesively due to the stress at the interface caused by thermal expansion differences.

Safety glass is manufactured by laminating sheets of glass together with an adhesive film such as polyvinyl butyral. Optical adhesives used for bonding glass lenses are usually styrene-modified polyesters and styrene monomer based adhesives. Thermosetting epoxies and acrylics are also used.

11.5 Nature of the Joint Design

11.5.1 Relation between stress and adhesive selection

In an ideal world, the optimum adhesive joint would be designed to handle loads in a mode that is consistent with the physical properties of any given adhesive. In other words, rigid adhesives would be selected for joints where the loading will be in shear or tension. Joints would not be designed so that they could be loaded in peel or cleavage. If such loading forces could not be avoided, flexible, tough adhesives would be chosen. However, in the real world, certain practicalities such as the geometry of mating members, cost of machining components, and placement of attachments may force the designer to come to a compromise.

The thickness and strength of the substrates are important factors in selecting an optimum adhesive. Flexible materials, such as rubbers or thin metal, plastic films, etc. are often subject to flexure in service and should not be bonded with rigid, brittle adhesives. A rigid bond may crack and cause a reduction in bond strength.

Differences in flexibility or thermal expansion between adherends can introduce internal stresses into the glue line. Such stresses can lead to premature bond failure. To a certain extent, stresses can be minimized through joint design, but the performance of the bond is still affected by them. Where minimum stress between adherends of the same materials is the objective, it is desirable to choose an adhesive that is similar with respect to rheological properties, thermal expansion, and chemical resistance. This assumes that all other adhesive requirements are met. Frequently, these other adhesive properties are more critical for the joint than ultimate bond strength.

Generally, rigid adhesives are selected for applications requiring high shear strength where the joint is designed so that the load is usually in shear. These resins are generally highly crosslinked thermosets such as epoxies, acrylics, phenolics and other such polymers. Often an elevated temperature cure is required. These adhesives have very high lap shear strengths but at the expense of low peel and impact strength.

Peel and impact strengths of adhesives are related in that adhesives with good peel properties generally will have good impact strength. Thermosetting adhesives tend to form rigid, relatively nonyielding bonds and, therefore, peel strength and impact strength will be lower than for tough rubber/elastomer based and thermoplastic adhesives. Of the epoxies, the polysulfide-epoxies, epoxy-nylons, and polyamide-epoxies have excellent peel properties with strengths ranging from 20 to 100 pwi for solvent wiped aluminum bonded to itself and nearly 150 pwi for a chromic acid etched aluminum bonded to itself. Urethanes also have very high peel strength and flexibility.

Soft mastic adhesives should be selected for applications where large areas are likely to move, warp, or twist. Mastics can include epoxies, urethanes, silicones, and rubber based materials that, depending on hardener and type, can cure to a flexible joint. The soft adhesive flows to relieve the stresses created when the substrates move.

11.5.2 Relationship between adhesive selection, bondline thickness, and viscosity

Thickness of the adhesive in the joint (i.e., bond-line thickness) can be a significant parameter in applications requiring a high strength requirement. The highest tensile and shear strength are obtained with high modulus adhesives when the film thickness is a minimum. As was indicated in Chapter 3, optimum strength is usually achieved when the bondline thickness is between 0.002 in. and 0.006 in.[2] For bond-line thicknesses above 0.006 in., the stress distribution is such that the adhesive may experience cleavage type forces. Below 0.001 in, bond strengths usually decrease, depending on the smoothness and gap tolerance of the substrates. Starvation of the adhesive in the joint is always a potential hazard when one tries for a very thin bond thickness, or the design of the joint and viscosity of the adhesive is such that the adhesive squeezes out from the joint before it cures.

For maximum peel strength and cleavage, elastic adhesives are chosen. Optimum strengths in these modes of stress are usually achieved with increased bond thickness (exceeding 0.005 in). This is due to the elastic distribution of stress over a larger bond area, as was shown in Fig. 3.3.

Adhesives can vary in form from thin liquids, to pastes, to solids such as supported film and powder. The choice of adhesive form will often determine the ease in which a bondline thickness can be maintained. Adhesives with a thin liquid consistency may tend to run out of the joint before it has a chance to cure. Excessive flow of the adhesive is usually related to the decrease in viscosity caused by elevated temperatures that the adhesive sees before cure, or when the substrates are heavy and place a load on the liquid adhesives as it cures. The main methods used to control adhesive are either to use an adhesive that is of heavier consistency, or to use "internal shims" within the adhesive.

Viscosity increase can be made by suitable changes to the formulation. Thixotropic fillers are usually required. This provides a formulation with the consistency of a mastic that is less likely to create adhesive starved regions. Automatic stops can be formulated into the adhesives by either using fillers such as glass beads or microballoons or by applying the liquid adhesives with a supported fabric or cloth. The fibers in the reinforcement provide a "stop" to maintain a constant separation distance between the substrates. A more conventional method is to incorporate mechanical stops into the joint design. In this case the geometry of the substrate does not permit bondline thickness to be less than a specified amount and the maximum thickness can be controlled by suitable "contact" pressure on the part as it cures. See Fig. 11.4 for a sketch of a part containing bondline stops.

Adhesive viscosity is also an important criterion in certain joint designs, such as bell and socket or slip ring joints, where one substrate must be slipped over another prior to adhesive cure. Close tolerances between substrates and high adhesive viscosity could result in the adhesive wiping itself off the substrate leaving a starved joint.

Adhesive tack properties are frequently important for assembly processes. Sometimes tack is a characteristic that is highly desirable, such as when substrates must be restrained from movement. However, there are also times when the parts must be capable of being adjusted prior to final cure. The tack range determines the time interval between the adhesive application and the assembly of components. Thermosetting adhesive types generally have little tack. Whereas, elastomeric and thermoplastic adhesives (i.e., dried latex dispersions and

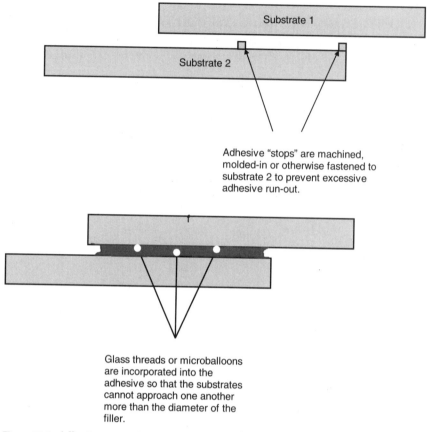

Figure 11.4 Adhesive starvation in a joint can be prevented by incorporating shims (top) in the substrate itself, or (bottom) directly within the adhesive.

solvent solutions) have considerable tack. Tack properties vary widely and are dependent on specific formulations.

11.6 Effect of the Part and Production Requirements on Adhesives Selection

The part or assembly being bonded and its production requirements are frequently important factors in determining which adhesive to use. These factors should be considered early in the bonding process, as they could determine the chemical type and form of adhesive most appropriate to the application. Especially when the strength or service environments are not an overriding concern, the part being assembled and the production tools available quickly become major parameters in the selection process.

11.6.1 Nature of the part or assembly

The assembly being bonded may consist of two simple parts; complicated, multipart subsections; or fully developed products with several subsections. They may be one-of-a-kind articles or mass production units. The nature of the part being bonded may dictate the type of adhesive or bonding process that could be used.

The method chosen for the application of the adhesive to the substrate is determined by the size and geometry of the parts to be assembled, the number of components to be bonded, and the dimensions of the part, as well as by the physical properties of the adhesives. Most adhesive types are available in forms ranging from thin liquids to pastes and solids. The method of application may depend on the viscosity of the adhesive. Table 11.5 lists available forms and processing requirements for common types of adhesives.

The shape of the components often favors the use of a particular form of adhesive. The joining of aluminum honeycomb structures to flat metal skins is best accomplished with thermosetting film adhesives that are supported on glass cloth. On the other hand, it is more convenient to use a paste adhesive for the construction of heat exchangers formed from aluminum tubes and spiral copper fins. A low viscosity, liquid adhesive is convenient to use for laminating parts that have a large surface area.

High volume production items usually require that the adhesive be of a form that can be handled by processing equipment designed for fast assembly. Hand assembled units often employ adhesives in physical forms unsuited to mass production machinery. The assembly of small, delicate parts is dependent on human skill, and the relatively low volume requirements makes it feasible to consider a wider range of adhesive types.

11.6.2 Production requirements

The production conditions under which the adhesive is to be applied are also important criteria for the selection of the adhesive. Particular production requirements could lead the user to choose a product that will bond at a certain temperature or within a certain time. In fact, such circumstances may be the major criterion in the selection process. This is especially true in high volume assembly of low strength parts, for example auto parts, clothing, and toys. Here, the method and time of bonding may be more important than the adhesive's strength or environmental resistance. Often the use of fast-setting expensive adhesives, such as cyanoacrylates, which avoid complex jigging of assembly parts, may be an economic choice over a higher strength adhesive that requires extended cure time.

TABLE 11.5 Selecting Adhesives with Respect to Form and Processing Factors

Adhesive type	Common forms available					Method of cure					Processing conditions		
	Solid	Film	Paste	Liquid	Solvent sol, emulsion	Solvent release	Fusion on heating	Pressure-sensitive	Chemical reaction	Room temp	Elevated temp	Bonding pressure required	Bonding pressure not required
Thermosetting Adhesives													
1. Cyanoacrylate	...	...	...	X	...	...	...	...	X	X	...	X	...
2. Polyester + isocyanate	...	X	...	...	X	X	...	...	X	X	X	X	...
3. Polyester + monomer	...	...	X	X	...	...	...	...	X	X	X	...	X
4. Urea formaldehyde	X	...	...	X	...	...	...	...	X	X	X	X	...
5. Melamine formaldehyde	X	...	...	X	...	...	...	...	X	...	X	X	...
6. Urea-melamine formaldehyde	...	...	...	X	...	...	...	...	X	X	X	X	...
7. Resorcinol formaldehyde	...	...	...	X	...	...	...	...	X	X	X	X	...
8. Phenol-resorcinol formaldehyde	...	...	...	X	...	...	...	...	X	X	X	X	...
9. Epoxy (+ polyamine)	...	...	X	X	...	...	...	...	X	X	X	...	X
10. Epoxy (+ polyanhydride)	X	X	X	X	...	...	...	...	X	X	X	...	X
11. Epoxy (+ polyamide)	...	...	X	X	...	...	...	...	X	...	X	...	X
12. Polyimide	...	X	...	...	X	...	...	...	X	...	X	X	...
13. Polybenzimidazole	...	X	...	...	...	...	...	...	X	...	X	X	...
14. Acrylic	...	...	X	X	...	...	...	...	X	X	...	...	...
15. Acrylate acid diester	...	...	X	X	...	...	...	...	X	X	X	X	X
Thermoplastic Adhesives													
16. Cellulose acetate	...	...	...	...	X	X	...	...	...	X	...	...	X
17. Cellulose acetate butyrate	...	...	...	...	X	X	...	...	...	X	...	...	X
18. Cellulose nitrate	...	...	...	...	X	X	...	...	...	X	...	...	X
19. Polyvinyl acetate	...	X	...	...	X	X	X*	...	...	X	X*	X*	X

	Adhesive	1	2	3	4	5	6	7	8	9	10
20.	Vinyl vinylidene		X								X
21.	Polyvinyl acetal			X			X	X	X	X*	X*
22.	Polyvinyl alcohol						X		X		
23.	Polyamide		X				X		X	X	X
24.	Acrylic							X		X	X
25.	Phenoxy		X				X			X	

Elastomer Adhesives

	Adhesive	1	2	3	4	5	6	7	8	9	10
26.	Natural rubber					X	X	X	X	X	X
27.	Reclaimed rubber					X	X	X	X		X
28.	Butyl					X	X	X	X	X	X
29.	Polyisobutylene					X	X	X	X	X	X
30.	Nitrile			X*		X	X	X	X	X*	X
31.	Styrene butadiene	X				X	X	X			
32.	Polyurethane		X					X			
33.	Polysulfide		X					X			
34.	Silicone (RTV)		X		X		X	X			
35.	Silicone resin				X	X	X	X	X		X
36.	Neoprene					X	X	X		X	X

Alloy Adhesives

	Adhesive	1	2	3	4	5	6	7	8	9	10
37.	Epoxy-phenolic		X	X				X	X	X	X
38.	Epoxy-polysulfide		X	X			X	X	X		
39.	Epoxy-nylon			X				X	X	X	X
40.	Phenolic-nitrile			X				X	X	X	X
41.	Phenolic-neoprene			X				X	X	X	X
42.	Phenolic-polyvinyl butyral			X				X	X	X	X
43.	Phenolic-polyvinyl formal			X				X	X	X	X

*Heat and pressure required for hot-melt types.

Other methods than adhesive bonding may also be considered at this stage in the process. Heat welding (e.g., ultrasonic assembly) methods for bonding thermoplastic parts may provide fast bonds and the elimination of adhesive mixing or application. Solvent welding is also a fast method of bonding that uses the resin in the substrate itself as the adhesive.

Factors that must be considered in selecting an adhesive relative to the assembly operation include: form of the adhesive; method of preparation and use; shelf (storage) life; working life; method or machinery necessary for bonding; and processing variables. The last factor includes the time allowed between coating and bonding; drying time and temperature; temperature required for application and curing; bonding pressure required; rate of strength development at various temperatures; and other properties such as odor, flammability, and toxicity of the adhesive which may call for extra equipment or precautionary measures on the shop floor. These safety and environmental factors are discussed more fully in Chapter 18.

Adhesives require time, pressure, and heat singly or in combination to achieve hardening. Curing conditions are often a severe restricting factor in the selection of an adhesive. In some applications, the curing temperature influences the selection of an adhesive. A number of thermosetting adhesives require heat and pressure to form the bond. These tend to be adhesives having high degrees of strength, temperature resistance, and environmental stability. Where such elevated temperature cures cannot be used, room temperature setting adhesives, such as two component epoxies or urethanes, can be used.

Several production factors must be taken into account when selecting an adhesive for an assembly operation. These factors include:

- The process and type of equipment necessary for metering and mixing of the adhesive
- The processes and equipment necessary for applying the adhesive to the substrate
- The processes required for curing the adhesive and associated equipment whether it be jigs and fixturing, ovens, or pressure devices
- The amount and type of pressure required to mate the substrates and to force the adhesive to flow as with the case of pressure sensitive or contact adhesives
- The availability of substrate cleaning and treatment facilities may also be a factor in selecting adhesives (certain adhesives such as the hybrid acrylics and epoxy formulations are more resistant to oily substrates than other adhesives

- The degree of inspection or quality control that is available or affordable may also factor into the selection of adhesive systems

11.6.3 Cost

Economic analysis of the bonding operation must consider not only the raw-materials cost, but also the processing equipment necessary, time and labor required, and cost incurred by wasted adhesive and rejected parts. The raw-materials cost for the adhesive should not be based on price per weight or volume because of variations in solids and density of adhesives. Cost per bonded part is a more realistic criterion for selection of an adhesive.

Satisfactory end-use properties should be more important in the selection of an adhesive than cost considerations. However, end-use properties should not be over-specified so that the cost becomes an unreasonable burden. Where the consideration of strength and durability requirements indicate that several adhesives may do the job, choice will then become a cost factor for the bonding process on the whole. The following cost factors must be considered in addition to the materials cost of the adhesive:

- Efficiency of coverage
- Ease of application and process equipment needed
- Processing time
- Cost of labor for assembly and for quality control and inspection of completed parts
- Amount of reject material due to improper storage conditions and production difficulties
- Poor bond quality
- Safety and environmental considerations including waste disposal

11.7 Service Conditions

Adhesives must also be selected with regard to the type of stress and environmental conditions to which they will be exposed. These factors may be the most important criteria in deciding which adhesive to use. Depending on the nature of the stress and the environments expected, these factors could drastically limit the number of candidate adhesives. The designer must be able to confidently predict all the conditions that the adhesive bond will see in service. He or she must also be able to determine the possible effects of these conditions on the bond — in the long term as well as the short term. Detailed information regarding the environmental resistance of various adhesive classifications appears in Chapter 17.

11.7.1 Service stress

The adhesive should be chosen with respect to the type of stress that will be encountered. Peel or cleavage types of stress dictate a tough, somewhat flexible adhesive to accommodate uneven stress distribution. Shear or tensile types of stress will result in more uniform stress distribution and will allow the selection of rigid adhesives that have higher cohesive strength. The chosen adhesive should have strength great enough to resist the maximum stress experienced at any time in service with reasonable safety factors. Over-specifying may result in certain adhesives being overlooked which can do the job at lower cost and with less demanding curing conditions. The adhesive must also be resistant to lower stresses that will occur during long continuous periods. Thermosetting adhesive systems are more likely to resist these creep stresses than are elastomeric or thermoplastic adhesives.

Adhesive types display wide differences in response to different stresses and strain rates. Thermoplastic and many elastomeric adhesives are not suitable for structural applications because they have a tendency to distort and creep under moderate load. They also soften on heating at moderate temperatures. Thermoplastic adhesives are unable to withstand vibratory stresses for long periods although they may exhibit greater strengths than thermosetting types for short duration tests. Thermoplastic and elastomeric types usually possess high peel strengths but relatively low tensile or shear strengths.

The thermosetting resins, on the other hand, are often used as the base material in structural adhesives. They provide rigid bonds that retain much of their strength at elevated temperatures. Tough thermosetting and hybrid adhesives also perform well under vibratory loads and have moderately good impact loading characteristics. Amine cured epoxies have a room temperature lap shear strength of 2,000 psi; the strength is approximately 500 psi greater when an anhydride or a polyamide curing agent is specified. Epoxy-phenolics have lap shear strengths of nearly 3,500 psi, and the epoxy-nylons reach lap shear strengths of 6,000 psi. Phenolic resins combined with polyvinyl butyral result in adhesives with excellent shear strength. Room temperature strength may be as high as 5,000 psi. The neoprene modified phenolics have lower shear strengths than most modified phenolics.

11.7.2 Service environment

Another important factor to be considered is the temperature range over which the adhesive will be used. At elevated temperatures, all organic adhesives lose some of their strength, and some types soften or decompose to the extent that they become useless. Up to 160°F, several types of thermosetting and thermoplastic adhesives may be used. However, at 250°F, while only a few thermoplastic adhesives will withstand intermittent exposures at low stress rates, most thermosetting adhesives will function satisfactorily under continuous exposure. Above 250°F, only the more thermally resistant thermosetting adhesives, such as phenolic-nitriles and epoxy-novolacs, will perform adequately for continuous duty. Some very special and expensive adhesives can withstand 400–500°F for continuous exposure and 600°F for short periods. Some adhesives have been noted to retain shear strengths exceeding 500 psi after 1 h exposure at 900°F. The adhesive engineer must realize that as the temperature service requirement becomes greater and greater, the number of candidate adhesives will become less and less. The high temperature resistant adhesives have onerous curing and production requirements, lack resistance to peel and cleavage, and are very expensive systems.

Low temperatures cause embrittlement and internal stresses on many adhesives so that premature cohesive failure of the bond may occur below −100°F. For subzero temperature conditions, it is preferable to choose flexible thermoplastic or elastomeric adhesive types that retain some resilience on exposure to cold. Flexible adhesives, such as urethanes and epoxy nylons, provide shear strengths in excess of 8,700 psi at cryogenic temperatures as low as −330°F.

Although performance varies considerably, most organic adhesives are adversely affected by moisture to some extent, especially when under conditions of stress. Natural adhesives are more vulnerable than the synthetic adhesives. Adhesives can fail in moist service environments either cohesively by hydrolysis of the adhesive resin or adhesively by preferential adsorption of moisture at the interface.

Considerations must also be given to such environmental influences as chemical reagents, hot and cold water, oils, solvents, hydraulic fluids, chemical atmospheres (e.g., ozone, acidic gases, salt-spray), exterior weathering and aging, and the effects of biological agents (mold growth, fungi, insect and rodent attack). Also to be considered are the effects of radiation (sunlight, x-rays, radioactivity, infrared), hard vacuum, etc. These environments are especially severe on natural adhesives. Such environments affect the thermoplastic and elastomeric adhesives and have lesser affect on the thermosetting and hybrid adhesives. However, all potential adhesive systems must be screened against the harmful effects of these environmental conditions. No adhesive system is resistant to all possible environments. Generally, adhesive systems are resistant to certain environments and less resistant or catastrophically damaged by other environments. So once again, the adhesive engineer must understand the environmental influences that will most likely occur and weigh these factors in the selection process.

11.8 Special Considerations

For the majority of applications, the considerations outlined above generally provide a sufficient basis on which to select a suitable adhesive. Other circumstances and requirements, however, can influence the selection process.

Special electrical, thermal, or optical properties, such as dielectric constant and loss, insulation resistance, or transparency, are sometimes required for assemblies in certain industries. Such properties are often found in the adhesive specifications for printed circuits, insulation components, optical lenses, etc. Adequate properties usually can be achieved with suitable formulation. However, the user must often make compromises. These special properties, if required, will take priority over other selection criteria.

For example, surface pretreatment may affect the electrical and thermal properties of the bonded joint. Often a specialty adhesive will not have the production capability or temperature resistance that are present in other, more standard adhesives, and the user will have to make provisions for using that adhesive. Transparent colorless glue lines of a given refractive index are usually required for optical systems. Clear adhesives can be formulated, but care must be given so that internal stressing does not cause optical imperfections. Some adhesives have permanent color properties that prevent their application to many optical assemblies.

Adhesives are sometimes called upon to perform functions other than joining. Adhesives may be required to act as sealants or as thermal and/or electrical insulators. Resistance against corrosion of metal joints, vibration, and fatigue are additional requirements that may be needed in an adhesive assembly. The secondary functions of adhesives are sometimes of major importance, and adhesive selection may be considerably influenced by them.

11.9 Verification of Reliability

It is always prudent to plan a detailed test program. This program will not only estimate the reliability of the adhesive bond but will also allow optimization of the parameters of the bonding processes. Check the joint design by bonding a simple assembly and testing it. Such testing also determines the optimum bond dimensions and thickness of the adhesive film. Specimens should be tested in a simulated operating environment. Accelerate the environmental conditions (moisture, temperature, etc.) making sure that the effect on bond life is consistent with the actual conditions expected in service. Test specimens should simulate the final assembly as closely as possible.

11.10 Sources of Information and Assistance

To select adhesive candidates for testing, it is necessary to have meaningful physical data for comparison. Literature from adhesive manufacturers and trade periodicals offers some adhesive design and test information. Usually the data presented in these documents are from reproducible ASTM test methods. General physical properties such as shear or peel strength can easily be compared. Unfortunately, information concerning specific stresses, environments, or adherends, is often lacking.

Adhesive information and property data can also be found in the technical literature. Many excellent texts also exist on adhesive technology. Various sources for such information are provided in Appendix B.

The adhesive supplier can help narrow the choice of adhesives, recommend a certain formulation, assist in designing the joint, and establish surface-preparation and bonding requirements. Before contacting the adhesive supplier or attempting to select candidates alone, the buyer must first clearly define his bonding problem. A questionnaire such as that illustrated in Fig. 9.8 is often used by suppliers to determine adhesive-bond requirements.

References

1. Polaski, G., "Putting Adhesives in Their Place", *Plastics Design Forum,* Sept./Oct. 1993.
2. Schields, J., "Adhesive Selection", Chapter 3 in *Adhesives Handbook,* 3rd ed., (London: Butterworths, 1984).
3. Bickerman, J. L., *The Science of Adhesive Joints,* 2nd ed., (New York: Academic Press, 1968).
4. Hemming, C. B., "Wood Gluing", in *Handbook of Adhesives,* I. Skiest ed., (New York: Reinhold, 1962).
5. Caster, D., Wood Adhesives—Research, Application and Needs, Symposium Proceedings (Madison, WI: USDA Forest Service FPL, 1980) at 168.
6. Moser, F., "Bonding Glass", in *Handbook of Adhesives,* I. Skiest ed., (New York: Reinhold, 1962).

Chapter 12

Sealant Classification and Composition

12.1 Introduction

Sealants are like adhesives in many ways. In fact, they are often considered together because many formulations can perform either as an adhesive or as a sealant. Usually, a sealant must effectively bond to a substrate in order for it to perform its function. However, a sealant is normally used to exclude or contain while an adhesive is used to hold. Sealants generally are used to exclude dirt, moisture, and chemicals or to contain a liquid or gas. Materials that are used as sealants generally have lower strength than materials that are used as adhesives. This is because sealant formulations contain large amounts of inert filler material, primarily for cost reduction and gap filling purposes. Sealant formulations also contain elastomeric resins to provide flexibility and elongation rather than high tensile strength.

In addition to adhering to the substrate and providing a certain degree of holding power, sealants can also be used to provide acoustical insulation, vibration damping, and electrical insulation or conduction depending on the conductivity of the fillers. Specialty sealants can also be used for thermal insulation or even as a fire barrier. Certain sealant formulations are also used for heavy surface coatings to protect against mechanical or chemical attack or to improve the appearance of a part. Therefore, like adhesives, sealants are often required to provide a multiplicity of functions in addition to their primary role.

This chapter describes methods of classifying sealant materials. The various types of sealant materials are characterized. The components used in common sealant formulations are identified and discussed in relation to the final performance requirements. Common families of sealant materials and formulations are identified in the next chapter. Chapter 14 will cover the selection of sealants for specific applications.

12.2 Basic Sealant Classifications

The large number and diversity of sealant formulations make their classification difficult. Often sealants are divided into the same product classes, chemical families, and industrial applications as are adhesives. However, there are also particular methods of classification that have been developed specifically for sealants. These unique methods of classifying sealants are by their:

- Hardening ability
- Cure
- End-use
- Performance

12.2.1 Hardening and non-hardening types

Sealants are often classified as either hardening or non-hardening. Hardening sealants are applied as a heavily bodied fluid, and they "set" to form either a rigid or a flexible solid seal. Non-hardening sealants do not "set". They stay "wet" or flowable after application and have physical properties similar to before they were applied. It should be noted, however, that although non-hardening sealants remain soft after application, they might still go through a chemical conversion, solvent evaporation, or some other "curing" mechanism.

Hardening sealants can be further sub-divided into rigid and flexible classifications. Rigid sealants have little or no resiliency and are characterized by their inability to flex. These types of sealants are commonly used when both sealing and joining functions are needed. Sealants that are representative of this group include the epoxies and acrylics. Table 12.1 lists the common sealants that are classified as rigid systems along with their more significant characteristics.

Flexible sealants remain flexible after curing and are generally elastomer-based products. Their range of flexibility and hardness varies considerably. Hardness values from Shore 10A to 85A durometer are common. Flexible sealants are either true elastomers and return to their original dimensions after compression or elongation (e.g., silicones and nitriles), or else they are made from materials that permanently deform on stress without tearing (e.g., asphalt and various gum stocks). Typical flexible sealants are described in Table 12.2.

Non-hardening sealants generally cannot be depended upon to perform a joining function although they may provide adequate sealing

TABLE 12.1 Hardening Sealants, Rigid Type[1]

Sealant base	Sealant form	Curing characteristics	Curing time and temp range	Pot life, hr	Application methods	Remarks
Epoxy	Two-part liquid	Cures at room temperature because of addition of curing agent (accelerator) or hardener. No fumes released during epoxy polymerization	1–8 hr at 75°F. Some formulations require 4–6 days. ½–3 hr at 150°F	¼–3	GM, G, K, B	Formulations available which are easy to apply because of low viscosity. Available in thixotropic form. Normally very brittle
Modified epoxy	Two-part liquid modified with polysulfide polymer	Thermosetting resin cures at room temperature because of addition of curing agent or hardener. No fumes released during cure	½–8 hr at 75°F	⅙–2	GM, GP, K, S	
Acrylic	Made in one-part emulsion, one-part solvent release, and two-part systems	....	Varies, depending on formulation and curing system	....	GM, B, S	More commonly used as adhesives, but have limited sealing applications
Oleoresin	One-part putty or paste	Thermoplastic-type material, normally cures by solvent evaporation	14 days or more at 75°F to reach specific hardness	Very long	GM, GP, K	Easy to extrude, requires long time to reach measurable physical properties. High degree of shrinkage
Asphalt and bituminous	One-part putty or paste	Thermoplastic-type material, normally cures by solvent release	Several weeks at 75°F	Very long	GM, GP, K	Easy to handle and extrude. Available in black only. Can cause staining and tendency to cold flow
	Two-part flow-type paste	Thermoplastic type with catalyst curing agent	1–8 hr at 75°F	4–8	GM, GP, K, B, S	

APPLICATION METHODS CODE: GM—manually operated gun for extruding sealant. GP—pressure-operated gun. B—brush. K—spatula, putty knife, trowel, squeegee. S—spray.

TABLE 12.2 Hardening Sealants, Nonrigid (flexible) Type[1]

Sealant base	Curing characteristics	Curing time and temp range	Pot life	Application methods	Toxicity during application	Remarks
Two-part Chemical reaction Systems (Catalyst-cured)						
Polysulfide	Thermosetting liquid polymer, cures at room temperature. Elevated temperature reduces cure time	16–24 hr at 75°F	1–6 hr	GM, GP, K, B, S	None	Economical, handles easily. Primers sometimes used for substrates in critical applications. Wide range of properties. Good flexibility. A versatile sealant
Polyurethane	Thermosetting elastomer, cures at room temperature	16–24 hr at 75°F, 6 hr at 180°F	3–6 hr	GM, GP, K, B, S	None	Economical, somewhat difficult to handle. Has high strength and good abrasion resistance. Primer system necessary
Silicone	Thermosetting elastomer, room-temperature curing. Should be postcured at high temperature if elevated-temperature service is anticipated	24 hr at 75°F	3–6 hr	GM, GP, K, B	None	Excellent for high-temperature service. Good thermal barrier. High in cost. Normally requires primer system
Modified epoxy	Thermosetting resin, room-temperature or heat-activated curing	2 hr at 75°F	50 min	GM, GP, K, B, S	Can affect allergic individuals	Normally modified with plasticizers or other ingredients to impart various degrees of flexibility. Polysulfides and polyamides often used. Economical, handle easily
Neoprene	Thermosetting. Room-temperature curing	24 hr at 75°F	3–4 hr	GM, GP, K, B, S	None	
Polysulfide	Thermosetting liquid polymer; cures at room temperatures. Curing accelerated by heat and/or moisture	14–21 days at 75°F	1 year at 80°F	GM, GP, K	None	Very easy to handle. Tacks out rapidly. Can be applied without primer. Supplied only in thixotropic form

Polyurethane	Thermosetting elastomer, cures at room temperature. Curing accelerated by heat and/or moisture	14–21 days at 75°F	3–9 months	GM, GP, K	None	Excess moisture can cause blowing or sponging. Normally requires primer
Silicone	Thermosetting elastomer at room temperature. Curing accelerated by heat and/or moisture	7–14 days at 75°F	6 months	GM, GP, K	None	Easy to extrude. Can be supplied in transparent color. Normally requires primer
Acrylic	Thermoplastic-type resin that cures at room temperature	21 days at 75°F	3–9 months	GM, GP, K	None, strong odor	Sometimes requires preheating to handle, low cohesive strength generally in the mastic or nonhardening family
Solvent-release Systems						
Neoprene	Thermoplastic elastomer. Sets at room temperature	7–14 days at 75°F	6–12 months	GM, GP, K, B, S	None	Fairly easy to apply. Available in colors. Good weathering characteristics: high shrinkage. Poor package stability
Hypalon	Thermoplastic elastomer. Sets at room temperature	7–14 days at 75°F	6–12 months	GM, GP, K, B, S	None	Similar to neoprene. Somewhat better weathering resistance
Butyl	Room-temperature-curing thermoplastic resin	...	6–12 months	GM, GP, K	None	Easy to apply. High degree of tenacity. Relatively costly. Requires long set time to achieve required strength.

APPLICATION METHODS CODE: GM—manually operated gun for extruding sealant. GP—pressure-operated gun. B—brush. K—spatula, putty knife, trowel, squeegee. S—spray.

capability in certain applications. These materials are available in many forms and are characterized by a "mastic" consistency. Soft setting natural resins and oil-based sealers are examples of non-hardening sealants. They are available in drying or non-drying types. The drying types remain soft inside the seal by developing a protective skin. The non-drying types remain tacky to the touch and are uniformly soft. Some non-hardening sealants can be formed into various shapes and applied as continuous beads or tape. They are usually supplied with a coated paper or plastic film release lining that is removed during application so that the sealant can be pressed into place by hand. Table 12.3 lists representative non-hardening sealants.

12.2.2 Cure type

Sealants are sometimes characterized by their method of cure. Curing parameters generally involve catalytic agents, time, temperature, or humidity. As with adhesives, curing characteristics can vary significantly within the various families and formulations. Depending on their curing method, sealants are more popularly classified as: two-part systems, single component sealants, or solvent or water release sealants.

12.2.2.1 Two-part systems. Two-part systems are usually faster reacting than the other types of curing sealants. Faster cures make it possible to handle prefabricated units and reduce the overall cycle time required for the sealing operation. Faster cures are also important when the sealant is used in outdoor applications. When applied outdoors, the time from application of the sealant until its first exposure to abrasion, weather conditions, or thermal cycle may be critical.

Similar to two-part adhesives, the two-part sealant is prepared and packaged as two or more components. One component contains the base resin component and compatible fillers, extenders, etc. Another component contains the catalyst or curing agent and appropriate additives. Perhaps a third component will contain a catalyst or accelerator. These components must be metered to the proper mixing proportions and mixed thoroughly prior to application. Some sealant systems, such as the polysulfides, are known as oxidatively or catalytically cured sealants. These are cured by the addition of an oxidizing agent such as a peroxide. Other two-part systems, such as the silicones, urethanes, and epoxy systems, are cured by chemical reaction.

Two-part sealants may cure at either ambient or elevated temperatures. Generally, sealants are applied and cured at either room or plant conditions or at outdoor ambient conditions. When sealants are

TABLE 12.3 Non-hardening Sealants[1]

Sealant base	Sealant form	Pot life	Toxicity during application	Application methods	Remarks
Butyl............	One-part mastic, putty or paste	Indefinite	None	GM, GP, K	High degree of tenacity. Relatively costly. Difficult to measure properties of strength. Takes many months for sealant to skin over
Acrylic............	One-part mastic, putty or paste	Indefinite	None, strong odor	GM, GP, K	Poor cohesive strength. At times require pre-heating material to extrude
Polybutene.......	One-part mastic, putty or paste	Indefinite	None	GM, GP, K	Similar to butyls
Asphalt and bi-tuminous	One-part mastic or paste	Indefinite	None, strong odor	GM, GP, K	Normally available only in black. Cannot be used on porous substrates where black color would counter the aesthetics involved. Easy to handle. Tendency to cold flow

APPLICATION METHODS CODE: GM—extruded by manual calking gun. GP—extruded by pressure gun. K—spatula, putty knife, trowel, or squeegee.

Non-hardening sealants usually have unlimited pot life since no cure is involved in application of these materials. However, where solvents or thinners are used to adjust the viscosity for production application purposes, the sealant will change consistency if left uncovered, because of evaporation of the thinner.

Generally the non-hardening sealants are not harmful; however, the odor from some materials might be objectionable, especially when they are used in confined quarters. In addition, some individuals may experience allergic reactions to contact with certain formulations.

Since most non-hardening sealants are heavy-consistency, they must be applied by gun extrusion or spatula. Some formulations can be thinned enough to permit brush application.

used in high volume manufacturing and speed of set is an important criterion, the designer may choose for production purposes heat reacting grades or room temperature grades that can be cured at elevated temperature.

12.2.2.2 Single component sealants. Single component sealants cure by a variety of methods including reaction with water vapor, oxidation, anaerobic reaction, or a latent catalyst. Single component sealants generally contain an integral catalyst or hardener that has some degree of latency. This permits long shelf life and working life, but this usually means that the sealant must be stored under refrigeration especially if ambient cures are possible.

Sometimes single-part systems are preferred to two-part sealants because they eliminate the problems of mixing and verifying that mixing is complete. However, at times a two-part sealant is needed for a faster, complete cure. One-part systems sometimes suffer from a disadvantage in that their curing characteristics and resulting properties are affected by the environmental conditions that they are exposed to during application including relative humidity, temperature, joint design allowing ingress of moisture, etc. This is of special concern when the sealants are applied and cured in a variable outdoor climate. Two-part systems are also affected but to a lesser extent depending on their reaction mechanism.

Many single component sealant formulations cure by a chemical change that is initiated by the presence of humidity. The application of heat will speed the curing reaction with some of these formulations. Sealants that cure via this reaction mechanism include many urethanes, polysulfides, and silicones. A popular representative of this class is the moisture activated silicone that can be identified by the odor of acetic acid produced during reaction with the moisture in the air. These particular sealants can also be formulated into two-part sealants by using catalysts to accelerate the reaction process.

Some single component sealants, such as certain mercaptan terminated sealants, cure by the absorption of oxygen. Oleoresinous sealants that are based on linseed oil, tung oil, fish oil, etc. and contain fillers of clay and fiber cure by surface oxidation. Additives of metallic napthenates are included in these formulations to obtain various degrees of cure depth. These sealants can be formulated to dry hard and, thus, classify as rigid hardening-type sealants, or else they can be formulated to stay pliable and, thus, fall into the non-hardening class. These materials are best recognized as the low cost caulks and glazing compounds that are commonly used in the construction industry.

Sealants formulated from combinations of hydrocarbon based plastics, such as polybutenes and polyisobutylenes, can be compounded with non-drying types of vegetable oils. These sealants are a permanent non-hardening type. They are often termed resinous or plastic sealants.

Anaerobic sealants are single component compounds that remain in the liquid state in the presence of oxygen. However, when oxygen is excluded (such as by placing the sealant into a joint) and metal ions are available (generally supplied by the substrate), the sealant cures. These sealants are best known as thread locking compounds for sealing fasteners or for use in pipe and tube joints.

12.2.2.3 Solvent and water release sealants. Some base polymers that are used in sealant formulations can be dissolved in an aqueous or organic solvent that will evaporate during the cure cycle and leave a tack free surface condition. Typical sealants that fall into this class are the butyls, acrylics, chloroprenes (neoprene), and chlorosulfonated polyethylenes. Solvent release sealants give good adhesion to a variety of substrates and perform satisfactorily in joints with small movement. Solvent release sealants vary considerably in cost, depending on the areas of application, the solvent content, and the movement capability. Consideration of environmental and health restrictions is necessary with solvent based sealant systems.

Water dispersed sealants can also be formulated with certain polymers, notably acrylics. The base resin is emulsified in water and then formulated with the required additives and extenders. Once applied, the sealant cures to a rubber like consistency when the water is driven off at either ambient or elevated temperatures.

12.2.3 Classification by end-use

Sealants are often classified according to their function or end-use. Interior sealants act to contain a medium; exterior sealants exclude contaminants. Some sealers perform both functions. These end-use classifications include concrete sealers, automobile sealers, insulated glass sealers, construction and architectural sealers, and airport and highway sealers. Table 12.4 shows typical application areas for generic types of sealants.

12.2.4 Performance classifications

There are several obvious parameters by which sealants can be classified for specification purposes. These include movement capabilities and recovery properties. Most of performance classifications have centered on construction applications where standards and specifications are commonly used with this type of product. From a performance

TABLE 12.4 Typical Application Areas for Generic Sealants[2]

Generic type	Cost range	Typical applications
Oil base	Low	Small wooden window sash, caulking for expansion and contraction of joints
Oil resin base	Low	Sealing of concrete joints, masonry copings, glazing, metal windows
Butyl mastics	Medium	Caulking expansion and contraction joints, sealing electrical conduit
Butyl curable	Medium	Home sealants, repair of gaskets, tapes, with resins for hot melts on insulating glass
Butyl/polybutylene	Medium	Metal buildings, slip joints, interlocking curtain wall joints, sound deadening tapes
Polyisobutylene	Medium	Primary seal on insulating glass, hot melt sealants
Emulsion acrylic (high modulus)	Medium	Water based for interior use joints on plywood
Emulsion acrylic (low modulus)	Medium	Water based caulks for exterior joints on low rise housing, with good movement capabilities, excellent weathering
Hypalon	Medium–High	Exterior joints on high rise construction, around doors and windows
Solvent acrylic	Medium	Exterior joints on high rise construction, around doors and windows with low movement, pipe joints, glazing, masonry and metal caulking, liquid gaskets and pipe dope
Block copolymer	Medium	Low rise buildings with good movement
One part polysulfide	High	High rise building joints, caulking, and glazing, sealing between dissimilar metals, deck caulking and sealing of refracting mirrors
Two part polysulfide	High	High rise building joints, aircraft fuel tanks, boating, insulting glass sealant, airport aprons, potting, molding
One part urethane	Medium–High	High rise building joints, sealing, caulking and glazing

TABLE 12.4 Typical Application Areas for Generic Sealants (*Continued*)

Generic type	Cost range	Typical applications
Two part urethane	Medium–High	High rise building joints, insulating glass sealant, waterproofing, potting and molding, caulking where compatibility with low temperatures required
Silicone structural	Very high	High rise building joints, sealing, caulking and glazing
Silicone (medium modulus)	Very high	High rise building joints, general construction, insulating glass, two part for potting and molding, electrical connectors
Silicone (low modulus)	Very high	Highways and difficult joints

standpoint, movement capability is one of the best parameters for qualifying sealants for certain applications such as in the construction industry. Table 12.5a lists application characteristics for generic classes of sealants. Table 12.5b provides performance capabilities for these sealants. Applicable Federal, ASTM, and Canadian standards are identified in Table 12.6.

One of the most comprehensive classification methods used in the construction industry is by movement capability. Usually this is expressed at +/− percentage value. The +% is the amount of movement that a sealant can take in extension, and the −% value is the amount of movement that it can take in compression of the same joint. In each case, the base is the original dimension of the joint at the time the sealant is installed. Most sealant standards today include a qualifying statement that specifies the joint movement capability of the sealants that are covered by the standard.

Recovery properties measure the degree to which a sealant exhibits elastomeric properties. Recovery is the degree to which the sealant returns to its original dimensions after being elongated. Recovery is essential in joints requiring good movement. The high modulus solvent release sealants and the unplasticized latex caulks exhibit poor recovery. Chemically cured sealants exhibit fair to excellent recovery. Silicone is the best in this category with up to 95% recovery even after one year in compression at elevated temperatures. Urethanes are next best, and polysulfides are a poor third.

12.3 Sealant Composition

Like adhesives, sealant formulations contain a base polymer that is modified by many possible extenders and additives. However, sealant

TABLE 12.5 Application Characteristics (a) and Performance Capabilities (b) of Generic Classes of Sealants[2,3]

(a) Application characteristics

Generic type	Cure type	Joint limits w × d, in.	Nonvolatiles, weight %	Shrinkage, %	Adhesion to common building materials (unprimed)
Oil base	oxidation	¼ × ¼	70–90	10–20	good
Oil resin base	oxidation	¼ × ¼	96–99	4	good
Butyl mastics	solvent rel.	⅜ × ⅜	74–99+	5–15	good
Butyl curable	oxidation	½ × ⅜	74–99+	5–15	good
Butyl/polybutylene	none	mastic	99+	1	good–excellent
Polyisobutylene	thermoplastic	thin beads	99+	1	good–excellent
Emulsion acrylic (high modulus)	water rel.	½ × ½	80–85	10–20	good
Emulsion acrylic (low modulus)	water rel.	½ × ½	80–85	10–20	good
Hypalon	solvent rel.	⅝ × ½	80–85	10–20	good–excellent
Solvent acrylic	solvent rel.	¾ × ⅜	80–85	10–20	excellent
Block copolymer	solvent rel.	½ × ½	60–70	20–30	fair
One part polysulfide	catalyst cure	¾ × ½	90–99+	10	good
Two part polysulfide	catalyst cure	¾ × ½	90–99+	10	good
One part urethane	moisture cure	1¼ × ⅜	94–99+	6	good
Two part urethane	catalyst cure	2 × ½	94–99+	6	good
Silicone structural	moisture cure	¾ × ¾	98+	2	fair–good
Silicone (medium modulus)	moisture cure	¾ × ¾	98+	2	fair–good
Silicone (low modulus)	moisture cure	¾ × ¾	98+	2	fair–good

(b) Performance capabilities

Generic type	Movement, +/-%	Recovery	Maximum service temperature, °F	Minimum service temperature, °F	Expected life, yrs	Heat aged weight loss, %	Resistance to: UV	Ozone	Heat aging
Oil base	2	poor	150	0	5–10	10	fair	fair	hardens
Oil resin base	5	poor	150	0	5–10	10	fair	fair	fair
Butyl mastics	5–10	poor	200	−20	5–10	5–20	fair	fair	stays soft
Butyl curable	10	poor	200	−20	5–10	10	fair	fair	good
Butyl/polybutylene	10	none	180	−40	20+	0	good	good	good
Polyisobutylene	7.5–10	none	180	−40	20+	0	excel	excel	excel
Emulsion acrylic (high modulus)	10	fair	180	0	10	10	poor	poor	hardens
Emulsion acrylic (low modulus)	25	good	180	0	20	30	good	good	good
Hypalon	12.5	fair	200	−40	20	10	good	good	toughens
Solvent acrylic	12.5	fair	180	0	20	10	good	good	toughens
Block copolymer	25	good	180	−10	20	30	good	good	good
One part polysulfide	12.5–25	fair	250	−40	10	12	crazes	crazes	crazes
Two part polysulfide	25	good	250	−40	20	10	crazes	crazes	toughens
One part urethane	<50	excellent	250	−40	20	10	good	good	good
Two part urethane	<50	excellent	250	−40	20+	10	super	super	super
Silicone structural	25 and 50	excellent	400	−90	30	10	supr	super	super
Silicone (medium modulus)	50	excellent	400	−90	30	10	super	super	super
Silicone (low modulus)	+100 to −50	excelent	400	−90	30	10	super	super	super

463

TABLE 12.6 Applicable Federal, ASTM, and Canadian Standards for Classes of Sealants[2]

Generic type	Standards		
	Federal	ASTM	Canadian
Oil base	TT-G-410E	none	19GP-1G, 19GP-6A
Oil resin base	TT-V-598	C-570	19GP-2B
Butyl mastics	none	none	none
Butyl curable	1657	none	19GP-14M
Butyl/polybutylene	none	none	none
Polyisobutylene	none	none	none
Emulsion acrylic (high modulus)	none	C-834	19GP-17
Emulsion acrylic (low modulus)	230C	C-920	19GP-24
Hypalon	none	none	19GP-26
Solvent acrylic	none	none	19GP-5M
Block copolymer	230C	C-920	19GP-24
One part polysulfide	230C	C-920	19GP-25
Two part polysulfide	227E	C-920	19GP-3M
One part urethane	230C	C-920	19GP-16A
Two part urethane	227E	C-920	19GP-15M
Silicone structural	230C	C-920	19GP-9M
Silicone (medium modulus)	230C	C-920	19GP-9M
Silicone (low modulus)	230C	C-920	19GP-9M

formulations are usually more complex than adhesive formulations because of the many compromises that must be made regarding application, curing, and performance characteristics. The sealant's material cost is also often an important criterion for certain industries because of the high volumes that are used.

Modern sealants are generally formulated from synthetic elastomeric polymers. Many of these modern sealants were commercially nonexistent before the early 1960s. Some of the polymers that have been used to formulate sealants include polysulfides, polyurethanes, chloroprenes, acrylic, butyl and chlorosulfonated polyethylene. Like adhesive systems, sealant formulations contain base polymer, fillers, plasticizers, thixotropic agents, adhesion promoters, catalysts, curing agents, and other ingredients. Various external accessory materials such as primers, bond-breaker tapes, back-up materials, and other aids may also be needed during the application stage to provide for an effective sealant joint.

12.3.1 Primary resin

The performance of a finished sealant formulation will depend heavily on the properties of the base resin material used. The base resin is chosen not as much for its high tensile strength as for its elongation, flexibility, and environmental resistance. The sealant base resin needs to be in a form that is convenient to formulate with many other compounds, depending on the specific application. Base resins for sealant formulations include a large spectrum of materials from modern high performance elastomers to low strength waxes and rosins.

The base polymer is generally supplied in the form of a solid rubber, a liquid polymer, or a latex dispersion. The solid rubber requires solvent to reduce it to a formulating liquid consistency. Major elastomer manufacturers can supply the rubber in a solution grade for small sealant manufacturers. Low viscosity liquid polymers, such as polysulfide, various urethanes, silicones, polybutene, and polyisobutylene, can be used to make sealants with little or no solvent. Often the compounding must be done at elevated temperatures to ensure uniform incorporation of all the additives. Latex emulsions, such as acrylic, polyvinyl chloride, polyvinyl acetate, and blends of such materials, are also easy to formulate since the starting base is a fluid water emulsion.

There is generally an optimal percentage range for each polymer base resin in a sealant formulation. From a performance standpoint, there is also an optimum balance of filler, plasticizer, and other additives to give the best performance possible with any system. Polymer content is usually measured as the weight percentage of base polymer in the formulation after the solvent or water has been evaporated from the system and after the sealant cures. Since the base polymer is generally the most expensive component in the formulation, it is the first item to be trimmed when cost is a concern. Table 12.7 shows typical solvent concentration (given as a weight percentage of the formulation before cure) and polymer content for typical sealants.

For most premium sealants, the polymer content is in the 30–50% range. However, there are perfectly adequate sealants at considerably lower solids content and lower cost. It is for this reason that one should develop functional performance specifications to best describe the end properties expected from the sealant.

TABLE 12.7 General Solvent Contents for Typical Sealants[2]

Base resin	Solvent content, %	Polymer content, %
Acrylic	20–40	35–45
Latex emulsions	35–45	35–45
Polysulfide	0	30–45
Silicone	0	60–85
Urethane	0	30–45

12.3.2 Solvents

Solvents are needed in large amounts to help liquefy the base resin so that the other additives may be easily incorporated into the formulation and to lower the viscosity of the system to make it easier to apply. In small amounts, solvents are used to improve the extrudability of the sealant.

Usually, the solvent type and concentration can be spotted quickly from the manufacturers' data sheet, the information on the sealant container, or the information on the Material Safety Data Sheet (MSDS). Total solvent concentration is usually quantitatively represented as 1.00 − solids content in weight percentage.

Normal solvent contents for typical sealants are shown in Table 12.7. It is possible to use too much solvent in a formulation. This may result in excessive shrinkage when the sealant cures. Volume shrinkage will always be greater than the weight percent of solvent because the density of the solvent is much lower than the other components in the sealant. Generally, for every 1% weight loss there is about 2% volume shrinkage.

The common solvents used in sealant formulations are toluene, xylene, petroleum spirits, water, and others. Solvent mixtures are carefully chosen so that rapid evaporation of one component does not create a problem. Care must be used in selecting the solvent mixture so that the sealant's skin does not dry so quickly that it prevents the escape of solvent from the body of the sealant. The solvent system is also an important determinant for the viscosity and application characteristic of the sealant. Often the solvent in which the resin is the most soluble is also the most volatile, and application problems are common with solvents having excessive volatility.

12.3.3 Fillers

Fillers are generally used in sealant formulations as additives to increase the viscosity of the sealant and provide better gap filling properties and to lower the cost of the formulation. For sealants, fillers do not give much reinforcement or added strength. However, fillers can significantly affect other performance factors such as water resistance and hardness.

The most common filler is calcium carbonate, because it is readily available and comes in various sizes. Other common fillers include various clays, silica, colorants (such as titanium dioxide), zinc sulfide, zinc oxide, carbon black, and various iron oxides. Table 12.8 provides examples of the fillers that are commonly used in sealant formulations and the beneficial effects that are derived from each.

Filler content can vary significantly depending on the base polymer and the requirements of the formulation. The volume percent of fillers

TABLE 12.8 Fillers Commonly Used in Sealant Formulations[2,4]

Filler	Benefit
Calcium carbonate	Viscosity adjustment
Silica	Common for silicone resins
Fumed silica	Thixotropic agent
Titanium dioxide	Used whenever white or light shades are required
Carbon black	Used when darker shades are required
Iron oxides	Colorants in certain sealants
Aluminum and other metal stearates, titanium esters	Thixotropic agents
Glass and polymeric fibers	Thixotropic agents

can be up to 90% in sealant putties and in oil and resin based caulks. Fillers in silicone sealants, however, may constitute only 10–20% of the total volume.

12.3.4 Plasticizers

Plasticizers are used to reduce the elastic modulus of some sealants. A general approach is to use a high performance base polymer that gives high modulus and high recovery and then incorporate a plasticizer to reduce the modulus and increase flexibility. Plasticizers are also used in chemically cured liquid polymers to assure softness and the ability to relax stress.

The best results are achieved by using relatively nonvolatile plasticizers. Volatile plasticizers are less expensive, but the beneficial properties that they provide will dissipate with time. Common construction sealant specifications will specify a maximum weight loss (10–12%) after heat aging samples for 2 weeks at 158°F. This will assure that the plasticizers in the sealant will not evaporate during normal service aging. Satisfactory plasticizers include the less volatile phthalates, adipates, sebacates, phosphates, and other ester types. Most sealants, except for silicones, use plasticizers in their formulations. Silicone sealants can be plasticized only by low molecular weight silicone oils.[2]

12.3.5 Additives to improve adhesion

Often the base polymer in the sealant formulation will provide adequate adhesion to many substrates. However when necessary, adhesion properties can be improved by using additives such as coupling agents, primers, or a combination of the two. Phenol-formaldehyde resins and mercaptan terminated silane coupling agents are probably the most widely used adhesion additives for sealants.

Integral coupling agents may greatly improve the durability of the joint by providing additional resistance to the penetration of water (see Chapter 7). It is assumed that the coupling agent will migrate through the sealant and preferentially adsorb at the adherend interface. This can only be true if the coupling agent is not already adsorbed at the far greater competing interface that exists between the base resin and filler materials. Thus, it may be necessary to add the coupling material as a primer directly to the substrate rather than as an integral ingredient in the sealant formulation.

Silane coupling agents are used both as additives and as primers for sealant systems. They are organofunctional monomers that possess dual reactivity. This characteristic enables groups at one end of the molecule to react with active hydroxy sites on glass and metals and even masonry while the other end reacts with the organic base resin. This chemical bond is more lasting against the water and UV. Marsden[5] describes the numerous silanes available and provides recommendations for a wide range of applications. Silanes are used in most sealants including latex caulks. Only a very small amount of silane additive is necessary to achieve improved adhesion. Usually less than several percent by weight is sufficient. Commercial sealants, with silane additives to improve adhesion to a variety of substrates, are sometimes marketed as "siliconized" sealants.

Primers are needed when the standard sealant formulation does not adhere to a surface. Conventional primers for sealants either are a monomolecular film or film formers. The monomolecular film is generally a dilute solution of an appropriate silane at about 5% solids content in anhydrous solvent such as toluene or xylene. The primer is wiped on the surface and dries in minutes. The sealant should be applied within 15 min of primer application; otherwise, the primer will react with moisture in the air. Silane primers should always be clear water white. Any milky appearance indicates that the silane has lost its potency. These primers may be used on steels, various forms of anodized aluminum, baked finishes, fluorocarbon surfaces, and any unusual surface in question.

Film forming primers are applied by brush or spray and actually form a film across the surface, sealing most porosity. These film formers are especially recommended on masonry and other porous surfaces. Film forming primers are often combinations of phenolic resins, epoxy resins, and polyester resins with silanes. They are formulated with solvent to a final product that is approximately 25% solids content. These primers generally dry within 15 min.

12.3.6 Other additives

Many other additives, similar to adhesive system additives, are used for sealant formulations to provide special performance, application, or storage properties. For example, antibacterial agents and freeze-thaw cycle suppressants are commonly used in water-based sealant formulations as well as with adhesives. Many sealant additives are necessary because of the unique characteristics required of a sealant such as long term exposure outdoors or color matching. Sealant formulations may also include UV absorbers and ozone inhibitors to prevent any undesirable sensitivity to outdoor environments. The linear hydrocarbon rubbers are especially sensitive to sunlight and ozone and are generally used only for indoor applications.

12.4 Sealant Formulation

Minor formulation changes in sealants will often result in major changes in application and performance properties. Because of this and because of the variety of base polymers available, it is difficult to provide an organized guide to sealant formulation. A basic understanding of how some of the more important properties are controlled through formulation will help in determining a correct sealant choice.

Prior to formulating a sealant, the desired property profile needs to be defined. Table 12.9 presents a list of properties to be considered when defining the performance requirements of a sealant in both its cured and uncured states.

12.4.1 Application properties

The speed at which a sealant cures is critically important in some industries. Catalytic primers, two component systems, or controlled environments are used to increase the rate of cure. Depth of cure and the rate at which cure is achieved throughout the joint are other important considerations for a sealant. Single component sealants such as silicones, urethanes, and polysulfides, which depend on the diffusion of atmospheric moisture or oxygen into the sealant for curing, can take days or even weeks to cure entirely throughout the sealant bead. These sealants will develop a skin that will then inhibit the diffusion of water vapor to the center regions of the seal. In fact, some of these systems are limited to very small cross-sectional beads or else they will never fully cure throughout.

Shrinkage on curing can be a critical factor for sealant systems as well as adhesives. Excessive shrinkage can result in internal stresses and voids in the joint. Low shrinkage products are those with very

TABLE 12.9 Property Consideration of Sealants[6]

	Typical units	Applicable ASTM test methods
Uncured properties		
Skin over time	minutes	
Tack free time	hours	C-679
Sag, weep	inches	C-713
Extrusion rate	grams/minute	C-603
Flow	seconds	C-639
Cured properties		
Durometer hardness	Shore A	C-661
Tensile at maximum elongation	psi	C-908
Elongation at maximum tensile	percent	
Modulus	psi	
Tear	lbs/inch	
Peel	piw	C-794
Application Properties		
UV resistance		C-793
Ozone resistance		C-793
Staining/dirt pick-up		C-510
Color change		C-771
Application temperature range		
Service temperature range		
Adhesion		C-719
Durability		C-973, C-732
Compatibility		
Compression set		C-972
Solvent release		C-733
Life expectancy		
Cost		
Material		
Installation		

high or 100% solids such as many of the two component systems. Medium shrinkage systems include hot melt sealants that shrink as they go from a molten state to a gelatinous state. High shrinkage systems are solvent or aqueous base sealants in which shrinkage is due to evaporation of solvent or water.

For improved gap filling characteristics, some sealants are formulated to expand prior to curing. These sealants include automotive sealants with foaming agents in the formulation similar to the foaming agents used in packaging materials. Hot-melt formulations are also prepared with gases dissolved into the resin that will expand when heated. All of these formulations will expand before cure to fill in the joint gap (Fig. 12.1) and maintain positive pressure on the substrate as the sealant cures.

Foam-in-place gaskets promised to eliminate production bottlenecks in mass produced automobiles, appliances, electronics, and office

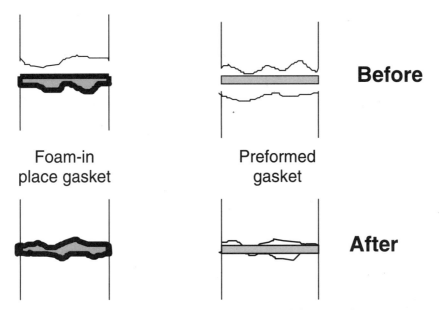

Figure 12.1 Foam-in-place gaskets provide complete fill of the joint volume.

equipment. These sealants are either hot-melt thermoplastic (e.g., ethylene vinyl acetate, butyl) or multicomponent curable thermosets (e.g., urethane). These materials can be applied directly to the joint, thus eliminating the need for die-cut gaskets and the resulting large quantities of scrap. Preformed solid gaskets conserve material but are labor intensive and require secondary adhesives to install. In addition, because it can be applied in three dimensions, foam-in-place gasketing provides greater flexibility in joint design.

The sealant must wet the substrate and easily flow into the joint; therefore, it must be a liquid or flowable during application. However, if it is to remain in a vertical joint it must behave as a solid. The former condition is met by using a liquid polymer or a solution of a solid polymer. In either case, the flow properties of the liquid must be such that flow is reasonable under a moderately applied stress, but ceases when the stress is small. This requirement is called thixotropy, flow control, or anti-slump. It is obtained in sealant formulations generally by fillers if the base polymer is a liquid or by solvents when the polymer is a solid.

Safety and toxicity effects are a concern when using certain types of sealants. Chemical reactions can cause the release of toxic fumes, heat, or possibly both. Solvent evaporation may contribute to a safety or health hazard in the working place. Metering and mixing of multiple component systems could bring personnel into direct contact with the materials resulting in dermatitis or other health related problems.

12.4.2 Performance properties

In some applications, strength may be more important than elasticity. Low strength or more precisely low tensile modulus may be the most important factor in a situation where a sealant joins one or more weak surfaces. Tensile strength is needed primarily to avoid cohesive failure under stress and not to transfer stress between substrates as the case with most adhesives.

Modulus sometimes can predict the extension or compression characteristics of a sealant. In general, low to medium modulus sealants are able to take significant movement without putting much stress on the sealant or the substrate materials. Some high performance sealants are formulated for a higher movement capability than a joint actually was designed to accommodate. The fact is that joints designed for about 25% extension/compression often must accommodate movement of 50% or more. Thus, higher performance sealants provide an added safety factor. A change in elasticity or hardness on aging may be an indication that further curing or degradation is taking place.

Compressive strength is the maximum compressive stress that a sealant can withstand without breaking down or experiencing excessive extrusion from the joint. Compression set is the inability of a sealant to return to its original dimension after being compressed. High compression set is usually caused by further curing or degradative cross-linking of the material while under compression. Compression set is undesirable in a joint that needs to expand and contract. Stress relaxation is a condition in which the stress decays as the strain remains constant. Some very low modulus sealants literally become pulled apart when held at low elongation.

Sealants may be exposed to scuffing and mechanical wear. Examples include the sealant used as an expansion joint in the highway, and the sealant used in preparing stone walkways. Thus, they must offer good abrasion, puncture, and tear resistance. Flexible sealants, which are available in either chemical curing or non-curing types, exhibit varying degrees of tear resistance. Urethanes have the highest tear resistance. Dynamic loads, shock, and rapid variations in stress also can cause seals to fail. Thus, tough, flexible elastomeric sealants that can stretch and then return to their original length in a short time are often selected for occilating mechanical loads.

Weatherability is the degree of resistance of a sealant to heat, moisture, cold, solar radiation, etc. The degree of weatherability is determined by the base polymer and the nature of the additives in the sealant formulation. Generally, sealants are formulated for maximum resistance to a single element such as moisture. Often, this chemistry will give it resistance to other elements as well. However, if a sealant's long term performance and weatherability are important to an application, contact the sealant manufacturers and ask for results of accelerated testing performed on specific substrates and under specific loading conditions. Such data are often available and can identify at least a family of sealants that may be candidates for the application. For more information on environmental resistance of sealants and adhesives, the reader is directed to Chapter 17.

In many situations, the appearance of the sealant is almost as important as its physical properties. Thus, most sealants are available in a variety of colors to match the environment in which they are used. Several questions must be considered when determining the appearance requirements of sealants:[7]

- Does a sealant cause discoloration of surrounding areas, initially or over a period of time?
- Does water runoff over the material cause unsightly residues?
- Does one product cause discoloration of another?
- Does the product itself change in appearance over time for any reason?

Sealants may also need to be compatible with a specific environment for certain applications. Examples of this may be a requirement for a sealant to have USDA or FDA acceptance because food or drugs are to be processed in the area near the sealant. It may be that in an installation such as a food processor or clean room, the sealant cannot outgas or liberate certain chemical components either during or after cure. Another end-use requirement could be that the sealant must meet certain fire resistance properties to meet code requirements in housing construction or in a particular area of use.

12.5 Commercial Products and Formulations

A wide range of materials and formulations are available for preparing sealant products that meet a specific need. Most sealants, however, are a compromise formulation. That is, manufacturers formulate their products to meet a variety of applications. These products, then, meet a majority of needs, while the final choice rests with the informed user. Modern demands for increased productivity, higher performance, and environmentally safer materials have led to a decline in the use of some of the older chemistries, particularly solvent-based sealants. Over the past several years, polyurethanes have become one of the mainstays for sealing of building joints. Silicones have had success in structural glazing applications due to their favorable performance in weathering, particularly UV resistance. During this time, polysulfides have been losing ground. Most sealants have the capability of meeting a variety of standards relative to joint movement.

It is important to seek the advice and help of knowledgeable people in all areas of sealant selection and formulation. Of equal importance is the timing of the questions. Determine property requirements early in the project so that tests can be run, development tasks scheduled in the proper sequence, and design changes can be made to assure that the optimal sealant is chosen or formulated for the specific application.

References

1. Amstock, J. S., "Sealants", Chapter 7 in *Handbook of Adhesive Bonding*, C. V. Cagle ed., (New York: McGraw-Hill, 1973).
2. Paneck, J. R., and Cook, J. P., "Sealant Classifications", Chapter 2 in *Construction Sealants and Adhesives,* 3rd ed. (New York: John Wiley & Sons, 1992).
3. Usmani, A. M., "Environmental Considerations Unique to Sealants", *Adhesives and Sealants,* vol. 3, Engineered Materials Handbook, ASM International, 1990.
4. Wake, W. C., "Adhesion and the Formulation of Adhesives", Chapter 12 in *Sealants* (London: Applied Science Publishers Limited, 1976).
5. Marsden, J. G., "Organofunctional Silane Coupling Agents", in *Handbook of Adhesives,* 3rd ed., I. Skiest ed., (New York: Van Nostrand Reinhold, 1990).
6. Elias, M., "Silicone Sealant Technology Markets Continue to Grow", *Adhesives Age,* May 31, 1986 at 9.
7. Brower, J., "Sealant Compatibility Issues Include Appearance and Chemical Effects", *Adhesives Age,* August 1996.

Chapter 13

Sealant Families

13.1 Introduction

The various generic families of sealant materials are reviewed in this chapter. Properties and performance characteristics are presented although they can vary considerably within each family. There are over 15 families of polymers that are used singly or in blends to achieve the storage characteristics, application properties, physical performance, and durability required for each application. Table 13.1 shows some of the strengths and weaknesses of the most popular synthetic based sealant families.

The following sections describe the characteristics of many of the sealants commonly in use today. The various sealant families are classified in this chapter primarily by their movement capability:

- **Low performance sealants**—typically 0–5% movement capability; 2–10 years service life; low cost
  - Oil- and resin-based sealants
  - Asphaltic and bituminous mastics
  - Polyvinyl acetate
  - Epoxy
  - Polyvinyl chloride plastisol
- **Medium performance sealants**—5–12% movement capability; 5–15 years service life; medium cost
  - Hydrocarbon rubber-based sealants
  - Acrylic
  - Chlorosulfonated polyethylene
  - Hot-melt sealants based on synthetic rubber
- **High performance sealants**—greater than 12% movement capability; 10–50 years service life; high cost

TABLE 13.1 Characteristics of Popular Sealant Families[1]

Chemical family and curing type	Strengths	Weaknesses
Silicones, one component (moisture-initiated condensation) and two components (condensation or addition)	Best weathering, highest flexibility, good adhesion, heat resistance	High MVTR, low depth of cure, slow curing[b]
Polyurethanes, one component (moisture-initiated condensation) and two components (condensation)	Good weathering, best adhesion, high flexibility	Weak UV resistance, weak heat resistance
Polysulfides, one component (moisture-initiated condensation) and two components (condensation)	Low MVTR[a], fuel resistant, good flexibility	Slow curing[b], low depth of cure
Acrylic latex (water evaporation)	Easy to use	High shrinkage, poor weathering, fair flexibility
Butyls (sulfur vulcanization)	Lowest MVTR, good flexibility	Fair weathering
Anaerobics (metal/peroxide initiated free radical)	Fast curing, chemical resistant, heat resistant	Brittleness, poor gap filling
Vinyl plastisols (heat fusion)	Good adhesion, low cost	Fair flexibility, fair weathering
Asphalts/coal tar resins (cooling oxidation)	Low cost, fuel resistant	Poor weathering
Polypropylene hot melts (cooling)	Low cost, expandable	Limited adhesion, fair flexibility

[a]MVTR, moisture vapor transmission rate. [b]Two-component version cures faster.

- Fluorosilicones and fluoropolymers
- Polysulfides
- Polyethers
- Polyurethane
- Silicone
- Styrene butadiene copolymer
- Neoprene

Table 13.2 and 13.3 summarizes the properties of several common sealants in each of these categories.

13.2 Low Performance Sealants

Sealants with movement capabilities of less than 5% include systems based on oils, resins, asphalt and bitumen products, and polyvinyl

TABLE 13.2 Summary of Sealant Properties for Common Low and Medium Performance Sealants[2]

Characteristic	Low performance	Medium performance		
	Oil base, one part	Latex (acrylic), one part	Butyl skimming, one part	Acrylic, solvent-release
Maximum recommended joint movement, percent of joint width	±3	±5, ±12.5	±7.5	±10 to 12.5
Life expectancy, years(a)	2–10	2–10	5–15	5–20
Service temperature range, °C (°F)	−29 to 66 (−20 to 150)	−20 to 82 (−20 to 180)	−40 to 82 (−40 to 180)	−29 to 82 (−20 to 180)
Recommended application temperature range, °C (°F)(b)	4–50 (40–120)	4–50 (40–120)	4–50 (40–120)	4–80 (40–180)
Cure time to a tack-free condition, h(c)	6	0.5–1	24	36
Cure time to specified performance, days(c)	Continues	5	Continues	14
Shrinkage, %	5	20	20	10–15
Hardness, new (1–6 mo), at 25°C (75°F), Shore A scale	...	15–40	10–30	10–25
Hardness, old (5 years) at 25°C (75°F), Shore A scale	...	30–45	30–50	30–55
Resistance to extension at low temperature	Low to moderate	Moderate to high	Moderate to high	High

TABLE 13.2 Summary of Sealant Properties for Common Low and Medium Performance Sealants[2] (*Continued*)

	Type of sealant			
	Low performance		Medium performance	
Characteristic	Oil base, one part	Latex (acrylic), one part	Butyl skinning, one part	Acrylic, solvent-release
Primer required for sealant bond to				
Masonry	No	Yes	No	No
Metal	No	Sometimes	No	No
Glass	No	No	No	No
Applicable specifications				
United States	TT-C-00593b	ASTM C 834, TT-S-00230	TT-S-001657	TT-S-00230
Canada	CAN/CGSB 19.6-N87		CGSB 19-GP-14M	CGSB 19-GP-5M

NOTE: Data from manufacture data sheets; U.S. made sealants are generally considered. (a) Affected by conditions of exposure. (b) Some sealants may require heating in low-temperature environments. (c) Affected by temperature and humidity. Source: Adapted from Ref 1

TABLE 13.3 Summary of Sealant Properties for Common High Performance Sealants[2]

Characteristic	Type of sealant					
	High performance					
	Polysulfide		Urethane		Silicone	
	One part	Two part	One part	Two part	One part	Two part
Maximum recommended joint movement, percent of joint width	±25	±25	±25	±25	±25 to +100/−50	±12.5 to ±50
Life expectancy, years(a)	10–20	10–20	10–20	10–20	10–50	10–50
Service temperature range, °C (°F)	−40 to 82 (−40 to 180)	−51 to 82 (−60 to 180)	−40 to 82 (−40 to 180)	−32 to 82 (−25 to 180)	−54 to 200 (−65 to 400)	−54 to 200 (−65 to 400)
Recommended application temperature range, °C (°F)(b)	4–50 (40–120)	4–50 (40–120)	4–50 (40–120)	4–80 (40–180)	−30 to 70 (−20 to 160)	−30 to 70 (−20 to 160)
Cure time to a tack-free condition, h(c)	24(c)	36–48(c)	12–36	24	1–3	0.5–2
Cure time to specified performance, days (c)	30–45	7	8–21	3–5	5–14	0.25–3
Shrinkage, %	8–12	0–10	0–5	0–5	0–5	0–5
Hardness, old (5 years) at 25°C (75°F), Shore A scale	30–55	20–55	30–55	20–60	15–40	15–50
Resistance to extension at low temperature	Low to high	Low to moderate	Low to high	Low to high	Low	Low

TABLE 13.3 Summary of Sealant Properties for Common High Performance Sealants[2] *(Continued)*

Characteristic	Type of sealant					
	Polysulfide		High performance			
			Urethane		Silicone	
	One part	Two part	One part	Two part	One part	Two part
Primer required for sealant bond to						
Masonry	Yes	Yes	Yes	Yes	No	Yes
Metal	Yes	Yes	No	No	...	No
Glass	No	No	No	No	No	No
Applicable specifications						
United States	ASTM C-920, TT-C-00230C	ASTM C 920, TT-S-00227E	ASTM C 920, TT-S-00230C	ASTM C 920, TT-S-00227C	ASTM C 920, TT-S-00230C TT-S-001543A	ASTM C 920, TT-S-001543A
Canada	CAN 2-19.13M		CAN 2-19.13M		CAN/CGSB 19.18	TT-S-00227E

NOTE: Data from manufacture data sheets; U.S. made sealants are generally considered. (a) Affected by conditions of exposure. (b) Some sealants may require heating in low-temperature environments. (c) Affected by temperature and humidity. Source: Adapted from Ref 1 acetate. They are noted chiefly for their very low cost. Some of these sealants have been in general use since the early 1900s. Low performance sealants function mainly as crack fillers to prevent ingress of dust, rain, and wind in substantially static conditions. However, even in this application excessive shrinkage, hardening on age, and poor adhesion severely limit the service. Compatibility issues and chemical effects must usually be considered with low performance sealants.[3] Low performance sealants see limited use in mainly the consumer markets.

13.2.1 Oil- and resin based sealants

Oil-based caulks are made from various unsaturated oils, including linseed and vegetable oil. They have been used for centuries as glazing sealants and putties. These sealants are 100% solids. The main cure mechanism is oxidation causing crosslinking of the oil in the presence of a catalyst. Typical components in an oil-based sealant formulation are linseed or soy oil, fibrous fillers, calcium carbonate filler, pigment, gelling agents, and catalyst.

Oil-based sealants with movement capability of only $\pm 2\%$ are more rigid than other sealants. Oil-based sealants continue to cure via oxidation while they are in service. After about two years they lose their pliability and can be considered hard and brittle. Shrinkage of the sealant in the joint also continues on aging. The service life is only considered to be several years depending on the application. They are generally not used in exterior applications where joint movement will occur. A hand applied linseed oil-based sealant is used as a "putty" for window sealing. Most oil-based sealants can be applied with an extrusion gun.

The rigidity associated with oil-based sealants is often overcome by the addition of elastomeric resins such as polyisobutylene. These more elastic formulations are called resin based sealants or caulks. They are limited in movement to $\pm 5\%$, and their service life is longer than oil-based sealants. A typical formulation for a resin based caulk is shown below.[4]

Composition	Parts by weight
Bodied vegetable oil	100
Kettle-bodied vegetable oil	70
Polyisobutylene	100
Cobalt carboxylate	0.20
Calcium carbonate	483
Thixotropic agent	as required for thixotropy
Titanium dioxide	17

13.2.2 Asphaltic and other bituminous mastics

Mastics are formulated from bituminous materials obtained as by-products of petroleum refining. These bitumens are hydrocarbon materials containing various polymeric materials of high molecular weight and complexity. They are usually solids or semi-solids that must be applied hot, or else the sealant is formulated from a solvent or water-based emulsion of the asphaltic product.

Because of their hydrocarbon nature, these bituminous materials have a low surface tension and wet most surfaces moderately well. However, bitumens can be displaced from the substrate by water, or they can be prevented from wetting a substrate by a film of moisture. To overcome this problem, a surfactant is usually added to the formulation, and the substrate must be completely dry before application of the sealant.

Bitumen based sealants can become brittle at low temperatures. This fault is overcome by the addition of rubber (e.g., scrap neoprene) to the formulation to act as a fexibilizing agent. Drying oils (about 5% by weight) are added to the formulations to prevent oxidation and resulting embrittlement and surface crazing. A tendency to cold flow is reduced by the incorporation of epoxy resins.

A major application for these sealants is in highway joints. Typically the highway sealant will contain 50–60% asphalt, 20–30% ground rubber (usually from scrap automobile tires), and 20% cyclic hydrocarbons. This sealant is applied at 250–400°F. For applications where fuel resistance is important, such as airfield joints, nitrile rubber is generally added to the asphaltic formulation. Asphaltic-based mastics are also used for automotive joints, construction sealing, pipe seals, and marine joints.

13.2.3 Polyvinyl acetate

Polyvinyl acetate sealants have the same base latex resin that is used for wood glues. These sealants are a low performance, inexpensive consumer product that dries to a hard, rigid seal. A typical sealant formulation consists of polyvinyl acetate latex, filler, plasticizer (for improved flexibility), and water. Since these sealants set by evaporation of water, their shrinkage is greater than that of 100% solids sealants.

Polyvinyl acetate-based systems are generally used for interior applications such as, spackling compounds, sealing wall framing, or bath tub caulk. Their advantages include ease of application, water clean-up, paintability, and good adhesion to most substrates. Disadvantages are low movement capability, high modulus, and shrinkage on curing. Polyvinyl acetate sealants have a modulus that is very temperature sensitive. Thus, these materials are rather weak and pliable at elevated temperatures, and hard and brittle at lower temperatures.

Latex caulks based on polyvinyl acetate plasticized with dibutyl phthalate or based on vinyl acetate-acrylic emulsion copolymer are noted for their ease of application. However, they are less flexible, harden on aging and have inferior exterior durability when compared to the acrylic emulsion caulks (Section 13.3.2).

13.2.4 Epoxy

Epoxy resins usually are considered to be brittle systems and not applicable to many sealant applications. However, the use of reactive flexibilizers, such as polyamide and polysulfide, can make epoxy sealants useful as bridge sealants, masonry adhesives, and monolithic heavy duty coatings. The compositions are resistant to many nonoxidizing acids, salts, and alkalines. A typical formulation for a corrosion resistant epoxy sealant is:[5]

Composition	Parts by weight
Liquid epoxy resin	100
Liquid polysulfide	75
Dimethyl aminopropylamine	10

Most epoxy sealant or caulk systems are limited to ±5% movement. The main advantage of an epoxy sealant is its high degree of adhesion. They are also easy to handle, have low shrinkage being 100% solids systems, and have good overall environmental and chemical resistance. These sealants are used in the construction, electrical, and transportation industries. They find diverse applications such as crack repair for concrete, machinery grouting material, and automotive body solder.

13.2.5 Polyvinyl chloride plastisol

Polyvinyl chloride (PVC) plastisols have been used widely as gap fillers and sealants in automobiles. These plastisols are dispersions of finely divided PVC resin in liquid plasticizers such as dioctyl phthalate. These liquid sealants set when heated at 250–400°F to form flexible compositions. Filled plastisols or plastigels may be used like putty.

13.3 Medium Performance Sealants

Medium performance sealants include solvent based and emulsion based acrylic systems, higher quality butyl and polyisobutylene sealants, neoprene, and styrene butadiene rubber (SBR) based compounds. They have a movement capability of approximately 5–12% and will accept low levels of thermally induced contraction/expansion. Medium performance sealants have a longer service life than the low performance sealants. Life of 2–5 years on most common substrates is possible if they are used within their movement limitations. The price is higher than the low performance grade sealants and much lower than the high performance grades.

These intermediate type systems generally cure by solvent evaporation and oxidation or a mixture of the two processes. As a result, they have 10–30% shrinkage after installation. Medium performance sealants are formulated with non-volatile extenders and fillers to minimize shrinkage and silane additives to enhance adhesion. The silane additive formulations are termed "siliconized". Siliconized butyl and siliconized acrylic systems are conventionally found in hardware stores for the consumer home repair markets.

13.3.1 Hydrocarbon rubber-based sealants

Hydrocarbon rubber-based sealants are formulated from polybutenes, butyl rubber, and polyisobutlyene. They are characterized by high elongation, but they are susceptible to atmospheric oxidation and attack by ozone. These systems are available as solvent release sealants, deformable preformed tapes, and hot applied sealants.

Polybutenes are used for the formulating of non-drying mastic sealants as well as rope caulks and extruded tapes. They are also used as plasticizing agents for oleoresinous caulks and certain elastomeric sealants. A typical dry rope caulk formulation is as follows:[4]

Composition	Parts by weight
Polybutene (1300–1500 MW)	27.4
Petrolatum	3.2
Tall oil fatty acid	0.6
Thixotropic agent	As required
Calcium carbonate	32.6
Titanium dioxide	1.8

Butyl elastomers are copolymers of isobutylene with sufficient isoprene to permit crosslinking by agents such as p-quinone dioxime and an oxidizer. Commercial sealants contain fillers such as carbon black, zinc oxide, or silica; tackifiers, such as pentaerythirotol esters of rosin; and solvents, such as cyclohexane. A typical butyl rubber caulking compound has the following composition:[4]

Composition	Parts by weight
Butyl rubber	175
Mineral spirits	270
Petroleum resins	34
Pentaerythritol esters of rosin	8
Bentone clay derivative	23
Finely divided silica	364
Fiber	91
Titanium dioxide	45

ASTM C 1085 is a standard specification for a butyl rubber-based solvent release sealant with a maximum 10% joint movement capability. Solvent release butyl sealants have a service life of 5–15 years depending on the application. Butyl sealants are known for excellent adhesion to most substrates. However, they can stain adjacent surfaces and have relatively poor recovery from extension.

Butyl rubber compositions are available as one and two component solvent release sealants, preformed tapes, and hot melt extruded sealants. One component butyl caulks are superior to oleoresinous caulks for outdoor exposure, and they are impermeable to gases and vapors. They do not harden or oxidize. They can be used in joints with movement up to 10–15%. Two component, chemically curing butyl systems are crosslinked with p-quinine dioxime activated with a peroxide. They have lower viscosity and are used in applications such as insulated glass and electrical components. Curing butyl sealants have generally poor adhesion and cohesion, but they can perform with ±10% movement.

Preformed butyl tapes are 100% solids so that there is no shrinkage on cure. Butyl tapes can also be applied as a hot melt at temperatures between 300–400°F with an appropriate heated applicator. The compound is heated and extruded into a joint opening to form a smooth joint that cools rapidly with excellent adhesion, even to oily and otherwise poorly prepared surfaces. Preformed butyl tapes are used extensively in the glazing industry between glass and aluminum. These tapes have 20–30% compression and a 10-year service life.

Polyisobutlyene as a homopolymer is permanently tacky. It has the characteristic of being self-healing. Polyisobutylene is a glass-clear polymer with elastomeric properties although showing much greater cold flow than most unvulcanized rubbers. Most of its use is as an additive for other sealants or tapes, primarily of the butyl type.

Polyisobutylene sealants are available as non-curing, solvent release systems and as preformed tape systems.

13.3.2 Acrylic

Emulsions and solutions of polymethyl methacrylate or its copolymers are used as caulking materials and sealants. They have moderate movement capability and excellent weatherability. Acrylics tend to have a strong odor and are usually applied outdoors.

As the solvent types gradually lose solvent, their hardness increases and their movement capability changes from ±12.5% to about ±7.5%. Recovery is poor, as the polymer is not an elastomer. The acrylic solvent sealant has excellent adhesion to common construction surfaces without primers. It is non-staining, and has excellent durability and color performance. Resistance to weathering and UV is excellent. Only the relatively expensive silicone sealants have better exterior durability.

Acrylic emulsions have approximately ±10% movement capability. They are water sensitive and will lose adhesion outdoors. However, they have low shrinkage, excellent flexibility, and fair recovery. Acrylic emulsion caulks are used indoors as bathtub caulks; for baseboard and trim seams; and outdoors for glazing, masonry joints, and roof and siding joints. Although most latex caulks have a movement capability of 5–12.5%, plasticized acrylic latex sealants have ±25% movement and excellent adhesion to a variety of substrates.[6] Primers are generally not necessary with acrylic sealants. They are relatively water resistant. They have found application in the home repair market for both indoor and outdoor projects. Acrylic emulsion sealants are considered to have better movement capability than polyvinyl acetate sealants that are sold for the same applications. These systems clean easily with water and can be painted over. The toxicity and odor are not as much a concern as with the acrylic solvent release sealants. As with most water-based sealants, one disadvantage is that the sealant must be applied at application temperatures greater than freezing. Acrylic sealants are subject to ASTM C 834, "Standard Specification for Latex Sealing Compounds".

Two-part acrylic sealant systems are also available that contain acrylic monomers, fillers, and initiators. They are 100% solids and polymerize at room temperature. They have been used as polymer concrete and for repairing potholes and cracks in concrete highways. Although these systems are solvent free, precautions need to be taken to avoid inhaling the strong acrylic monomer vapors.

13.3.3 Chlorosulfonated polyethylene (Hypalon)

Hypalon is the DuPont tradename for chlorosulfonated polyethylene resin, and it is used as the base for a one part sealant. A typical formulation for a sealant of this type is as follows:[5]

Composition	Parts by weight
Blend of chlorosulfonated polyethylene	17.5
Chlorinated paraffins	17.5
Thixotropic fillers	20
Titanium dioxide	14
Curing agents, e.g., litharge	7.5
Dibutyl phthalate plasticizer	19
Solvents, e.g., isopropyl alcohol	4.5

These compositions are cured by cross-linking at ambient temperatures to produce flexible sealants with good adhesion.

Hypalon-based sealants have excellent weathering properties, good resistance to ozone, oxygen, and ultraviolet light, and excellent resistance to strong acids. These systems generally have limited shelf stability. They also require extended cure time to reach the ultimate elastomeric properties. Hypalon sealants are used in exterior joints where movement is moderate. Construction applications include doors, windows, and roofing.

13.3.4 Hot-melt sealants

In addition to butyl rubber and bitumens, several other polymers are used as hot-melt sealants. Among these are the copolymers of ethylene and vinyl acetate (EVA), atactic polypropylene, and mixtures of paraffin wax and polyolefins. Being 100% solids, there is very little shrinkage.

The set time of hot-melt sealants is much shorter than the conventional sealant system. Thus, hot-melt sealants are finding their way into high volume, production operations such as car manufacture. The introduction of UV stable styrenic thermoplastic elastomers, such as styrene-ethylene/butylene-styrene (SEBS), and thermally stable tackifying resins have accelerated the development and commercialization of hot-melt sealants.

13.4 High Performance Sealants

High performance sealants have movement capability of greater than 12% and generally excellent resistance to environmental extremes.

They are often used in exterior, high movement applications. Sealants of this type generally cure by the chemical crosslinking of elastomeric-base resins. They are available as either single part or two part systems. The single part systems generally require atmospheric moisture to cause cure either by inducing oxidation or condensation of a reactive end group.

High performance sealant families include fluoropolymers, fluorocarbons, polysulfides, urethanes, and silicones. By far the polysulfides, polyurethanes, and silicones make up the bulk of the high performance sealant market. Table 13.4 compares properties of these three sealants. High performance sealants have a higher price than other sealants and due to their chemically reactive nature tend to be more difficult to apply.

13.4.1 Fluorosilicone and fluoropolymer sealants

Fluorosilicone polymers are intermediate in cost and performance between the fluoropolymers and silicones. They have unusually high resistance to chemicals and maintain elastomeric properties over a very wide temperature range (-80 to $+450°F$). Fluorosilicone sealants are available as single component sealants or as two component systems that are peroxide cured. They are used as sealants for aircraft, automobiles, and oil well heads in cold climates.

Fluoropolymers are used in small volume, high temperature and fuel resistant applications. They are also available in single- and two-part systems. Cured fluoroelastomers can withstand prolonged expo-

TABLE 13.4 Comparison of Properties of Polysulfide, Silicone, and Urethane Sealants[7]

Property	Polysulfide	Silicone	Urethane
Recovery from stress	−	++	++
Ultraviolet resistance	−	++	+
Cure rate[a]	−	++	− to ++
Cure rate, two component[b]	+		++
Cure rate, latent hardener	NA	NA	++
Low temperature gunnability	−	++	−
Tear resistance	−	−	++
Cost	−	−	++
Paintability	++	− −	++
Available in colors	+	−	++
Unprimed adhesion to concrete	−	−	++
Resistance to hydrolysis			
Non-bubbling	++	++	−
Self levelling available[c]	++	−	++

[a]One component sealant.
[b]Silicone is available only as one component
[c]Desirable for plaza decks and pavements.

sure to high temperatures and retain their elastomeric, mechanical, chemical, and electrical properties better than other elastomers. Their temperature rating is 450°F for continuous service and 600°F for intermittent exposure. Although resistance to compression set is excellent at upper temperatures, low temperature properties are poor compared to the silicones and fluorosilicones.

13.4.2 Polysulfides

Polysulfide was one of the first synthetic elastomers. They entered the construction market in the 1950s as the first high performance sealant. Polysulfide sealants are 100% solids systems that are available as one-part and two-part formulations. Compounded products can vary significantly in processing and performance. Table 13.5 lists five types of polysulfide sealant formulations that are useful in several end-applications.

Mercaptan terminated liquid polysulfide polymers are polymerized by being oxidized by metallic oxides such as lead dioxide or activated manganese dioxide or by epoxy resins in-situ. They cure at ambient temperatures to produce flexible caulking compounds. Additives are generally fillers (carbon black, calcium carbonate, or clay), plasticizers (dibutyl phthalate), retarders (stearic acid) or accelerators (amino compounds), and adhesion promoters (epoxy resins). Polysulfide sealants are commercially designated as LP-2, LP-32, LP-3, LP-33, etc. for different average molecular weights and crosslink densities. (LP designates a liquid polymer.)

Silane adhesion promoters are often utilized in polysulfide sealant formulations. A thin layer of silane as a primer provides excellent adhesion of polysulfide to metals, glass, and ceramic substrates. A film forming primer is required for porous surfaces. Masonry primers generally contain a chlorinated rubber or phenolic resin alone or in combination with a plasticizer.

Two component polysulfide sealants cure to 80-90% of their final properties in 16–24 hrs at room temperature. Cure rate can be accelerated by elevated temperatures. Metering and mixing of the compo-

TABLE 13.5 Polysulfide Sealant Formulations for Various Applications[8]

	Sealant				
Ingredients	One-part	Building	Insulating glass	Aircraft	Casting compound
Polysulfide polymer	20	35	30	65	35
Fillers	50	40	50	25	35
Plasticizers	25	20	15	5	27
Adhesion additives	2	2	2	2	—
Curing agents	3	3	3	3	3

nents is critical and must be carefully controlled at the job site. Surfaces must be very clean to achieve the best adhesion.

Single component polysulfides are an advantage in that no metering or mixing is required. The formulation is similar to two part systems except for the curing agent, an alkali oxide, which reacts with moisture to initiate the polymer chain extension and crosslinking. Single component polysulfides skin rapidly at high humidities. Cure is slow reaching only about 50% of the ultimate properties in 7 days depending on the relative humidity.

Polysulfide sealants have a movement capability of ±25%. Hardness can be varied from that of soft rubber (Shore A of 20) to hard rubbers (Shore A of >50) for applications that must withstand penetration. Properly formulated, polysulfide sealants can have a service life of 20 years or more. Most of the polysulfide sealants have excellent resistance to solvents and fuels along with good weatherability.

Polysulfide sealants have been widely used in such applications as potting and molding of electrical connectors, glazing of windows, and sealing of windshields of automobiles. They have also been used with asphalt or coal tar as sealants for roadways, bridges, and airport runways.

Table 13.6 lists some of the advantages and disadvantages with polysulfide sealants over other common sealants. The principle advantages of polysulfide sealants are ease of application, good adhesion, good resistance to weathering and solvents, negligible shrinkage, and good moisture vapor transmission.[9]

13.4.3 Polyethers

Silyl terminated polyether sealants, known as MS sealants, have become the most used type of construction sealant in Japan since their introduction some 20 years ago. In the U.S., MS sealants have been primarily used for industrial application; construction use is still in the developmental stages.[10,11] The MS sealants have high performance capabilities with many of the same characteristics as a urethane sealant. However, the one-part MS sealants cure much faster than a one-part urethane sealant and find use in the construction industry where property development speed is important.

One component and two component MS sealant formulations are available. The MS polymer is silyl terminated polyether prepared from high molecular weight polypropylene oxide. It is end capped with allyl groups, followed by hydrosilylation to produce a polyether end capped with methyldimethoxysilane groups. In the presence of appropriate catalyst, the methoxysilane group can be cured by moisture. The water

TABLE 13.6 Advantages and Disadvantages of Polysulfide Sealants[9]

Advantages:
- Ease of application, both one and two part compounds
- Excellent adhesion retention after ultraviolet and water exposure
- Excellent resistance to ozone and weathering
- Outstanding resistance to fuels
- Good color retention
- Negligible shrinkage
- Good moisture vapor transmission
- Excellent adhesion to a variety of substrates
- Excellent rheological properties

Disadvantages:
- Poor puncture resistance
- Generally not recommended for pedestrian traffic because of puncture and abrasion resistance
- One part systems provide a slow cure
- High compression set
- Difficult to handle below 40°F
- Increase in Shore durometer on aging

reacts with the group to liberate methanol and produce a silanol. Further reaction of the silanol with either another silanol or methoxysilane produces siloxane linkages.

The raise in MS sealant popularity has been due to its versatility and well balanced properties. MS sealants are suitable for a wide range of applications, except for glazing. The initial adhesion of MS sealants to glass is good, but its exposure to sunlight or UV over a long period of time causes deterioration. MS sealants, like polyurethane sealants, should not be used at temperatures greater than 180°F. Table 13.7 compares the performance of silicone rubber, polyurethane, and MS sealants. Compared with these other sealants, MS sealants have well balanced properties and performance. Some of the unique properties of MS sealants are:[12]

- Environmental friendliness: MS sealants are solvent-free
- Low temperature gunnability: the viscosity of MS sealants is less than others depending on temperature
- Storage stability: shelf life is excellent although MS sealants must be protected from moisture

TABLE 13.7 Construction Sealant Performance Comparison[12]

Property	Silyl terminated polyether MS sealants	Polyurethane sealants	Silicone rubber sealants
Environmental friendliness	10	5	9
Non-bubbling	10	6	10
Low temperature gunnability	10	8	10
Slump resistance	10	10	10
Quick cure	10	7	10
Storage stability	10	7	9
Body (tooling)	8	10	8
Weather resistance	8	6	10
Adhesion to various substrates	10	5	8
Mechanical properties	10	10	10
Heat resistance, mechanical stability (cure time/temperature)	9	8	10
Non-dirt pickup	10	10	5
Stain resistance	8	8	5
Paintability with water based paint	10	10	3

Rating: 10 = high; 1 = low.

- Weather resistance and durability: MS sealants show no cracking, splitting, discoloration, or adhesion failure after 7 years testing in an Arizona climate
- Stain resistance: MS sealants do not stain as some silicone sealants do because of low molecular weight silicon materials that bleed from the surface of sealed joints
- Paintability: MS sealants provide good paintability unlike silicone sealants
- Adhesion: MS sealants provide adhesion to various substrates including metals, plastics, wood, and ceramics

13.4.4 Polyurethane

Polyurethane sealants are usually two part systems, but one part moisture curing systems are also available. Raw materials that are used in formulating polyurethanes include isocyanates, polyols, chain extenders, catalysts, and solvents and additives. The two part systems are based on diisocyanate. Polyurethene sealants also contain antioxidants, pigments, fillers, cross-linking agents, and organometallic catalysts.

The two-part polyurethane sealants have a faster curing rate than the single component varieties. Two-part systems are generally preferred for outdoor application where property development must be fast. Working life once mixed is 1–4 hrs, and properties are nearly fully developed in 24–48 hrs. Component mixing in the field must be very thorough. Since polyurethanes cure by reacting with moisture, single component sealants have joint design restrictions and delayed cure characteristics due to the skin effect. Reaction mechanisms of the one and two component polyurethane sealants are similar to those described for polyurethane adhesive formulations in Chapter 10.

Early polyurethane formulations suffered from high modulus, and they had a higher cohesive strength than adhesive strength resulting in adhesive failure in many applications. Modern polyurethane sealants have lower modulus and greater movement capabilities. There are a broad variety of formulations available with differing performance property profiles. Polyurethane sealants are characterized by a wide range of hardness and flexibility. They have good sag resistance, good thixotropy, and good adhesion to various substrates. Urethane sealants have excellent elongation and recovery. Tear and abrasion resistance are outstanding, making polyurethane sealants the choice for floor joints and highway applications. Low temperature performance is better than that of the polysulfides. However, UV exposure will cause the sealant to age, crack, and chalk. Urethane sealants are not recommended for continuous temperatures over 180°F. Typical characteristics of urethane sealants are presented in Table 13.8.

Polyurethane sealants are normally formulated for single part, gun-grade application. They form waterproof seals on concrete, masonry, glass, plastic, wood, aluminum, and most other metals. Properly formulated polyurethane sealants can have a service life of 10–20 years, equal to the best sealants. These sealants have movement capabilities of $\pm 25\%$ and $\pm 50\%$. Polyurethanes form tough, flexible, abrasion and cold flow resistant sealants for use over a temperature range of $-127°F$ to $+400°F$.

Polyurethane sealants are often used for tasks such as installing skylight and windows; mounting fixtures; sealing gutters, ductwork and pipe flanges; and forming expansion joints. When compared to RTV silicone sealants, urethanes offer greater adhesion but lower thermal resistance. However, urethanes have very good cold temperature properties. Urethanes can be painted and generally offer better chemical resistance and twice the elongation of RTV silicone sealants. Table 13.9 lists the advantages and disadvantages of polyurethane sealants.

TABLE 13.8 Typical Characteristics of Urethane Sealant[13]

Characteristic	Value or comment
Tensile strength, MPa (ksi)	1.4–5.5 (0.2–0.8)
Elongation (ultimate), %	200–1200
Hardness, Shore A	25–60
Weatherometer rating	No elastomeric property change after 3000 h
Low-temperature flexibility, °C (°F)	−40 (−40)
Service temperature range, °C (°F)	−40 to 80 (−40 to 180)
Expected life, years	≤20
Dynamic movement, %	±25
Adhesion peel strength, kN/m (lbf/in.)	3.5–8.8 (20–50)
Cure time, days	1–7
Durability	Excellent
Shrinkage, %	0–2
Gun dispensability at 5°C (40°F)	Excellent
Water immersion	Yes(a)
Sag resistance	Excellent
Objectionable odor	No
Primer required	No(a)
Water permeability, [kg/Pa · s · in.] × 10^{-12} (perm · in., 23°C)	0.15–1.2 (0.1–0.8)

(a) Primer is required for water immersion conditions.

Silyated urethane sealants have been developed to extend the property range of conventional sealant formulations. Silane endcapped urethane polymers undergo crosslinking reactions in the presence of catalyst and moisture to form a stable siloxane linked network. This mechanism is an alternative to existing urethane technology. It allows the formulator to produce fast room temperature cure construction sealants of low to medium modulus that are free of residual isocyanate. These sealants exhibit generally superior physical properties to conventional urethane sealants. Silyated polyurethane sealants have excellent adhesion to conventional construction substrates and good durability.[14,15] These sealants have excellent adhesion characteristics and bond well to plastic surfaces including PVC, ABS, polystyrene, and acrylic. Figure 13.1 shows that silyated polyurethane provides superior adhesive peel strength on all but acrylic substrates.

One component urethane foam sealants have been developed to fill and seal gaps in cars, trucks, and trailers; as vibration dampening mounts for electrical equipment; as well as for innumerable applica-

TABLE 13.9 Advantages and Disadvantages of Polyurethane Sealants[7]

Advantages	Disadvantages
Excellent recovery	Light colors discolor
Excellent UV resistance	Some require priming
Fast cure for multicomponent	Relatively slow cure for one component sealant
Fast cure for latent hardeners	
Negligible shrinkage	
Excellent tear resistance	
Excellent chemical resistance	
Meets ASTM C920	
±40% Movement capability	
Unprimed adhesion to concrete	
Paintable: available in colors	

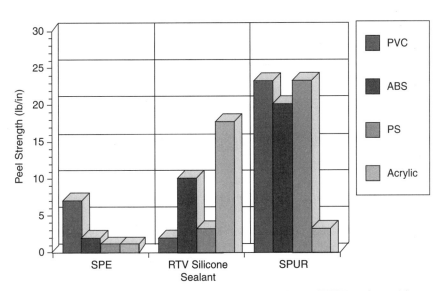

Figure 13.1 Performance comparison of silyated polyurethane (SPUR) sealant with one part conventional polyurethane (SPE) and RTV silicone sealants.[16]

tions in building construction.[17] Once the material reaches a surface it expands to two to three times its original volume, filling nearby space as it sets to a flexible or semi-rigid form. The foam becomes fully cured with exposure to moisture in the air.

13.4.5 Silicone

Silicone-based sealants have been formulated as single and two-part systems that cure either with heat or at room temperature (i.e., room temperature vulcanizing, RTV). Two-part room temperature vulcanizing systems cure via the silicone molecule's silanol ends. One-package silicon sealants, cure on exposure to moisture in the air. Primers are required for optimal adhesion of the silicone sealants although often primers are not found to be necessary.

Silicone sealants are based on mixtures of fillers (e.g., silica), silicone polymers, cross linking components, and catalysts. The polymer has a siloxane backbone, i.e., Si-O-Si, with alkyl and alkoxy or acetoxy pendant groups. The latter are readily hydrolyzed to silanol groups (SiOH) which form larger chains by condensation and loss of alcohol or acetic acid. Trifunctional silanes are used as crosslinking agents with metal carboxylates as catalysts.

Silicone sealants are solventless and have a low shrinkage characteristic. They can be applied over a wide temperature range of -40 to $150°F$. The single-part systems require access to moisture for cure, and therefore cannot be cured in deep sections. Usually the depth of adequate cure is about 1 in. through the bond line. Therefore, joints greater than 2 in. deep are not recommended. The two-part systems do not require access to moisture to cure.

The cured products are used as flexible space fillers or sealants. Silicone sealants can bond, waterproof, and electrically insulate as well as seal. They provide generally excellent adhesion to most metals, wood, glass, rubber, some plastics, and concrete and mortar. Paint does not adhere to silicone sealants. This class of sealants has a high range of movement capability, from $+100\%$ to -50%. Their resistance to heat, water, and UV is excellent. Silicones offer high temperature resistance up to typically $450°F$. Special grades provide properties similar to standard silicone sealants with temperature resistance up to $600°F$ and superior solvent and oil resistance. Silicone sealants have very few drawbacks other than their relatively high cost. They are expected to perform for extended periods of 20–30 years or more. Table 13.10 lists the advantages and disadvantages of silicone sealants.

Room temperature vulcanizing silicones have been used for formed-in-place gaskets and potting compounds. They are also used in sealing of exterior joints, metal-to-glass sealing, expansion joints, and

TABLE 13.10 Advantages and Disadvantages of Silicone Sealants[7]

Advantages	Disadvantages
Low temperature gunnability	Poor unprimed adhesion to masonry
Glass adhesion	Dirt pickup
UV and ozone resistance	Poor tear resistance
Fast cure	Short tooling time
No shrinkage	Liable to stain substrate due to low modulus weep out
20 year durability	
+100% to −50% Movement	Poor resistance to hydrolysis

bathtub sealing. One component materials also go into electronics, automotive sealing, and marine and appliance applications. Similar to the other moisture cured sealants, the single component silicones, must be protected from moisture during storage.

The release of acetic acid as a byproduct of the condensation curing reaction could be of concern in certain applications where the acid may corrode metallic components that are near to the sealant. For these applications, alternative silicone formulations have been developed that liberate methanol rather than acetic acid during cure.

Two component silicone sealants are used in aircraft, automotive, and electronic industries. They are favored when the joint design offers little access to ambient humidity or when out-gassing products are not allowed during cure. Two component silicones are used as sealants and also as electrical potting and encapsulation materials.

13.4.6 Styrene butadiene copolymer sealants

Solvent based block copolymer sealants based on styrene butadiene resins have two levels of movement capability: ±10% and ±25%. Present formulations include large quantities of various oils, plasticizers, and resins, since the base polymer may be quite tough. Tackifying agents and UV stabilizers are also commonly used in styrene butadiene copolymer sealants. If these modifiers are either volatile or reactive, the result can be toughening and lowering of the movement capability with age. The construction industry uses these sealants for ceramic and tile bonding compounds. They have also been used for bonding floor boards to studs to reduce the amount of nailing required and to provide vibration damping.

Hot-melt sealants have also been developed based on styrene butadiene copolymers.[18] Since the thermoplastic sealant can be applied as a hot melt, the set time is much shorter than the conventional chemically cured or solvent and water release sealants.

13.4.7 Chloroprenes

Polychloroprene (neoprene) sealants can be compounded as either one or two component systems. For the two-part systems, pot life and curing rates can be varied over a wide range. A two-part system consists of the base resins and one or more fast acting accelerators such as polyalkyl polyamine or aldehyde amines, lead peroxide, and some tertiary amines. A combination of various accelerators is often used to control the work life. Intensive mixing is necessary in the field. Some two package systems are capable of curing at room temperature.

A typical single component system has the following composition:[5]

Composition	Parts by weight
Neoprene AG	110
Magnesia	4
Zinc oxide	5
Heat reactive resin	45
Solvent (methyl ethyl ketone, naphtha, or toluene)	600

The single component systems set by evaporation of solvent and cure at elevated temperatures.

Cured neoprene compositions have excellent elastomeric properties as well as good oil, chemical, ozone, oxidation, and heat resistance. Applications are usually related to oil resistance and chemical resistance. Neoprene sealants are compatible with asphaltic concrete and bitumen and, thus, are used often in roadway applications.

13.5 Other Specialty Sealants

A family of sulfur containing sealants has been developed based on thioethers. The thioether structure is thermally more stable than the polysulfide. The operating temperature of these sealants is 60–70°F higher than polysulfides. The thioethers also show better hydrolytic stability and chemical resistance than the polysulfides. These sealants were specially developed for composite application in the aircraft industry.[19]

Nitrile rubbers require high solvent levels to produce extrudable grades. Therefore, nitrile sealants are not common except in applications requiring very high solvent and oil resistance.

Closed cell polyvinyl chloride and polyolefin foam adhesive tapes have also been used to displace conventional sealant in many applications. These tapes can be used to form compression seals that are resistant to the flow of liquids or gas. They are generally easier to use and provide a neater appearance than extrudable sealants especially for thicker seals.

Rigid polymeric systems have been used as grouting materials or crack sealers where there is no movement. These systems include highly filled polyester, phenolic, urea, and furane resins. They have also been used as sealants and coatings for porous substrates such as brick and concrete.

References

1. Dunn, D. J. "Sealants and Sealant Technology", in *Adhesives and Sealants,* vol. 3 Engineered Materials Handbook, vol. 3, ASM International, 1990.
2. Carbary, L. D., "Types of Sealants", in *Adhesives and Sealants,* vol. 3 Engineered Materials Handbook, vol. 3, ASM International, 1990.
3. Brower, J., "Sealant Compatibility Issues Include Appearance and Chemical Effects", *Adhesives Age,* August 1996.
4. Kirk-Othmer Encyclopedia of Chemical Technology, J. I. Kroschwitz, ed. (New York: John Wiley and Sons, 1997).
5. Prane, J. W., "Sealants and Caulks", in *Handbook of Adhesives,* 3rd ed., I. Skeist, ed. (New York: Van Nostrand Reinhold Company, 1992).
6. Lomax, J., "Acrylic Polymer Caulks and Sealants", paper presented at the Sealant Short Course, Adhesives and Sealants Council, January, 1990.
7. Evans, R. M., "Polyurethane Sealants", Chapter 1 in *Urethane Construction Sealants* (Lancaster, PA: Technomic Publishing, 1993).
8. Paneck, J. R., "Polysulfide Sealants", in *Handbook of Adhesives,* I. Skiest, ed. (New York: Van Nostrand Reinhold, 1992).
9. Amstock, J. S., "Sealants", in *Handbook of Adhesive Bonding,* C. V. Cagle, ed. (New York: McGraw Hill, 1973).
10. Schuman, M. A., and Yazujian, A. D., "Technology of Polysulfide Sealants", in *Plastic Mortars, Sealants, and Caulking Compounds,* R. B. Seymour, ed. (Washington, D.C.: ACS Symposium Series 113, 1979).
11. Hutchinson, A., and Pagliuca, A., "MS Sealants: A Comparative Study of Performance Properties", *Adhesives Age,* April 1996.
12. Hashimoto, K., "Silyl-Terminated Polyethers for Sealant Use: Performance Updates", *Adhesives Age,* August 1998.
13. Regan, J. F., "Urethanes", in *Adhesives and Sealants,* vol. 3 Engineered Materials Handbook, ASM International, 1990.
14. Landon, S. J., et. al., "Silyated Urethane Polymers for Sealants", *European Adhesives and Sealants,* December 1995.
15. Ta-Min Feng and Waldman, B. A., "Silyated Urethane Polymers Enhance Properties of Construction Sealants", *Adhesives Age,* April 1995.
16. Landon, S. J., "Hybrid Sealants Demonstrate Enhanced Adhesion to Plastic Substrates", *Adhesives and Sealants Industry,* March 1997.
17. "One Component Urethane Foam is Easy to Use", *Product Engineering,* Nov. 1976 at 12.
18. Gun Chu, S., "Hot Melt Sealants Based on Thermoplastic Elastomers", *Proceedings of the ACS Division of Polymeric Materials Science and Engineering,* Denver CO, 1987.
19. Silverman, B., and Norrbom, A., "Use of Polysulfide Sealants in Aircraft Composite Structures:", *Adhesives Age,* July 1983.

Chapter 14

Selecting and Using Sealants

14.1 Introduction

This chapter will review the necessary steps that must be taken to provide the proper sealing system for specific applications. Whereas the previous chapter gave descriptions of the various materials that are commonly used as sealants, this chapter will provide guidance for both the selection of the sealant and the proper methodology for joint design and application.

The design of a sealing system involves more than merely choosing a material with the right physical and chemical properties. The following considerations are also essential to obtain optimal performance:

1. Joint design, including back-up or bedding materials
2. Type and nature of the substrates
3. Application, curing, and performance characteristics of the sealant

In this chapter, we will first try to understand the requirements imposed by the design of the joint and by the substrates, and then we will try to define the application and performance considerations that are necessary for selecting the proper sealant.

As with adhesives, a well developed methodology is necessary in the design of sealing systems because there is no ideal sealant. If there were an ideal sealant, it would be one that would stick to any unprimed substrate, yet would easily squeeze out of the caulking gun and flow into the joint. It would become tack-free quickly to minimize dust pickup, and it would cure quickly to become impervious to the environment. The ideal sealant would form smooth beads that are easily cleaned. It would shrink little on curing, and once cured, it would remain plastic and rubbery for life. The ideal sealant would be capable of being painted over, but this would not likely be necessary because of its outstanding resistance to weathering. The sealant would last as long as the substrates that it seals. Finally, the ideal sealant would be very inexpensive so that it could be used economically in large volumes. Since there is no single sealant that will meet all of these qualifications, usually compromises are necessary with regard to the selection of the material and to the design of the joint.

Many factors must be taken into account when designing a sealing system. Included in the decision process are considerations such as the type of base resin, the physical form of the sealant, applicable specifications, method of applying, and performance properties. The sealant and the joint design must be decided on in unison. The sealant's ability to expand and contract is the most important factor in choosing a sealant. This characteristic, along with the properties of the substrates, will determine the nature of the joint design. In selecting the proper sealant for a specific situation, consideration must also be given to such characteristics as its service life, hardness, cure time, abrasion resistance, application parameters, adhesion, weatherability, chemical resistance, and rate of recovery.

A significant difference between adhesives and sealants is that sealants are designed to allow rather than to restrict movement. The sealant must provide the physical properties required to move with the joint and the adhesion properties required to remain attached to the substrates. It must have the durability required to resist the external conditions during time of application and cure and maintain its sealing function during the life of the product.

14.2 Nature of the Joint Design

The design of a sealant joint depends to a great extent on its function or on the industry that is using the sealant. For example, there are numerous types of construction joints including expansion and contraction, overlap, perimeter, and butt. They can be horizontal, vertical, or inclined. Threaded joints are commonly used in the machinery and equipment industries. In this use, the sealant could be semipermanent and capable of being removed for maintenance operations. Sealants can be used for the function of providing a gasket. Gasketing sealants usually must have good chemical compatibility, and they must completely fill the entire volume of the joint. Sealants can also be used for sealing porosity in metal castings and powdered metal parts. In these applications, the sealant must be capable of being applied over the entire area of the part since the location of the porosity is often not known.

Typical sealant joints that are common in most industries are the butt, lap, and angle joints. These were generally discussed in Chapter 3 and are shown again in Fig. 14.1. There are many modifications and variations of these configurations. Joints should be designed to accommodate the expansion characteristics of the sealant and to allow for easy application.

14.2.1 Common butt and lap joints

For common butt joints, there must be room enough in the joint design to place a sufficient volume of sealant so that it can elongate and recover properly within the limits of the material selected. However, to understand the design principles involved one needs to examine the stresses and modes of failure in a sealant joint. In practice, joints usually fail because of peeling stresses at the sealant-to-substrate interface. Most sealant joints and especially the high performance sealant joints rarely fail cohesively.

The stress at the bond line is a function of the elongation of the free sealant surface as shown in Fig. 14.2. A thinner sealant produces less

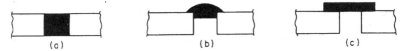

Butt joint—Use sealant if thickness of plate is sufficient (a), or bead seal if plates are thin (b). Tape can also be used. If joint moves due to dynamic loads or thermal expansion and contraction, a flexible sealant with good adhesion must be selected (c). Select flexible tape for butt joint if movement is anticipated.

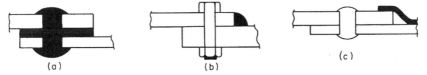

Lap joint—Sandwich sealant between mating surfaces and rivet, bolt, or spot weld seam to secure joint (a). Thick plates can be sealed with a bead of sealant (b); and tape can also be used if sufficient overlap is provided as a surface to which the tape can adhere (c).

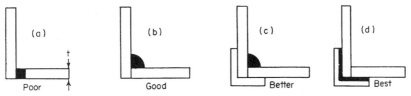

Angle joint—Simple butt joint can be sealed as shown (a) if material thickness t is sufficient. But a better choice is the bead of sealant (b), which is independent of material thickness. Supported angle joints with bead (c) or sandwich seal (d) are better choices.

Figure 14.1 Common methods of sealing joints.[1]

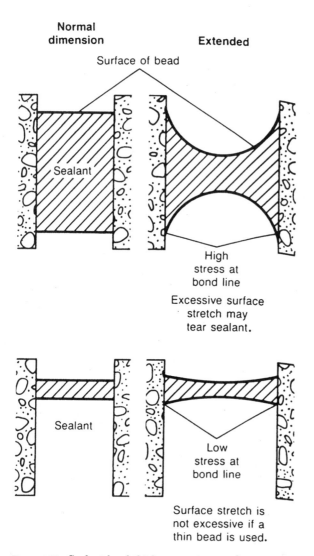

Figure 14.2 Sealant bead thickness and stress.[2]

stress at the bond line and has greater capability for movement than a thicker sealant. To maintain stress within limits, some common guidelines have been established regarding butt joint design.[2,3]

- In joints up to ½ in. wide, the depth of the sealant should be the same as its width.

- In joints greater than ½ in. wide and up to 1 in. wide, the depth of the sealant should not be greater than ½ in.

- In joints greater than 1 in. wide, the depth of sealant should not be greater than 50% of the width.

Construction joints that are subject to wear from travel are recommended not to exceed 1 in. in width.

Joints that are too deep cause severe stresses and should be avoided. Joints can be made shallower with a backing material. In Fig. 14.3 a compressible back-up material is used which should be under compression even at the maximum joint volume. Where back-up or bedding materials are used, care must be taken to assure that these materials are compatible with the sealant used. They must be free of oil and asphalt. The ingredients used in compounding a bedding material must be of a nonmigrating nature and exhibit no cold flow.

The sealant must be capable of expanding and contracting along with the substrates to which it is joined. To allow for complete freedom for expansion and contraction, a release agent is sometimes placed on the inner surface of the joint so that the sealant does not bond to this surface. By not bonding to the bottom surface of the joint, the sealant is free to expand and contract.

Application of such a release agent is shown in Fig. 14.4. A release agent, either chemical or film, prevents the sealant from bonding to the bottom and allows the sealant material to stretch normally (Fig. 14.4d). If a tight bond were to occur to the inner surface, the sealant

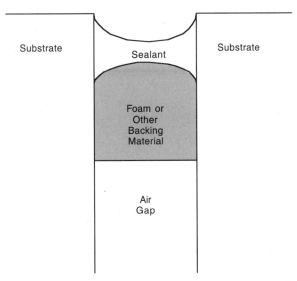

Figure 14.3 Compound butt joint has back-up material, which should be under compression even at maximum joint volume.

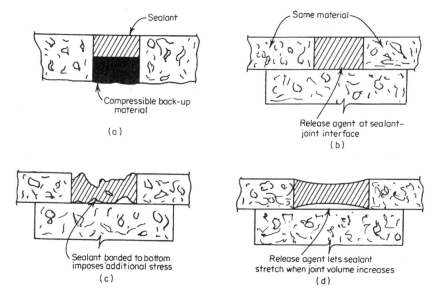

Figure 14.4 Compound butt joint showing compressible back-up (*a*) and release agent at sealant-joint interface (*b*). The release agent, either chemical or film, prevents sealant from bonding to bottom and allows materials to stretch normally if necessary. (*d*). Sealant failure may be caused by expansion and contraction, if sealant is bonded to bottom layer (*c*).[1]

could cohesively fail during periods when there was a large joint gap such as during cold weather conditions (Fig. 14.4*c*).

Lap joint designs are shown in Fig. 14.1. Lap joints can effectively be sealed by sandwiching sealant between mating surfaces that are secured with mechanical fasteners. Lap joints undergoing shearing forces can develop significant strains depending on the movement and on the joint thickness. Using simple geometrical relationships, the strain on a joint undergoing shear deformation can be approximated for most joint applications as shown in Fig. 14.5.

Some sealant materials are capable of being welded through with spot welding equipment to achieve mechanical fastening. These were developed primarily for the automobile industry where the sound deadening characteristic of an elastomeric sealant is desired, but the production line can not wait for the sealant to set before the subassembly is moved. Thus, by spot welding through the sealant, the substrates can be held together while the sealant sets. Lap joints can also be sealed by applying a bead of sealant to the edges of a joint. If lap joints cannot be sandwich-sealed, the joint should be designed with a built-in receptacle to receive the sealant.

Figure 14.6 shows various seam joint designs where the substrates cannot be overlapped. Seam joints that are applied to hold in a fluid,

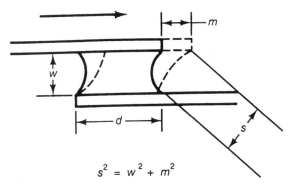

Figure 14.5 Strain, s, on a lap joint with width, w, and movement, m; $s^2 = w^2 + m^2$.[2]

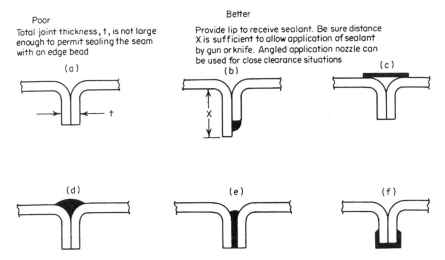

Figure 14.6 Various seam sealing joints. Edge (a) is not thick enough to support bead. Lip joint (b) provides receptacle for sealant. Seam can be inside sealed by tape (c), putty (d), or sandwich sealed with putty or anaerobic (e), tape can also be used on the outside of seam (f).[1]

such as on storage tanks, should be sealed from the inside so that the seal is augmented by the fluid pressure. Outside sealants are not recommended because the force of the fluid pressure will be working to cleave the sealant from the substrate.

Sometimes tape can be used to replace a sealant. Examples of tape sealed joints are shown in Figs. 14.1 (lap joint, c) and 14.6 (c and f). The adhesion of pressure sensitive tape is very dependent on the surface condition of the substrate, and sufficient overlap area must be provided to which the tape can adhere. The tape must be resistant to the environment and fluids that the joint will be exposed to during service. Of course, pressure sensitive tapes only provide sealing in two dimensions. Preformed rubbery mastic tapes or continuous extrusion beads are also available. These are applied and pressed into place usually with simple finger pressure. They provide three dimensional sealing capabilities and generally offer better adhesion and sealing characteristics than pressure sensitive tapes.

14.2.2 Backup materials

The primary purpose of the backup material in a joint is to control the depth of the sealant, thus assuring the proper shape factor. Another purpose is to provide support or reinforcement for the sealant in horizontal joints, such as floors. Improper selection of a backup material could cause premature loss of adhesion. The backup materials must be unaffected by solvent or other material contained in the sealant. Backup materials containing asphalt, coal tar, or polyisobutylene should never be used. These extrude oils, could cause staining, and are incompatible with some sealants.

Neoprene, polyurethane, polyethylene foam, cork board, fiber board, and cotton rope have all been used as backing materials. The foam materials are often used because they are compressible and are conventionally available in continuous tapes or strips. Rubber tubing made from various elastomers also provide an excellent backup material. It is desirable that the backup material offer a water barrier to provide a secondary seal if the primary sealant loses adhesion.

14.2.3 Threaded joints

Threaded fittings come with tapered or parallel thread gaps or a combination of the two. The common types of thread designs used for threaded fittings are shown in Fig. 14.7. The spiral leak paths that could occur when they are tightened are indicated. Regardless of thread design, such fittings will leak unless some form of a sealant is used.

Polytetrafluoroethylene (PTFE) or Teflon tape has been used for many years to lubricate and seal threaded assemblies. This tape cold flows and forms a seal when pressure is exerted on the tape as the threads are engaged. However, PTFE tape does not stop the parts from loosening because of vibration, and the tape tends to shred and clog hydraulic lines.

Conventional elastomer and latex sealants that can be applied to the male threads are also very effective in sealing threaded joints. However, with large fittings, shrinkage of the sealant on drying can be a problem, and the application can be a time consuming process.

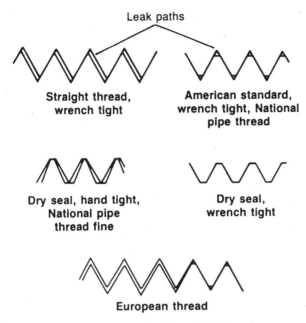

Figure 14.7 Thread designs for threaded fittings.[4]

Anaerobic sealants are perhaps the most versatile sealants for threaded fittings. They lubricate the threads for assembly and cure to a hard, solvent resistant sealant. They also adhere well and act as a threadlocker to prevent looseness due to vibration. However, anaerobic sealants are brittle materials and cannot be reused on disassembly. Since they harden by the catalytic effect of metal substrates, there are limitations regarding cure. Pipe joints over 1 in. in diameter may have gaps that exceed 1 mm in the crest of the threads. Large gaps between threads will either extend the cure time necessary or cause incomplete cure depending on the type of anaerobic sealant used.

There are also anaerobic and epoxy sealants that can be preapplied on bolts in which the sealant material is microencapsulated in a polymeric shell. The sealant material is released upon assembly when the microcapsules are broken due to the threading action. These materials offer indefinite shelf-life so that the sealant can be preapplied to the substrate at the fastener manufacturer. Examples of microencapsulated sealant applied to hex head bolts are shown in Fig. 14.8.

14.2.4 Gasketing

Gasketing materials are generally preformed elastomers that are available as o-rings, rope, and other convenient shapes. Gaskets may

Figure 14.8 Examples of thread locking adhesive. Microspheres encapsulate the resin. (Photo courtesy of ND Industries, Inc.)

also be cut from stock material such as thin sheet. However, sealants can be used to make formed-in-place gaskets.[5] A ribbon of uncured sealant is applied to the flange area of the components to be sealed. Upon assembly of the components, the sealant is compressed to conform exactly to the mating surfaces.

The formed-in-place gasket functions differently than a cut gasket. The cut gasket maintains a seal by virtue of compression that is maintained in service by mechanical fasteners. The cut gasket must remain under compression to function. Any loss of load due to relaxation of the gasket material or loosening of the fasteners could cause the gasket to leak. Formed-in-place gaskets, on the other hand, do not require compression. Instead, they rely on adhesion for maintaining the seal. In a formed-in-place gasket joint design, there is metal-to-metal contact between flanges so that the gasket cures to a predetermined constant thickness after assembly. If tension is lost in the mechanical fasteners, the gasket would still function by virtue of the adhesion of the sealant to the metal surfaces. It is very important that the flange systems be specifically designed for formed-in-place gaskets. Joint thickness especially will depend on the physical properties expected of the cured gasket material. However, because of its joint filling capacity, greater irregularities in either mating surface can be tolerated with formed-in-place gaskets than with conventional gaskets. Components whose sealing surfaces are scratched or nicked and would be impossible to seal with preformed gaskets can be sealed directly using formed-in-place gasketing.

Formed-in-place gasketing alleviates many of the problems associated with conventional cut gaskets. In addition, formed-in-place gasketing lowers manufacturing costs. Because each gasket is dispensed to the desired shape, there is no waste and only one formulation is needed to meet many different gasketing needs. Formed-in-place gaskets also allow the gasket profile to be changed as needed to ensure a proper seal for the surfaces being mated. The challenge created by formed-in-place gaskets is that the sealant must be applied uniformly with no knit line at the start/stop point. However, development of dispensing valves and robotic application techniques have minimized these concerns.[6] Figure 14.9 illustrates formed-in-place gasket knit lines of various quality.

Two types of sealants are conventionally used for formed-in-place applications. These are room temperature curing silicones and anaerobics. Anaerobic sealants will cure to a rigid system. As a result, they require rigid flanges and a joint arrangement with a small gap that is designed to prevent any joint movement. RTV (room temperature vulcanizing) silicones are generally used for formed-in-place gaskets where longer cure times are possible. These sealants are generally formulated from medium modulus silicone elastomers. They are used in joints where movement is required. With this material, it is important that the flange design allow a sufficient gap to provide for the movement capability. With RTV silicone sealants, it is also necessary to provide a joint design that permits atmospheric moisture to cure the silicone material.

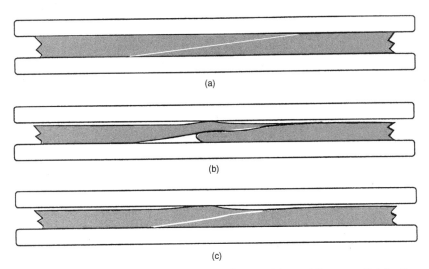

Figure 14.9 Cross-section of zero knit line application (*a*), providing even distribution of sealant and compression of lid to housing provide a void free seal. Cross-section of knit line with void (*b*), creating potential leak paths on top and bottom of gasket. Cross-section of knit line with too much sealant on top for proper compression (*c*), creating potential leak path between gasket and lid.[6]

Another type of formed-in-place gasket can be made by applying a multi-component or a UV cured sealant directly to only one flange, allowing the sealant to cure, and then mating the surfaces together. A low durometer, single component silicone sealant has been developed that can be applied robotically and cures in place within 30 secs after passing beneath a UV lamp.[7] Only very soft types of sealants will work well with this method. Formed-in-place gaskets can also use foamed material. These foamed-in-place sealants are described in a later section of this chapter.

General difficulties in working with formed-in-place gaskets include non-flat or bowed flanges, excessive joint thickness variation, and gaps that exceed specification. It is also very important that the applied formed-in-place gasket be a continuous material filling the entire joint cavity. Gaps in the joint could be caused by erratic application techniques or air bubbles entrapped in the sealant. Such gaps could easily lead to leaking gaskets.

14.2.5 Porosity sealing

Low viscosity sealants are also used to seal against micro-porosity in metal casting and powder metal parts. Such parts are commonly found with air conditioner housings and other metal castings that must contain pressurized gas. These parts must be non-porous to fluids and/or gases. Usually in the machining of these parts from stock, cavities are uncovered that could lead to a leakage path through the entire thickness of the part. Sealing of these parts can be achieved by impregnation with liquid sealants having very low viscosity. The sealant is capable of wicking into the pores and filling the leak path due to its capillary effects. Pressure and/or vacuum can also be used to insure that the sealant penetrates the pores before curing.

The most common types of porosity sealing materials are water solutions of sodium silicates, styrene based polyesters, low viscosity methacrylate resins, and anaerobic sealants. The anaerobic type is available as either a heat cure or room temperature curing version. They are used for the most critical sealing operations where resin bleed-out cannot be tolerated, and chemical or heat resistance is required. Table 14.1 describes some of the materials that can be used for porosity sealing.

Factors that should be considered when selecting the appropriate resin for sealing porosity are:

- Curing process (heat cure vs. room temperature cure)
- Chemical, solvent, and temperature resistance

TABLE 14.1 Porosity Sealing Materials[4]

Material	Advantages	Disadvantages
Water solution of sodium silicate	Temperature resistance over 260°C	Slow production and cure cycles and large shrinkage on cure
Styrene based unsaturated polyesters	Effective for large degrees of porosity	Hazardous vapor emissions; requires heat cure for 1–2 hrs at up to 140°C; cure can produce resin bleed out
Low viscosity methacrylate resins	Less hazardous than polyesters; low temperature or room temperature cure	Cost

- Pore size sealing capability
- Surface quality requirements
- Yield requirements

MIL-STD-276 and MIL-I-17563B have been developed by the U.S. military to cover the requirements for impregnation of military and aerospace hardware.

The inherent porosity that is a natural result of powdered metal parts makes resin impregnation almost a necessity for sealing these parts. Vacuum impregnation with room temperature curing anaerobic resins is widely utilized for sealing the pores. Heat curing sealants usually are not satisfactory in powdered metal parts because of problems related to the bleedout of resin with heat curing and extensive porosity.

A variety of processing methods may be used to seal the porosity in parts. The actual method selected depends on the sealant to be used and the requirements of the part. Table 14.2 outlines the processing procedures for three of the most popular impregnation processes:

- Dry vacuum/pressure
- Wet vacuum/pressure
- Wet vacuum

14.3 Sealant Substrates

As with adhesives, the substrates must be carefully cleaned prior to the application of sealants. Weak boundary layers that could cause premature failure must be removed or strengthened. Some substrates

TABLE 14.2 Processing Steps for Impregnating Parts to Seal Porosity[8]

Dry vacuum/pressure

- Slowest and most complicated vacuum impregnation method. The vacuum impregnation cycle requires two tanks, one which holds the sealant, and one in which the parts are processed.
- Traditionally used with high viscosity sealants. This application method is sometimes specified with modern sealants where porosity is very small and sealing requirements are unusually good.
- Processing steps:
  1. Place parts in process basket and load into process tank.
  2. Draw vacuum in process tank to remove air from pores of parts.
  3. Transfer sealant from storage tank to process tank and submerge parts still under vacuum.
  4. Release vacuum and pressurize process tank with shop pressure. This helps to drive the sealant into the pores.
  5. Release pressure and transfer sealant back to storage tank.
  6. Remove parts. Wash and complete other process steps.

Wet vacuum/pressure

- This process method requires only one tank. Parts are submerged in the sealant which remains in the process tank at all times. The vacuum is applied to parts and sealant together, followed by pressurization with shop air.
- This method is sometimes recommended for production processing of castings with very fine porosity and for high density powder metal parts.
- Processing steps:
  1. Place parts in process basket and load into process tank.
  2. Draw vacuum in process tank to remove air from pores of parts.
  3. Release vacuum and pressurize then with shop air.
  4. Release pressure.
  5. Remove part. Wash and complete other process steps.

Wet vacuum

- This is the simplest and fastest of the vacuum impregnation methods. It is similar to the wet vacuum/pressure method, except that the tank is not pressurized. Instead, the tank is simply vented to atmosphere after establishing the vacuum. Penetration of the sealant into the parts takes place at atmospheric pressure. The resin flows in to fill the vacuum created inside the porosity of the parts.
- This is the most widely used application method by far. The simplicity and rapid processing along with lower equipment cost, make this the method of choice in most impregnation systems installations.
- Processing steps:
  1. Place parts in baskets and load into process tank.
  2. Draw vacuum to remove air from porosity.
  3. Release vacuum, vent tank to atmospheric pressure.
  4. Allow parts to soak briefly while sealant penetrates.
  5. Remove parts. Wash and complete other process steps.

may have to be specially treated in order for the sealant to provide adequate adhesion. Listed here are a number of treatment methods, surfaces, and common concerns regarding the substrate's role in assuring adequate sealant performance.

14.3.1 Surface preparation

Generally, cleaning is an adequate surface treatment for many sealant joints. Cleaning can be accomplished by simply washing with detergent or by wiping the surface with solvent. However, there are certain precautions that must be noted. Cleaners and solvents will only remove organic contamination and very loose surface material. More significant contamination may require wire brushing, sandblasting, or other mechanical abrasion methods for removal. Low energy surfaces, such as many thermoplastics, may require special chemical treatment to raise the surface energy to provide adequate adhesion.

Surface preparation methods for sealants are identical to those used with adhesives. These are fully described in Chapter 6. Often the surface preparation is less critical with sealants in that a high degree of adhesion is not the primary criterion for success. Thus, solvent or detergent cleaning is usually all that is necessary to achieve an adequate sealant system.

14.3.2 Primers

Primers play an important role in applications where adhesion of the sealant to the substrate material is important. The selection of a primer is based on the substrate, sealant material, and performance considerations. Primers in common use are based on chlorinated rubber, polyvinyl chloride solutions, aminosilicones, polyurethanes, two-part epoxy/amide and epoxy/polysulfide, and other low viscosity polymeric resins. Primers can be used to provide improved adhesion, provide a barrier to ingress of fluids and water to the joint's interface, and even improve the cohesive strength of some substrate surfaces. The sealant manufacturer should always be consulted as to the need of a primer. If a primer is required, the sealant manufacturer can usually recommend the primer of choice for the specific application along with the primer's application and curing requirements.

14.3.3 Common substrate surfaces

Below are characteristics of common substrate surfaces encountered by sealants. Many of the surfaces that sealants must bond to are construction materials. If a surface treatment is required it often must be completed out-of-doors and in uncontrolled conditions. For more common substrate surfaces, refer to the surface preparation processes and methodologies described in Chapter 6.

14.3.3.1 Concrete and masonry. Factors that can affect sealant adhesion to concrete and masonry surfaces include, but are not limited to: concrete formulation and additives, concrete curing conditions, the presence of moisture, finishing techniques, and contamination from release materials. Concrete or masonry substrates that have a weak surface layer should be cleaned by sandblasting, grinding, or wire brushing until a sound, strong cohesive surface is obvious. Dust must be removed by repeated brushing with a soft bristled brush or, more effectively, by blasting with oil-free compressed air.

Surfaces that are contaminated by hardeners, curing compounds, or releasing agents, for example, can be cleaned by sandblasting or wire brushing. Normally, a solvent is not very effective on concrete, since it may only drive contaminants deeper into the concrete or spread them over the surface. Moisture in or on the surface of the concrete or masonry can be hard to detect. It usually is noticeable by a surface darkening effect. In some cases, the outer surface may look and feel dry, but there may be enough water in the concrete to inhibit adhesion. In this case, it is best to wait until the substrate dries or to force-dry the substrate with a hot air blower. Moisture removal should be the last step in any concrete surface preparation process.

Some sealant manufacturers recommend a primer to provide a better concrete or masonry surface for adhesion. Primers are normally used to either provide a barrier to the penetration of moisture, chemicals, and other environmental fluids into the concrete substrate. They are also used to strengthen a weak substrate surface. Loose or flaking concrete surfaces can be reinforced by the application of a low viscosity epoxy-based primer.

In certain applications, where a high degree of adhesion is required and when grit blasting is not possible, concrete surfaces can be acid etched with a solution of 15% by weight of concentrated hydrochloric acid in water. One gallon of the solution will treat 5 sq. yds. of concrete. It is important that all of the acid be removed from the surface before application of the sealant. This is usually accomplished by washing the surface with a mild alkaline solution and then clean water.

14.3.3.2 Stone. Stone surfaces are normally good substrates for sealant adhesion. They must be cohesively sound, dry and free of contaminants. Weak, dusty, or contaminated stone surfaces may be cleaned by water blasting, sand blasting, or wire brushing. Surfaces must be dry and free of dust before primer or sealant application. On dense and uncontaminated surfaces of granite or marble, a wipe with methyl

ethyl ketone solvent may be all that is needed for optimum sealant adhesion.

Many sealant suppliers recommend primers for optimum adhesion to stone surfaces such as granite, limestone, marble, and sandstone. The main purpose of the primer is to reinforce and seal cracks and fissures in the stone substrate.

14.3.3.3 Glass or porcelain. Because of their high surface energy, glass, porcelain, or vitrified substrates are usually excellent surfaces for sealants to bond. This, of course, assumes that they are clean, dry, and free of contamination. Silicone sealants, in particular, exhibit excellent adhesion without a primer. Some other sealants may require primers for optimum adhesion, so consult the manufacturer's literature.

Glass surfaces should be cleaned with methyl ethyl ketone (MEK) or alcohol. The effect of these solvents on adjacent surfaces should be examined before use. Oily contaminants should be removed by wiping the surface with a solvent such as MEK or xylene. Contaminants, such as paint or old sealant, should be removed from the surface with a knife before solvent cleaning.

14.3.3.4 Painted surfaces. In general, sealants, should not be applied over paint that has not adhered well to the substrate. Paint that is peeling and flaking before sealant installation may continue to do so after the sealant is applied, and adhesion problems will result. If the paint is well adhered to the substrate, then consult the sealant manufacturer's literature for recommendations on whether or not a primer is required between the old paint and the sealant.

If the sealant to be used is compatible with the paint and if the coating is sound, the surface of the paint should be cleaned by wiping with a solvent such as naphtha to remove contaminants, chalk, and dust. In all cases, the possible deleterious effect of the solvent on the paint film should be determined.

14.3.3.5 Unpainted metal. In attempting to seal metal surfaces, the proper surface preparation and solvent cleaning steps are especially important. Mill-finished aluminum surfaces may contain an invisible oil film or oxide. Clean this substrate with a degreasing solvent such as xylene or trichloroethylene. In some cases, abrasion of the surface with steel wool or fine emery paper may be required for proper adhesion. Many sealant suppliers recommend the use of a primer to metal surfaces usually for corrosion protection.

Anodized aluminum substrates usually provide an excellent surface for adhesion, but the substrate may be contaminated by dirt, oil, etc. Normally, a methyl ethyl ketone (MEK) or xylene cleaning step is all that is required. Certain sealant suppliers may recommend a primer for anodized aluminum surfaces, but silicone sealants may be used without a primer on clean anodized surfaces.

Copper may contain a heavy oxide film, which should be removed by sanding or by use of steel wool. Once mechanically abraded the sealant should be applied to the surface as quickly as possible. MEK or xylene may be used for cleaning. Copper is not compatible with some sealants such as acetic acid liberating RTV silicones. Consult the sealant manufacturer if sealant compatibility is in question.

Most steel surfaces are painted where sealants are used. Should the need arise to seal unpainted steel, it should be dry and free of rust, oil, and other contamination. Sandblasting may be required to provide a sound, clean surface. Solvent cleaning should be completed with xylene or naphtha.

14.4 Application Requirements

Different types of sealants require different application techniques. Sealants can be supplied to the job site in various forms, each of which has advantages and disadvantages and will require special techniques and considerations. The conventionally available sealant forms are bulk materials, extruded tapes, preformed gaskets, and foams.

Application procedures have a very important bearing on sealant selection. Factors to be considered in addition to the form of the sealant include curing characteristics, handling characteristics, application methods, surface preparation, and health hazards. The uncured properties of the adhesive that need to be considered relative to their application are shown in Table 14.3.

TABLE 14.3 Property Considerations in the Application of Sealants

Property	Units
Skin-over time	Minutes
Tack-free time	Hours
Sag, weep	Inches
Extrusion rate	Grams/minute
Flow	Seconds
Rate of cure	Hours
Depth of cure	Inches
Shrinkage during cure	Percentage

14.4.1 Bulk materials

Bulk sealant materials are either single or multiple component high viscosity liquids or pastes. The easiest way to characterize bulk sealant materials is through their rheological properties once they are mixed and ready to apply. Bulk sealants can be divided into two general groups by means of rheology: self-leveling and non-sagging (thixotropic). Self-leveling sealants have some degree of flow so that they will tend to seek their own level. Generally, sealants are considered liquids if their viscosity is less than 500,000 centipoise and putties or pastes if above 500,000. Thixotropic sealants have mastic-like consistency and require some degree of force to cause them to flow. Once applied they will not sag and, thus, can be applied to vertical joints. Depending on their viscosity, sealants can be applied by brush or roller, trowel, spatula, putty knife, or by extrusion with a caulking gun or pneumatic equipment.

Bulk sealant materials are available in tubes or cartridges (single component sealants) and in cans or drums (single or multicomponent). The cartridges are typically 10.7 oz–1 qt. The sealant can be extruded from the cartridge and into the joint by hand pressure or power activated caulking guns. The cans or drums are sold in units of 1–50 gal. They are measured and mixed, if multicomponent, and either put into a caulking tube or applied by knife directly to the joint. Because of limited pot life when mixed or because of possible contamination once the cans or drums are open, these bulk sealants are usually applied immediately.

There is some skill required in applying a sealant from a caulking gun or tube. There is no universally proper way to use a caulking gun. There are those who prefer to push, and those who prefer to pull. Either method is satisfactory as long as the sealant is properly applied. More important is the cut angle on the nozzle and the angle of the gun to the joint. When properly applied, the sealant is forced against the three sides of the joint, and the surface is smooth and curved. Tooling is desirable to force out small air pockets as shown in Fig. 14.10. Tooling also becomes necessary when the edges of the joint are irregular, as with some porous stone or precast concrete (Fig. 14.11).

The properties of the uncured sealant are extremely important factors in certain applications. For construction type sealants, the resistance of the uncured sealant to possible weather conditions during the time of application is an important factor. Some bulk sealants cure very slowly at temperatures below 45°F. In many cases, standing water accumulates in horizontal joints, thereby making sealing impossible.

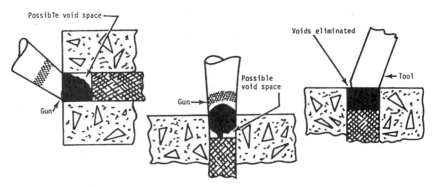

Figure 14.10 Voids in gunned sealant can be removed by tooling.[9]

Bulk sealant materials are available as one component and multiple component types. They basically use the same type of caulking equipment. The advantages of single component sealants are that they do not require on-site metering and mixing, and they maintain a longer working life. This longer working life is of value especially when working with larger containers that must be used over longer periods of time. However, they may not be appropriate where the joint will experience abrasion, dirt, or adverse weather conditions a short time after the sealant is applied.

Multiple component systems must be thoroughly mixed prior to application. The time between mixing and application will depend on the speed of cure. The availability and type of metering and mixing equipment at the job site are other factors to consider. Mixing is usually by hand with a paddle or with a slow speed electric drill with a mixing blade attachment. Care must be exercised so as not to mix too quickly and entrap air into the sealant. There are also excellent bulk mixers that use rotating screws to mix the materials. Chapter 18 provides a detailed description of such equipment.

In some applications, multiple component sealants may be premixed, loaded into polyethylene cartridges and then quick-frozen to delay the curing process. The freezing process consists of immersion in a dry ice/acetone bath of $-90°F$. The cartridges are then thawed at the job site and used. Not all sealants are applicable to this freezing process; it is best to check with the sealant manufacturer before it is attempted. This process is used when it is desirable to have consistency and control in the mixing, metering, and cartridge filling processes.

Multiple component sealants are preferred when a short cure time is required, such as when sealing joints in heavy traffic areas. Working life varies from about 1–8 hrs and will vary with temperature and humidity depending on the type of sealant. Working life could be a

Selecting and Using Sealants 521

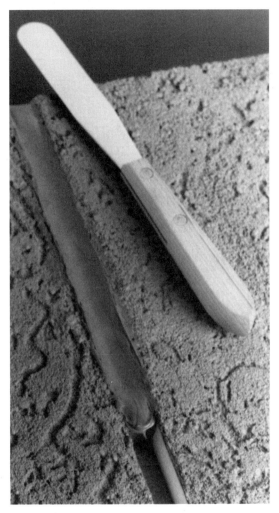

Figure 14.11 Tooling is necessary when sealing edges of irregular surfaces. (Photo courtesy of Polymeric Systems Inc.)

consideration on how fast a curing system you can use. Sealants with very short curing times also have very short working lives so that automated metering, mixing, and application equipment is often necessary. In warm climates, the working life could also be a limiting factor in that the ambient temperature may significantly decrease the working life.

Bulk materials come in several types, depending on the formulation and application requirements of the job. The viscosity may require heating of the sealant either before it is applied, before it is loaded into a cartridge, or as it is applied. Heating of the bulk materiel could be done in ovens, heat chests, or boiling water baths to the manufacturer's recommended temperatures and specifications. Specially designed electrically heated cartridge guns are available to maintain a heated cartridge at temperature and to assure that the sealant is heated as it is applied to the joints. Bulk sealants and other forms of sealing materials are listed in Table 14.4.

14.4.2 Tape sealants

Tape sealants are elastomeric materials that have been manufactured in bulk and are of sufficient viscosity that they can be furnished as a semi-solid tape, ribbon, or rope in various cross sectional shapes. In certain applications such as glazing, tapes can be specified to very

TABLE 14.4 Forms of Sealing Materials[10]

Type	Description
Hot-pour sealants	Melt to required temperature for proper handling and performance, specifically adhesion
Cold-pour, multicomponent	Cure by cross linking of polymers. Sealant base, curative, and color pack, if required, are thoroughly mix blended. Blended material can be poured into joints
Nonsag, noncured sealants	Apply by knife into joint; may require warming for proper application
Nonsag, one-component, chemically cured sealants	Cure by chemical reactions of polymers. Applied by knife, caulking cartridge, or bulk dispensing equipment
Nonsag, multicomponent, chemically cured sealants	Cure by chemical reactions of polymers. Sealant base, curative and color pack must be mix blended. Filled into cartridges and dispensed by automatic dispensers or applied with a knife.
Heat-softened, nonsag sealants	Require heating for dispensing and application
Strip sealants, cold applied	Apply strip to the joining surface, remove backing paper, and place second joining surface for sealing
Strip sealants, hot applied	Cut, remove backing paper, and fit into slot to be sealed. Heat, making sure that heat-softened material has continuous contact with sides of forming surfaces to be sealed
Compression seals	Place preform material in compression between the forming surfaces

accurate dimensions. They are generally supplied on a paper backing with a release lining that is removed prior to application. Usually, a suitable length of the tape sealant is cut from a roll and applied to one of the joining surfaces with the backing paper still in place. The sealant is pressed into place by hand, and then the paper is removed. Butt splices and corner splices are easily formed by pushing the tape into place. Extruded tapes of polybutene, polyisobutylene, and butyl are available in many states of cure, and in various combinations with other resins. These tapes are available in continuous lengths and in a variety of forms such as bead, half-bead, ribbon, and sheet.

Heat-applied preformed sealants are also available. These sealants are also supplied with backing paper. Usually the backing paper is removed and the solid, relatively nontacky sealant is fitted into the joint. The sealant is then heated to the manufacturer's recommended temperature with a hot air gun or oven. Caution must be taken that the heat softened material achieves continuous contact with the sides of the joining surfaces and that wetting and adhesion occur. These preformed sealants can also be applied through a heated gun.

14.4.3 Preformed gaskets

Preformed seals are supplied specifically for the application in usually a multitude of available shapes and materials. An example of such a gasket would be a continuous length of fully cured silicone tape with pressure sensitive adhesive on both sides. Other materials and forms are commonly available.

Common to all these shapes is the fact that they are installed under compression. In this application, adhesion is not a problem since seal integrity is dependent on the gasket remaining under compression. Thus, joint cleaning consists of simply blowing or brushing loose dirt and contamination out of the joint. Preformed gaskets can have a pressure sensitive adhesive on one side of the sealant material to hold the gasket in place until it is mated with its other substrate. In certain applications, it may be desirable to have adhesive on both sides of the gasket.

The critical factors to be considered in gasket installation are stretching and splicing. The gasket must be designed for the joint so that it is under compression at all times. Most gaskets can leak at corners unless properly spliced or molded. Molded 90 degree picture-frame gaskets prevent leakage at the corners.

14.4.4 Foam sealants

Foam sealants are available as: (1) solid, preformed closed cell gaskets (a type of preformed gasket) and as (2) extrudable hot melt or multicomponent bulk material that will foam in place within the joint and form a seal.

Preformed foam seals are compression type seals and require compression of up to 25% to form a positive seal. Foam sealants are supplied in a wide variety of densities from 6.5–16 lbs per cubic foot, and a wide range of hardnesses from 15–40 on the Shore 00 scale. Preformed foam sealants can also be supplied with adhesive on one or both sides.

Equipment and materials have also been developed that produce closed cell foam gaskets from a range of pumpable sealants and caulks. The equipment mixes the gasketing material with an inert gas that expands during dispensing forming a resilient, reusable foam gasket. Foam-in-place systems are more economical than conventional gasketing and provide more consistent quality and higher production rates. Sealant materials used for foam-in-place gaskets include hot-melt formulations, silicone RTVs, and moisture cure urethanes.

A patented "transfer molding" process has also been developed for applying foam-in-place gaskets in high volume applications.[12] In the first step, the foam sealant is dispensed similar to a hot melt into a groove of a special treated, water cooled mold. The mold, filled with unexpanded sealant, is then pressed onto the part to be gasketed. Five seconds later, the mold is removed leaving the expanded gasket adhering to the part. Depending on the size and shape of the gasket, total processing time is 7–20 secs. The proprietary sealant reacts quickly with ambient moisture causing the outer skin of the gasket to cure. Full curing will occur as the gasket continues to absorb ambient moisture.

These transfer molded gaskets can be molded easily to various cross sections. Height of the gasket can exceed its width, and molding provides superior consistency in gasket size and finish. Through speed of application and the use of about 30% less material, overall gasket installation costs can be cut 55% on average.[13]

14.4.5 Application properties

Application parameters are critical in the selection of the sealant. Consideration must be given to such characteristics as the sealant's rheological properties, temperature at time of application, cure rate, shrinkage during cure, and depth of cure.

14.4.5.1 Rheological properties. Rheological or flow properties determine how a sealant can be best applied: by brush, trowel, spatula, and putty knife, and by extrusion via caulking gun. Sealants can be classified as either self-leveling or non-sagging (thixotropic). Self-leveling sealants have sufficient flow to form a smooth, level surface when applied to a horizontal joint at 40°F. Non-sag sealants can be applied to vertical surface without sagging or slumping. ASTM C 639 describes test methods to determine whether a sealant can be considered self-leveling or non-sagging and the degree of sag-resistance. ASTM D 2202 can also be used to determine the degree of sag resistance.

True liquid sealants have a viscosity of less than 500,000 centipoise. Sealants with greater viscosities are considered to be mastics. Sealant viscosity can greatly affect application procedures. Whether a sealant can be brushed or troweled or extruded onto a substrate depends on its viscosity and rheological characteristics. The viscosity of bulk sealants may be adjusted by incorporated fillers or thinners. This is usually done in the factory under controlled conditions, and consideration must be given to the effect of the final formulation on ultimate physical properties.

14.4.5.2 Ambient temperature on application. Generally two problems arise due to low temperatures at the time of sealant application. With water-based emulsions or latex sealants, the temperature at the time of the application must be greater than about 40°F to prevent freezing of the adhesive. Regardless of the setting mechanism, cooler temperatures usually result in longer times until the sealant reaches a condition to which it can resist its service environment. In construction sealants especially, long cure times may require that the setting sealant be protected from abrasion, weather, and other external influences.

Application at cold temperatures generally means that the sealant will be applied when the joint is at its most "open" configuration (i.e., dimension of greatest gap expansion). At higher temperatures, the substrates have expanded, and the sealant will be in compression. This could force the sealant out of the joint and create a portion of the sealant to be exposed directly to the environment such as traffic.

14.4.5.3 Cure rate. The speed at which the sealant develops strength can be critically important in some applications. Although many elastomeric sealants require long curing times, most set to a tack free state in several minutes to 24 hrs.

In general, all adhesives have a defined time that the material can remain in a form that is ready to apply. This time is called the sealant's pot life or working life. The pot life is related to the mechanism that controls the curing or setting reaction. Some chemical reactive sealants can be frozen to extend the pot life once the components are mixed.

Depth of cure is another important consideration for a sealant. Depth of cure is also related to the mechanism by which the sealant cures or sets. Certain moisture curing, single component sealants relay on moisture transmission through the sealant to effect cure. If the skin of the sealant sets first, it may inhibit the passage of moisture to the interior center section of the sealant bead. For these types of sealants, a joint design is required where moisture can easily reach the material. With solvent-based systems, a skinning effect can also inhibit curing of the adhesive in the center of the bead. In this case, the skin inhibits the volatilization of the solvent, and the sealant retains a "wet" condition.

Tack free time can also be an important characteristic of the sealant. As long as the surface of the adhesive retains a degree of tack, it is subject to possible contamination by dirt or other air borne particles. With long tack free times, the sealant can easily be disturbed if a tool or workman's body accidentally touches the bead before it reaches handling strength.

Shrinkage during cure is another important characteristic. Excessive shrinkage can cause voids in the sealant joint and uneven distribution of stress. Shrinkage can lead to finished joints that are not completely filled with sealant. This could cause dirt and other contaminants to build up in the recess of the joint over time affecting appearance and perhaps durability. Dirt trapped in the surface of the sealant could cause abrasion effects as the sealant is cycled over time and thus lead to early cohesive failure. Shrinkage can also place added stress on the adhesion characteristics of the sealant. Shrinkage is most significant with sealants that set by release of solvent or water from the formulation.

14.5 Performance Requirements

No mater what the application, the sealant must perform three basic functions:

- Fill the space to create a seal
- Form an impervious barrier to fluid flow or particulate transfer
- Maintain the tightness of the seal in the operating environment

The key performance properties that determine how well a sealant accomplishes these main functions include: movement capability, mechanical properties, adhesion, and durability. In certain applications, appearance is also an important performance requirement. Of course, cost must be a factor in all of these considerations. No single characteristic, but a proper balance of several essential properties, leads to selection of the proper sealant. Table 14.5 describes some of the more conventional sealants as a function of several common performance characteristics.

TABLE 14.5 Comparison of Various Sealants by Performance Properties[11]

Properties	Asphalt, bituminous	Oleo-resinous	Poly-butene	Butyl	Hypalon	Neoprene	Styrene-butadiene	Acrylic-solution	Acrylic-emulsion	Polyvinyl acetate	Poly-sulfide	Poly-urethane	Silicone	Fluoro-polymer
NV, % (by weight)	70–90	96–99+	99+	74–99+	85–90	80–85	60–70	80–85	80–85	70–75	96–99+	94–99+	98+	99+
Wt (lb/gal)	9–12	14–20	13–14	10–13	10–11	11–12	9–10	12–13	12–14	12–13	12–13	11–12	10–12	14–16
Max. design joint movement, % (±)	<5	<5	5–10	10–15	10–15	10–15	5–10	10–15	5–10	<5	25	20–25	20–25	10–20
Recovery—after joint movement[1]	P	P	NA[2]	F–G	F	P–F	P–F	F	F	P–F	F–G	G	E	F–G
Shrinkage, %	10–20	<4	<1	<1–20	10–15	10–20	20–30	10–20	10–20	20–25	<4	<6	<2	<1
Life expectancy, yr (exterior)[3]	1–2	2–10	5–10	5–15	5–15	5–15	3–10	5–20	2–20	1–3	10–20	5–15	20+	10–20
Cure type	Evap.	Oxidtn	Non-Cure	Evap.	Chem.	Chem.	Evap.	Evap.	Evap.	Evap.	Chem.	Chem.	Chem.	Chem.
Practical service temp. range— max. °F	150	150	180	200	225	210	180	180	180	150	250	250	400	500
min. °F	0	0	–40	–20	–25	–25	–10	0	0	0	–40	–40	–90	–10
Adhesion—unprimed[4] (to common bldg. materials)	G	G	G–E	G	G–G	F	F	E	G	F–G	G	F	P–F	F
Components—one-, two-part, tape	1	1	1, T	1, T	1	1, 2	1	1	1	1	1, 2	1, 2	1, 2	2
Applications:														
Glazing		X	X	X			X	X	X		X	X	X	
Exterior caulking		X	X	X	X	X	X	X	X		X	X	X	
Curtain walls														
Metal		X	X	X		X		X	X		X	X	X	
Masonry		X	X	X		X		X	X		X	X	X	
Expansion joint				X	X	X		X	X		X	X	X	
Interior caulking	X	X								X	X			
Floor joints											X	X		

[1] E = less than 10% compression set; G = 10–20%; F = 20–30%; P = >30%.
[2] NA = Not applicable.
[3] Mild to severe conditions.
[4] Adhesion of all materials improved with suitable primers.

14.5.1 Movement capability

The most important performance requirement of any sealant is its movement capability expressed in percentage extension (+%) and percentage compression (−%). Movement capability is expressed as a ±value. Most sealant standards include a specification for the movement capability of the sealants. Some standards include several classes, such as Class 12.5 and Class 25 in ASTM C-920 for ±12.5% and ±25% joint movement.

Movement capability should be determined by the ability of the joint to move either through external forces (i.e., mechanical separation) or through natural forces such as thermal expansion and contraction. Coefficients of expansion of common materials were given in Fig. 2.12 earlier in this Handbook. Coefficients of expansion of common building materials that often provide the substrates for sealants are given in Table 14.6.

Sealants that set to a tough, inflexible compound, such as oil and resin-based sealants, have a low movement capacity. The development of synthetic sealants has improved the movement capabilities of sealant formulations to new stature. Low modulus silicones for example have been developed with movement capability of +100% and −50%. Sealants with large movement capabilities are often used for difficult building joints and highway joints. Table 14.5 describes a range of sealants and their movement capabilities.

Sealants that have low movement capabilities are polybutene, polyisobutene, and oil and resin-based caulks. These are low cost sealants with an approximate joint movement capability of only ±5%. Generally, they require a minimal amount of surface treatment, have good adhesion, and are supplied in one component cartridges. Low movement sealants are primarily utilized in static joints having minimal exterior exposure.

Medium movement sealants include latexes and butyls (±7.5%) and Hypalon, neoprene, and solvent release acrylics (±12.5%). These sealants offer a longer life than the low movement caulks and significantly better performance characteristics, but at a higher price.

High movement elastomeric sealants include polysulfides (±25%), polyurethanes (+40/−25%), and silicones (+100/−50%). The service lives of these sealants are 20 years or more. They have excellent performance characteristics, but their price is significantly higher than the low and medium movement capability sealants.

ASTM C 719 can be used to measure the ability of the sealant to withstand repeated movement (expansion and contraction). This method incorporates a conditioning period followed by testing of specific movement capability requirements.

TABLE 14.6 Coefficients of Linear Expansion of Common Building Materials[11]

Clay, masonry (Brick, clay or shale)	
Brick, fire clay	3.6
Tile, clay or shale	3.3
Tile, fire clay	2.5
Concrete	
Gravel aggregate	6.0
Lightweight structural	4.5
Concrete, masonry	
Cinder aggregate	3.1
Dense aggregate	5.2
Expanded-shale aggregate	4.3
Expanded-slag aggregate	4.6
Volcanic pumice and aggregate	4.1
Metals	
Aluminum	13.0
Brass, red 230	10.4
Bronze, arch. 385	11.6
Copper, 110	9.8
Iron	
Cast gray	5.9
Wrought	7.4
Lead, common	16.3
Monel	7.8
Stainless steel	
Type 302	9.6
Type 304	9.6
Structural steel	6.7
Zinc	19.3
Glass, plate	5.1
Plaster	
Gypsum aggregate	7.6
Perlite	5.2
Vermiculite aggregate	5.9
Plastics	
Acrylics	40–50
Lexan®	37.5
Phenolics	25–66
Plexiglas®	39
Polyesters, glass reinforced	10–14
PVC	33
Vinyls	24–40
Stone	
Granite	6.2
Limestone	3.5
Marble	7.3

14.5.2 Other mechanical properties

Physical properties other than movement capability of the sealant could be a significant consideration in their selection, depending on the nature of the application. Properties that are often of interest include hardness, modulus, tensile strength, tear resistance, and compressive strength and set.

Hardness is the relative measurement of the flexibility of the sealant. As the hardness values increase, flexibility decreases. Hardness is often used as a quality control mechanism or as a measurement of aging or change within the sealant. A change of hardness upon aging is an indication that further curing or degradation is taking place. Hardness, by itself, is not an adequate property to use to select a sealant. However, most sealants have Shore A durometer hardness of 15–70. ASTM D 2240 and ASTM C 661 are used to measure hardness.

Modulus is often related to the movement capability of the adhesive. It is the ratio of the force (stress) required to elongate (strain) a sealant. Modulus has an important bearing on the ability of the sealant to handle joint movement. A superior sealant offers minimum resistance to such movement and, therefore, would have a lower modulus rating. In general, low or medium modulus sealants are able to accommodate much greater joint movement without putting large stress on either the sealant or the substrate surface. Elongation when considered with modulus sometimes can predict the extension and compression characteristics of a sealant. Dynamic loads, shock, and rapid variations in stress can also cause certain sealants to fail. More flexible sealants, that can stretch and return to their original shape in a short time, are preferred for such dynamic loading conditions. ASTM D 412 is used to measure modulus of sealants.

Tensile strength is needed to some degree to avoid cohesive failure of the sealant due to movement. A high degree of tensile strength is usually not necessary for most sealants. However, the sealant should have the cohesive strength necessary so that it does not become the mode of failure. ASTM D 412 is used to measure tensile strength in addition to elongation and modulus.

Tear or abrasion resistance is important because sealants are often exposed to abrasion and mechanical wear. They must have good puncture, abrasion, and tear resistance. Sealants exhibit varying degrees of tear resistance, with urethanes having the best overall properties. Tear resistance of sealants is measured by ASTM D 624.

Compressive strength and compression set are important parameters for many applications. Compressive strength is the maximum compressive force that the sealant can withstand before physically deteriorating. Compression set is the inability of the sealant to recover after prolonged compression. Compression set is usually caused by further curing or degradation of the sealant by the service environment. Compression set is undesirable in a joint that must expand and contract. In general, silicones and urethanes have the lowest compression sets, followed by polysulfides, butyls and acrylics.

Stress relaxation is the reduction in stress as the elongation remains constant. Sealants having a high degree of stress relaxation may fail cohesively when held in an elongated period for a long duration of time. Stress relaxation is more or less the reverse of creep, where elongation increases under constant stress. Sealants that have high creep do not support continuous loads very well.

14.5.3 Adhesion

Adhesion is a measure of the ability of a sealant to adhere to a substrate. Adhesion must be maintained even while the sealant material is being stretched. Adhesion is often a critical requirement for most sealing applications. Loss of adhesion in a threaded or overlap joint can result in the fluid wicking through the sealant-substrate interface, leading to increased degradation of the sealant or corrosion of the substrate. Loss of adhesion can also allow vibration loosening of a threaded joint or loss of acoustical coupling in a sound deadening sealant. Joint design, surface preparation, and primers are used to assure good adhesion.

The adhesion requirement will depend upon the nature of the substrate and the end-use for the product that contains the seal. Since the sealant that provides the best balance of physical properties may not provide a high degree of adhesion to the substrate, it is frequently necessary to resort to the use of primers or sealants that use adhesion promoters as additives. Of course, the importance of bond preparation and cleanliness cannot be overemphasized. A surface preparation that is more than simple solvent wiping or abrasion may be required in many of the industries that use sealants. Surface shape, surface smoothness, and joint design are also important factors that can immensely affect the degree of adhesion in a sealant joint.

With sealants, often the most important stresses affecting adhesion are not external stresses but those from within the sealant material itself. Many sealants change elastic modulus or shrink significantly as they age, and this could drastically affect the ability of the sealant to maintain adhesion.

The adhesion quality of sealants is often measured by ASTM test methods that have been designed for adhesives. (See Chapter 4.) MIL-S-8802 can be also be used to indicate the adhesive qualities of a sealant. Since peel stress is a prominent type of stress in a sealant joint, ASTM C 794 has been designed to test a sealant's adhesion in peel.

14.5.4 Durability

The broad requirement of durability can be broken down into properties that are more specifically related to the particular sealant application, such as thermal capability, chemical resistance, chemical compatibility, permeability, and weatherability. Table 14.7 shows some of the advantages and disadvantages of common sealant materials with regard to their durability in standard environments.

Commercial sealants have a wide variation of thermal capabilities. Two factors must be considered with regard to temperature resistance: extremes of temperatures during the service life and the range and frequency of temperature fluctuations. Temperature affects sealants in different ways. These changes could include strength loss and embrittlement. When testing for heat resistance ASTM D 573 can be used. Silicones have proved to be an excellent high temperature resistant sealant with long term resistance to 400°F and short term resistance to 500°F. Fluorosilicones are good for continuous exposure up to 500°F. Acrylic latex and certain polysulfides are good up to 200°F.

Cyclic temperature variations are important to many construction sealants. These variations can occur by daily excursions in temperatures especially when the sealant is in a region that sees direct exposure to both sunlight and shade depending on the time of day. Such temperature variation could be as much as 60°F. These daily variations are likely to occur within a few hours after the sealant has been first applied. Another temperature cycle is due to the seasonal variation. A joint will commonly see extension in the winter and compression in the summer. Consequently most test methods for movement capability require a measurement of the effect of compression at high temperatures and of tension at low temperatures.

Chemical properties of concern in selecting sealants usually consist of resistance to fluids and other chemicals. Continuous exposure can cause sealants to disintegrate, contract, swell, embrittle, or become permeable. As might be expected, chemical resistance varies considerably between sealant families and within families depending on the exact formulations.

Resistance to oil, water, and chemicals is best measured by ASTM D 471. In oil, chlorosulfonated polyethylenes, acrylics, and polychloroprene swell moderately. Polysulfides, fluorosilicones, and polyurethanes have the best resistance. For years, polysulfides have been used as integral fuel tank sealants. For general chemical resistance, silicones, polychloroprenes, and butyls are useful. Urethanes retain

TABLE 14.7 Advantages and Disadvantages of Common Sealant Materials with Respect to Their Durability (Based on Reference 10)

Sealant	Advantages	Disadvantages
Oleoresinous	Low cost Easy applications	Poor joint movement Limited adhesion Cracks on aging
Acrylic, solvent based	Good adhesion Good UV resistance	Limited joint movement Odor Slow setting
Butyl, solvent based	Low cost Broad adhesion range	Joint shrinkage Flammability Stringy when applied
Vinyl Acrylic, latex	Water cleanup Good shelf stability	High joint shrinkage Limited water resistance Poor extrudability at low temperatures
Polyurethane	Excellent joint movement Excellent adhesion range No shrinkage Cold temperature resistance	Limited UV resistance Limited package stability Prone to hydrolysis Low cure (one component)
Polysulfide	Excellent joint movement Excellent adhesion range No shrinkage Good fuel and solvent resistance	Limited package stability Limited UV resistance Bad sulfur type odor
Silicone, acetoxy	Superior UV resistance Excellent joint movement Rapid curing No shrinkage Excellent water resistance High temperature resistance	High cost Poor adhesion to concrete Cannot paint sealant surface
Silicone, modified	Excellent joint movement Good UV resistance Good adhesion range Paintable Odorless High temperature resistance	Medium cost Limited adhesions to concrete and some plastic substrates
Styrene Block Copolymer	Excellent joint movement Excellent UV resistance	Flammability High joint shrinkage

good properties after being exposed to alkalis. Polysulfides also have good chemical resistance; silicones are fair.

Polysulfide, polyurethanes, silicones, acrylics, polychloroprenes and chlorosulfonated polyethylenes exhibit excellent water resistance. For measuring vapor transmission, ASTM E 96 can be used. Generally, butyls have the lowest moisture transmission rates and are used as the standard for all other types of elastomers.

A sealant's resistance to the weather must include its endurance to water, heat, cold, cycling, ultraviolet, and solar radiation. In industrial areas, resistance to acid fumes and ozone must be taken into account. In shore areas, saltwater spray and algae must also be carefully considered. UV radiation has an detrimental effect on certain sealant materials such as urethanes. Some sealants, such as silicones are inherently UV resistant. Where UV resistance is a concern, UV absorbers and inhibitors can be added to the sealant formulation.

14.5.5 Appearance

Appearance may be an important selection criterion in choosing a sealant, whereas it is not generally used in choosing adhesives. It must be determined whether the sealant causes discoloration of surrounding substrate areas and whether the sealant itself changes in appearance or color over time. Staining of substrate surfaces is usually caused by some chemical extruding from the sealant over time. Polysulfides and silicones have been notable for staining in the past, but newer formulations have been developed to correct the problem. ASTM has developed a test, C-510, which can determine staining on questionable surfaces.

Color plays an important part in the selection of a sealant. There are at least 10 standard and 30 semi-standard colors widely available for sealants. Color matches are commonly required for many sealant materials. One-part silicones are limited to about 9 shades, but a wide variety of colors are available for the two component silicone sealants. It should be noted that the sealant's color could change on aging for a variety of reasons. For example, silicone sealants can acquire dirt by electrostatic attraction; urethanes and certain other sealants will darken and change color from UV exposure.

14.6 Sources of Information and Assistance

The leading supplier of information and assistance regarding the selection and use of sealant materials is the manufacturer of the sealant itself. Although this information will be specific to the manufacturer's product, it will provide guidance as to the appropriateness of the sealant to the requirements in question, and it may provide information that may be useful to compare one sealant to another.

The performance data that are offered by the manufacturer must be quantifiable and comparable to other materials and to other applications. Test methods and specification have been established by ASTM, U.S. Bureau of Standards, Adhesive and Sealant Council and other agencies. Appendix C lists important specification for sealants and caulks. Appendix C also lists ASTM test methods for seals and sealants.

The National Research Council of Canada has developed an expert computer-based system for the selection of sealants. This work has been developed into a Microsoft Windows based system called "Sealant System" that is relatively easy to understand and use.[14] The program is capable of accepting a joint design and substrate material and will calculate thermal and moisture movements and provide the joint width dimensions or sealant movement range. The program selects from its database of sealants those materials that meet the joint movement and input sealant characteristic properties. Computer selection of sealants and adhesives are discussed in Chapter 19.

References

1. Amstock, J. S., "Sealants", in *Handbook of Adhesive Bonding,* C. V. Cagle, ed. (New York: McGraw Hill, 1973).
2. Thompson, J. E., "Design Considerations Unique to Sealants", *Adhesives and Sealants,* vol. 3, Engineered Materials Handbook, ASM International, 1990.
3. Zakim, J. and Shihadeh, M., "A Comprehensive Guide to Sealants and Caulking Compounds", *Adhesives Age,* August 1965.
4. Dunn, D. J., "Sealants and Sealant Technology", *Adhesives and Sealants,* vol. 3, Engineered Materials Handbook, ASM International, 1990.
5. "Form in Place Gaskets Are Leak-Free, Low-Cost", *Materials Engineering,* January, 1973.
6. Beals, J., "Novel Dispensing Methods for Zero Knit Line Liquid Gaskets", *Adhesives and Sealants Industry,* September 1997.
7. Loctite Corporation, FastGasket 5950. Also, "Bonding, Sealing Materials Enhance Reliability of Auto Electric Systems", *Adhesives Age,* August 1997.
8. *Loctite Worldwide Design Handbook,* Loctite Corporation, 1996–97.
9. Amstock, J. S., "Reclassification of Joint Sealants", *Adhesives Age,* February, 1969.
10. Usmani, A. M., "Environmental Considerations Unique to Sealants", *Adhesives and Sealants,* vol. 3, Engineered Materials Handbook, ASM International, 1990.
11. Prane, J. W., Elias, M., and Redman, R., "Sealants and Caulks", in *Handbook of Adhesives,* 3rd ed. I. Skiest, ed. (New York: Van Nostrand Reinhold Publishing, 1992).
12. Hoover, S., "Foam-In-Place Gasketing System Offers Benefits", *Adhesives Age,* August 1994.
13. Dynafoam Gasketing, Norton Performance Plastics Corp., "New Gasketing Technique Cuts Costs", *Plastics Technology,* April 1997.
14. O'Connor, T. F. and Myers, J. C., "Black Magic and Sealant Joints: Very Little in Common Anymore", *Adhesives Age,* August 1997.

Chapter 15

Methods of Joining Plastics Other Than With Adhesives

15.1 Introduction

Plastics are common substrates found in all industries. Plastic materials range from inexpensive, commodity-type products, such as those used for packaging, to a group of resins that have become known as "engineering plastics" by virtue of their outstanding mechanical and chemical properties and durability. Engineering plastics are often used as an alternative to metals in applications requiring high specific strength, low cost, and high volume production methods.

To make large plastic assemblies, the most cost effective method often involves molding smaller subsections and joining them together. In such cases, the manufacturer has a variety of joining options available including adhesives, thermal welding, solvent cementing, and mechanical fastening.

The final plastic assembly relies heavily on the characteristics of the joint. In most applications, the joint must be nearly as strong as the substrate throughout the expected life of the product. The method used to join plastics should be carefully considered by the designer. In addition to strength and permanence requirements, selection criteria for an assembly method should include tooling cost, labor and energy cost, production time, appearance of the final part, and disassembly requirements. Some plastic materials will be more suited for certain joining processes than others due to their physical and chemical characteristics.

A consideration in plastic assembly that is usually not dominant when joining other substrates is the speed of the joining operation. Plastic products generally require very high volume assembly processes in industries such as consumer products, automotive, and packaging. Speed, simplicity, and reliability are key concerns in most of these high volume assembly processes. Speed, simplicity, and reliability are also often the order of priority in selecting a joining method. Because of the nature of the substrate and the application, exceedingly high strengths and durability in exotic environments are not generally necessary, and these properties are gladly sacrificed for faster, low cost production methods. Often, there is not enough time for critical surface preparation or non-destructive testing of every part. In certain industries such as the automotive industry, plastic materials may be chosen because of their fast joining ability. Thus, thermoplastics are often preferred over thermosets because they can be joined via thermal welding processes in a few seconds, compared to possibly several hours for adhesive bonding.

Although adhesive bonding often proves to be effective, there are various other ways of joining plastics to themselves or to other materials. The principal methods of joining plastics are adhesive bonding, thermal welding, solvent cementing, and mechanical fastening. The non-adhesive methods of joining will be the subject of this chapter. The sections that follow will describe these various methods of joining. Information will be provided regarding how to choose the most appropriate process for a specific substrate and application. The plastic materials best suited for each will be identified. Important process parameters and test results are reviewed. Recommendations regarding joining methods for specific types of plastic materials (e.g., polyethylene, glass reinforced epoxy, polysulfones) will be given in Chapter 16.

15.2 Plastic Materials

There are many types of plastic materials with a wide range of properties depending on the base polymer and the additives used. There are excellent sources of information available regarding plastics materials, design practices, manufacturing methods, and applications. References 1–4 provide broad tutorials on these materials. Plastics are used routinely in many commodity items such as packaging, pipe, clothing, appliances, and electronics. They are also increasingly being used in structural and engineering applications such as in the aerospace, building, and automotive sectors.

All plastics can be classified into two categories: thermoplastics and thermosets. Thermoplastics are not crosslinked, and the polymeric molecules making up the thermoplastic can easily slip by one another. This slip or flow can be caused by thermal energy, by solvents or other chemicals, and by the application of continuous stress. Thermoplastics can be repeatedly softened by heating and hardened on cooling. Hence, they can be welded by the application of heat. Thermoplastics can also be dissolved in solvents so that it is also possible to join thermoplastic parts by solvent cementing. Typical thermoplastics are polyethylene, polyvinyl chloride, polystyrene, polypropylene, nylon, and acrylic.

"Engineering thermoplastics" are a class of thermoplastic materials that have exceptionally high temperature and chemical resistance. They have properties very similar to the thermoset plastics and to metals. As a result, they are not as easy to heat weld or to solvent cement as are the more conventional thermoplastics. Typical engineering thermoplastics are polysulfone, acetal, amide-imide, and thermoplastic polyimide.

Thermoset plastics are crosslinked by chemical reaction so that their molecules cannot slip by one another. They are rigid when cool and cannot be softened by heat. If excessive heat is applied, thermoset plastics will only degrade. Consequently, they are not heat weldable. Because of their chemical resistance, they are also not capable of being solvent cemented. Thermoset plastics are usually joined by either adhesive bonding or mechanical fastening. Typical thermosetting plastics are epoxy, urethane, phenolic, and melamine formaldehyde.

Several common thermosetting plastics, their tradenames, and suppliers are presented in Table 15.1. Similar information for thermoplastics is offered in Table 15.2. The suppliers represented are only a few of the base resin suppliers. In addition, there are many companies which formulate filled plastic systems from these resins. These smaller companies are quite often the ultimate supplier to the end-product fabricator. Both the resin manufacturer and the formulator have considerable influence on the bondability of the final material. This is discussed in other parts of this Handbook, and the influence of surface properties is specifically addressed in Chapters 6 and 16.

Plastics offer several important advantages compared with traditional materials, and their joining process should not detract from these benefits. The greatest advantage is low processing cost. Low weight is also a major advantage. Relative densities of most unfilled plastics materials range from 0.9 to 1.4, compared with 2.7 for aluminum and 7.8 for steel. Relative densities for highly filled plastic compounds can rise to 3.5.

Other advantages of plastics are their low frictional resistance, good corrosion resistance, insulation properties, and ease at which they can be fabricated into various shapes. The chemical resistance of plastics is dependent on the type of plastic. Some plastics have chemical resistance comparable with metals, others are attacked by acids, and others are attacked by solvents and oils. Certain plastics are also attacked by moisture especially at moderately elevated temperatures. The nature of this attack is generally first a swelling of the substrate

TABLE 15.1 Typical Tradenames and Suppliers of Common Thermosetting Plastics[5]

Plastic	Typical trade names and suppliers
Alkyd	Plaskon (Allied Chemical) Durez (Hooker Chemical) Glaskyd (American Cyanamid)
Diallyl phthalates	Dapon (FMC) Diall (Allied Chemical) Durez (Hooker Chemical)
Epoxies	Epon (Shell Chemical) Epi-Rez (Celanese) D.E.R. (Dow Chemical) Araldite (Ciba) ERL (Union Carbide)
Melamines	Cymel (American Cyanamid) Plaskon (Allied Chemical)
Phenolics	Bakelite (Union Carbide) Durez (Hooker Chemical) Genal (General Electric)
Polybutadienes	Dienite (Firestone) Ricon (Colorado Chemical Specialties)
Polyesters	Laminae (American Cyanamid) Paraplex (Rohm and Haas) Selectron (PPG)
Silicones	DC (Dow Corning)
Ureas	Plaskon (Allied Chemical) Beetle (American Cyanamid)

and then finally dissolution. Some plastics are transparent and others are translucent. These plastics can pass UV radiation so that the adhesive or joint could be affected by UV. In certain cases, the plastic itself can be degraded by UV. Resistance to degradation can be improved by the addition of UV absorbers such as carbon black. All plastics can be colored by the addition of pigments. This generally eliminates the need for painting.

The most serious disadvantage of plastics is that, compared to metals, they have low stiffness and strength and are not suitable for use at high temperatures. Some plastics are unusable at temperatures above 120°F. Others are capable of resisting temperatures up to 600°F. Surface hardness of plastics is generally low. This can lead to indentation and compression under local stress. The elastic modulus of plastic materials is relatively low compared to metals. They can be elongated and compressed with applied stress. The thermal expansion coefficient of plastics is generally an order of magnitude greater than metals. This results in internal joint stress when joined to a substrate having a much lower coefficient of expansion.

TABLE 15.2 Typical Tradenames and Suppliers of Common Thermoplastics[5]

Thermoplastic	Typical trade names and suppliers
ABS	Marbon Cycolac (Borg-Warner) Abson (B. F. Goodrich) Lustran (Monsanto)
Acetals	Delrin (E. I. du Pont) Celcon (Celanese)
Acrylics	Plexiglas (Rohm and Haas) Lucite (E. I. du Pont)
Aramids	Nomex (E. I. du Pont)
Cellulosics	Tenite (Eastman Chemical) Ethocel (Dow Chemical) Forticel (Celanese)
Ionomers	Surlyn A (E. I. du Pont) Bakelite (Union Carbide)
Low-permeability thermoplastics	Barex (Vistron/Sohio) NR-16 (E. I. du Pont) LPT (Imperial Chemical Industries)
Nylons (see also Aramids)	Zytel (E. I. du Pont) Plaskon (Allied Chemical) Bakelite (Union Carbide)
Parylenes	Parylene (Union Carbide)
Polyaryl ether	Arylon T (Uniroyal)
Polyaryl sulfone	Astrel (3M)
Polycarbonates	Lexan (General Electric) Merlon (Mobay Chemical)
Polyesters	Valox (General Electric) Celanex (Celanese) Celanar Film (Celanese) Mylar Film (E. I. du Pont) Tenite (Eastman Chemical)
Polyethersulfone	Polyethersulphone (Imperial Chemical Industries)
Polyethylenes, polypropylenes, and polyallomers	Alathom Polyethylene (E. I. du Pont) Petrothene Polyethylene (U.S.I.) Hi-Fax Polyethylene (Hercules) Pro-Fax Polypropylene (Hercules) Bakelite Polyethylene and Polypropylene (Union Carbide) Tenite Polyethylene and Polypropylene (Eastman) Irradiated Polyolefin (Raychem)
Polyimides and polyamide-imides	Vespel SP Polyimides (E. I. du Pont) Kapton Film (E. I. du Pont) Pyralin Laminates (E. I. du Pont) Keramid/Kinel (Rhodia) P13N (Ciba-Geigy) Torlon Polyamide-Imide (Amoco)
Polymethyl pentene	TPX (Imperial Chemical Industries)
Polyphenylene oxides	Noryl (General Electric)
Polyphenylene sulfides	Ryton (Phillips Petroleum)

TABLE 15.2 Typical Tradenames and Suppliers of Common Thermoplastics[5] (*Continued*)

Thermoplastic	Typical trade names and suppliers
Polystyrenes	Styron (Dow Chemical)
	Lustrex (Monsanto)
	Dylene (Koppers)
	Rexolite (American Enka)
Polysulfones	Ucardel (Union Carbide)
Vinyls	Pliovic (Goodyear Chemical)
	Diamond PVC (Diamond Alkali)
	Geon (B. F. Goodrich)
	Bakelite (Union Carbide)

15.3 Joining Methods for Plastics

The joining of plastics with adhesives is generally made difficult because of their low surface energy, poor wettability, and presence of weak boundary layers. Adhesive bonding is a relatively slow process that could be a significant bottleneck in many industries that produce high volume plastic assemblies. However, with plastic substrates, the designer has a greater choice of joining techniques than with many other substrate materials. Thermoset plastics must usually be bonded with adhesives or mechanically fastened, but many thermoplastics can be joined by solvent or heat welding as well as by adhesives or with mechanical fasteners.

In the thermal or solvent welding processes, the plastic resin that makes up the substrate itself acts as the adhesive. These processes require that the surface region of the substrate be made fluid, so that it can wet the mating substrate. If the mating substrate is also a polymer, both substrate surfaces can be made fluid so that the resin can molecularly diffuse into the opposite interface. This fluid interface region is usually achieved by thermally heating the surface areas of one or both substrates, or by dissolving the surfaces in an appropriate solvent. Once the substrate surfaces are in a fluid condition, they are then brought together and held in place with moderate pressure. At this point, the molecules of substrate A and substrate B will diffuse into one another and form a very tight bond. The fluid polymer mix then returns to the solid state by the dissipation of solvent or by cooling from the molten condition.

Thermal welding can also be used to bond one polymeric substrate to another substrate that is not polymeric. In these cases, the molten polymer surface wets the other substrate and acts as a hot melt type of adhesive. Internal stresses that occur on cooling the interface from the molten condition are the greatest detriment to this method of bonding. Solvent welding is also appropriate when the non-polymeric substrate is porous. If it is not porous, the solvent may become entrapped at the interface and cause very weak joints.

Table 15.3 indicates the most common joining methods for various plastic materials. Descriptions of these joining techniques are summarized both in Table 15.4 and in the sections to follow. In general, joining methods for plastics can be classified as either:

1. Welding by direct heating (heated tool, hot gas, resistance wire, infrared, laser, extrusion)
2. Induced heating (induction, electrofusion, dielectric)
3. Frictional heating (ultrasonic, vibration, spin)
4. Solvent welding
5. Mechanical fastening

Equipment costs for each method vary considerably, as does the amount of labor involved and the speed of the operation. Most techniques have limitations regarding the design of the joint and the types of plastic materials that can be joined.

15.4 Direct Heat Welding

Welding by the direct application of heat provides an advantageous method of joining many thermoplastic materials that do not degrade rapidly at their melt temperatures. The principal methods of direct welding are:

- Heated tool
- Hot gas
- Resistance wire
- Laser
- Infrared

These methods are generally capable of joining thermoplastics to themselves and other thermoplastics. In certain cases, they may also be used to weld thermoplastics to non-plastic substrates.

15.4.1 Heated tool welding

Fusion or heated tool welding is an excellent method of joining many thermoplastics. In this method, the surfaces to be fused are heated by holding them against a hot surface. When the plastic surface becomes visibly molten, the parts are removed from the hot surface. They are then immediately joined under slight pressure (5–15 psi) and allowed to cool and harden. The molten polymer acts as a hot-melt adhesive providing a bond between the substrates.

TABLE 15.3 Assembly Methods for Plastics[6]

Plastic	Adhesives	Dielectric welding	Induction bonding	Mechanical fastening	Solvent welding	Spin welding	Thermal welding	Ultrasonic welding
Thermoplastics								
ABS	X	. .	X	X	X	X	X	X
Acetals	X	. .	X	X	X	X	X	X
Acrylics	X	. .	X	X	X	X	. .	X
Cellulosics	X	. .	. .	X	X	X	. .	. .
Chlorinated polyether	X	X	. .	. .	. .	. .	. .	. .
Ethylene copolymers	. .	. .	. .	. .	. .	. .	X	. .
Fluoroplastics	X	. .	. .	. .	. .	. .	. .	. .
Ionomer	. .	X	. .	. .	. .	. .	X	. .
Methylpentene	X	. .	. .	X	. .	. .	. .	X
Nylons	X	. .	X	X	X	X	X	X
Phenylene oxide-based materials	X	. .	. .	X	X	X	X	X
Polyesters	X	. .	. .	X	X	. .	. .	X
Polyamide-imide	X	. .	. .	X	. .	. .	. .	. .
Polyaryl ether	X	. .	. .	X	. .	. .	. .	X
Polyaryl sulfone	X	. .	. .	X	. .	. .	. .	X
Polybutylene	. .	. .	. .	. .	. .	. .	. .	X
Polycarbonate	X	. .	X	X	X	X	X	X
Polycarbonate/ABS	X	. .	. .	X	X	X	X	X
Polyethylenes	X	X	X	X	. .	X	X	X
Polyimide	X	. .	. .	X	. .	. .	. .	. .
Polyphenylene sulfide	X	. .	. .	X	. .	. .	. .	. .
Polypropylenes	X	X	X	X	. .	X	. .	X
Polystyrenes	X	. .	X	X	X	X	X	X
Polysulfone	X	. .	. .	X	X	. .	. .	. .
Propylene copolymers	X	X	X	X	. .	X	X	. .

PVC/acrylic alloy	X		X					X
PVC/ABS alloys	X	X			X	X	X	X
Styrene acrylonitrile	X		X	X	X	X	X	
Vinyls	X	X	X	X			X	
Thermosets								
Alkyds	X							
Allyl diglycerol carbonate	X			X				
Diallyl phthalate	X			X				
Epoxies	X			X				
Melamines	X							
Phenolics	X			X				
Polybutadienes	X							
Polyesters	X			X				
Silicones	X							
Ureas	X			X				
Urethanes	X							

TABLE 15.4 Bonding or Joining Plastics; What Techniques are Available and What Do They Have to Offer?[7]

Technique	Description	Advantages	Limitations	Processing considerations
Solvent cementing and dopes	Solvent softens the surface of an amorphous thermoplastic; mating takes place when the solvent has completely evaporated. Bodied cement with small percentage of parent material can give more workable cement; fill in voids in bond area. Cannot be used for polyolefins and acetal homopolymers	Strength, up to 100% of parent materials, easily and economically obtained with minimum equipment requirements	Long evaporation times required; solvent may be hazardous; may cause crazing in some resins	Equipment ranges from hypodermic needle or just a wiping media to tanks for dip and soak. Clamping devices are necessary, and air dryer is usually required. Solvent-recovery apparatus may be necessary or required. Processing speeds are relatively slow because of drying times. Equipment costs are low to medium
Thermal Bonding				
Ultrasonics	High-frequency sound vibrations transmitted by a metal horn generate friction at the bond area of a thermoplastic part, melting plastics just enough to permit a bond. Materials most readily weldable are acetal, ABS, acrylic, nylon, PC, polyimide, PS, SAN, phenoxy	Strong bonds for most thermoplastics; fast, often less than 1 s. Strong bonds obtainable in most thermal techniques if complete fusion is obtained	Size and shape limited. Limited applications to PVCs, polyolefins	Converter to change 20 kHz electrical into 20 kHz mechanical energy is required along with stand and horn to transmit energy to part. Rotary tables and high-speed feeder can be incorporated.
Hot-plate and hot-tool welding	Mating surfaces are heated against a hot surface, allowed to soften sufficiently to produce a good bond, then clamped together while bond sets. Applicable to rigid thermoplastics	Can be very fast, e.g., 4–10 s in some cases; strong	Stresses may occur in bond area	Use simple soldering guns and hot irons, relatively simple hot plates attached to heating elements up to semiautomatic hot-plate equipment. Clamps needed in all cases

Method	Description	Notes	Equipment
Hot-gas welding	Welding rod of the same material being joined (largest application is vinyl) is softened by hot air or nitrogen as it is fed through a gun that is softening part surface simultaneously. Rod fills in joint area and cools to effect a bond	Strong bonds, especially for large structural shapes	Relatively slow; not an "appearance" weld	Requires a hand gun, special welding tips, an air source and welding rod. Regular hand-gun speeds run 6 in./min; high-speed hand-held tool boosts this to 48–60 in./min
Spin welding	Parts to be bonded are spun at high speed developing friction at the bond area; when spinning stops, parts cool in fixture under pressure to set bond. Applicable to most rigid thermoplastics	Very fast (as low as 1–2 s); strong bonds	Bond area must be circular	Basic apparatus is a spinning device, but sophisticated feeding and handling devices are generally incorporated to take advantage of high-speed operation
Dielectrics	High-frequency voltage applied to film or sheet causes material to melt at bonding surfaces. Material cools rapidly to effect a bond. Most widely used with vinyls	Fast seal with minimum heat applied	Only for film and sheet	Requires rf generator, dies, and press. Operation can range from hand-fed to semiautomatic with speeds depending on thickness and type of product being handled. 3–25 kW units are most common
Induction	A metal insert or screen is placed between the parts to be welded, and energized with an electromagnetic field. As the insert heats up, the parts around it melt, and when cooled form a bond. For most thermoplastics	Provides rapid heating of solid sections to reduce chance of degradation	Since metal is embedded in plastic, stress may be caused at bond	High-frequency generator, heating coil, and inserts (generally 0.02–0.04 in. thick). Hooked up to automated devices, speeds are high. Work coils, water cooling for electronics, automatic timers, multiple-position stations may also be required

TABLE 15.4 Bonding or Joining Plastics; What Techniques are Available and What Do They Have to Offer? (Continued)

Technique	Description	Advantages	Limitations	Processing considerations
		Adhesives*		
Liquids solvent, water base, anaerobics	Solvent-and-water-based liquid adhesives, available in a wide number of bases — e.g., polyester, vinyl — in one- or two-part form fill bonding needs ranging from high-speed lamination to one-of-a-kind joining of dissimilar plastics parts. Solvents provide more bite, but cost much more than similar base water-type adhesive. Anaerobics are a group of adhesives that cure in the absence of air	Easy to apply; adhesives available to fit most applications	Shelf and pot life often limited. Solvents may cause pollution problems; water-base not as strong; anaerobics toxic	Application techniques range from simply brushing on to spraying and roller coating-lamination for very high production. Adhesive application techniques, often similar to decorating equipment, from hundreds to thousands of dollars with sophisticated laminating equipment costing in the tens of thousands of dollars. Anaerobics are generally applied a drop at a time from a special bottle or dispenser
Pastes, mastics	Highly viscous single- or two-component materials which cure to a very hard or flexible joint depending on adhesive type	Does not run when applied	Shelf and pot life often limited	Often applied via a trowel, knife or gun-type dispenser; one-component systems can be applied directly from a tube. Various types of roller coaters are also used. Metering-type dispensing equipment in the $2,500 range has been used to some extent

Hot melts	100% solids adhesives that become flowable when heat is applied. Often used to bond continuous flat surfaces	Fast application; clean operation	Virtually no structural hot melts for plastics	Hot melts are applied at high speeds via heating the adhesive, then extruding (actually squirting) it onto a substrate, roller coating, using a special dispenser or roll to apply dots or simply dipping
Film	Available in several forms including hot melts, these are sheets of solid adhesive. Mostly used to bond film or sheet to a substrate	Clean, efficient	High cost	Film adhesive is reactivated by a heat source; production costs are in the medium-high range depending on heat source used
Pressure-sensitive	Tacky adhesives used in a variety of commercial applications (e.g., cellophane tool). Often used with polyolefins	Flexible	Bonds not very strong	Generally applied by spray with bonding effected by light pressure
Mechanical fasteners (staples, screws, molded-in inserts, snap fits and variety of proprietary fasteners)	Typical mechanical fasteners are listed on the left. Devices are made of metal or plastic. Type selected will depend on how strong the end product must be, appearance factors. Often used to joint dissimilar plastics or plastics to nonplastics	Adaptable to many materials; low to medium costs; can be used for parts that must be disassembled	Some have limited pull-out strength; molded-in inserts may result in stresses	Nails and staples are applied by simply hammering or stapling. Other fasteners may be inserted by drill press, ultrasonics, air or electric gun, hand tool. Special molding — i.e., molded-in-hole — may be required

*Typical adhesives in each class are: Liquids: 1. Solvent—polyester, vinyl, phenolics, acrylics, rubbers, epoxies, polyamide; 2. Water—acrylics, rubber-casein; 3. Anaerobics—cyanoacrylate; mastics—rubbers, epoxies; hot melts—polyamides, PE, PS, PVA; film—epoxies, polyamide, phenolics; pressure, sensitive—rubbers.

This method is often used in high volume operations where adhesive bonding is objectionably long. It is also often used to join low surface energy materials such as polypropylene where the cost and complexity required for substrate treatment and adhesive bonding cannot be tolerated. Surface treatment, other than simple cleaning, is not required for thermal welding. Heated tool welding is a simple, economical technique in which high strength joints can be achieved with large and small parts. Hermetic seals can also be achieved. Heated tool welding does not introduce foreign materials into the part and as a result plastic parts are more easily recycled.

Success in heated tool welding depends primarily on having the proper temperature at the heating surface and on the timing of the various steps in the process. These periods include time for application of heat, the time between removal of heat and joining of parts, and the time the parts are under pressure. The tool should be hot enough to produce sufficient flow for fusion within 10 secs. The parts are generally pressed against the heated tool with a certain degree of pressure. However, to avoid strain, the pressure on the parts should be released for a period of time before they are removed from contact with the heated tool. While some rules of thumb can apply, the final process settings for temperature, duration of heating and cooling times, and pressures will depend on the polymer. Adjustments will be required until the desired bond quality is achieved. The thickness of the molten layer is an important determinant of weld strength. Dimensions are usually controlled through the incorporation of displacement stops at both the heating and mating steps in the process. If welds are wider than ¼ in., the heated parts should be glided across each other during the mating step to prevent air entrapment in the joint.

Heated tool welding is suitable for almost any thermoplastic, but it is most often used for softer, semicrystalline thermoplastics. Common plastic substrates that are suitable for heated tool welding include polyethylene, polypropylene, polystyrene, ABS, PVC, and acetals. Usually, heated tool welding is not suitable for nylon or other materials that have long molecular chains. Dissimilar yet chemically compatible materials that have different melting temperatures can be welded by using two platens each heated to the melting temperature of the separate part.

Heated tool welding can be accomplished with either no surface treatment or very minor surface preparation (degreasing and removal of mold release) depending on the strength and reliability dictated by the application. Generally, surface degreasing to remove mold release or other organic contaminants is the only prebond treatment necessary. Mechanical roughening or chemical treatment of the surface provides no advantage since the surface will be melted and a new surface will be formed. Plastic parts that have a significant degree of internal moisture may have to be dried before heated tool welding, or else the moisture will tend to escape the molten surface in the form of vapor bubbles.

Electric-strip heaters, soldering irons, hot plates, and resistance blades are common methods of providing heat locally. Usually, the heating platen is coated with a fluorocarbon such as PTFE to prevent sticking. A simple hot plate has been used extensively with many plastics. Fig. 15.1 illustrates an arrangement for direct heat welding consisting of heated platens and fixturing. The parts are held against the heated platen until sufficient fusible material has been developed. Once the plastic is softened, the blade is raised, and the sheets are pressed together and fused. Table 15.5 lists typical hot-plate temperatures for a variety of plastics.

The direct heat welding operation can be completely manual, as in the case of producing a few prototypes, or it can be semi- or fully-automatic for fast high volume production. For automated assembly, rotary machines are often used where there is an independent station for each process: clamping into fixtures, heating, joining and cooling, and unloading.

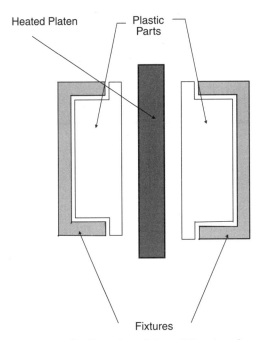

Figure 15.1 In direct hot plate welding, two fixtures press components into a hot moving platen, causing the plastic to melt at the interface.

TABLE 15.5 Hot Plate Temperatures to Weld Plastics[8]

Plastic	Temp, °F
ABS	450
Acetal	500
Phenoxy	550
Polyethylene LD	360
HD	390
Polycarbonate	650
PPO	650
Noryl*	525
Polypropylene	400
Polystyrene	420
SAN	450
Nylon 6, 6	475
PVC	450

*Trademark of General Electric Co.

Heated wheels or continuously moving heated bands are common tools used to bond thin plastic sheet and film. Such devices are commonly used for sealing purposes such as packaging of food. Care must be taken, especially with thin film, not to apply excessive pressure or heat. This could result in melting through the plastic. Table 15.6 provides heat-sealing temperature ranges for common plastic films.

Heated tool welding is commonly used in medium to high volume industries that can make use of the simplicity and speed of this joining process. The appliance and automotive industries, among others, commonly use this fastening method. Welding times range from 10–20 secs for small parts to up to 30 mins for larger parts such as heavy duty pipe. However, typical cycle times are less than 60 secs. Although heated tool welding is faster than adhesive bonding, it is not as fast as other welding methods such as ultrasonic or induction welding.

TABLE 15.6 Heat Sealing Temperatures for Plastic Films[9]

Film	Temp, °F
Coated cellophane	200–350
Cellulose acetate	400–500
Coated polyester	490
Poly(chlorotrifluoroethylene)	415–450
Polyethylene	250–375
Polystyrene (oriented)	220–300
Poly(vinyl alcohol)	300–400
Poly(vinyl chloride) and copolymers (nonrigid)	200–400
Poly(vinyl chloride) and copolymers (rigid)	260–400
Poly(vinyl chloride)–nitrile rubber blend	220–350
Poly(vinylidene chloride)	285
Rubber hydrochloride	225–350
Fluorinated ethylene–propylene copolymer	600–750

The direct heat welding process is extremely useful for pipe and duct work, rods and bars, or for continuous seals in films. However, irregular surfaces are difficult to heat unless complicated tools are provided. Special tooling configuration can be used for bonding any structural profile to a flat surface. In certain applications, the directed heat can also be used to shape the joint. With pipe, for example, a technique called groove welding is often employed. Groove welding involves two heating elements. One element melts a groove in one substrate that is the exact shape of the mating part, and the other element heats the edge of the mating substrate. The heated male part is quickly placed into the heated groove, and the joint is allowed to cool.

15.4.2 Hot gas welding

A welding gun can be used to bond many thermoplastic materials. An electrical heating element in the welding gun is capable of heating either compressed air or an inert gas to 425–700°F and forcing the heated gas onto the substrate surface. The pieces to be joined are beveled and positioned with a small gap between them. A welding rod made of the same plastic that is being bonded is laid in the joint with a steady pressure. The heat from the gun is directed to the tip of the rod, where it fills the gap, as shown in Fig. 15.2. Several passes may be necessary with the rod to fill the pocket. Thin sheets that are to be butt welded together, as in the case of tank linings, use a flat strip instead of a rod. The strip is laid over the joint and welded in place in a single pass (Fig. 15.3). Usually, the parts to be joined are held by fixtures so that they do not move during welding or while the weld is cooling. Alternatively, the parts can be first tacked together using a tool similar to a soldering iron.

Hot gas welding is usually a manual operation where the quality of the joint corresponds to the skill and experience of the operator. However, automatic welding machines are available and are used for overlap welding of seams or membranes. In either case, bond strength at least 85% of the strength of the bulk material can be achieved. Hot gas welding is a relatively fast operation. It can be used to weld a 1 in. wide tank seam at rates up to 60 in./min. It can also be used to do temporary tack work and to repair faulty or damaged joints that are made by gas welding or other joining processes.

Hot gas welding can be used to join most thermoplastics including polypropylene, polyethylene, acrylonitrile butadiene styrene, polyvinyl chloride, thermoplastic polyurethane, high density polyethylene, polyamide, polycarbonate, and polymethylmethacrylate. For polyolefins and other plastics that are easily oxidized, the heated gas must be inert (e.g., nitrogen or argon) since hot air will oxidize the surface of the plastic.

554   Chapter Fifteen

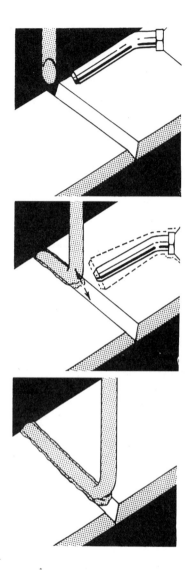

THERMOPLASTIC WELDING CHART

	PVC	H.D. Poly-ethylene	Poly-pro-pylene	Penton	ABS	Plexi-glass
Welding Temperature	525	550	575	600	500	575
Forming Temperature	300	300	350	350	300	350
Welding Gas	Air	WP* Nitro-gen	WP* Nitro-gen	Air	WP* Nitro-gen	Air
*W.P.—water pumped nitrogen						

Figure 15.2 Hot gas welding apparatus, method of application, and thermoplastic welding parameters.[10]

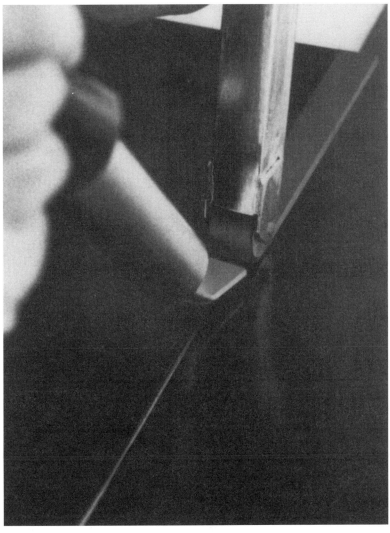

Figure 15.3 Thin thermoplastic sheets can be hot gas welded using a butt joint and a thin strip of welding stock. (Photo courtesy of Seelye Inc.)

Process parameters that are responsible for the strength of hot gas weld include the type of plastic being welded, the temperature and type of gas, the pressure on the rod during welding, the preparation of the material before welding, and the skill of the welder. After welding, the joint should not be stressed for several hours. This is particularly true for polyolefins, nylons, and polyformaldehyde. Hot gas welding is not recommended for filled materials or substrates that are less than $1/16$ in. thick. Conventional hot gas welding joint designs are shown in Fig. 15.4.

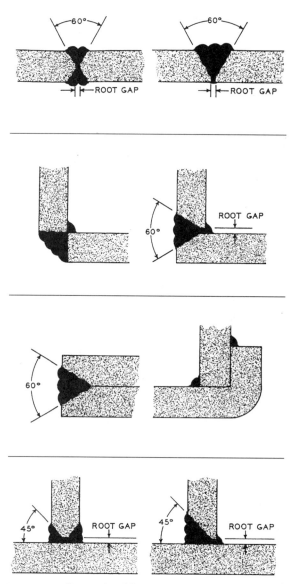

Figure 15.4 Conventional hot gas welding joint designs.[10]

Ideally, the welding rod should have a triangular cross section to match the bevel in the joint. A joint can be filled in one pass using triangular rod, saving time and material. Plastic welding rod of various types and cross sections are commercially available. However, it is also possible to cut the welding rod from the sheet of plastic that is being joined. Although this may require multiple passes for filling and the chance of air pockets is greater, the cost of the welding rod is very low, and the user is guaranteed of material compatibility between the rod and the plastic being joined.

Hot gas welding can be used in a wide variety of welding, sealing and repair applications. The most common applications are usually large structural assemblies. Hot gas welding is used very often in industrial jobs such as chemical storage tank repair and assembly of pipe fittings. It is an ideal system for a small fabricator or for anyone looking for an inexpensive joining system. The weld may not be as cosmetically attractive as other joining methods, but fast processing and high joint strength can easily be obtained.

Another form of hot gas welding is extrusion welding. In this process, an extruder is used instead of a hot gas gun. The molten welding material is expelled continuously from the extruder and fills a groove in a preheated weld area. A welding shoe follows the application of the hot extrudate and actually molds the seam in place. The main advantage with extrusion welding is the pressure that can be applied to the joint. This adds to the quality and consistency of the joint.

15.4.3 Resistance wire welding

The resistance wire method of welding employs an electrical resistance heating element laid between mating substrates to generate the needed heat of fusion. Once the element is heated, the surrounding plastic melts and flows together. Heating elements can be anything that conducts current and can be heated through Joule heating. This includes nichrome wire, carbon fiber, woven graphite fabric, and stainless steel foil. Fig. 15.5 shows an example of such a joint where a nichrome wire is used as the heating element. After the bond has been made, the resistance element that is exterior to the joint is cut off. Implant materials should be compatible with the intended application, since they will remain in the bond-line for the life of the product.

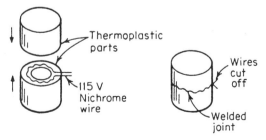

Figure 15.5 Resistance wire welding of thermoplastic joints.[11]

Like hot plate welding, resistance welding has three steps: heating, pressing, and maintaining contact pressure as the joint gels and cures. The entire cycle takes 30 secs to several minutes. Resistance welders can be automated or manually operated. Processing parameters include power (voltage and current); weld pressure; peak temperature; dwell time at temperature; and cooling time.

With resistance wire welding, surface preparation is only necessary when one of the substrates cannot be melted (e.g., thermosets and metals). Standard surface preparation processes such as those suggested in Chapter 6 can be used on these substrates.

The resistance heating process can be performed at either constant power or at constant temperature. When using constant power, a particular voltage and current is applied and held for a specified period of time. The actual temperatures are not controlled, and they are difficult to predict. In constant temperature resistance wire welding, temperature sensors monitor the temperature of the weld and automatically adjust the current and voltage to maintain a predefined temperature. Accurate control of heating and cooling rates is important when welding some plastics and when welding substrates having significantly different melt temperatures or thermal expansion coefficients. This heating and cooling control can be used to minimize internal stresses in the joint.

Resistance wire welding can be used to weld dissimilar material including thermoplastics, thermoplastic composites, thermosets, and metal in many combinations. When one of the substrates cannot be melted, such as when bonding two aluminum strips together, then a thermoplastic film with an embedded heating element can be used as the adhesive. A similar type of process can be used to cure thermosetting adhesives when the heat generated by the resistance wire is used to advance the cure.

Large resistance welded parts can require considerable power requirements. Resistance welding has been applied to complex joints in automotive applications, including vehicle bumpers and panels, joints in plastic pipe, and in medical devices. This type of welding is not restricted to flat surfaces. If access to the heating element is possible, repair of badly bonded joints is possible, and joints can be disassembled in a reverse process to which they were made.

15.4.4 Laser welding

Laser welding of plastic parts has been possible for the last 30 years. However, only recently has the technology and cost allowed this joining technique to be considered broadly.[12] Laser welders produce small beams of photons or electrons. The beams are focused onto the workpiece. Power density varies from a few to several thousand W/mm^2, but generally low power lasers (less than 50 W/mm^2) are used for plastic parts.

Laser welding is a high speed, non-contact process for welding thermoplastics. It is expected to find applications in the packaging and medical products industries.[13] Thermal radiation absorbed by the workpiece forms the weld. Solid state Nd:YAG and CO_2 lasers are most commonly used for welding. Laser radiation, in the normal mode of operation, is so intense and focused that it very quickly degrades thermoplastics. However, lasers have been used to butt weld polyethylene by pressing the unwelded parts together and tracking a defocused laser beam along the joint area. High speed laser welding of polyethylene films has been demonstrated at weld speeds of 164 ft/min using carbon dioxide and Nd:YAG lasers. Weld strengths are very near to the strength of the parent substrate.

Processing parameters that have been studied in laser welding are the power level of the laser, shielding gas flow rate, offset of the laser beam from a focal point on the top surface of the weld interface, travel speed of the beam along the interface, and welding pressure.[14] Butt and lap joint designs are most commonly welded by directing the laser beam at the edges of the mated joint.

Lasers have been used primarily for welding polyethylene and polypropylene. Usually laser welding is applied only to films or thin walled components. The least powerful beams, of approximately 50 W/mm^2, with the widest weld spots are used for fear of degrading the polymer substrate. The primary goal in laser welding is to reach a melt temperature where the parts can be joined quickly before the plastic degrades. To avoid material degradation, accurate temperature measurement of the weld surfaces and temperature control by varying laser strength is essential.

Lasers have been primarily used for joining delicate components that cannot stand the pressure of heated tool or other thermal welding methods. Applications exist in the medical, automotive, and chemical industries. Perhaps the greatest opportunity for this process will be for the high speed joining of films.

Laser welding has also been used for filament winding of fiber reinforced composite materials using a thermoplastic prepreg. A defocused laser beam is directed on the area where the prepreg meets the winding as it is being built-up. With suitable control over the winding speed, applied pressure, and the temperature of the laser, excellent reinforced structures of relatively complex shape can be achieved.

Laser welding requires high investment cost in equipment and the need for a ventilation system to remove hazardous gaseous and particulate materials resulting from polymer degradation. Of course, suitable precautions must also be taken to protect the eyesight of anyone in the vicinity of laser welding operation.

15.4.5 Infrared welding

Infrared radiation is a non-contact alternative to hot plate welding. Infrared is particularly promising for higher melting polymers, since the parts do not contact and stick to the heat source. Infrared radiation can penetrate into a polymer and create a melt zone quickly. By contrast, hot plate welding involves heating the polymer surface and relying on thermal conduction to create the required melt zone.

Infrared welding is at least 30% faster than heated tool welding. High reproducibility and bond quality can be obtained. Infrared welding can be easily automated, and it can be used for continuous joining. Often heated tool welding equipment can be modified to accept infrared heating elements.

Infrared radiation can be supplied by high intensity quartz heat lamps. The lamps are removed after melting the polymer, and the parts are forced together, as with hot plate welding. The depth of the melt zone depends on many factors including minor changes in polymer formulation. For example, colorants and pigments will change a polymer's infrared absorption properties and will affect the quality of the welding process. Generally, the darker the polymer the less infrared energy is transferred down through a melt zone, and more likely will surface degradation occur due to overheating.

15.5 Indirect Heating

Many plastic parts may be joined by indirect heating. With these methods, the materials are heated by external energy sources. The heat is induced in the polymer or at the interface. Indirect methods of heating consist primarily of induction and dielectric welding. For induction welding, the source is an electromagnetic field, and for dielectric welding the source is an electric field of high frequency.

Indirect heat joining is possible for almost all thermoplastics; however, it is most often used with the newer engineering thermoplastics. The engineering thermoplastics generally have greater heat and chemical resistance than the more conventional plastics. In many applications, they are reinforced to improve structural characteristics. They are generally stronger and have excellent strength-to-weight ratios. However, many of these plastics are not well suited to joining by direct heat because of the high melt temperatures and pressures required. Indirect heating methods and frictional heating methods may be used to obtain fast, high quality bonds with these useful plastic materials.

15.5.1 Induction welding

Induction welding is very much like resistance wire welding. An implant is heated to melt the surrounding polymer. Rather than heating the implant resistively, in induction welding the implant is heated with an electromagnetic field.

The electromagnetic induction field can be used to heat a metal grid or insert placed between mating thermoplastic substrates. Energy from the electromagnetic field induces eddy currents in the conductive material, and the material's resistance to these currents produces heat. When the joint is positioned between induction coils, the hot insert causes the plastic to melt and fuse together. Slight pressure is maintained as the induction field is turned off and the joint hardens. The main advantage of induction welding is that heating only occurs where the electromagnetic insert is applied. The bulk substrate remains at room temperature, avoiding degradation or distortion.

More popular forms of induction welding have been developed which use a bonding agent consisting of a thermoplastic resin filled with metal or ferromagnetic particles. This bonding agent melts in the induction field and forms the adhesive joint. The advantage of this method is that stresses caused by large metal inserts are avoided.

The bonding agent should be similar to the substrates. When joining polyethylene, for example, the bonding agent may be a polyethylene resin containing 0.5–0.6% by volume magnetic iron oxide powder. Electromagnetic adhesives can be made from many types of filled thermoplastics. These adhesives are commonly shaped into gaskets or film that melt in the induction field. Induction welding of a plastic nozzle to a hose with an electromagnetic adhesive gasket is illustrated in Fig. 15.6.

Four basic components comprise the electromagnetic welding process.

- An induction generator converts 60 Hz electrical supply to 3–40 MHz output frequency and output power from 1–5 kW.
- The induction heating coil, consisting of water cooled copper tubing, is usually formed into hairpin shaped loops.
- Fixturing is used to hold parts in place.
- The bonding material is in the form of molded or extruded preforms which becomes an integral part of the welded product.

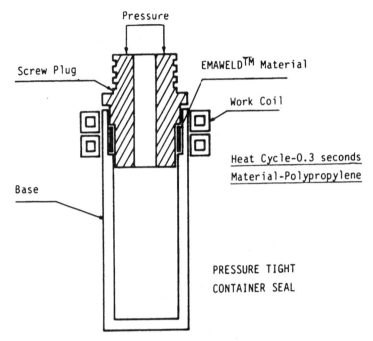

Figure 15.6 Schematic of induction welded plastic nozzle; adhesive is supplied by Emaweld.[15]

Induction heating coils should be placed as close as possible to the joint. For complex designs, coils can be contoured to the joint geometry. Electromagnetic welding systems can be designed for semi-automatic or completely automatic operation. With automated equipment, a sealing rate of up to 150 parts/min can be achieved. Equipment costs depend on the degree of automation required.

The bonding agent is usually produced for the particular application to ensure compatibility with the materials being joined. However, induction welding equipment suppliers also offer proprietary compounds for joining dissimilar materials. The bonding agent is often shaped into a profile to match the joint design (i.e., gaskets, rings, ribbon). The fillers used in the bonding agents are micron-sized ferromagnetic powders. They can be metallic, such as iron or stainless steel, or a ceramic ferrite material.

Quick bonding rates are generally obtainable because heating occurs only at the interface. Heat does not have to flow from an outside source or through the substrate material to the point of need. Polyethylene joints can be made in as fast as 3 secs with electromagnetic welding. Depending on the weld area, most plastics can be joined by electromagnetic welding in 3–12 sec cycle times.

Plastics that are readily bonded with induction welding include all grades of ABS, nylon, polyester, polyethylene, polypropylene, and polystyrene, as well as those materials often considered more difficult to bond such as acetals, modified polyphenylene oxide, and polycarbonate. Reinforced thermoplastics with filler levels up to 65% have been joined successfully.[16] Many combinations of dissimilar materials can be bonded with induction welding processes. Table 15.7 shows compatible plastic combinations for electromagnetic adhesives. Thermoset and other non-metallic substrates can also be electromagnetically bonded. In these applications, the bonding agent acts as a hot-melt adhesive.

Advantages of induction welding include the following factors.

TABLE 15.7 Compatible Plastic Combinations for Bonding with Electromagnetic Adhesives[17]

● = Compatible

Row \ Column	ABS	Acetals	Acrylic	Cellulosics	Ionomer (Surlyn)	Nylon 6,6 11, 12	Polybutylene	Polycarbonate	Polyethylene	Polypropylene	Polyphenylene Oxide (Noryl)	Polystyrene	Polysulfone	Polyvinyl Chloride	Polyurethane	SAN	Thermoplastic Polyester	Copolyester	Styrene Bl. Copolymer	Olefin Type
ABS	●	●						●								●				
Acetals		●																		
Acrylic	●	●						●				●				●				
Cellulosics				●																
Ionomer (Surlyn)					●															
Nylon 6,6, 11, 12						●														
Polybutylene							●													
Polycarbonate	●	●						●				●	●			●				
Polyethylene									●											●
Polyphenylene Oxide (Noryl)									●	●										
Polypropylene										●										●
Polystyrene			●					●	●		●	●				●				
Polysulfone								●					●							
Polyvinyl Chloride														●						
Polyurethane															●					
SAN	●	●						●				●				●				
Thermoplastic Polyester																	●			
Thermo-Plastic Elastomers — Copolyester																		●		●
Thermo-Plastic Elastomers — Styrene Bl. Copolymer								●											●	
Thermo-Plastic Elastomers — Olefin Type										●									●	●

- Heat damage, distortion, and over-softening of the parts are reduced.
- Squeeze-out of fused material from the bond line is limited.
- Hermetic seals are possible.
- Control is easily maintained by adjusting the output of the power supply.
- No pretreatment of the substrates is required.
- Bonding agents have unlimited storage life.

The ability to produce hermetic seals is cited as one of the prime advantages in certain applications such as medical equipment. Welds can also be disassembled by placing the bonded article in an electromagnetic field and remelting the joint. There are few limitations on part size or geometry. The only requirement is that the induction coils be designed to apply a uniform field. Significant development has occurred over the years relative to coil geometry.

The primary disadvantages of electromagnetic bonding are that the metal inserts remain in the finished product, and they represent an added cost. The cost of induction welding equipment is high. The weld is generally not as strong as those obtained by other welding methods.

Induction welding is frequently used for high speed bonding of many plastic parts. Production cycles are generally faster than with other bonding methods. It is especially useful on plastics that have a high melt temperature such as the modern engineering plastics. Thus, induction welding is used in many under-the-hood automotive applications. It is also frequently used for welding large or irregularly shaped parts.

Electromagnetic induction methods have also been used to quickly cure thermosetting adhesives such as epoxies. Metal particle fillers, wire, or mesh inserts are used to provide the heat source. These systems generally have to be formulated so that they cure with a low internal exotherm or else the joint will overheat, and the adhesive will thermally degrade.

15.5.2 Dielectric heating

Dielectric heating can be used on most thermoplastics except those that are relatively transparent to high frequency electric fields. This method is used mostly to seal vinyl sheeting such as automobile upholstery, swimming pool liners, and rainwear. An alternating electric field is imposed on the joint, which causes rapid reorientation of polar molecules. As a result, heat is generated within the polymer by molecular friction. The heat causes the polymer to melt and pressure is applied to the joint. The field is then removed, and the joint is held until the weld cools. The main difficulty in using dielectric heating as a bonding method is in directing the heat to the interface. Generally, heating occurs in the entire volume of the polymer that is exposed to the electric field.

Variables in the bonding operation are the field frequency, dielectric loss of the plastic, the power applied, pressure, and time. The materials most suitable for dielectric welding are those that have strong dipoles. These can often be identified by high electrical dissipation or power factors. Materials most commonly welded by this process include polyvinyl chloride, polyurethane, polyamide, and thermoplastic polyester. Since the field intensity decreases with distance from the source, this process is normally used with thin polymer films.

Dielectric heating can also be used to generate the heat necessary for curing polar, thermosetting adhesives. It can also be used to quickly evaporate water from a water-based adhesive formulation. Dielectric processing water-based adhesives is commonly used in the furniture industry for very fast drying of wood joints in small furniture parts. Common white glues, such as polyvinyl acetate emulsions, can be dried in seconds using dielectric heating processes.

There are basically two forms of dielectric welding: radio frequency welding and microwave welding. Radio frequency welding uses high frequency (13–100 MHz) to generate heat in polar materials, resulting in melting and weld formation after cooling. The electrodes are usually designed into the platens of a press. Microwave welding uses high frequency (2–20 GHz) electromagnetic radiation to heat a susceptor material located at the joint interface. The generated heat melts thermoplastic materials at the joint interface, producing a weld upon cooling. Heat generation occurs in microwave welding through absorption of electrical energy similar to radio frequency welding.

Polyaniline doped with an aqueous acid such as HCl is used as a susceptor in microwave welding. This introduces polar groups and a degree of conductivity into the molecular structure. The polar groups preferentially generate heat when exposed to microwave energy. These doped materials are used to produce gaskets which can be used as an adhesive in dielectric welding.

Dielectric welding is commonly used for sealing thin films such as polyvinyl chloride for lawn waste bags, inflatable articles, liners, and clothing. It is used to produce high volume stationary items such as loose-leaf notebooks and checkbook covers. Because of the cost of the equipment and the nature of the process, the commodity industries are the major users of dielectric welding.

15.6 Friction Welding

In friction welding, the interface alone is heated due to mechanical friction caused by one substrate contacting and sliding over another substrate surface. The frictional heat generated is sufficient to create a melt zone at the interface. Once a melt zone is created the relative movement is stopped, and the parts are held together under slight pressure until the melt zone cools and sets. Common friction welding processes include:

- Spin welding
- Ultrasonic welding
- Vibration welding

15.6.1 Spin welding

Spin welding uses the heat of friction to provide the heat of fusion at the interface. One substrate is rotated very rapidly while in touch with the other substrate which is fixed in a stationary position. The surfaces melt by frictional heating without heating or otherwise damaging the entire part. Sufficient pressure is applied during the process to force out a slight amount of flash along with excess air bubbles. Once the rotation is stopped, position and pressure are maintained until the weld sets. The rotation speed and pressure are dependent on the thermoplastics being joined.

Spin welding is an old and uncomplicated technique. The equipment required can be as simple as a lathe or modified drill. Spin welding has a lower capital cost than other welding methods. The base equipment required is comparatively inexpensive; however, auxiliary equipment such as fixtures, part feeders, and unloaders can drive up the cost of the system. Depending on the geometry and size of the part, the fixture that attaches the part to the rotating motor may be complex. A production rate of 300 parts/min is possible on simple circular joints with an automated system containing multiple heads.

The main advantages of spin welding are its simplicity, high weld quality, and the wide range of possible materials that can be joined. Spin welding is capable of very high throughput. Heavy welds are possible with spin welding. Actual welding times for most parts are only several seconds. A strong hermetic seal can be obtained which is frequently stronger than the material substrate itself. No foreign materials are introduced into the weld, and no environmental considerations are necessary. The main disadvantage of this process is that spin welding is used primarily on parts where at least one substrate is circular.

When considering a part as a candidate for spin welding, there are several factors that must be considered: the type of material and the temperature at which it starts to become tacky; the diameter of the parts; and how much flash will develop and what to do with the flash.

The parts that are to be welded must be structurally stiff enough to resist the pressure required. Joint areas must be circular, and a shallow matching groove is desirable in one of the parts in order to index the parts and provide a uniform bearing surface. In addition, the tongue and groove type joint is useful in hiding the flash that is generated during the welding process. However, a flash "trap" will usually lower the ultimate bond strength. It is generally more desirable to either remove the flash or to design the part so that the flash accumulates on the inside of the joint and hidden from view. Figure 15.7 shows conventional joints used in spin welding.

Since the heating which is generated at the interface depends on the relative surface velocity, the outside edges of circular components

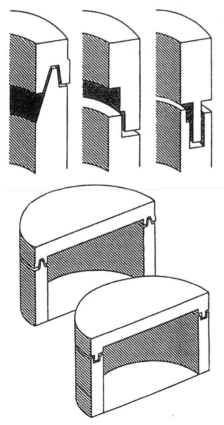

Figure 15.7 Common joints used in the spin welding process.[18]

will see higher temperatures by virtue of their greater diameter and surface velocity. This will cause a thermal differential that could result in internal stress in the joint. To alleviate this effect, joints with hollow section and thin walls are preferred.

The larger the part, the larger the motor required to spin the part as more torque is required to spin the part and obtain sufficient friction. Parts with diameters of 1–15 in. have been spin welded using motors from ¼–3 horsepower.[19] The weld can be controlled by the RPM of the motor and somewhat by the pressure on the piece being joined, the timing of the pressure during spin and during joining, and the cooling time and pressure. In commercial rotation welding machines, speed can range from 200 to 14,000 RPM. Welding times range from tenths of a second to several seconds. Cool down times are in the range of ½ sec. A typical complete process time is 2 secs.[18] Axial pressure on the part ranges from 150 to 1,000 psi. A prototype appraisal is usually completed to determine the optimum parameters of the process for a particular material and joint design.

Table 15.8 shows the temperature at which tackiness starts for most thermoplastics that can be spin welded. This data is useful for all forms of thermal welding not only spin welding. The tackiness temperature can be used as a guide to determine RPM or SFM (surface feet per min) required at the joint. An unfilled 1 in. diameter polyethylene part may be spun at 1,000 RPM (260 SFM) to reach the tackiness temperature of 280°F. As the amount of inert filler increases in

TABLE 15.8 Tack Temperatures of Common Thermoplastics[19]

Plastic	Tackiness temperature, °F
Ethylene, Vinyl, Acetate	150
PVC	170
Polystyrene, High Impact	180
ABS, High Impact	200
Acetal	240
Polyurethane, Thermoplastic	245
SAN, CAB	250
Polypropylene, Noryl	260
Cellulose Acetate	270
Polycarbonate	275
Polyethylene	280
Acetal	290
Acrylic	320
Polysulfone	325
PET	350
PES	430
Fluorcarbon, Melt Processable	630

the part, the RPM needs to be increased. Effects of increasing RPM are similar to those of increased pressure.

Typical applications include small parts such as fuel filters, check valves, aerosol cylinders, tubes, and containers. Spin welding is also a popular method of joining large-volume products such as packaging and toys. Spin welding can also be used for attaching studs to plastic parts.

15.6.2 Ultrasonic welding

Ultrasonic welding is also a frictional process that can be used on many thermoplastic parts. Frictional heat in this form of welding is generated by high-frequency linear vibration.

The basic parts of a standard ultrasonic welding device are shown in Fig. 15.8. During ultrasonic welding, a high-frequency electrodynamic field is generated which resonates a metal horn that is in contact with one substrate. The horn vibrates the substrate sufficiently fast relative to a fixed substrate that significant heat is generated at the interface. With pressure and subsequent cooling, a strong bond can be obtained. The stages of the ultrasonic welding process are shown in Fig. 15.9.

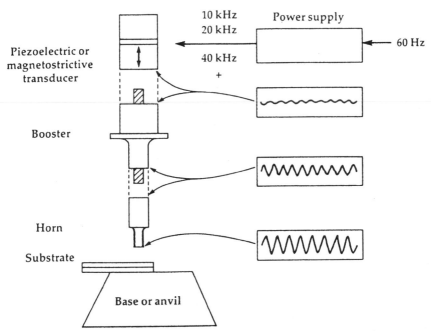

Figure 15.8 Equipment used in a standard ultrasonic welding process.[20]

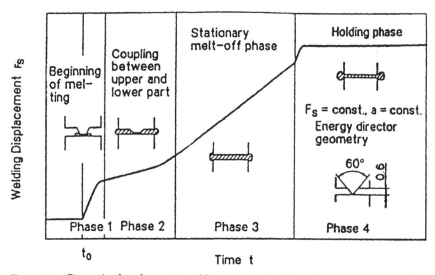

Figure 15.9 Stages in the ultrasonic welding process. In Phase 1, the horn is placed in contact with the part, pressure is applied, and vibratory motion is started. Heat generation due to friction melts the energy director, and it flows into the joint interface. The weld displacement begins to increase as the distance between the parts decreases. In Phase 2, the melting rate increases, resulting in increased weld displacement, and the part surfaces meet. Steady-state melting occurs in Phase 3, as a constant melt layer thickness is maintained in the weld. In Phase 4, the holding phase, vibrations cease. Maximum displacement is reached, and intermolecular diffusion occurs as the weld cools and solidifies.[21]

The frequency generally used in ultrasonic assembly is 20 kHz because the vibration amplitude and power necessary to melt thermoplastics are easy to achieve. However, this power can produce a great deal of mechanical vibration which is difficult to control, and fixturing becomes large. Higher frequencies (40 kHz) that produce less vibration are possible and are generally used for welding engineering thermoplastics and reinforced polymers. Higher frequencies are also more appropriate for smaller parts and for parts where less material degradation is required.

Ultrasonic welding is clean, fast (20–30 parts per min), and usually results in a joint that is as strong as the parent material. Materials handling equipment can be easily interfaced with the ultrasonic system to further improve rapid assembly. Ultrasonic welding can provide hermetically sealed components if the entire joint can be welded at one time. Large parts generally are too massive to be joined with one continuous bond, and spot welding is necessary. It is difficult to obtain completely sealed joints with spot welding.

Rigid plastics with a high modulus are best for ultrasonic welding. Rigid plastics readily transmit ultrasonic energy, whereas softer plastics tend to dampen the energy before it reaches the critical joint area. Excellent welds are generally obtainable with polystyrene, SAN, ABS, polycarbonate, and acrylic plastics. PVC and the cellulosics tend to attenuate energy and deform or degrade at the surfaces. Figure 15.10 shows an index for the ultrasonic weldability of conventional thermoplastics. Dissimilar plastics may be joined if they have similar melt temperatures and are chemically compatible. A plastic compatibility chart for ultrasonic welding is shown in Table 15.9. Materials such as polycarbonate and nylon must be dried before welding, otherwise their high level of internal moisture will cause foaming and interfere with the joint.

Common ultrasonic welding joint designs are shown in Fig. 15.11. The most common design is a butt joint that uses an "energy director". This design is appropriate for most amorphous plastic materials. The wedge design concentrates the vibrational energy at the tip of the energy director. A uniform melt then develops from the volume of material formed by the energy director. This becomes the material that is consumed in the joint. Without the energy director, a butt joint would produce voids along the interface, resulting in stress and a low

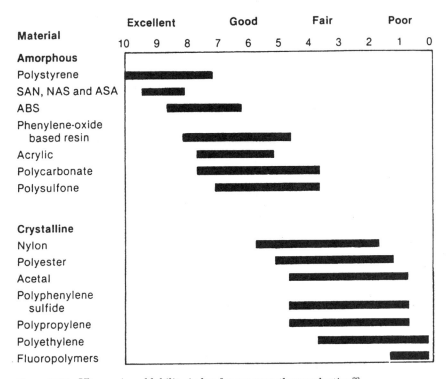

Figure 15.10 Ultrasonic weldability index for common thermoplastics.[22]

TABLE 15.9 Compatibility of Plastics for Ultrasonic Welding[20]

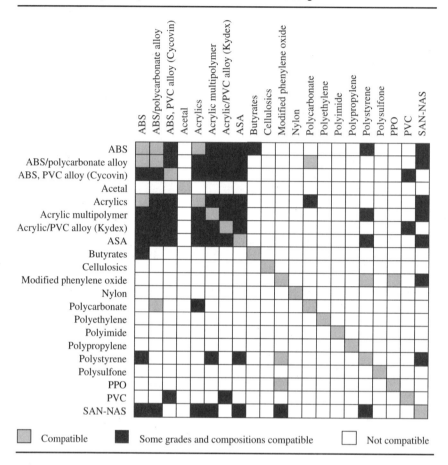

strength joint. Shear and scarf joints are employed for crystalline polymeric materials. They are usually formed by designing an interference fit.

Ultrasonic welding can also be used to stake plastics to other substrates, for inserting metal parts, and for spot welding two plastic components. Figure 15.12 illustrates ultrasonic insertion, swaging, staking, and spot welding operations. In ultrasonic spot welding, the horn tip passes through the top sheet to be welded. The molten plastic forms a neat raised ring on the surface that is shaped by the horn tip. Energy is also released at the interface of the two sheets producing frictional heat. As the tip penetrates the bottom substrate, displaced molten plastic flows between the sheets into the preheated area and forms a permanent bond.

Methods of Joining Plastics Other Than With Adhesives 573

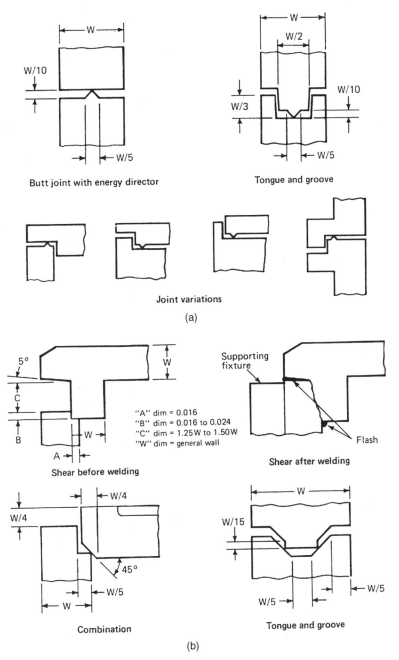

Figure 15.11 Ultrasonic welding joints for (a) amorphous and (b) crystalline polymeric materials.[23]

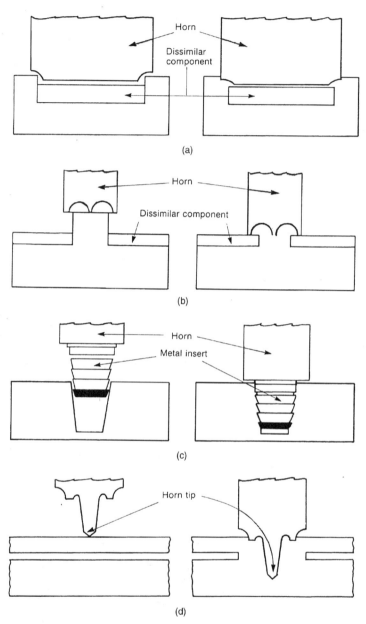

Figure 15.12 Ultrasonic joining operations. (*a*) Swaging: the plastic ridge is melted and reshaped (left) by ultrasonic vibrations in order to lock another part into place. (*b*) Staking: ultrasonic vibrations melt and reform a plastic stud (left) in order to lock a dissimilar component into place (right). (*c*) Insertion: a metal insert (left) is embedded in a preformed hole in a plastic part by ultrasonic vibration (right). (*d*) Spot welding: two plastic components (left) are joined at localized points (right).[24]

Ultrasonic heating is also applicable to hot-melt and thermosetting adhesives.[25] In these cases, the frictional energy is generated by the substrate contacting an adhesive film placed between the two substrates. The frictional energy generated is sufficient either to melt the hot-melt adhesive or to cure the thermosetting adhesive.

15.6.3 Vibration welding

Vibration welding is similar to ultrasonic welding in that it uses the heat generated at the surface of two parts rubbing together. This frictional heating produces melting in the interfacial area of the joint. Vibration welding is different from ultrasonic welding, however, in that it uses lower frequencies of vibration, 120–240 Hz, rather than the 20–40 kHz used for ultrasonic welding. With lower frequencies, much larger parts can be bonded because of less reliance on the output of the power supply. Figure 15.13 shows the joining and sealing of a two-part plastic tank design of different sizes using vibration welding.

There are two types of vibration welding: linear and axial. Linear vibration welding is most commonly used. Friction is generated by a linear, back and forth motion. Axial or orbital vibration welding allows irregularly shaped plastic parts to be vibration welded. In axial welding, one component is clamped to a stationary structure; the other component is oscillated using orbital motion.

Vibration welding fills a gap in the spectrum of thermoplastic welding processes in that it is suitable for large, irregularly shaped parts. Vibration welding has been used successfully on large thermoplastic parts such as canisters, pipe sections, and other geometries that are too large to be excited with an ultrasonic generator and ultrasonically

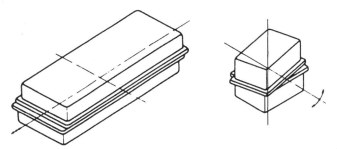

Linear vibration (left) is employed where the length to width ratio precludes the use of axial welding (right) where the axial shift is still within the width of the welded edge.

Figure 15.13 Linear and axial vibration welding of a two part container.[26]

welded. Vibration welding is also capable of producing strong, pressure tight joints at rapid rates. The major advantage is its application to large parts and to non-circular joints, provided that a small relative motion between the parts in the welding plane is possible.

Usually, the same manufacturers of ultrasonic welding equipment also provide vibration welding equipment. Vibration welding equipment can be driven either electrically (variable frequency) or hydraulically (constant frequency). Capital cost is generally higher than with ultrasonic welding.

The main process parameters to control in vibration welding are the amplitude and frequency of motion, weld pressure, and weld time. Most industrial vibration welding machines operate at frequencies of 120–240 Hz. The amplitude of vibration is usually less than 0.2 in. Lower weld amplitudes are used with higher frequencies, and are necessary when welding parts into recessed cavities. Lower amplitudes (0.020 in.) are used for high temperature thermoplastics. Joint pressure is held in the range of 200–250 psi, although at times much higher pressures are required. Vibration welding equipment has been designed to vary the pressure during the welding cycle in order to improve weld quality and decrease cycle times. This also allows more of the melted polymer to remain in the bond area, producing a wider weld zone.

Vibration times depend on the melt temperature of the resin and range from 1–10 secs. Solidification times are usually less than several secs. Total cycle times typically range from 3–15 secs. This is slightly longer than typical spin welding and ultrasonic welding cycles, but shorter than hot plate welding and solvent cementing.

A number of factors must be considered when vibration welding larger parts. Clearances must be maintained between the parts to allow for movement between the halves. The fixture must support the entire joint area, and the parts must not flex during welding. Vibration welding is applicable to a variety of thermoplastic parts with planar or slightly curved surfaces. The basic joint is a butt joint, but unless parts have thick walls, a heavy flange is generally required to provide rigidity and an adequate welding surface. Typical joint designs for vibration welds are shown in Fig. 15.14.

Vibration welding is ideally suited to parts made from engineering thermoplastics as well as acetal, nylon, polyethylene, ionomer, and acrylic resins. Almost any thermoplastic can be vibration welded. Unlike other welding methods, vibration welding is applicable to crystalline or amorphous materials and to filled, reinforced, or pigmented materials. Vibration welding also can be utilized with certain fluoropolymers and polyester elastomers, which cannot be joined by ultrasonic welding. By optimizing welding parameters and glass fiber loadings, nylon 6 and nylon 66 butt joints can be produced having up to

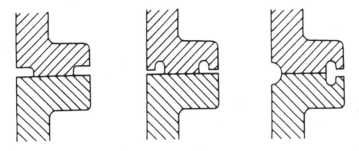

Unless the parts have thick walls it is necessary to form a flange for the butt weld. In practice melt traps are usually formed so that the molten resin does not create a flashing.

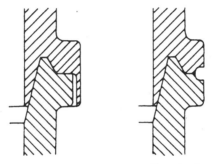

Circular parts can be joined more effectively with tongue-in-groove joints which retain most of the melt and create a joint of high integrity.

Figure 15.14 Typical vibration welded joint designs.[26]

17% higher strength than the base resin.[27] Any pair of dissimilar materials that can be ultrasonically joined can also be vibration welded.

Vibration welding techniques have found several applications in the automobile industry, including emission control canisters; fuel pumps and tanks; head and tail light assemblies; heater valves; air intake filters; water pump housings; and bumper assemblies. They have also been used for joining pressure vessels and for batteries, motor housings, and butane gas lighter tanks.

15.7 Solvent Cementing

Solvent cementing is the simplest and most economical method of joining non-crystalline thermoplastics. In solvent cementing, the application of solvent softens the substrate surfaces being bonded. The solvent diffuses into the surface allowing increased freedom of movement of the polymer chains. As the parts are then brought together under pressure, the solvent softened plastic flows. Van der Waals attractive forces are formed between molecules from each part, and polymer chains from each part intermingle and diffuse into one another. The parts are then held in place until the solvent evaporates from the joint area.

Solvent-cemented joints of similar materials are less sensitive to thermal cycling than joints bonded with adhesives because there is no stress at the interface due to differences in thermal expansion between the adhesive and the substrate. When two dissimilar plastics are to be joined, adhesive bonding is generally desirable because of solvent and polymer compatibility problems. Solvent cemented joints are as resistant to degrading environments as the parent plastic. Bond strength greater than 85% of the parent plastic can generally be obtained. Solvents provide high strength bonds quickly due to rapid evaporation rates.

Solvent bonding is suitable for all amorphous plastics. It is used primarily on ABS, acrylics, cellulosics, polycarbonates, polystyrene, polyphenylene oxide, and vinyls. Solvent welding is not suitable for crystalline thermoplastics. It is not affective on polyolefins, fluorocarbons, or other solvent resistant polymers. Solvent welding is moderately effective on nylon and acetal polymers. Solvent welding cannot be used to bond thermosets. It can be used to bond soluble plastics to unlike porous surfaces, including wood and paper, through impregnation and encapsulation of the fibrous surface.

The major disadvantage of solvent cementing is the possibility of stress cracking in certain plastic substrates. Stress cracking or *crazing* is the formation of microcracks on the surface of a plastic part that has residual internal stresses due to its molding process. The contact with a solvent will cause the stresses to release uncontrollably resulting in stress cracking of the part. When this is a problem, annealing of the plastic part at a temperature slightly below its glass transition temperature will usually relieve the internal stresses and reduce the stress cracking probability. Annealing time must be sufficiently long to allow the entire part to come up to the annealing temperature. Another disadvantage of solvent welding is that many solvents are flammable and/or toxic and must be handled accordingly. Proper ventilation must be provided when bonding large areas or with high volume production.

Solvent cements should be chosen with approximately the same solubility parameter as the plastic to be bonded. Table 15.10 lists typical solvents used to bond major plastics. Solvents used for bonding can be a single pure solvent, a combination of solvents, or a solvent(s) mixed

TABLE 15.10 Typical Solvents for Solvent Cementing of Plastics[28]

with resin. It is common to use a mixture of a fast-drying solvent with a less volatile solvent to prevent crazing or to extend tack time. The solvent cement can be bodied up to 25% by weight with the parent plastic to increase viscosity. These bodied solvent cements can fill gaps, and they provide less shrinkage and internal stress than if only pure solvent is used.

The parts to be bonded should be unstressed and annealed if necessary. For solvent bonding, surfaces should be clean and should fit together uniformly throughout the joint. Close-fitting edges are necessary for good bonding. The solvent cement is generally applied to both substrates with a syringe or brush. In some cases, the surface may be immersed in the solvent. However, solvent application generally must be carefully controlled since a small difference in the amount of applied solvent greatly affects joint strength. After the area to be bonded softens, the parts are mated and held under light pressure until dry. Pressure should be low and uniform so that the joint will not be stressed. After the joint hardens, the pressure is released. An elevated-temperature cure may by necessary depending on the plastic and desired joint strength. Exact processing parameters for solvent welding are usually determined by trial and error. They will be dependent on the exact polymer, ambient conditions, and type of solvent used.

The bonded part should not be packaged or stressed until the solvent has adequate time to escape from the joint. Complete evaporation of solvent may not occur for hours or even days. Some solvent joined assemblies may have to be "cured" at elevated temperatures to encourage the release of solvent prior to packaging.

15.8 Methods of Mechanical Joining

There are instances when adhesive bonding, thermal welding, or solvent cementing are not practical joining methods for plastic assembly. This usually occurs because the optimum joint design is not possible, the cost and complexity is too great, or the skill and resources are not present to attempt these forms of fastening. Another common reason for foregoing bonding or welding is when repeated disassembly of the product is required. Fortunately, when these situations occur, the designer can still turn to mechanical fastening as a possible solution.

There are basically two methods of mechanical assembly for plastic parts. The first uses fasteners, such as screws or bolts, the second uses interference fit such as *press-fit* or *snap-fit* generally used in thermoplastic applications. This latter method of fastening is also called "design for assembly" of "self-fastening." If possible, the designer should try to design the entire product as a one-part molding or with the capability of being press-fit or snap-fit together because it will eliminate the need for a secondary assembly operation. However, mechanical limitations often will make it necessary to join one part to another using a fastening device. There are a number of mechanical fasteners designed for metals that are also generally suitable with plastics, and there are many other fasteners specifically designed for plastics. Typical of these plastic fasteners are thread-forming screws, rivets, threaded inserts, and spring clips.

As in adhesive bonding or welding, special considerations must be given to mechanical fastening because of the nature of the plastic material. Care must be taken to avoid overstressing the parts. Mechanical creep can result in loss of preload in poorly designed systems. Reliable mechanically fastened plastic joints require:

- A firm strong connection
- Materials that are stable in the environment
- Stable geometry
- Appropriate stresses in the parts including a correct clamping force

In addition to providing joint strength, mechanically fastened joints should prevent slip, separation, vibration, misalignment, and wear of the parts. Well designed joints provide the above without being excessively large or heavy, or burdening assemblers with bulky tools. Designing plastic parts for mechanical fastening will depend primarily on the particular plastic being joined and the functional requirements of the application.

15.8.1 Mechanical fasteners

A large variety of mechanical fasteners can be used for joining plastic parts to themselves and to other materials. These include:

- Machine screws and bolts
- Self-threading screws
- Rivets
- Spring fasteners and clips

In general, when repeated disassembly of the product is anticipated, mechanical fasteners are used. Metal fasteners of high strength can overstress plastic parts, so torque controlled tightening or special design provisions are required. Where torque cannot be controlled, various types of washers can be used to spread the compression force over larger areas.

15.8.1.1 Machine screws and bolts. Parts molded of thermoplastic resin are sometimes assembled with machine screws or with bolts, nuts, and washers especially if it is a very strong plastic. Machine screws are generally used with threaded inserts, nuts, and clips. They rarely are used in pre-tapped holes. Figure 15.15 shows correct and incorrect methods of mechanical fastening of plastic parts using this hardware.

Inserts into the plastic part can be effectively used to provide the female part of the fastener. Inserts that are used for plastic assembly consist of molded-in inserts and post-molded inserts.

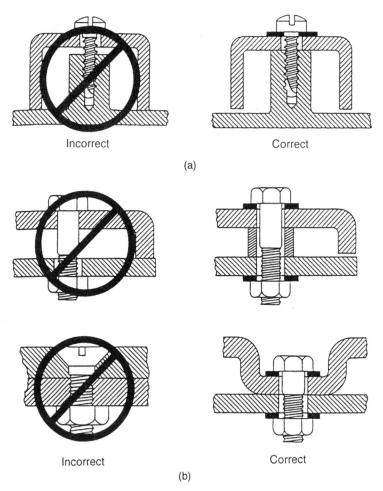

Figure 15.15 Mechanical fastening with self taping screws (a) and with bolts, nuts, and washers (b).[29]

Molded-in inserts represent inserts that are placed in the mold before the plastic resin is injected. The resin is then shaped to the part geometry and locks the insert into its body. Molded-in inserts provide very high strength assemblies and relatively low unit cost. However, molded-in inserts could increase part cycle time while the inserts are manually placed in the mold.

When the application involves infrequent disassembly, molded-in threads can be used successfully. Coarse threads can also be molded into most materials. Threads of 32 or finer pitch should be avoided, along with tapered threads, because of excessive stress on the part. If the mating connector is metal, over-torque will result in part failure.

Post-molded inserts come in four types: press-in, expansion, self-tapping, and thread forming. Metal inserts are available in a wide range of shapes and sizes for permanent installation. Inserts are typically installed in molded bosses and designed with holes to suit the insert to be used. Some inserts are pressed into place and others are installed by methods designed to limit the stress and increase strength. Inserts are commonly installed in thermoplastic parts using ultrasonic welding equipment (Fig. 15.11c). Generally, the outside of the insert is provided with projections of various configurations that penetrate the plastic and prevent movement under normal forces of assembly.

Whatever mechanical fastener is used, particular attention should be paid to the head of the fastener. Conical heads, called flat heads, produce undesirable tensile stress and should not be used. Bolt or screw heads with a flat underside, such as pan heads and round heads (Fig. 15.16) are preferred because the stress produced is more compressive. Flat washers are also suggested. and should be used under both the nut and the fastener head. Sufficient diametrical clearance

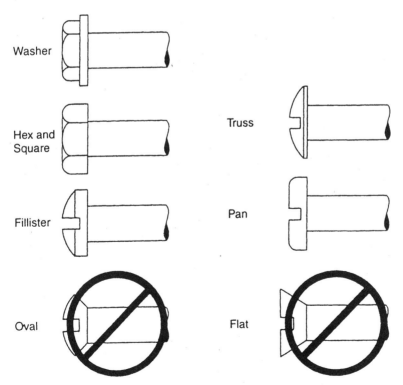

Figure 15.16 Common head systems of screws and bolts. Flat underside of head is preferred.[29]

for the body of the fastener should always be provided in the parts to be joined. This clearance can nominally be 0.25 mm (0.010 in.).

15.8.1.2 Self-threading screws. Self-threading screws can either be thread cutting or thread forming. To select the correct screw, the designer must know which plastic will be used and its modulus of elasticity. The advantage of using these types of screws are:

- Off-the-shelf availability
- Low cost
- High production rates
- Minimum tooling investment

The principal disadvantage of these screws is limited reuse; after repeated disassembly and assembly, these screws will cut or form new threads in the hole, eventually destroying the integrity of the assembly.

Thread forming screws are used in the softer, more ductile plastics with moduli below 200,000 psi. There are a number of fasteners specially designed for use with plastics (Fig. 15.17). Thread forming screws displace plastic material during the threading operation. This

Blunt-tip fasteners are suitable for most commercial plastics. Harder plastics require a fastener with a cutting tip. Hardest plastics require both a piercing and drilling tip, as in these fasteners.

BLUNT

CUTTING

PIERCING

Twin lead fastener seats in two revolutions.

TWIN LEAD

For rapid installation on lightly loaded joints, some fasteners have a thread configuration that allows the screws to be pushed into place. Typical is this design. Suitable for ductile plastics, this fastener relies on plastics relaxation around the shank to form threads. The thread is helical so that it can be unscrewed, but reuse is limited.

Reverse saw-tooth edges bite into the walls of the plastic.

 MILFORD

Triangular configuration is another technique for capturing large amounts of plastic. After insertion, the plastic cold-flows or relaxes back into the area between lobes. The Trilobe design also creates a vent along the length of the fastener during insertion, eliminating the "ram" effect. In some ductile plastics, pressure builds up in the hole under the fastener as it is inserted, shattering or cracking the material.

 TRILOBE

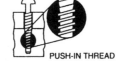 PUSH-IN THREAD

Dual-height thread design boosts holding power by increasing the amount of plastic captured between threads.

 HI-LO

Some specials have thread angles smaller than the 60° common on most standard screws. Included angles of 30 or 45° make sharper threads that can be forced into ductile plastics more readily, creating deeper mating threads and reducing stress. With smaller thread angles, boss size can sometimes be reduced.

SHARP THREAD

Barbs provide holding power.

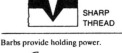 BARBED

Pushtite fastener is pushed into place and can be screwed out.

 PUSHTITE

Figure 15.17 Thread forming fasteners for plastics.[30]

type of screw induces high stress levels in the part and is not recommended for parts made of weak resins.

Assembly strength with thread forming screws can be increased by reducing hole diameter in the more ductile plastics, by increasing screw thread engagement, or by going to a larger diameter screw when space permits. The most common problem encountered with these types of screws is boss cracking. This can be minimized or eliminated by increasing the size of the boss; increasing the diameter of the hole; decreasing the size of the screw; changing the thread configuration of the screw; or changing the part to a more ductile plastic.

Thread cutting screws are used in harder, less ductile plastics, and remove material as they are installed, thereby avoiding high stress. However, these screws should not be installed and removed repeatedly.

15.8.1.3 Rivets. Rivets provide permanent assembly at very low cost. Clamp load must be limited to low levels to prevent distortion of the part. To distribute the load, rivets with large heads should be used with washers under the flared end of the rivet. The heads should be three times the shank diameter.

Riveted composite joints should be designed to avoid loading the rivet in tension. Generally, a hole $1/64$ in. larger than the rivet shank is satisfactory for composite joints. A number of patented rivet designs are commercially available for joining aircraft or aerospace structural composites.

15.8.1.4 Spring steel fasteners. Push-on spring steel fasteners (Fig. 15.18) can be used for holding light loads. Spring steel fasteners are simply pushed on over a molded stud. The stud should have a minimum 0.38 mm (0.015 in.) radius at its base. Too large a radius could create a thick section, resulting in sinks or voids in the plastic molding.

15.8.2 Design for self-fastening

It is often possible and desirable to incorporate fastening mechanisms in the design of the molded part itself. The two most common methods of doing this are by interference-fit (including press-fit or shrink-fit) and by snap-fit. Whether these methods can be used will depend heavily on the nature of the plastic material and the freedom one has in part design.

15.8.2.1 Press-fit. In press- or interference-fits, a shaft of one material is joined with the hub of another material by a dimensional interference between the shaft's outside diameter and the hub's inside diameter. This simple, fast assembly method provides joints with high

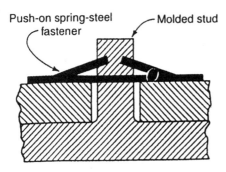

Figure 15.18 Push-on spring steel fasteners.[29]

strength and low cost. Press-fitting is applicable to parts that must be joined to themselves or to other plastic and non-plastic parts. The advisability of its use will depend on the relative properties of the two materials being assembled. When two different materials are being assembled, the harder material should be forced into the softer. For example, a metal shaft can be press-fitted into plastic hubs. Press-fit joints can be made by simple application of force or by heating or cooling one part relative to the other.

Press-fitting produces very high stresses in the plastic parts. With brittle plastics, such as thermosets, press-fit assembly may cause the plastic to crack if conditions are not carefully controlled.

Where press-fits are used, the designer generally seeks the maximum pullout force using the greatest allowable interference between parts that is consistent with the strength of the plastic. Figure 15.19 provides general equations for interference fits (when the hub and shaft are made of the same materials and for when they are a metal shaft and a plastic hub). Safety factors of 1.5–2.0 are used in most applications.

For a press-fit joint, the effect of thermal cycling, stress relaxation, and environmental conditioning must be carefully evaluated. Testing of the factory assembled parts is obviously indicated under expected

General equation for interference

$$I = \frac{S_d D_s}{W}\left[\frac{W\mu_h}{E_h} + \frac{1-\mu_s}{E_s}\right]$$

in which

$$W = \frac{1+\left(\dfrac{D_s}{D_h}\right)^2}{1-\left(\dfrac{D_s}{D_h}\right)^2}$$

I = Diametral interference, mm (in.)

S_d = Design stress limit or yield strength of the polymer, generally in the hub, MPa (psi) (A typical design limit for an interference fit with thermoplastics is 0.5% strain at 73°C.)

D_h = Outside diameter of hub, mm (in.)

D_s = Diameter of shaft, mm (in.)

E_h = Modulus of elasticity of hub, MPa (psi)

E_s = Elasticity of shaft, MPa (psi)

μ_h = Poisson's ratio of hub material

μ_s = Poisson's ratio of shaft material

W = Geometric factor

If the shaft and hub are of the same material, $E_h = E_s$ and $\mu_h = \mu_s$. The above equation simplifies to:

Shaft and hub of same material

$$I = \frac{S_d D_s}{W} \times \frac{W+1}{E_h}$$

If the shaft is a high modulus metal or other material, with $E_s > 34.4 \times 10^3$ MPa, the last term in the general interference equation is negligible, and the equation simplifies to:

Metal shaft, plastic hub

$$I = \frac{S_d D_s}{W} \times \frac{W+\mu_h}{E_h}$$

Figure 15.19 General calculation of interference fit between a shaft and hub.[31]

temperature cycles, or under any condition that can cause changes to the dimensions or modulus of the parts. Differences in coefficient of thermal expansion can result in reduced interference due to one material shrinking or expanding away from the other, or it can cause thermal stresses as the temperature changes. Since plastic materials will creep or stress-relieve under continued loading, loosening of the press-fit, at least to some extent, can be expected during service. To counteract this, the designer can knurl or groove the parts. The plastic will then tend to flow into the grooves and retain the holding power of the joint.

15.8.2.2 Snap-fit. In all types of snap-fit joints, a protruding part of one component, such as a hook, stud, or bead, is briefly deflected during the joining operation, and it is made to catch in a depression (undercut) in the mating component. This method of assembly is uniquely suited to thermoplastic materials due to their flexibility, high elongation, and ability to be molded into complex shapes. However, snap-fit joints cannot carry a load in excess of the force necessary to make or break the snap-fit. Snap-fit assemblies are usually employed to attach lids or covers which are meant to be disassembled or which will be lightly loaded. The design should be such that after the assembly, the joint will return to a stress free condition.

The two most common types of snap-fits are those with flexible cantilevered lugs (Fig. 15.20) and those with a full cylindrical undercut and mating lip (Fig. 15.21). Cylindrical snap-fits are generally stronger, but require deformation for removal from the mold. Materials with good recovery characteristics are required.

In order to obtain satisfactory results, the undercut design must fulfill certain requirements:

- The wall thickness should be kept uniform.

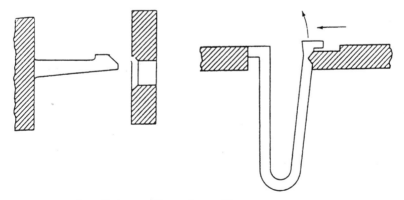

Figure 15.20 Snap fitting cantilevered arms.[29]

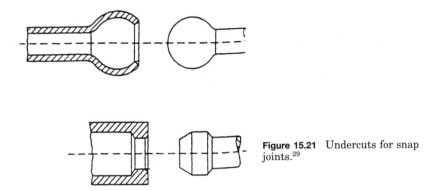

Figure 15.21 Undercuts for snap joints.[29]

- The snap fit must be placed in an area where the undercut section can expand freely.
- The ideal geometric shape is circular.
- Ejection of an undercut core from the mold is assisted by the fact that the resin is still at relatively high temperatures.
- Weld lines should be avoided in the area of the undercut.

In the cantilevered snap-fit design, the retaining force is essentially a function of the bending stiffness of the resin. Cantilevered lugs should be designed in a way so as not to exceed allowable stresses during assembly. Cantilevered snap-fits should be dimensioned to develop constant stress distribution over their length. This can be achieved by providing a slightly tapered section or by adding a rib. Special care must be taken to avoid sharp corners and other possible stress concentrations. Cantilever design equations have been recently developed to allow for both the part and the snap-fit to flex.[32] Many more designs and configurations can be used than only cantilever or snap-fit joints. Individual plastic resin suppliers are suggested for design rules and guidance on specific applications.

15.9 More Information on Joining Plastics

Additional details on joining plastics by adhesive bonding, direct heat welding, indirect heat welding, frictional welding, solvent cementing, or mechanical fastening can be found in numerous places. The best source of information is often the plastic resin manufacturers who often have recipes and processes. These are usually freely offered because it is in the manufacturers' interest that their materials successfully see their way into assembled products.

Another source of information is the equipment manufacturers. The manufacturers of induction bonding, ultrasonic bonding, spin welding, vibration welding, etc. equipment will often provide guidance on the correct parameters to be used for specific materials and joint designs. Many of these equipment suppliers will have customer service laboratories where prototype parts can be tried and guidance provided regarding optimum processing parameters.

Of course, the adhesive supplier and the mechanical fastener supplier can also provide detailed information on their products and which substrates are most appropriate. They can generally provide complete processes and specifications relative to the assembly operation. They usually also have moderate amounts of test data to provide an indication of strength and durability.

Finally, a very useful source of information is the technical literature, conference publications, books, and handbooks devoted to the subject of joining plastics. The following works are especially recommended for anyone requiring detailed information in this area. These are specifically devoted to plastic joining. Sources of information related to general joining are identified in Appendix B.

- *Handbook of Plastics Joining;* Plastics Design Laboratory (1997)
- *Joining of Composite Materials* by K. T. Kedward; ASTM (1980)
- *Joining of Fibre-Reinforced Plastics* by F. L. Matthews (ed.), Chapman and Hall (1987)
- *Joining of Composite Matrix Materials* by M. M. Schwartz; ASM International (1994)
- *Joining of Advanced Composites,* Engineering Design Handbook, Pamphlet DARCOM-OP 706-316, U.S. Government
- *Plastics Design Forum* (periodical); Edgell Communications
- *Plastics Packaging* (periodical); Edgell Communications

References

1. Harper, C. A., ed., *Handbook of Plastics, Elastomers, and Composites,* 3rd ed. (New York: McGraw Hill, 1996).
2. "Modern Plastics Encyclopedia '99", *Modern Plastics,* Mid-November 1998.
3. Ruben, I. I., ed., *Handbook of Plastic Materials and Technology,* (New York: John Wiley & Sons, 1990).
4. Bardson, J. R., *Plastic Materials,* 5th ed. (London: Butterworths, 1989).
5. Meier, J. F., "Fundamentals of Plastics and Elastomers", *Handbook of Plastics, Elastomers, and Composites,* 3rd ed., C. A. Harper, ed. (New York: McGraw Hill, 1996).
6. "Engineer's Guide to Plastics", *Materials Engineering,* May 1972.
7. Trauenicht, J. O., "Bonding and Joining, Weigh the Alternatives; Part 1: Solvent Cement, Thermal Welding", *Plastics Technology,* August 1970.
8. Gentle, D. F., "Bonding Systems for Plastics", *Aspects of Adhesion,* vol. 5, D. J. Almer, ed. (London: University of London Press, 1969).
9. Mark, H. F., Gaylord, N. G., and Bihales, N. M., eds., "Encyclopedia of Polymer Science and Technology", vol. 1 (New York: Wiley, 1964) at 536.
10. "All About Welding of Plastics", Seelye Inc., Minneapolis, MN.

Methods of Joining Plastics Other Than With Adhesives 591

11. "How to Fasten and Join Plastics", *Materials Engineering,* March 1971.
12. Spooner, S. A., "Designing for Electron Beam and Laser Welding", *Design News,* September 23, 1985.
13. Troughton, M., "Lasers and Other New Processes Promise Future Welding Benefits", *Modern Plastics,* mid-November, 1997.
14. "Laser Welding", Chapter 13 at *Handbook of Plastics Joining* (Norwich, NY.: Plastics Design Library, 1997).
15. Chookazian, M., "Design Criteria for Electromagnetic Welding of Thermoplastics" (Emabond Corporation, Norwod, NJ).
16. Leatherman, A., "Induction Bonding Finds a Niche in an Evolving Plastics Industry", *Plastics Engineering,* April 1981.
17. "Electromagnetic Welding System for Assembling Thermoplastic Parts" (Emabond Corporation, Norwood, NJ).
18. "Spin Welding", Chapter 4 of *Handbook of Plastics Joining* (Norwich NY: Plastics Design Library, 1997).
19. LaBounty, T. J., "Spin Welding Up-Dating and Old Technique", SPE ANTEC, 1985, at 855–856.
20. Grimm, R. A., "Welding Process for Plastics", *Advanced Materials and Processes,* March 1995.
21. "Ultrasonic Welding", Chapter 5 of *Handbook of Plastics Joining,* (Norwich, NY: Plastics Design Library, 1997).
22. Branson Sonic Power Co., Danbury, CT
23. "Ultrasonic Joining Gains Favor With Better Equipment and Knowhow", *Product Engineer,* January 1977.
24. Mainolfi, S. J., "Designing Component Parts for Ultrasonic Assembly", *Plastics Engineering,* December 1984.
25. Hauser, R. L., "Ultra Adhesives for Ultrasonic Bonding", *Adhesives Age,* 1969.
26. Scherer, R., "Vibration Welding Could Make the Impossible Design Possible", *Plastics World,* September 1976.
27. Kagan, V. A., et. al., "Optimizing the Vibration Welding of Glass Reinforced Nylon Joints", *Plastics Engineering,* September 1996.
28. Raia, D. C., "Adhesives—the King of Fasteners", *Plastics World,* June 17, 1975.
29. "Engineering Plastics", *Engineered Materials Handbook,* vol. 2, ASM International, Metals Park, OH, 1988.
30. Fastening, Joining, and Assembly Reference Issue, *Machine Design,* November 17, 1988.
31. "Mechanical Fastening", Chapter 14 of *Handbook of Plastics Joining* (Norwich, NY: Product Design Library, 1997).
32. McMaster, W., and Lee, C., "New Equations Make Fastening Plastic Components a Snap", *Machine Design,* September 10, 1998.

Chapter 16

Bonding and Sealing Specific Substrates

16.1 Introduction

This chapter identifies various materials and processes that have been used to successfully bond and seal specific substrates. Methods, other than adhesive bonding, are also included if they have been recommended for use on the particular substrate. Recommended substrate surface treatments are usually the same for both adhesives and sealants, although surface treatment might not be as critical with sealants.

The examples provided in this chapter will be grouped by substrate material. The substrates are broadly identified as:

- Metals
- Plastics (thermosets and thermoplastics)
- Composites
- Foams
- Elastomers
- Wood and wood products
- Glass and ceramics
- Sandwich and honeycomb structures

Specific substrates are described under each classification. This chapter certainly does not consider all possible substrates. However, the guidance that is presented and the lessons provided throughout the rest of this Handbook should be sufficient for the user to wisely select "candidate" joining processes and materials no matter what substrates are involved. This initial list of candidates may require further modification or refinement relative to parameters unique to the user's application or because of the results of prototype testing. However, they provide a "first-cut" as to the best possible assembly process. Along with the information contained in this chapter, the reader will want to review Chapter 6, Surface Preparation; Chapter 11, Selection of Adhesives; and Chapter 14, Selecting and Using Sealants. Together, these chapters should provide the information necessary to choose a candidate joining process.

16.2 Metal Bonding

All metals have a relatively high surface energy and are generally considered easy to bond. However, there are several problems that could occur when working with metallic substrates. One difficulty in bonding metals is the durability of the joint. It is not so much a problem of making a strong joint as one of keeping it that way throughout its expected service life. A weld may have a strength of only 600 lbs, but it is likely to remain that strong for 5–10 years afterward. An adhesive, on the other hand, may have three to four times the initial strength of a weld, but it could weaken when exposed to high humidities, cycled between hot and cold temperatures, or immersed in salt water and then dried. By definition, a structural adhesive must be able to withstand such conditions without significant deterioration.

A second difficulty in bonding metals is understanding the nature of the surface. One of the important points to consider when bonding metals is that only the surfaces are involved. Adhesives and sealants are active only on that top molecular surface layer and on any surfaces contained in the porosity of the metal itself. Thus, if unprepared steel is being bonded, it is not the bulk iron/carbon alloy that is being bonded, but the iron oxide layer on the surface (presuming that the metal surface was cleaned of organic contaminants). Similarly with aluminum, the actual bond is to aluminum oxide rather than the pure metal.

The base metal is highly reactive in most cases and forms various oxides, sulfides, and hydrates when exposed to the atmosphere. As a result, it is necessary to consider not only the bulk metal but also the ability to bond to its hydrated oxide. One must also consider the inherent nature of the adhesive force existing between the base metal and its oxide. The final joint will be no stronger than its weakest link:

1. The cohesive strength of all the materials involved
2. The forces between the adhesive and the metal oxide
3. The strength of the metal oxide bond to the base metal

The actual metal surface that takes part in bonding was illustrated in Fig. 6.3. Adhesives recommended for metal bonding are in reality used for metal-oxide bonding. They must be compatible with the firmly bound layer of water attached to surface metal-oxide crystals. Even materials such as stainless steel and nickel or chromium are coated with transparent metal oxides that tenaciously bind at least one layer of water.

The nature and characteristic of these oxide layers depend on the base metal and the conditions that were present during its formation. With steel, for example, the oxide adhesion to the base metal is very weak. In the case of aluminum, however, the oxide is extremely stable and clings tightly to the base metal. In fact, it adheres so well that it serves as a protective coating for the aluminum. This is one reason why aluminum is a corrosion resistant metal. Certain metals will possess surfaces that interact more effectively with a particular type of adhesive than with another. This is the reason why adhesive formulators need to know as much as possible about the surfaces being assembled.

General property data for common structural metal adhesives were presented in Table 11.2. This table may prove useful in making preliminary selections or for eliminating obviously unsuitable adhesives for bonding metals. Once the candidate adhesives have been limited to only a few types, the designer can search more efficiently for the best bonding system. However, selecting a specific adhesive from a table of general properties is difficult because formulations within one class of adhesive may vary widely in physical properties, curing conditions, and environmental resistance.

Non-structural adhesives for metals include elastomeric and thermoplastic resins. These are generally used as contact, pressure-sensitive, or hot-melt adhesives. They are noted for fast production, low cost, and low to medium shear strength. Typical adhesives for non-structural-bonding applications were described in Chapter 11. Most pressure-sensitive and hot-melt adhesives can be used on any clean metal surface and on many plastics and elastomers.

16.2.1 Aluminum and its alloys

Aluminum is an almost ideal substrate for adhesives and sealants. It has high surface energy and is very resistant to most environments. It is also a material with good formability and high strength-to-weight ratio that can benefit greatly from properties offered by adhesive joints. As a result, adhesive bonded aluminum joints are commonly used in the aircraft and automotive industries. Aluminum joints are also commonly used in adhesive studies and for comparison of different adhesive materials and processes. Adhesive manufacturers' literature generally describe the properties of bonded aluminum joints.

The oxide layer which forms on aluminum, however, is more complex than with other metal substrates. Aluminum is a very reactive surface, and oxide forms almost instantaneously when a freshly machined aluminum surface is exposed to the atmosphere. The oxide is extremely stable and adheres to the base metal with strengths higher than could be provided by most adhesives. The oxide is also cohesively strong and electrically nonconductive. These surface characteristics make aluminum a desirable metal for adhesive bonding, and they are the reasons why many adhesive comparisons and studies are done with aluminum substrates.

Adhesive wetting may be enhanced by converting the existing surface to a new surface of higher surface energy and improved topology. In the case of aluminum surfaces, chemical conversion can also protect the base metal from corrosion and enhance the durability of the bonded joint in various service environments. The common surface preparations that have been used for bonding aluminum can be generally segregated into three groups:

- Simple cleaning and abrading
- Chemical etching
- Primers and conversion coatings

Included in the simple cleaning and abrading category are: solvent wiping, vapor degreasing, and either of these methods combined with mechanical abrading. Descriptions of these processes were provided in Chapter 6. In each instance, care must be taken to assure that the cleaning materials themselves do not become unknowingly contaminated, thus providing ineffectual cleaning or cross-contamination resulting in poor bond performance.

Sandblasting is commonly used for treating aluminum surfaces prior to adhesive bonding because of its simplicity and economics. However, chemical treatments, such as the etchant processes, produce higher reliability and longer service life in a bonded assembly. If aluminum adherends are first cleaned, then sandblasted, and finally chemically treated, the surface area is increased, the contaminants are removed, and the initial and long term bond strengths are generally excellent. However, this three-step process is often not necessary when only low strengths (500–2,000 psi) are required. Useful bonds in these applications can be achieved simply with cleaning and/or abrasion.

When bonding aluminum to itself or to other materials, the optimal surface preparation should be determined for the application based on the initial strength and durability required, and then the process should be rigidly followed. Over-specifying the strength requirements must be avoided since it could result in the selection of a surface preparation process that is time consuming and expensive. Table 6.4 serves as a guideline for selecting the pretreatment to try first.

In selecting a pretreatment process for aluminum or any other substrate, both the initial strength and the permanence in a specific operating environment must be considered. Mechanical abrasion is a useful pretreatment in that it removes the oxide and exposes bare aluminum. When this is done, however, many of the benefits of the protective oxide layer are lost. For example, if bare abraded aluminum is bonded, the reactive metal at the joint interface can potentially become hydrolyzed and oxidized, which will displace the adhesive. Hence, this bonded joint may initially be much stronger than one made with unabraded metal, but it will deteriorate rapidly when exposed to harsh environments such as high heat or humidity. This is why pretreatments that modify the oxide layer or create a new, stable oxide layer are especially desirable when permanence is a primary consideration. They improve bondability *and* maintain protection.

To protect the aluminum joint from the effects of the environment, especially water and corrosion, an artificially thickened oxide layer is formed on the surface. Historically, chemical etching as a surface preparation has provided the surest way of obtaining durable adhesive bonds with aluminum. While various acidic or caustic procedures can be employed with or without vapor degreasing, the best recognized etching pretreatment has been the sulfuric acid-dichromate solution used by the aircraft industry and described in ASTM D 2651. This process is sometimes known as the FPL etch, named after its developers, Forest Products Laboratories.[1] The first step in this process is vapor degreasing, followed by alkaline cleaning, and then chemical immersion. The substrates are finally forced dried. (The exact recipe for the FPL etch process can be found in Appendix D-1.) There are many modifications of this treatment including a paste-like etching solution (ASTM D 2641 Method F) to allow for parts that cannot be immersed in the acid solution, and a chromate free etching process (designated PT2) for improved environmental and occupational health and safety perspectives.[2]

Other important methods of pretreating aluminum for adhesive bonding include anodizing and chromate conversion coating.[3] In anodizing, the aluminum is immersed in various concentrations of acids (usually phosphoric or chromic) while an electrostatic charge is applied. The oxide reacts with the etchant to form a compound that protects the surface and is compatible with the adhesive. In this way the aluminum oxide is retained, but it is rendered more receptive to bonding. It has been shown that an anodized surface can provide a very durable surface for adhesive bonding with excellent resistance to seacoast or saltwater exposure. Examples of widely used anodizing processes are the Boeing phosphoric acid anodize (PAA) process[4] and the chromic acid anodize (CAA) process.[5]

The landmark U.S. Air Force Primary Adhesively Bonded Structure Technology (PABST) Program in the late 1970s demonstrated that properly designed and manufactured bonded fuselage panels made from the correct materials can actually be operated safely at higher stress levels than comparable rivet joined aluminum structures.[6,7] The results of this program show an optimal way to achieve durable aluminum bonds:

- PAA is the most durable pretreatment for aluminum that is processable within reasonable production tolerances.
- PAA with a corrosion resistant primer provides the best corrosion resistance.
- The adhesive should be selected on the basis of durability as defined by slow cyclic testing in a hot/humid environment.

The chemical conversion coating method is also commonly used to treat aluminum substrates prior to bonding. Chemical conversion coating is an amorphous phosphatization process where the aluminum is treated with a solution containing phosphoric acid, chromic acid, and fluorides. Chromate conversion coatings on aluminum constitute an effective way to enhance the surface bond and also improve the corrosion resistance of the bondline.[8] However, the resulting durability observed with chemical conversion coatings are variable and depend on the particular processing conditions. By contrast, anodizing and etching processes produce consistent and generally durable aluminum joints.

The durability of bonded aluminum joints immersed in water are shown in Fig. 16.1. The anodized and grit blasted surface treatments, although giving different initial joint strengths, showed no deterioration after two years' exposure. Both the vapor degreased and conversion coating treatments were significantly degraded by the moist environment. Exposure of similarly prepared specimens to a more aggressive soak-freeze-thaw cycle gave rise to even greater differences in performance with only the anodized aluminum joint showing a high percentage of joint strength after a two-year period.[9]

However, the recognition that chrome is carcinogenic is forcing alterations in surface treatment processes and may eventually require changes in primers. Chromic acid anodizing and FPL etching are being phased out in many locations. Development of sulfuric acid ano-

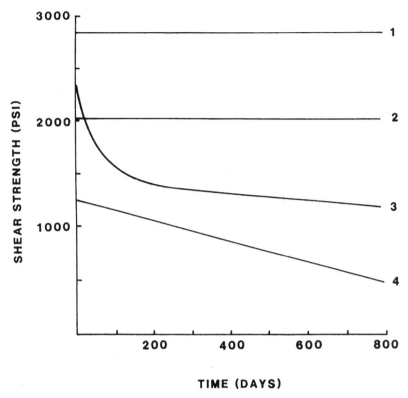

Figure 16.1 Effect of surface treatment on the durability of epoxy/aluminum joints exposed to room temperature water immersion. (1) Anodized, (2) grit blasted plus vapor degrease, (3) vapor degrease, (4) chromate conversion coating.[9]

dizing and sulfuric boric acid anodizing is now in progress.[10] In the automotive industry, a pretreatment has been developed for aluminum coil that is nontoxic and compatible with weld bonding. This proprietary treatment is claimed to be as effective as chromium-based pretreatment processes on exposure to salt spray.[2]

The usual approach to good bonding practice is to prepare the aluminum surface as thoroughly as possible, then wet it with the adhesive as soon afterwards as practical. In any event, aluminum parts should ordinarily be bonded within 48 hrs after surface preparation. However, in certain applications this may not be practical, and primers are used to protect the surface between the time of treatment and the time of bonding. Primers are also applied as a low viscosity solution which enables them to chemically wet a metal surface more effectively than more viscous, higher solid content adhesives.

The chemical cleaning methods used for aluminum vary in effectiveness with different aluminum alloys. The permanence of these bonds also depend on the type of alloys used because of their different corrosion degradation rates under extreme environmental conditions. The yield strength of the alloy also has an influence on bond strength when the joint is stressed in shear. The peel test is usually considered a more meaningful test method for measuring the surface treatment effectiveness. The yield factor of the adherend is never approached because of relatively poor peel strength of the adhesive.

Table 16.1 shows the effect of surface treatments on bonded aluminum alloy joints using a heat cured epoxy adhesive. The initial shear strength and permanence depend on the type of alloy and the pretreatment used. It should be noted that the data presented here only show the relative differences in joint strength for a single adhesive and is not representative of other adhesive formulations. There was no attempt to maximize any of these values through choice of the adhesive.

Once the surface considerations are taken care of, there are many adhesives that can bond well to aluminum. The selection will depend on the strength needed, the type of stress involved (e.g., peel or shear, static or dynamic), and the operating environment. Reynolds Metals Company[12] offers some general rules in selecting an adhesive for aluminum bonding.

- Bonds to aluminum are generally stronger than bonds to steel.
- A chromic-sulfuric acid etch gives the best resistance to weathering and salt water environments.
- Room temperature curing epoxies offer the best salt water resistance.
- Higher strengths are usually obtained with heat curing epoxies than with room temperature curing epoxies.
- Modified phenolic films give the highest peel and shear strength combinations.
- The most severe adhesive environment is a hot, humid climate (temperature 90–120°F; humidity +90%).
- Structural adhesives are strong in shear; weakest in peel and cleavage.
- Heat curing adhesives are less sensitive to surface preparation than room temperature curing adhesives.

Table 16.2 presents a concise summary of adhesives for bonding aluminum to itself and to a wide variety of other materials. Sell[13] has also ranked a number of aluminum adhesives in order of decreasing durability as follows: nitrile-phenolics, high temperature epoxies,

TABLE 16.1 Effect of Surface Treatment on Bonded Joints Using EC-3443 (3M Co.) Epoxy Heat Cured Adhesive[11]

	Alloy 2036			Alloy 6151			Alloy X5085		
	A	B	C	A	B	C	A	B	C
Initial shear strength[2] (psi)	1930	1850	2200	2550	2690	2530	2250	2270	2110
85% humidity at 75°F[3]	1390	1050	2400	1350	1550	2410	1860	1890	1910
100% humidity at 125°F[3]	420	350	1110	920	1050	1100	1260	1520	1230
5% salt spray[4]	0	0	1410	80	150	2090	690	530	1210

[1]A—Mill Finish; B—Vapor Degreased; C—Alodine 401–45.
[2]All averages based on triplicate specimens.
[3]Three months exposure.
[4]Three weeks exposure.

TABLE 16.2 Selection Guide for General Aluminum Bonding Applications[12]

Material	Adhesive	Type & form	Service temperature (°F)
Aluminum to itself Cryogenic (very low temp. appl.)	Metlbond 305	Glass-supported mod. phenolic tape	−423 to +500
	Metlbond 329	Modified epoxy supported film	−423 to +500
	Epon 422	Epoxy-phenolic glass cloth	−423 to +700
Aluminum to itself Electrical conductive appl.	544-29	2-part epoxy silver paste	−50 to +150
	Eccobond 56-C	2-part epoxy silver paste	up to +350
	AC-2	1-part mod. epoxy paste	−50 to +200
	#3012	1-part epoxy silver paste	up to +250
Aluminum to itself General purpose	Epon 907	2-part epoxy paste	−67 to +180
	EC-1838	2-part epoxy paste	−67 to +140
	M-688	2-part epoxy paste	−67 to +200
	EC-2086	1-part epoxy paste	−70 to +200
	EC-2214	1-part epoxy paste	−40 to +250
Aluminum to itself (High temp. appl.)	#5523	2-part system Part A paste Part B powder	up to +600
	Epon 422	Epoxy-phenolic glass cloth	up to +700
	#5524	2-part system Part A liquid Part B liquid	long time service to +600
	A-701	1-part epoxy paste	up to +450
Aluminum to itself (low temp. appl. below 70°F)	Epibond 8061	Not available	Not available
	A-32	2-part epoxy paste	−60 to +200
Aluminum to itself (where fast cures are necessary)	Epon 911F	2-part epoxy	−67 to +150
	Ankpoxy 973-2/974-1	2-part epoxy syrup	up to +120
	Mereco X-305	2-part epoxy liquid	up to +120
Aluminum to itself (where moisture is present on metal)	Metagrip 303	2-part paste	up to +180
	Concresive #1078	2-part epoxy paste	−50 to +150
	Epocast 546-9514A	2-part epoxy paste	−50 to +150

TABLE 16.2 Selection Guide for General Aluminum Bonding Applications[12] (*Continued*)

Material	Adhesive	Type & form	Service temperature (°F)
Aluminum to itself (where surface may by oily)	Epibond 8083	2-part epoxy	−50 to +200
Concrete to Aluminum	Epon 936	2-part epoxy paste	up to +120
	Epocast 546 with Hardener 9514-A	2-part epoxy paste	−20 to +180
	Concresive 1078	2-part epoxy paste	+35 to +180
Fiberglass (flexible) to aluminum	EC-1357	1-part neoprene phenolic	−30 to +300
Fiberglass (rigid) to aluminum	EC-2158	2-part paste	−67 to +180
	EC-2214	1-part paste	−40 to +250
Foams to aluminum (styrofoam, urethane, etc.)	Tygoweld 56—Act. 63	2-part epoxy light paste	−40 to +200
	EC-321	1-part reclaimed rubber	−35 to +300
	M-666	2-part epoxy paste	−67 to +200
Glass to aluminum	EC-1294/EC-1295	2-part epoxy thin syrup	−67 to +200
	DC-780	silicone rubber paste	−80 to +350
	EC-2216	2-part epoxy amine paste	−67 to +200
Nickel to aluminum	Epibond 8277	Not available	Not available
Nickel steel to aluminum	Metagrip 303	2-part paste	up to +180
Plastics to Aluminum	Metagrip 303	2-part paste	up to +180
Delrin to Aluminum	Epibond 8277	Not available	Not available
Mylar to Aluminum	G414-22	1-part synthetic rubber resin	−50 to +250
Nylon to Aluminum	Metagrip 303	2-part paste	up to +180
	Eastman 910	Cyanoacrylate 1-part liquid	−65 to +125
	Loctite 404	Cyanoacrylate 1-part liquid	up to +175
Penton to Aluminum	Epibond 8277	Not available	Not available
PVC to Aluminum	Prime with G414-22	1-part synthetic rubber resin	−50 to +250
	Bond with M-666	2-part epoxy paste	−67 to +200

TABLE 16.2 Selection Guide for General Aluminum Bonding Applications[12] **(Continued)**

Material	Adhesive	Type & form	Service temperature (°F)
Teflon to Aluminum	Metagrip 303	2-part paste	up to +180
	Chemgrip Adhesives	1 & 2-part epoxy paste systems	up to +500
Rubber to Aluminum	Eastman 910	Cyanoacrylate 1-part liquid	−65 to +200
General (For high strengths epoxies may be used. Any adhesive manufacturer can recommend a suitable one for application)	Loctite 404	Cyanoacrylate 1-part liquid	up to +175
Rubber to Aluminum Buna N (nitrile), hyperlon, natural, filled gum stock	Ad-E-Bond #148	1-part synthetic rubber liquid	−30 to +230
Rubber to Aluminum GR-S, Kel-F, buna N, neoprene, red sheet, sponge, pure gum	Ad-E-Bond #228	1-part synthetic rubber liquid	−30 to +230
Silicone Rubber to Aluminum	RTV-109 (1)	Silicone rubber 1-part paste	−75 to +300
	Silastic 732	Silicone rubber 1-part paste	−85 to +450
Wood to Aluminum	Epon 907	2-part epoxy paste	−67 to +180
	EC-1838	2-part epoxy paste	−67 to +140
	EC-2214	1-part epoxy paste	−40 to +250
	EC-1595	1-part epoxy paste	−67 to +350

This table is purely a general guide and the recommended adhesives may not necessarily work for every application. Each application should be thoroughly tested.

Notes: (1) For a satisfactory bond to aluminum with this adhesive the metal should be primed.

Suggested primers:

    For service under 400°F, Chemlok 607
    For service over 400°F, Dow Corning 796

250°F curing epoxies, 250°F curing rubber modified epoxies, vinyl epoxies, two-part room temperature curing epoxy paste with amine cure, and two-part urethanes.

16.2.2 Beryllium and its alloys

Beryllium and its alloys (e.g., beryllium copper) have gained interest in the aerospace industry and specialty sports equipment industry in recent years. Brazing or riveting can be used for joining, but these methods are expensive and distortion or highly stressed areas may be encountered. The metal must be handled with care when the processing produces dust, chips, scale, slivers, mists, or fumes, since air-borne particles of beryllium and beryllium oxide are toxic with latent health effects. Abrasives and chemicals used with beryllium must be disposed of properly.

The gain in the popularity of beryllium has in large part been due to adhesive bonding. Bonding permits one to take advantage of the excellent combination of physical and mechanical properties of beryllium and minimizes the inherent problems of high notch sensitivity and low ductility. Many bonded beryllium products are found in applications where high strength-to-weight ratios are important.

Beryllium is a very reactive metal, and it reacts quickly with methyl alcohol, Freon, perchloroethylene, and blends of methyl ethyl ketone and Freon. Beryllium can be pitted by long exposure to tap water containing chlorides or sulfates.[14] Proprietary coatings are used to provide a corrosion-resistant barrier.[15]

A number of prebonding surface preparations for beryllium and its alloys have been suggested. One procedure is to degrease the substrate with trichloroethylene, followed by immersion in the solution listed below for 5–10 mins at 68°F.

Sodium hydroxide	20–30 pbw
Distilled water	170–180 pbw

The substrate should then be washed in tap water and rinsed with distilled water. The final step in the process is an oven-dry of 10 mins at 250–350°F.[16] Several other surface preparation procedures for beryllium have been reported to have merit. Fig. 16.2 shows results obtained with a paste etch and various adhesives.

Many of the adhesives and surface preparations developed for beryllium have been used to exploit the metal's high temperature properties in aerospace applications. Since beryllium retains significant strength at temperatures up to 1,000°F, many of the adhesives used with beryllium are high temperature adhesives such as polyimides and polybenzimidizoles. The more conventional adhesive systems will

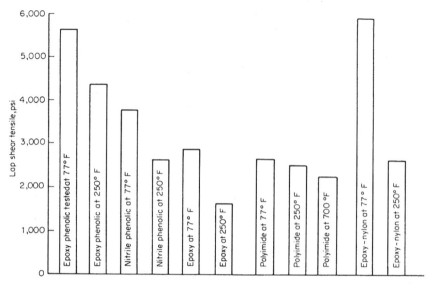

Figure 16.2 Bond strength for beryllium to beryllium joints using various adhesives and a sulfuric acid—sodium dichromate etch.[17]

offer higher shear strength at room temperature, but will have properties that begin to degrade rapidly once their glass transition temperature is exceeded.

Beryllium copper alloys have been bonded with an adhesive consisting of polyarylsulfone, a high temperature thermoplastic resin.[14] The suggested pretreatment process consisted of scrubbing with a pumice and water mixture, rinsing with distilled water and drying in an oven at 200°F. The prepared adherends were then coated with the resin in solution and dried. The polyarylsulfone coated beryllium copper specimens were then bonded at 570°F at a pressure of 55–90 psi. Useful adhesive values were obtained over a wide temperature range of −65 to +400°F. A linear relationship exists between lap shear strength and temperature over the range of 75–500°F. The joints exhibited excellent resistance to various chemicals and water at room temperature. However, immersion in boiling water reduced lap shear strength by approximately 50%.

16.2.3 Cadmium plating

Cadmium had once been a relatively common coating on steel. Health and safety concerns, however, have significantly reduced its use. Cadmium can best be made bondable by electroplating with silver or nickel. Another procedure is to degrease the substrate with solvent, scour it with a non-chlorinated abrasive cleaner, and then rinse with distilled water and dry.

Some adhesive bonded cadmium joints show rapid degradation when subjected to environmental conditions, especially humidity. Weathering properties may be improved by the use of a sealant or primer prior to adhesive bonding.

16.2.4 Copper and copper alloys

Copper substrates are commonly bonded with adhesives in the microelectonics and marine industries. Like most metals, they generally form acceptable initial bond strengths, but the durability is poor especially in moist environments.

Copper is used in three basic forms: pure, alloyed with zinc (brass), and alloyed with tin (bronze). Copper and copper alloys are difficult to bond satisfactorily especially if high shear and peel strengths are desired. The prime reason for this difficulty is that the oxide that forms on copper develops rapidly (although not as fast as the rate of oxide development on aluminum). The copper oxide layer is weakly attached to the base metal under usual conditions. Thus, if clean, bare copper substrates were bonded, the initial strength of the joint would be high, but with environmental exposure an oxide layer could develop which would reduce the durability of the joint.

Cleaning and mechanical abrasion are often used as pretreatments for copper when low to medium bond strengths are acceptable. For optimum bond strength and permanence, the oxide layer must be specifically "engineered" for adhesive bonding. This is done through what has been called the "black oxide" coating process or through chromate conversion coatings.

The black oxide processes form micro-rough morphologies on the copper surface. One such process uses a commercially available product[18] to form a strong black oxide coating on the copper surface. This process requires alkaline cleaning and acid immersion before the copper surface is immersed at elevated temperatures in a solution containing the proprietary product. The process steps are critical, and time and temperature must be carefully controlled. Parts should be joined within 4 hrs after chemical treating. Another black oxide coating process, Method E in ASTM D 2651, is intended for relatively pure copper alloys containing 95% copper.

The sodium dichromate–sulfuric acid process has been found by some to be superior to ferric chloride methods for the prebond treatment of copper. This method is also defined by ASTM D 2651. Nitric acid and ferric chloride etching processes have also been found to be useful for copper, brass, and bronze substrates in certain applications.

Copper substrates can be bonded with a number of adhesives. Generally, epoxy compounds are used for structural joints. Adhesives and sealants which liberate acidic by-products during cure should not be used since these will attack delicate copper parts. Certain RTV silicone adhesives have been formulated so that they cure by liberating methanol rather than acidic acid. Copper may also react with certain chemicals in the adhesive or sealant formulation at elevated temperatures.

Copper has a tendency to form brittle amine compounds with some curing agents. Many room temperature curing adhesives will give very good adhesion to copper and copper alloy substrates. For elevated temperature cures, however, only heat cured epoxies containing dicyandiamide or melamine should be used. Dicyandiamide and melamine curing reactions have been shown to be beneficial in both epoxy-based adhesives and coatings when used on copper substrates. These formulations show significantly increased time to adhesive failure on either bare or alkaline permanganate-treated copper.[19]

Brass is an alloy of copper and zinc, and bronze is an alloy of copper and tin. Sandblasting or other mechanical means of surface preparation may be used for both of these alloys. Surface treatments combining mechanical and chemical treatment with a solution of zinc oxide, sulfuric acid, and nitric acid is recommended for maximum adhesion properties. Adhesives similar to those recommended for copper may be used on brass and bronze substrates.

16.2.5 Gold, silver, platinum, and other precious metals

Most adhesives bond readily to precious metal substrates. Either vapor degreasing and light abrasion (180–240 grit), or scrubbing with non-chlorinated abrasive cleanser are suitable pretreatments for these precious metals. Chromate conversion coatings and primers have also been used with silver substrates.

16.2.6 Lead

Lead is a commonly bonded substrate for sheet laminations (x-ray shielding, sound barriers, etc.), decorative trim, chemical resistant foil tape, and balancing shapes. There are four grades of lead designated as "common", "corroding", "copper bearing" (chemical or acid), and "hard" (antimonial). The first three grades are similar, although copper bearing grades have advantages in corrosive exposure and better fatigue properties. Hard lead, generally with 4–8% antimony, has better mechanical properties than the other grades up to about 250°F.

Lead depends upon an inert surface film for corrosion resistance. Thus, it cannot be used with chemicals that form soluble lead compounds. Acetic, nitric, and sulfuric acids; most common alkalis; and chloride bleaches will corrode lead. Strong agitation, especially with suspended solids or aeration, will result in abrasion and destroy lead's protective film.

When exposed to moist atmosphere in the absence of chemical pollutants, the lead surface is usually a basic lead carbonate. Under conditions such as high humidity or water spray and moderately warm temperatures, oxide films may be formed. In some cases, chemical reaction between an adhesive and lead oxides in the film has been suspected in bond failures. The lead oxide film can also be a weak boundary layer under conditions of cyclic movement such as due to thermal expansion and contraction. It makes good sense to remove lead oxides by either mechanical or chemical cleaning prior to bonding.

Many adhesives can be used to bond lead. High strength or alloyed adhesives are seldom justified for bonding lead because of the lead's ductility and relatively low strength. Low cost and easy to use adhesives are generally preferred.

16.2.7 Magnesium and magnesium-based alloys

The surface preparation and bonding methods developed for magnesium and magnesium-based alloys are closely associated with corrosion prevention because magnesium is one of the most reactive of all metals. Numerous processes have been developed for both painting and bonding of magnesium. The choice of a bonding process will be determined by criteria such as high bond strength, high corrosion resistance, or both.

The position of magnesium in the electrochemical or the galvanic series indicates that it has a potential for reacting chemically with a great number of other metals when there is electrical connection between the two metals. The conductive path could be caused by direct metal-to-metal contact, an aqueous solution in which there is an electrolyte (e.g., chloride ions in solution), or by other ways. The key in any assembly involving magnesium is to design and assemble the parts in such a manner that the conductive path is eliminated.

Magnesium will react with moisture to form a magnesium hydroxide that, in itself, provides some corrosion protection. However, this coating is not quite uniform and may contain some carbonates and sulfates or other ionic materials that are present in the surrounding environment. Magnesium is quite resistant to alkaline conditions, and under controlled conditions this factor is used to advantage in adhesive surface preparation. Organic compounds, in general, do not react with magnesium as do the aqueous electrolytes. Anhydrous methyl alcohol is an exception—it can react readily with magnesium.

Minor constituents found in magnesium alloys can play a significant part in adhesive bonding. Magnesium alloys do not all react to protective surface treatments in the same way. Knowledge of the alloy composition must be coupled with a careful surface preparation selection process. Grain structure of the metal can also influence the nature of the surface preparation. The temper of the alloy is usually indicated as strain hardened (W), annealed (O), or as fabricated (F), and the various ways of treating are specified (T1 through T10). Different surfaces due to temper and treatment could have a bearing on the effectiveness of the surface preparation process.

Removal of light oil or light chromate coatings (used to protect the magnesium during shipment and storage), mill scale, lubricants, welding fluxes, etc. must occur before the desired surface preparation is applied. Three cleaning methods, either used separately or in combination, constitute the necessary cleaning before the final surface preparation: solvent degreasing, mechanical cleaning (galvanic reactions from the grit must be avoided), and chemical treatment (acid pickling using chromic, nitric, or phosphoric acid solutions). It is not uncommon for the acid pickling or cleaning procedure to suffice as the final surface preparation process. The alkaline detergent cleaning process is described in ASTM D 2651 Method A. A hot chromic acid cleaning process that can be combined with the alkaline cleaning process is given by ASTM D 2651 Method B.

Light anodic treatment and various corrosion preventive treatments produce good surfaces on magnesium for adhesive bonding (ASTM D 2651 Method D). These treatments were developed by magnesium alloy producers, such as Dow Chemical Company (Dow 17 and Dow 7) and others. Details are available from the ASM Metals Handbook[20] and in MIL-M-45202, Type I, Class 1, 2, and 3.[21] Some surface dichromate conversion coatings and wash primers designed for corrosion protection can also be used for adhesive bonding of magnesium (ASTM D 2651 Method E). Details are found in the ASM Metals Handbook[20] and MIL-M-3171[22].

The thickness of the surface coating will influence the relative degree of protection for the magnesium against corrosive elements. Heavy anodic coatings of about 0.001 in. are used for maximum corrosion resistance. However, such thick coatings are generally not optimal for adhesive bonding because cohesive failure occurs in the coating. Thinner coatings of 0.0001–0.0003 in. obtained from the Dow 17 anodic surface treatment is better for adhesive bonding. Thin films also provide more coating flexibility with less likelihood of cracking upon flexure.

A problem unique to magnesium substrates is the formation of loose particles on the surface of the metal. The source of the smut can be varied but is usually related to the cleaning process or to the environment during the cleaning process. Such loose, gray, black, or brown smut particles are believed responsible, in some cases, for erratic adhesive strength results. These loose particles generally can be rubbed off with a solvent dampened pad.

High humidity and salt spray environments have been found to cause the greatest decrease in bond strength of bonded magnesium joints. Of the several surface treatments that have been evaluated, the Dow 17 surface preparation provides the best overall performance with all adhesives under these types of environmental conditions.[23]

Almost any adhesive can be used on magnesium provided that proper corrosion protection is maintained. Epoxy adhesives have generally been used. In view of the corrosion potential, water-based adhesives or adhesives that allow water permeation may be expected to cause problems with magnesium substrates.

16.2.8 Nickel and nickel alloys

Nickel and its common alloys such as Monel (nickel-copper), Inconel (nickel-iron-chromium) and Duranickel (primarily nickel) can be bonded with procedures that are recommended for stainless steels.[24] A simple nitric acid process has also been used consisting of solvent cleaning, immersion for 4–6 secs at room temperature in concentrated nitric acid, rinsing with cold deionized water, and finally drying. Also, a chromium trioxide–hydrochloric acid process consisting of a 60–80 sec immersion in acid solution has been suggested. If immersion is impossible, this solution may be applied with a cheesecloth after solvent cleaning. The solution is applied to approximately 1 sq ft of the substrate surface at a time, and it remains on the part for approximately 1 min.

Epoxies are commonly used to bond nickel substrates. However, the nickel alloys generally find applications at temperatures higher than most organic adhesives are capable of resisting. Thus, there is relatively little development work done on bonded nickel alloy substrates.

16.2.9 Plated parts (zinc, chrome, and galvanized)

One of the major problems associated with bonding plated parts is the different surface conditions that can be caused by variations in plating equipment, process methods, and solution concentrations. These variables result in plated surfaces with broad conditions of surface finish and inconsistent metallurgical and adhesion properties.

Nickel plated parts should not be heavily etched or sanded. Roughening surfaces to obtain good bonds is unnecessary, as excellent bonds can be made to the smoothest of surfaces as long as they are properly cleaned. The likely cleaning candidates are solvent cleaning, vapor degreasing, or soap cleaning. A recommended practice for surface treating plated parts is lightly scouring with a non-chlorinated commercial cleaner, rinsing with distilled water, and drying at temperatures under 120°F. The substrate is then primed or bonded as soon as possible after surface preparation.

Chromium and chrome plated alloys can be etched in a 50% solution of concentrated hydrochloric acid for 2–5 mins at 190°F. Zinc and galvanized metal parts can be similarly immersed for 2-4 mins at room temperature in such a solution at 15% concentration. In both cases, the part should be primed or the adhesive applied as soon as possible after surface treatments.

Another metal for which adhesive bonding is widely utilized is galvanized steel. Adhesive bonding is a preferred method of joining this material because it is difficult to weld. Galvanized steel is nominally zinc coated steel, but in reality it is much more complex. Aluminum, lead, tin, and magnesium are also present, and these contribute to the difficulty in bonding this material. Magnesium oxide, in particular, has virtually no bond strength. The most common way to prepare galvanized steel for bonding is through repeated detergent washings. These remove the magnesium oxide progressively with the aluminum oxide, so that eventually only zinc oxide remains, which is relatively easily bonded.

16.2.10 Steel and iron

Because of their widespread use in industry, steel and iron are frequently bonded. Like the surfaces of most metals, their surfaces actually exist as a complex mixture of hydrated oxides and absorbed water. Unfortunately, iron oxides are often not the best surface for adhesives because the oxides may continue to react with the atmosphere after an adhesive has been applied, forming weak crystal layers. Iron oxides are more difficult to "engineer" than aluminum, copper, or titanium oxides. As a result, grit blasting is often used. Although it provides adequate adhesion and durability for many applications, grit blasting does not provide great durability in severe environments. Conversion coatings are often used in these cases.

Bonding operations frequently require the mechanical or chemical removal of loose oxide layers from iron and steel surfaces before adhesives are applied. In order to guard against slow reaction with environmental moisture after the bond has formed, iron and steel surfaces are often phosphated prior to bonding. This process converts the relatively reactive iron atoms to a much more passive, "chemically stable" form that is coated with zinc or iron phosphate crystals. Such coatings are applied in an effort to convert a reactive and largely unknown surface to a relatively inert one whose structure and properties are reasonably well understood.

Corrosion protection is critical in bonding steel—even more critical than for many other metallic adherends. The initial adhesion to steel is usually good, but deteriorates rapidly during environmental conditioning. Thus, corrosion preventing primers are usually recommended because they protect the surface against change after bonding. Steel alloys will form surface oxides in a very short time. Drying cycles after cleaning can be critical. During these processes, an alcohol rinse after a water rinse tends to accelerate drying and reduce undesirable surface layers.

Mild steel (carbon steel) may require no more extensive treatment than degreasing and abrasion to give adequate adhesive bonds. Tests should be carried out with the actual adhesive to be used to determine whether a chemical etch or other treatment is essential.[17,25,26] Figure 16.3 illustrates the initial and residual strength obtained with joints of mild steel employing different adhesives and different surface treatments. From these results it may be concluded that the durability of steel bonds produced with a hot curing epoxy adhesive is better than that of specimens produced with room temperature curing adhesive systems.

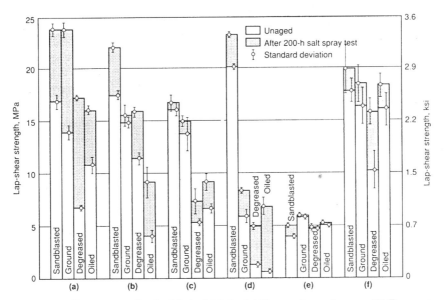

Figure 16.3 Shear strength of mild steel joints. (*a*) Two part cured epoxy. (*b*) Two part acrylic. (*c*) Anaerobic acrylic. (*d*) Cyanoacrylate. (*e*) PVC plastisol. (*f*) One part heat cured epoxy.[27]

The general sequence of surface preparation for ferrous surfaces consist of the following methods: degreasing, acid etch or alkaline clean, rinse, dry, chemical surface treatment, and priming. The chemical surface treatment step is not considered a standard procedure, but it is sometimes used when optimum joint strength is required. It consists of the formation of a corrosion preventing film of controlled chemical composition and thickness. These films are a complex mixture of phosphates, fluorides, chromates, sulfates, nitrates, etc. The composition of the film may be the important factor that controls the strength of the bonded joint.

Stainless steel, or corrosion resistant steel, has a high chromium content (11% or higher) as the primary alloying element. The stable oxide film that exists on stainless steel appears to tightly bind ions from prior manufacturing steps. Rinsing seems to be especially critical in assuring the successful bonding of stainless steel surfaces, but even bond quality on well-rinsed surfaces may not be comparable with that on other, more easily bonded materials.

There have been a large number of surface preparation methods reported to give excellent bonds with stainless steels. In addition to mechanical methods, strong acids and strong alkalis are used. A wet abrasive blast with a 200 girt abrasive followed by thorough rinsing to remove the residue is an acceptable procedure for some uses, but does not produce high bond strengths. Strong acid treatments are usually used to produce strong bonds with most adhesives. Passivation in nitric acid solution and concentrated sulfuric acid saturated sodium dichromate solution both produce high bond strength but with low or marginal peel strength. Such joints may fail under vibration stress, particularly when a thin stainless steel sheet is bonded with a low peel strength adhesive. Acid etching can be used to treat types 301 and 302 stainless. These processes result in a heavy black smut formation on the surface. This material must be removed if maximum adhesion is to be obtained. The acid etch process produces bonds with high peel and shear strengths. The 400 series of straight chromium stainless steels should be handled in the same manner as the plain carbon steels. The precipitation hardening (pH) stainless steels each present an individual problem. Processes must be adopted or developed for each type.

Once properly treated, there are practically thousands of organic adhesive compounds that are available for bonding steel alloys. The most common of the structural adhesives for bonding steel consist of thermosetting adhesives such as epoxies, urethanes, and phenolics. In high volume industries such as automotive and appliances, there is a desire not to provide a surface preparation process for steel. Special adhesive systems have been developed that will bond well to steel coming directly from the mill. This substrate is usually coated with light oil for protection against corrosion. Toughened epoxy adhesives, thermosetting acrylic adhesives, and certain other systems have been found suitable for this application. These adhesives usually work with oily surfaces through either thermodynamic displacement of the oil at the surface by the adhesive, or absorption of the oil into the body of the adhesive as a plasticizer.

16.2.11 Tin

Tin surfaces provide no unusual problems for adhesive bonding. Solvent cleaning is recommended prior to surface abrasion by scraping, fine sanding, or scouring. The abrasion process should be followed with a second solvent cleaning. Acid etching is usually not used on tin substrates.

Recommended adhesives include epoxies, acrylic, polyurethane, styrene butadiene rubber, and polyisobutylene. Usually flexibilized adhesives are recommended because of the tendency of thin tin parts to undergo peel or cleavage stresses.

16.2.12 Titanium and titanium-based alloys

Titanium is widely used in aerospace applications that require high strength-to-weight ratios at elevated temperatures. As a result, a number of different prebonding surface preparation processes have been developed for titanium. These follow the same sequence for steel and other major industrial metal substrates: degrease; acid etch or alkaline clean; rinse; dry; chemical surface treatment; rinse and dry; and finally prime or bond. Mechanical abrasion is generally not recommended for titanium surfaces.

If chlorinated solvents are used with titanium surfaces, they must be completely removed prior to bonding. Chlorinated solvents give rise to stress corrosion cracking in the vicinity of welds. Welding of titanium often occurs in the same manufacturing area as adhesive bonding, and it is sometimes done on the same parts. Therefore, the best practice is to avoid the use of chlorinated solvents completely. Several airframe manufacturers who fabricate titanium alloys no longer permit the use of chlorinated solvents.

Different titanium alloys are attacked by acid etching solutions at different rates. Titanium containing lower percentages of alloy elements are generally more resistant; so minimum treating times will differ. Extreme caution must be used when treating titanium with acid etchants that evolve hydrogen. In strongly acidic etching solutions, and particularly in sulfuric acid pickling solutions, there can be appreciable hydrogen pickup during treatment. Hydrogen pickup on surfaces of titanium can cause embrittlement. Immersion times must be closely controlled and minimized.

Shaffer, et al., provide a excellent review of adhesive bonding titanium including surface preparation processes.[28] Of 31 surface preparation procedures for adhesive bonding of titanium alloy studied by General Dynamics,[29] a phosphate-fluoride (PF) process was selected as optimal. The process was modified by Picatinny Arsenal and is described in MIL-A-9067. This method gives excellent bond durability for both 6,4 titanium and chemically pure (CP) titanium. The former, however, shows a loss of lap shear strength after 5 years outdoor weathering. The CP titanium does not show this effect.

Alkaline peroxide (AP) surface treatments of titanium have led to bonded structures that possess high adhesive strength and improved resistance at elevated temperatures.[30,31] The use of this process, however, has declined in recent years due to instability of its components and long treatment times at room temperature (up to 36 hrs).[28] Titanium surfaces produced by electrochemical anodization typically provide the best initial strength and long term durability. This treatment provides a porous cellular oxide morphology that is very compatible with adhesives. The most commonly used anodic process is chromic acid anodization (CAA).[32-34] Recently, a sodium hydroxide anodization (SHA) treatment has also gained favor because of bond durabilities equal or greater than provided by the CAA process,[35,36] and it is a more environmentally friendly process

A proprietary alkaline cleaner, Prebond 700, appears to be satisfactory for a number of metal adherends including titanium and is recommended as a versatile one step surface preparation process for a variety of metals, including titanium.[37] A proprietary alkaline etch solution, Turco 5578, is available form Turco Product Division, Purex Corporation. When combined with vapor degreasing and alkaline cleaning, this process offers very high strength bonds with titanium.

Plasa Jell is a proprietary chemical marketed by Semco Division, Products Research and Chemical Corp. This formulation is available either as a thixotropic paste suitable for brush application or as an immersion solution for tank treatment. The VAST process was developed by Vought Systems of LTV Aerospace Corporation; VAST is an acronym for Vought Abrasive Surface Treatment. In this process, the titanium is blasted in a specially designed chamber with a slurry of fine abrasive containing fluorosilicic acid under high pressure. The aluminum oxide particles are about 280 mesh, and the acid concentration is maintained at 2%. The process produces a gray smut on the surface of 6,4 titanium alloy that must be removed by a rinse of 5%

nitric acid. The resulting joint strength is claimed to be superior to that provided by the unmodified phosphate fluoride process, but is slightly lower than that after the Turco 5578 alkaline etch.

Adhesives recommended for bonding titanium substrates include epoxies, nitrile-epoxy, nitrile-phenolic, polyimide, and epoxy-phenolic. Epoxy adhesives have generally been selected when a combination of high initial strength and durability is required. Representative data are shown in Fig. 16.4 for an epoxy adhesive and several proprietary and non-proprietary treating processes. Keith[39] has covered all aspects of titanium adhesive bonding, including adhesive selection.

16.2.13 Zinc and zinc alloys

The most common use of zinc is in galvanized metals. Pure zinc surfaces are usually prepared mechanically. Acid etchings (hydrochloric acid immersion at room temperature for 2–4 mins and sulfuric acid/ dichromate immersion for 3–6 mins at 100°F) have also been suggested as surface treatments for zinc substrates. Phosphate and chromate conversion coatings are available for zinc from commercial sources as proprietary materials. Adhesives recommended for zinc substrates include nitrile-epoxies, epoxies, silicones, cyanoacrylates, and rubber-based adhesives.[40]

16.3 Plastics Substrates in General

Plastics are usually more difficult to bond with adhesives than are metal substrates. Plastic surfaces can be unstable and thermodynamically incompatible with the adhesive. The actual bonding surface may be far different than the expected surface. The plastic part may have physical properties that can cause excessive stress in the joint. The operating environment can change the adhesive-plastic interface, the base plastic, the adhesive, or all three.

However, even with these potential difficulties, adhesive bonding can be an easy and reliable method of fastening one type of plastic to itself, to another plastic, or to a non-plastic substrate. Pocius, et. al., provides an excellent treatise on the use of adhesives in joining plastics.[41] Many thermoplastic materials can also be solvent or thermally welded. In these cases, the resin from the plastic part being bonded acts as the adhesive.

The physical and chemical properties of both the solidified adhesive and the plastic substrate affect the quality of the bonded joint. Major elements of concern are the thermal expansion coefficient, modulus, and glass translation temperature of the substrate relative to the ad-

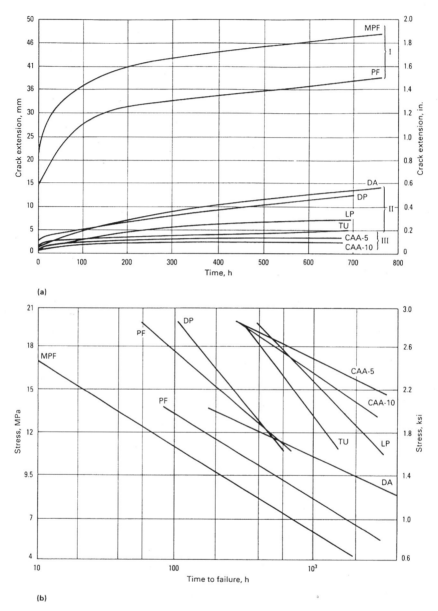

Figure 16.4 Typical bond durability data for Ti-6Al-4V adherends bonded with an epoxy adhesive and aging at 60°C and 100% relative humidity. (*a*) Crack propagation versus time for the wedge crack propagation test. (*b*) Applied stress versus time to failure for the lap shear geometry. PF—phosphate fluoride; MPF—modified phosphate fluoride; DP—PasaJell 109 dry hone; LP—Pasa Jell 107 liquid hone; CAA5—5% solution; CAA10—10% solution; TU—Turco 5578 etch; DA-Dapcotreat.[38]

hesive. Special consideration is also required of polymeric surfaces than can change during normal aging or on exposure to operating environments.

Significant differences in thermal expansion coefficient between the substrate and the adhesive can cause severe stress at the interface. This is common when plastics are bonded to metals because of the order of magnitude difference in thermal expansion coefficients. Residual stresses are compounded by thermal cycling and low temperature service. Selection of a resilient adhesive or adjustments in the adhesive's thermal expansion coefficient via fillers or additives can reduce such stress.

Bonded plastic substrates are commonly stressed in peel because the part thickness is usually small and the modulus of the plastic is low. Tough adhesives with high peel and cleavage strength are usually recommended for bonding plastics.

Structural adhesives must have a glass transition temperature, T_g, higher than the operating temperature or preferably higher than the T_g of the part being bonded. Modern engineering plastics, such as polyimide or polyphenylene sulfide, have very high glass transition temperatures. Common adhesives have a relatively low glass transition temperature, so that the weakest thermal link in the joint may often be the adhesive. The use of an adhesive too far *below* the glass transition temperature could result in low peel or cleavage strength. Brittleness of the adhesive at very low temperatures could also manifest itself in poor impact strength.

Plastic substrates could be chemically active, even when isolated from the operating environment. Many polymeric surfaces slowly undergo chemical and physical change. The plastic surface, at the time of bonding, may be well suited to the adhesive process. However, after aging, undesirable surface conditions may present themselves at the interface, displace the adhesive, and result in bond failure. These weak boundary layers could come from the environment or from within the plastic substrate itself. Plasticizer migration and degradation of the interface through UV radiation are common examples of weak boundary layers that can develop with time at the interface.

Moisture, solvent, plasticizers, and various gases and ions can compete with the cured adhesive for bonding sites. The process by which a weak boundary layer preferentially displaces the adhesive at the interface is called desorption. Moisture is the most common desorbing substance, being present both in the environment and within many polymeric substrates. Solutions to the desorption problem consist of eliminating the source of the weak boundary layer or selecting an adhesive that is compatible with the desorbing material. Excessive moisture can be eliminated from a plastic part by postcuring or drying the part before bonding. Additives that can migrate to the surface can possibly be eliminated by reformulating the plastic resin. Also, certain adhesives are more compatible with oils and plasticizers than others. For example, the migration of plasticizer from flexible polyvinyl chloride can be counteracted by using a nitrile-based adhesive. Nitrile rubber adhesives are capable of absorbing the plasticizer without degrading.

16.4 Thermosetting Plastic Substrates

Thermosetting plastics cannot be dissolved in solution and do not have a melting temperature since these materials are crosslinked. Therefore, they cannot be heat or solvent welded. In some cases, solvent solutions or heat welding techniques can be used to join thermoplastics to thermoset materials. However, most thermosetting plastics are not particularly difficult to bond with adhesive systems. They are generally bonded with many different types of adhesives such as epoxies, thermosetting acrylics, and urethanes. Since thermosetting parts are often highly filled and rigid, a flexible adhesive is not so important as one that can resist the service environment and provide practical joining processes.

A list of typical tradenames and suppliers was given in Table 15.1 for the thermoset plastics discussed in this section. In general, unfilled thermosetting plastics tend to be harder, more brittle, and not as tough as thermoplastics. Thus, it is common practice to add filler to thermosetting resins. These fillers can affect the nature of the adhesive bond (either positively or negatively) and are a possible source of lot-to-lot and supplier-to-supplier variability.

Thermosetting plastics, being chemically crosslinked, shrink during cure. Sometimes the cure is not entirely complete when the part is bonded. In these cases, continued cure of the part during the bonding operation or even on aging will result in shrinkage and residual stresses in the joint. Depending on the nature of the crosslinking reaction, volatile byproducts could also generate due to post curing of the part and provide materials for a weak boundary layer.

The surface of thermoset materials may be of slightly different chemical character than the material beneath the surface because of surface inhibition during cure or reaction of the surface with oxygen and/or humidity in the surroundings. By abrading the surface, a more consistent material is available for the adhesive to bond.

Abrasion and solvent cleaning are generally recommended as a surface treatment for thermosetting plastics. Frequently, a mold release agent is present on thermoset materials and must be removed before adhesive bonding. Mold release agents are usually removed by a detergent wash, solvent wash, or solvent wipe. Clean lint free cloth or paper tissue is commonly used, and steps must be taken to ensure that the cleaning materials do not become contaminated. Cleaning solvents used for thermosetting materials are acetone, toluene, trichloroethylene, methyl ethyl ketone (MEK), low boiling petroleum ether, and isopropanol.

Surface abrasion processes can be applied on all thermosetting plastics. Mechanical abrasion methods consist of abrasion by fine sandpaper, carborundum or alumina abrasives, metal wools, or steel shot. The following surface treatment procedure is usually recommended for thermosetting plastics:

1. Solvent degrease with MEK or acetone
2. Grit or vapor blast, or abrade with 100–300 grit emery cloth
3. Wash with solvent

The roughness of the abrasion media can vary with the hardness of the plastic. Usually, this is not a critical parameter except where decorative surfaces are important.

Adhesives commonly used on thermosetting materials include epoxies, urethanes, cyanoacrylates, thermosetting acrylics, and a variety of non-structural adhesive systems. The following discussion includes a brief description of the thermosetting substrate material, the properties that are critical relative to adhesion, and any special processes that should be noted for the particular substrate.

16.4.1 Alkyds

Alkyd resins consist of a combination of unsaturated polyester resins, a monomer, and fillers. Alkyd compounds generally contain glass fiber filler, but they may also include clay, calcium carbonate, alumina, and other fillers. Alkyds have good heat, chemical, and water resistance, and they have good arc resistance and electrical properties. Alkyds are easy to mold and economical to use. Postmolding shrinkage is small. Their greatest limitation is extremes of temperature (above 350°F) and humidity.

Alkyd parts are generally very rigid, and the surfaces are hard and stiff. Surface preparation for alkyd parts consists of simple solvent cleaning and mechanical abrasion. Epoxies, urethanes, cyanoacrylates, and thermosetting acrylics are commonly used as adhesives. Elastomeric adhesives are also used in non-structural applications.

16.4.2 Diallyl phthalate

Diallyl phthalates are among the best of the thermosetting plastics with respect to high insulation resistance and low electrical loss.

These properties are maintained up to 400°F or higher and in the presence of high humidity environments. Also, diallyl phthalate resins are easily molded and fabricated. There are several chemical variations of diallyl phthalate resins, but the two most commonly used are diallyl phthalate (DAP) and diallyl isophthalate (DAIP). The primary difference is that DAIP will withstand somewhat higher temperatures than will DAP. Both DAP and DAIP have excellent dimensional stability and low shrinkage after molding. Surfaces are hard and tough, and they pick-up very little moisture. DAP parts are ordinarily molded or laminated with glass fibers. Only filled molding resins are commercially available.

Typical surface preparation calls for cleaning with acetone, MEK, or other common solvent. Once clean, the substrate is then mechanically abraded with sand, grit or vapor blast, or steel wool. The surface is again wiped clean with fresh solvent. Typical adhesives that are employed include epoxies, urethanes, and cyanoacrylates. Polysulfides, furanes, and polyester adhesives have also been suggested.

16.4.3 Epoxy

Epoxy resins are one of the most commonly used thermosetting material. They offer a wide variety of properties depending on base resin, curing agent, modifiers, fillers, and additives. Epoxies show good dimensional stability, electrical properties, and mechanical strength. They have good creep resistance and will operate over a wide temperature range. However, high temperatures tend to oxidize epoxies after long periods of time. Both filled and unfilled grades are available. Common fillers include minerals, glass, silica, and glass or plastic microballoons. Epoxy composites are available with continuous or discontinuous reinforcing materials of several types.

The common surface preparation treatment for epoxy resins is to wipe with solvent, mechanical abrasion, and final solvent cleaning. Epoxy parts can be most easily bonded with an epoxy adhesive similar to the substrate being bonded. Urethanes, cyanoacrylates, and thermosetting acrylics have also been used when certain properties or processing parameters are required. Elastomeric adhesives can be used for semi-structural applications.

16.4.4 Phenolic, melamine, and urea resins

The phenolics are heavily commercialized thermosetting materials that find their way into many applications. They have an excellent combination of physical strength and high temperature resistance. They have good electrical properties and dimensional stability. Like epoxies and diallyl phthalate, phenolic resins are often found to contain fillers and reinforcement.

Phenolics are formed by a condensation reaction. This results in the formation of water during cure. If a phenolic substrate is not completely cured, it may continue to cure during processing or when exposed to elevated temperatures in service and liberate additional water vapor. This water vapor could form a weak boundary layer and drastically reduce the strength of a bonded joint.

Phenolics can be surface treated by any of the standard processes listed above for thermosetting materials. Solvent cleaning and mechanical abrasion are commonly employed for high joint strengths. The parts must be completely dry and cleaned of any mold release.

Phenolic parts can be bonded with a wide variety of adhesives. With many adhesives it is possible that the bond strength of the joint will be greater than the strength of the adherend. However, phenolics are used in many high temperature applications. Adhesives with good high temperature properties are required when these parts are bonded. For this reason, high temperature epoxy, nitrile-phenolic, and urea-formaldehyde adhesives are commonly used. Neoprenes and elastomeric contact adhesives are used for bonding decorative laminate.

Melamine (melamine formaldehyde) resins and urea (urea formaldehyde) resins are similar to the phenolics. They are hard, rigid materials that have excellent electrical and mechanical characteristics. Melamine parts are noted for high impact resistance and resistance to water and solvents. Only filled melamine resins are available. As a result of their properties, melamines are often used as decorative laminates. The melamine resins cure via an addition reaction mechanism, so no reaction byproducts can be produced on post cure as with the phenolic resins.

The specific surface preparation for adhesive bonding and the preferred adhesives for bonding melamine and urea parts are similar to those suggested for phenolic resins.

16.4.5 Polyimides

Polyimide plastics have exceptional thermal stability. Some polyimides are able to withstand temperatures up to 900°F for short periods. Polyimides are available both as thermosetting and thermoplastic materials. Thermosetting polyimides cure by either a condensation reaction or by an addition reaction mechanism depending on their chemistry. Polyimides are available in many forms such as fabricated parts, molding resins, films, and coatings. Polyimides are often filled for greater physical properties and dimensional stability. In certain bearing products, polyimides are filled with low friction materials such as graphite, polytetrafluoroethylene (PTFE), and molybdenum disulfide. These low surface energy fillers could interfere with adhesion when present at the interface.

Polyimide parts can either be bonded using a standard thermosetting substrate surface treatment listed above, or one of a number of specialty processes developed for higher adhesive strength and permanence. DuPont recommends that Vespel polyimide parts be bonded by a process consisting of solvent cleaning in trichloroethylene (Freon TF), perchloroethylene, or trichloroethylene. Because of the high temperature and solvent resistance of this plastic, parts have also been cleaned by refluxing in the solvent and by ultrasonic agitation. Once clean, the parts are then abraded with a wet or dry abrasive blast. The part is then solvent cleaned again and dried before bonding.

A sodium hydroxide etch process has been developed for polyimide parts that require maximum adhesive strength.[42] The parts are first degreased, and then etched for 1 min at 140–200°F in 5% solution of sodium hydroxide in water. After etching, the parts are rinsed in cold water and air dried.

Thermosetting polyimide materials can be bonded with any structural adhesive. Unfortunately, the high temperature strength of the adhesive is generally not as good as the high temperature characteristics of the polyimide. Where the maximum strength is needed at elevated temperatures, high temperature curing epoxy, phenolics, or polyimide adhesive formulations must be applied. Where high strengths are not crucial, silicone adhesives are commonly employed.

16.4.6 Polyesters

Polyesters are also common thermosetting resins that are used in many applications such as automobile and boat parts, fiberglass structures, and molding compounds. Polyesters are generally heavily filled and/or contain reinforcing fibers such as glass. Polyester resins are somewhat lower in cost than epoxy resins and can be formulated to have very short molding times. They have good resistance to water, good electrical properties, good resistance to oils and solvents, and high strength-to-weight ratios. However, polyester parts exhibit a high shrinkage rate.

Adhesive bonding of polyester parts could be complicated by the addition of a variety of additives in the resin to enhance curing, mold release, and surface gloss. In polyester resin formulations that are cured in contact with the atmosphere, waxes are included to shield the free radical polymerization process from inhibition by oxygen. Certain polyester mold release agents are formulated directly into the resin (internal mold release). For a glossy finished part appearance, thermoplastic polyolefins are sometimes incorporated into the polyester formulation. Surface preparation must take into consideration these possible weak boundary layers. The recommended surface preparation is simple solvent cleaning and mechanical abrasion. Generally, the same surface preparations as recommended for epoxies and phenolics are also recommended for polyesters.

Polyester parts are frequently bonded with epoxy, polyester, or polyurethane adhesives. Polyester adhesives, however, form a rigid bond and exhibit a high shrinkage on cure resulting in internally stressed joints. A primer is usually not necessary for polyester parts. Good results have been obtained in the automotive industry with two-part epoxies and one- and two-part urethanes.[43] Outdoor weather resistance of polyester fiberglass/aluminum bonds have been reported for epoxy, acrylic, and silicone adhesives.[44]

16.4.7 Silicones

Silicones are a family of unique polymers that are partly organic and partly inorganic. Silicones have outstanding thermal stability. Silicone polymers may be filled or unfilled. They can be cured by several different mechanisms and are available as both flexible and rigid resins. They are low surface energy materials and are generally difficult to bond with adhesives. In fact, the poor bonding characteristics of silicone surfaces are often used profitably for baked-on release coating and graffiti resistant paints. Silicone resins are also used for molding compounds, laminates, impregnating varnishes, high temperature paints, and encapsulating materials.

Flexible silicone resins can be cured either at room temperature (room temperature vulcanizing, RTV) or at elevated temperatures. They are available as single-part and two-part systems. The two-part systems have very little exotherm and cure without formation of by-products. The single-part systems, however, generally cure by reaction with the moisture in the air and form as by-products acidic acid or methanol. If not completely reacted before bonding, these by-products could form a weak boundary layer after the bond is made. The acid by-product could also cause corrosion of nearby metal substrates.

Flexible resins find extensive use in electrical and electronic applications where stable dielectric properties and resistance to harsh environments are important. They are also used in many industries to make rubber molds and patterns. Flexible silicone resins are also commonly used themselves as high temperature sealants and adhesives.

Medium to high strength bonds to cured silicone elastomers are best obtained by using room temperature curing, one-part silicone sealant/adhesives (RTV silicone). In applications where possible corrosion problems can occur with metal substrates, a methanol producing silicone has been developed and is suggested over the silicone formulations that produce acidic acid during cure. Where a closed joint design prevents moist air from migrating to the center of the joint or where long bond overlaps make the escape of the acid or methanol byproducts impossible, two-part elastomeric silicone adhesives must be used.

Surface abrasion is not recommended for flexible silicone parts. For maximum strength a compatible primer may be used. The primers are available from the same source as the silicone adhesives.

Rigid silicone resins are used as organic coatings, electrical varnishes, laminates, and circuit board coatings. The non-coating products are generally filled or reinforced with mineral fillers or glass fibers. Thermosetting molding compounds made with silicone resins are finding wide application in the electronic industry as encapsulants for semiconductor devices.

Rigid silicone resins also have a low surface energy and are difficult materials to bond with adhesives or sealants. Parts should be clean and dry. The best bonding candidate is another silicone resin (either rigid or flexible) plus a compatible primer.

16.4.8 Thermosetting polyurethanes

Polyurethanes can be furnished as either thermosetting or thermoplastic material. The thermosetting variety can be obtained in a number of different densities, rigidities, and forms, depending on the curing agents and reaction mechanism. They are excellent for low temperatures including cryogenic applications and have good chemical resistance, skid resistance, and electrical properties. Polyurethanes are limited to 250°F maximum temperature applications. Polyurethanes have good electrical properties and are often used in electrical applications. Polyurethanes are elastomeric materials that deform under pressure. This could cause internal strains to be frozen into the joint if the part is bonded under pressure.

Substrate cleaning usually involves the light sanding of a clean, dry bonding surface. A primer (urethane or silane) is sometimes used to improve adhesion. Urethanes are generally bonded with a flexible epoxy or a urethane adhesive system.

16.5 Thermoplastic Substrates

Unlike the thermosetting resins, the thermoplastic resins will soften on heating or on contact with solvents. They will then harden on cooling or on evaporation of the solvent from the material. This is a result of the non-crosslinked chemical structure of thermoplastic molecules. Tradenames of common thermoplastic materials can be found in Table 15.2. The following are important characteristics of thermoplastic resins that can affect their joining capability:

- Many thermoplastic compounds are alloyed and really consist of two or more resins.
- Additives and mold release agents are commonly employed in the formulation.
- Thermoplastics exhibit a relatively high degree of water absorption and mold shrinkage.
- Dimensional changes due to moisture migration, thermal expansion, etc. are generally greater than with other materials.
- The properties of the surface, such as surface energy and crystallinity, may be different from the bulk, (this is especially true for thermoplastics that are molded at very high temperatures).
- Many grades of the same material are available (e.g., high flow, high density).

Thermoplastic materials usually have a lower surface energy than thermosetting materials. Thus, physical or chemical modification of the surface is sometimes necessary to achieve acceptable bonding. This is especially true of the crystalline thermoplastics such as polyolefins, linear polyesters, and fluoropolymers. Methods used to increase the surface energy and improve wettability and adhesion include:

- Oxidation by chemical means or by flame treatment
- Roughening of the surface by electrical discharge
- Gas plasma treatment

As with metal substrates, the effect of plastic surface treatments decrease with time, so it is important to carry out priming or bonding as soon as possible after surface preparation. The surface preparation methods suggested in the following sections are recommended for conventional adhesive bonding. Greater care must be taken in cleaning thermoplastics than with thermosets. Thermoplastic parts can be attacked or swell on contact with certain solvents. Therefore, cleaning solvent selection must be made depending on the materials being joined.

Unlike thermosets, many thermoplastics can also be joined by solvent cementing or thermal welding methods. Solvent cementing and thermal welding do not require abrasion or chemical treatment of the plastic surfaces. The surfaces must be clean, however, and free of impurities that could cause a weak boundary layer. Bond strengths achieved by solvent or thermal welding are generally as high or higher than adhesive bonding. Bond strengths are often greater than 80% of the strength of the substrate material.

A universal joining process is not possible for thermoplastic parts. The process ultimately chosen must provide the strength and durability necessary as well as the production cycle required. In many high volume industries such as the automotive and appliance industries, the method and speed of assembly may be the most important criteria in selecting a material for a specific part. The most common types of thermoplastics and their suggested joining procedures are described below. Included are methods for adhesive bonding and sealing, solvent cementing, and thermal welding.

16.5.1 Acrylonitrile-butadiene-styrene (ABS)

ABS plastics are derived from acrylonitrile, butadiene, and styrene. ABS materials have a good balance of physical properties. There are many ABS modifications and many blends of ABS with other thermoplastics that can affect adhesion properties. ABS resins can be bonded to themselves and other materials with adhesives, by solvent cementing, or by thermally welding.

Common solvents such as methyl ethyl ketone, methyl isobutyl ketone, or tetrahydrofuran have been used to solvent cement ABS. These solvents can be used by themselves as single component systems. Open or working time may be increased by dissolving ABS resin into the solvent or by using blends of solvents having different volatility. Solvents can be bodied up to 25% with resin. These bodied adhesives also have good gap filling properties. Because of the rapid softening action of ABS solvents, the pressure and the amount of solvent applied to the part should be minimal. It may be several days at room temperature before the solvent dissipates from the joint entirely, even though holding strength occurs in a few hours. Therefore packaging of the joined part may have to be delayed until the solvent has evaporated, or a forced drying step may need to be included as a final step in the solvent cementing process.

ABS resins can also be joined by a number of thermal welding techniques. Heated tool welding, vibration welding, spin welding, and ultrasonic welding have all been effectively used. Table 16.3 shows the effect of thermal aging and boiling water aging on the properties of hot plate welded ABS.

TABLE 16.3 The Effect of Thermal Air Aging and Boiling Water on the Properties of Hot Plate Welded ABS[45]

ABS sample	Exposure type	Aging temp. °C	Aging time days	Ultimate tensile strength		Elongation at break	
				MPa	% retained	mm	% retained
Hot Plate Weld	thermal air aging	120	3	19.7	61	0.9	42
Parent	thermal air aging	120	3	46.1	107	2.8	78
Hot Plate Weld	thermal air aging	120	7	20.5	63	0.9	41
Parent	thermal air aging	120	7	47.2	109	3.1	85
Hot Plate Weld	thermal air aging	120	14	12.8	40	0.7	33
Parent	thermal air aging	120	14	39.1	91	2.5	67
Hot Plate Weld	boiling water	100	3	9.3	29	0.6	29
Parent	boiling water	100	3	41.4	96	6.5	181
Hot Plate Weld	boiling water	100	7	12.6	38	0.7	31
Parent	boiling water	100	7	40.5	94	5.4	150

When solvent welding or thermal welding is not practical or desired, adhesive systems can be used. Adhesive types such as epoxies, urethanes, thermosetting acrylics, nitrile-phenolics, and cyanoacrylates permit ABS to be bonded to itself and to other substrates. The best adhesives have shown strength greater than that of ABS; however these adhesives provide very rigid bonds.

For low to medium strength bonds, simple mechanical abrasion is a suitable surface preparation if the substrates are first cleaned. This surface preparation has found success in most ABS applications. A silane primer such as Dow Corning A-4094 or General Electric SS-4101 may be used for higher strength.[46] For maximum joint strength, a warm chromic acid etch of the ABS substrate is suggested.[47]

Electroplated ABS is used extensively for many electrical and mechanical products in many forms and shapes. Adhesion of the electroplated metal to the ABS resin is excellent; therefore, the electroplating does not generally have to be removed to have a good quality joint.

16.5.2 Acetal

Acetals are among the group of high performance engineering thermoplastics that resemble nylon somewhat in appearance but not in properties. Acetals are strong and tough and have good moisture, heat, and chemical resistance. There are two basic types of acetals: the homopolymers by Du Pont and the copolymers by Celanese. The homopolymers are more rigid and stronger, and have greater resistance to fatigue. The copolymers are more stable in long term high temperature service and are more resistant to hot water. Both types of acetal resins are degraded by UV light.

Because acetals absorb a small amount of water, the dimensions of molded parts will vary slightly with the relative humidity of the environment. The absorbed water should be considered when bonding because it could migrate to the interface during exposure to elevated temperatures either from the joining process or from the service environment. This released moisture could create a weak boundary layer. Dimensional changes must also be considered when acetal parts are bonded to other substrates and then exposed to changing temperature and humidity conditions.

The acetal surface is generally hard, smooth, glossy, and is not easily bonded with adhesives. The non-stick and or solvent resistant nature of acetal requires that the surface be specially prepared before adhesive bonding can occur. Once prepared, the surface can then be bonded with adhesives to adhere to like substrates or others.

Because of the solvent and chemical resistance of acetal *copolymer*, special etching treatments have been developed for surface preparation prior to adhesive bonding. A chromic acid etch and a hydrochloric acid etch have been suggested. Acetal parts that have been formed by heat treatment or machining should be stress relieved before etching.

Acetal *homopolymer* can be effectively bonded with various adhesives after the surface is either sanded, etched with a chromic acid solution at elevated temperature, or "satinized". Satinizing is a patented process developed by Du Pont for preparing Delrin acetal homopolymer for painting, metallizing, and adhesive bonding. In this process, a mild acid solution produces uniformly distributed anchor points on the adherend surface. Adhesives bond mechanically to these anchor points, resulting in strong adhesion to the surface of the homopolymer. Details of the satinizing process can be found in Appendix D-2. Performance of various adhesives on acetal homopolymer is shown in Table 16.4. Oxygen plasma and corona discharge surface treatments have also shown to be effective on acetal substrates.[48]

Epoxies, isocyanate cured polyester, and cyanoacrylates are used to bond acetal copolymer. Epoxies have shown 150–500 psi shear strength on sanded surfaces and 500–1000 psi on chemically treated surfaces. For structural applications, cyanoacrylate adhesives will bond acetal to rubber, metal, leather, and other plastics. Two component epoxies give slightly lower bond strengths. However, they provide better temperature and chemical resistance. Urethane adhesives also work well with acetal copolymer. Epoxies, nitrile, and nitrile-phenolics can be used as adhesives with treated acetal homopolymer substrates.

Solvent cementing is generally not recommended as a method of joining acetal homopolymer (Delrin) parts because of the material's outstanding solvent resistance. A room-temperature-solvent cement for bonding acetal copolymer (Celcon) to itself, nylon, and ABS is hexafluoroacetone sesquihydrate (Allied Chemical Corp.). Lower-strength bonds can also be made to polyester, cellulosics, and rubber. This solvent will not bond acetal to metal.

Acetal parts can also be joined by various thermal welding methods. Heated tool welding, hot gas welding, spin welding, and ultrasonic wielding have all been used. These are effective on both the homopolymer and copolymer varieties.

16.5.3 Acrylics

Acrylic resins (polymethyl methacrylate) have exceptional optical clarity and good weather resistance, strength, electrical properties, and chemical resistance. They have low water absorption characteristics. However, acrylics are attacked by strong solvents, gasoline, acetone, and similar organic fluids.

Solvent cementing is the usual method for bonding acrylic to itself. Acrylics must be annealed before solvent welding to avoid stress cracking and to achieve maximum joint strength. Typical annealing tem-

TABLE 16.4 Performance of Adhesives with Acetal Homopolymer[49]

Adhesive	Base	Color of adhesive	Tensile shear strength, lb/in.^{2}						Adhesive curing conditions	
			Acetal to acetal		Acetal to aluminum		Acetal to steel		Room temp	Oven, °F
			Sanded	"Satinized"	Sanded	"Satinized"	Sanded	"Satinized"		
F&F 46950- and 2% RC 805[a]	Polyester	Amber	300	450	200	...	150	600	...	1 h at 275
F&F 4684[a]	Rubber	Amber	300	250	300	100	300	300	...	1 h at 275
F&F 46950[a]	Polyester	Amber	300	700	250	300	350	700	...	1 h at 275
Chemlock 220–203[b]	Rubber	Black	...	300	...	150	...	150	...	1 h at 250
Chemlock 220[b]	Rubber	Black	...	300	...	150	...	250	...	1 h at 250
3M-EC711[c]	Rubber	Tan	500	400	450	200	500	350	...	1 h at 250
Phenoweld No. 8[d]	Phenolic	Red	100	600	...	400	250	400	...	30 min at 300
Narmco 3135A[e] 3135B	Modified epoxy	Yellow	150	600	...	450	300	850	24 h or	1 h at 200
Narmco 3134[e] 3134C	Modified epoxy	Gray	150	450	...	400	250	830	24 h or	1 h at 200
Resiweld No. 7004[f]	Epoxy	Yellow	150	850	...	600	200	600	24–48 h	50 min at 200
Biggs R-363[g]	Epoxy	Amber	...	550	...	...	550	...	8 h or	30 min at 150
Biggs R-313[g]	Epoxy	Gray	500	600	...	60	300	50	8 h or	30 min at 150
Cycleweld C-14[h]	Epoxy	Straw yellow	...	400	...	150	...	200	...	1 h at 150
Cycleweld C-6[h]	Epoxy	Straw yellow	...	500	...	...	...	...	...	1 h at 180
CD 200[i]	Vinyl	Amber	300	500	...	550	400	700	...	1 h at 180
EA VI[j]	Epoxy	Red	500	500	...	600	500	600	48 h or	2 h at 165

[a] E. I. du Pont de Nemours & Co., Inc., Industrial Finishes Dept., Wilmington, Del.
[b] Hughson Chemical Company, Erie, Pa.
[c] 3M Corporation, 900 Bush Avenue, St. Paul, Minn.
[d] H. V. Hardman Co., 571 Cortland St., Belleville, N. J.
[e] Narmco Resins & Coatings Company, 600 Victoria St. Costa Mesa, Calif.
[f] H. B. Fuller Co., St. Paul, Minn.
[g] Carl H. Biggs, Inc., 1547 15 St., Santa Monica, Calif.
[h] Cycleweld Cement Product Co., 5437 W. Jefferson St., Trenton, Mich.
[i] Chem. Develop. Corp., Danvers, Mass.
[j] Hysol Corp., Olean, N.Y.

peratures are about 10°F below the plastic's heat distortion temperature. The following annealing conditions have been found suitable in some applications. The part should be annealed for a sufficient time so that the entire part is uniformly at the annealing temperature.

Annealing temperature		
Acrylic (easy flow).....	140F	2h*
Acrylic (hard flow)	170F	2h*

*Approximate time for thin sections; longer times required for heavy sections.

Typical solvents for cementing acrylic are methylene chloride and ethylene dichloride. A polymerizable solvent cement for acrylics can be made from a mixture of 60 parts by weight methylene chloride, 40 parts methyl methacrylate monomer, 0.2 parts benzoyl peroxide, and sufficient resin for body. This is a very good solvent cement with gap-filling properties; but once mixed, the cement has a limited working life.

Amorphous noncrystalline acrylic parts can be welded with thermal welding processes easier than semicrystalline plastics such as acetal or nylon. Differences in weldability have been noticed with different flow grades of acrylic resin. Welded joints can be transparent if the joint is properly designed. Ultrasonic welding is the most popular method of thermally welding acrylic parts. Infrared welding offers a promising way to join materials in a continuous process where access to only one side of the weld zone is needed.[50]

Adhesive bonded acrylic joints usually give lower strength than solvent or heat-welded joints. Cyanoacrylate, epoxy, and thermosetting acrylic adhesives offer good adhesion but poor resistance to thermal aging. Table 16.5 shows the bond strength of acrylic bonded to other plastic substrates with various adhesives.

Surface preparation for bonding can be accomplished by wiping with methanol, acetone, MEK, trichloroethylene, isopropanol, or detergent;

TABLE 16.5 Lap Shear Values (MPa) for Acrylic Bonded to Various Substrates[51]

Substrate	Adhesive			
	Acrylic	Epoxy	Urethane	Elastomeric
Acrylic	7.5	2.8	0.62	0.35
ABS	6.7	3.4	1.6	0.55
PVC	9.5	3.5	2.1	0.17
PPO	5.5	4.2	2.2	0.59
FRP	9.0	4.3	2.8	0.41
CRS	10.3	4.2	2.5	0.38
Aluminum	9.1	6.9	1.9	0.66

or abrading with fine abrasive media. The precautionary note relative to stress cracking (crazing) caused by contact of highly stressed areas of the part with solvent cements also applies to solvent cleaning prior to adhesive bonding.

16.5.4 Cellulosics (cellulose acetate, cellulose acetate butyrate, cellulose propionate, ethyl cellulose, and cellulose nitrate)

Cellulosics are among the toughest of plastics. However, they are temperature limited and are not as resistant to extreme environments as other thermoplastics. The four most prominent industrial cellulosics are cellulose acetate, cellulose acetate butyrate, cellulose propionate, and ethyl cellulose. A fifth member of this group is cellulose nitrate.

Cellulosic resins are formulated with a wide range of plasticizers for specific properties. The extent of plasticizer migration should be determined before cementing cellulose acetate (and, to a much lesser extent, butyrate and propionate) to cellulose nitrate, polystyrene, acrylic, or polyvinyl chloride. Plasticizer migration will, in some cases, cause crazing or softening of the mating material or degradation of the adhesive.

Cellulosics are normally solvent cemented unless they are to be joined to another substrate. In such cases, conventional adhesive bonding is employed. Either methyl ethyl ketone, or acetone, or blends of the two are generally appropriate for solvent cementing cellulose-based plastics. Solvent cements are generally bodied by dissolving 10% of the cellulosic resin into the solvent. This provides good gap filling characteristics and extends the time for solvent evaporation. Ethyl cellulose is usually bonded with ethyl acetate solvent.

Polyurethane, epoxy, and cyanoacrylate adhesives are commonly used to bond cellulosics. Surface treatment generally consists of solvent cleaning and abrasion. A recommended surface cleaner is isopropyl alcohol. Cellulosics can be stress cracked by certain solvents, uncured cyanoacrylate adhesives, and some components of acrylic adhesives.

16.5.5 Chlorinated polyether

Chlorinated polyether resists most solvents and is attacked only by nitric acid and fuming sulfuric acids. Thus, it is not capable of being solvent cemented. Chlorinated polyether parts can be bonded with epoxy, polyurethane, and polysulfide-epoxy adhesives after treatment with a hot chromic acid solution. Tensile-shear strength of 1,270 psi has been achieved with an epoxy-polysulfide adhesive.

16.5.6 Fluorocarbons

There are eight types of common fluorocarbons. Chemical structure, suppliers, and tradenames are given in Table 16.6. Like other plastics, each type of fluorocarbon is available in several different grades. They

TABLE 16.6 Structure, Tradenames, and Suppliers of Fluorocarbon Material[52]

Fluorocarbon	Structure	Typical trade names and suppliers
TFE (tetrafluoroethylene)	$\left[\begin{array}{c} F\ \ F \\ \|\ \ \| \\ -C-C- \\ \|\ \ \| \\ F\ \ F \end{array}\right]_n$	Teflon TFE (E. I. du Pont) Halon TFE (Allied Chemical)
FEP (fluorinated ethylenepropylene)	$\left[\begin{array}{c} F\ F\ F\ F \\ \|\ \|\ \|\ \| \\ C-C-C-C \\ \|\ \|\ \|\ \| \\ F\ F\ F-C-F \\ \| \\ F \end{array}\right]_n$	Teflon FEP (E. I. du Pont)
ETFE (ethylene-tetra-fluoroethylene copolymer)	Copolymer of ethylene and TFE	Tefzel (E. I. du Pont)
PFA (perfluoroalkoxy)	$\left[\begin{array}{c} F\ F\ F\ F\ F \\ \|\ \|\ \|\ \|\ \| \\ C-C-C-C-C \\ \|\ \|\ \|\ \|\ \| \\ F\ F\ O\ F\ F \\ \| \\ R_l^* \end{array}\right]_n$	Teflon PFA (E. I. du Pont)
CTFE (chlorotrifluoro-ethylene)	$\left[\begin{array}{c} Cl\ \ F \\ \|\ \ \| \\ -C-C- \\ \|\ \ \| \\ F\ \ F \end{array}\right]_n$	Kel-F (3M)
E-CTFE (ethylene-chloro-trifluoroethylene copolymer)	Copolymer of ethylene and CTFE	Halar E-CTFE (Allied Chemical)
PVF_2 (vinylidene fluoride)	$\left[\begin{array}{c} H\ \ F \\ \|\ \ \| \\ -C-C- \\ \|\ \ \| \\ H\ \ F \end{array}\right]_n$	Kynar (Pennsalt Chemicals)
PVF (polyvinyl fluoride)	$\left[\begin{array}{c} H\ \ H \\ \|\ \ \| \\ -C-C- \\ \|\ \ \| \\ H\ \ F \end{array}\right]_n$	Tedlar (E. I. du Pont)

$R_f = C_n F_{2n-1}$.

differ principally in the way they are processed and formed and in the way their properties vary over the useful temperature range.

The original basic fluorocarbon, and perhaps the most widely known one, is tetrafluoroethylene (TFE). It has the optimum electrical and thermal properties and almost complete moisture resistance and chemical inertness. However, TFE does cold flow or creep at moderate loading and temperatures. Filled modifications of TFE resins are available, and these are generally stronger than unfilled resins. Fluorinated ethylenepropylene (FEP) is similar to TFE except that its operating temperature is limited to 400°F. FEP is more easily processed, and it can be molded which is not possible with TFE. Ethylene-tetrafluoroethylene copolymer (ETFE) is readily processed by conventional methods including extrusion and injection molding. Perfluoroalkoxy (PFA) is a class of melt processable fluoroplastics, that perform successfully in the 500°F area. Chlorotrifluoroethylene (CTFE) resins are also melt processable. They have greater tensile and compressive strength than TFE within their service temperature range. Ethylene-chlorotrifluoroethylene copolymer (E-CTFE) is a strong, highly impact resistant material. E-CTFE retains its strength and impact resistance down to cryogenic temperatures. Vinylidene fluoride (PVF_2) is another melt processable fluorocarbon with 20% lower specific gravity compared to that of TFE and CTFE. Thus, PVF_2 is found to be economical to use in parts requiring a useful temperature range from −80 to 300°F. Polyvinyl fluoride (PVF) is manufactured as film and has excellent weathering and fabrication properties. It is widely used for surfacing industrial, architectural, and decorative building materials.

Because of their high thermal stability and excellent resistance to solvent, fluorocarbons cannot be joined by solvent cementing, and they are difficult to join by thermal welding methods. Because of their inertness and low surface energy, they also tend to be difficult materials to join by adhesive bonding. Surface treatment is necessary for any practical bond strength to the fluorocarbon parts.

Surface preparation consists of wiping with acetone, treating with commercially available sodium naphthalene solutions, washing again with acetone and then with distilled or deionized water, and drying in a forced air oven at 100°F. Because of the hazardous nature of these surface treatment chemicals, the user must provide caution and follow all of the manufacturers' recommendations. The sodium naphthalene etching process removes fluorine atoms from the surface of the substrate, leaving it with a higher surface energy and a surface that is more wettable by conventional adhesives. The sodium naphthalene treated fluorocarbon surface is degraded by UV light and should be protected from direct exposure.[53]

Epoxies and polyurethane adhesives give good bond strengths to treated fluorocarbon surfaces. Table 16.7 shows the effect of surface treatments on the bondability of epoxy adhesives to polytetrafluoroethylene (Teflon) and polychlorotrifluoroethylene (Kel-F). Plasma treatment can also be used to increase adhesion to fluorocarbon surfaces,[55] but the joint strength is not as high as with the sodium naphthalene etching treatment.

16.5.7 Nylon (polyamide)

Nylons, also known as polyamides, are strong, tough thermoplastics with good impact, tensile, and flexural strengths at temperatures up to 300°F. They provide excellent low friction properties and good electrical resistivities. Four common varieties of nylon are identified by the number of carbon atoms in the diamine and dibasic acid that are used to produce that particular grade. They are referred to as nylon 6, nylon 6/6, nylon 6/10 and nylon 11. These materials generally vary by their processing characteristics and their dimensional stability.

All nylons absorb some moisture from environmental humidity. Moisture absorption characteristics must be considered in designing and joining these materials. They will absorb from 0.5 to 8% by weight of moisture after 24 hrs of water immersion. Freshly molded objects will contain less than 0.3 % moisture since only dry molding powder can be successfully molded. Once molded, these objects will then absorb moisture when they are exposed to humid air or water. The amount of absorbed moisture will increase until an equilibrium condition is reached based on the relative humidity of the environment. Equilibrium moisture contents of two commercial nylon resins for two humidity levels are as follows:

	Zytel 101	Zytel 31
50% RH air	2.5%	1.4%
100% RH air (or water)	8.5%	3.5%

The absorbed water can create a weak boundary layer under certain conditions. Generally, parts are dried to less than 0.5% moisture before bonding. Caution must be observed when mating the nylon to another substrate. The nylon part will grow and shrink due to ingress and egress of moisture from within the substrate. This will lead to stresses at the interface that may cause warpage and degradation of the bond strength.

Various commercial adhesives have been used to provide bond strength with nylon on the order of 250–1,000 psi. However, adhesive

TABLE 16.7 Effect of Surface Treatment on Bondability of Epoxy Adhesives to Teflon and Kel-F[54]

Adhesive	Ratio	Cure	Fluorinated material	Tensile shear strength, lb/in.^{2}		
				No treatment	No treatment, abraded	Treated*	Abraded and treated
Armstrong A2/Act E	100:6	4 h/165°F	Teflon	Fell apart	...	1,570	1,920
Epon 828/Versamid 125	60:40	16 h/RT, 4 h/165°F	Kel-F 270	380	1,120	2,820	3,010
Epon 828/Versamid 125	60:40	16 h/RT, 4 h/165°F	Kel-F 300	550	1,250	2,030	2,910
Epon 828/Versamid 125	60:40	16 h/RT, 4h/165°F	Kel-F 500	780	1,580	1,780	2,840
Epon 828/Versamid 125	60:40	16 h/RT, 4 h/165°F	Teflon*	...	...	1,150	

*Treated with sodium naphthalene etch solution.

bonding is usually considered inferior to heat welding or solvent cementing. Priming of nylon adherends with a composition based on resorcinol formaldehyde, isocynanate modified rubber, and cationic surfactants have been reported to provide improved joint strength when adhesive bonding. Elastomeric (nitrile, urethane), hot-melt (polyamide, polyester), and reactive (epoxy, urethane, acrylic, and cyanoacrylate) adhesives have been used for bonding nylon. Some epoxy, resorcinol formaldehyde, phenol-resorcinol, and rubber-based adhesives have been found to produce satisfactory joints between nylon and metal, wood, glass, and leather. Table 16.8 shows bond strength of various adhesives in joining nylon to metal. Exposure of nylon 6 to oxygen and helium plasmas for 30 sec to 1 min greatly improves adhesion of two-part epoxy adhesives.[57]

Solvents generally recommended for bonding nylon consist of aqueous phenol, solutions of resorcinol in alcohol, and solutions of calcium chloride in alcohol. These solvents are sometimes bodied by adding nylon resin. Solvent solutions are recommended only when the mating surfaces are smooth and very close fitting. Where gap-filling is required because of loose joint tolerances, solvent cementing should not be used, and an adhesive should be selected for joining.

Aqueous phenol containing 12% water is the most generally used solvent cement for nylon. After the solvent is applied, the parts are clamped together under pressure. After air drying, the joined part is placed in boiling water for 5 min to remove excess phenol. In cases where immersion is impractical, the joints may be cured by prolonged air exposure at ambient condition or at 150°F. Handling strength is achieved in 48 hrs, and maximum strength is realized in 3–4 days. Bond strengths as high as 2,400 psi in shear can be obtained with aqueous phenol solvent cement.

A solution of resorcinol in ethanol provides equivalent bond strength, and the solvent cement is less corrosive and hazardous than

TABLE 16.8 Tensile Shear Strength of Nylon to Metal Joints Bonded With Various Adhesives.[56]

Adhesive	Tensile-shear strength, lb/in.^{2}			
	Aluminum	Brass	Stainless steel	Iron
Phenoweld No. 7*............	900	1,200	1,500	1,600
Nt 422† (primer) + Penacolite G-1124‡....................	1,000	1,100	1,300	1,200
Cycleweld 6§	500	500	500	500
Cycleweld 14§	800	800	900	800

*H. V. Hardman Co., Belleville, N.J.
†Har-Lee Fishing Rod Co., Jersey City, N.J.
‡Koppers Co., Pittsburgh, Pa.
§Cycleweld Chemical Products Div., Trenton, Mich.

aqueous phenol. Equal parts by weight resorcinol and 95% ethanol make up the cement. The solution is brushed on both surfaces and allowed to dry for 30 sec. The parts are then assembled and held firmly. The joint will cure at room temperature. Handling strength is achieved after 90 min, and maximum strength occurs after 4–5 days. If desired, the cure can be accelerated by a 30 min, 150°F oven exposure.

Calcium chloride (22.5 pbw) and ethanol (67.5 pbw) bodied with nylon resin produces bonds as strong as those obtained with phenol or resorcinol except when used with high-molecular-weight nylons. This adhesive also forms a less brittle bond. The cement is painted on both surfaces and after 60 secs, the parts are assembled and clamped. Bond strength develops at a rate similar to the resorcinol-ethanol cement.

Nylon parts can also be joined by various thermal welding techniques including vibration, ultrasonic, and spin welding. High frequency dielectric welding may also be used in some cases, mainly for heat sealing polyamide film. The mechanical properties of the welded joint depend to a great extent on the moisture trapped in the interface. Formation of bubbles in the welding zone is something that will degrade the joint. Fortunately, this is visually evident by the appearance of bubbles or frothing of the interface.

16.5.8 Polycarbonates

Polycarbonates are among those plastic materials that are grouped as engineering thermoplastics because of their high performance characteristics in engineering designs. Polycarbonates are especially outstanding in impact strength, having strengths several times higher than other engineering thermoplastics. Polycarbonates are tough, rigid, and dimensionally stable. Polycarbonate resins are available in either transparent or colored grades. An important molding characteristic is low and predictable mold shrinkage, which sometimes gives polycarbonates an advantage over nylons and acetals for close tolerance parts. Polycarbonates are also alloyed with other plastics in order to increase strength and rigidity. Polycarbonates are somewhat hygroscope, like nylon; so it is important to control and keep the humidity low before bonding.

Polycarbonate plastics are generally joined by solvent cementing or thermal welding methods. However, caution needs to be used when solvent cementing because these plastics can stress crack in the presence of certain types of solvents. When joining polycarbonate parts to metal parts, a room temperature curing adhesive is suggested to avoid stress in the interface caused by differences in thermal expansion.

Methylene chloride is a very fast solvent cement for polycarbonate. This solvent is recommended only for temperate climate zones and when bonding small areas. A mixture of 60% methylene chloride and 40% ethylene chloride is slower drying and the most common solvent cement used. Ethylene chloride is recommended alone in very hot climates. These solvents may be bodied with 1–5% by weight with polycarbonate resin where gap-filling properties are important. A joining pressure of 200 psi is recommended.

Polycarbonate materials can be joined by many thermal welding techniques. Because of its high degree of rigidity and amorphous structure, polycarbonate can be readily welded with ultrasonic and vibration methods. Heated tool welding has also been used successfully to join polycarbonate parts. It is recommended that the parts be dried prior to being joined by thermal welding. The preferred drying conditions are 250°F for an amount of time which is dependent on part thickness.

When adhesives are used to join polycarbonate, bonding processes based on epoxies, urethanes, and cyanoacrylates are generally chosen. Table 16.9 shows bond strength of solvent welded polycarbonate joints compared with adhesive bonded joints. Adhesive bond strengths with polycarbonate are generally 1,000–2,000 psi. Cyanoacrylates, however, are claimed to provide over 3,000 psi when bonding polycarbonate to itself. Elastomeric RTV (room temperature vulcanizing) silicone adhesives are recommended for polycarbonate joining applications requiring low to moderate bond strength, high service temperature, and good thermal expansion. Generally, a silicone primer is also used with the RTV silicone adhesive.[59]

Recommended solvents for cleaning are methanol, isopropanol, petroleum ether, heptane, and white kerosene. Ketones, toluene, trichloroethylene, benzene, and a number of other solvents including paint thinners can cause crazing or cracking of the polycarbonate surface. After cleaning, possible surface preparations include flame treatment or abrasion. In flame treatment, the part is passed through the oxidizing portion of a propane flame. Treatment is complete when all surfaces are polished to a high gloss, and then the part must be cooled before bonding.

16.5.9 Polyolefins (polyethylene, polypropylene, and polymethylpentene)

This large group of polymers, classified as polyolefins, is basically wax-like in appearance and extremely inert chemically. They exhibit decrease in strength at lower temperatures than most other thermoplastics. All of these materials come in various processing grades. Polyethylenes are relatively soft with thermal stability in the 190–250°F range depending on grade. Polypropylenes are chemically similar to polyethylene, but they have somewhat better physical strength

TABLE 16.9 Bonding Data for Polycarbonate Plastics[58]

Method or materials	For bonds to			Hardener	Proportions, pph	Temp, °F	Time, h	Adhesion quality	Tensile-shear strength, lb/in.²	Service range, °F	Color of adhesive
	Polycarbonate	Metals	Rubber, wood, plastics								
Solvent cementing:											
Methylene chloride	X	...	...	...	...	Room	48	Excellent	4,500–6,500	−40 + 270	Clear
Ethylene dichloride	X	...	...	...	...	Room	48	Excellent	4,500–6,500	−40 + 270	Clear
One-part systems:											
Room temp. curing:											
Eastman 910	X	X	X	...	...	Room	2–24	Excellent	...	70–212	Clear
Central Coil HB-56	...	X	X	...	...	Room	4–24	Good	...	70–1752	Red
Vulcalock R-266-T	...	...	X	...	...	Room	Air dry	Good	...	...	Amber
G-E RTV-102	X	X	X	...	...	Room	2–24	Excellent	350	−75–350	White
Hot-melt adhesive:											
Terrell Corp. 10 × 104	X	X	X	...	...	Melts at 350	Instantaneous	Excellent	...	32–175	Dark
Two-part systems:											
Room temp curing:											
Uralane 8089	X	X	X	Part B	100:40	Room	24–48	Very good	150	...	Translucent amber
Epon 828	X	X	X	Diethylene triamine	100:8	Room	2–48	Excellent	22,500	−40 + 250	Clear
Epon 828	...	...	...	Versamid 115	100:100	Room	24–48	Fair*	600	32–160	Dark amber
Epon 828–Thiokol LP-3	...	X	X	DMP-30	100:100:10	Room	48	Fair*	...	70–100	Dark amber
EA IV	...	X	X	Cur. Ag. A	100:6	Room	24–6 days	Very good	1,800	−70 + 190	Red–tan

Adhesive	One-part system	Two-part system	Part B	Ratio	Cure temp, °F	Cure time	Flexibility	Strength, psi	Temp range, °F	Color
EA 913	...	X	Part B	100:12	Room	3–4 days	Very good	2,900	−60 + 250	Gray–black
Bondmaster M688	...	X	CII-8	100:16	Room	3–4	Very good	2,000	−40 + 212	Tan
Elevated temp curing: Epon 828	...	X	Methyl nadic anhydride DMP-30	100:84:2	175	8 (min)	Very good	1,200	32–250	Light amber
Epon 828	...	X	Phthalic anhydride	100:40	250	8 (min)	Very good	1,200	−40 + 250	Light amber
EA 913	...	X	Part B	100:12	180	3	Very good	2,900	−60 + 250	Gray–black
Bondmaster M688	...	X	CII-8	100:16	200	15 min	Very good	2,000	−40 + 250	Tan
Flexicast 4002	X	X	Part B	100:25	212	3	Very good	600	−90 + 250	Gray–black
Pressure and Heat-sensitive:										
Lepage 4050	...	X	...	...	Room	...	...	25 lb/in.² at point of bond	Room	Clear
Nalco No. 101	X	X	...	...	Room	...	Excellent		−40–400	Clear
Kleen–Stik 295	X	X	...	...	120–140†	<1 min	Good	...	−40 + 200	Amber
Kleen–Stik 298	X	X	...	...	120–140†	<1 min	Good	...	−40 + 200	Amber
G-P Transmount	X	X	...	...	150–250†	10–15 min	Excellent	...	−60 + 250	Amber

Adhesive Sources

One-part system	Two-part systems	Pressure and heat-sensitive
Eastman Chemical Products, Kinsport, Tenn	Furane Plastics Inc., Los Angeles, Calif.	Lepage's Permacel, Inc., New Brunswick, N.J.
Central Coil Co., Indianapolis, Ind.	Hysol Corp., Olean, N.Y.	Wesrep Corp., Los Angeles, Calif
B. F. Goodrich, Akron, Ohio	National Starch Corp., Bloomfield, N.J.	Kleen-Stik Products, Inc., Newark, N.J.
Terrell Corp., Waltham, Mass.	Industrial Tires, Ltd., Toronto, Canada	Girder Process, Rutherford, N.J.
G.E. Co., Silicone Products Dept., Waterford, N.Y.		

*Both combinations make flexible joints, with the Epon-Thiokol system the more flexible. Both fail in hot water.
†Also cures at room temperature.

at lower density. Mold shrinkage is less of a problem and more predictable than with the other polyolefins. Polymethylpentene is a rigid, chemically resistant polyolefin that has greater thermal stability than the other polyolefins. It is used in microwave compatible containers, autoclavable medical products, and the like.

Because of their excellent chemical resistance, polyolefins are impossible to join by solvent cementing. Because of their very low surface energy, polyolefins can only be adhesively bonded after surface treatment processes. The most common way of joining polyolefins is by thermal welding techniques.

Polyolefin materials can be thermally welded by any of the techniques described in Chapter 15. The only thermal welding technique that is not appropriate is high frequency dielectric welding. The electrical characteristics of polyolefins (low dielectric constant and dissipation factor) are such that the polymer will not readily heat in a dielectric field.

To obtain a useable adhesive bond with polyolefins, the surface must be treated in one of the ways listed in Chapter 6 or Appendix D-2. A number of surface preparation methods, including flame, chemical, plasma, and primer treatments are in use. The chromic acid etch method, similar to the FPL etch developed for treating aluminum, had been recognized as one of the most effective ways of surface treating polyolefin parts. Recently, plasma treatment has been recognized as the optimum surface treatment for polyolefins when high bond strength is the only criterion. Plasma surface treatment provides the strongest adhesive joints with conventional adhesives. Difficulties with plasma treatment are that it is a batch process and requires significant investment in equipment. Gas plasma treatment is effective on geometrically complex parts. Epoxy and nitrile-phenolic adhesives have been used to bond polyolefin plastics after plasma surface preparation. Shear strengths in excess of 3000 psi have been reported on polyethylene treated for 10 min in an oxygen plasma and bonded with an epoxy adhesive.[57] Table 16.10 presents the tensile strength of epoxy-bonded polyethylene and polypropylene joints after various surface treatments.

16.5.10 Polyphenylene oxides

The polyphenylene oxide (PPO) family of engineering thermoplastics is characterized by outstanding dimensional stability at elevated temperatures, broad service temperature range, outstanding hydrolytic stability, and excellent dielectric properties over a wide range of frequencies and temperatures. Several grades are available for a wide range of engineering applications. The PPO materials that are com-

TABLE 16.10 Tensile Strength of Polyethylene and Polypropylene Bonded Joints[54]

Adhesive	Ratio	Cure	Adherend materials	Tensile shear strength; lb/in.^{2}			
				No treatment	No treatment, abraded	Unabraded, treated	Abraded, treated
Epon 828/Versamid 125............	50:50	2 h RT + 6 h	Aluminum to polypropylene	Fell apart (3)	393 (3)	760 (3)	695 (3)
Epoxy 1001/Versamid 125............	50:50	165°F 16 h	Aluminum to polyethylene	...	366 (4)	1,000 (4)	
Epon 828/Versamid 125............	60:40	RT 7 days	Aluminum to polyethylene	83 (5)	...	580 (5)	

The number of specimens tested is indicated in parentheses. Polyolefin treatment was a sulfuric acid–sodium dichromate etch.

mercially available today are polystyrene modified. PPOs are usually joined by solvent cementing or thermal welding.

Polyphenylene oxide joints must mate almost perfectly; otherwise solvent welding provides a weak bond. Excellent solvents for cementing are ethylene dichloride, toluene, chloroform, or a blend of 95% ethylene dichloride and 5% carbon tetrachloride. A mixture of 95% chloroform and 5% carbon tetrachloride is the best solvent system for general-purpose bonding, but very good ventilation is necessary. Ethylene dichloride offers a slower rate of evaporation for large structures or hot climates.

Best results have been obtained by applying the solvent cement to only one substrate. An even film of solvent should be applied. Very little solvent cement is needed. An open time of 10 secs should be allowed before mating the parts. Optimum holding time has been found to be 4 min, with an applied pressure of approximately 400 psi. Table 16.11 lists properties of solvent-welded PPO bonds.

Thermal welding techniques found suitable for modified polyphenylene oxide parts include: heat sealing, spin welding, vibration welding, resistance wire welding, electromagnetic bonding, and ultrasonics. Excellent bonds are possible because the low thermal conductivity of the modified polyphenylene oxide prevents dissipation of heat from the bonding surfaces.

Various adhesives can be used to bond polyphenylene oxide to itself or to other substrates. Parts must be prepared by sanding or by chromic acid etching at elevated temperature. Methyl alcohol is a suitable solvent for surface cleaning. The prime adhesive candidates are ep-

TABLE 16.11 Tensile Shear Strength of Solvent Cemented Polyphenylene Oxide Bonds[60]

Solvent	Cure temp	Cure time, h	Tensile-shear strength, lb/in.^{2}
95% chloroform + 5% carbon tetrachloride	100°C	2	4,000
	RT	100	4,100
Chloroform	100°C	1	2,350
	RT	48	1,150
	RT	200	2,650
Ethylene dichloride	100°C	1	2,000
	RT	48	1,000
	RT	200	2,700
Ethylene dichloride + 2% PPO	100°C	2	2,150
	RT	200	2,700
Methylene chloride + 2% PPO	100°C	3	2,100
	RT	48	1,050
	RT	200	2,650

oxies, modified epoxies, nitrile-phenolics, and polyurethanes. Table 16.12 lists tensile-shear strength of PPO bonded with a variety of adhesives.

16.5.11 Thermoplastic polyesters (polyethylene terephthalate, polybutylene terephthalate, and polytetramethylene terephthalate)

Thermoplastic polyesters have achieved significant application in film and fiber form. For years, polyethylene terephthalate (PET) was the primary thermoplastic polyester available. This material is best known in its film form as Mylar®. A new class of high performance molding and extrusion grade of thermoplastic polyester has been made available and is becoming increasingly competitive among engineering plastics. These polymers are denoted chemically as polybutylene terephthalate (PBT) and polytetramethylene terephthalate. These newer thermoplastic polyesters are highly crystalline with melting points above 430°F.

Thermoplastic polyesters are quite inert; thus, compatibility with other substrates does not pose major problems. The terephthalates have high tensile and tear strengths, excellent chemical resistance, good electrical properties, and an operating temperature range from −67 to +400°F. These materials are generally joined with adhesives, and surface treatments are used to enhance adhesion. Commonly used adhesives for both PET and PBT substrates are isocyanate cured polyesters, epoxies, and urethanes.

Surface treatments recommended specifically for PBT include mechanical abrasion and solvent cleaning with toluene. Gas plasma surface treatments and chemical etch have been used where maximum strength is necessary. PBT can be heat welded but not solvent cemented. Vibration welding methods have been successfully applied to both filled and unfilled PBT parts.[61]

Polyethylene terephthalate cannot be solvent cemented or heat welded. Adhesives are the prime way of joining PET to itself and to other substrates. Solvent cleaning of PET surfaces is recommended. The linear film of polyethylene terephthalate (Mylar®) provides a surface that can be pretreated by alkaline etching or plasma for maximum adhesion, but often a special treatment such as this is not necessary.

16.5.12 Polyimide (PI), polyetherimide (PEI), and polyamide-imide (PAI)

Polyimide, polyetherimide and polyamide-imide aromatic resins are a group of high temperature engineering thermoplastics. They are avail-

TABLE 16.12 **Tensile Shear Strength of Polyphenylene Oxide Bonded with Various Adhesives**[60]

Adhesive type and manufacturer	Breaking strength, lb/in.²	
	Sanded surface	Etched surface*
Epoxy:		
J-1151 (E-9) (Armstrong Cork)	880 (1,180)	1,315 (2,447)
Armstrong A-701 (Armstrong Prod.)	1,355 (3,628)	1,398 (3,013)
Chemlock 305 (Hughson Chem.)	704 (940)	2,145 (2,817)
Ray-Bond R-86009 (Raybestos-Manhatten)	600 (1,052)	1,545 (3,060)
EA 907 (Hysol Corp.)	1,322 (722)	2,167 (2,186)
Ecco Bond 285 (HV) (Emerson-Cuming)	. . .	1,450
Ecco Bond 285 (HVA) (Emerson-Cuming)	. . .	1,340
Polyester:		
46930 (6% RC-805) (du Pont)	1,340 (1,180)	1,594 (1,557)
46950 (9% RC-805) (du Pont)	1,086 (1,182)	903 (1,127)
46971 (6% RC-805) (du Pont)	924 (1,460)	1,225 (1,190)
46960 (8% RC-805) (du Pont)	738 (387)	951 (432)
Silicone:		
RTV 102 (General Electric)	49 (72)	82 (148)
RTV 108 (General Electric)	46 (73)	106 (132)
RTV 103 (General Electric)	53 (84)	59 (97)
Elastomer and resin-modified elastomer:		
Pliobond 20 (Goodyear Tire & Rubber)	545 (1,108)	470 (1,285)
Pliobond HT 30 (Goodyear Tire & Rubber)	280 (375)	255 (440)
D-220 (Armstrong Cork)	225 (645)	208 (500)
Ray-Bond R-81005 (Raybestos-Manhatten)	475 (700)	935 (1,703)
Ray-Bond R-82015 (Raybestos-Manhatten)	184 (140)	140 (145)
Ray-Bond R-82016 (Raybestos-Manhattan)	343 (507)	330 (488)
Cyanoacrylate:		
Eastman 910 (Eastman Chem. Products)	1,160	. . .
Polyurethane:		
Uralane 8089 (Furarne Plastics)	. . .	748 (1,036)

Values in parentheses have been obtained on samples exposed for 1 week to 100°C.
*Chromic acid etch, conditions not specified.

able in a variety of forms including molded parts, coatings, and film. These are commercially available plastics that provide the highest service temperatures. They are also one of the strongest and most rigid plastics available. Parts made from these resins have thermal expansion coefficients very similar to metals, and generally there is no problem in bonding them to metals because of differences in thermal expansion rates.

Polyimide parts can be either thermosetting or thermoplastic. Even the thermoplastic variety must be considered like a thermoset because of its high temperature resistance and resistance to solvents. Polyimide (PI) parts are generally joined with adhesives or thermal welding techniques (if thermoplastic), and they are not commonly joined by solvent cementing. Bonding is generally accomplished with moderate surface preparation processes and high temperature adhesives. There are certain grades of polyamide-imide that are used as low friction bearing material and have inherent lubricity. These are more difficult to bond. Polyimide parts can be bonded with epoxy adhesives. Only solvent cleaning and abrasion are necessary to treat the substrate prior to bonding. Selection of an adhesive for high temperature service could be a problem since the plastic substrate will generally have a higher thermal rating than the adhesive.

Parts molded from thermoplastic polyetherimide (PEI) can be assembled with all common thermoplastic assembly methods. Thermal welding techniques that have been successfully used include heated tool welding, vibration welding, and ultrasonic welding. Solvent welding can be accomplished using methylene chloride with a 1–5% addition of PEI for body. Adhesives that are recommended include epoxy, urethane, and cyanoacrylate. However, service temperature must be taken into consideration in choosing an adhesive because PEI parts are generally used for high temperature applications. Good adhesion can be affected by simple solvent wipe, but surface treatment by corona discharge, flame treatment, or chromic acid etch will provide the highest bond strengths.

Polyamide-imide (PAI) parts can be joined mechanically or with adhesives. They have too great a resistance to solvents and too high a thermal stability to be solvent cemented or thermally welded. A variety of adhesives including amide-imide, epoxy, and cyanoacrylate can be used to bond polyamide-imide parts. Cyanoacrylates have poor environmental resistance and are not recommended in structural applications. Silicone, acrylic, and urethane adhesives are also generally not recommended unless environmental conditions preclude other options. Polyamide-imide parts are relatively easy to bond and only solvent cleaning and mechanical abrasion are necessary as a surface preparation for good bonds. Plasma surface preparation has also been shown to provide excellent bonds. Polyamide-imide parts should be dried before bonding for at least 24 hrs at 300°F in a desiccant oven. Thick parts (over ¼ in.) may require longer drying times to dispel casual moisture prior to bonding.

16.5.13 Polyetheretherketone (PEEK), polyaryletherketone (PAEK), and polyetherketone (PEK)

Polyetheretherketone (PEEK) is a high performance material developed primarily as a coating, but it is also available as film and molded parts. PEEK can be either adhesive bonded or thermally welded using ultrasonic, friction, or hot plate welding techniques. When welding PEEK, it must be remembered that the melting point is very high, and considerable amounts of energy must be put into the polymer during welding to achieve a good bond.

PEEK parts can be bonded with epoxy, cyanoacrylate, silicone, and urethane adhesives. The epoxies give the strongest bond. High temperature epoxy or silicone adhesives may be necessary in high temperature applications. Surfaces must be clean, dry, and free of grease. Isopropanol, toluene, and trichloroethylene can be used to clean PEEK surfaces prior to mechanical abrasion. Surface roughening and flame treatment or etching may improve bond strengths. PEEK composites that have been surface treated by plasma activation show excellent bond strength with epoxy adhesives.[62]

Polyaryletherketone (PAEK) and polyetherketone (PEK) parts can also be joined with adhesives by the methods described above. These materials can also be thermally welded. Heated tool, vibration, spin, and ultrasonic techniques have been found to be very successful. With all thermal welding techniques, the part to be assembled must be dried before bonding.

16.5.14 Polystyrenes

Polystyrene homopolymer is characterized by its rigidity, sparkling clarity, and ease of processability; however, it tends to be brittle. Polystyrenes have good dimensional stability and low mold shrinkage. They are easily processed at low costs. They have poor weatherability, and they are chemically attacked by oils and organic solvents. Resistance is good, however, to water, inorganic chemicals, and alcohol. Impact properties are improved by copolymerization or grafting polystyrene chains to unsaturated rubber such as polybutadiene (SBR) or acrylonitrile (SAN). Rubber levels typically range from 3–12% by weight. Commercially available, impact-modified polystyrene is not as transparent as the homopolymers, but it has a marked increase in toughness. Properties of polystyrene can be varied extensively through the polymerization process. They can even be crosslinked to produce higher temperature material.

Polystyrene resins are subject to stresses during their fabrication and forming operations, and polystyrene parts often require annealing. Parts can usually be annealed by exposing them to an elevated temperature that is approximately 5–10°F lower than the temperature at which the greatest tolerable distortion occurs.

Polystyrene is ordinarily bonded to itself by solvent cementing, although conventional adhesive bonding, thermal welding, and electromagnetic bonding have been used. When bonding polystyrene to other surfaces, conventional adhesive bonding is usually employed. However in certain applications, solvent cementing can also be used to bond polystyrene to a dissimilar material.

Many solvents can be used for solvent cementing of polystyrene parts. The selection is usually based on the time required for the joint to harden. The faster drying solvents, as shown in Table 16.13, can cause crazing. Slower drying solvents are often mixed with these faster solvents for optimum properties. A 50:50 mixture of ethyl acetate and toluene bodied with polystyrene is an excellent general-purpose adhesive. With solvent cementing, bond strengths up to 100% of the strength of the parent material are common.

Polyurethane or cyanoacrylate adhesives offer bonds to polystyrene that are often stronger than the substrate itself. Unsaturated poly-

TABLE 16.13 Solvent Cements for Polystyrene[63]

Solvent	Boiling point, °C	Tensile strength of joint, lb/in.^{2}
Fast-drying:		
Methylene chloride	40	1,800
Carbon tetrachloride	77	1,350
Ethyl acetate	77	1,500
Benzene	80	
Methyl ethyl ketone	80	1,600
Ethylene dichloride	84	1,800
Trichloroethylene	87	1,800
Medium-drying:		
Toluene	111	1,700
Perchloroethylene	121	1,700
Ethyl benzene	136	1,650
Xylenes	138–144	1,450
Diethyl benzene	185	1,400
Slow-drying:		
Mono-amyl benzene	202	1,300
Ethyl naphthalene	257	1,300

ester and epoxy adhesives give lower bond strengths, but are acceptable in many applications. Hot-melt adhesives are used in the furniture industry. Generally, only solvent cleaning and abrasion are necessary for surface preparation of polystyrene parts. Methanol or isopropanol are acceptable solvents for solvent cleaning of polystyrene. For maximum bond strength, the substrates can be etched with a sodium dichromate/sulfuric acid solution at elevated temperature. Table 16.14 shows the results of a study on the durability of joints formed between polystyrene and aluminum with different types of adhesives.

Polystyrene can be joined by any number of thermal welding techniques including ultrasonic, hot-plate, vibration, spin, and electromagnetic welding. The quality of the welds can be significantly influenced by selection of the polystyrene product grade and by the welding parameters.

16.5.15 Polysulfone

Polysulfone is a rigid, strong engineering thermoplastic which can be molded, extruded, or thermoformed into a wide variety of shapes. Characteristics of special significance are their high heat deflection temperature, 340°F at 264 psi, and their long term resistance to temperatures in the 300 to 340°F range. This material can be joined by either adhesive bonding, solvent cementing, or thermal welding.

Solvent cementing, using a 5% solution of the resin in methylene chloride, can be used to bond polysulfone to itself. A minimum amount of solvent should be applied to the mating surfaces of the joint. The assembled pieces should be held for 5 min under a pressure of about 500 psi. The strength of the joint will improve over a period of several weeks as the residual solvent evaporates. Strength of solvent-welded joints can surpass the strength of the polysulfone part.

A number of adhesives have been found suitable for joining polysulfone to itself or other materials. No special surface treatments are

TABLE 16.14 Joint Strengths for Polystyrene/Aluminum Overlap Shear Specimens[64]

Adhesive	Initial, MPa	Percent strength retention after exposure		
		52°C/100% RH	Salt fog	Seacoast
Epoxy	2.3	100	18	30
Acrylic	2.3	100	57	71
Urethane	1.7	23	14	22
Silicone	0.14	25	—	—

generally required other than simple solvent cleaning, although an elevated temperature sodium dichromate/sulfuric acid etch has been used at times for maximum joint strength. An epoxy adhesive, EC-2216, and a pressure-sensitive adhesive, EC-880 (both from 3M Co.) offer good bonds at temperatures to 180°F. EC-880 is not considered a structural adhesive. For higher temperature applications, BR 89 and BR 92 (American Cyanamid) are recommended. These adhesives require heat cure but are useful to 300–350°F.

Conventional thermal welding techniques are suitable for the various polysulfone products. It is essential to dry the parts before welding in order to avoid foaming in the zone of the weld during the melting phase. Direct heat sealing requires a hot plate or other suitable heat source capable of attaining 700°F. Polysulfone, considered a good adhesive resin on its own, can also be joined to metal using a direct heat seal method.

16.5.16 Polyethersulfone (PES)

Polyethersulfone is a high temperature engineering thermoplastic with outstanding long term resistance to creep. It is capable of being used continuously under load at temperatures of up to about 360°F and in some low stress applications up to 400°F. Certain grades are capable of operating at temperatures above 400°F. Some polyethersulfones have been used as high temperature adhesives. PES is especially resistant to acids, alkalis, oils, greases, aliphatic hydrocarbons, and alcohol. It is attacked by ketones, esters, and some halogenated and aromatic hydrocarbons.

PES can be bonded to itself using solvent cementing or ultrasonic welding techniques. Solvents such as N-methy-2-pyrrolidone (NMP), N,N-dimethylforamide, and dichloromethane can be used. Allowance must be made for the fact that these solvents may give rise to stress cracking in parts that are under mechanical stress. The viscosity of the solvent cement can be increased by adding 3–15% by weight of PES resin. If sufficiently designed, these solvent welded joints can transmit more than 60% of the assembly's strength.

Successful thermal welding methods include heated tool, hot gas, vibration, spin, and ultrasonic. Heated tool welding occurs at temperatures ranging from 715–1020°F, with 10–90 secs of heating, and a joining pressure of 300–700 psi.

For adhesive bonding of PES to itself or to other materials, epoxy adhesives are generally used. Cyanoacrylates provide good bond strength if environmental resistance of the bonded joint is not a factor. Parts made from PES can be cleaned using ethanol, methanol, isopropanol, or low boiling petroleum ether. Solvents that should not be used are acetone, MEK, perchloroethylene, tetrahydrofuran, toluene, and methylene chloride.

16.5.17 Polyphenylene sulfide (PPS)

Polyphenylene sulfide is a semicrystalline polymer with high melting point (550°F), outstanding chemical resistance, thermal stability, and nonflammability. There are no known solvents below 375–400°F. This engineering plastic is characterized by high stiffness and good retention of mechanical properties at elevated temperatures. PPS resins are available as filled and unfilled compounds and as coatings. Polyphenylene sulfide resin itself offers good adhesion to aluminum, steel, titanium, and bronze and is used as a non-stick coating that requires a baking operation of near 700°F.

Polyphenylene sulfide parts are commonly bonded together with adhesives. Adhesives recommended for polyphenylene sulfide include epoxies and urethanes. Joint strengths in excess of 1,000 psi have been reported for abraded and solvent cleaned surfaces. A suggested surface preparation method is to solvent degrease the substrate in acetone, sandblast, and then repeat the degreasing step with fresh solvent. Somewhat better adhesion has been reported for machined surfaces over as-molded surfaces. It is suspected that the molecular nature of the as-molded surface is affected by the PPS resin's contact with the high temperature metal mold. The high heat and chemical resistance of polyphenylene sulfide plastics make them inappropriate for either solvent cementing or heat welding.

16.5.18 Polyvinyl chloride (PVC)

Polyvinyl chloride (PVC) is perhaps the most widely used type of plastic in the vinyl family. PVC is a material with a wide range of modulus, from very rigid to very flexible. One of its basic advantages is the way it accepts compounding ingredients. For instance, PVC can be elasticized with a variety of plasticizers to produce soft, yielding materials to almost any desired degree of flexibility. Without plasticizers, PVC is a strong, rigid material that can be machined, heat formed, or welded by solvents or heat. It is a tough material, with high resistance to acids, alcohol, alkalis, oils, and many other hydrocarbons. It is available in a wide range of forms and colors. Typical uses include profile extrusions, wire and cable insulation, and various foam applications. It is also made into both rigid and flexible film and sheet.

PVC and its acetate copolymers can be bonded using solvent solutions containing ketones. Methyl ethyl ketone is commonly used as a solvent cement for low and medium molecular weight polyvinyl chloride copolymers. High molecular weight copolymers require cyclohexane or tetrahydrofuran as solvent cements. The following cement formulation has worked satisfactorily to bond flexible PVC to itself:

Parts by weight	Material
100	Polyvinyl chloride resin (medium molecular weight)
100	Tetrahydrofuran
200	Methyl ethyl ketone
1.5	Tin organic stabilizer
20	Dioctylphthalate (a plasticizer)
25	Methyl isobutyl ketone

This cement also works well on rigid PVC.

Plasticizer migration from the vinyl substrate into the adhesive bond-line can degrade the strength of the joint. Adhesives must be tested for their ability to resist the plasticizer. PVC can be made with a variety of plasticizers. An adhesive suitable for a certain flexible PVC formulation may not be compatible with a PVC from another supplier. Nitrile rubber adhesives have been found to be very resistant to plasticizers and are often the preferred adhesive for flexible PVC films. Polyurethane and neoprene adhesives are also commonly used to bond PVC parts. A comparison of the performance of several adhesives bonding PVC to itself and to other materials is given in Table 16.15.

16.6 Composites

Modern structural composites are a blend of two or more materials. One component is generally made of reinforcing fibers, either polymeric or ceramic. The other component is generally made up of a resinous binder or matrix that is polymeric in nature, at least for polymeric composites. The fibers are strong and stiff relative to the matrix.

TABLE 16.15 Lap-Shear Values (MPa) for PVC Bonded to Various Substrates[51]

Substrate	Adhesive			
	Acrylic	Epoxy	Urethane	Elastomeric
PVC	11.0	6.3	5.6	0.28
Fiberglass	9.9	9.3	5.4	0.41
Steel	9.9	9.1	5.1	0.28
Aluminum	9.5	8.5	4.6	0.35
ABS	6.9	3.4	1.3	0.62
PPO	6.4	4.3	2.0	0.52
Acrylic	9.5	3.5	2.1	0.17

Composites are generally orthotropic (having different properties in two different directions) materials. When the fiber and matrix are joined to form a composite, they both retain their individual identities and both directly influence the composite's final properties. The resulting composite is composed of layers (laminates) of the fibers and matrix stacked to achieve the desired properties in one or more directions. The reinforcing fiber can be either continuous or discontinuous in length. The common commercially available reinforcing fibers are:

- Glass
- Polyester
- Graphite
- Aramide
- Polyethylene
- Boron
- Silicon carbide
- Silicon nitride, silica, alumina, and alumina silica

Glass fiber composites are the most common type of composite. However, graphite, aramid, and other reinforcements are finding applications in demanding aerospace functions and in premium sporting equipment such as fishing rods, tennis rackets, and golf clubs.

The resin matrix can be either thermosetting or thermoplastic. Thermosetting resins such as epoxy, polyimide, polyester, phenolic, etc. are used in applications where physical properties and environmental resistance are important. Polyester and epoxy composites make the bulk of the thermoset composite market. Of these two, polyesters dominate by far. Reinforced with glass fiber, these are known as "fiberglass" reinforced plastics (FRP). FRPs are molded by lay-up and spray-up methods or by compression molding from either a preform or from sheet molding compound (SMC). Thermoplastic matrix composites are generally employed where there are high volume and economic considerations such as in the automotive and decorative paneling industries. Thermoplastic resin based composites range from high priced polyimide, polyethersulfone and polyetheretherketone to the more affordable nylon, acetal, and polycarbonate resins. Practically all thermoplastics are available in glass reinforced grades.

Resin based composites are usually defined as either "conventional" or "advanced". Conventional composites usually contain glass or mineral fiber reinforcement, and sometimes carbon fiber, either alone or in combination with others. The fiber often is randomly oriented. Conventional composites are usually produced in stock shapes such as sheet, rod, and tube. There are many methods of processing composite materials. These include filament winding, lay-up, cut fiber spraying, resin transfer molding, and pultrusion.

Advanced composites is a term that has come to describe materials that are used most often for the most demanding applications, such as aircraft, having properties considerably superior to those of conventional composites and much like metals. These materials are "engineered" from high performance resins and fibers. The construction and orientation of the fibers are selected to meet specific design requirements. Advanced composite structures are usually manufactured in specific shapes. An advanced composite can be tailored so that the directional dependence of strength and stiffness matches that of the loading environment.

Thermoset composites are joined by either adhesive bonding or mechanical fasteners. Thermoplastic composites offer the possibility of thermal welding techniques, adhesives, or mechanical fasteners for joining. Composites are also often joined with a combination of mechanical fasteners and adhesives. Many manufacturers distrust adhesive bonds in applications where joints undergo large amounts of stress (e.g., aircraft structures). Mechanical fasteners must be sized to avoid fiber crushing and delamination; adhesives must balance strength and flexibility. The joining of composite materials involves some special problems not faced with other materials.

16.6.1 Adhesive bonding of composites

Adhesives have certain advantages and disadvantages in joining composite materials. These pros and cons are summarized in Table 16.16. A significant advantage is that adhesive bonding does not require the composite to be drilled or machined. These processes cut through the reinforcing fibers and drastically weaken the composite. A cut edge will also allow moisture or other chemicals to wick deep into the composite along the fiber-matrix interface, thereby further weakening the structure.

Much of what we know about bonding to composite materials has come through the aerospace industry. The early studies on adhesives, surface preparation, test specimen preparation, and design of bonded composite joints reported for the PABST Program[66] gave credibility to the concept of a bonded aircraft and provides reliable methods of transferring loads between composites and metals or other composites. The Navy F-18 Hornet is a dramatic example of adhesive bonding. The aircraft wing is bonded to the fuselage, the ultimate in primary structure dependent on a bonded attachment. Figure 16.5 diagrams the wing joint: carbon fiber composite top and bottom wing cover skins

TABLE 16.16 Reasons for and Against Adhesive Bonding of Composite Materials[65]

For	Against
Higher strength-to-weight ratio	Sometimes difficult surface preparation techniques cannot be verified 100% effective
Manufacturing cost is lower	Changes in formulation
Better distribution of stresses	May require heat and pressure
Electrically isolated components	Must track shelf life and out-time
Minimized strength reduction of composite	Adhesives change values with temperature
Reduced maintenance costs	May be attacked by solvents or cleaners
Corrosion of metal adherend is reduced (no drilled holes)	Common statement: "I won't ride in a glued-together airplane"
Better sonic fatigue resistance	

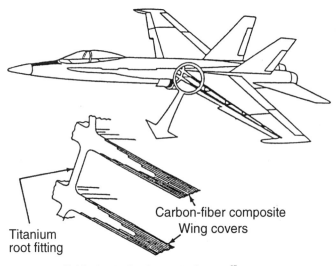

Figure 16.5 F-18 wing to fuselage attachment.[67]

bonded in a stepped lap shear configuration to titanium fittings at the sides of the fuselage. The configuration removes the load gradually in a series of shear areas from the carbon composite.

16.6.1.1 Types of adhesives for composites. Adhesives that give satisfactory results on the resin matrix alone may also be used to bond composites. The three adhesives most often used to bond composites are epoxies, acrylics, and urethanes. Polyester adhesives are also often used with polyester-based composites. Table 16.17 describes common

TABLE 16.17 Common Adhesives that Have Been Used to Bond Composite Parts in Various Applications[68]

Adhesive	Form	Cure	Properties
Epoxy	Liquid, pastes, solid, supported film	Room or elevated temperature	■ Short term strength retention to 225°C ■ Limited to 175°C long term ■ High temperature versions have 80% strength retention after 3000 hrs at 215°C
Epoxy Phenolic	Paste, supported film	Elevated temperature with light pressure	■ Excellent for short term high temperature exposure
Condensation Reaction Polyimide	Solvent solutions, supported film	High temperature cure with pressure and vacuum in certain cases	■ Excellent strength retention after 1200 hrs aging at 260°C ■ Short term resistance somewhat better than epoxy phenolics
Addition Reaction Polyimide	Support film	High temperature cure with pressure. Pressure also required during post cure	■ Similar to condensation polyimides
Bismalimide Resins	Supported film, pastes	Elevated temperature cure under light pressure	■ Short term strength retention is good up to 300°C ■ Long term temperature resistance limited to 200–225°C

commercially available adhesives that have been used to bond composite parts. There are numerous adhesive products available. Suppliers are continually working to improve temperature ranges and properties.

Epoxies are especially reliable when used with epoxy-based composites because they have similar chemical characteristics and physical properties. Room temperature curing adhesives are often used to bond large composite structures to eliminate expensive fixturing tools and curing equipment required of higher temperature cure adhesives. However, room temperature epoxies require long cure times, so they are not suitable for high speed production runs. Some of the lower temperature composite materials are sensitive to the heat required to cure many epoxies. Epoxies are too stiff and brittle to use with flexible composites.

Polyurethane adhesives are also used to bond composite materials. They are suitable for joining flexible composites because polyurethanes make flexible bonds. They also cure quickly at room temperature. In fact, certain formulations will cure so quickly that they may require automated metering, mixing, and dispensing equipment. The cost of such equipment cannot be easily justified for low production applications. Urethane adhesives are sensitive to moisture in both their uncured and cured states.

Thermosetting acrylic adhesives are also used to bond composites, but these adhesives are somewhat rigid. Acrylics cure quickly at room temperature; bond unprepared substrates; can be mixed, metered, and dispensed in a variety of ways; and have a long pot life. However, acrylics have a noticeable acrylic monomer odor; are flammable; have poor impact resistance at low temperatures; and may cause internal stress due to high shrinkage on cure.

Structural adhesives that are commonly used for composites are supplied in two basic forms: semisolid B-stage film and thixotropic pastes. The film adhesives are cast or extruded onto carrier fabrics or films and partially cured to a semisolid. They can easily be handled, cut, and applied to the joint area. There is no need for mixing, metering, or dispensing of liquid components. In use, these adhesive systems are activated by heat and pressure. The semisolid B-stage film liquefies briefly on application of heat and then cures to an insoluble state. Epoxy, polyimides, epoxy-nylons, epoxy-phenolic, and nitrile-phenolic adhesives are available as B-stage film.

Paste adhesives are supplied as either one or two component adhesive systems. They can be used in applications where pressure cannot be applied. Some two-part pastes cure at room temperature after the appropriate proportions are mixed. Epoxy, urethane, and acrylic adhesives are all available as paste adhesives.

Non-structural adhesives that are commonly used for bonding composites are rubber-based and hot-melt adhesives. Rubber-based adhesives are usually chosen where high shear strength is not required, but where high peel strength is important such as bonding of decorative laminates to a countertop. Generally, these adhesives are available in solvent form as a "contact" adhesive. The adhesive is coated on both substrates to be bonded, allowed to dry to a tacky state, and then the coated parts are carefully mated together with a slight rolling pressure. Hot-melt adhesives are employed in high volume production such as the furniture industry. A wide range of application processes are possible since hot-melt adhesives are available in powder, granule, blocks or film.

16.6.1.2 Surface preparation for adhesive bonding. Surface preparation of composite parts for adhesive bonding will depend on the specific adherend and adhesive. Recommended preparations of many composite adherends simply consist of a solvent wipe to remove loose dirt, oil, and mold release followed by a mechanical abrading operation. Abrasion should be done carefully to avoid damaging the composite's surface fibers. A degree of abrasion is desired so that the glaze on the resin surface is removed, but the reinforcing fibers are not exposed. Many surface abrasion methods have been applied to composite parts, and all of these have some merit. These abrasion processes include light sanding, grit honing, vapor honing, Scotch-briting (an abrasive product supplied by 3M Company), and other methods. In some cases, a primer may be used to coat the composite before the adhesive is applied.

Reinforced thermoplastic parts are generally abraded and cleaned prior to adhesive bonding. However, special surface treatment such as used on the thermoplastic resin matrix may be necessary for optimum strength. Care must be taken so that the treatment chemicals do not wick into the composite material and cause degradation. It may not be a good idea to use chemical surface treatment without first verifying that the treatment improves the final joint strength of the assembly and does not adversely affect the parts.

One surface preparation method that is unique to composites employs a peel ply.[42] Utilization of the peel ply is illustrated in Chapter 6 (Fig. 6.6). With this technique, a closely woven nylon or polyester cloth is incorporated as the outer layer of the composite during its production lay-up. This outer ply is then torn or peeled away just before bonding. The tearing or peeling process fractures the resin matrix coating and exposes a clean, fresh, roughened surface for the adhesive. This method is fast and eliminates the need for solvent cleaning and mechanical abrasion.

16.6.2 Thermoplastic composites

Although the fabrication and joining technology that has been developed for thermoset composites is well advanced and established, the associated enabling technology for thermoplastic composites is still in the early stages. However, since thermoplastic composites can be softened by the action of heat and solvent, many techniques that are unsuitable for use with thermosets can be used for joining thermoplastic composites.

The percentage of filler in the thermoplastic composite substrate must be limited or else the bond will be starved of resin. Generally, solvent cementing or thermal welding becomes difficult when the filler content in thermoplastic composites is greater than 10–25% by weight. With higher filler loading, adhesive bonding or mechanical fastening are the only alternatives for thermoplastic composites.

The thermal welding processes available can be divided into two groups:

1. Processes involving mechanical movement—these include ultrasonic welding, friction welding (spin, ultrasonic, vibration); and

2. Processes involving external heating—these include hot plate welding, hot gas welding and resistive and inductive welding.

Figure 16.6 compares some of the thermal welding methods with adhesive bonding for thermoplastic composites. A capsule assessment of these common methods for joining thermoplastic composites is shown in Table 16.18. With the appropriate modifications to the process parameters, all of these thermal welding techniques can be used to join thermoplastic composites.[69]

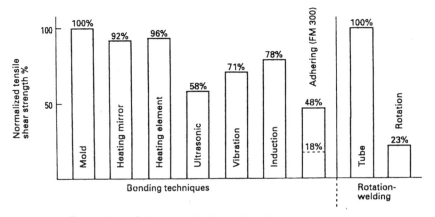

Figure 16.6 Comparison of the normalized tensile yield strength of thermoplastic advanced composite bonds produced by selected techniques.[68]

TABLE 16.18 Assessment of Current Methods for Joining Thermoplastic Composites[68]

Technique	Assessment
Resistance welding	Ideal for short bond lines in nonconductive composites. Requires conductive metal or carbon strips that remain in the joint after the application of resistance current; this may weaken the bond by producing a discontinuity in the integrity of the joint. A drawback of conductive thermoplastic composites is that the resistance heating element must be insulated from the conductive constituents of the composite. Ohm's law restricts the length or width of bond line in resistance welding.
Ultrasonic bonding	Ideal for spot welding in thicknesses less than 1 mm (0.040 in.) where disruption of the fiber is not a critical factor. Generally effective in layers of less than 1 mm (0.040 in.) and suitable for short bond lines under 254 mm (10 in.) long; high-frequency vibrating of the materials may dislocate and abrade the fiber structure of the composite.
Vibrational bonding	Ideal for relatively small, flat parts and some three-dimensional parts. Requires the rapid frictional abrading of the two surfaces to be bonded, which may disrupt the fiber embedded in the matrix. Generally successful with smaller parts.
Induction bonding	Ideal for nonconductive composites and where introduction of a susceptor as a permanent bond line element is acceptable. Requires the use of ferrous metal powders or particles that are left in the joint area of the interface.
High-frequency welding	Not suitable for conductive composites because high-frequency welding uses the material as a buffer. If the reinforcing fibers are dielectrically conductive, then transfer of heat to the interior of the composite may result in overheating, melting, thinning, and weakening.
Traditional infrared	With traditional approaches to the use of infrared, electromagnetic energy is absorbed and dissipated by the entire body of material. The bonding surface, when exposed to so much unfocused radiant energy, melts, deforms, and weakens. When removed from the heat the material cools before it can be brought together.
Heated air	Tends to be limited in bond line area, and sometimes ambient air conditions make temperature control difficult.
Hot plate	Ideal for lower-temperature thermoplastics. High-temperature thermoplastics require longer heating cycles during which heated metal is in physical contact with the resin of adhered interfaces. This often results in disruption of the material and difficulty in controlling welding temperatures.
Mechanical fasteners	Often unsatisfactory because the integrity of the composite sheet is interrupted by the puncture necessary to affix the fasteners; mechanical joints often leak and create weakness of the joints.
Hot-melt, room-cure adhesives	Difficult to apply uniformly and, because of the difference in composition and lack of molecular continuity, create fracture lines. Fast coding time of hot-melt adhesives may create difficulties.

Solvent welding can also be used on thermoplastic parts having relatively low filler loading. However, effort must be exerted not to immerse edges of cut fibers into the solvent. This could lead to degradation of the composite caused by solvent wicking into the fiber-resin interface.

16.7 Plastic Foams

Plastic foams are manufactured from thermoplastic and thermoset resins in various forms. The main pitfalls in joining plastic foam are (1) causing the foam to swell or collapse by contacting it with a solvent or monomer, and (2) having the adhesive change the properties of the foam through its absorption into the foam. Adhesion is usually not a serious problem because of the porous nature of the foam.

It should be noted that there are closed cell and open cell foams. Adhesives may spread or wick deeply into the open cell variety, thereby affecting the resulting mechanical properties of the foam and perhaps even weakening the foam. When bonding foam to another less porous substrate, the adhesive should be applied to the non-foam substrate to minimize the wicking and ingress of the adhesive into the body of the foam. With the closed cell variety, the adhesive cannot wick deeply into the foam, but usually the foam's skin must be machined or abraded to allow for some surface roughness for the adhesive to mechanically attach.

There are also low surface energy foams, such as polyethylene, that require either surface treatment or special adhesives for bonding. Fortunately, extremely strong bonds are generally not required because the foam has a relatively low cohesive strength. Therefore, simple cleaning is generally the only surface preparation required. The surface treatments that are recommended are those that are described in the previous section for the parent plastics. However, the possibility of wicking of the chemical compounds into the foam and degradation of the foam must be considered.

Some solvent cements, solvent containing adhesives, and adhesives with certain monomers will swell or collapse foams. Water-based adhesives based on styrene butadiene rubber (SBR), neoprene, or polyvinyl acetate, and 100% solids adhesives are less reactive and are often used on thermoplastic structural foams.

Recommendations for adhesives for specific types of thermoplastic and thermosetting foams are shown in Table 16.19. Generally, a flexible adhesive should be used for flexible foams. These adhesives are available in the following forms:[70]

TABLE 16.19 Recommended Adhesives for Thermoplastic and Thermosetting Foams[70]

Thermoplastic foam	Recommended adhesives
Phenylene oxide based resins (Noryl®)	Epoxy, polyisocyanate polyvinyl butyral, nitrile rubber, neoprene, polyurethane, polyvinylidene chloride, and acrylic
Polyethylene	Nitrile rubber, polyisobutylene rubber, flexible epoxy, nitrile-phenolic, and water based (emulsion) adhesives
Polystyrene	100% solids adhesives based on urea formaldehyde, epoxy, polyester isocyanate, polyvinyl acetate, vinyl chloride-vinyl acetate copolymer, and reclaimed rubber, water based adhesives based on SBR or polyvinyl acetate
Polyvinyl chloride	Urethane, epoxy, rubber based adhesives
Polycarbonate	Urethane, epoxy, rubber based adhesives
Thermoplastic polyester	Urethane, epoxy
Thermosetting foam	**Recommended adhesives**
Epoxy	Epoxy (including syntactic foams), heat cured epoxies
Phenolic	Epoxy, polyester isocyanate, polyvinyl acetate, vinyl chloride acetate copolymer, polyvinyl formal phenolic, nitrile rubber, nitrile-phenolic, reclaimed rubber, neoprene rubber, polyurethane, butyl, melamine formaldehyde, neoprene-phenolic, and polyvinyl formal phenolic
Polyurethane	Epoxy, polyester, polyacrylate, nitrile rubber, butyl, water based emulsion, polyurethane, neoprene, SBR, melamine formaldehyde and resorcinol formaldehyde
Silicone	Silicone rubber
Urea-formaldehyde	Urea-formaldehyde, resorcinol formaldehyde

- **Water based emulsion**—best for bonding polystyrene foam to porous surface; one surface needs to be porous to allow the adhesive to dry.

- **Contact adhesives**—give optimum initial strength; both water-based and solvent based available. Brush on both surfaces, let dry, and then mate with slight pressure. May need auxiliary heating system for faster drying. Solvent types are recommended for bonding to metal or other structural surfaces.

- **Pressure sensitive adhesives**—will bond to almost any substrate; both water and solvent based materials are used. They cannot be used in applications requiring long term exposure to stress or resistance to high levels of heat. Synthetic elastomer adhesives with fast tack characteristics are available in spray cans; however, the resistance of the foam to the solvent needs to be determined.

- **100% solids** — standard epoxy, polyurethane, silicone paste type adhesive; generally room temperature curing for foams. They form an extremely strong, heat and environmentally resistant bond.

16.8 Elastomers

There are over 30 broad groupings of chemical types of elastic polymers. These are arranged by ASTM D 1418 into categories of materials having similar chemical chain structures. These elastomer categories are shown in Table 16.20. Included is information regarding their surface energy.

There are several problems with joining elastomer materials. One problem is the significant variation that can exist within a given chemical type. This is due to differences in average molecular weight, molecular weight distribution, polymerization processes, variation of structural arrangement, copolymer or terpolymer ratios, etc. There also can be significant differences in the compounding recipe for a given product. This is due to differences and latitude of choice in fillers and reinforcing agents, liquid plasticizers, compounding agents, and the like. Thus, there is almost an infinite number of compounds possible for a given generic elastomer like "neoprene" or "nitrile", and each variation can affect the adhesion properties of the material.

Elastomer materials specifications usually do not focus on the adhesive properties, but mainly address the chemical and physical properties of the rubber. Thus, the supplier has wide latitude within the specification to make changes in the compound formulation that could be disastrous to the adhesive bond. One solution is to qualify every new lot of material for adhesion as well as the more standard properties.

Besides curing systems, fillers, and plasticizers, an elastomeric compound may contain protective chemicals such as antioxidants, antiozone agents, waxes, and fungicides. Some of these are purposefully designed to "bloom" or to come to the surface of the elastomer either during processing or on aging. These weak boundary layers often cannot be removed prior to bonding because their supply to the interface is relatively unlimited by the capacity of the bulk elastomer.

Another problem in joining elastomers with adhesives is that the substrate is a deformable material. It is easy to develop internal stresses at the bond interface. These stresses could adversely affect

TABLE 16.20 ASTM Designation of Common Elastomers

ASTM D 1418 Nomenclature	Elastomer—generic name	Common name	Polymer source	Critical surface energy, mJ/m^2	Water contact angle, degrees
ABR	Acrylate-butadiene rubber	None	Uniroyal B. F. Goodrich Polymer Corporation	—	—
BR	Butadiene rubber	None	B. F. Goodrich Goodyear Chemical Phillips Petroleum General Tire Polymer Corporation	32	—
CR	Chloroprene rubber	Neoprene	DuPont	38	—
IIR	Isobutylene-isoprene rubber	Butyl	Enjay B. F. Goodrich Polymer Corporation	27	115
IR	Isoprene, synthetic	Synthetic natural rubber	Goodyear Chemical Shell Chemical B. F. Goodrich	31	106
NR	Cis-polyisoprene	Natural rubber	Various	—	120
NBR	Nitrile butadiene rubber	Bune-N or nitrile	Firestone Goodyear Chemical B. F. Goodrich Uniroyal Polymer Corporation	37	97

TABLE 16.20 ASTM Designation of Common Elastomers *(Continued)*

ASTM D 1418 Nomenclature	Elastomer—generic name	Common name	Polymer source	Critical surface energy, mJ/m^2	Water contact angle, degrees
SBR	Styrene butadiene rubber	SBR	B. F. Goodrich Firestone Uniroyal Phillips Petroleum Goodyear Chemical Shell Chemical	33	95
CSM	Chlorosulfonated polyethylene	Hypalon	DuPont	37	96
EPDM	Ethylene-propylene-unsaturated dieneterpolymer	EPDM	Enjay DuPont Uniroyal	—	—
EPM	Ethylene propylene copolymer	EPM	Enjay	28	105
FKM	Vinylidine fluoride hexafluoropropylene copolymer	Viton Fluorel Kel-F	DuPont 3M	—	—
CO	Epichlorohydrin elastomer	Hydrin	B. F. Goodrich Chemical Hercules, Inc.	35	95
AU	Polyester urethane rubber	Ester urethane	General Tire Uniroyal	—	—
EU	Polyether urethane rubber	Ether urethane	DuPont	—	—
None	Polysulfide rubber	Polysulfide	Thiokol	—	—
MQ or Si	Dimethyl silicone	Silicone rubber	Dow Corning General Electric	—	—

the bond strength and permanence of the joint. Minimal pressure to achieve close substrate contact with the adhesive is all that is necessary when bonding elastomers.

Elastomeric bonding can be divided into two main categories: bonding to vulcanized elastomers, and bonding with unvulcanized elastomers. When bonding to vulcanized elastomers, one is working with an elastomeric substrate that is completely cured and formed. In unvulcanized bonding, the elastomer is in its liquid form, and one tries to cure the elastomer and achieve the bond in the same step.

16.8.1 Unvulcanized bonding

It is generally recognized that the best bond between an elastomer and a substrate is usually obtained from a bond accomplished simultaneously with the cure of the elastomer itself. Examples of products that are produced in this manner include roller wheels, shock mounts, rubber covered conveyor rollers, seals, etc.

The substrate in these cases (e.g., metal insert, center shaft) must usually be specially designed so that it is not affected by the conditions of the molding operation such as high temperatures and pressures. The substrate design must also be compatible with the flow pattern of the elastomer when it is in the liquid state and under pressure.

The surface of the substrate must be prepared prior to vulcanization by techniques similar to typical adhesive bonding practices. Usually, a commercially available unvulcanized bonding system or "tie coat" is then applied to the substrate. This primer-like material provides both good adhesion to the substrate and a surface to which the vulcanized rubber can readily bond. It is necessary to either air or heat dry the adhesive tie coat prior to vulcanizing. Table 16.21 offers recommended adhesives for unvulcanized bonding of different elastomers. These fully formulated, proprietary compounds are based on resins similar to the rubber being vulcanized.

16.8.2 Bonding vulcanized elastomers

Bonding of already vulcanized elastomers to themselves and other materials is generally completed by using a pressure sensitive adhesive derived from an elastomer similar to the one being bonded. Flexible thermosetting adhesives such as epoxy-polyamide or polyurethane also offer excellent bond strength to most elastomers. Figure 16.7 illustrates a bondability index of common and specialty elastomers.

Surface treatments prior to bonding consist of washing with solvent, abrading, or in the most demanding applications, cyclizing with acid as described in Appendix D-3. In general adhesives that are commonly

TABLE 16.21 Commercial Adhesives for Unvulcanized Bonding[71]

Type	Product (adhesive/primer)	Elastomer
Proprietary Metal Primers—single coat systems for metal inserts	Chemlok 205 Ty Ply UP Ty Ply T Thixon P5 Thixon P6 Thixon P7 Thixon P9	NBR CR NBR NBR, CR, CO NBR NBR NBR, CR
Proprietary Cover Coats—better environmental resistance than metal primer alone	Chemlok 220/205 Chemlok 233/205 Chemlok 234B/205 Chemlok 236/205 Chemlok 238/205 Ty Ply BC/UP Ty Ply Q/T Ty Ply RC/UP Ty Ply S/T Ty Ply TL/UP Thixon 807/P7 Thixon 814/any primer	CR, BR, IR, SBR NBR, IR, SBR, AU, EU IIR, CSM, NBR, BR, IR, SBR IIR, CR, EPDM, EPM, NR IIR, EPDM, EPM IIR NR, SBR, BR, IR, Reclaim NR, SBR, BR, IR, Reclaim CR, NBR NR, IR, BR, SBR IIR, EPM, EPDM IIR, EPM, EPDM
General Purpose Elastomer Single Coat Bonding Agents—single coat systems developed for specific elastomers	Chemlok 216 Chemlok 217 Thixon 511 Thixon 705 Thixon 710 Thixon 750 Ty Ply BN Ty Ply UP Ty Ply S Ty Ply D Ty Ply BC	Vinyl polymers NBR EPM, EPDM Carboxylated NBR NBR NBR NBR CR CR, NBR NR, IR, SBR (specifically to brass) IIR (specifically to brass)
Broad Spectrum General Purpose Elastomer Single Coat Bonding Agents—single coat systems formulated for a variety of elastomers	Chemlok 250 Chemlok 252 Ty Ply 1863 Thixon OSN-2 Thixon OSN-3	IIR, EPDM, EPM, NR, BR, IR, NBR IIR, EPDM, EPM, NR, BR, IR NR, CR, SBR, BR, CSM NR, SBR, CR, BR, IR, CSM NR, SBR, CR, BR, IR, IIR, CSM
Specific Specialty Elastomer Bonding Agents—developed specifically for specialty elastomers	Chemlok 218 Chemlok 307 Chemlok 608 Thixon 405 Thixon 408 Thixon 300/301 Thixon 302 Thixon 305	AU, EU Si Si AU, EU (castable) AU, EU (millable) FKM Si Si

TABLE 16.21 Commercial Adhesives for Unvulcanized Bonding[71] (*Continued*)

Type	Product (adhesive/primer)	Elastomer
Splice Cements and Post Vulcanizing Bonding Agents— used mainly for splicing	Chemlok 234B Chemlok 236 Chemlok 238	NR, SBR, CR, NBR, EPM, EPDM NR, CR, NBR, IIR, EPDM General purpose
Vulcanized Elastomeric Parts	Chemlok 252 Ty Ply BC Ty Ply S Thixon OSN-2 Thixon 508 Thixon 552 Thixon 403/404 Thixon 405 Thixon 412/413 Thixon 814	General purpose IIR CR, NBR Automotive engine elastomers General purpose Water borne AU, EU to metal AU, EU to metal EU, EU to themselves or others EPDM, IIR
Elastomer to Fiber—used as a primer on fibers	Chemlok 402 Chemlok 607 Chemlok 252	NR, SBR, CR, IIR, NBR General purpose General purpose
Elastomer to Fiber—used as a primer on fibers	Chemlok 402 Chemlok 607 Chemlok 252	NR, SBR, CR, IIR, NBR General purpose General purpose
Elastomer to Plastics—used as a primer on plastics for insert bonding	Chemlok 216 Chemlok 205/220 Chemlok 220 Chemlok 252 Chemlok 607 Ty Ply BN Ty Ply W	NBR, AU and PU (millable), CSM General to glass reinforced plastic General to glass reinforced plastic General to a variety of plastics General to nylon, epoxy, phenolic NBR to phenol formald and nylon General to a variety of plastics

Guide to Suppliers: Chemlok: Hughson Chemicals, Lord Corporation, Cary, NC; Ty Ply: Hughson Chemicals, Lord Corporation, Cary, NC; Thixon: Dayton Coatings and Chemicals, Whittaker Corp., West Alexandria, OH

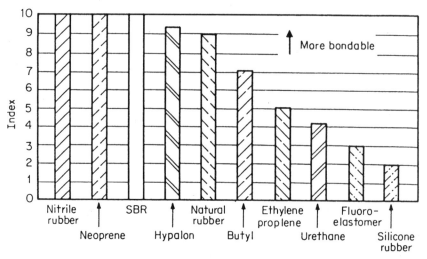

Figure 16.7 Bondability index of common and specialty elastomers.[72]

used for bonding rubber include natural, chlorinated, reclaim, butyl, nitrile, butadiene styrene, urethane, polysulfide, and neoprene-based elastomers. Also commonly used are adhesives based on acrylics, cyanoacrylates, polyester isocyanates, resorcinol-formaldehyde, silicone, epoxy, polyisocyanate, furane, phenolic-nitrile, phenolic-neoprene, phenolic polyvinyl formal, and flexible epoxy-polyamides.

The bonding of vulcanized rubber to itself can be accomplished with an unvulcanized rubber that acts as an adhesive. Here the unvulcanized elastomer is chemically similar to the rubbers being joined, and the process is similar to the one described above for unvulcanized bonding. Typical applications are seals and o-rings. The cured rubber ends are butted or skewed at an angle that is appropriate for the final article. The faces are primed with an adhesive solution that is chemically identical or similar to the rubbers being bonded. Often the adhesives recommended in Table 16.21 can be used. Using local clamping and heating, the splice is cured. Usually the bond strength will be 50–100% that of the parent rubber depending on the elastomer system and the difficulty of the geometry.

The most common elastomers to be bonded in this way include nitrile, neoprene, urethane, natural rubber, SBR, and butyl rubber. It is more difficult to achieve good bonds with silicones, fluorocarbons, chlorosulfonated polyethylene, and polyacrylate. However, recently developed adhesive systems improve the bond of these unvulcanized elastomers to metal.

16.9 Wood and Wood Products

Wood is an important structural material consisting of a cellulosic composition with a highly porous nature. Adhesives are commonly employed to bond wood in the furniture industry. They are increasingly being used in laminating and veneering of wood-based products onto composite panels. Sealants are commonly applied to wood framing members in the construction industry. There are several properties that are unique to these materials that affect their ability to be joined.

Wood is an anisotropic material. That is, wood and other cellulosic materials have different properties when measured along different axis. Tensile and shear strengths of wood are greater along the longitudinal direction, parallel with the wood fibers. Wood is also a hygroscope material because of its cellulosic nature. Generally, dimensional stability as a result of changing moisture content is highest in the longitudinal direction, somewhat less in the tangential direction, and least in the radial direction. Maximum performance is achieved if the wood has a moisture content during bonding that is close to the average moisture content anticipated during service. In this way, the internal stresses induced in the joint as a result of moisture change in service will be at a minimum. Wood changes little in dimensions as a result of temperature change, unless such change also affects the internal moisture content.

Wood of different species will vary considerably in density or specific gravity. Higher density wood of any species is usually stronger than wood of a lower density. Higher density wood will swell and shrink more for a given moisture content than will wood of lower density. This means that internal stresses on high density wood joints will be greater as a result of swelling and shrinkage effects than between two pieces of wood of lower density with the same joint design. Such internal stresses may be sufficiently large to pull the joint apart. Therefore, higher quality joints must be developed when bonding higher density wood. It is desirable to have the adhesive joint as strong as or stronger than the wood itself, particularly for structural applications.

Several important rules need to be remembered when joining wood. Good fitting joints are critical. A thin, even layer of adhesive will form a strong bond between two pieces of wood, but a thick cushion of adhesive weakens the joint. To achieve a successful edge joint, the long mating surfaces must be perfectly tight all along their length. One should not rely on clamps and pressure to pull warped or deformed parts together. The fit of a mortise and tenon joint should also be precise. If the parts must be forced together, there will be no room for the glue between the pieces, and the joint will be starved. If the gap is too large, the adhesive layer will be too thick and stresses will develop. The adhesive layer should be approximately the thickness of a sheet of notebook paper. In order to ensure that there is sufficient glue in a joint, spread a thin layer on both mating surfaces.

The principal objective in surface preparation of wood for bonding is to provide a clean, undamaged surface that is flat and smooth and will permit the two pieces to mate properly when the liquid adhesive is applied between them. It is important not to damage the wood in cutting or sawing prior to bonding. Damaged wood fibers provide a cohesively weak area near to the adhesive, and they will become a failure mode under stress.

Thin glue lines are preferred in bonded wood joints. The prevailing theory is that penetration of the adhesive into the wood is not necessary for good adhesion. Good bonding need occur only to the outside cellulose fibers. Thick bond-lines tend to give erratic joint quality because of uneven stresses introduced in the joint. Thus, the best surface finish is one that does not tear wood fibers and provides sufficiently close tolerances so that the parts fit together uniformly. Usually, such surface preparation can be accomplished by standard wood machining operations.

The bonding of wood is perhaps one of the easiest applications for adhesives. Most adhesives wet clean wood very well and form bonds that are stronger than the substrate. Adhesives used for wood include natural occurring products such as animal or hide glues, starch, casein, soybean, blood glues, and fish glues. These have been used for many years, even before the advent of modern synthetic polymeric adhesives. Table 16.22 provides a description of the common terms used for wood adhesives. Table 16.23 describes properties of several polymeric adhesives that have been used for bonding wood.

Hide or animal glues are useful for applications, like musical instruments requiring disassembly to make repairs. With the application of heat and humidity, it is a relatively easy task to separate bonded pieces without damaging them. Hide glues also cure slowly. Thus, it is a good candidate for joints that require repositioning as the product is assembled. However, in many wood bonding applications, these characteristics are disadvantages, and fortunately there are many other adhesives that can be used on wood substrates.

Resorcinol-formaldehyde resins are cold setting adhesives that are used for wood structures. Urea formaldehyde adhesives, commonly modified with melamine-formaldehyde, are used in the production of plywood and in wood veneering for interior applications. Phenol-formaldehyde and resorcinol-formaldehyde adhesive systems have the best heat and weather resistance.

Polyvinyl acetates are quick drying, water-based adhesives commonly used for the assembly of furniture. These water-based emulsion

TABLE 16.22 Description and Use of the Most Common Wood Adhesives

Adhesive	Informal Name	Description	Use
Casein		Powder to be mixed with water, working life of several hours	Good gap filling characteristics for furniture with loose joints that must be filled. Stains wood and may attract mold. Good on high moisture and oily woods such as teak
Clear cement	Duco cement	Clear liquid adhesive (various resins in solvent solution)	For lighter wood and paper or other porous materials. Somewhat water resistant
Animal glues	Hide glues	One part liquid or flakes that must be mixed with water	Traditional furniture glue, for both construction and repair, though not waterproof and with a longer set time than white or yellow glue
Hot melt	Glue stick	Preformed hot melt adhesive cylinders to easily heat in glue gun	General purpose, fast setting adhesive particularly suited for complicated fits without clamping
Plastic resin	Urea resin, urea formaldehyde	Powdered material mixed with water	Any kind of wood. Colorless adhesive. Mainly for interior applications such as counter tops
Waterproof	Resorcinol, resorcinol formaldehyde	Two part system (liquid and powder)	For exterior applications
White glue	Polyvinyl acetate, polyvinyl alcohol	Resin emulsion that dries by evaporation of water	General use adhesive. Not moisture or heat resistant. Not a gap filler. Medium joint strength
Yellow glue	Aliphatic glue, carpenter's glue, woodworker's glue	Resin in water solution	Strong joint strength. Used for general indoor applications. Medium heat and moisture resistance. Good gap filler

TABLE 16.23 Characteristics of Common Wood Adhesives[73]

Resin type used	Resin solids in glue mix, %	Principal use	Method of application	Principal property	Principal limitation
Urea formaldehyde	23–30	Wood to wood interior	Spreader rolls	Bleed-throughfree; good adhesion	Poor durability
Phenol formaldehyde	23–27	Plywood exterior	Spreader rolls	Durability	Comparatively long cure times
Melamine formaldehyde	68–72	Wood to wood, splicing, patching, scarfing	Sprayed, combed	Adhesion, color, durability	Relative cost; poor washability; needs heat to cure
Melamine urea 2/1	55–60	End and edge gluing exterior	Applicator	Colorless, durability and speed	Cost
Resorcinol formaldehyde	50–56	Wood to wood exterior (laminating)	Spreader rolls	Cold set durability	Cost, odor
Phenol-resorcinol 10/90	50–56	Wood to wood exterior (laminating)	Spreader rolls	Warm-set durability	Cost, odor
Polyvinyl-acetate emulsion	45–55	Wood to wood interior	Brushed, sprayed, spreader rolls	Handy	Lack of H_2O and heat resistance

adhesives are known as white glues. They set by the evaporation of moisture from the adhesive. Polyvinyl acetate adhesives produce very strong bonds, but are not resistant to moisture or high temperatures. This drying process is often accelerated in the furniture industry by the application of radio frequency electric fields. Where resistance to moisture is required, there are crosslinking polyvinyl acetate adhesives. They cure through chemical reaction as well as evaporation. Unlike the more conventional white glues, crosslinked polyvinyl acetate glues cannot be cleaned with water after it cures.

Aliphatic resin glues, also known as yellow glue or carpenter's glue, are similar to polyvinyl acetate adhesives. They are specifically formulated to use for interior woodworking applications and should not be used for exterior parts. Yellow glues sand cleanly, without clogging and leave an invisible glue line if the joint is tight.

Epoxies have been used for certain specialized wood joining applications such as when wood is bonded to metal substrates. Rubber-based contact adhesives have also been used for bonding wood-especially in decorative laminating. Mastic adhesives and sealants have been used in construction work and are usually applied in caulking guns. These are based on elastomers, including reclaimed rubber, neoprene, butadiene styrene, polyurethane and butyl rubber. Neoprene based mastics have excellent overall characteristics and good resistance to many environments. These adhesives are commonly used to bond wall or floor boards to wood studs. They provide a method of fastening as well as vibration damping.

16.10 Glass and Ceramics

Glass and ceramic substrates are generally high surface energy materials, and most adhesives wet them readily. One problem in bonding optically clear glass is to select an adhesive that is optically clear and does not change the optical characteristics of the glass. Another problem is that shrinkage of the adhesive or differences in thermal expansion coefficients could provide high internal stresses that cause catastrophic failure of the brittle glass substrate.

Adhesives used for glass substrates are generally transparent, heat setting resins that are water- and UV-resistant to meet the requirement of outdoor applications. They are usually flexibilized systems in order not to place stress on the glass substrate either after cure or during thermal cycling. These adhesives include polyvinyl butyl phenolic butyral, phenolic nitrile, neoprene, polysulfide, silicone, vinyl acetate, and flexible epoxy adhesives. Polyvinyl butyral adhesives are commonly used to manufacture safety glass. Commercial adhesives normally used to bond glasses are described in Table 16.24. Optical adhesives used for bonding glass lenses are usually styrene modified polyesters and styrene monomer-based adhesives. Epoxies are also beginning to be used in this application.

Adhesives used for bonding glass and glazed ceramics should have minimum volatile content, since these substrates are non-porous. Epoxy adhesives, being 100% solid materials, are probably best for these applications. Thermosetting acrylic is good for bonding glass and ceramics to thermoplastic polymers and to metal surfaces. Where large areas and materials with greatly differing coefficients of thermal expansion are being bonded, polysulfides should be considered, since these adhesives have very high elongation.

TABLE 16.24 Commercial Adhesives Most Desirable for Glass[74]

Trade name	Chemical type	Bond characteristics		
		Strength, lb/in.^{2}	Type of failure	Weathering quality
Butacite, Butvar	Polyvinyl butyral	2,000–4,000	Adhesive	Fair
Bostik 7026, FM-45, FM-46	Phenolic butyral	2,000–5,500	Glass	Excellent
EC826, EC776	...	...	Adhesion and glass	
N-199, Scotchweld	Phenolic nitrile	1,000–1,200	...	Excellent
Pliobond M-20, EC847	Vinyl nitrile	1,200–3,000	Adhesion and glass	Fair
EC711, EC882	...	...	...	...
EC870	Neoprene	800–1,200	Adhesion and cohesive	Fair
EC801, EC612	Polysulfide	200–400	Cohesive	Excellent
EC526, R660T, EC669	Rubber base	200–800	Adhesive	Fair to poor
Silastic	Silicone	200–300	Cohesive	Excellent
Rez-N-Glue, du Pont 5459	Cellulose vinyl	1,000–1,200	Adhesive	Fair
Vinylite AYAF, 28-18	Vinyl acetate	1,500–2,000	Adhesive	Poor
Araldite, Epon L-1372, ERL-2774, R-313, C-14, SH-1, J-1152	Epoxy	600–2,000	Adhesive	Fair to good

Usually solvent cleaning is the only surface preparation needed for bonding glass. However, solvent cleaning coupled with light sandblasting, or a chromium trioxide etch followed by ultrasonic detergent bath has been suggested for highest bond strength. Mechanical or chemical surface treatment can degrade the optical transmission quality of glass.

Most ceramic surfaces, like metal surfaces, have an adsorbed layer of water that may cause difficulty in bonding. If this water is partially removed by heating prior to bond formation, atmospheric moisture may diffuse into the bond during service and displace adhesive molecules from the interface since they will be preferentially adsorbed on the glass interface. To overcome this problem, coupling agents, typically silanols, have been developed to enhance the attraction between the adhesive and the glass or ceramic surface. Coupling agents can also provide a degree of stress release at the interface and reduce the thermal expansion-related stresses between the substrate and the adhesive.

Adhesives for glass or ceramic to metal seals should never become fully rigid because thermal expansion rates for adhesives are much higher than for the substrate. Temperature changes can lead to high stresses, which can cause cracking and joint failure. Nitrile rubber epoxies, phenolic resin blends, silicones, cyanoacrylates and anaerobics are some of the important organic adhesives that can be used to make glass or ceramic to metal seals. The silicones can be used from -100 to $450°F$ while maintaining good flexibility, but they are relatively costly. The cyanoacrylates are especially useful in bonding glass to metals, but they are less useful for joining porous ceramics because the porosity will absorb most of the adhesive and inhibit its cure.

16.11 Honeycomb and Other Structural Sandwich Panels

Structural sandwich panels represent what is probably one of the most common applications for structural adhesives. Facing materials can be made to adhere to core materials, such as aluminum or paper honeycomb, to give a high composite strength to weight ratio.

Adhesives with filleting properties are required for honeycomb cores. A modified phenolic is often used with aluminum honeycomb for high strength, while a neoprene or nitrile based organic solvent type of adhesive is often used with impregnated paper honeycomb. Epoxy adhesives are also commonly used in the fabrication of honeycomb sandwich panels.

Facings of metal, paper, or plastic may be combined with a variety of solid core materials. The adhesive must transmit shear stress from the facing to the core. It must also be able to resist the tendency of the thinner facing to buckle under design loads or under thermal stresses caused by variations in the panel components' coefficients of thermal expansion.

Adhesives for sandwich structures such as honeycomb and faced cores must have different properties than other adhesives. In order to achieve a good attachment to an open cell core, such as honeycomb, the adhesive must have a unique combination of surface wetting and controlled flow during its early stages of cure. This controlled flow prevents the adhesive from flowing down the cell wall and leaving a low strength top-skin attachment and an overweight bottom-skin attachment. The adhesive must resist being squeezed out from between faying (close fitting) surfaces when excessive pressure is applied during cure to a local area of the part. Many adhesives are formulated to achieve good core filleting and are subsequently given controlled flow by adding an open weave cloth or fibrous web, cast within a thicker film of adhesive. This cloth, called a scrim cloth, then prevents the faying surfaces from squeezing out all of the adhesive, which would result in an area of low bond strength.

The adhesive for sandwich construction must also have good toughness. It must be resistant to loads that act to separate the facings from the core under either static or dynamic conditions. Experience has shown that greater toughness in the bond line usually equates to greater durability and longer service life. If the facing is flexible enough so that it can be rolled on a drum, the climbing drum peel test (ASTM D 1781) is the most common test to measure toughness in sandwich construction. The resulting value of peel strength will vary considerably depending on the properties of the facing, toughness of the adhesive, amount of adhesive used, density of the core, cell size of the core, direction of peel, adequacy of surface preparation, and degradation of the joint.

Common adhesives for sandwich construction include nitrile-phenolic films, modified epoxy pastes and films, epoxy-nylon films, polyimide films, and modified urethane liquid and pastes. Accessory materials that are commonly used in the joining process are core splicing adhesives, syntactic foams, tapes, and corrosion inhibiting primers.[76]

References

1. Eichner, H. W., and Schowalter, W. E., Forest Products Laboratory Report 1813, 1950.
2. Wilson, I., et al., "Pretreatment for Bonded Aluminum Structures", *Advanced Materials and Processes,* August, 1997.
3. Minford, J. D., "Surface Preparations and Their Effect on Adhesive Bonding", *Adhesives Age,* July 1974.
4. Kabayashi, G. S., and Donnelly, Boeing Company Report DG-41517, February 1974.
5. Bijlmer, P. F. A., "Influence of Chemical Pretreatment on Surface Morphology and Bondability of Aluminum", *Journal of Adhesion,* 5 (1973).
6. Thrall, E., "Bonded Joints and Preparation for Bonding", AGARD Lecture Series 102 5.1, 1979.
7. Shannon, et al, "Primary Adhesively Bonded Structure Technology (PABST) General Materials Property Data", Douglas Aircraft Co., McDonnell Douglas Corp., Air Force Flight Dynamics Laboratory, Technical Report AFFDL-TR-77-107, September 1977.
8. Kim, G., and Ajersch, F., "Surface Energy and Chemical Characteristics of Interfaces of Adhesively Bonded Aluminum Joints", *Journal of Materials Science,* vol. 24, 1994 at 676–681.
9. Hartshorn, S. J., "Durability of Adhesive Joints", in *Structural Adhesives,* S. R. Hartshorn, ed. (New York: Plenum Press, 1986). Also in Minford, J. D., *Treatise on Adhesion and Adhesives,* vol. 5, R. L. Patrick, ed. (New York: Marcel Dekker, 1981) at 45.
10. Browne, J., "Aerospace Adhesives in the 90s", 38th International SAMPE Symposium, May 10–13, 1993.
11. *Adhesive Bonding of Aluminum Automotive Body Sheet Alloys,* The Aluminum Assn. Inc., 750 Third Ave., New York, NY, November 1975.
12. *Adhesive Bonding Aluminum,* Reynolds Metals Co., Richmond, VA, 1966.

13. Sell, W. D., "Some Analytical Techniques for Durability Testing of Structural Adhesives", Proceedings 19th National SAMPE Symposium and Exhibition, vol. 19, New Industries and Applications for Advanced Materials Technology, April 1974.
14. Magariello, E. and Hannon, M., "Bonding Beryllium Copper Alloys with High Temperature Adhesives", *Adhesives Age,* March 1971.
15. Berylcoat D available from Brush Wellman, Inc., Cleveland, OH.
16. Shields, J., *Adhesives Handbook,* 3 ed., Chapter 5 (London: Butterworths, 1984).
17. Cagle, C. V., "Surface Preparation for Bonding Beryllium and Other Adherends", in *Handbook of Adhesives Bonding,* C. V. Cagle, ed. (New York: McGraw Hill, 1973).
18. Ebanol C Special, Enthane Company, New Haven, CT.
19. Bolger, J. C., et al., "A New Theory of Improving the Adhesion of Polymers to Copper", Final Report, INCRA Project No. 172, International Copper Research Assn., New York, August 1971.
20. American Society for Metals (ASM), "Properties and Selections: Nonferrous Alloys and Pure Metals", in *Metal Handbook,* 9th ed., vol. 2., 1979.
21. Military Specification, MIL-M-45202C, Magnesium Alloy, Anodic Treatment of, April 1981.
22. Military Specification, MIL-M-3171C, Magnesium Alloy, Processes for Pretreatment and Prevention of Corrosion on, March 1974.
23. Eickner, H. W., "Adhesive Bonding Properties of Various Metals as Affected by Chemical and Anodizing Treatments of the Surfaces", Forest Products Laboratory Reports 1842 and 1842-A, February 1955 and August 1960, Forest Products Laboratory, Madison, WI.
24. Keith, R. E., et al., "Adhesive Bonding of Nickel and Nickel Base Alloys", NASA TMX-63428, October 1965.
25. Landrock, A. H., "Processing Handbook on Surface Preparation for Adhesive Bonding", *Picatinny Arsenal Technical Report* 4883 (Dover, NJ: Picatinny Arsenal, December 1975).
26. Devine, A. T., "Adhesive Bonded Steel: Bond Durability as Related to Selected Surface Treatments", U.S. Army Armament Research and Development Command, Large Caliber Weapon Systems Laboratory, Technical Report ARLCD-TR-77027, December 1977.
27. Brockmann, W., "Durability Assessment and Life Prediction for Adhesive Joints", in *Adhesives and Sealants,* vol. 3, Engineered Materials Handbook, ASM International, 1990.
28. Shaffer, D. K. et al., "Titanium as an Adherend", Chapter 4 in *Treatise on Adhesion and Adhesives,* vol. 7, J. D. Minford, ed. (New York: Marcel Dekker, 1991).
29. Johnson, W. E., et al., "Titanium Sandwich Panel Research and Development", General Dynamics, Fort Worth, TX, Final Report AMC-59-7-618, vol. 3, Contract AF 33(600)-34392, November 1959.
30. Allen, K. W., et al., "Bonding of Titanium Alloys", *J. Adhesion* 6, 153, 1974.
31. Cotter, J. L., and Mahoon, A., "Development of New Surface Pretreatments Based on Alkaline Hydrogen Peroxide Solutions for Adhesive Bonding of Titanium", *Int. J. Adhesion Adhesives,* 2:47, 1982.
32. Ditchek, B. M., et al., Proc. 12th SAMPE Tech. Conf., 882, 1980.
33. Brown, S. R., Proc. 27th Natl. SAMPE Symp., 363, 1982.
34. Clearfield, H. M., et al., "Adhesion Tensile Testing of Environmentally Exposed Ti-6Al-4V Adherends", *J. Adhesion,* 23:83, 1987.
35. Fibey, J. A., et al., "Sodium Hydroxide Anodization of Ti-6Al-4V Adherends", *J. Adhesion,* 20:283, 1987.
36. Kennedy, A. C., et al., "A Sodium Hydroxide Anodize Surface Treatment for the Adhesive Bonding of Titanium Alloys", *Int. J. Adhesion Adhesives,* 3:133, 1983.
37. Morita, W. H., "Titanium Tankage Program — Titanium Tankage Development", North American Aviation, Inc., Space and Information Systems Division, Technical Documentary Report AFRPL-TR-64-154M, November 1964.
38. Clearfield, H. M., et al., "Surface Preparation of Metals", in *Adhesives and Sealants,* vol. 3, Engineered Materials Handbook, ASM International, 1990.

39. Keith, R. E., "Adhesive Bonding of Titanium and Its Alloys", *Handbook of Adhesives Bonding*, C. V. Cagel, ed. (New York: McGraw Hill, 1973).
40. McIntrye, R. T., "Adhesive Bonding to Cadmium and Zinc Plated Steel Substrates", Applied Polymer Symposia, no. 29, M. J. Bodnar, ed. (New York: Wiley, Interscience, 1972).
41. Pocius, A. V., et al., "The Use of Adhesives in the Joining of Plastics", in *Treatise on Adhesion and Adhesives*, J. D. Minford, ed. (New York: Marcel Dekker, 1991).
42. Shields, J., *Adhesives Handbook*, 3rd ed. (London: Butterworths, 1984) at 102.
43. Lupton, D. C., "Selection of an Adhesive for Bonding FRP Automotive Panels", *Int. J. Adhesion Adhesives*, 3(2):155–158, 1983.
44. Minford, J. D., *Proc. Int. Symp. Physiochem. Aspects Polym. Surfactants* 2:1139–1160, 1983.
45. Tavakioli, S. M., "The Effect of Aging on Structure Properties of Hot Plate Welds in Acrylonitrile Butadiene Styrene", ANTEC Conference Proceedings, Society of Plastic Engineers, 1994.
46. Landrock, A., "Surface Preparation of Adherends", in *Adhesives Technology Handbook* (Park Ridge, NJ: Noyes Publishing Co., 1985).
47. *Adhesive Bonding*, MIL-HDBK-691B, U.S. Dept. of Defense, 1987.
48. Snogren, R. C., *Handbook of Surface Preparation,* Chapter 12 (New York: Palmerton Publishing Co., 1974).
49. *Delrin Acetal Resins Design Handbook,* E. I. DuPont de Nemours & Co., Technical Bulletin.
50. Grimm, R. A., "Through Transmission Infrared Welding of Polymers", ANTEC Conference Proceedings, Society of Plastics Engineers, 1996.
51. Toy, L. E., "Plastics/Metals: Can They Be United?", *Adhesive Age,* 17(10):19–24, 1974.
52. Meier, J. F., "Fundamentals of Plastics and Elastomers", in *Handbook of Plastics, Elastomers, and Composites,* 3rd ed., C. A. Harper, ed., (New York: McGraw Hill, 1996).
53. Meier, J. F., and Petrie, E. M., "The Effect of Ultraviolet Radiation on Sodium-Etched Polytetrafluoroethylene Bonded to Polyurethane Elastomer", *Journal of Appl. Poly. Sci.,* 17, 1973, at 1007–1017.
54. DeLollis, N. J., and Montoya, O., "Surface Treatment for Difficult to Bond Plastics", *Adhesives Age,* January 1963.
55. Tung, B., et al., "Surface Modification of Fluoropolymers with Plasma Generated via Microwave Discharge", *SAMPE Quarterly,* April 1988.
56. King, A. F., *Adhesives for Bonding DuPont Plastics,* E. I. duPont de Nemours & Co., TR 152, August 1966.
57. Bresin, R. L., "How to Obtain Strong Adhesive Bonds Via Plasma Treatment", *Adhesives Age,* March 1972.
58. Lexan Polycarbonate Resins Fabrication Data, General Electric Plastics Dept., TR CDC-446.
59. Tanner, R. C., "Manufacturing Processes with Adhesive Bonding", Applied Polymer Symposia no. 19, *Processing for Adhesive Bonded Structures,* M. J. Bodnar, ed. (New York: Interscience Publishers, 1972).
60. Abolins, V., and Eickert, J., "Adhesive Bonding and Solvent Cementing of Polyphenylene Oxide", *Adhesives Age,* July 1967.
61. Stokes, V. K., "The Effects of Fillers on the Vibration Welding of Polybutylene Terephthalate", ANTEC Conference Proceedings, Society of Plastics Engineers, 1993.
62. Ballmann, A., et al., "Surface Treatment of Polyetheretherketone Composite by Plasma Activation", *Journal of Adhesion,* vol. 46, 1994.
63. "Cementing, Welding, and Assembly of Plastics", Chapter 17 of *Plastics Engineering Handbook* 3rd ed., A. F. Randolph, ed., (New York: Reinhold, 1960).
64. Minford, J. D., *Proc. Int. Symp. Physiochem. Aspects Polym. Surfactants,* 2:1161–1180, 1983.
65. Peters, S. T., "Advanced Composite Materials and Processes", *Handbook of Plastics Elastomers and Composites,* 3rd. ed., C. A. Harper, ed., (New York: McGraw Hill, 1996).

66. Potter, D. L., *Primary Adhesively Bonded Structure Technology (PABST) Design Handbook for Adhesive Bonding* (Long Beach, CA: Douglas Aircraft Co., McDonnell Douglas Corp., January 1979).
67. Albee, N., "Adhesives for Structural Applications", *Advanced Composites*, November/December 1989.
68. Schwartz, M. M., "Joining of Composite Matrix Materials", *ASM International* (Materials Park, OH, 1994).
69. Watson, M. N., "Techniques for Joining Thermoplastic Composites", Automated Composites, First International Conf., University of Nottingham, U.K., September 10–12, 1986.
70. Landrock, A., "Adhesives for Specific Adherends", Chapter 6 of *Adhesives Technology Handbook* (Park Ridge, NJ: Noyes Publications, 1985).
71. Symes, T., and Oldfield, D., "Technology of Bonding Elastomers", Chapter 2 in *Treatise on Adhesion and Adhesives,* vol. 7, J. D. Minford, ed. (New York: Marcel Dekker, 1991).
72. Cox, D. R., "Some Aspects of Rubber to Metal Bonding", *Rubber Journal,* April/May 1969.
73. Hemming, C. B., "Wood Gluing", *Handbook of Adhesives,* I. Skeist, ed., (New York: Reinhold, 1962).
74. Moser, F., "Bonding Glass", *Handbook of Adhesives,* I. Skeist, ed. (New York: Reinhold, 1962).
75. TSB 124, "Bonded Honeycomb Sandwich Construction", Hexcel Corp., Dublin CA. (also in *Advanced Composites,* March/April 1993).

Chapter 17

Effect of the Environment

17.1 Introduction

For an adhesive or sealant bond to be useful, it not only must withstand the mechanical forces that are acting on it, but it must also resist the elements to which it is exposed during service. Thus, one of the most important characteristics of an adhesive joint or sealant is its endurance to the operating environment. This endurance is also referred to as the joint's *permanence* or *durability*. Strength and permanence are influenced by many common environmental elements. These include high and low temperatures, moisture or relative humidity, chemical fluids, and outdoor weathering. Table 17.1 summarizes the relative resistance of various adhesive types to common operating environments.

The effect of simultaneous exposure to both mechanical stress and a chemical environment is often more severe than the sum of each factor taken separately. Mechanical stress, elevated temperatures, and high relative humidity can be a fatal combination for certain adhesives and sealants if all occur at the same time. Add to this the possible cyclic effects of each factor, and one can easily see why it is important to understand the effects of environment on the joint.

Environmental consequences are so severe that it is usually necessary to test preproduction joints, both in the laboratory and in the field, and under conditions as close to the actual service environment as possible. The parameters that will likely affect the durability of a given joint are:

- Maximum stress level
- Average constant stress level

TABLE 17.1 Relative Resistance of Synthetic Adhesives to Common Service Environments[1]

Adhesive type	Shear	Peel	Heat	Cold	Water	Hot water	Acid	Alkali	Oil, grease	Fuels	Alcohols	Ketones	Esters	Aromatics	Chlorinated solvents
Thermosetting Adhesives															
1. Cyanoacrylate	2	6	5	...	6	6	6	6	3	3	5	5	5	4	4
2. Polyester + isocyanate	2	2	3	2	2	1	3	3	2	2	3	3	5	6	2
3. Polyester + monomer	2	6	5	3	3	3	6	6	2	2	2	6	6	6	6
4. Urea formaldehyde	2	6	3	3	2	2	3	2	2	2	2	2	2	2	2
5. Melamine formaldehyde	2	6	2	2	2	5	2	2	2	2	2	2	2	2	2
6. Urea-melamine formaldehyde	2	6	2	2	2	2	1	1	2	2	2	2	2	2	2
7. Resorcinol formaldehyde	2	6	2	2	2	2	2	2	2	2	2	2	2	2	2
8. Phenol-resorcinol formaldehyde	2	6	2	2	2	2	2	2	2	2	2	2	2	2	2
9. Epoxy (+ polyamine)	2	5	3	5	2	2	2	2	2	3	1	6	6	1	
10. Epoxy (+ polyanhydride)	2	5	1	4	3	3	2	2	...	2	2	6	6	2	
11. Epoxy (+ polyamide)	2	2	6	2	2	3	3	6	2	2	1	6	6	3	
12. Polyimide	2	4	1	1	2	4	2	2	2	2	2	2	2	2	2
13. Polybenzimidazole	2	4	1	1	2	4	2	2	2	2	2	2	2	2	2
14. Acrylic	2	6	5	3	1	3	2	2	2	2	2	2	2	2	2
15. Acrylate acid diester	2	5	3	3	4	6	6	6	3	3	5	5	5	4	4
Thermoplastic Adhesives															
16. Cellulose acetate	2	6	2	3	1	6	1	2	2	...	4	6	6	6	6
17. Cellulose acetate butyrate	2	3	3	3	2	...	3	2	...	...	6	6	6	6	6
18. Cellulose nitrate	2	6	3	3	3	3	3	6	2	2	6	6	6	6	6
19. Polyvinyl acetate	2	6	6	...	3	6	3	3	2	2	6	6	6	6	6
20. Vinyl vinylidene	2	3	3	3	3	3	...	...	2	2	2	2	2	2	
21. Polyvinyl acetal	2	6	5	3	2	...	6	3	2	2	3	3	6	3	2
22. Polyvinyl alcohol	...	2	3	...	6	6	5	5	2	1	3	3	1	1	1
23. Polyamide	2	3	5	...	5	6	6	2	2	2	6	2	2	2	6
24. Acrylic	2	2	4	3	3	3	...	...	2	...	...	4	4	...	4
25. Phenoxy	2	3	4	3	3	4	3	2	3	3	5	...	...	6	4

Elastomer Adhesives

No.	Adhesive	1	2	3	4	5	6	7	8	9	10	11	12	13
26.	Natural rubber	2	3	3	...	3	3	3	6	2	4	4	6	6
27.	Reclaimed rubber	2	3	3	...	2	3	3	6	2	4	4	6	6
28.	Butyl	3	6	6	3	2	2	2	6	2	2	2	6	6
29.	Polyisobutylene	6	6	6	3	2	2	2	6	2	2	2	6	6
30.	Nitrile	2	3	3	3	2	6	6	2	3	6	6	3	6
31.	Styrene butadiene	3	6	3	3	2	5	2	...	2	6	6	6	6
32.	Polyurethane	2	3	2	2	2	3	3	2	2	6	5	...	5
33.	Polysulfide	3	2	6	2	1	6	2	2	2	6	6	2	6
34.	Silicone (RTV)	3	5	1	1	2	2	3	2	3	3	3	3	3
35.	Silicone resin	2	2	1	2	2	2	...	2	2	4	4	3	6
36.	Neoprene	2	3	3	3	2	2	2	2	3	6	6	6	6

Alloy Adhesives

No.	Adhesive	1	2	3	4	5	6	7	8	9	10	11	12	13
37.	Epoxy-phenolic	1	6	3	2	2	2	3	3	2	6	6	2	6
38.	Epoxy-polysulfide	2	2	2	1	6	2	...	2	2	6	6	2	6
39.	Epoxy-nylon	1	1	6	2	6	...	...	2	3	6	6	6	6
40.	Phenolic-nitrile	2	2	2	2	2	2	2	2	2	6	6	6	6
41.	Phenolic-neoprene	2	3	3	2	...	3	2	2	3	6	6	6	6
42.	Phenolic-polyvinyl butyral	2	3	3	2	3	4	2	2	4	6	6	6	6
43.	Phenolic-polyvinyl formal	2	3	6	2	6	4	2	2	4	6	6	6	6

Key:
1. Excellent
2. Good
3. Fair
4. Poor
5. Very poor
6. Extremely poor

- Nature and type of environment
- Cyclic effects of stress and environment (rate and period)
- Time of exposure

In applications where possible degrading elements exist, candidate adhesives must be tested under simulated service conditions. Standard lap shear tests, such as ASTM D 1002, which use a single rate of loading and a standard laboratory environment, do not yield optimal information on the service life of the joint. Important information such as the maximum load that the adhesive joint will withstand for extended periods of time and the degrading effects of various environments are addressed by several test methods. Table 17.2 lists common ASTM environmental tests that are often reported in the literature. A more complete list of standard ASTM test methods may be found in Appendix C.

Time and economics generally allow only short term tests to verify the selection of the adhesive system relative to the environment. It is tempting to try to accelerate service life in the laboratory by increasing temperature or humidity, for example, and then to extrapolate the results to actual conditions. However, often too many interdependent variables and modes of strength degradation are in operation, and a

TABLE 17.2 ASTM Test Method for Determining Environmental Resistance of Adhesives and Sealants

Title	Method
Atmospheric Exposure of Adhesive Bonded Joints and Structures, Recommended Practice for	ASTM D 1828
Exposure of Adhesive Specimens to High Energy Radiation, Recommended Practice	ASTM D 1879
Integrity of Glue Joint in Structural Laminated Wood Products for Exterior Use, Test for	ASTM D 1101
Resistance of Adhesives to Cyclic Laboratory Aging Conditions, Tests for	ASTM D 1183
Effect of Bacteria Contamination on Permanence of Adhesive Preparations and Adhesive Bonds	ASTM D 1174
Effect of Moisture and Temperature on Adhesive Bonds, Tests for	ASTM D 1151
Effect of Mold Contamination on Permanence of Adhesive Preparations and Adhesive Bonds, Test for	ASTM D 1286
Resistance of Adhesive Bonds to Chemical Reagents, Test for	ASTM D 896
Strength Properties of Adhesives in Shear by Tension Loading in the Temperature Range of -450 to $-57°F$, Test for	ASTM D 2557
Strength Properties of Adhesives in Shear by Tension Loading at Elevated Temperatures, Test for	ASTM D 2295

reliable estimate of life using simple extrapolation techniques cannot be achieved.

For example, elevated temperature exposure could cause oxidation or pyrolysis and change the rheological characteristics of the adhesive. Thus, not only is the cohesive strength of the adhesive weakened by temperature, but its ability to absorb stresses due to thermal expansion or impact are also degraded by thermal aging. Chemical environments cause corrosion at the interface; however, the adhesive may become more flexible and be better able to withstand cyclic stress because of the softening effect of the chemical. Exposure to a chemical environment may also result in unexpected elements replacing the adhesive at the interface and resulting in a weak boundary layer. They are dependent not only on the type and degree of environment, but also on the time the bond is in service. What makes life estimation difficult is that these effects could all occur at various times.

If there is only one parameter that changes due to environmental exposure, then the application of accelerated test techniques and analysis may yield useful information with regard to service life. In the electrical insulation industry, for example, Arrhenius plots are often used to predict the end-of-life of insulating materials by simple extrapolation. This can be accomplished because insulation life is dependent mainly on temperature and other factors are relatively minor. However, there are a multiplicity of consequences that can occur within the adhesive/interface/adherend, and each consequence could possibly affect the other. The formidable task of determining the end-of-life is one of the most difficult challenges in adhesives science and possibly the single item that most inhibits the use of adhesives and sealants in structural applications.

In this chapter, we will examine both the individual effects of various environments on the bond, and we will also consider the significance of certain combined environments. The root cause of environmental degradation will be studied so that better judgments about the durability of bonded joints can be made by those seeking to design, manufacture, and use adhesives and sealants. This chapter will point to common problem areas, suggest changes in the bonding system to provide greater bond durability, and act as groundwork for further investigation into this important area.

17.2 High Temperatures

17.2.1 Factors affecting temperature resistance

All polymeric materials are degraded to some extent by exposure to elevated temperatures. Not only are short term physical properties lowered by elevated temperature exposure, but these properties will also likely degrade with prolonged thermal aging. Several important questions need to be asked of an adhesive or sealant if high service temperatures are expected. What is the maximum temperature that the bond will be exposed to during its life? What is the average temperature to which the bond will be exposed? Ideally one would like to have a definition of the entire temperature–time relationship representing the adhesive's expected service history. This data would include time at various temperatures and rates of temperature change.

Certain polymers have excellent resistance to high temperatures over short duration (e.g., several minutes or hours). The short term effect of elevated temperature is primarily one of increasing the molecular mobility of the adhesive or sealant. Thus, depending on the adhesive, the bond may actually show increased toughness but lower shear strength. Certain products will show softness and a high degree of creep on exposure to elevated temperatures. However, prolonged exposure to elevated temperatures may cause several reactions to occur in the adhesive or sealant. These mechanisms can weaken the bond both cohesively and adhesively. The reactions that affect the bulk material are oxidation and pyrolysis. Thermal aging can also affect the interface by causing changes (chemical and/or physical).

If heating brings a non-crosslinked adhesive far above its glass transition temperature, the molecules will become so flexible that their cohesive strength will drastically decrease. In this flexible, mobile condition, the adhesive is susceptible to creep and greater chemical or moisture penetration. Generally, prolonged heating at an excessive temperature will have the following effects on a crosslinked adhesive:

- Split polymer molecules (chain scission) causing lower molecular weight, degraded cohesive strength, and byproducts
- Continued crosslinking resulting in bond embrittlement and shrinkage
- Evaporation of plasticizer resulting in bond embrittlement
- Oxidation (if oxygen or a metal oxide interface is present) resulting in lower cohesive strength and weak boundary layers

Most synthetic adhesives and sealants degrade rapidly at service temperatures greater than 300°F. However, several polymeric materials have been found to withstand up to 500–600°F continuously and even higher temperatures on a short term basis. To use these materials, one must pay a premium in adhesive cost and also be able to provide long, high-temperature cures, often with pressure. Long term temperature resistance greater than 500–600°F can only be accomplished with ceramic, metal, or other non-organic materials. Lee[2] has divided high temperature adhesives into three temperature ranges as shown in Table 17.3.

For an adhesive to withstand elevated temperature exposure, it must have a high melting or softening point and resistance to oxidation. Materials with a low melting point, such as many of the thermoplastic adhesives, may prove excellent adhesives at room temperature. However, once the service temperature approaches the glass transition temperature of the adhesive, plastic flow results in deformation of the bond and degradation of cohesive strength. Thermosetting adhesives, exhibiting no melting point, consist of highly crosslinked networks of macromolecules. Because of this crosslinked structure, they show relatively little creep at elevated temperatures and exhibit relatively little loss of mechanical function when exposed to either elevated temperatures or other degrading environments. Many of these materials are suitable for high-temperature applications. When considering thermosets, the critical factor is the rate of strength reduction due to thermal oxidation and pyrolysis.

Thermal oxidation initiates a progressive chain scission of molecules resulting in losses of weight, strength, elongation, and toughness within the bulk adhesive or sealant. Figure 17.1 illustrates the effect

TABLE 17.3 Classification of Adhesives and Sealants According to Use Temperature and Time[2]

Range	Use temperature, °C	Use time
I	538–760	Seconds to minutes
II	288–371	Hundreds of hours
III	177–232	Thousands of hours

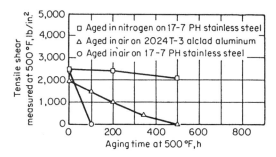

Figure 17.1 The effect of 500°F aging in air and nitrogen on an epoxy phenolic adhesive (HT-424).[3]

of oxidation by comparing adhesive joints aged in both high-temperature air and inert-gas environments. The rate of bond strength degradation in air depends on the temperature, the adhesive, the rate of airflow, and even the type of adherend. Some metal-adhesive interfaces are chemically capable of accelerating the rate of oxidation. For example, it has been found that nearly all types of structural adhesives exhibit better thermal stability when bonded to aluminum than when bonded to stainless steel or titanium (Fig. 17.1).

Adhesive and sealant formulators can extend the thermal endurance of their products by incorporating additives into the formulations. Chelating agents and antioxidants such as antimony trioxide and others are commonly found in the formulations of high temperature adhesives to forestall as best as possible the effects of oxidation. Usually, concentrations of less than 1% by weight are effective. Catalyst poisons may also be incorporated into the system to react directly with the metallic substrate, thereby inhibiting its catalytic effect on oxidation.

Pyrolysis is simple thermal destruction of the molecular chain of the base polymer in the adhesive or sealant formulation. Pyrolysis causes chain scission and decreased molecular weight of the bulk polymer. This results in both reduced cohesive strength and brittleness. Resistance to pyrolysis is predominantly a function of the intrinsic heat resistance of the polymers used in the adhesive formulation. As a result, many of the base resins used in high temperature adhesive formulations are rigidly crosslinked or are made-up of a molecular backbone referred to as a "ladder structure". The ladder structure is made from aromatic or heterocyclic rings in the main polymer structure. The rigidity of the molecular chain decreases the possibility of chain scission by preventing thermally agitated vibration of the chemical bonds. The ladder structure provides high bond dissociation energy and acts as an energy sink to its environment. Notice in Fig. 17.2 that to have

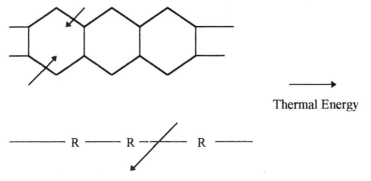

Figure 17.2 Degradation of ladder polymer and straight chain polymer due to thermal aging.

a complete chain scission resulting in a decrease in molecular weight, two bonds must be broken in a ladder polymer, whereas only one needs to be broken in a more conventional linear straight chain structure.

In order to be considered as a promising candidate for high temperature applications, an adhesive must provide all of the usual functions necessary for good adhesion (wettability, low shrinkage on cure, thermal expansion coefficient matched to the substrates, etc.), and it must also possess:

1. A high melting or softening point
2. Resistance to oxidative degradation
3. Resistance to thermally induced chain scission

17.2.2 Adhesives and sealants for high temperature service

High temperature adhesives are usually characterized by a rigid polymeric structure, high softening temperature, and stable chemical groups. The same factors also make these adhesives very difficult to process. Only epoxy-phenolic-, bismaleimide-, polyimide-, and polybenzimidazole-based adhesives can withstand long-term service temperatures greater than 350°F. However, modified epoxy and even certain cyanoacrylate adhesives have moderately high short term temperature resistance. Silicone adhesives also have excellent high temperature permanence, but exhibit low shear strength. Properties of these adhesives are compared in Table 17.4. Figure 17.3 compares various high temperature adhesives as a function of heat resistance and thermal aging.

There are even fewer sealants suitable for long term, high temperature service. The thermal endurance requirement of high crosslinking density generally are counter to the requirement that a sealant be flexible in nature. Silicone based elastomers and some very special elevated temperature elastomers are the only products that will provide both thermal endurance and significant flexibility.

17.2.2.1 Epoxy.
Epoxy adhesives are generally limited to applications below 250°F. Figure 17.4 illustrates the aging characteristics of a typical epoxy adhesive at elevated temperatures. The epoxy adhesives using aliphatic polyamine hardening agents are not serviceable above 150°F. The aromatic diamine and monoanhydride cured products are usable at temperatures of 250–300°F. Certain epoxy adhesive formulations, however, have been able to withstand short terms at 500°F and long term service at 300–350°F. These systems were formulated specifically for thermal environments by incorporation of stable epoxy

TABLE 17.4 Properties of High Temperature Adhesives[4]

Property	Mod. epoxy	Mod. epoxy	Mod. epoxy	Epoxy-phenolic	Epoxy-phenolic	Cyano-acrylate	Cyano-acrylate	Polyimide	Vulcanizing silicone rubber	Pressure sensitive silicone
Temperature range, °F	−67 to +420	−67 to +350	350	−423 to +500	−400 to +500	475	475	400–600	−100 to +450	500
Optimum cure condition										
Time, min	60	60	60	60	30	seconds	seconds	90	24 h	5–10
Temp, °F	350	350	350	325–350	340 ± 10	68	68	550–700	68	212
Pressure, psi	10–50	10–50	15–50	10–50	0–100	contact	contact	40	—	—
Tensile shear, psi										
68°F	3700[a]	2500[a]	4331[b]	2100[a]	3800[c]	3120[d], 950[e]	3130[d], 1800[e]	3300	275[f]	Adhesion per ASTM D-1000
350°F	2800	2200	1742	1970	2500	970, 400	1070, 640	—		
500°F	—	—	—	1600	2000	430[f], —	0, —	2300		
Environmental resistance of bond: tensile shear, psi after 18 h at 365°F (30 days salt spray test at 365°F)	1800	—	—	—	—	990[g]	1000[g]	—	—	—
30 days, 100% rel hum, 120°F, test at 350°F	—	1570	—	—	2050	3800[h]	3670[h]	—	—	—

694

Applicable specification	Remarks	Manufacturer	Tradename
MMM-A132 MIL-A-25463	Also useful in bonding titanium alloys; supported by synthetic knitted carrier	Narmco	Metbond 329-7
MMM-A-132 MIL-A-25463	For load bearing & sandwich applic. light nylon carrier	Narmco	Metbond 1510
MMM-A-132	Glass supported	Narmco	Metbond 306
MIL-A-5090D MIL-A-25463	—	American Cyanamid	FM96 FM96U
MM-A-132 MIL-A-25463	Load bearing & sandwich aluminum-filled on cloth	American Cyanamid	HT424
—	Low viscosity, fast setting	Eastman	910 MHT
—	Medium viscosity thickened grade	Eastman	910THT
MMM-A-132 MIL-A-25463	Primer recommended; for load bearing and sandwich 112 glass cloth carrier	American Cyanamid	FM34
MIL-A-46106A	Ready-to-use adhesive-sealant	Dow Corning	732RTV
—	Experimental, pressure sensitive adhesive, highly volatile	Dow Corning	X30494

[a] Unclad 2024-T81 aluminum alloy; after 18 h at 365°F.
[b] MIL-A-5090D.
[c] Clad 2024-T3 aluminum alloy; cured 40 min at 335 F.
[d] Steel-to-steel after 7 days.
[e] Al-to-Al after 7 days.
[f] 482°F.
[g] Steel-to-steel; 30 days in water.
[h] Steel-to-steel; 30 days in gasoline.
[i] ASTM D412.

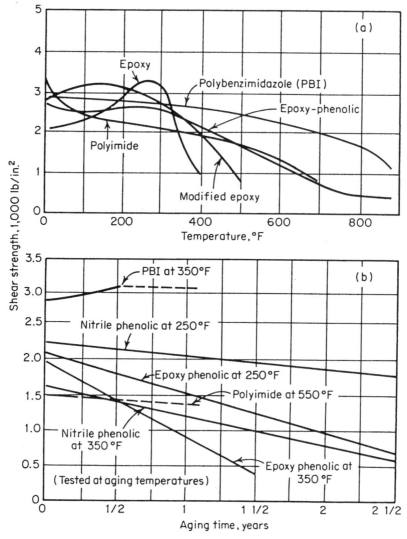

Figure 17.3 Comparison of (a) heat resistance and (b) thermal aging of several high temperature structural adhesives.[5]

coreactants or high temperature curing agents into the adhesive. Epoxy adhesives based on multifunctional resins are available that will provide strength retention to about 450°F. However, where long term service is required, epoxies are limited to temperatures no higher than 350°F.

Anhydride curing agents give unmodified epoxy adhesives somewhat greater thermal stability than most other epoxy curing agents.

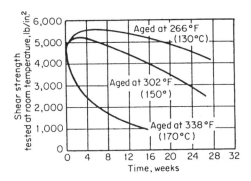

Figure 17.4 Effect of temperature aging on typical epoxy adhesive in air. Strength measured at room temperature after aging.[6]

However, benzophenonetetracarboxylic dianhydride (BTDA), phthalic anhydride, pyromellitic dianhydride, and chlorendic anhydride allow greater cross linking and result in short term heat resistance to 450°F. Long term thermal endurance, however, is limited to 300°F. Table 17.5 shows the high temperature properties of a typical epoxy formulation cured with pyromellitic dianhydride.

One of the most successful epoxy coreactant systems developed thus far is an epoxy-phenolic alloy. The excellent thermal stability of the phenolic resins is coupled with the valuable adhesion properties of epoxies to provide an adhesive capable of 700°F short term operation and continuous use at 350°F. The heat-resistance and thermal-aging properties of an epoxy-phenolic adhesive are compared with those of other high temperature adhesives in Fig. 17.3. Epoxy-phenolic adhesives are generally preferred over other high temperature adhesives, such as the polyimides, because of their lower cost and ease of processing.

Advantages of epoxy-based systems include relatively low cure temperatures, no volatiles formed during cure, low cost, and a variety of formulating and application possibilities. The higher-temperature ad-

TABLE 17.5 Tensile Shear Strength of Anhydride Cured Epoxy Adhesives at Elevated Temperatures[7]

Substrate	Treatment	Tensile-shear strength, lb/in.², at		
		75°F	300°F	450°F
Aluminum	Etched	2,300	2,600	900
Aluminum	Untreated	1,800	1,400	500
Steel, cold-rolled....	Etched	1,700	2,000	1,100
Steel, cold-rolled....	Untreated	1,200	1,200	900

hesives lose many of these advantages in favor of improved thermal-aging characteristics.

17.2.2.2 Modified phenolics. The phenolic resins exhibit reasonably good resistance to elevated temperatures and have been used in high temperature adhesive systems for many years. To overcome their inherent brittleness and low peel strength, phenolic resins are alloyed with modifiers such as nylons, acetals, and elastomers (neoprene and nitrile) as well as epoxies. These systems do not have exceptionally high tensile shear strength, but their peel strength is relatively high (10–25 piw) and for this reason they find use in applications where both peel strength and high temperature resistance are important. For example, nitrile-phenolics have been used to bond break linings in automobiles and for high temperature sandwich constructions.

Of the common modified phenolic adhesives, the nitrile-phenolic blend has the best resistance to elevated temperatures. As shown in Fig. 17.3, nitrile-phenolic adhesives have high shear strength up to 250–350°F, and the strength retention on aging at these temperatures is very good. The nitrile-phenolic adhesives are also extremely tough and provide high peel strength.

17.2.2.3 Silicone. Silicone resins have long been known for their thermal stability as both adhesives and sealants. However, they have relatively weak cohesive strength. Silicone-based adhesives and sealants are good for continuous service up to 600°F. Because of their relatively low cohesive strength, their chief application as an adhesive is in nonstructural uses such as high-temperature pressure-sensitive tape.

Flexible silicone sealants have been used in ovens and furnaces, electrical connectors, and aerospace parts such as ablative shielding. Attempts have been made to incorporate silicones with other resins such as epoxies and phenolics, but long cure times and low strength have limited their use.

Single part, room temperature vulcanizing silicone rubber may be used for extended periods up to 450°F and for longer periods up to 550°F. When bonded to glass, metal, or wood, this adhesive has a peel strength at 446°F of 24 piw. The peel strength drops sharply at 482°F (see Fig. 10.9). Typical applications are sealing windows in oven doors, flues on gas appliances, and bonding gaskets in heating units.

Silicone adhesives are also used in high temperature pressure sensitive tapes. These systems usually include a peroxide catalyst for improved cohesive strength. Peel strength is relatively constant from 212°F to 482°F (Table 10.19).

17.2.2.4 Polyimides and other aromatic resins. The polyaromatic resins, such as polyimide, polybenzimidazole, and bismaleimide, offer greater thermal resistance than any other commercially available adhesive. The rigidity of their molecular chains decreases the possibility of chain scission caused by thermally agitated chemical bonds. The aromaticity of these structures provides high bond dissociation energy and acts as an "energy sink" to the thermal environment. The rigid nature of these polymers usually excludes their use as sealants.

17.2.2.4.1 Polyimide. The strength retention of polyimide adhesives for short exposures to 1000°F is slightly better than that of an epoxy-phenolic alloy. However, the long term thermal endurance of polyimides at temperatures greater than 500°F is unmatched by other commercially available adhesives. Because of the inherent rigidity of this polymer, peel strength is low. Two types of polyimides are currently in use. These are cured by condensation or addition reaction mechanisms.

The condensation polyimides were the first high temperature adhesives developed. Polyimide adhesives are usually supplied as a glass-fabric-reinforced film having a limited shelf life. However, liquid solution forms of condensation cured polyimide resins are also available. Because of their condensation curing reaction and the high temperature solvents used in their processing, these polyimides give off volatiles during cure. Condensation cured polyimides are often cured under vacuum to facilitate the removal of these by-products. Even so, the high-boiling volatiles that are released during cure cause a somewhat porous adhesive layer. A cure of 90 min at 500–600°F and 15–200 psi pressure is necessary for optimum properties. The processing conditions for a commercially available condensation polyimide adhesive demonstrate the tedious nature of the curing process.[8]

1. Apply vacuum at room temperature.
2. Raise temperature to approximately 365°F.
3. Release vacuum and apply pressure of 100 psi.
4. Maintain temperature and pressure for 1–1.5 h.
5. Increase temperature to 650°F and maintain pressure and temperature for 1–1.5 h.

Such processing requirements are especially cumbersome for large or complex substrates.

FM 34 (American Cyanamid) had long been a popular polyimide adhesive for high temperature structural applications. An arsenic free derivative, FM 34B-18 shows excellent strength retention after 1200 h aging at 500°F, but degrades rapidly during the next 2000 h. In applications where transition metals are not present in the substrate (e.g., polyimide composites), FM 34 and FM 34B-18 have equivalent long term heat aging properties.[9]

Addition reaction polyimides have somewhat less stringent curing conditions and are less likely to have volatiles affect the bondline, although they still require high temperature cures for processing. Typical processing conditions require heating from room temperature to 400°F in 1 h under light contact pressure; holding 30 min at 400°F; then increasing pressure to 100 psi and temperature to 550°F; and finally holding at 550°F for 2 h.[8] This sequenced processing method results in very low void content in the bonds. Addition reaction polyimides must be held under pressure during the entire curing and post curing operations. Addition reaction polyimides are usually supplied as supported film containing sufficient solvent to impart tack and drape to the adhesive.

17.2.2.4.2 Polybenzimidazole. As illustrated in Fig. 17.3, polybenzimidazole (PBI) adhesives offer the best short term performance at elevated temperatures. However, PBI resins oxidize rapidly and are not recommended for continuous use at temperature over 450°F.

PBI adhesives require a cure at 600°F. Release of volatiles during cure contributes to a porous adhesive bond. Supplied as a very stiff, glass-fabric-reinforced film, this adhesive is also very expensive. Applications are limited by a long, high temperature curing cycle and the cost of the adhesive.

17.2.2.4.3 Bismaleimide. Bismaleimide adhesives are suitable for long term exposure to temperatures up to 400°F and short term exposure up to 450°F. They have excellent electrical properties, making them especially suitable for high energy radomes where low loss transparency to microwave electrical frequencies are important. Like epoxies, almost limitless formulations are possible with bismaleimide adhesives.

Bismaleimide adhesives cure by addition reaction, so no volatiles are given off during cure. This simplifies processing and produces nonporous bondlines with low bonding pressures. These systems are usually cured for several hours at 350°F under contact pressure and then post cured at 400–450°F with or without pressure.

The strength retention of bismaleimide adhesives is very good up to about 570°F, but long term resistance suffers at temperatures above 430°F. Bismaleimide systems are very rigid, and for this reason have low metal peel and sandwich peel properties.

17.2.2.5 Polysulfone thermoplastics.
Polysulfone resins, although thermoplastic, undergo no abrupt changes in mechanical properties between −150°F and 375°F and feature excellent creep resistance at temperatures to 300°F. Because they are thermoplastic and have good thermal stability, high bonding temperatures are required to obtain optimum joints. Polysulfone adhesive resins may be processed at temperatures up to 800°F. The adhesives have good resistance to strong acids, alkalis, oils, greases, and aliphatic hydrocarbons. However, resistance to aromatic hydrocarbons, ketones, esters, and solvents is poor.

Good substrate wetting is required for sulfone adhesives to obtain optimum bonding. Highest bond strengths are obtained if the adhesive is applied as a free film, and the bond is formed under pressure at a temperature of 700°F. A press cycle of 3–7 min at 50–100 psi is generally required.

The polysulfone resin also can be applied from solution. For best results, the adhesive resin is coated on both substrates to a dry film thickness of 1–2 mils and then dried at 400–500°F to remove any solvent. Bonding is then done by heat sealing the coated surfaces together in a press at 400–700°F. Bond strengths are generally lower with the solvent application approach than with the free film method because of possibility of entrapped solvent in the glue line.

17.2.2.6 High temperature cyanoacrylates.
Conventionally cured cyanoacrylate adhesives are thermoplastic, and they will creep on exposure to high temperature and stress. Cyanoacrylates are usually employed in applications requiring service temperatures below 170°F. However, several variations are now available that are useful for short terms up to 475°F. High temperature service has been made possible by the use of crosslinking agents and heat resistant monomers.

Examples of these high temperature cyanoacrylate formulations are 910MHT and 910THT.[10] The former is a low viscosity fast setting version and the latter is medium viscosity and thickened for easier handling. As with regular cyanoacrylate, these variants will bond virtually all types of metals and alloys, plastics, and rubber to each other or to other materials. Shear strength and elevated temperature resistance vary considerably with the type of adhesive and the adherend material. These products have shown shear strengths comparable to low grade epoxies at room temperature and several hundred psi at temperatures to 400°F. Very long term aging is not suggested. However, several days at temperatures to 400°F are possible.

One promising application for these high temperature cyanoacrylate adhesives is the bonding of strain gauges to metal surfaces where high temperatures are expected. The cyanoacrylate bonds well to the substrates and provides high temperature resistance for the short term. Usually strain gauge testing only requires that the strain gauge be attached to the substrate for a short time period at elevated temperatures.

17.2.2.7 Specialty elastomers. Several specialty elastomers have been developed for high temperature service. These sealants have been used primarily in the aerospace industries. However, they have never been widely used because of their high cost and relatively unique applications.

Cyanosilicone sealants have been synthesized for the earliest space shuttles.[11] The polymer is resistant to hot hydrocarbon fuels and has an operating range of −60 to +450°F. Fluoroalkylarylenesiloxanylene (FASIL) sealants have been developed by the Air Force Materials Laboratory for service at temperatures from −65 to +480°F. The elastomer bonds well to titanium and aluminum. Another fluorine containing polymer, phosphonitrilic fluoroelastomer (PNF) has been developed for service from −90 to 350°F.[12,13] The unique mechanical properties provided by these systems include low compression set, high modulus, excellent fatigue, and good abrasion resistance. This product has been commercially developed as an advanced sealant for the aerospace industry.[12] Flexible polyimide resins have been developed for high temperature fuel tank sealants.[14] In addition to excellent resistance to jet fuels, these sealants provide sealing capability from −50 to 500°F.

17.3 Low Temperature

Many applications for adhesives and sealants require service life at very low temperatures. For example, adhesives and sealants for superconducting components in advanced machines should maintain their mechanical performance in the range of −450°F (liquid helium) to −320°F (liquid nitrogen).

There are also certain applications that require the adhesive or sealant not only to resist extremely low temperatures but also to resist extremely high temperatures. These applications exist primarily in the aerospace industry. Several examples are readily evident. Bonded structures on space vehicles could be exposed to cold when oriented away from the sun and to heat when oriented directly at the sun. Ablative tiles on re-entry vehicles require resistance to low temperatures during normal flight and extremely high temperatures when penetrating the atmosphere.

17.3.1 Factors affecting low temperature performance

The factors that determine the strength of an adhesive when used over wide temperature ranges are:

1. The difference in coefficient of thermal expansion between adhesive and adherend
2. The elastic modulus of the adhesive at the service temperature
3. The thermal conductivity of the adhesive

At low service temperatures, the difference in thermal expansion is very important, especially since the elastic modulus of the adhesive generally increases with falling temperature. It is necessary that the adhesive retain some resiliency if the thermal expansion coefficients of the adhesive and adherend cannot be closely matched. The adhesive's thermal conductivity is also important in minimizing transient stresses during cooling. This is why thinner bond thicknesses and adhesives or sealants with higher levels of thermal conductivity have better cryogenic properties than thicker ones.

In addition to external stress, the final bonding strength at a given temperature is affected by internal stress arising mainly from shrinkage during cure and differences in thermal expansion and contraction. Residual stress resulting from thermal expansion or contraction are due to the differences in the thermal expansion coefficient between adhesive and adherend and also due to temperature distribution in the joint due to differences in thermal conductivity.

Other opportunities for stress concentration in bonded joints that may be aggravated by low temperature service include trapped gases or volatiles evolved during bonding and residual stresses in adherends as a result of the release of bonding pressure. These internal stresses are magnified when the adhesive or adherend is not capable of deforming to help release the stress. At room temperature, a standard low modulus adhesive may readily relieve stress concentration by deformation. At cryogenic temperatures, however, the modulus of elasticity may increase to a point where the adhesive can no longer effectively release the concentrated stresses.

17.3.2 Low temperature adhesives and sealants

Most conventional low modulus adhesives and sealants, such as polysulfides, flexible epoxies, silicones, polyurethanes, and toughened acrylics are flexible enough for use at intermediate low temperatures such as −40°F to −100°F. Low temperature properties of common structural adhesives used for applications down to −400°F are illustrated in Fig. 17.5.

17.3.2.1 Modified epoxies. Modified epoxies are often selected for low temperature applications whenever urethanes and silicones are not suitable. Conventional room temperature cured epoxy-polyamide adhesive systems can be made serviceable at very low temperatures by the addition of appropriate fillers to control thermal expansion. However, the epoxy-based systems are not as attractive for low temperature applications as some others because of their brittleness and corresponding low peel and impact strength at cryogenic temperatures. On a basis of lap shear strength at low temperatures (below −67°F), the epoxy formulations are ranked in decreasing order as follows: epoxy-nylon, epoxy-polysulfide, epoxy-phenolic, epoxy-polyamide, and amine cured and anhydride cured epoxy.[15]

Epoxy-nylon adhesives are among the toughest and strongest adhesives and are usually produced as a dry B-staged film. Epoxy-nylon adhesives retain their flexibility and provide 5,000 psi shear strength in the cryogenic temperature range. They also have useful impact resistance properties to −250°F. Peel strengths can be as high as 40 piw, and resistance to vibration and fatigue is excellent. Limitations of the epoxy-nylon adhesives include poor moisture and chemical resistance and the need to provide pressures up to 20 psi and temperatures of up to 350°F during cure.

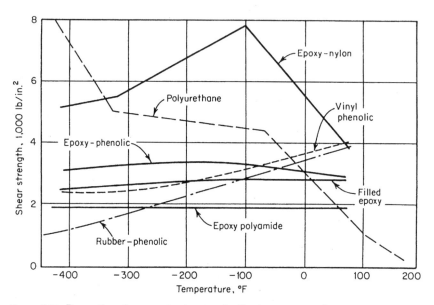

Figure 17.5 Properties of cryogenic structural adhesives systems.[5]

Other useful epoxies for lower temperature service include epoxy-polysulfide, epoxies modified with nitrile butadiene rubber, and epoxy-phenolic. Epoxy-phenolic adhesives are exceptional in that they have good adhesive properties at both elevated and low temperatures.

17.3.2.2 Modified phenolics. These adhesives are generally available as supported and unsupported films, solvent solutions, and solutions with powder. Vinyl acetal phenolic and nitrile-phenolic are examples of two modified phenolics that have been used in low temperature applications.

Vinyl acetal phenolic adhesives maintain fair shear and peel strength at cryogenic temperatures, but strength falls with decreasing temperature because of brittleness in the thermoplastic constituent. Nitrile-phenolic adhesives do not have high strength at low service temperatures because of their relative rigidity. However, these adhesives may find use in certain low temperature applications because of their high peel strengths.

17.3.2.3 Polyurethanes. Polyurethanes have the best low temperature properties of all commercially available adhesives. Some grades of polyurethane offer outstanding cryogenic properties. Polyurethane adhesives are easily processable and bond well to many substrates.

The lap shear strength of a polyurethane bonded aluminum joint at 77°F is 1,800 psi, but this increases to 4,500 psi at −100°F, and increases further to 8,000 psi at −423°F. Elongation of the urethane adhesive reaches nearly 600% under some conditions, and this flexibility is preserved at low temperatures. Excellent peel strength at room temperature is also maintained at low temperatures. Peel strengths range from 22 piw at 75°F to 26 piw at −423°F.[15]

17.3.2.4 Silicones. Silicone adhesives and sealants retain excellent properties over a temperature range from nearly 500°F down to the cryogenic range. In general, the silicone resins are used where organic materials cannot withstand harsh environmental conditions, where superior reliability is required, or where the durability of silicone gives an economic advantage. These adhesives and sealants can form tough bonds at −40°F and do not become brittle at −100°F.

Silicone joints are designed to utilize the good peel strength of the silicone elastomer rather than its tensile or shear properties. Peel strength in bonding rubber to aluminum is from 17–20 piw. However, the lap shear strength of metal to metal bonds with a typical silicone adhesive is only 250–500 psi. Both peel and shear strength increase with decreasing temperatures for silicone adhesives and sealants that have been specifically developed for low temperature applications.

Commercially available room temperature vulcanizing silicone adhesives and sealants have been used at temperatures as low as $-260°F$.

17.3.2.5 Modified acrylics. Thermosetting acrylic resins are considered excellent structural adhesives for metals, thermoplastics, and thermosets at temperatures down to $-40°F$ and under a wide range of severe environmental conditions. Thermosetting acrylics that have been modified for flexibility show excellent properties down to $-100°F$ along with good impact and peel properties.

17.3.2.6 Rubber-based adhesives. There are several rubber-based adhesives and sealants that have useful properties at low temperatures. Included in this group are butyl, neoprene, and polysulfide rubbers. The polysulfides are used where high strength is not mandatory. They are used more frequently as sealants. They have good resistance to oxygen, oil, and solvents and are impermeable to many gases. The polysulfides retain their flexibility to $-80°F$.

Two-part neoprene adhesives can be used for bonding neoprene, Hypalon, urethane, rigid polyvinyl chloride, and other rubber used in low temperature service. Neoprene adhesives are useful to $-60°F$, and it can be used for bonding many substrates including textiles.

Certain butyl rubber compounds have useful flexibility down to $-65°F$. They are typical of the rubber-based adhesives in that they are more suitable as sealants because they lack the necessary cohesive strength for adhesives. However, adhesion is excellent, and elongation values are on the order of 300–500%.

17.3.2.7 Polyaromatic resins. Heat resistant polyaromatic adhesives, such as the polyimides and polybenzimidazoles, also show excellent low temperature properties. The shear strength of a polybenzimidazole adhesive on stainless-steel substrates is 5,690 psi at a test temperature of $-423°F$. Polyimide adhesives have exhibited shear strength of 4,100 psi at $-320°F$. In particular, the polyimide polymers indicate good resistance to thermal shock at cryogenic temperatures. This is thought to be due to a secondary glass transition temperature in the cryogenic range.

17.3.2.8 Sealants. Most conventional sealants such as polysulfides, flexible epoxies, silicones, polyurethanes, and toughened acrylics, do not require extreme low temperature flexibility. They are generally flexible enough for use at temperatures to $-20°F$. For aerospace applications, the sealants often need to operate both at high and low temperatures. In these applications, the high temperature specialty sealant materials described earlier (cyanosilicones, fluoroalkylarylenesiloxanylene, phosphonitrilic fluoroelastomer and flexible polyimides) have served the purpose. However, the cost and special nature of these sealants preclude their use in more general purpose applications.

17.4 Humidity and Water

Water is the substance that gives the greatest problems in terms of environmental stability for many adhesive joints. Water is a problem because it is very polar and permeates most polymers. Other common fluids, such as lubricants and fuels, are of low or zero polarity and do not permeate and weaken thermosetting adhesive or sealant joints significantly. Moisture can affect an adhesive bond in three distinctive ways.

- Moisture can degrade the properties of the bulk adhesive or sealant itself.
- Moisture can degrade the adhesion properties at the interface.
- Moisture can also degrade the properties and cause dimension changes of certain adherends.

17.4.1 Effect on the bulk materials

Internal degradation within the bulk adhesive or sealant occurs primarily by absorption of water molecules into the polymer structure. All polymers will absorb water to some extent. Moisture can also enter by wicking along the adhesive–adherend interface or by wicking along the interfaces caused by reinforcing fibers and the resin. Moisture can alter the properties of the bulk material by reducing its glass transition temperature, inducing cracks, or by chemically reacting with the polymer—a process called hydrolysis.

The water ingress properties of various polymers can be predicted by values of their permeability coefficient and the diffusion constant of water (Table 17.6). The permeability coefficient is defined as the amount of vapor at standard conditions permeating a sample that is 1 cm sq. and 1 cm thickness within 1 sec with a pressure difference of 1 cm Hg across the polymer. The diffusion coefficient is a measure of the ease at which a water molecule can travel in a polymer. Both epoxies and phenolics show relatively low diffusion rates and are less susceptible to moisture attack than most other polymers. As a result, these materials are often used in adhesive and sealant formulations where resistance to moisture is essential.

Water permeation in polymers generally lowers the glass transition temperature of the polymer by reducing the forces between molecules. Data for certain epoxy adhesives are given in Table 17.7. The effect of

TABLE 17.6 Permeability Coefficients (P) and Diffusion Constants (D) of Water in Various Polymers[16]

Polymer	Temperature °C	Temperature °F	$P \times 10^{-9}$	$D \times 10^{-9}$
Vinylidene chloride/acrylonitrile copolymer	25	75	1.66	0.32
Polyisobutylene	30	85	7–22	...
Phenolic	25	75	166	0.2–10
Epoxy	25	75	10–40	2–8
Epoxy/tertiary amine	20	70	...	2.4
	40	105	...	6.5
	60	140	...	18.1
	90	195	...	60.7
Polyvinyl chloride	30	85	15	16
Polymethyl methacrylate	50	120	250	130
Polyethylene (low density)	25	75	9	230
Polystyrene	25	75	97	...
Polyvinyl acetate	40	105	600	150

TABLE 17.7 Effect of Water Immersion on Glass Transition Temperatures of Epoxy Adhesives Based on DGEBA[17]

Hardener	Dry	T_g(°C) After initial uptake	After 10 months
DAPEE	67	37	49
TETA	99	86	111
DAB	161	143	157
DDM	119	110	130

absorbed water on the mechanical properties of cured adhesives is shown in Table 17.8. Water lowers tensile strength and modulus but increases elongation at break. These properties generally recover fully when the polymer is dried unless irreversible hydrolysis has taken place.

Some polymeric materials, notably ester-based polyurethanes, will chemically change or "revert" on continuous exposure to moisture. Reversion or hydrolysis causes the adhesive or sealant to lose hardness, strength, and in the worst cases transform to a fluid during exposure to warm, humid air. Figure 17.6 illustrates degradation of polymer chains by hydrolytic reaction with water. The rate of reversion or hydrolytic instability depends on the chemical structure of the base adhesive, the type and amount of catalyst used, and the permeability of the adhesive. Certain chemical linkages such as ester, urethane, amide, and urea can be hydrolyzed. The rate of attack is fastest for ester-based linkages. Ester linkages are present in certain types of polyur-

TABLE 17.8 Effect of Water on Mechanical Properties of Structural Adhesives[17]

Exposure conditions	Weight gain (%)	Tensile strength (MPa)	Elongation at break (%)	Modulus (MPa)	Failure mode
FM1000 epoxide–polyamide					
None		73	5.0	1880	brittle
3 months at 65% r.h.	2.9	52	263	623	ductile
5 d in water at 50°C	9.4	19	260	3.0	rubbery
5 d in water at 50°C, then dry at 60°C for 2 d	3.3	76	5.7	1980	brittle
DGEBA/DAPEE epoxide					
None		41	7.1	1700	ductile
24 h in water at 100°C		24	37	1020	ductile
24 h in water at 100°C, then dry at 65°C for 2 d		52	6.8	1560	ductile

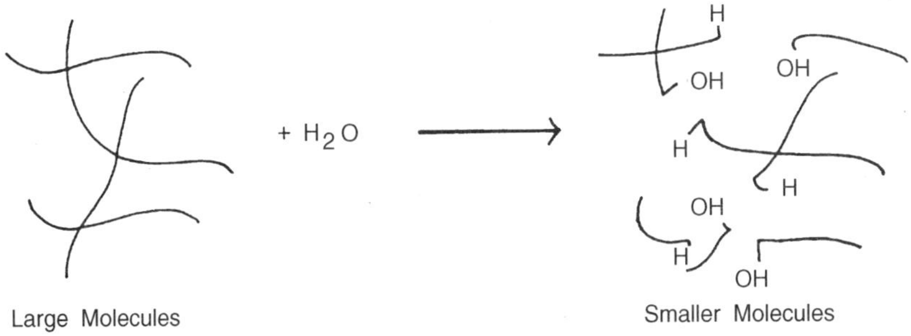

Large Molecules   Smaller Molecules

Figure 17.6 The degradation of polymer chains by reaction with water hydrolysis.[18]

ethanes and anhydride-cured epoxies. Generally, amine-cured epoxies offer better hydrolytic stability than anhydride-cured types. The reversion rate also depends on the amount of catalyst used in the formulations. Best hydrolytic properties are obtained when the proper stoichiometric ratio of base material to catalyst is used. Reversion is usually much faster in flexible materials because water permeates it more easily.

Figure 17.7 illustrates the hydrolytic stability of various polymeric materials determined by a hardness measurement after exposure to high relative humidity. This technique for comparing hydrolytic stability is known as the *Gahimer and Nieske procedure*.[20] A time period of 30 days in the 100°C (212°F), 95% RH test environment corresponds roughly to a period from 2–4 years in a hot, humid climate such as Vietnam. The hydrolytic stability of urethane potting compounds was not believed to be a problem until it resulted in the failure of many potted electronic devices during the 1960s military action in Vietnam.

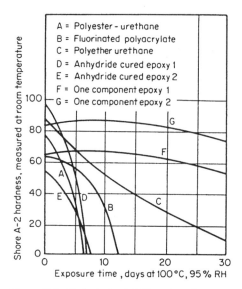

Figure 17.7 Hydrolytic stability of potting compounds. Materials showing rapid loss of hardness in this test soften similarly after 2–4 years in high temperature, high humidity climate zones.[19]

Hydrolysis reaction mechanisms have been recognized for epoxy, polyurethane, and cyanoacrylate adhesives. In the case of conventional construction sealants, polysulfides, polyurethanes, epoxies, and acrylics have all shown sensitivity to moisture. Hydrolysis causes the breaking of molecular bonds within the sealant. Thus, the strength decreases and cohesive failure results. However, before this occurs the sealant usually swells and may cause deformation or bond failure before hydrolysis can completely act. For electronic sealants, it is highly desirable to keep moisture from penetrating into critical areas. Hydrophobic polymers have been developed to accomplish this task. Such polymers "repel" moisture. They are siloxyimides, fluorosilicones, fluoroacrylics, phenylated silicone, and silastyrene.

17.4.2 Effect on the Interface

Water can also permeate the adhesive and preferentially migrate to the interface region displacing the adhesive at the bond interface. This mechanism is illustrated in Fig. 17.8. It is the most common cause of adhesive strength reduction in moist environments. Thus, structural adhesives not susceptible to the reversion phenomenon may also lose adhesive strength when exposed to moisture. The degradation curves shown in Fig. 17.9 are typical for an adhesive exposed to moist,

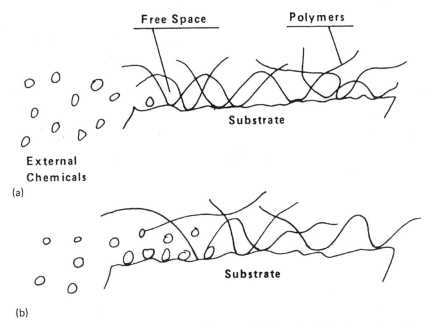

Figure 17.8 Competition between an adhesive and other chemicals for surface sites leading to displacement of the adhesive from the surface. (*a*) Adhesive adsorbed at surface sites. (*b*) Adhesive displaced from the surface sites.[18]

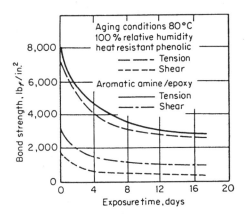

Figure 17.9 Effect of humidity on adhesion of two structural adhesives to stainless steel.[21]

high-temperature environments. The mode of failure in the initial stages of aging usually is truly cohesive. After 5–7 days, the failure becomes one of adhesion. It is expected that water vapor permeates the adhesive through its exposed edges. The water molecules are preferentially adsorbed and concentrate on the metal adherend, thereby, displacing the adhesive at the interface. This effect is greatly dependent on the type of adhesive and the adherend material.

Certain adhesive systems are more resistant to interfacial degradation by moist environments than other adhesives. Table 17.9 illustrates that a nitrile-phenolic adhesive does not succumb to failure through the mechanism of preferential displacement at the interface. Failures occurred cohesively within the adhesive even when tested after 24 months of immersion in water. A nylon-epoxy adhesive, however, degraded rapidly under the same conditioning owing to its permeability and preferential displacement by moisture. Adhesive strength deteriorates more quickly in a 100% RH environment than in liquid water because of more rapid permeation of the vapor.

The shape of the curve in Fig. 17.9 is common for most adhesives being weakened on exposure to wet surroundings. Strength falls most rapidly at the beginning of aging and then slows down to a low or zero rate of weakening. The initial rate and overall percent degradation will vary with the adhesive and surface treatment.

There also appears to exist a critical water concentration within the adhesive below which water induced damage of the joint may not occur. This also infers that there is a critical humidity for weakening of an adhesive joint. For an epoxy system, it is estimated that the critical water concentration is about 1.35–1.45%, and that the critical humidity is 50–65%.[23,24]

Another way moisture can degrade the strength of adhesive joints is through hydration or corrosion of the metal oxide layer at the interface. Common metal oxides, such as aluminum and iron oxides, undergo hydration. The resulting metal hydrates become gelatinous, and they act as a weak boundary layer because they are poorly bonded

TABLE 17.9 Effect of Humidity and Water Immersion on the Shear Strength (psi) of Two Structural Adhesives[22]

Exposure time, months	Nylon/epoxy adhesive		Nitrile/phenolic adhesive	
	Humidity cycle*	Water immersion	Humidity cycle*	Water immersion
0	4,370	4,370	3,052	3,052
2	1,170	2,890	2,180	2,740
6	950	1,700	2,370	2,280
12	795	500	2,380	2,380
18	1,025	200	2,350	2,640
24	850	120	2,440	2,390

Substrate aluminum treated with a sulfuric acid-dichromate etch.
*Humidity cycle of 93% RH between 149 and 89°F with a cycle time of 48 h.

to their base metals. Resistance of adhesive joints to moisture can be improved either by preventing water from reaching the interface or by improving the durability of the interface itself. The first method of preventing degradation can be achieved through formulation of the adhesive system. These possibilities include:

1. Proper selection of a base polymer having low water permeability
2. Incorporating inert fillers into the adhesive to lower the volume that can be affected by moisture
3. Coating the exposed edges of the joint with sealants having very low permeabilities to inhibit wicking of moisture along the interface
4. Chemically modifying the adhesive to reduce water permeation

Methods that are available to improve the durability of the interface center on surface preparation and use of primers or coupling agents. These are most beneficial on metallic surfaces that are prone to degradation by corrosion. Primers and surface treatments tend to hinder adhesive strength degradation in moist environments by providing corrosion protection to the adherend surface. A fluid primer that easily wets the interface presumably also tends to fill-in minor discontinuities on the surface. Organosilane, organotitanate, and phenolic primers have been found to improve the bond strength and permanence of many adhesive systems.

Approaches for the development of water resistant surface treatments include application of inhibitors to retard the hydration of oxides or the development of highly crystalline oxides as opposed to more amorphous oxides. Standard chemical etching procedures, which remove surface flaws, also result in improved resistance to high humidity.

17.4.3 Combined effects of stress, moisture, and temperature

Mechanical stress accelerates the effect of the environment on the adhesive joint. A great amount of data is not available on this phenomenon for specific adhesive systems because of the time and expense associated with stress-aging tests. However, it is known that moisture, as an environmental burden, markedly decreases the ability of certain adhesive to bear prolonged stress especially at slightly elevated temperatures. The interaction of temperature and moisture causes greater degradation than can be attributed to either environment by itself.

This effect was noted in the 1960s, when stressed and non-stressed aluminum lap shear joints were aged in a natural weathering environment in Florida.[25] Stress was applied by flexural bending of lap shear samples and keeping them in that state during the aging period. Depending on the type of adhesive, there was a significant degradation after 1–2 years due to stress-weathering, whereas the joints that were aged in the non-stressed condition showed little degradation.

In high humidity environments, stressed joints weaken more rapidly than unstressed joints (Fig. 17.10). Joints made with a flexible adhesive having low glass transition temperature failed by creep of the adhesive at relatively short service times. Figure 17.11 illustrates the effect of stress aging on specimens exposed to humidity cycling from 90–100% R.H. and simultaneous temperature cycling from 80–120°F. The loss of load-bearing ability of a flexibilized epoxy adhesive (Fig. 17.11b) is exceptionally great. The stress on this particular adhesive had to be reduced to 13% of its original static strength in order for the joint to last a little more than 44 days in the high temperature, high humidity environment.

Table 17.10 shows the adverse effect of stress and tropical climate on aluminum joints bonded with various adhesives. However, it also shows that accelerated exposure can be very misleading. If a 30-day

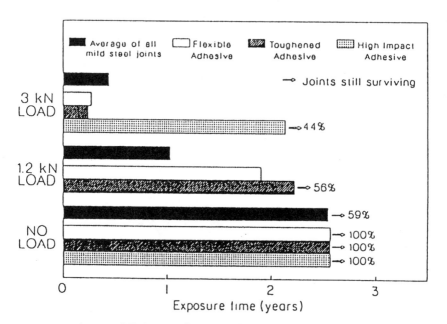

Figure 17.10 Average failure times for stressed and unstressed zinc-nickel coated steel joints in a tropical environment.[26]

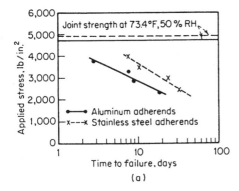

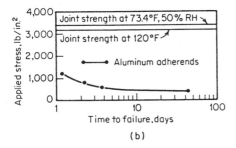

Figure 17.11 Time to failure vs. stress for two adhesives in a warm, high humidity environment. (a) One part, heat cured modified epoxy adhesive. (b) Flexibilized amine cured epoxy adhesive.[27]

salt water spray is used as an accelerated test to determine the long term performance of these adhesives in a tropical climate, it could lead to the wrong conclusions. Salt water spray had very little effect on the strength of stressed or unstressed joints with the exception of one acrylic adhesive. However, the stressed specimens in Florida almost all went to zero strength except for the vinyl phenolics. Panama was not nearly as severe. These data illustrate that permanence or durability must be tested in the specific intended environment.

Several sources seem to come to general agreement as to the relative durabilities of structural adhesives. Results of sustained load durability testing and outdoor weathering studies provide the same order for the relative durabilities of different adhesive classes. These are summarized in Table 17.11. However, such a ranking of performance should only be taken as an approximate guide, since it is difficult to make reliable predictions about the performance of an adhesive in any general way.

TABLE 17.10 Comparison of Long Term and Accelerated Exposures[28]

Designation	Chemical type	Cure time and temperature	Strength retention, percent					
			Florida (3 years)		Panama (3 years)		Salt water spray (30 days)	
			Stress	No stress	Stress	No stress	Stress	No stress
Redux 775	Vinyl phenolic	30 min @ 340°F	60	97	87	83	97	95
FM-47	Vinyl phenolic	125 min @ 350°F	0	78	95	97	100	103
EC-1471	Vinyl phenolic	60 min @ 350°F	0	62	75	96	97	100
Swedlow 371 W	Acrylic	30 min @ 340°F	19	79	72	105	104	103
Griffabond P262A	Acrylic	30 hr @ 77°F	0	24	15	54	16	94
EC 1469	One part epoxy	60 min @ 350°F	0	57	0	79	106	120

8 × 9 in. panels stressed by mounting on a steel bending frame to get 0.25 in. deflection at center of a 6 in. span; ½ in. overlap is at center of span.

TABLE 17.11 Relative Durabilities of Structural Adhesives[29]

Most durable	350°F-cure film
	nitrile-phenolic, vinyl-phenolic, novolac-epoxy
	250°F-cure film
	modified epoxy, nitrile-epoxy, nylon-epoxy
	Heat-cure paste
	nitrile-epoxy, nylon-epoxy, vinyl-epoxy
Least durable	RT-cure paste
	epoxy/polyamide, epoxy/anhydride

17.5 Outdoor Weathering

Adhesives and sealants are generally not significantly affected by moderate outdoor weathering. However, there are certain circumstances that could affect the permanence of joints exposed to outdoor service. It is important that these be considered early in the design of the adhesives joint and selection of materials.

By far the most detrimental factors influencing adhesives aged in a non-sea coast environment are heat and humidity. The reasons why warm, moist climates degrade many adhesive joints were presented in the last section. Near the sea coast, however, salt water and salt spray must also be considered when designing an adhesive joint. Thermal cycling, oxygen, ultraviolet radiation, and cold are relatively minor factors with most structural adhesives.

17.5.1 Non-sea coast environment

When exposed to weather, structural adhesives may rapidly lose strength during the first 6 months to 1 year. After 2–3 years, however, the rate of decline usually levels off at strength that is 25–50% of the initial joint strength depending on the climate zone, adherend, adhesive, and stress level. Adhesive systems that are formulated specifically for outdoor applications show little strength degradation over time in a moderate environment. Figure 17.12 shows the weathering characteristics of unstressed epoxy adhesives to the Richmond, VA climate region.

The following generalizations are of importance in designing a joint for outdoor service.[30]

1. The most severe locations are those with high humidity and warm temperatures.

2. Stressed panels deteriorate more rapidly than unstressed panels.

3. Stainless-steel panels are more resistant than aluminum panels because of corrosion.

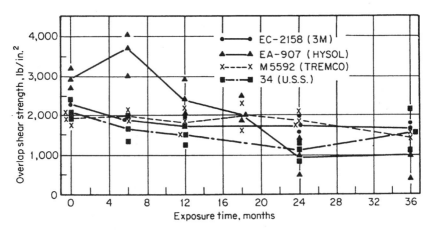

Figure 17.12 Effect of outdoor weathering on typical aluminum joints made with four different two part epoxies cured at room temperature.[30]

4. Heat-cured adhesive systems are generally more resistant than room-temperature-cured systems.

5. For the better adhesives, unstressed bonds are relatively resistant to severe outdoor weathering, although all joints will eventually exhibit some strength loss.

MIL-STD-304 is a commonly used accelerated-exposure test method to determine the effect of weathering and high humidity on adhesive specimens.[31] However, only relative comparisons can be made with this type of test; it is not possible to extrapolate the results to actual service life. In this test procedure, bonded panels are exposed to alternating cold (−65°F) and heat and humidity (160°F, 95% RH) for 30 days. The effect of MIL-STD-304 conditioning on the joint strength of common structural adhesives is presented in Table 17.12.

17.5.2 Sea coast environment

For most bonded metal joints that see outdoor service, corrosive environments are a more serious problem than the influence of moisture. The degradation mechanism is corrosion of the metal interface resulting in a weak boundary layer. Surface preparation methods and primers that make the adherend less corrosive are commonly employed to retard the degradation of adhesive joints in these environments.

The quickest method to measure corrosion dominated degradation in bonded metal joints is to store the bonded parts in a salt spray chamber with a continuous 5% NaCl salt solution spray in 95% RH at 95°F. This is a procedure described in ASTM B 117. Usually only

TABLE 17.12 Effect of MIL-STD-304 on Bonded Aluminum Joints[32]

Adhesive	Shear, lb/in.^{2}, 73°F		Shear, lb/in.^{2}, 160°F	
	Control	Aged	Control	Aged
Room-temp. cured:				
Epoxy-polyamide....................	1,800	2,100	2,700	1,800
Epoxy-polysulfide	1,900	1,640	1,700	6,070
Epoxy-aromatic amine...............	2,000	Failed	720	Failed
Epoxy-nylon........................	2,600	1,730	220	80
Resorcinol epoxy–polyamide	3,500	3,120	3,300	2,720
Epoxy-anhydride....................	3,000	920	3,300	1,330
Polyurethane.......................	2,600	1,970	1,600	1,560
Cured 45 min at 330°F, epoxy–phenolic..	2,900	2,350	2,900	2,190
Cured 1 h at 350°F:				
Modified epoxy	4,900	3,400	4,100	3,200
Nylon–phenolic.....................	4,600	3,900	3,070	2,900

several hundred hours of exposure are needed to show significant differences in corrosion resistance of various adhesive joint systems. Figure 16.3 showed the shear strength of mild steel joints with various adhesives before and after exposure for 200 h in a salt spray test.

Resistance of the adhesive joint to salt climates depends not only on the type of adhesive, but on the method of surface preparation and on the type of primer used. Durability of bonded aluminum alloy having different types of anodize treatment and aged in 3.5% salt spray is shown in Table 17.13. The good bond durability of the anodized surface pretreatment has been shown by several studies.[34]

17.6 Chemicals and Solvents

Some organic adhesives tend to be susceptible to chemicals and solvents, especially at elevated temperatures. Most standard tests to determine chemical resistance of adhesive joints last only 30 days or so. Practically all adhesives are resistant to these fluids over short time periods and at room temperatures. Some epoxy adhesives even show an increase in strength during aging in fuel or oil over these time periods. This effect is possibly due either to postcuring or plasticizing of the epoxy by the oil.

There are two properties of adhesive joints that protect them from exposure to chemical or solvent environments: high degree of crosslinking and low exposure area.

1. Most crosslinked thermosetting adhesives such as epoxies, phenolics, polyurethanes, and modified acrylics are highly resistant to

TABLE 17.13 Effect of Surface Treatment and Exposure to 3.5% Salt Water Intermittent Spray on the Durability of 6061-T6 Aluminum Alloy Joints Exposed in the Unstressed Condition (nitrile modified epoxy paste adhesive)[33]

Surface treatment	Initial shear strength		Average percent retained bond strength after indicated exposure time (months)*			
	psi	MPa	3	6	9	12
Chromic acid anodize	5513	38.0	>50	>50	>50	82.0
Phosphoric acid anodize (30 volts, 30 min)	5700	39.3	>50	>50	>50	89.8
Phosphoric acid anodize (60 volts, 18.5 min)	6030	41.6	>50	>50	>50	92.9
Phosphoric acid anodize (110 volts, 5 min)	6070	41.9	>50	>50	>50	95.9
Phosphoric acid anodize (Boeing procedure)	6480	44.7	>50	>50	>50	79.3
Sulfuric acid anodize (12 asf, 60 min unsealed)	3940	27.2	>50	>50	>50	75.1
Sulfuric acid anodize (12 asf, 60 min, boil-water seal)	3550	24.5	>50	>50	>50	49.6

*In this procedure the specimens were stressed weekly to 50% of initial shear strength and then returned to the exposure conditions, providing no bond failure occurred. After 52 weeks testing, the joints were deliberately failed for quantitative determination of the actual bond strength as shown.

many chemicals at least at temperatures below their glass transition temperature.

2. The adhesive bond-line is usually very thin and well protected from the chemical itself. This is especially true if the adherends are non-porous and non-permeable to the chemical environments in question.

Adhesive joint designers will take advantage of this latter effect by designing the joint configuration for maximum protection or by specifying a protective coating and/or sealant around the exposed edges of the adhesive.

Figure 17.13 shows the long term effect to various chemical environments of aluminum joints bonded with a heat cured epoxy adhesive. As can be seen, the temperature of the immersion medium is a significant factor in the aging of the adhesive. As the temperature increases, more fluid is generally absorbed by the adhesive, and the degradation rate increases.

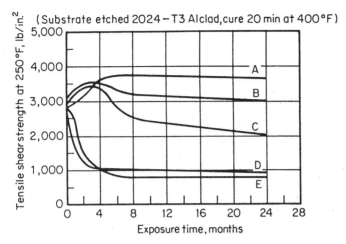

Figure 17.13 Effect of immersion in various chemical environments on a one part heat curing epoxy adhesive (EA929, Hysol Corp.). (a) gasoline at 75°F, (b) gear oil at 250°F, (c) distilled water at 75°F, (d) tapwater at 212°F, (e) 38% Shellzone at 250°F.[35]

Epoxy adhesives are generally more resistant to a wide variety of liquid environments than other structural adhesives. However, the resistance to a specific environment is greatly dependent on the type of epoxy curing agent used. Aromatic amine (e.g., metaphenylene diamine) cured systems are frequently preferred for long term chemical resistance. Urethane adhesives generally show good resistance to most chemicals, solvents, oils, and greases. There is no universal "best adhesive" for all chemical environments. As an example, maximum resistance to bases almost axiomatically means poor resistance to acids. It is relatively easy to find an adhesive that is resistant to one particular chemical environment. It becomes more difficult to find an adhesive that will not degrade in two widely differing chemical environments. Generally, adhesives that are most resistant to high temperatures have good resistance to chemicals and solvents because of their dense crosslinked molecular structure.

With regard to aging of adhesive joints in chemical environments, it can be summarized that:

1. Chemical resistance tests are not uniform, in concentrations, temperature, time, or properties measured.
2. Generally, chlorinated solvents and ketones are severe environments.
3. High boiling point solvents, such as dimethylforamide and dimethyl sulfoxide, are severe environments.

4. Acetic acid is a severe environment.

5. Amine curing agents for epoxies are poor in oxidizing acids. Anhydride curing agents are poor in caustics.

ASTM D 896, "Standard Test Method for Resistance of Adhesive Bonds to Chemical Reagents", specifies the testing of adhesive joints for resistance to solvents and chemicals. Standard chemical reagents are listed in ASTM D 543, and standard oils and fuels are identified in ASTM D 471. Standard test fluids and immersion conditions that are used by many adhesive suppliers are also specified in MMM-A-13[36] (Table 17.14).

17.7 Vacuum and Outgassing

Adhesives or sealants may be composed of low molecular weight constituents that can be extracted from the bulk when exposed to a vacuum. If these low molecular weight constituents also have a low vapor pressure, they may migrate out of the bulk on exposure to elevated temperatures with or without the presence of a vacuum. This results in an overall weight loss and possible degradation of the adhesive or sealant. The ability for an adhesive to withstand long periods of exposure to a vacuum is of primary importance for materials used in space applications or in the fabrication of equipment that requires a vacuum for operation. The outgassed constituents can also become a source of contamination and be highly objectionable in certain applications such as with electronic products, optical equipment, and solar arrays.

The degree of adhesive evaporation is a function of the vapor pressure of its constituents at a given temperature. Loss of low-molecular-weight constituents such as plasticizers or diluents could result in hardening and porosity of adhesives or sealants. Since most structural adhesives are relatively high-molecular-weight polymers, exposure to pressures as low as 10^{-9} torr is not harmful to the base resin. How-

TABLE 17.14 Standard Test Fluids and Immersion Conditions for Adhesive Evaluation per MMM-A-132[36]

Conditions, 77°F after
30 days in RT water
30 days in 110°F humid cabinet (100% RH)
30 days in 95°F salt-spray cabinet (5% salt)
7 days in JP-4 jet-engine fuel
7 days in anti-icing fluid (isopropyl alcohol)
7 days in hydraulic oil (MIL-H-5606)
7 days in HC test fluid (70/30 v/v isooctane/toluene)

ever, high temperatures, radiation, or other degrading environments may cause the formation of low molecular weight fragments that tend to bleed-out of the adhesive in a vacuum.

Epoxy and polyurethane adhesives are not appreciably effected by 10^{-9} torr for 7 days at room temperature. However polyurethane adhesives exhibit significant outgassing when aged under 10^{-9} torr at 225°F.[33] Table 17.15 shows results from a study that indicates that under room temperature conditions a high vacuum does not cause significant weight loss in commercial adhesive and sealant materials.

17.8 Radiation

High-energy particulate and electromagnetic radiation including neutron, electron, and gamma radiation have similar effects on organic adhesives. Radiation of sufficient energy causes molecular chain scission of polymers used in adhesives and sealants. This results in weakening and embrittlement of the bond. The degradation is worsened when the adhesive is simultaneously exposed to both elevated temperatures and radiation. ASTM D 1879, "Standard Practice for Exposure of Adhesive Specimens to High Energy Radiation" specifies methods to evaluate resistance to radiation.

Figure 17.14 illustrates the effect of radiation dosage on the tensile shear strength of several structural adhesives. Generally, heavily crosslinked, heat-resistant adhesives have been found to resist radiation better than less thermally stable systems. Fibrous reinforcement, fillers, curing agents, and reactive diluents will also affect the radiation resistance of adhesive systems. In epoxy-based adhesives, aromatic curing agents offer greater radiation resistance than aliphatic-type curing agents.

Polyester resins and anaerobic adhesives and sealants have high radiation resistance. Anaerobic adhesives have several years of long term exposure in radiation environments due to their use as thread locking sealants in nuclear reactors and accessory equipment. Thread

TABLE 17.15 Effect of 10^{-7} torr on Commercial Adhesives and Sealants[37]

Adhesive	Type	Weight change (%)	Moisture change (%)
Lefkoweld 109	Modified epoxy	−0.03	+0.60
EC2216 B/A	Flexibilized epoxy	−0.06	+0.61
Adiprene L-100 + MOCA	Polyurethane	+0.01	+0.38
PR 1535	Polyurethane	+0.01	+0.44
EC 1605	Polysulfide	−0.23	+0.39

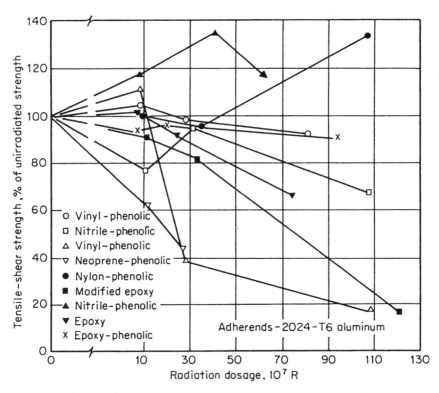

Figure 17.14 Percent change in initial tensile shear strength caused by nuclear radiation dosage.[38]

locking grades of anaerobic adhesives have sustained over 2×10^7 rads without molecular change or loss of locking torque.[39]

In an early study by McCrudy and Rambosek,[40] radiation does not appear to have serious effect on the tensile shear strength of highly crosslinked adhesives. Nitrile-phenolic adhesives are more resistant to radiation damage than epoxy based adhesives. The study also showed that thick adhesive layers retain useful strength better than thin glue lines in a radiation environment. Ten mils is recommended as the minimum glue line thickness when radiation is an environmental factor.

References

1. Weggemans, D. M., "Adhesives Charts", in *Adhesion and Adhesives,* vol. 2 (Amsterdam: Elsevier, 1967).
2. Lee, L. H., "Adhesives and Sealants for Severe Environments", *Int. J. of Adhesion and Adhesives,* April 1987.
3. Krieger, R. B., and Politi, R. E., "High Temperature Structural Adhesives", in *Aspects of Adhesion,* vol. 3, D. J. Alner, ed. (London: University of London Press, 1967).

4. "High Temperature Adhesives are Easy to Apply, Usable Up to 700F", *Materials Engineering,* May 1973.
5. Kausen, R. C., "Adhesives for High and Low Temperatures", *Materials Engineering,* August–September 1964.
6. Burgman, H. A., "The Trend in Structural Adhesives", *Machine Design,* November 21, 1963.
7. Licari, J. J., "High Temperature Adhesives", *Production Engineering,* December 7, 1964.
8. Shaw, S. J., "High Temperature Polymers for Adhesive and Composite Applications", *Materials Science and Technology,* August 1987.
9. Schwartz, M. M., *Joining of Composite Matrix Materials,* Chapter 2, (Materials Park, OH: ASM International, 1994).
10. Eastman Chemical Co., Kingsport, TN.
11. Singh, H., "Cyanosilicone Sealants Used for Space Shuttle Orbiter", *Adhesives Age,* 21, 4, 1978.
12. "PNF Elastomer by Firestone", Technical Publication, Firestone Tire and Rubber Co., October 1979.
13. Lohr, D. F., and Becham, J. A., "PNF Phosphonitrilic Fluoroelastomer: Properties and Applications", Technical Publication, Firestone Tire and Rubber Co.
14. Jones, R. F., and Casey, H. N., "Flexible Polyimide Fuel Tank Sealants", *Adhesives Age,* 22, 11, 1979.
15. Miska, K. H., "Which Low Temperature Adhesive is Best for You?", *Materials Engineering,* May 1975.
16. Sung, N. H., "Moisture Effect on Adhesive Joints", *Adhesives and Sealants,* vol. 3, Engineered Materials Handbook, ASM International, 1990.
17. Comyn, J. *Adhesion Science,* Chapter 10, (Cambridge, UK: Royal Society of Chemistry, 1977).
18. Schenberger, G. L., "Polymer Structure and Adhesive Behavior", in *Adhesives in Manufacturing,* G. L. Schneberger, ed. (New York: Marcel Dekker, 1983).
19. Bolger, J. C., "New One Part Epoxies are Flexible and Reversion Resistant", *Insulation,* October 1969.
20. Gahimer, F. H., and Nieske, F. W., "Navy Investigates Reversion Phenomena of Two Elastomers", *Insulation,* August 1968.
21. Falconer, D. J., et al., "The Effect of High Humidity Environments on the Strength of Adhesive Joints", *Chem. Ind.,* July 4, 1964.
22. DeLollis, N. J., and Montoya, O., " Mode of Failure in Structural Adhesive Bonds", *J. Appl. Polymer Science,* vol. 11, at 983–989, 1967.
23. Brewis, D. M., et al., *Int. J. of Adhesion and Adhesives,* 10, 247, 1990.
24. Kinloch, A. J., "Interfacial Fracture: Mechanical Aspects of Adhesion Bonded Joints", Review, *Journal of Adhesion,* 10, 1979, at 193.
25. Carter, G. F., "Outdoor Durability of Adhesive Joints Under Stress", *Adhesives Age,* 10, 32, 1967.
26. Davis, R. E., *Int. J. Adhesion Adhesives,* 13, 97, 1993.
27. Sharpe, L. H., "Aspects of the Permanence of Adhesive Joints", in *Structural Adhesive Bonding,* M. J. Bodnar, ed. (New York: Interscience, 1966).
28. Olson, W. Z., et al., "Resistance of Adhesive Bonded Metal Lap Joints to Environmental Exposure", Report No NADC TR59-564, October 1962. Also in DeLollis, N. J., "Durability of Structural Adhesive Bonds: A Review", *Adhesives Age,* September 1977.
29. Hartshorn, S. R. "The Durability of Structural Adhesive Joints", in *Structural Adhesives,* S. R. Hartshorn, ed. (New York: Plenum, 1986).
30. *Adhesive Bonding Aluminum,* Reynolds Metals Co., 1966.
31. MIL-STD-304, Department of Defense, Washington, DC.
32. Tanner, W. C., "Adhesives and Adhesion in Structural Bonding for Military Material", in *Structural Adhesive Bonding,* M. J. Bodnar, ed. (New York: Interscience, 1966).
33. Landrock, A. H., "Effect of Environment on Durability of Adhesive Joints", Chapter 9 of *Adhesives Technology Handbook* (Park Ridge, NJ: Noyes Publications, 1985).

34. Minford, J. D., "Comparison of Aluminum Adhesive Joint Durability as Influenced by Etching and Anodizing Testaments of Bonded Surfaces", *J. of Applied Polymer Science,* Applied Polymer Symposia, 32, 1977, at 91–103.
35. Aerospace Adhesive EA 929, Hysol Division, Dexter Corp. Tech. Bull. A5-129.
36. MMM-A-132, "Adhesives, Heat Resistant, Airframe Structural, Metal to Metal", Department of Defense, Washington, DC.
37. Resealed, L. M., "Structural Adhesives in Space Applications", *Structural Adhesive Bonding,* M. J. Bodnar, ed. (New York: Interscience, 1966).
38. Arlook, R. S., and Harvey, D. G., "Effect of Nuclear Radiation on Structural Adhesives Bonds", Wright Air Development Center, Report WADC-TR-46-467, February 1957.
39. Pearce, M. B., "How to Use Anaerobics Successfully", Applied Polymer Symposia No. 19, *Symposium on Processing for Adhesive Bonded Structures*, M. J. Bodnar, ed. (New York: Interscience Publisher, 1972) at 207–230.
40. McCrudy, R. M., and Rambosek, G. M., "The Effect of Gamma Radiation on Structural Adhesive Joints", SAMPE National Symposium on the Effects of Space Environments on Materials, St. Louis, MO, May 1962.

Chapter 18

Production Processes and Equipment

18.1 Introduction

The production processes that are used to convert the liquid adhesive or sealant into a practical working joint are important considerations in the overall design of the bonding system. These processes include storage of the materials, metering and mixing, application, fixturing of parts, and setting of the adhesive or sealant material. The multiple methods available can represent both an advantage and a hindrance to the end-user. There is a great likelihood that the user will find a cost effective process for the specific application. However, choosing the wrong process can have just as dire consequences as choosing the wrong adhesive.

Deliberations concerning the exact processes and equipment to be employed are, unfortunately, often left to the end of the decision making process. Selection of the substrate, adhesive, joint design, and surface pre-treatment are often considered first. However, considerations regarding the application and processing of the adhesive or sealant should be made as early in the process as possible because they alone may:

- Restrict the degrees of freedom in designing the end product
- Widen or narrow the types of adhesives that can be considered
- Affect the quality and reproducibility of the joint
- Affect the total assembly cost

The production methods that are at the disposal of the user may affect the production rate, simplicity, economy, and overall bond quality. The simpler the processes, the less training the operators will require and the more reliable will be the results. Several generic factors must be considered in the selection of the methods used in the bonding process:[1,2]

- The size and shape of the part to be bonded
- The specific areas to which the adhesive is to be applied
- The number of assemblies to be produced
- The required production speed
- The viscosity and other working characteristics of the adhesive
- The form of the adhesive (liquid, paste, powder, film, hot melt)

In this chapter, manufacturing processes and equipment related to the adhesive bonding processes will be reviewed. Sealing processes and equipment were described in Chapter 14. However, many of the factors (e.g., storage, metering, and mixing) that are discussed here are also directly applicable to sealants. This chapter will provide an overview to the various production methods that are generally available and discuss the implications of each on the overall bonding operation. The various processes will be presented in order of their sequence as shown in Fig. 18.1.

18.2 Storage

The method in which the adhesive or sealant is stored can affect the quality of the bond even before the containers are open. The manufacturer's storage directions are usually found in the technical bulletins describing the particular adhesive or sealant. As a general rule, all adhesives and sealants should be stored in a cool, dry place until it is time for their use. Certain adhesives are very tolerant to storage conditions, but certain types may have to be stored at low temperature or under special conditions. Some adhesive systems are affected by light or moisture, and some will require periodic agitation to make sure that the components do not settle irreversibly.

As organic resins age in storage due to exposure to temperature, moisture, light, gravity, or other environments, the viscosity of the system will usually undergo noticeable change. Once the storage life is exceeded, the system will generally be too viscous to use or else it will be unusable from separation of its components. Figure 18.2 shows the increase in viscosity of a paste adhesive as a function of storage time at room temperature. Ordinarily, the shelf life is chosen so that the majority of the viscosity change will occur just after the expiration of the adhesive's shelf life.

Production Processes and Equipment 729

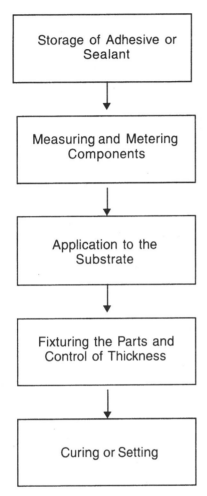

Figure 18.1 Major processes in producing adhesive or sealant joints.

Quite often the containers of an adhesive or sealant will bear a special label stating the safe upper limit of temperature during storage. Containers should be checked when received to see if such a label is present. As soon as the container is received, the user should place a second label on it stating the date received and the date when the shelf life will expire. This will provide useful information when it comes to disposal of old product or maintaining the freshest product in inventory. A record should also be kept of the manufacturer's lot number for possible future reference. Containers should be kept tightly closed, and labels should be kept clean for proper identification.

Three levels of storage temperature are common with adhesives and sealants: room temperature (60–80°F); refrigerated (35–40°F); and fro-

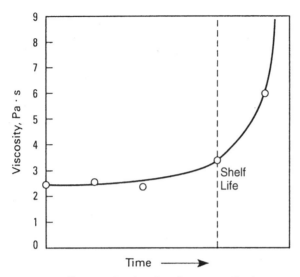

Figure 18.2 Increase in viscosity of a paste adhesive as a function of time at room temperature.[3]

zen (0°F). Generally adhesives or sealants that are delivered as multiple component systems have long shelf lives and are storable for very long times. They can usually be stored for 6–12 months at room temperature and often longer. Single component systems, where the catalyst and base resin are combined in the same product, generally require refrigerated or frozen storage depending on the type of catalyst used and the reactivity of the system. These systems may only have storage lives of several weeks or months at the conditions specified.

Extended exposure of the uncured material to temperatures or conditions outside those recommended by the manufacturer will cause change of physical properties to the uncured material and will likely reduce its resulting cohesive and adhesive strength once used. Figure 18.3 shows the effect of aging conditions on the tensile shear strength of an epoxy film adhesive. ASTM D 1337 describes a standard test method for measuring storage life of adhesives by consistency and bond strength.

The safe storage of adhesives and sealants that contain flammable, corrosive, or hazardous substances must also be considered. Many organic solvents are flammable, and certain adhesives will generate copious amounts of heat when reacted. For this reason, different containers used for two or three part materials should be stored separately to prevent accidental mixing of the components in the case of spillage.

The more hazardous adhesives and sealants should be stored in a small building or shed separate from the main building or storage area

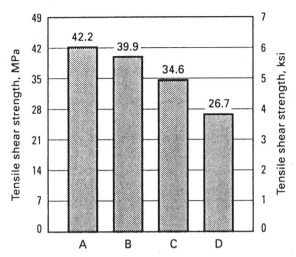

Figure 18.3 Effect of aging conditions on epoxy film adhesive: (a) fresh adhesives, (b) cured after aging 90 days at 24°C, (c) cured after 90 days at 32°C, (d) cured after 1 hr at condensing humidity.[3]

in the plant. An outside storage building is recommended if several drums of flammable materials are to be stored. The storage facility should be built of fire resistant materials and a fire extinguishing system installed. For storage of just a few drums of flammable materials, insulated metal storage cabinets are available that will protect the drums by closing their doors automatically in the event of a fire. An electrical ground should also be provided to all drums so that a potentially explosive spark cannot occur. The most appropriate internal location for a flammable storage area is a room with brick or masonry walls and fire doors that open outward. Some adhesive products contain corrosive compounds that require special containers and need to be separated from other materials to avoid interaction.

When removed from refrigerated storage, the adhesive must be brought to its application temperature, usually room temperature. The containers should remain closed until they are at temperature. A material is most vulnerable to moisture collection when it is removed from cold storage and the package or container is not properly sealed. Moisture condenses on the adhesive and quickly reduces its capabilities. Once opened, the desired amount of material should be removed from the container, and then it should be resealed as quickly as possible. Containers that are not completely used when originally opened should be sealed immediately after use to prevent loss of volatile ingredients or contamination from moisture in the air. Equipment used to remove the contents from the containers, such as spatulas, flasks, cups, etc. should be cleaned between each use.

18.3 Preparing the Adhesive or Sealant for Application

The process of getting the adhesive ready for application to the substrate includes:

- Removal of the product from the container
- Metering
- Mixing the adhesive or sealant
- Distributing the adhesive to the bonding area

These are also simple processes that can have consequences on the resulting bond quality.

18.3.1 Transferring the product

Bulk adhesives and sealants are normally supplied in containers that range in volume from 1 qt. to 55 gal. drums. The method used to extract the adhesive from the container should be carefully considered.

If a commercially filled resin is used, it will be advantageous to first place the unopened can on a paint shaker or roller and vibrate or roll the contents for a period of time. If the resin is in a drum, drum rollers may be used to assure that the components within the drum are uniformly dispersed before the drum is opened. The least aggressive method should always be used so as to obtain uniform distribution of the components with minimal addition of air into the product.

Once the contents of the container are evenly distributed, the adhesive or sealant material can be removed. The use of pumps for transferring the adhesive or sealant has several drawbacks. Filled compounds have an abrasive action on the moving pump parts. This is particularly true with fillers such as silica. Compounds that polymerize without the presence of a catalyst or heat will tend to begin their polymerization due to friction within the pump. This will gum the moving parts and make them inoperable. Therefore, transfer is best done under some type of pressure rather than mechanical pumping.

Although it may require special designs, pressure transfer does eliminate the problems that occur with gear pumping methods. Many resinous systems may be moved from their original drums by air pumps. Transfer distances should be short and lines should be smooth and non-corrosive. In some instances, barrel warmers and heated lines may be required to reduce pressures required to transport materials. Under no circumstances should the same transfer equipment be used on different resins without complete and thorough cleaning. This will avoid the contamination of one resin system with the other.

18.3.2 Metering

The components of multiple part adhesive or sealant systems must be measured out carefully. The concentration ratios may have a significant effect on the quality of the joint. Strength differences caused by varying curing-agent concentration are most noticeable when the joints are tested at elevated temperatures or after exposure to water or solvents.

Exact proportions of resin and hardener must be weighted out on an accurate balance or in a measuring container for best reproducibility. Possible problems that can occur by not adhering to the proper mixing proportions can include incomplete polymerization (too little catalyst), brittleness (excessive catalyst), and corrosion of metallic adherends (excessive catalyst).

Certain two component adhesives (e.g., fatty polyamides used with epoxies) have less critical mixing ratios, and component volumes may often be measured volumetrically by eye without too adverse an effect on the ultimate bond strength. In fact, with certain types of adhesives, the mixing ratio is used as a parameter to vary the final properties of the cured adhesive. For epoxy systems cured at room temperature with a polyamide hardener, flexibility and toughness can be improved within limits by increasing the polyamide resin concentration. Temperature and chemical resistance of the adhesive is generally improved by decreasing the amount of polyamide hardener used, but at the sake of decreasing toughness.

If the components require weighing, the weighing equipment should be clean and calibrated. Containers used for weighing should have their tare weights plainly marked on the outside. Containers for weighing adhesive must not be interchanged with those for weighing water or any other ingredients.

Many adhesives and sealant manufacturers supply products in preproportioned containers designed to simplify the metering and mixing processes. These innovative containers include:

- Side-by-side syringes with attachable static mixer
- Bubble pack or divider pack with a removable hinge that can be used for spreading and mixing
- Barrier and injection cartridges

Examples of these are shown in Fig. 18.4. These packaging systems store metered amounts of liquid adhesive components and provide means for mixing and dispensing, all within the same package. An example is the familiar flexible epoxy pouches with removable clamps that separate the base resin and hardener. Other forms consist of a

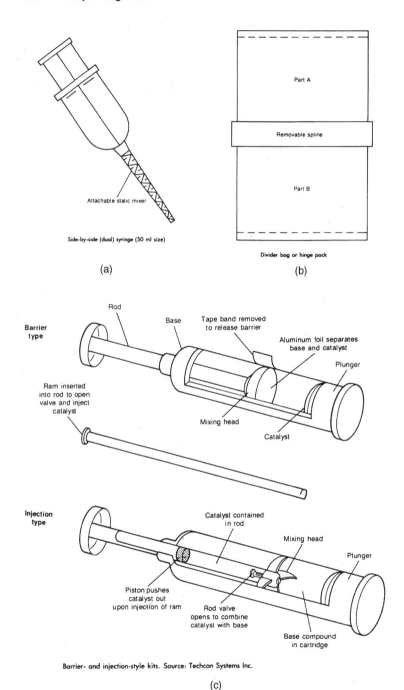

Figure 18.4 Several varieties of preproportioned adhesive containers: (*a*) side-by-side syringe with static mixer, (*b*) bubble pack with hinge, and (*c*) barrier style cartridge.[4]

plastic syringe for storing the adhesive and a disposable plastic static mixer for mixing the components and applying them to the substrate. These useful applicators are available with capacity from several grams to large industrial cartridge size (quart).

Another "package" is a semi-solid two component epoxy ribbon with hardener and base resin identified by different colors. Suitable lengths of ribbon are cut off and kneaded together by hand until the colors combine to form a third color. Once this is completed, the paste-like material is applied to the joint in a conventional manner. Single part adhesives such as cyanoacrylates and anaerobics do not require metering and mixing, yet they can cure at room temperature. "Honeymoon" cure concepts have been developed for thermosetting acrylic systems. With these adhesives the catalyst can be applied as a primer to the substrates. The catalyst can also be applied to substrate "A" and the base resin to substrate "B". When the joint is made, the catalyst and resin contact one another to begin the curing process. There have also been adhesive formulations developed with encapsulated curatives. These systems have 1–3 years shelf life at room temperature and can be cured using conventional processes.

18.3.3 Mixing

The measured components must be mixed thoroughly. Mixing should be continued until no color streaks or density stratifications are noticeable. Caution should be taken to prevent air from being mixed into the adhesive through over agitation. This can cause foaming of the adhesive during cure, resulting in porous bonds. If air does become mixed into the adhesive, vacuum degassing may be necessary before application.

In mixing, cleanliness is almost as important as accuracy. Scoops, weighing pans, mixing spatulas, and other accessories should be kept clean. Adhesives that appear to be similar may be ruined if one is contaminated with the other. Water will spoil many adhesives and sealants, and organic solvents are recommended for cleaning.

Only enough adhesive should be mixed to work with before the adhesive begins to cure. Working life of an adhesive is defined as the period of time during which a mixed adhesive remains suitable for use. Working life is decreased as the ambient temperature increases and the batch size becomes larger. One-part and some heat curing, two-part adhesives have very long working lives at room temperature, and application and assembly speed or batch size is not critical.

For mechanical mixing often a variable speed mixer, preferably air driven, and a container are all that are required. An air driven mixer is safer, less costly, and more easily regulated than an electrically driven mixer. For the most reliable mixing operation, the temperature of the compound in the mixing vessel should be uniform. For adhesives or sealants that are stored at cold temperature, adequate mixing viscosity can be reached quickly by the use of a hot liquid jacket around the vessel.

For a large scale bonding operation, hand mixing is costly, messy, and slow; repeatability is entirely dependent on the operator. Equipment is available that can meter, mix, and dispense multicomponent adhesives on either a continuous or one-shot basis. Figure 18.5 illustrates a standard adhesive meter-mix-dispensing machine for multicomponent systems. An important advantage for using such equipment is that it allows the use of short working life materials by mixing only the smallest practical quantity continuously. This offers shorter curing cycles, resulting in faster assembly times. A disadvantage of this type of equipment is that it should be used almost continuously for best results. When the equipment must be shut down, it requires thorough cleaning. Often the equipment is supplied with an automatic solvent flush to make sure that all resin is purged from the lines and from internal mixing devices. Automated mixing, metering, and dis-

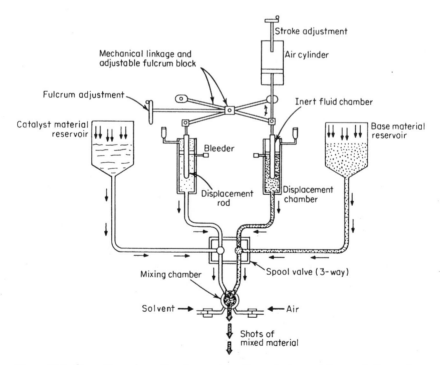

Figure 18.5 Operating schematic of an automatic metering, mixing, and dispensing machine for adhesive systems.[5]

pensing equipment can be mounted on a cart for easy access to the bonding area (Figure 18.6). These systems are also ideal for robotic application of adhesives on a high volume production line.

For maximum strength glue lines, it is necessary to remove all air from the adhesive before using it. Entrapped air represents a source of voids in the cured bond-line and, hence, the possibility of a failure. Trapped air is especially noticeable when the adhesive requires an elevated temperature cure. Significant amounts of trapped air can cause the adhesive to foam when exposed to elevated temperatures.

When hand mixing adhesives or sealants, two methods are commonly employed to degas the mixed material. These are centrifuging, and subjecting the mixed system to a vacuum. The latter is the more popular method and consists of placing the mixed material, in the container that it was mixed, into a vacuum chamber. The level of the mixture in the container should be low enough to allow for a foam

Figure 18.6 Cart-mounted meter, mixing, and dispensing machine. (Photo courtesy of Sealant Equipment and Engineering, Inc.)

head to rise and break during application of a vacuum. If the compound level is too high, the foam head will not break. It is desirable to know when the foam head has broken so the vacuum tank should be provided with two sight windows: one for illumination, and one for observation. It is best to repeatedly break the vacuum during the degassing operation so that multiple foam heads rise and break. The user can determine when all of the air is removed from the adhesives, because it will then become very difficult to cause a foam head to rise.

Static mixing nozzles, such as shown in Figure 18.7, offer a method to automatically mix two components on-line. The nature of the static mixing process avoids entraining air into the mixed adhesive or sealant. However, available mixing ratios are limited with static mixers.

18.3.4 Transferring to the application area

The adhesive or sealant should be mixed and furnished for application in clean containers. The containers should be small enough so that no adhesive or sealant will be left beyond the time when the resin be-

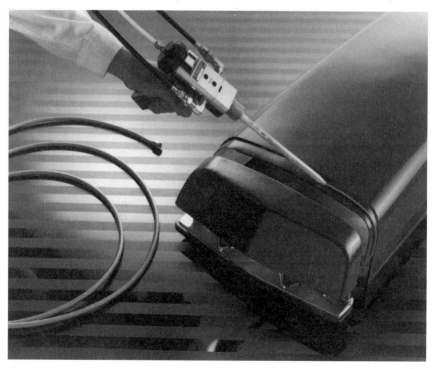

Figure 18.7 Structural bonding of air intake on a tractor. Adhesive is metered and mixed in an on-line static mixing nozzel. (Photo courtesy of Sealant Equipment and Engineering, Inc.)

comes unusable due to its working life. Disposable waterproof cardboard cups or aluminum containers work very well. Containers may be marked to show the hour at which the contents were mixed. Any unused adhesive should then be collected at the end of the pot life period recommended by the manufacturer.

Header systems with manual or automatic dispensing valves have been designed to carry adhesives and sealants over long distances within the plant. Distances of as much as 300 ft are not uncommon with drops at points of application. Figure 18.8 shows a modular header system. Adhesives and sealants are nearly ideal materials for application by robots. Robotic application is commonly used in the automobile industry to increase quality and to reduce labor and material cost.[7,8]

18.4 Application

The method used to apply the adhesive or sealant can have as much to do with the success of the joining operation as the kind of material applied. The selection of an application or dispensing method depends primarily on the form of the adhesive: liquid, paste, powder, or film. Another important consideration is the shape of the parts to be bonded.

Table 18.1 describes the advantages and limitations that are realized in using each of these four basic forms. Other factors influencing

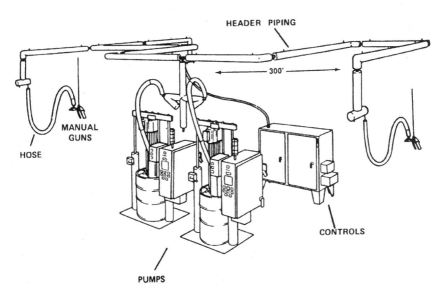

Figure 18.8 Modular header system.[6]

TABLE 18.1 Characteristics of Various Adhesive Application Methods[9]

Application method	Viscosity	Operator skill	Production rate	Equipment cost	Coating uniformity	Material loss
Liquid						
Manual brush or roller	Low to medium	Little	Low	Low	Poor	Low
Roll coating, reverse, automatic, gravure	Low	Moderate	High	High	Good	Low
Spray, manual, automatic, airless, or external mix	Low to high	Moderate to high	Moderate to high	Moderate to high	Good	Low to high
Curtain coating	Low	Moderate	High	High	Good to excellent	Low
Bulk						
Paste and mastic	High	Little	Low to moderate	Low	Fair	Low to high
Powder						
Dry or liquid primed	....	Moderate	Low	High	Poor to fair	Low
Dry Film	....	Moderate to high	Low to high	Low to high	Excellent	Lowest

the application method are the size and shape of parts to be bonded, the areas where the adhesive is to be applied, and the production volume and rate required.

18.4.1 Liquids

Liquids are the most common form of adhesive, and they can be applied by a variety of methods. They have an advantage in that they are relatively easy to transfer, meter, and mix. Liquids also tend to wet the substrate easily and provide uniform bond-line thickness. However, they have the disadvantages of sometimes being messy, requiring clean-up, and providing a relatively high degree of waste.

Brush, simple rollers, syringes, squeeze bottles, and pressurized glue guns are manual methods that provide simplicity, low cost, and versatility. These methods are probably the most widely used because of their simplicity. Manual dispensing methods allow application of adhesive to only a small segment of a surface and are particularly effective for small or irregular parts or low production rates.

It is possible to mechanize brush methods by using hollow brushes through which the liquid adhesive is fed. This improves production rates, coating uniformity, and cuts waste. Uniform films are generally difficult to achieve, but operators can become proficient with time.

Squeeze bottles, oil cans, and pressurized glue guns permit precise and speedy adhesive application. By adjusting the pressure, the rate of adhesive flow can be matched to the production rate. These devices can apply the adhesive or sealant inside a blind hole or limited access area. The tip of the applicator can be used to deliver multiple spots of adhesive.

Silk screen application is often used when the adhesive has to be applied to specific controlled areas. The liquid adhesive is forced through pores in a cloth or screen. It is possible to coat only selected areas by masking parts of the screen so that adhesive does not pass through in the unwanted areas. Adhesives generally must be specifically formulated for silk screen processing. Very low viscosity adhesives, with flow characteristics similar to coatings, are best for silk screening operations.

Spray, dipping, and mechanical-roll coaters are generally used on large production runs. Dipping is also used to cover large areas quickly. Areas that do not require adhesive can be masked-off. Manual dipping is often effective; and baskets, screens, and perforated drums are available to facilitate immersion of the part into the adhesive bath.

Spraying usually is evaluated against other methods on the basis of cost. Spraying applies the liquid resin fast, reduces the required drying time through better evaporation of solvent, and is capable of reaching areas that are inaccessible to other manual application tools. When automated, spraying is particularly useful for coating long runs of identical or similar products. Even hot melt adhesives can be applied by spraying (Fig. 18.9).

Several spraying methods can be used to apply adhesives, including conventional air spray, hydraulic cold airless spray, hot spray, and hot airless spray. The main types of spray equipment are described below.

Air spray—Utilizes air pressure to spray a fine mist of atomized adhesive on the substrate. Generally unsuitable for small parts. Maintenance is high due to nozzle clogging and overspray. Several passes are necessary for thick coating build-up.

Hot spray—Adhesives are first heated, permitting the use of high viscosity materials. Heavier coatings are possible than with air spray. Drying is accelerated due to evaporation of solvent in the spray.

Airless spray—Uses hydraulic pressure rather than air pressure. Saves energy and overspray. Can be applied both hot and cold.

Heating the adhesive before atomization allows heavier adhesive buildup, reduces overspray losses, and minimizes contamination from atmospheric water vapor. Spray methods can be used on both small

Figure 18.9 Hot melt adhesives can be applied by spray application for fast bonding of most foams, plastics, fabrics, particle board, and light gauge metals. (Photo courtesy of 3M Company.)

and large production runs. The liquid to be sprayed is generally in solvent solution. Sizable amounts of product may be lost from overspray. Two component adhesives are usually mixed prior to placing in the spray-gun reservoir. Application systems are available, however, that meter and mix the adhesive within the spray-gun barrel. This is ideal for fast-reacting systems, but guns must be thoroughly cleaned to avoid build-up of polymerized product.

When large flat surfaces and webs of materials are to be coated and when production reaches rates of 200-300 pieces per day, machine methods of application should be considered over the manual methods described above. Mechanical-roller methods are commonly used to apply a uniform layer of adhesive via a continuous roll. Such automated systems are used with adhesives that have a long working life and low viscosity. Various machine methods of adhesive application are illustrated in Fig. 18.10, and several are described below.

Bench coaters are roughly 20-35% faster than hand brushing and cut waste by up to 20–35%. They start to be useful at rates of 200 pieces per day and can reach 12,000 pieces per day. Roller sizes normally range from 4–26 in. and they come with various surfaces from smooth for thin applications to increasingly coarse for heavier adhesive layers.

Floor mounted roll coaters are noted for their extreme efficiency with large size webs and flat sheets. Their waste is very low, less than 2% in some cases. Several types are available because no single model or type of roll coater is best for all of the variables of the material surface, coating formulation, and end use. The following roll coaters are commonly used:

Kiss type—No back-up roll; coater is in light contact with substrate; contact is controlled by web tension. Best with adhesives having good flow characteristics and relatively slow drying rates.

Pressure or squeeze type—Also called nip coaters; can provide greater penetration into open web structures. Often used to apply coatings to both sides simultaneously.

Reverse roll coater—Popular for web precision coating because of better control over thickness.

Two-roll reverse coater—For additional control over adhesive coating; can handle wet film thicknesses from 1-20 mils.

Dip roll coaters—Simplest of the basic types of coaters; used mostly with low viscosity adhesives; works well where impregnation or saturation is required.

Engraved roll coaters—Provides accurate control over coating uniformity; best for extremely low adhesive weights.

Kiss Roll Coater

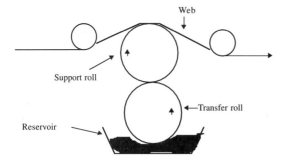

Knife-Over-Roll Coater

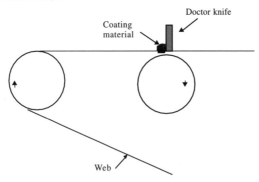

Squeeze Roll Coater

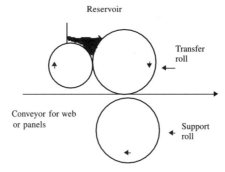

Figure 18.10 Methods of machine application of adhesive coatings.[10]

Reverse Roll Coater

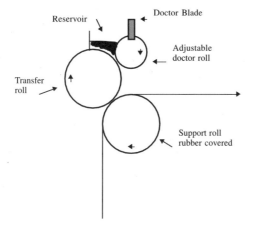

Engraved Roll Coater

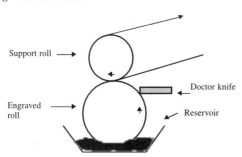

Figure 18.10 (*Continued*)

Air blade coater—Similar to the kiss coater in that the adhesive is applied where the web touches it; possible to apply coating uniformly at high speeds; limited use with solvent-based adhesives.

Bead coaters—Two rollers set at specific clearance; a rod between the bottom roll and the web makes a bead of adhesive that coats the web passing above it.

Knife spreader—Usually used to apply high viscosity adhesives to flexible webs; web is drawn between a precision knife edge and a rubber roller or flat bed plate.

Flow or curtain coaters—Uses a head with a slot in the bottom through which the adhesive is poured or lowered onto the parts or surfaces as they pass beneath the head by a conveyor; provides materials savings and ability to coat irregular surfaces uniformly.

18.4.2 Pastes and mastics

Bulk adhesives such as pastes or mastics are the simplest and most reproducible adhesive to apply. They produce heavy coatings that fill voids, bridge gaps, or seal joints. However, the process is limited to high viscosity materials. They can be in the form of high viscosity extrudable liquid or a trowelable mastic.

These systems can be troweled on or extruded through a caulking gun. Little operator skill is required. Since the thixotropic nature of the paste prevents it from flowing excessively, application is usually clean, and not much waste is generated. With trowel application, the depth of notch and the spaces between the notches help regulate the amount of adhesive applied. Shallow, rounded and closely spaced notches are often used with lower viscosity adhesives to allow the lines of adhesive to flow together and form a continuous, unbroken film.

18.4.3 Powders

Powdered adhesive systems are not common although powder resin coatings are ordinarily used for protective and decorative coating. B-staged powder adhesives may be applied dry to a natural or primed surface. They can be applied electrostatically like powder coatings or they can be applied gravimetrically by being dispersed onto a hot substrate surface. Heat and pressure must be applied for full curing, and considerable care must be taken to obtain uniform thickness over large areas. If coating uniformity is poor, large variations of joint strength may result with powder adhesives. The main advantages of using powder adhesives are their cleanliness and minimal waste. Spills and overspray may be collected and reused if not contaminated. The main disadvantage is the cost of the application equipment and the lack of manual methods of application.

Powder adhesives are generally one-part, epoxy-based systems that do not require metering and mixing but often must be refrigerated for extended shelf life. Elastomeric types are generally not available in powder form. Systems that need a primer require an additional coating step prior to applying the powder. Equipment cost for powder application tends to be high.

Powder adhesives can be applied in three ways: sifting or spraying the powder on the substrate; dipping the substrate into the powder; or melting the powder to liquid form for more conventional application. When sifted or sprayed onto a substrate, the powder falls onto the preheated substrate, melts, and adheres. Electrostatic spraying of adhesive powder has also been used to apply a B-staged powder to various substrates. A preheated substrate could also be dipped into the powder and then extracted with an attached coating of adhesive.

Lastly, the powder can be melted into a paste or liquid and applied by conventional means. Once the powder is adhered to the substrate, the assembly is mated and cured under heat and pressure according to recommended processes.

18.4.4 Films

Both structural and non-structural adhesives are commonly available in film form. Adhesives applied in the form of dry films offer a clean, hazard free operation with minimum waste and excellent control of film thickness (Figure 18.11). However, the method is generally limited to parts with flat surfaces or simple curves. Optimum bond strength requires curing under heat and pressure, which may involve considerable equipment and floor space, particularly for large parts.

Figure 18.11 AF-32 nitrile-phenolic adhesive film bonds doubler stacks in aerospace application. Adhesive provides vibration damping. (Photo courtesy of 3M Company.)

Film material cost is high in comparison to liquids, but waste or material loss is the lowest of any application method.

Practically any solid or B-stageable adhesive that has adequate stability at room temperature can be supplied in the form of a film. This film may be pressure sensitive, thermoplastic or, in applications where the greatest structural strength is required, a thermosetting type. Epoxy, epoxy hybrid, phenolic and other thermosetting film adhesives are commonly used in many industrial applications. Elastomeric adhesives are available in dry films in both solvent and heat reacting types. Both elastomer and resin based adhesives are also supplied as pressure sensitive adhesives, but they have lower strength than the solvent or heat reacting types.

The use of dry adhesive films is expanding more rapidly than other forms because of their following advantages:

1. High repeatability—no mixing or metering, constant thickness.
2. Easy to handle—low equipment cost, relatively hardware-free; clean operating.
3. Very little waste—preforms can be cut to size.
4. Excellent physical properties—wide variety of adhesive types available.

Application requires a relatively high degree of care to ensure non-wrinkling and removal of separator sheets. Films are often supported on scrim which distributes stress in the cured joint and assures a uniform bondline thickness throughout the bonded area.

Characteristics of typical film adhesives (Table 18.2) vary widely depending on the type of adhesive used. Film adhesives are made in both unsupported and supported types. The carrier for supported films is generally fibrous fabric or mat. Film adhesives are supplied as heat-activated thermosetting or thermoplastic hot melt, pressure-sensitive, or solvent-activated forms. The type of dry film adhesive used will usually determine the technique by which it is applied.

- Pressure sensitive adhesives, available in the form of single coated or double coated films, are tacky and bond instantly when hand pressure is applied.
- Thermoplastic adhesive films are normally positioned between the mating surfaces, and heat and pressure are applied to make the bonded joint; this joint can be processed very quickly.
- Thermosetting adhesive films are applied between the two mating surfaces and the assembly is subjected to heat and pressure for a predetermined period of time.

TABLE 18.2 Characteristics of Adhesive Films[11]

Adhesive type	Carrier	Thickness, mils.	Minimum bonding temp., °F	Characteristics and typical applications
Thermosetting	Unsupported	2.2	325	Heat reactivatable type for nameplates and lightweight flat materials. Peel strength, 22 to 28 lb/in.
Thermosetting	Unsupported	2.2	325	Heat or solvent reactivatable type (standard, blends, or chlorinated) for nameplates.
Thermosetting	Glass mat	11	250	General purpose metal-to-metal bonding film. Provides oil, fuel and water resistant bonds. Peel strength, 25 to 30 lb/in. Good exterior exposure properties. Used for exterior trim on cars.
Thermosetting	Fibrous mat	10	250	Can be used on great variety of surfaces because low pressure is required in bonding. Employs thin carrier with heavy adhesive coating. Can be bonded at slightly lower temperatures and shorter dwell times than those required for other types. Peel strength 25 to 30 lb/in.
Thermosetting	Glass mat	16 to 18	325	Features good heat and moisture resistance. Excellent low temperature flexibility and high bond strength (aluminium to aluminium peel values up to 50 lb/in.). Thickness is sufficient to compensate for minor irregularities on surface of substrates.
Pressure sensitive	Supported	3 to 6	R.T.	Good ageing; retains tack indefinitely. Good ultraviolet resistance. Not good for use on pvc (flexible).
Pressure sensitive	Supported	Approx. 11	R.T.	Good adhesion to wide variety of surfaces, including vinyls. Excellent heat resistance. Resists plasticiser exudation. Heavy film with good void filling characteristics makes this suitable for use on poorly mated surfaces.
Pressure sensitive	Unsupported	2.5	R.T	High performance type of film specifically formulated for bonding nameplates, escutcheons, and similar parts. Peel values in excess of 10 lb/in. immediately after bonding are possible on high impact polystyrene, stainless steel, and backed enamel. These values can be increased to over 13 lb/in. after 24 h cure.

- Solvent reactivatable adhesive films are pressure sensitive adhesives that are reactivated by wiping with a cloth that has been dampened with solvent.

Solvent-reactivated adhesives are made tacky and pressure-sensitive by wiping with solvent. They are not as strong as other types of adhesives but are well suited for contoured, curved, or irregularly shaped parts. Manual solvent reactivation methods should be closely monitored so that excessive solvent is not used. Chemical adhesive formulations that are available in solvent-activated films include neoprene, nitrile, and butyral phenolics. Decorative trim and nameplates are usually fastened to a product with solvent-reactivated adhesives.

18.4.5 Hot-melt adhesives

Thermoplastic hot-melt adhesives are generally used in non-structural applications where convenience and production speed are important. Hot-melt equipment consists of hot-spray devices (air and airless), wheel and extrusion dispensers, and heated handguns to handle preforms of various shapes. Hot-melt materials and equipment are usually integrated into high volume production lines. The equipment is very compatible with robotic application, and hot-melt materials can be applied from any direction.

Hot-melt adhesives are available in a variety of solid forms. Usually they are supplied as billets and applied to the substrate through a hot-melt glue gun. The hot-melt application system is equipped with a temperature-regulated glue reservoir and insulated hosing to the gun. Once out of the gun, the adhesive solidifies quickly. A high degree of operator skill is necessary to make the assembly as quickly as possible. Hot-melt adhesives that are stored in molten baths may require a blanket of nitrogen above the adhesive to eliminate the possibility of oxidation.

A wheel applicator can dispense hot melts continuously in a variety of configurations. This is especially advantageous for bag or carton sealing. Nozzle applied hot melts are usually chosen for spot application or for robotics. Hot melt adhesives can also be applied to continuously laminated products (Figure 18.12). The main advantages of this process is that it is fast and solvent-free.

Preformed, shaped hot-melt adhesives and heated guns provide an alternative to the nitrogen blanketed hot-melt storage tank. Certain hot melt systems function by feeding a configured coil, rope, or rod of solid adhesive into a melt chamber of a heated glue gun. A piston forces the hot melt adhesive or sealant through a heating chamber. The product is dispensed as a continuous bead or individual dots. An advantage of this technique is that a limited volume of adhesive is

Figure 18.12 Hot melt laminating system can produce 66 inch wide product. (Photo courtesy of Harlan Machinery Company.)

molten at any given time, so that thermal degradation of the material is minimized. However, the cost of the shaped adhesive will generally exceed that of bulk adhesives.

18.5 Bonding Equipment

After the adhesive is applied, the assembly must be mated as quickly as possible to prevent contamination of the adhesive surface. The substrates are held together under pressure and heated if necessary until cure is achieved. The equipment required to perform these functions must provide adequate heat and pressure, maintain constant pressure during the entire cure cycle, and distribute pressure uniformly over the bond area. Of course, many adhesives cure with simple contact pressure at room temperature, and extensive bonding equipment is not necessary.

It is important when selecting equipment to give consideration to life cycle costs which include: original investment cost as well as operation and maintenance cost such as energy, repair, trouble shooting, etc. A checklist of minimal equipment considerations should include:

- Simplicity of operation
- Compatibility with adhesive or sealant
- Ease of disassembly and cleanup
- Trouble detection devices and self-diagnostics
- Service backup relative to spare parts, local resources, etc.
- Knowledge of life expectancy of the system and its components
- Clearly defined warranty coverage
- Knowledge of the equipment production rates

- Ease of pressure regulation
- A means of checking the proportions of different adhesive components
- Temperature controls and indicators[12]

18.5.1 Pressure equipment

Pressure devices should be designed to maintain constant pressure on the bond during the entire cure cycle. They must compensate for bondline thickness reduction from adhesive flow or thermal expansion of assembly parts. Thus, screw-actuated devices like C-clamps and bolted fixtures are not acceptable when constant pressure is important. However, clamp fixtures perform well on assemblies machined to distribute mechanical loads. Spring pressure should supplement the clamps during heat curing to compensate for variation in pressure when the adhesive begins to flow and adherends expand. Spring pressure can often be used to supplement clamps and compensate for thickness variations. Dead-weight loading may also be applied in many instances. However, this method is sometimes impractical, especially when a heat cure within an oven is necessary.

Pneumatic and hydraulic presses are excellent tools for applying constant pressure. Steam or electrically heated platen presses with hydraulic rams are often used for adhesive bonding. Some units have multiple platens, thereby permitting the bonding of several assemblies at one time.

Large bonded areas, such as aircraft parts, are usually cured in an autoclave. The parts are mated first and covered with a rubber blanket to provide uniform pressure distribution. The assembly is then placed in an autoclave, which can be pressurized and heated. This method requires heavy capital-equipment investment.

Vacuum-bagging techniques can be a less expensive method of applying pressure to large parts. A film or plastic bag is used to enclose the assembly, and the edges of the film are sealed airtight. A vacuum is drawn on the bag, enabling atmospheric pressure to force the adherends together. Vacuum bags are especially effective on large areas because the size is not limited by pressure equipment. Pressures, of course, are limited to atmospheric pressure.

18.5.2 Heating equipment

Many structural adhesives require heat as well as pressure to cure. Even with conventional room temperature curing systems, most often the strongest bonds are achieved by an elevated temperature cure. With many adhesives, trade-offs between cure times and temperature are permissible. Generally, the manufacturer will recommend a certain curing schedule for optimum properties.

If a cure of 60 min at 300°F is recommended, this does not mean that the assembly should be simply placed in a 300°F oven for 60 min. The temperature is to be measured at the adhesive bond-line. A large part will act as a heat sink and may require substantial time for an adhesive in the bond-line to reach the necessary temperature. In this example, total oven time would be 60 min in addition to whatever time is required to bring the adhesive up to 300°F. Bond-line temperatures are best measured by thermocouples placed very close to the adhesive. In some cases, it may be desirable to place the thermocouple directly in the adhesive joint for the first few assemblies being cured.

Oven heating is the most common source of heat for bonded parts, even though it involves long curing cycles because of the heat-sink nature of large assemblies. Ovens may be heated with gas, oil, electricity, or infrared units. Good air circulation within the oven is mandatory for uniform heating. Temperature distribution within an oven should always be checked before items are placed in the oven. Many ovens will have significant temperature distributions and dead-spaces in corners where air circulation is not uniform. The number of items placed in the oven will also affect the time for the bond to get to temperature. The geometry and size of the part may also affect air circulation and cause variations in temperature distribution.

Heated-platen presses are good for bonding flat or moderately contoured panels when faster cure cycles are desired. Platens are heated with steam, hot oil, or electricity and are easily adapted with cooling-water connections to further speed the processing cycle. Strip or cartridge heaters are useful for localized heating in small areas. They adapt easily to flat dies, pressure bars, specially shaped dies, and other pressure fixtures. They are also used for tack-bonding small areas to facilitate handling prior to complete cure.

Induction and dielectric heating are fast heating methods because they focus heat at or near the adhesive bond-line. Workpiece-heating rates greater than 100°F/sec are possible with induction heating. For induction heating to work, the adhesive must be filled with metal or ferromagnetic particles or the adherend must be capable of being heated by an electromagnetic field as shown in Fig. 18.13. Dielectric heating is also an effective way of curing adhesives if at least one substrate is a non-conductor. However, metal-to-metal joints tend to break down the microwave field necessary for dielectric heating. This heating method makes use of the polar characteristics of the adhesive materials. Dielectric heating is used in the furniture industry to drive off water and harden water-based adhesives. Dielectric heating is also being explored as a rapid method of curing adhesives on substrates

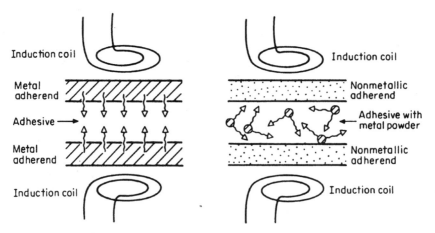

Figure 18.13 Left—induction heating metal parts heat adhesive by conduction. Right—Parts not heated by induced electric currents are bonded with an adhesive that can be heated because it contains metal particles.[14]

such as glass, plastics, and composites.[13] Both induction and dielectric heating involve relatively expensive capital equipment outlays, and the bond area is limited. Their most important advantages are assembly speed and the fact that an entire assembly does not have to be heated to cure only a few grams of adhesive.

An interesting development that minimizes the fixturing and heating equipment needed for curing the adhesive joint is "weldbonding". Weldbonding is adhesive bonding combined with metallurgical spot welding to make fast, strong, reliable joints. The process produces joints that are stronger, more durable, and more fatigue resistant than when either method is used by itself. This process has found its way into the automotive industry for use on aluminum and steel substrates. Table 18.3 shows that a variety of adhesives and substrates are compatible with this method.

With weldbonding the adhesive must be capable of being welded through the adherends. The spot welds hold the adherends together while the majority of the adhesive reacts in a conventional manner. Weldbonding adhesives generally must be specially formulated to allow for the localized heat generated by conventional spot welders. By-products from the decomposition of the organic adhesive must not affect the remaining joint area. A spot welded lap-shear joint is shown in Fig. 18.14.

18.6 Environmental and Safety Concerns

Four primary safety factors must be considered in all adhesive bonding or sealing operations: toxicity, flammability, hazardous incompat-

TABLE 18.3 Strength of Weldbonded Adhesive Joints[15]

Joint alloy	Adhesive	Curing condition	Spotweld lb	Joint breaking strength in shear Adhesive bond lb	Weldbond lb
2036-T4	None		770		
2036-T4	Polysulfide-epoxy	Ambient		700	1175
2036-T4	High peel epoxy	Ambient		1385	1090
2036-T4	Polyamide-epoxy	Ambient		735	875
2036-T4	Vinyl plastisol	Not cured	690		
2036-T4	Vinyl plastisol	2 hr @ 350 F		910	1450
2036-T4	One part epoxy	Not cured	750		
2036-T4	One part epoxy	2 hr @ 350 F		1610	1270
X5085-H111	None		700		
X5085-H111	Vinyl plastisol	Not cured	740		
X5085-H111	Vinyl plastisol	2 hr @ 350 F		860	1510
X5085-H111	One part epoxy	Not cured	770		
X5085-H111	One part epoxy	2 hr @ 350 F		1220	1210
2036-T4 to X5085-H111	None		880		
2036-T4 to X4085-H111	Vinyl plastisol	Not cured	840		
2036-T4 to X4085-H111	Vinyl plastisol	2 hr @ 350 F		930	1300
2036-T4 to X5085-H111	One part epoxy	Not cured	510		
2036-T4 to X5085-H111	One part epoxy	2 hr @ 350 F		1430	1180
Steel	None		1440		
Steel	Vinyl plastisol	Not cured	1380		
Steel	Vinyl plastisol	2 hr @ 350 F		900	1800
Steel	One part epoxy	Not cured	1430		
Steel	One part epoxy	2 hr @ 350 F		2000	1930

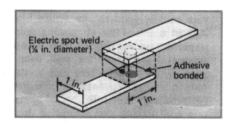

Figure 18.14 Weldbonded lap-shear joint. Adhesive flows beyond point of weld, then carbonizes when heat is applied.[15]

ibility, and the equipment. The adhesive or sealant must be carefully considered because it not only can provide health and safety issues within the factory, but it can also provide issues within the community relative to release of volatiles and waste disposal.

All adhesives, solvents, and chemical treatments must be handled in a manner preventing toxic exposure to the workforce. Methods and facilities must be provided to ensure that the maximum acceptable concentrations of hazardous materials are never exceeded. These values are prominently displayed on the material manufacturer's Material Safety Data Sheet (MSDS), which must be maintained and available for the workforce.

Cotton, leather, or rubber gloves should be worn to protect the hands from repeated contact with the materials. There is generally little danger with many of these systems, but repeated contact over long periods of time can sensitize the skin and produce unpleasant reactions such as itchiness, redness, swelling, and blisters.

Open use of adhesive and sealants, especially solvent-based sprays, should be limited to well ventilated areas. Spraying should be done in hoods with proper air circulation and safety equipment. The operator should wear a fume mask. Goggles and other safety clothing are also required in the event of splashing or spilling of liquid materials.

Where flammable solvents and adhesives are used, they must be stored, handled, and used in a manner preventing any possibility of ignition. Proper safety containers, storage areas, and well ventilated workplaces are required.

Certain adhesive materials are hazardous when mixed together. Epoxy and polyester catalysts, especially, must be well understood, and the user should not depart from the manufacturer's recommended procedure for handling and mixing. Certain unstabilized solvents, such as trichloroethylene and perchloroethylene, are subject to chemical reaction on contact with oxygen or moisture. Only stabilized grades of solvents should be used.

Certain adhesive systems, such as those based on heat curing epoxy and room temperature curing polyester resins, can develop very large exothermic reactions on mixing. The temperature generated during this exotherm is dependent on the mass of the materials being mixed. Exotherm temperatures can get so high that the adhesive will catch fire and burn. Mixing volume should not be greater than that recommended by the manufacturer. Adhesive products should always be applied in thin bond-lines to minimize the exotherm. Never use elevated temperature curing sealants or adhesives for casting or for application in excessively thick cross-sections without first consulting the manufacturer. These materials are formulated to be cured in thin cross-sections so that the heat generated by exotherm can easily be dissipated.

Safe equipment and its proper operation are, of course, crucial to a workplace. Sufficient training and safety precautions must be installed in the factory before any bonding process is established.

References

1. Landrock, A. H., "Adhesive Bonding Process", Chapter 7 of *Adhesives Technology* (Park Ridge: Noyes Publishing Co., 1985).
2. Bruno, E. J., "Methods of Bonding", Chapter 5 of *Adhesives in Modern Manufacturing* (Dearborn, MI: Society of Manufacturing Engineers, 1970).
3. Van Twisk, J., and Aker, S. C., "Storing Adhesive and Sealant Materials", in *Adhesives and Sealants,* vol. 3, Engineered Materials Handbook, ASM International, 1990.
4. Devlin, W., "Metering and Mixing Equipment", in *Adhesives and Sealants,* vol. 3, Engineered Materials Handbook, ASM International, 1990.
5. Plyes Industries, Inc., Form 1-810-100.
6. Killick, B. R., "Dispensing High Viscosity Adhesives and Sealants", in *Adhesives in Manufacturing,* G. L. Schneberger, ed. (New York: Marcel Dekker, 1983).
7. "Robotic Dispensing of Adhesives and Sealants", *Handbook of Adhesives,* 3rd ed., I. Skiest, ed. (New York: Van Nostrand Reinhold, 1990).
8. "Developing a Robotic Systems", *Adhesives Age,* April 1983.
9. Carroll, K. W., "How to Apply Adhesives", *Production Engineering,* November 22, 1965.
10. Martin, R. A., "How to Apply Adhesives", *Materials Engineering,* July 1970.
11. Baker, A., "Dry Film Adhesives Simplify Application", *Assembly and Fastener Methods,* February 1968
12. Chestain, C. E. and Berry, N., "Application Methods", in Adhesives Digest, 1995 Edition (Englewood, CO: DATA Business Publishing, 1994).
13. Raulauskas, F. L., "Adhesive Bonding/Joining Via Exposure to Microwave Radiation", 27th International SAMPE Technical Conference, October 9–12, 1995.
14. Bolger, J. C., and Lysaght, M. J., "New Heating Methods and Cures Expand Uses for Epoxy Bonding", *Assembly Engineering,* March 1971.
15. "Spot Welding Teams Up with Adhesives for Stronger Metal to Metal Bonds", (no author) *Product Engineering,* May 1975.

Chapter 19

Information Technology

19.1 Introduction

The computer age has made significant and somewhat unexpected impacts on the adhesives and sealants industries in recent years. Computers and analytical programs have, of course, greatly assisted both adhesive developers and manufacturers. As examples, computer analysis capabilities have been able to help us understand the stress distributions within complex joints and to simulate the effect of chemical or moisture penetration within the adhesive joint. Such advances are expected of a mature industry and technology.

However, what was somewhat unexpected is the assistance that information technology provides to the end user of adhesives and sealants. These advances can primarily be classified as:

- Information access
- Search and selection capabilities

There are several examples given in this chapter for each of these areas.

It is apparent that information handling technologies are just now making inroads into this rather conservative industry. The future use of computers and information science in providing assistance to the end users appears to have significant potential. It may be appropriate to add "information sciences" to the sciences shown in Fig. 1.3 that make-up what we know today about adhesives and sealants technology.

19.2 Information Access

There are, of course, a proliferation of Internet web sites belonging to manufacturers of adhesives and sealants. These range from those that provide simply a contact person, to those that supply complete product descriptions, instructions on how to use the product, tutorials on surface preparation and joint design, and links to related web sites. The web sites that are related to the adhesives and sealants industry are identified in a number of publications and directories. A good directory is in the Annual Buyers' Guide (Supplier Section) for the periodical "Adhesives & Sealants Industry". Several trade magazines have also begun periodic review of industry related web sites.

It is not the intention in this chapter to review these web sites, but to highlight several that are good examples of the potential capabilities and that have generally provided useful information to the user of adhesives and sealants. The following web sites are in that category and worthy of a bookmark.

The Adhesives and Sealant Council, Inc.	www.ascouncil.org
Adhesives and Sealants Online	www.adhesivesandsealants.com
Assembly Magazine Online	www.assemblymag.com
Adhesives for Industry Tech Center	ww.gluguru.com

19.3 Adhesive and Sealant Selection

Several web sites have been set-up to not only inform the user of products and services offered by the manufacturers of adhesives and sealants, but also to provide real knowledge that can be used in the every day selection and utilization of these materials. Several of these web sites will be reviewed here. The list is definitely not meant to be all encompassing. There are new sites coming on-stream daily. Those that are identified in this chapter represent the opportunities that such sites provide to both the user and manufacturer of adhesives and sealants.

Users should be cautious of working with the programs embedded in these websites, however. Their ease of use can create a false sense of security. If possible, time should be spent learning the underlying assumptions of the program. As with use of any computer intelligence program, one should apply experience, common sense, and outside opinions to the results that are provided to check their accuracy and to help in avoiding catastrophic mistakes. These computer systems can be used profitably as a "front end engine" to make the initial gathering of information and sources more efficient. One should never "blindly use" the result provided from such a program.

Because the adhesives and sealants industry is very fragmented (several very large companies, but thousands of very small companies), locating the right adhesive has been very much of a hit or miss proposition. On the part of the adhesive or sealant user, it requires contacting dozens of manufacturers, listing requirements, and discussion trade-offs. Also, the prospective adhesive user doesn't always know what questions to ask the manufacturer.

From the manufacturer's side, it is also a difficult process. The response to potential customers' inquiries takes up much of their technical marketing staff's time, and volumes may not be there to justify significant development or follow-through. So frequently, both user and manufacturer spend much time on a query only to find that what the manufacturer offers and what the user wants do not match. The web sites listed below attempt to make this process much more efficient.

In addition to selecting adhesives for a specific application, selection programs also provide a means to comparison shop, in order to find a less expensive product or a second source of supply. It is also possible to search for alternative products that conform to regulatory requirements which the current product may not fulfill.

19.3.1 AdhesivesMart

A new industrial adhesive web site at www.AdhesivesMart.com already contains information on over 1,000 adhesives.[1] Developed by AdhesivesMart Inc., located in West Newton, MA, the web site is designed to make life easier for both users and manufacturers of industrial adhesives. In essence, AdhesivesMart.com acts as a middleman and matchmaker by efficiently connecting the two sides of the market.

AdhesivesMart.com is the Internet's first example of providing a complete and sophisticated selection process for an industrial product. It has combined a search engine with a questionnaire that helps define the characteristics of the adhesives that are required. After the questionnaire is completed, the prospective user receives a list of products that meets the "specifications" provided. If the list is too long, the user is asked to refine the questionnaire to narrow down the search and be more focused.

Searchers working through AdhesivesMart.com can also request pricing details, technical specifications, Material Safety Data Sheets, and other information from the manufacturers through the web site. The web site forwards these requests directly to the manufacturer who responds to the site user via e-mail, surface mail, or a phone call.

AdhesivesMart does not take orders or represent specific manufacturers. Instead, the company receives a commission from suppliers on sales that occur as a result of the web site connection.

19.3.2 Adhesive Selector

Another selector tool, Adhesive Selector, can be found on www.assemblymag.com/toolbox/adhesive.[2] Similar to AdhesivesMart.

com, this tool allows a user to enter a number of variables, and the selector will suggest the best possible materials to use. The variables include:

- Substrate to be bonded
- Upper limit of temperature to which the adhesive can be exposed
- Lower limit of temperature to which the adhesive can be exposed
- Maximum tensile strength
- Viscosity
- Cure time
- Additional cure needs
- Environmental conditions to which the adhesive may be subjected
- Chemicals to which the adhesive may be subjected
- Special requirements and traits, such as thermal conductivity and FDA clearance

The Adhesives Selector is very intuitive. The user follows a clear and logical path to get results that let him or her speak intelligently to adhesive suppliers. The user first enters the basic material types that are to be joined. From there, the selector allows a more detailed description of the materials, and the user can enter more detailed process and joining information. The results are provided to the user in a weighted manner, indicating which adhesive type works best in the conditions the user has selected. Clicking on the selected adhesive types provides a detailed description of adhesive properties.

The Adhesive Selector is the result of a cooperative effort between the editors of *Assembly* magazine and the Edison Welding Institute (EWI). Future additions to the selector include links to treatises on adhesive use and technical papers.

19.3.3 Sealant System

The National Research Council of Canada began developing an expert computer-based system for sealant joint design.[3] Innovative Technology, Inc. of Ottawa Ontario, has taken this work and developed it into a Microsoft Windows-based program called Sealant System.

The following data is provided as input to the Sealant System:

- Project characteristics
- Joint description
- Joint design
- Joint characteristics
- Sealant characteristics

The user first inputs the geographical location, construction tolerances, and units of measure that are applicable to the sealant project. The joint descriptions are provided by various pull-down menus. They feature parameters such as the joint location in a building, and whether or not to use a primer. For the joint design, the user selects the joint shape as either butt or fillet. From various pull-down menus, the substrate materials and characteristics are also chosen. Dimensions are added to the joint design at this stage. The user then chooses various desirable sealant properties, such as mildew resistance, paintability, or abrasion resistance. Finally, a decision is made to specify the joint width either directly or by indicating the movement range of interest.

Once this data are placed into the system, the Sealant System designs the joint and provides the following output:

- All calculations
- Possible commercial sealants
- All relevant specifications

Joint design calculations for thermal and moisture movements are provided as well as joint width dimensions and sealant movement range. The program selects from its database of sealants those materials that meet the joint movement and input sealant characteristics. The program can also print out sealant specifications that are relevant to the project.

Sealant System is a stand alone computer program. Future versions of the program will incorporate user comments, the latest contribution to sealant joint design testing and theory. Other joint design methodologies, such as tension and compression combined with shear movement, will also be incorporated into the program.

19.3.4 ADHESIVES

ADHESIVES is a database access program whose main focus is facilitating access to basic adhesive bonding technology.[4] It is sponsored by the U.S. Army Material Command and the Army Aviation Systems Command (AVS-COM). Prototype adhesives databases were constructed and have been reported in the literature, and these have led to the more advanced ADHESIVES database

The ADHESIVES database includes not only the properties of specific adhesives, but also additional supporting information on adherends, surface preparations, physical/chemical test methods, and repair techniques and procedures. In addition, major sections of the database are devoted to information on the design and manufacturing of bonded structures and lessons learned from past adhesive bonding problems. The database also provides the technical results from adhesive evaluations performed by DOD agencies, independent test labs, and adhesive manufacturers.

The information contained in the database is unclassified, but a significant portion is categorized for government use only. Program access is therefore restricted to government and contractor personnel strictly on a "need to know" basis. In an effort to assure maximum program use, company proprietary information has been excluded.

The information that is provided and maintained in the ADHESIVES database is the following:

- Laboratory test data
- Trade identifications
- Adhesive materials
- Surface preparations
- Adherends
- Test methods
- Documents
- Glossary

One of the most interesting sections of the ADHESIVES database is the "Lessons Learned" section. One could argue that the best available source of information is the historical knowledge gained from experience with actual adhesive bonding systems. The Lessons Learned files can be accessed in three ways: type of material, hardware item, or category of failure. Submenus then narrow the possible choices to a small list of fields. The Category of Failures files classifies the origin of failure to either the design, manufacturing, or service phases of the hardware life cycle. In each case, a record appears that describes the hardware, the problem, its solution, and any lesson derived from the incident.

References

1. "Web Site Speeds Search for Industrial Adhesives", (no author), *Adhesives Age,* May 1998.
2. Williams, T. A., "Adhesive Selector Debuts", *Assembly,* November 1998.
3. O'Connor, T. F., and Myers, J. C., "Black Magic and Sealant Joints: Very Little in Common Anymore", *Adhesives Age,* 1997.
4. Desmond, A. T., et al., "Database Helps Engineers Choose Proper Bonding Method", *Adhesives Age,* May 1992.

Appendix A

Glossary

A.1 Standard Definitions of Terms Relating to ADHESIVES

A-stage — an early stage in the reaction of certain thermosetting resins in which the material is fusible and still soluble in certain liquids. Sometimes referred to as Resol. (See also **B-stage** and **C-stage**)

abhesive — a material which is adhesive resistant and applicable as a nonsticking surface coating; release agent

acceptance test — a test, or series of tests conducted by the procuring agency, or an agent thereof, upon receipt to determine whether an individual list of materials conforms to the purchase order or contract or to determine the degree of uniformity of the materials supplied by the vendor, or both. Note: specifications usually state sampling technique, test procedures, and minimum requirements for acceptance.

accelerated aging — see **aging, accelerated.**

adhere — to cause two surfaces to be held together by adhesion.

The following sources were used in compiling this Appendix:
ASTM D 907, "Standard Definitions of Terms Relating to Adhesives," American Society for Testing and Materials, Conshohocken, PA.
ASTM C717, "Terminology of Building Seals and Sealants," American Society for Testing and Materials, Conshohocken, PA.
MIL-HDBK-691B, Military Standardization Handbook, *Adhesive Bonding,* Department of Defense, Washington, DC.
Panek, J. R. and Cook, J. P., *Construction Sealants and Adhesives,* John Wiley & Sons, New York, 1991.
Sharpe, L. H., "Fundamentals of Adhesives and Sealants Technology," *Adhesives and Sealants,* vol. 3, Engineered Materials Handbook Series, ASM International, 1990.

adherend—a body which is held to another body by an adhesive. (See also **substrate**)

adherend preparation—see **surface preparation.**

adhesion—the state in which two surfaces are held together by interfacial forces which may consist of valence forces or interlocking action, or both. (See also **adhesion, mechanical** and **adhesion, specific**)

adhesion, mechanical—adhesion between surfaces in which the adhesive holds the parts together by interlocking action. (See also **adhesion, specific**)

adhesion, specific—adhesion between surfaces which are held together by valence forces of the same type as those which give rise to cohesion. (See also **adhesion, mechanical**)

adhesive—a substance capable of holding materials together by surface attachment. Note: Adhesive is a general term and includes among others cement, glue, mucilage, and paste. All of these terms are loosely used interchangeably. Various descriptive adjectives are applied to the term adhesive to indicate certain characteristics as follows: (1) physical form, that is, liquid adhesive, tape adhesive, etc.; (2) chemical type, that is, silicate adhesive, resin adhesive, etc.; (3) materials bonded, that is, paper adhesive, metal-plastic adhesive, can label adhesive, etc.; (4) condition of use, that is, hot setting adhesive, room temperature setting adhesive, etc.

adhesive, assembly—an adhesive that can be used for bonding parts together, such as in the manufacture of a boat, airplane, furniture, and the like. Note: The term assembly adhesive is commonly used in the wood industry to distinguish such adhesives (formerly called "joint glues") from those used in making plywood (sometimes called "veneer glues"). It is applied to adhesives used in fabricating finished structures or goods, or subassemblies thereof, as differentiated from adhesive used in the production of sheet materials for sale as such, for example, plywood or laminates.

adhesive, anaerobic—an adhesive that cures spontaneously in the absence of oxygen, the curing being inhibited by the presence of oxygen and catalyzed by metallic ions.

adhesive, cellular—see **adhesive, foamed**

adhesive, cold setting—an adhesive that sets at temperatures below 68°F. (See also **adhesive, hot setting; adhesive, intermediate temperature setting;** and **adhesive, room temperature setting**)

adhesive contact—an adhesive that is apparently dry to the touch and which will adhere to itself instantaneously upon contact; also called contact bond adhesive or dry bond adhesive.

adhesive, dispersion (or emulsion) — a two phase system with one phase (the adhesive material) in a liquid suspension.

adhesive, encapsulated — an adhesive in which the particles or droplets of one of the reactive components are enclosed in a protective film (microcapsules) to prevent cure until the film is destroyed by suitable means.

adhesive, film — an adhesive in film form, with or without a carrier, usually set by means of heat and/or pressure. The main advantage is uniformity of glueline thickness.

adhesive, foamed — an adhesive, the apparent density of which has been decreased substantially by the presence of numerous gaseous cells dispersed through its mass.

adhesive, gap filling — an adhesive subject to low shrinkage in setting, can be employed as a sealant.

adhesive, heat activated — a dry adhesive film that is rendered tacky or fluid by application of heat or heat and pressure to the assembly.

adhesive, heat sealing — a thermoplastic film adhesive which is melted between the adherend surfaces by heat application to one or both of the adjacent adherend surfaces.

adhesive, hot melt — an adhesive that is applied in a molten state and forms a bond on cooling to a solid state.

adhesive, hot setting — an adhesive that requires a temperature at or above 212°F to set it. (See also **adhesive, cold setting; adhesive, intermediate temperature setting;** and **adhesive, room temperature setting**)

adhesive, intermediate temperature setting — an adhesive that sets in the temperature range from 90° to 210°F.

adhesive, latex — an emulsion of rubber or thermoplastic rubber to water.

adhesive, multiple layer — a film adhesive usually supported with a different adhesive composition on each side; designed to bond dissimilar materials such as the core to face bond of a sandwich composite.

adhesive, one component — an adhesive material incorporating a latent hardener or catalyst activated by heat. Usually refers to thermosetting materials, but also describes anaerobic, hot melt adhesive, or those dependent on solvent loss for adherence. Thermosetting one component adhesives require heat to cure.

adhesive, pressure sensitive — a viscoelastic material which in solvent free form remains permanently tacky. Such materials will adhere instantaneously to most solid surfaces with the application of very slight pressure.

adhesive, room temperature setting—an adhesive that sets in the temperature range from 68° to 86°F.

adhesive, separate application—a term used to describe an adhesive consisting of two parts, one part being applied to one adherend and the other part to the other adherend and the two brought together to form a joint. Also known as honeymoon adhesives.

adhesive, solvent—an adhesive having a volatile organic liquid as a vehicle. Note: This term excludes water based adhesive.

adhesive, solvent activated—a dry adhesive film that is rendered tacky just prior to use by application of a solvent.

adhesive, structural—an adhesive of proven reliability in engineering structural applications in which the bond can be stressed to a high proportion of its maximum failing load for long periods without failure.

adhesive, two component—an adhesive supplied in two parts which are mixed before application. Such adhesives usually cure at room temperature.

adhesive, warm setting—a term that is sometimes used as a synonym for Intermediate Temperature Setting Adhesive.

adsorption—the action of a body in condensing and holding gases and other materials at its surface.

aging, accelerated—a set of laboratory conditions designed to produce in a short time the results of normal aging. Usual factors include temperature, light, oxygen, water, and other environments as needed.

aging time—see **time, joint conditioning.**

aggressive tack—see **tack, dry.**

ambient temperature—see **temperature, ambient.**

amorphous phase—noncrystalline; most plastics are amorphous at processing temperature. Many retain this strength under normal temperatures.

amylaceous—pertaining to, or of the nature of, starch; starchy.

assembly—a group of materials or parts including adhesive, which has been placed together for bonding or which has been bonded together.

assembly adhesive—see **adhesive, assembly.**

assembly glue—see **adhesive, assembly.**

assembly time—see **time, assembly.**

autoclave — a closed container which provides controlled heat and pressure conditions.

B-stage — an intermediate stage in the reaction of certain thermosetting resins in which the materials soften when heated and swell when in contact with certain liquids, but may not entirely fuse or dissolve. The resin in an uncured thermosetting adhesive is usually in this stage. Sometimes referred to as Resitol. (See also **A-stage** and **C-stage**)

backing — the flexible supporting materials for an adhesive. Pressure sensitive adhesives are commonly backed with paper, plastic films, fabric, or metal foil; heat curing thermosetting adhesives are often supported on glass cloth backing.

batch — the manufactured unit or a blend of two or more units of the same formulation and processing.

bag molding — a method of molding or bonding involving the application of fluid pressure, usually by means of air, steam, water, or vacuum, to a flexible cover which, sometimes in conjunction with the rigid die, completlely encloses the materials to be bonded.

bag, vacuum — a flexible bag by which pressure may be applied to an assembly inside the bag by means of evacuation of the bag.

binder — a component of an adhesive composition that is primarily responsible for the adhesive forces which hold two bodies together. (See also **extender** and **filler**)

bite — the penetration or dissolution of adherend surfaces by an adhesive.

blister — an elevation of the surface of an adherend, somewhat resembling in shape a blister on the human skin; its boundaries may be indefinitely outlined and it may have burst and become flattened. Note: A blister may be caused by insufficient adhesive; inadequate curing time, temperature, or pressure; trapped air, water, or solvent vapor.

blocked curing agent — a curing agent or hardener rendered unreactive, which can be reactivated as desired by physical or chemical means.

blocking — an undesired adhesion between touching layers of material, such as occurs under moderate pressure during storage or use.

body — the consistency of an adhesive which is a function of viscosity, plasticity, and rheological factors.

bond, n — the union of materials by adhesives.

bond, v — to unite materials by means of an adhesive. (See also **adhere**)

bond line — see **glue line.**

bond strength — the unit load applied in tension, compression, flexure, peel, impact, cleavage, or shear, required to break an adhesive assembly with failure occurring in or near the plane of the bond. Note: The term adherence is frequently used in place of bond strength.

bond, structural — see **structural bond.**

caul — a sheet of materials employed singly or in pairs in hot or cold pressing of assembles being bonded. Note 1: A caul is used to protect either the faces of the assembly, of the press platens, or both, against marring and staining; to prevent sticking; to facilitate press loading; to impart a desired surface texture or finish; and to provide uniform pressure distribution. Note 2: A caul may be made of any suitable materials such as aluminum, stainless steel, hardboard, fiberboard, or plastic; the length and width dimensions being generally the same as those of the platen of the press where it is used.

C-stage — the final stage in the reaction of certain thermosetting resins in which the material is relatively insoluble and infusible. Certain thermosetting resins in a fully cured adhesive layer are in this stage. Sometimes referred to as Resite. (See also **A-stage** and **B-stage**)

catalyst — a substance that markedly speeds up the cure of an adhesive when added in minor quantity as compared to the amounts of the primary reactants. (See also **hardener** and **inhibitor**)

cement, n — see **adhesive.**

cement, v — see **bond, v.**

closed assembly time — see **time, assembly.**

cohesion — the state in which the particles of a single substance are held together by primary or secondary valence forces. As used in the adhesive field, the state in which the particles of the adhesive (or the adherend) are held together.

cold flow — see **creep.**

cold pressing — a bonding operation in which an assembly is subjected to pressure without the application of heat.

cold setting adhesive — see **adhesive, cold setting.**

collagen — the protein derived from bone and skin used to prepare animal glue and gelatin.

colophony — the resin obtained from various species of pine trees.

condensation — a chemical reaction in which two or more molecules combine with the separation of water or some other simple substance. If a polymer is formed, the process is called polycondensation. (See also **polymerization**).

conditioning time — see **time, joint conditioning.**

consistency—that property of a liquid adhesive by virtue of which it tends to resist deformation. Note: Consistency is not a fundamental property but is comprised of viscosity, plasticity, and other phenomena. (See also **viscosity** and **viscosity coefficient**)

contact bonding—the deposition of cohesive materials on both adherend surfaces and their assembly under pressure.

copolymer—see **polymer.**

copolymerization—see **polymerization.**

core—the honeycomb structure used in sandwich panel construction.

corrosion—the chemical reaction between the adhesive or contamination and the adherend surfaces, due to reactive compounds in the adhesive film, leading to deterioration of the bond strength.

coverage—the spreading power of an adhesive over the surface area of the adherend

crazing—fine cracks that may extend in a network on or under the surface of or through a layer of adhesive.

creep—the dimensional change with time of a material under load, following the initial instantaneous elastic or rapid deformation. Creep at room temperature is sometimes called cold flow.

cross laminated—see **laminated, cross.**

crosslinking—the union of adjacent molecules of uncured adhesive (often existing as long polymer chains) by catalytic or curing agents.

crystallinity—a state of molecular structure in some polymers denoting uniformity and compactness of the molecular chains.

cure—to change the physical properties of an adhesive by chemical reaction, which may be condensation, polymerization, or vulcanization; usually accomplished by the action of heat and catalyst, alone or in combination with or without pressure.

curing agent—see **hardener.**

curing temperature—see **temperature, curing.**

curing time—see **time, curing.**

degrease—to remove oil and grease from adherend surfaces.

delamination—the separation of layers in a laminate because of failure of the adhesive, either in the adhesive itself or at the interface between the adhesive and the adherend.

dextrin—a water based product derived from the acidification and/or roasting of starch.

dielectric curing—the use of a high frequency electric field through a joint to cure a synthetic thermosetting adhesive. A curing process for wood and other nonconductive joint materials.

diluent—an ingredient usually added to an adhesive to reduce the concentration of bonding materials. (See also **extender** and **thinner**)

doctor-bar or blade—a scraper mechanism that regulates the amount of adhesive on the spreader rolls or on the surface being coated.

doctor roll—a roller mechanism that is revolving at different surface speed, or in opposite directions, resulting in a wiping action for regulating the adhesive supplied to the spreader roll.

double spread—see **spread.**

dry—to change the physical state of an adhesive or an adherend by the loss of solvent constituents by evaporation or absorption, or both. (See also **cure** and **set**)

dry strength—see **strength, dry.**

dry tack—see **tack, dry.**

drying temperature—see **temperature, drying.**

drying time—see **time, drying.**

elasticity, modulus of—the ratio of stress to strain in elastically deformed materials.

elastomer—a macromolecular material which, at room temperature, is capable of recovering substantially in size and shape after removal of a deforming force.

epoxy—a resin formed by combining epichlorohydrin and bisphenols. Requires a curing agent for conversion to a plastic-like solid. Has outstanding adhesion and excellent chemical resistance.

exothermic—a chemical reaction that gives off heat.

extender—a substance, generally having some adhesive action, added to an adhesive to reduce the amount of the primary binder required per unit area. (See also **binder, diluent, filler,** and **thinner**)

failure, adherend—joint failure by cohesive failure of the adherend.

failure, adhesive—rupture of an adhesive bond, such that the separation appears to be at the adhesive-adherend interface. Note: Sometimes termed failure in adhesion.

failure, cohesive—rupture of an adhesive bond, such that the separation appears to be within the adhesive.

failure, contact—the failure of an adhesive joint as a result of incomplete contact during assembly, between adherend and adhesive surfaces or between adhesive surfaces.

faying surface—the surface of an adherend which makes contact with another adherend.

feathering—the tapering of an adherend on one side to form a wedge section, as used in a scarf joint.

filler—a relatively nonadhesive substance added to an adhesive to improve its working properties, permanence, strength, or other qualities. (See also **binder** and **extender**)

filler sheet—a sheet of deformable or resilient materials that when placed between the assembly to be bonded and the pressure applicator, or when distributed within a stack of assemblies, aids in providing uniform application of pressure over the area to be bonded.

fillet—that portion of an adhesive which fills the corner or angle formed where two adherends are joined.

flow—movement of an adhesive during the bonding process, before the adhesive is set.

gel—a semisolid system consisting of a network of solid aggregates in which liquid is held.

gelation—formation of a gel.

glue, n—originally, a hard gelatin obtained from hides, tendons, cartilage, bones, etc. of animals. Also, an adhesive prepared from this substance by heating with water. Through general use the term is synonymous with the term adhesive. (See also **adhesive, mucilage, paste,** and **sizing**)

glue, v—see **bond,** v.

glue line (bond line)—the layer of adhesive which attaches two adherends.

green strength—the ability of an adhesive to hold two surfaces together when brought into contact and before the adhesive develops its ultimate bond properties when fully cured.

gum—any class of colloidal substances exuded by or prepared from plants, sticky when moist, composed of complex carbohydrates and organic acids, which are soluble or swell in water (See also **adhesive, glue, resin**). Note: The term gum is sometimes used loosely to denote various materials that exhibit gummy characteristics under certain conditions, for example, gum balata, gum benzoin, and gum asphaltum. Gums are included by some in the category of natural resins.

hardener—a substance or mixture of substances added to an adhesive to promote or control the curing reaction by taking part in it. The term is also used to designate a substance added to control the degree of hardness of the cured film. (See also **catalyst**)

heat reactivation—the use of heat to effect adhesive activity, e.g., hot melt adhesive; completion of the curing process of a B-staged resin.

honey comb core—a sheet of material, which may be metal, formed into cells (usually hexagonal) and used for sandwich construction in structural assemblies, especially in aircraft construction.

hot setting adhesive—see **adhesive, hot setting.**

hydrolysis—decomposition of a substrate or adhesive by the reaction with water.

hygroscopic—material capable of absorbing and retaining environmental moisture.

inhibitor—a substance that slows down chemical reaction. Inhibitors are sometimes used in certain types of adhesives to prolong storage or working life.

interface—the contact area between adherend and adhesive surfaces.

intermediate temperature setting adhesive—see **adhesive, intermediate temperature setting.**

jig—an apparatus used to hold a bonded assembly until the adhesive has cured.

joint—the location at which two adherends are held together with a layer of adhesive. (See also **bond,** n)

joint aging time—see **time, joint conditioning.**

joint conditioning time—see **time, joint conditioning.**

joint, lap—a joint made by placing one adherend partly over another and bonding together the overlapped portions. (See also **joint, scarf**)

joint, scarf—a joint made by cutting away similar angular segments of two adherends and bonding the adherends with the cut areas fitted together. (See also **joint, lap**)

joint, starved—a joint that has an insufficient amount of adhesive to produce a satisfactory bond. Note: This condition may result from too thin a spread to fill the gap between the adherend, excessive penetration of the adhesive into the adherend, too short an assembly time, or the use of excessive pressure.

laminate, n—a product made by bonding together two or more layers of material or materials. (see also **laminated, cross** and **laminated, parallel**)

laminate, v—to unite layers of materials with adhesive.

lamination—the process of preparing a laminate. Also, a layer in a laminate.

laminated, cross—a laminate in which some of the layers of materials are oriented at right angles to the remaining layers with respect to the grain or strongest direction in tension. (See also **laminated, parallel**) Note: Balanced construction of the laminations about the center line of the thickness of the laminate in normally assumed.

laminated, parallel—a laminate in which all the layers of materials are oriented approximately parallel with respect to the grain or strongest direction in tension (See also **laminated, cross**)

lap joint—see **joint, lap.**

legging—the drawing of filaments or strings when adhesive bonded substrates are separated.

manufactured unit—a quantity of finished adhesive or finished adhesive component, processed at one time. Note: The manufactured unit may be a batch or a part thereof.

matrix—the part of an adhesive which surrounds or engulfs embedded filler or reinforcing particles and filaments.

mechanical adhesion—see **adhesion, mechanical** and **adhesion, specific.**

modifier—any chemically inert ingredient added to an adhesive formulation that changes its properties. (See also **filler, plasticizer,** and **extender**)

modulus—see **elasticity, modulus of.**

monomer—a relatively simple compound which can react to form a polymer. (See also **polymer**)

mucilage—an adhesive prepared from a gum and water. Also in more general sense, a liquid adhesive which has a low order of bonding strength. (See also **adhesive, glue, paste,** and **sizing**)

Newtonian fluid—a fluid in which the shearing rate is directly proportional to the applied torque.

novolak—a phenolic-aldehydic resin that, unless a source of methylene groups is added, remains permanently thermoplastic. (See also **reinoid** and **thermoplastic**)

open assembly time—see **time, assembly.**

parallel laminated—see **laminated, parallel.**

paste—an adhesive composition having a characteristic plastic type consistency, that is, a high order of yield value, such as that of a paste prepared by heating a mixture of starch and water and subsequently cooling the hydrolyzed product. (See also **adhesive, glue, mucilage,** and **sizing**)

penetration—the entering of an adhesive into an adherend. Note: This property of a system is measured by the depth of penetration of the adhesive into the adherend.

permanence — the resistance of an adhesive bond to deteriorating influences.

pick-up roll — a spreading device where the roll for picking up the adhesive runs in a reservoir of adhesive.

plasticity — a property of adhesives that allows the materials to be deformed continuously and permanently without rupture upon the application of a force that exceeds the yield value of the material.

plasticizer — a material incorporated in an adhesive to increase its flexibility, workability, or distensibility. The addition of the plasticizer may cause a reduction in melt viscosity, lower the temperature of the second order transition, or lower the elastic modulus of the solidified adhesive.

plywood — a cross-bonded assembly made of layers of veneer or veneer in combination with a lumber core or plies joined with an adhesive. Two types of plywood are recognized, namely (1) veneer plywood and (2) lumber core plywood. Note: Generally the grain of one or more plies is approximately at right angles to the other plies, and almost always an odd number of plies are used.

polycondensation — see **condensation.**

polymer — a compound formed by the reaction of simple molecules having functional groups which permit their combination to proceed to high molecular weights under suitable conditions. Polymers may be formed by polymerization (addition polymer) or polycondensation (condensation polymer). When two or more monomers are involved, the product is called a copolymer.

polymerization — a chemical reaction in which the molecules of a monomer are linked together to form large molecules whose molecular width is a multiple of that of the original substance. When two or more monomers are involved, the process is called copolymerization or heteropolymerization. (See also **condensation**)

porosity — the ability of an adherend to absorb and adhere.

post cure, n — a treatment (normally involving heat) applied to an adhesives assembly following the initial cure, to modify specific properties.

post cure, v — to expose an adhesive assembly to an additional cure, following the initial cure, for the purpose of modifying specific properties.

post vulcanization bonding — conventional adhesive bonding of previously vulcanized elastomeric adherends.

pot life — see **working life.**

prebond treatment — see **surface preparation.**

preproduction test—a test or series of tests conducted by (1) an adhesive manufacturer to determine conformity of an adhesive batch to established production standards, (2) a fabricator to determine the quality of an adhesive before parts are produced, or (3) an adhesive specification custodian to determine conformance of an adhesive to the requirements of a specification not requiring qualification tests.

pressure sensitive adhesive—see **adhesive, pressure sensitive.**

primer—a coating applied to a surface, prior to the application of an adhesive, to improve the performance of the bond.

qualification test—a series of tests conducted by the procuring activity, or an agent thereof, to determine conformance of materials, or materials system, to the requirements of a specification which normally results in a qualified products list under the specification.

release agent—an adhesive material which prevents bond formation.

release paper—a sheet, serving as a protectant or carrier, or both, for an adhesive film or mass, which is easily removed from the film or mass prior to use.

resin—a solid, semisolid, or pseudosolid organic material that has an indefinite and often high molecular weight, exhibits a tendency to flow when subjected to stress, usually has a softening or melting range, and usually fractures conchoidally. Note: A "liquid resin" is an organic polymeric liquid which when converted to its final state becomes a resin.

resinoid—any of the class of thermosetting synthetic resins, either in their initial temporary fusible state or in their final infusible state. (See also **novolak** and **thermosetting**)

resite—an alternative term for C-stage. (See also **C-stage**)

resitol—an alternative term for B-stage (See also **B-stage**)

resol—an alternative term for A-stage (See also **A-stage**)

retarder—see **inhibitor.**

retrogradation—a change of starch pastes from low to high consistency on aging.

room temperature setting adhesive—see **adhesive, room temperature setting.**

rosin—a resin obtained as a residue in the distillation of crude turpentine from the sap of the pine tree (gum rosin) or from an extract of the stumps and other parts of the tree (wood rosin).

sagging—run or flow-off of adhesive from an adherend surface due to application of excess or low viscosity material.

sandwich panel—an assembly composed of metal skins (facings) bonded to both sides of a lightweight core.

sealant—a gap filling material to prevent excessive absorption of adhesive, or penetration of liquid or gaseous substances.

scarf joint—see **joint, scarf.**

self-curing—see **self-vulcanizing.**

self-vulcanizing—pertaining to an adhesive that undergoes vulcanization without the application of heat.

separate application adhesives—see **adhesive, separate application.**

service conditions—the environmental conditions to which a bonded structure is exposed, e.g., heat, cold, humidity, radiation, vibration, etc.

set—to convert an adhesive into a fixed or hardened state by chemical or physical action, such as condensation, polymerization, oxidation, vulcanization, gelation, hydration, or evaporation of volatile constituents. (See also **cure** and **dry**)

setting temperature—see **temperature, setting.**

setting time—see **time, setting.**

shear, tensile—the apparent stress applied to an adhesive in a lap joint.

shelf life—see **storage life.**

shortness—a qualitative term that describes an adhesive that does not string on cotton, or otherwise form filaments or threads during application.

shrinkage—the volume reduction occurring during adhesive curing, sometimes expressed as percentage volume or linear shrinkage; size reduction of adhesive layer due to solvent loss or catalytic reaction.

single spread—see **spread.**

size—see **sizing.**

sizing—the process of applying a material on a surface in order to fill pores and thus reduce the absorption of the subsequently applied adhesive or coating or to otherwise modify the surface properties of the substrate to improve the adhesion. Also, the materials used for this purpose. The latter is sometimes called size.

slippage—the movement of the adherends with respect to each other during the bonding process.

solids content—the percentage of weight of the nonvolatile matter in an adhesive. Note: The actual percentage of the nonvolatile matter in an adhesive will vary considerably according to the analytical procedure that is used. A standard test method must be used to obtain consistent results.

solvent bonding—see **solvent welding**.

solvent cement—an adhesive utilizing an organic solvent as the means of depositing the adhesive constituent.

solvent cementing—see **solvent welding**.

solvent reactivating—the application of solvent to a dry adhesive layer to regenerate its wetting properties.

solvent welding—the process of joining articles made of thermoplastic resins by applying a solvent capable of softening the surfaces to be joined and pressing the softened surfaces together. Adhesion is attained by means of evaporation of the solvent, absorption of the solvent into the adjacent materials and/or polymerization of the solvent cement.

specific adhesion—see **adhesion, specific** and **adhesion, mechanical**.

spread—the quantity of adhesives per unit joint area applied to an adherend, usually expressed in points of adhesive per thousand square feet of joint area. (1) Single spread refers to application of adhesive to only one adherend of a joint. (2) Double spread refers to application of adhesive to both adherends of a joint.

squeeze-out—excess adhesive pressed out at the bond line due to pressure applied in the adherends.

stabilizer—an adhesive additive which prevents or minimizes change in properties, e.g., by adherend absorption, demulsification, or rapid chemical reaction.

storage life—the period of time during which a packaged adhesive can be stored under specified temperature conditions and remain suitable for use. Sometimes called Shelf Life. (See also **working life**)

starved joint—see **joint, starved**.

strength, cleavage—the tensile load expressed in force per unit of width of bond required to cause cleavage separation of a test specimen of unit length.

strength, dry—the strength of an adhesive joint determined immediately after drying under specified conditions or after a period of conditioning in the standard laboratory atmosphere.

strength, fatigue—the maximum load expressed in force per unit of width of bond required to cause cleavage separation of a test specimen of unit length.

strength, fatigue—the maximum load that a joint will sustain when subjected to repeated stress application after drying, or after a conditioning period under specified conditions.

strength, impact—ability of an adhesive material to resist shock by a sudden physical blow direct against it. Impact shock is the transmission of stress to an adhesive interface by sudden vibration or jarring blow of the assembly, measured in work units per unit area.

strength, lap joint—the force necessary to rupture an adhesive joint by means of stress applied parallel to the plane of the bond. Also referred to as tensile-shear strength.

strength, peel—the force per unit width necessary to bring an adhesive to the point of failure and/or to maintain a specified rate of failure by means of a stress applied in a peeling mode.

strength, shear—the resistance of an adhesive joint to shearing stresses; the force per unit areas sheared, at failure.

strength, tensile—the resistance of an adhesive joint to tensile stress; the force per unit area under tension, at failure.

strength, wet—the strength of an adhesive joint determined immediately after removal from a liquid in which it has been immersed under specified conditions of time, temperature, and pressure. Note: The term is commonly used also to designate strength after immersion in water. In the latex adhesive, the term is also used to describe the joint strength when the adherends are brought together with the adhesive still in the wet state.

stringiness—the property of an adhesive that results in the formation of filaments or threads when adhesive transfer surfaces are separated. (See also **webbing**) Note: Transfer surfaces may be rolls, picker plates, stencils, etc.

structural adhesive—see **adhesive, structural.**

structural bond—a bond which stresses the adherend to the yield point, thereby taking full advantage of the strength of the adherend. On the basis of this definition, a dextrin adhesive used with paper (e.g., postage stamps, envelopes, etc.) and which cause failure of the paper, forms a structural bond. The stronger the adherend, the greater the demands placed on the adhesive. Thus, few adhesives qualify as "structural" for metals. A further requirement for a structural adhesive is that it be able to stress the adherend to its yield point after exposure in its intended environment.

substrate—a material upon the surface of which an adhesive containing substance is spread for any purpose, such as bonding or coating. A broader term than adherend. (See also **adherend**)

surface preparation—a physical or chemical preparation, or both, of an adherend to render it suitable for adhesive joining.

synersis—the exudation of small amounts of liquid by gels on standing.

tack—the property of an adhesive that enables it to form a bond of measurable strength immediately after adhesive and adherend are brought into contact under low pressure.

tack, dry—the property of certain adhesives, particularly nonvulcanizing rubber adhesives, to adhere on contact to themselves at a stage in the evaporation of volatile constituents, even though they seem dry to the touch. Sometimes called Aggressive Tack.

tack range—the period of time in which an adhesive will remain in the tacky dry condition after application to an adherend, under specified conditions of temperature and humidity.

tacky-dry—pertaining to the condition of an adhesive when the volatile constituents have evaporated or been absorbed sufficiently to leave it in a desired tacky state.

tackifier—an additive intended to improve the stickiness of a cast adhesive film; usually a constituent of rubber based and synthetic resin adhesives.

tape—a film form of adhesive which may be supported on carrier material.

teeth—the resultant surface irregularities or projections formed by the beading of filaments or strings which may form when adhesive bonded substrates are separated.

telegraphing—a condition in a laminate or other type of composite construction in which irregularities, imperfections, or patterns of an inner layer are visibly transmitted to the surface.

temperature, ambient—temperature of the air surrounding the object under construction or test.

temperature, curing—the temperature to which an adhesive or an assembly is subjected to cure the adhesive (See also **temperature, drying and temperature, setting**) Note: The temperature attained by the adhesive in the process of curing it (adhesive curing temperature) may differ from the temperature of the atmosphere surrounding the assembly (assembly curing temperature).

temperature, drying—the temperature to which an adhesive on an adherend or in an assembly or the assembly itself is subjected to dry the adhesive. (See also **temperature, curing,** and **temperature, drying**)

temperature, maturing—the temperature, as a function of time and bonding condition, that produces desired characteristics in bonded components.

temperature, setting—the temperature to which an adhesive or an assembly is subjected to set the adhesive (See also **temperature, curing** and **temperature, drying**)

tests, destructive—tests involving the destruction of assemblies in order to evaluate the maximum performance of the adhesive bond.

tests, nondestructive—inspection tests for the evaluation of bond quality without damaging the assembly, e.g., ultrasonic, visual inspection, etc.

thermoplastic, adj—capable of being repeatedly softened by heat and hardened by cooling.

thermoplastic, n—a material that will repeatedly soften when heated and harden when cooled.

thermoset, adj—pertaining to the state of a resin in which it is relatively infusible.

thermoset, n—a material that will undergo or has undergone a chemical reaction by the action of heat, catalyst, ultraviolet light, etc., leading to a relatively infusible state.

thermosetting, adj—having the property of undergoing a chemical reaction by the action of heat, catalysts, ultraviolet light, etc., leading to a relatively infusible state.

thinner—a volatile liquid added to an adhesive to modify the consistency or other properties. (See also **diluent** and **extender**)

thixotropy—a property of adhesives systems to thin upon isothermal agitation and to thicken upon subsequent rest.

time, assembly—the time interval between the spreading of the adhesive on the adherend and the application of pressure or heat, or both, to the assembly. (1) Open assembly time is the time interval between the spreading of the adhesive on the adherend and the completion of the assembly of the parts for bonding. (2) Closed assembly time is the interval between completion of assembly of the parts for bonding and the application of pressure or heat, or both, to the assembly.

time, curing—the period of time during which an assembly is subjected to heat or pressure, or both, to cure the adhesive. (See also **time, joint conditioning** and **time, setting**)

time, drying—the period of time during which an adhesive on an adherend or an assembly is allowed to dry with or without the application of heat or pressure, or both. (See also **time, curing; time, joint conditioning;** and **time, setting**)

time, joint conditioning—the time interval between the removal of the joint from the conditions of heat or pressure, or both, used to accomplish bonding and the attainment of approximately maximum bond strength. Sometimes called joint aging time.

time, setting—the period of time during which an assembly is subjected to heat or pressure, or both, to set the adhesive. (See also **time, curing; time, joint conditioning;** and **time, drying**)

vehicle—the carrier medium (liquid) for an adhesive material which improves its ease of application to adherends; solvent component of an adhesive.

viscosity—the ratio of the shear stress existing between laminae of moving fluid and the rate of shear between these laminae. Note: A fluid is said to exhibit Newtonian behavior when the rate of shear is proportional to the shear stress. A fluid is said to exhibit non-Newtonian behavior when an increase or decrease in the rate of shear is not accompanied by proportional increase or decrease in the shear stress.

viscosity coefficient—the shearing stress tangentially applied that will induce a velocity gradient. A material has a viscosity of one poise when a shearing stress of one dyne per square centimeter produces a velocity gradient of (1 cm/s)/cm. (See also **viscosity**)

vulcanization—a chemical reaction in which the physical properties of a rubber are changed in the direction of decreased plastic flow, less surface tackiness, and increased tensile strength by reacting it with sulfur or other suitable agents. (See also **self-vulcanizing**)

vulcanize—to subject to vulcanization.

warm setting adhesives—see **adhesive, warm setting.**

warp—a significant variation from the original, true, or plane surface.

webbing—filaments or threads that may form when adhesive transfer surfaces are separated. (See also **stringiness**)

weld bonding—a process in which a joint is formed by spot welding through an uncured adhesive bond line, or by flowing an adhesive into a spot-welded joint.

wet strength—see **strength, wet.**

wetting—a surface is said to be completely wet by a liquid if the contact angle is zero, and incompletely wet if it is a finite angle. Surfaces are commonly regarded as unwettable if the angle exceeds 90 degrees.

wood build-up, laminated — an assembly made by bonding layers of veneer or lumber with an adhesive so that the grain of all laminations is essentially parallel.

wood failure — the rupturing of wood fibers in strength tests on bonded specimens, usually expressed as the percentage of the total area involved which shows such failure.

wood veneer — a thin sheet of wood, generally within the thickness range from 0.01 to 0.25 in. to be used in a laminate.

working life — the period of time during which an adhesive, after mixing with catalyst, solvent or other compounding ingredients, remains suitable for use. (See also **storage life**)

yield value — the stress (either normal or shear) at which a marked increase in deformation occurs without an increase in load.

A.2 Standard Definitions of Terms Relating to SEALANTS

abrasion resistance—resistance to wear resulting from mechanical action of a surface.

accelerated aging—a set of laboratory conditions designed to produce in a short time the results of normal aging (usual factors include temperature, light, oxygen, water, and other environments as needed).

accelerator—an ingredient used in small amounts to speed up the action of a curing agent (sometimes used as a synonym for curing agent).

acrylic—a group of thermoplastics polymers or resins formed from acrylic acid.

activator—a material that speeds up normal curing mechanisms.

adhesive failure—type of failure characterized by pulling the adhesive or sealant loose from the adherend.

adsorption—the action of a body in condensing and holding gases and other materials at its surface.

aging—the progressive change in the chemical and physical properties of a sealant or adhesive.

alligatoring—cracking of a surface into segments so that it resembles the hide of an alligator.

ambient temperature—temperature of the air surrounding the object under construction or test.

asphalt—naturally occurring mineral pitch or bitumen.

back-up—a compressible material used at the base of a joint opening to provide the proper shape factor in a sealant design; this material can also act as a bond breaker.

band-aid or **bandage sealant design**—a sealant placed above an opening designed to accommodate greater movement.

batten plate—a thin metal plate separated by sealant beads to bridge poorly designed joints.

bead—a sealant or compound after application in a joint irrespective of the method of application, such as caulking bead, glazing bead, and so on.

bedding compounds—any materials into which another material, such as a plate of glass or a panel may be embedded for close fit.

birefringence—the refraction of light in two slightly different directions to form two rays; the phenomenon can be used to locate stress in a transparent material.

bond breaker — thin layer of material such as tape used to prevent the sealant from bonding to the bottom of a joint.

bond durability — test cycle in ASTM C-920 for measuring the bond strength after repeated weather and extension cycling.

bulk compounds — any sealant or caulk that has no defined shape and is stored in a container.

butt joint — a joint in which the structural units are jointed to place the sealant into tension or compression.

carbon black — finely divided carbon used as a reinforcing filler in sealants.

cap bead — a bead placed above a gasket in a glazing design.

catalyst — substance used in small quantities to promote a reaction, while remaining unchanged itself; sometimes referred to as the curing agent for sealants.

caulk, n — an old term used to describe materials used in joints. Generally applied to oil based compounds, but later applied to materials with low movement capability. The word is also used as a substitute for sealant. ASTM C-24 has adopted the position that all caulks are sealants.

caulk, v — to fill joints with a material.

chain stopper — a material added during the polymerization process to terminate or control the degree of the reaction. This could result in soft sealants or higher elongation.

chalking — formation of slight breaks or cracks in the surface of a sealant.

chemical cure — curing by a chemical reaction. Usually involves the crosslinking of a polymer.

closed-cell foam — a foam that will not absorb water because all the cells have complete walls.

cohesion — the molecular attraction that holds the body of a sealant or adhesive together. The internal strength of an adhesive or sealant.

cohesive failure — the failure characterized by pulling the body of a sealant or adhesive apart.

compatible — the ability of two or more substances to be mixed or blended together without separating, reacting, or affecting the material adversely. However, two materials such as a sealant and a tape gasket are compatible if there is no interaction between them, and materials from one do not migrate into the other.

compression seal — a preformed seal that is installed by being compressed and inserted into the joint.

cone penetrometer—an instrument for measuring the relative hardness of soft deformable materials.

crazing—a series of fine cracks that may extend through the body of a layer of sealant or adhesive.

creep—the deformation of a body with time under constant load.

crosslinked—molecules that are joined side by side as well as end to end.

cure—to set or harden by means of a chemical reaction.

cure time—time required to effect a cure at a given temperature.

drying agent—a component of paint or a sealant that accelerated the oxidation of oils or unsaturated polymers.

depolymerization—separation of a complex molecule into simple molecules; also softening of a sealant by the same action.

durometer—an instrument used to measure hardness or Shore A hardness; may also refer to the hardness rather than the instrument.

elasticity—the ability of a material to return to its original shape after removal of a load.

elastomer—a rubbery material that returns to approximately its original dimensions in a short time after a relatively large amount of deformation.

epoxy—a resin formed by combining epichlorohydrin and bisphenols. Requires a curing agent for conversion to a plastic like solid. Has outstanding adhesion and excellent chemical resistance.

exothermic—a chemical reaction that gives off heat.

extender—an organic material used as a substitute for part of the polymer in a sealant or adhesive.

extensibility—the ability of a sealant to stretch under tensile load.

extrusion failure—failure that occurs when a sealant is forced too far out of a joint by compression forces. The sealant may be abraded by dirt or folded over by traffic.

face clearance—the distance between a glass plate and the edge of the stop.

fatigue failure—failure of a material as a result of rapid cyclic deformation.

filler—finely ground material added to a sealant or adhesive that either improves certain properties or, if used in excess, cheapens the compound.

flashing—strips, usually of sheet metal, rubber, or plastic, used to waterproof the junctions of building surfaces, such as roof peaks and valleys, or around windows.

gasket—a cured elastic but deformable material placed between two surfaces to seal the union.

gunability—the ability of a sealant to extrude out of a cartridge in a caulking gun.

gypsum wallboard—a sandwich type material. Gypsum plaster with a heavy paper coating on both sides. When fastened directly to studs, it forms a wall surface.

hardboard—fine pieces of wood bound together with an adhesive and pressed into sheets. Thermosetting resins are used as the adhesive binder.

hardener—a substance added to control the reaction of a curing agent in a sealant or adhesive. Sometimes used as a synonym for curing agent or catalyst.

hardness—the resistance of a material to indentation measured on an instrument such as a durometer. The value is an artificial number. On a Shore A Durometer scale, numbers range from 0 to 100 for rubber like materials.

head—the top member of a window or door frame.

heel bead—sealant applied at the base of a channel. This sealant bridges the gap between the glass and the frame.

Hochman test cycle—the bond durability test cycle used in ASTM C 920.

Hypalon—a chlorosulfonated polyethylene synthetic polymer that is a base for making solvent based sealants.

joint—the point at which two substrates are joined, or the opening between component parts of a structure.

laitance—a thin, weak coating that sometimes forms on the surface of concrete caused by water migration to the surface.

lap joint—a joint in which the component parts overlap so that the sealant or adhesive is placed into shear action.

latex—a colloidal dispersion of a rubber (synthetic or natural) in water, which is the base for a sealant.

latex caulks or sealants—a caulking compound or sealant using latex as the base raw material. The most common latex caulks are polyvinyl acetate and vinyl acrylic.

load transfer device—any device embedded in the concrete on both sides of a pavement joint to prevent relative vertical movement of slab edges.

mastic—a thick pasty coating or sealant.

mercaptan — an organic compound containing -SH groups; a main curing agent for polysulfide adhesives and sealants.

modulus — the ratio of stress to strain. Also the tensile strength at a given elongation.

monomer — a material composed of a single type of molecule. A building block in the manufacture of polymers.

mullion — external structural member in a curtain wall building. Usually vertical. May be placed between two opaque panels, between two window frames, or between a panel and a window frame.

MVT — moisture vapor transmission, usually expressed in terms of grams of water per square meter per 24 hours.

neck down — the change in the cross section areas of a sealant as it is extended.

needle glazing — the application of a small bead of sealant using a nozzle not exceeding 1/4 inch in diameter.

oil (drying) — an oil that dries to a hard, varnish like film. Linseed oil is a common example.

open cell foam — a foam that will absorb water and air because the walls are not complete and run together.

open time — time interval from when an adhesive is applied to when in becomes unworkable.

particle board — same as hardboard, except that larger wood chips are used as the filler.

pavement growth — an increase in the length of a pavement caused by incompressibles working into the joints.

parapet — extended upward portion of a wall above the roof line.

peel test — a test of an adhesive or sealant using one rigid and one flexible substrate. The flexible material is folded back (usually 180 degrees) and the substrates are peeled apart. Strength is usually measured in pounds per inch of width.

permanent set — the amount of deformation that remains in a sealant or adhesive after removal of a load.

phenolic resin — a thermosetting resin; usually formed by the reaction of phenol with formaldehyde.

pitch — the residue that remains after the distillation of oil and other substances from raw petroleum.

plasticizer — a material that softens a sealant or adhesive by solvent action, but is relatively permanent.

plastisol — a physical mixture of resin (usually vinyl) compatible plasticizers and pigments. Mixture requires fusion at elevated temperatures in order to convert the plastisol to a homogeneous plastic material.

poise — the cgs unit of viscosity.

polyester — resins manufactured by reacting a dicarboxylic acid and a dihydroxy alcohol. They may be used in one part and two part systems for coatings and molding compounds.

polyethylene — a straight chain polymer of ethylene.

polymer — a compound consisting of long chain like molecules.

pot life — the same as working life.

preformed sealant — a sealant that is preshaped by the manufacturer before being shipped to the job site.

reflection crack — a crack through a bituminous overlay on Portland cement concrete pavement. The crack occurs above any working joint in the base pavement.

reinforcement — in rubber or sealants, this is the increase in modulus, toughness, tensile strength, and so forth, by the addition of selected fillers.

resilience — a measure of energy stored and recovered during a loading cycle. It is expressed in a percentage.

retarder — a substance added to slow down the cure rate of a sealant.

routing — removing old sealant from a joint by means of a rotating bit or saw blade.

seal — ASTM definition is "a material applied to a joint or on a surface to prevent the passage of liquids, solids, or gases."

sealant — ASTM definition is "in building construction, a material which has the adhesive and cohesive properties to form a seal." Sometimes defined as an elastomeric material with movement capability greater than 10%, but this definition has been expanded to include all sealants covered by ASTM Committee C-24.

sealer — a surface coating generally applied to fill cracks, pores, or voids in a surface.

sealing tape — ASTM definition is "a preformed, uncured, or partially cured material which when placed in a joint, has the necessary adhesive and cohesive properties to form a seal".

self-leveling sealant — a sealant that is fluid enough to be poured into horizontal joints. It forms a smooth, level surface without tooling.

shape factor—the width-to-depth proportions of a field molded sealant.

shear test—a method of deforming a sealed or bonded joint by forcing the substrates to slide over each other. Shear strength is reported in units of force per units area (psi).

shelf life—the length of time a sealant or adhesive can be stored under specific conditions and still maintain its properties.

shore A hardness—the measurement of firmness of a rubbery compound or sealant by means of a Durometer hardness gauge.

silane—any monomeric tetrafunctional derivative of silicone, such as vinyl trichlorosilane.

skewed joints—transverse joints in a pavement slab, which are placed at an angle and not perpendicular to the direction of traffic.

spalling—a surface failure of concrete, usually occurring at the joint. It may be caused by incompressibles in the joint, by overworking the concrete, or by sawing joints too soon.

stopless glazing—the use of a sealant as a glass adhesive to keep glass in permanent position without the use of exterior stops.

strain—deformation per unit length. The change in length divided by the original length of a test specimen and expressed as a percentage.

stress—force per unit area, usually expressed in pounds per square inch.

stress relaxation—reduction in stress in a material that is held in constant deformation for an extended period of time.

structural glazing gaskets—a synthetic rubber section designed to engage the edge of glass or other sheet material in a surrounding frame by forcing an interlocking filler strip into a grooved recess in the face of the gasket.

substrate—an adherend; the surface to which a sealant or adhesive is bonded.

tackiness—the stickiness of the surface of a sealant or adhesive.

tear strength—the load required to tear apart a sealant specimen. ASTM test method D 624 expresses tear strength in pounds per inch of width.

tensile strength—resistance of a material to a tensile force and expressed in psi.

thermoplastic—a material that can be repeatedly softened by heating.

thermosetting—a material that hardens by chemical reaction that may be activated by heat or that may give off heat if the reaction is rapid.

thixotropic (nonsagging)—a material that maintains shape unless agitated. A thixotropic sealant can be placed in a joint in a vertical wall and will maintain its shape or position without sagging during the curing process.

tooling—the act of compacting and shaping a material in a joint, which may also assist in expelling any entrapped air.

transverse joint—a joint perpendicular to the direction of traffic in a highway pavement.

ultraviolet light (UV)—part of the light spectrum which can cause chemical changes in sealant materials.

viscosity—a measure of the flow properties of a liquid or paste, usually expressed in poises.

vulcanization—improving the elastic properties of a sealant or rubber by chemical change, usually by heat.

working life—period of time after mixing during which a sealant or adhesive can be used; it may be expressed in minutes or hours.

Appendix B

Other Sources of Information

Books

Adams, R. D., and Wake, W. C.,
Structural Adhesive Joints in Engineering
Elsevier Applied Science Publishers Ltd. (1984)

Allen, K. W. (ed.)
Adhesion 1 through Adhesion 13
The City University (1977–1989)

Anderson, G. P., Bennett, S. J., and DeVries, K. L.
Analysis and Testing of Adhesive Bonds
Academic Press, Inc. (1977)

Ash, M. and Ash, I.
Handbook of Adhesive Chemical and Compounding Ingredients
Synapac Information Resources, Inc. (1999)

Benedek, I. and Heymans, L. J.
Pressure Sensitive Adhesives Technology
Marcel Dekker, Inc. (1997)

Bickerman, J. J.
The Science of Adhesive Joints (2nd ed.)
Academic Press (1968)

Bikales, N. B.
Adhesion and Bonding
John Wiley & Sons (1971)

Bodnar, M. J. (ed.)
Structural Adhesives Bonding
Interscience Publishers (1966)

Bodnar, M. J. (ed.)
Processing for Adhesive Bonded Structures
Interscience Publishers (1972)

Bodnar, M. J. (ed.)
Durability of Adhesive Bonded Structures
Interscience Publishers (1976)

Bodnar, M. J. and Wentworth, S. E. (eds.)
Proceedings of the Joint Military / Government – Industry Symposium on Structural Adhesive Bonding
American Defense Preparedness Association (1987)

Bruno, E. J.
Adhesives in Modern Manufacturing
Society of Manufacturing Engineers (1970)

Brewis, D., and Comyn, J.
Advances in Adhesives: Applications, Materials and Safety
T/C Press (1983)

Buckley, J. D. and Stein, B. A. (eds.)
Joining Technologies for the 1990s – Welding, Brazing, Soldering, Mechanical, Explosive, Solid State, Adhesive
Noyes Data Corporation (1986)

Cagle, C. V. (ed.)
Handbook of Adhesive Bonding
McGraw Hill Book Co. (1973)

Comyn, J.
Adhesion Science
The Royal Society of Chemistry (1997)

DeFrayne, G. (ed.)
High Performance Adhesive Bonding
Society of Manufacturing Engineers (1983)

DeLollis, N. S.
Adhesives, Adherends, Adhesion
Industrial Press (1980)

Ellis, B. (ed.)
Chemistry and Technology of Epoxy Resins
Blackie Chapman and Hall (1993)

Evans, R. M.
Polyurethane Sealants
Technomic Publishing, Inc. (1993)

Flick, E. W.,
Construction and Structural Adhesives and Sealants — An Industrial Guide
Noyes Publications (1988)

Flick, E. W.
Handbook of Adhesive Raw Materials
Noyes Publications (1982)

Flick, E. W.
Adhesives, Sealants, and Coatings for the Electronics Industry
Noyes Publications (1986)

Flick, E. W.
Adhesives and Sealant Compound Formulations, (2nd ed.)
Noyes Publications (1984)

Gillespie, R. H. (ed.)
Adhesives for Wood — Research, Applications, and Needs
Noyes Publications (1984)

Gutcho, M.
Adhesives Technology — Developments Since 1979
Noyes Publications (1983)

Hartshorn, S. A. (ed.)
Structural Adhesives — Chemistry and Technology
Plenum Press (1986)

Haviland, G. S.
Machinery Adhesives for Locking, Retaining and Sealing
Marcel Dekker, Inc. (1986)

Houwink, R. and Salomon, G.
Adhesion and Adhesives (2nd ed.)
Volume 1: Adhesives (1965)
Volume 2: Applications (1967)
Elsevier Publishing Company

Johnson, W. S.
Adhesively Bonded Joints: Testing, Analysis and Design
ASTM Special Technical Publication STP 981
American Society of Testing and Materials (1988)

Kedward, K. T.
Joining of Composite Materials
American Society of Testing and Materials (1980)

Kinloch, A. J.
Adhesion and Adhesives — Science and Technology
Chapman and Hall (1987)

Kinloch, A. J.
Durability of Structural Adhesives
Applied Science Publisher Ltd. (1983)

Kinloch, A. J.
Structural Adhesives
Chapman and Hall (1986)

Klosowski, J. W.
Sealants in Construction
Marcel Dekker, Inc. (1989)

Koch, G. S., Klareich, F., and Exstrum, B.
Adhesives for the Composite Wood Panel Industry
Noyes Publications (1987)

Landrock, A. H.
Adhesives Technology Handbook
Noyes Publications (1985)

Lee, H. and Nevel, K.
Handbook of Epoxy Resins
McGraw Hill (1982)

Lee, Lieng Huang
Adhesion Science and Technology
Plenum Publishing Press (1975)

Matthews, F. L. (ed.)
Joining Fibre-Reinforced Plastics
Chapman and Hall (1987)

Mays, G. C. and Hutchinson, A. R.
Adhesives in Civil Engineering
Cambridge University Press (1992)

Minford, J. D.
Handbook of Aluminum Bonding Technology and Data
Marcel Dekker, Inc. (1993)

Mittal, K. L., (ed.)
Adhesives Joints — Formation, Characteristics, and Testing
Plenum Press (1984)

Panek, J. R. and Cook, J. P.
Construction Sealants and Adhesives (2nd ed.)
John Wiley & Sons, Inc. (1984)

Patrick, R. L. (ed.)
Treatise on Adhesion and Adhesives (6 volumes)
Marcel Dekker, Inc. (1967–1989)

Pizzi, A. (ed.)
Wood Adhesives — Chemistry and Technology
Vol. 1, 1983; Vol. 2, 1989
Marcel Dekker, Inc.

Pizzi A., and Mittal, K. L. (eds.)
Handbook of Adhesive Technology
Marcel Dekker, Inc. (1994)

Pocius, A.
Adhesion and Adhesives Technology
Hanser Gardner Publications (1997)

Sakek, M. M.
Industrial Application of Adhesive Bonding
Elsevier Applied Science (1987)

Satas, D.
Handbook of Pressure Sensitive Adhesive Technology
Van Nostrand Reinhold (1989)

Schneberger, G. L.
Adhesives in Manufacturing
Marcel Dekker, Inc. (1983)

Schwartz, M. M.
Joining of Composite Matrix Materials
ASM International (1994)

Skeist, I. (ed.)
Handbook of Adhesives (3rd ed.)
Van Nostrand Reinhold (1990)

Semerdjiev, S.
Metal to Metal Adhesive Bonding
Business Books Ltd. (1970)

Shields, J.
Adhesives Handbook (3rd ed.)
Butterworths (1984)

Snogren, R. C.
Handbook of Surface Preparation
Palmerton Publishing Co. (1974)

Thrall, E. W. and Shannon, R. W.
Adhesive Bonding of Aluminum Alloys
Marcel Dekker, Inc. (1985)

Wake, W. C.
Adhesion and the Formulation of Adhesives
Applied Science Publishers Ltd. (1982)

Wegman, R. F.
Surface Preparation Techniques for Adhesive Bonding
Noyes Publications (1989)

Collective Handbooks and Data Collections

Adhesives and Sealants
Volume 3, Engineered Materials Handbook
ASM International (1990)

Handbook of Plastics Joining
Plastics Design Library (1997)

Structural Adhesives Directory and Databook
Chapman and Hall (1996)

Adhesives, Sealants and Primers
D.A.T.A. Digest (5th ed.)
International Plastics Selector (1989)

American Society of Testing and Materials (ASTM)
STP 1054: *Science and Technology of Glazing Systems*
STP 1069: *Building Sealants: Materials, Properties, and Performance*
STP 1168: *Science and Technology of Building Seals, Sealants Glazing, and Waterproofing*
STP 1243: *Science and Technology of Building Seals, Sealants, Glazing and Waterproofing*
STP 1286: *Science and Technology of Building Seals, Sealants, Glazing and Waterproofing*

Government Handbooks and Publications

Military Handbook MIL-HDBK-691B
Adhesive Bonding
Naval Publications and Forms Center

Military Handbook MIL-HDBK-337
Adhesive Bonded Aerospace Structure Repair
Naval Publications and Forms Center

"Joining of Advanced Composites"
Engineering Design Handbook
Pamphlet DARCOM-OP 706-316
National Technical Information Service (NTIS)

Journals

Adhesion (in German)
Bertelsmann Fachzeitschrifen GmbH, Munich

Adhesion and Adhesives (in Japanese)
High Polymer Publishing Association, Kyoto

Adhesives Abstracts Journal
Elsevier Science Ltd., Oxford, UK

International Journal of Adhesion and Adhesives
Elsevier Science Ltd., Oxford, UK

Journal of Adhesion
Gordon Beach Science Publishers

The Journal of Adhesion Science and Technology
VSP Publishing (Netherlands)

SAMPE Journal and Quarterly
Society for the Advancement of Materials and Processing Engineering (SAMPE)

Journal of the Adhesives and Sealant Council
Adhesive and Sealant Council

Periodicals

Adhesives Age
Chemical Week Associates

Adhesives & Sealants Industry
Business News Publishing Company

The Adhesives & Sealants Newsletter
Adhesives and Sealants Consultants

The Composites & Adhesives Newsletter
T/C Press

Advanced Materials & Processes
ASM International

Materials Engineering
Penton Publishing Company

Modern Plastics
McGraw-Hill, Inc.

Packaging
Cahners Publishing Company

Plastics Compounding
Resin Publications, Inc.

Plastics Design Forum
Edgell Communications

Plastics Packaging
Edgell Communications

Plastics Technology
Bill Communications

Plastics World
Cahners Publishing Company

Technical, Trade, and Educational Organizations

United States

The Adhesives and Sealant Council (ASC)
Washington, DC
www.ascoucil.org

Adhesives Manufacturers Association (AMA)
Chicago, IL

Adhesion Society
Virginia Tech, Blacksburg, VA
www.mse.uc.edu/adsoc.ht

American Chemical Society
Washington, DC

American Institute of Chemical Engineers
New York, NY
www.aiche.org

ASM International
Materials Park, OH
www.asm-intl.org

American Society of Testing and Materials
ASTM Committee C-24 on Building Seals and Sealants
ASTM Committee D-14 on Adhesives
Conshohocken, PA
www.astm.org

Chemical Manufacturers Association (CMA)
Arlington, VA

Gordon Research Conferences, Annual Adhesion Program
University of Rhode Island
ww.grc.uri.edu

Institute of Packaging Professionals
Hendon, VA

Laminating Materials Association
Hillsdale, NJ
www.ima.org

Packaging Machinery Manufacturers Institute
Arlington, VA

Pressure Sensitive Tape Council (PSTC)
Chicago, IL
www.pstc.org

Sealant, Waterproofing & Restoration Institute (SWRI)
Kansas City, MO
www.swrionline.org

Society for the Advancement of Material and Process Engineering (SAMPE)
Covina, CA
www.et.byu.edu/sampe

Society of Plastics Engineers (SPE)
Brookfield, CT
www.4spe.org

Technical Association of the Pulp and Paper Industry (TAPPI)
Atlanta, GA

International

Adhesives & Sealants Manufacturers Association of Canada
Scarborough, ON, Canada

Association of European Adhesives Manufacturers (FEICA)
Dusseldorf, Germany
www.feica.com

British Adhesives and Sealants Association (BASS)
Stevenage Herts, UK

Japan Adhesive Industry Association
Tokyo, Japan

Japan Sealant Industry Association
Tokyo, Japan

Oxford Brookes University
School of Engineering, Joining Technical Research Centre
Headington, Oxford, UK

Appendix C

Specification and Standards

C.1 American Society for Testing and Materials (ASTM)

Adhesives and Adhesion

ASTM Standard	
B 117	Practice for Operating Salt Spray (Fog) Apparatus
C 297	Flatwise Tensile Strength of Metal to Honeycomb Core Bonds
C 557	Specification for Adhesives for Fastening Gypsum Wallboard to Wood Framing
D 69	Test Methods for Friction Tapes
D 229	Test Method for Shear Strength and Shear Modulus of Structural Adhesives (Napkin Ring Style Test Piece)
D 411	Methods for Testing Shellac Used for Electrical Insulation
D 570	Test Method for Water Absorption of Plastics
D 696	Test Method for Coefficient of Linear Thermal Expansion of Plastics Between $-30°C$ and $+30°C$ with a Vitreous Silica Dilatometer
D 816	Test Methods for Testing Rubber Cements
D 896	Test Method for Resistance of Adhesive Bonds to Chemical Reagents
D 897	Test Method for Tensile Properties of Adhesive Bonds
D 898	Test Method for Applied Weight per Unit Area of Dried Adhesive Solids
D 899	Test Method for Applied Weight per Unit Area of Liquid Adhesive

D 903	Test Method for Peel or Stripping Strength of Adhesive Bonds
D 904	Practice for Exposure of Adhesive Specimens to Artificial (Carbon Arc Type) and Natural Light
D 905	Test Method for Strength Properties of Adhesive Bonds in Shear by Compression Loading
D 906	Test Method for Strength Properties of Adhesives in Plywood Type Construction in Shear by Tension Loading
D 907	Terminology of Adhesives
D 950	Test Method for Impact Strength of Adhesive Bonds
D 997	Test Method for Tensile Properties of Adhesive Bonds
D 1000	Test Methods for Pressure Sensitive Adhesive Coated Tapes Used for Electrical and Electronic Applications
D 1002	Test Method for Strength Properties of Adhesives in Shear by Tension Loading (Metal-to-Metal)
D 1062	Test Method for Cleavage Strength of Metal-to-Metal Adhesive Bonds
D 1084	Test Methods for Viscosity of Adhesives
D 1101	Test Methods for Integrity of Glue Joints in Structural Laminated Wood Products for Exterior Use
D 1144	Practice for Determining Strength Development of Adhesive Bonds
D 1146	Test Method for Blocking Point of Potentially Adhesive Layers
D 1151	Test Method for Effect of Moisture and Temperature on Adhesive Bonds
D 1183	Test Methods for Resistance of Adhesive to Cyclic Laboratory Aging Conditions
D 1184	Test Method for Flexural Strength of Adhesive Bonded Laminated Assemblies
D 1304	Methods of Testing Adhesives Relative to Their Use as Electrical Insulation
D 1337	Test Method for Storage Life of Adhesive by Consistency and Bond Strength
D 1338	Test Method for Working Life of Liquid or Paste Adhesives by Consistency and Bond Strength
D 1344	Method of Testing Cross-Lap Specimens for Tensile Properties of Adhesives (Discontinued in 1987)
D 1382	Test Method for Susceptibility of Dry Adhesive Films to Attack by Roaches

Specification and Standards 805

D 1383	Test Method for Susceptibility of Dry Adhesive Films to Attack by Laboratory Rats
D 1488	Test Method for Amlaceous Matter in Adhesives
D 1489	Test Method for Non-Volatile Content Aqueous Adhesives
D 1490	Test Method for Non-Volatile Content of Urea Formaldehyde of Resin Solutions
D 1579	Test Method for Filler Content of Phenol, Resorcinol, and Melamine Adhesives
D 1580	Specification for Liquid Adhesives for Automatic Machine Labeling of Glass Bottles
D 1581	Test Method for Bonding Permanency of Water- or Solvent-Soluble Liquid Adhesives for Labeling Glass Bottles (Intent to Withdraw)
D 1582	Test Method for Non-Volatile Content of Phenol, Resorcinol, and Melamine Adhesives
D 1583	Test Method for Hydrogen Ion Concentration of Dry Adhesive Films
D 1584	Test Method for Water Absorptiveness of Paper Labels (Intent to Withdraw)
D 1713	Test Method for Bonding Permanency of Water- or Solvent-Soluble Liquid Adhesives for Automatic Machine Sealing Top Flaps of Fiberboard Specimens (Intent to Withdraw)
D 1714	Test Method for Water Absorptiveness of Fiberboard Specimens of Adhesives (Intent to Withdraw)
D 1779	Specification for Adhesive for Acoustical Materials
D 1780	Practice for Conducting Creep Tests of Metal-to-Metal Adhesives
D 1781	Method for Climbing Drum Peel Test for Adhesives
D 1828	Practice for Atmospheric Exposure of Adhesive Bonded Joints and Structures
D 1874	Specification for Water- or Solvent-Soluble Liquid Adhesives for Automatic Machine Scaling of Top Flaps of Fiberboard Shipping Cases
D 1875	Test Method for Density of Adhesives in Fluid Form
D 1876	Test Method for Peel Resistance of Adhesives (T-Peel Test)
D 1879	Practice for Exposure of Adhesive Specimens to High Energy Radiation
D 1916	Test Method for Penetration of Adhesives

D 1994	Test Method for Determination of Acid Numbers of Hot Melt Adhesives
D 1995	Test Method for Multi-Modal Strength of Autoadhesives (Contact Adhesives)
D 2093	Practice for Preparation of Surfaces of Plastics Prior to Adhesive Bonding
D 2094	Practice of Preparation of Bar and Rod Specimens for Adhesion Tests
D 2095	Test Method for Tensile Strength of Adhesives by Means of Bar and Rod Specimens
D 2182	Test Method for Strength Properties of Metal to Metal Adhesion by Compressive Load (Disk Shear)
D 2183	Test Method for Flow Properties of Adhesives in Shear by Compression Loading (Metal-to-Metal)
D 2235	Specification for Solvent Cementing of Acrilonitrile Butadiene Styrene (ABS) Plastic Pipe and Fittings
D 2293	Test Method for Creep Properties of Adhesives in Shear by Compression Loading (Metal-to-Metal)
D 2294	Test Method for Creep Properties of Adhesives in Shear by Tension Loading (Metal-to-Metal)
D 2295	Test Method for Strength Properties of Adhesives in Shear by Tension Loading at Elevated Temperatures (Metal-to-Metal)
D 2301	Specification for Vinyl Chloride Plastic Pressure Sensitive Electrical Insulating Tape
D 2339	Test Method for Strength Properties of Adhesives in Two-Ply Wood Construction in Shear by Tension Loading
D 2484	Specification for Polyester Film Pressure Sensitive Electrical Insulating Tape
D 2556	Test Method for Apparent Viscosity of Adhesives Having Shear Rate Dependent Flow Properties
D 2557	Test Method for Strength Properties of Adhesive in Shear by Tension Loading in the Temperature Range from -267.8 to $-55C$ (-450 to $-67F$)
D 2558	Test Method for Evaluation Peel Strength of Shoe Sole Attaching Adhesives (Intent to Withdraw)
D 2559	Specification for Adhesive for Structural Laminated Wood Products for Use Under Exterior (Wet Use) Exposure Conditions
D 2651	Practice for Preparation of Metal Surfaces for Adhesive Bonding
D 2674	Methods of Analysis of Sulfochromate Etch Solution Used in Surface Preparation of Aluminum

Specification and Standards

D 2686	Specification for Polytetrafluoroethylene Backed Pressure Sensitive Electrical Insulating Tape
D 2739	Test Method for Volume Resistivity of Conductive Adhesives
D 2754	Specification for High Temperature Glass Cloth Pressure Sensitive Electrical Insulating Tape
D 2851	Specification for Liquid Optical Adhesive
D 2918	Practice for Determining Durability of Adhesive Joints Stressed in Peel
D 2919	Test Method for Determining Durability of Adhesive Joints Stressed in Shear by Tension Loading
D 2979	Test Method for Pressure Sensitive Tack of Adhesive Using an Inverted Probe Machine
D 3005	Specification of Low Temperature Resistant Vinyl Chloride Plastic Pressure Sensitive Electrical Insulating Tape
D 3006	Specification of Polyethylene Plastic Pressure Sensitive Electrical Insulating Tape
D 3024	Specification for Protein Base Adhesive for Structural Laminated Wood Products for Use Under Interior (Dry Use) Exposure Conditions (Discontinued 1991)
D 3110	Specification for Adhesive Used in Nonstructural Glued Lumber Products
D 3111	Test Method for Flexibility Determination of Hot Melt Adhesives by Mandrel Bend Test Method
D 3121	Test Method for Tack of Pressure Sensitive Adhesives by Rolling Ball
D 3163	Test Method for Determining the Strength of Adhesively Bonded Rigid Plastic Lap Shear Joints in Shear by Tension Loading
D 3164	Test Method for Determining the Strength of Adhesively Bonded Plastic Lap Shear Sandwich Joints in Shear by Tension Loading
D 3165	Test Method for Strength Properties of Adhesives in Shear by Tension Loading of Laminated Assemblies
D 3166	Test Method for Fatigue Properties of Adhesives in Shear by Tension Loading (Metal-to-Metal)
D 3167	Test Method for Floating Roller Peel Resistance of Adhesives
D 3236	Test Method for Apparent Viscosity of Hot Melt Adhesives and Coating Materials

D 3310	Test Method for Determining Corrosivity of Adhesive Materials
D 3418	relating to Glass Transition Temperature
D 3433	Practice for Fracture Strength in Cleavage of Adhesives in Bonded Joints
D 3434	Practice for Multiple Cycle Accelerated Aging Test (Automatic Boil Test) for Exterior Wet Use Wood Adhesives
D 3482	Test Method for Determining Electrolytic Corrosion of Copper by Adhesives
D 3498	Specification for Adhesives for Field Gluing Plywood to Lumber Framing for Wood Systems
D 3528	Test Method for Strength Properties of Double Lap Shear Adhesive Joints by Tension Loading
D 3535	Test Method for Resistance to Deformation Under Static Loading for Structural Wood Laminating Adhesives Used Under Exterior (Wet Use) Exposure Conditions
D 3632	Practice for Accelerated Aging of Adhesive Joints by the Oxygen Pressure Method
D 3658	Test Method for Determining the Torque Strength of Ultraviolet (UV) Light Cured Glass-to-Metal Adhesive Joints
D 3706	Test Method for Hot Tack of Wax Polymer Blends by the Flat Spring Test
D 3747	Specification for Emulsified Asphalt Adhesive for Adhering Roof Insulation
D 3762	Test Method for Adhesive Bonded Surface Durability of Aluminum (Wedge Tests)
D 3807	Test Method for Strength Properties of Adhesive in Cleavage Peel by Tension Loading (Engineering Plastics-to-Engineering Plastics)
D 3808	Practice for Qualitative Determination of Adhesion of Adhesives to Substrate by Spot Adhesion Test Method
D 3929	Practice for Evaluating the Stress Cracking of Plastics by Adhesives Using the Bent-Beam Method
D 3930	Specification for Adhesives for Wood Based Materials for Construction of Manufactured Homes
D 3931	Test Method for Determining Strength of Gap Filling Adhesive Bonds in Shear by Compression Loading

D 3932	Practice for the Control of the Application of Structural Fasteners When Attached by Hot Melt Adhesives
D 3933	Practice for Preparation of Aluminum Surface for Structural Adhesive Bonding (Phosphoric Acid Anodizing)
D 3983	Test Method for Measuring the Strength and Shear Modulus of Non-Rigid Adhesives by the Thick Adherend Tensile Lap Specimen
D 4027	Test Method for Measuring Shear Properties of Structural Adhesive by the Modified Rail Test
D 4299	Test Methods for Effect of Bacterial Contamination on Permanence of Adhesive Preparations and Adhesive Films (Discontinued 1990. Replaced by D 4300 and D 4783)
D 4300	Test Methods for the Ability of Adhesive Films to Support or Resist the Growth of Fungi
D 4317	Specification for Polyvinyl Acetate Based Emulsion Adhesives
D 4338	Test Method for Flexibility Determination of Supported Adhesive Films by Mandrel Bend Test Method
D 4339	Test Method for the Determination of Odor of Adhesives
D 4426	Test Method for Determination of Percent Nonvolatile Content of Liquid Phenolic Resins Used for Wood Laminating
D 4497	Test Method for Determining the Open Time of Hot Melt Adhesives (Manual Method)
D 4498	Test Method for Heat-Fail Temperature in Shear of Hot Melt Adhesives
D 4499	Test Method for Heat Stability of Hot Melt Adhesives
D 4500	Test Method for Determining Grit, Lumps, or Undissolved Matter in Water Borne Adhesives
D 4501	Test Method for Shear Strength of Adhesive Bonds Between Rigid Substrates by the Block-Shear Method
D 4502	Test Method for Heat and Moisture Resistance of Wood Adhesives Joints
D 4562	Test Method for Shear Strength of Adhesives Using Pin-and-Collar Specimen
D 4680	Test Method for Creep and Time to Failure of Adhesives in Static Shear by Compression Loading (Wood-to-Wood)

D 4688	Test Methods for Evaluating Structural Adhesives for Fingerjointing Lumber
D 4689	Specification for Adhesives, Casein Type
D 4690	Specification for Urea-Formaldehyde Resin Adhesives
D 4783	Test Methods for Resistance of Adhesives Preparation in Container to Attack by Bacteria, Yeast and Fungi
D 4800	Guide for Classifying and Specifying Adhesives
D 4896	Guide for the Use of Adhesive Bonded Single Lap-Joint Specimen Test Results
D 5040	Test Methods for Ash Content of Adhesives
D 5041	Test Method for Fracture Strength in Cleavage of Adhesives in Bonded Joints
D 5113	Test Method for Determining Adhesive Attack on Rigid Cellular Polystyrene Foam
D 5215	Test Method for Instrumental Evaluation of Staining of Vinyl Flooring by Adhesives
D 5266	Practice for Estimating the Percentage of Wood Failure in Adhesive Bonded Joints
D 5267	Test Method for Determination of Extrudability of Cartridge Adhesives
D 5330	Specification for Tape, Pressure Sensitive, Packaging, Filament Reinforced
D 5375	Test Methods for Liner Removal at High Speeds from Pressure Sensitive Label Stock
D 5570	Test Method for Water Resistance of Tape and Adhesives Used as a Box Closure
D 5574	Test Methods for Establishing Allowable Mechanical Properties of Wood Bonding Adhesives for Design of Structural Joints
D 5677	Specification for Fiberglass (Glass Fiber Reinforced Thermosetting Resin) Pipe and Pipe Fittings, Adhesive Bonded Joint Type, for Aviation Jet Turbine Fuel Lines
D 5686	Standard Specification for Fiberglass (Glass Fiber Reinforced Thermosetting Resin) Pipe and Pipe Fittings, Adhesive Bonded Joint Type Epoxy Resin, For Condensate Return Lines
D 5749	Standard Specification for Reinforced and Plain Gummed Tape for Sealing and Securing
D 5751	Standard Specification for Adhesives Used for Laminate Joints in Nonstructural Lumber Products

D 5793	Standard Test Method for Binding Sites Per Unit Length or Width of Pile Yarn Floor Coverings
D 5824	Determining Resistance to Delamination of Adhesive Bonds in Overlay Wood Core Laminates Exposed to Heat and Water
D 5999	Test Method for Noninterference of Adhesives in Repulping
D 6004	Test Method for Determining Adhesive Shear Strength of Carpet Adhesives
D 6005	Test Method for Determining Slump Resistance of Carpet Adhesives
D 6105	Practice for Application of Electrical Discharge Surface Treatment (Activation) of Plastics for Adhesive Bonding
E 229	Test Method for Shear Strength and Shear Modulus of Structural Adhesives
E 864	Practice for Surface Preparation of Aluminum Alloys to be Adhesively Bonded in Honeycomb Shelter Panels
E 866	Specification for Corrosion Inhibiting Adhesive Primer for Aluminum Alloys to be Adhesively Bonded in Honeycomb Shelter Panels
E 874	Practice for Adhesive Bonding of Aluminum Facings to Nonmetallic Honeycomb Core for Shelter Panels
E 900	Specification for Core Splice Adhesive for Honeycomb Sandwich Structural Panels
E 1307	Surface Preparation and Structural Adhesive Bonding of Precured, Nonmetallic Composite Facings to Structural Core for Flat Shelter Panels
E 1512	Test Methods for Testing Bond Performance of Adhesive Bonded Anchors
E 1555	Specification for Structural Pate Adhesive For Sandwich Panel Repair
E 1793	Standard Practice for Preparation of Aluminum Alloy for Bonding in Foam and Beam Type Transportable Shelters
E 1794	Standard Specification for Adhesive for Bonding Foam Cored Sandwich Panels (200F Elevated Humidity Service), Type II Panels
E 1800	Specification for Adhesive for Bonding Foam Cored Sandwich Panels (160F Elevated Humidity Service), Type I Panels

E 1801	Practice for Adhesive Bonding of Aluminum Facings in Foam and Beam Type Shelters
E 1826	Specification for Low Volatile Organic Compound (VOC) Corrosion Inhibiting Adhesive Primer for Aluminum Alloys to be Adhesively Bonded
F 607	Test Method for Adhesion of Gasket Material to Metal Surfaces
G 85	Practice for Modified Salt Spray (Fog Testing)

Sealants and Sealing

ASTM Standard

C 510	Test Method for Staining and Color Change of Single or Multicomponent Joint Sealants
C 603	Test Method for Extrusion Rate and Application Life of Elastomeric Sealants
C 639	Test Method for Rheological (Flow) Properties of Elastomeric Sealants
C 661	Test Method for Indentation Hardness of Elastomeric Type Sealant by Means of a Durometer
C 679	Test Method for Tack Free Time of Elastomeric Sealants
C 711	Test Method for Low Temperature Flexibility and Tenacity of One Part Elastomeric Solvent Release Type Sealants
C 712	Test Method for Bubbling of One Part Elastomeric Solvent Release Type Sealants
C 717	Terminology of Building Seals and Sealants
C 719	Test Method for Adhesion and Cohesion of Elastomeric Joint Sealants Under Cyclic Movement (Hockman Cycle)
C 771	Test Method for Weight Loss After Heat Aging of Preformed Tape Sealants
C 792	Test Method for Effect of Heat Aging on Weight Loss, Cracking, and Chalking of Elastomeric Sealants
C 794	Test Method for Adhesion in Peel of Elastomeric Joint Sealants
C 804	Practices for Use of Solvent Related Type Sealants
C 811	Recommended Practice for Surface Preparation of Concrete for Application of Chemical Resistance Resin Monolithic Surfaces
C 879	Method of Testing Release Papers Used with Preformed Tape Sealants

C 906	Test Method for T-Peel Strength of Hot Applied Sealants
C 907	Test Method for Tensile Adhesive Strength of Preformed Tape Sealants by Disk Method
C 920	Specification for Elastomeric Joint Sealants
C 961	Test Method for Lap-Shear Strength for Hot Applied Sealants
C 962	Guide for Use of Elastomeric Joint Sealant
C 972	Test Method of Compression Recovery of Tape Sealant
C 1016	Standard Test Method for Determination of Water Absorption of Sealant Backing (Joint Filler Material
C 1021	Practice for Laboratories Engaged in Testing Building Sealants
C 1087	Standard Test Method for Determining Compatibility of Liquid Applied Sealants with Accessories Used in Structural Glazing Systems
C 1135	Determining Tensile Adhesion Properties of Structural Sealants
C 1184	Specification for Structural Silicone Sealants
C 1247	Test Method for Durability of Sealants Exposed to Continuous Immersion in Liquids
C 1248	Test Method for Staining of Porous Substances by Joint Sealants
C 1249	Guide for Secondary Seal for Sealed Insulating Glass Units for Structural Sealant Glazing Applications
C 1253	Test Method for Determining the Outgassing Potential of Sealant Backing
C 1257	Test Method for Accelerated Weathering of Solvent Release Type Sealants
C 1265	Standard Test Method for Determining the Tensile Properties of An Insulating Glass Edge Seal for Structural Glazing Applications
C 1294	Standard Test Method for Compatibility of Insulating Glass Edge Sealants with Liquid Applied Glazing Materials
C 1299	Guide for Use in Selection of Liquid Applied Sealant
C 1311	Specification of Solvent Release Sealants
C 1330	Specification for Cylindrical Sealant Backing for Use with Cold Liquid Applied Sealants

C 1369	Specification for Secondary Edge Sealants for Structurally Glazed Insulating Glass Units
C 1375	Guide for Substrates Used In Testing Building Seals and Sealants
C 1392	Guide for Evaluating Failure of Structural Sealant Glazing
C 1394	Guide of In-Situ Structural Silicone Glazing Evaluation
D 471	Test Method for Rubber Property—Effect of Liquids
D 1546	Practice for Testing the Performance of Clear Floor Sealers
D 1985	Practice for Preparing Concrete Blocks for Testing Sealant for Joints and Cracks
D 2202	Test Method for Slump of Sealants
D 2203	Test Method for Staining from Sealants
D 2377	Test Method for Tack Free Time of Caulking Compounds and Sealants
D 2828	Specification for Nonbituminous Inserts for Contraction Joints in Portland Cement Concrete Airfield Pavements, Sawable Type
D 3406	Specification for Joint Sealant, Hot Poured, Elastomeric Type, for Portland Cement Concrete Pavements
D 3538	Standard Test Method for Strength Properties of Double Lap Shear Adhesive Joints by Tension Loading
D 3569	Specification for Joint Sealant, Hot Applied, Elastomeric, Jet Fuel Resistant Type for Portland Cement Concrete Pavements
D 3581	Specification for Joint Sealant, Hot Applied, Jet Fuel Resistant Type, for Portland Cement Concrete & Tar-Concrete Pavements
D 3910	Practice for Design, Testing, and Construction of Slurry Seal
D 4070	Specification for Adhesive Lubricant for Installation of Preformed Elastomeric Bridge Compression Seals in Concrete Structures
D 4259	Practice for Abrading Concrete
D 4260	Practice for Acid Etching Concrete
D 5167	Practice for Melting of Hot Applied Joint and Crack Sealant and Filler for Evaluation
D 5249	Specification for Backer Materials for Use with Cold and Hot Applied Joint Sealants in Portland Cement Concrete and Asphalt Joints

D 5657	Fluid Tightness Ability of Adhesives Used on Threaded Fasteners
D 5749	Standard Specification for Reinforced and Plain Gummed Tape for Sealing and Securing
D 5893	Specification for Cold Applied, Single Component, Chemically Curing Silicone Joint Sealant for Portland Cement Concrete Pavements
E 773	Test Method for Accelerated Weathering of Sealed Insulating Glass Units
E 1068	Test Method for Testing Non-Metallic Seal Materials by Immersion in a Simulated Geothermal Test Fluid
E 1069	Test Method for Testing Polymeric Seal Materials for Geothermal and/or High Temperature Service Under Sealing Stress
F 37	Sealability of Gasketing Material
F 88	Test Method for Seal Strength of Flexible Barrier Materials

C.2 U.S. Federal Specifications and Standards

Specification	
A-A-272	Caulking Compounds
A-A-373	Glazing Compound
DOD-C-24176	Cement Epoxy, Metal Repair
HH-C-536	Caulking Compound for Gasket Connections
MIL-R-46082	Retaining Compounds, Single Component
MIL-A-101	Adhesive, Water Resistant
MIL-A-1154	Adhesive-Bonds Vulcanized Synthetic Rubber Parts
MIL-A-13883	Adhesive, Synthetic Rubber
MIL-A-14042	Adhesive Epoxy
MIL-A-21016	Linoleum and Tile Adhesive
MIL-A-22010	Adhesive, Solvent Type Polyvinylchloride
MIL-A-22397	Phenol and Resorcinol Resin Base
MIL-A-22895	Adhesives, Metal Identification Plate
MIL-A-24179	Adhesive, Flexible
MIL-A-25463	Adhesive Film Form, Metallic Structure Sandwich Construction
MIL-A-3167	Adhesives for Plastic Inhibitors
MIL-A-3316	Fire Resistant Adhesive
MIL-A-374	Adhesive, Paste for Demolition Charges
MIL-A-3920	Adhesive, Optical, Thermosetting
MIL-A-3941	Water Resistant Label Adhesives
MIL-A-43316	Adhesive, Patching
MIL-A-43365	Adhesive, Repair of Radome, Air
MIL-A-46050	Adhesive, Special Rapid Room Temperature
MIL-A-46051	Adhesives, Room Temperature and Intermediate Temperature
MIL-A-46091	Adhesive, Brake Lining to Metal
MIL-A-46106	Adhesive, Sealant Silicone, General Purpose
MIL-A-46146	Adhesive, Sealants, Silicone, RTV, Noncorrosive
MIL-A-46864	Adhesives, Epoxy, Modified
MIL-A-47040	Adhesive-Sealant, Silicone, RTV, High Temperature
MIL-A-4833	Adhesive, Cellulose Nitrate Base
MIL-A-48611	Adhesives Synthetic Epoxy-Elastomeric for Glass to Metal
MIL-A-5090	Adhesive, Heat Resistant Metal to Metal
MIL-A-5092	Light Stress Bonding Only
MIL-A-52194	Adhesive, Epoxy
MIL-A-5540	Polychloroprene Adhesive

Specification and Standards

MIL-A-60091	Adhesive for Bonding Demolition Charges
MIL-A-81236	Adhesive, Epoxy Resins
MIL-A-82484	Adhesives and Sealing Compounds
MIL-A-83376	Adhesives Bonded Metal Faced Sandwich Structures
MIL-A-83377	Adhesive Bonding (Structural) for Aerospace and Other Systems, Requirements for
MIL-A-8576	Adhesive, Acrylic Monomer Base
MIL-A-87135	Adhesive, Non-Conductive for Electrical Applications
MIL-A-9117	Synthetic Elastomeric Sealant
MIL-A-12850	Natural Liquid Rubber Cement
MIL-C-14064	Grinding Disk Cement
MIL-C-15705	Caulking Compound
MIL-C-18255	Caulking Compound with Synthetic Rubber Base
MIL-C-18969	Caulking Compound—Watertight Exterior Hull Seams of Vessels
MIL-C-23092	Cement, Natural Rubber
MIL-C-27315	Coating Systems, Elastomeric
MIL-C-27725	Coating, Corrosion, Preventative, Air Fuel Tanks
MIL-C-5539	Natural Rubber Cement
MIL-C-7438	Core Material, Aluminum, for Sandwich Construction
MIL-C-8073	Core Material, Plastic Honeycomb, Laminated Glass Fabric Base, for Aircraft Structural Applications
MIL-C-81986	Core Material, Plastic Honeycomb, Nylon Paper Base, for Aircraft Structural Applications
MIL-C-83019	Protective Coating, Integral Fuel Tank
MIL-C-83231	Coating, Polyurethane, Rain, Erosion Resistant
MIL-C-8514	Resin-Acid Metal Pretreatment Compound for Aircraft
MIL-C-897	Cement, Rubber
MIL-D-17951	Sealing Compound for Use with Deck Covering
MIL-G-413	Marine Glue
MIL-G-46030	Glue, Animal (Protective Colloid)
MIL-P-20628	Putty, Sealing
MIL-P-23236	Paint Coating, Ship, Fuel, Salt Tanks
MIL-P-46276	Primer, Bonding
MIL-P-47279	Primer, Silicone Adhesive
MIL-R-17882	Epoxy Resin for Pipe Repair
MIL-S-11030	Non-curing, Polysulfide Base Sealing Compound
MIL-S-11031	Adhesive, Curing Compound for Bonding Metal to Metal

MIL-S-12935	Synthetic Resin Lumber Knot Sealing Compound
MIL-S-15204	High Temperature Sealing Compound
MIL-S-17377	Compound Composed of Plastics or Resinous Binders
MIL-S-20541	Liquid Rubber Cement
MIL-S-22473	Anaerobic Single Component Sealing Compound
MIL-S-23498	Sealing Compound, Bearing Preservation
MIL-S-24340	Sealing, Ship Deck Polyurethane
MIL-S-3927	Sealing Compound, Thread
MIL-S-4383	Sealing Compound, Synthetic Rubber Base
MIL-S-45180	Thread and Gasket Sealing Compounds
MIL-S-46163	Sealing, Lubricating and Wicking Compounds Thread Locking
MIL-S-46897	Sealing Compound, Polyurethane Foam
MIL-S-7502	High Adhesion Sealant
MIL-S-7916	Thread and Gasket Sealing Compound
MIL-S-81732	Sealing Compound, Electrical
MIL-S-81733	Sealing, Coating, Corrosion Inhibitive
MIL-S-83315	Sealing, Aluminum Structure
MIL-S-83318	Sealant, Quick Repair Integral Fuel Tank
MIL-S-83430	Sealing Compound, Integral Fuel Tanks and Fuel Cell Cavities
MIL-S-8516	Sealing Compound—Protects Electrical Components
MIL-S-8784	Sealing Compound for Aircraft Fuel Tank
MIL-S-8802	High Temperature Sealant Component, for Integral Fuel Tanks and Fuel Cell Cavities
MIL-T-5542	Thread Compound, Antiseize and Sealing
MIL-T-83483	Thread Compound, Antiseize, Molybdenum Disulfide Petrolatum
MMM-A-001058	Adhesive, Rubber Base Pressurized Dispensers
MMM-A-001993	Adhesive, Epoxy, Flexible, Filled
MMM-A-100	Animal Glue
MMM-A-115	Asphalt Tile Cement
MMM-A-121	Adhesive, Binding
MMM-A-122	High Strength Adhesive
MMM-A-125	Casein Glue in Powder Form
MMM-A-132	Adhesive, Heat Resistant, Airframe Structural, Metal to Metal
MMM-A-134	Adhesive, Epoxy Resin, Metal to Metal Structural Bonding
MMM-A-137	Strong Linoleum Cement
MMM-A-138	Adhesive, Metal to Wood
MMM-A-139	Adhesive, Natural or Synthetic

MMM-A-150	Paste for Fibrous Acoustical Materials
MMM-A-1617	Adhesive, Rubber Base, General Purpose
MMM-A-1754	Adhesive and Sealing Compound, Epoxy, Metal Filled
MMM-A-177	Adhesive Paste
MMM-A-178	Adhesive, Paper Label
MMM-A-179	Adhesive, Paper Label
MMM-A-180	Thermoplastic Synthetic Resin
MMM-A-181	Phenol, Melamine or Resorcinol Resin Base
MMM-A-182	Adhesive, Innertube Repair
MMM-A-187	Epoxy Resin Paste
MMM-A-188	Thermosetting Urea Resin
MMM-A-189	Synthetic Rubber for Hot and Cold Bonding
MMM-A-193	Adhesive, Vinyl Acetate Resin Emulsion
MMM-A-1931	Adhesive, Epoxy, Silver Filled, Conductive
MMM-A-250	Water Resistant Rubber Base Liquid Adhesive
MMM-A-260	Water Resistant Liquid Adhesive
MMM-B-00350	Binder Adhesive, Epoxy Resin—Flexible
SS-S-1996	Sealer, Water and Weather Resistant
TT-C-00598	Caulking Compound
TT-C-1796	Caulking Compounds, Metal Seam and Wood Seam
TT-F-320	Filler for Cracks in Wood, Metal, Concrete and Cement
TT-F-322	Metal Surface Dent Filler
TT-F-336	Filler for Wood
TT-F-340	Plastic Wood Filler
TT-P-1536	Plumbing Fixture Setting Compound
TT-P-781	Putty and Elastic Compound
TT-P-791	Wood Sash Glazing Putty
TT-S-00230	Single Component Synthetic Rubber Base Joint Sealant
TT-S-0227	Sealing Compound
TT-S-1732	Sealing Compound, Pipe Joint and Thread
VV-S-190	Solid Form Sealing Compound for Overseas Shipments
MIL-STD-401	Sandwich Construction and Core Materials: General Test Methods
Federal Test Standard 175	Adhesives, Method for Testing

C.3 Other Industry Specification and Standards

Society of Automotive Engineers Specifications and Standards

ARP 1524	Surface Preparation and Priming of Aluminum Alloy Parts for High Durability Structural Adhesive Bonding, Phosphoric Acid Anodizing
APR 1575	Surface Preparation and Priming of Aluminum Alloy Parts for High Durability Structural Adhesive Bonding, Hand Applied Phosphoric Acid Anodizing
ARP 1843	Surface Preparation for Structural Adhesive Bonding Titanium Alloy Parts
ARP 4069	Aerospace Recommended Practice for Sealing Integral Fuel Tanks
AMS 1320	Decal Adhesive Remover
AMS 3374	Sealing Compound, One Part Silicone, Aircraft Firewall
AMS 3375	Adhesive/Sealant, Fluorosilicone Aromatic Fuel Resistant, One Part Room Temperature Vulcanizing
AMS 3376	Sealing Compound, Noncuring, Groove Injection, Temperature and Fuel Resistant
AMS 3491	Surface Treatment of Polytetrafluoroethylene
AMS 3106	Primer, Adhesive, Corrosion Inhibiting, -67 to 200F
AMS 3107	Primer, Adhesive, Corrosion-Inhibiting, for High Durability Structural Adhesive Bonding
AMS 3681	Adhesive, Electrically Conductive, Silver Organic Base
AMS 3685	Adhesive, Synthetic Rubber, Buna N Type
AMS 3686	Adhesive Polyimide Resin, Film and Paste, High Temperature Resistant, 315C or 600F
AMS 3687	Adhesive Film, Humidity Resistant, for Sandwich Panels
AMS 3688	Adhesive, Foaming, Honeycomb Core Splice, Structural, -55 to 82C
AMS 3689	Adhesive, Foaming, Honeycomb Core Splice, Structural, -55 to 177C
AMS 3690	Adhesive Compound, Epoxy, Room Temperature Curing Adhesive Compound, Epoxy, Room Temperature Curing
AMS 3691	Adhesive Compound, Epoxy, Medium Temperature Application

AMS 3692   Adhesive Compound, Epoxy, High Temperature Application
AMS 3693   Adhesive, Modified Epoxy, Moderate Heat Resistant, 120C Curing, Film Type
AMS 3695   Adhesive Film, Epoxy Base for High Durability Structural Adhesive Bonding
AMS 3696   Aerodynamic Fairing Compound, -55 to 85C
AMS 3697   Aerodynamic Fairing Compound, -55 to 150C
AMS 3698   Adhesive Film, Hot Melt, Addition Type Polyimide, for Foam Sandwich Structure, -67 to 450F
AMS 3704   Adhesive, Contact Chloroprene, Resin Modified
J 1523   Recommended Practice for Metal to Metal Overlap Shear Strength Test for Automotive Type Adhesives
J 1525   Recommended Practice for Lap Shear for Automotive Type Adhesives for Fiber Reinforced Plastic (FRP) Bonding

American Architectural Manufacturers Association

AAMA 1407.1   Voluntary Specification for a Single Component Sealant For Residential Sheet Products
AAMA 850   Penetration Sealants Guide Manual
AAMA CW-13   Curtain Wall Manual No. 13—Structural Sealant Glazing Systems

American Concrete Institute

ACI 504R   Guide to Joint Sealants for Concrete Structures

C.4 Other Standards Organizations

International Standard Organizations

Acronym	Publisher	Example
AS	Australian Standards—Australia	AS 2990
BS	British Standard Institute—UK	BS 3924
CGSB	Canadian General Standards Board—Canada	CGSB 25.14
DEF	UK Ministry of Defense Standards—UK	DEF 103
DIN	Deutsche Institut fur Normung—Germany	DIN 52451
EN	European Committed for Standardization—Europe	EN 60454
JIS	Japanese Industrial Standards—Japan	JIS Z9900
ISO	International Standard Organization—Switzerland	ISO 9653
NEN	Netherlands Normalisatie Instutuut	NEN 10244-7
NF	Norme Francaise (AFNOR)—France	NF T76-103
NS	Norges Standardiseringsforbund—Norway	NS 4828
SEN, SIS, SMS	Standardiserigns-Kommissionen I Sverige—Sweden	SEN 01 03 45

Other Industrial and Professional Organizations

Acronym	Publisher	Example
AA	Aluminum Association	AA53
ACI	American Concrete Institute	ACI 211.1
AISC	American Institute of Steel Construction	AISC M011
AMS	Aerospace Materials Specifications, Society of Automotive Engineers	AMS 1374
ANSI	American National Standards Institute	ANSI X12.27
ASCE	American Society of Civil Engineers	ASCE 10-90
ASME	American Society of Mechanical Engineers	ASME B31.3

ASQ	American Society for Quality Control	ASQ Q90
NACE	National Association of Corrosion Engineers	NACE MR 01 75
SPI	Society of Plastics Institute	SPI B 151.21
TAPPI	Technical Association of the Pulp and Paper Industry	TAPPI 207
UL	Underwriters Laboratory	UL 94

Appendix D

Surface Preparation Methods for Common Substrate Materials

The following sources were used in compiling this Appendix:

Wegman, R. R., *Surface Preparation Techniques for Adhesive Bonding,* Noyes Publications, Park Ridge, NJ, 1989.

Cagle, C. V., *Adhesive Bonding Techniques and Applications,* McGraw Hill, New York, 1968.

DeLollis, N. J., *Adhesives for Metals Theory and Technology,* Industrial Press, New York, 1970.

Schields, J., *Adhesives Handbook,* 3rd ed, Butterworths, London, 1984.

Guttman, W. H. *Concise Guide to Structural Adhesives,* Reinhold, New York, 1961.

"Preparing the Surface for Adhesive Bonding," Hysol Division, Dexter Corp., Bulletin Gl-600.

ASTM D 2093, "Preparation of Surfaces of Plastics Prior to Adhesive Bonding," American Society for Testing and Materials, Conshohocken, PA.

ASTM D 2651, "Preparation of Metal Surfaces for Adhesive Bonding," American Society for Testing and Materials, Conshohocken, PA.

Adhesives and Sealants, vol. 3, Engineered Materials Handbook Series, ASM International, 1990.

Handbook of Plastics Joining, Plastics Design Library, Norwich, NY, 1997.

Note: All formulations in the following tables of this Appendix are presented on a parts by weight basis unless otherwise indicated.

D.1 Surface Preparation Methods for Metal Substrates

Substrates	Cleaning solvents	Substrate treatments	Comments
Aluminum and aluminum alloys	Trichloroethylene	1. Sandblast or 100 grit emery cloth followed by solvent degreasing	Medium to high strength bonds, suitable for non critical applications
		2. Immerse for 10 min at 70–82°C in a commercial alkaline cleaner or Sodium metasilicate 3.0 Sodium hydroxide 1.5 Sodium dodecylbenzene sulfonate, such as Nacconol 90G (Stephan Co., Nothfield, IL) 1.5 Water (distilled) 128.0 Wash in water below 65°C and etch for 12–15 min at 66–71°C in Sodium dichromate 1.0 Sulfuric acid (96%) 10.0 Water (distilled) 30.0 Rinse in distilled water after washing in tap water and dry in air	Optimum bond strength per FPL etch process. Specified in ASTM D 2651 and MIL-A-9067. Solvent degrease may replace alkaline cleaning
		3. Alkaline clean as described in 2 above. Then immerse for 10–12 mins at 60-65°C in the following solution: Sulfuric acid (6.5 to 9.5N) 7–36% by wt. Ferric sulfate 135 to 165 g/l Rinse as described in 2 above	P-2 etch specified in ASTM D 2651

	Procedure	Remarks
	4. Degrease with solvent. Abrade lightly with mildly abrasive cleaner. Rinse in deionized water; wipe; or air dry. Etch 20 min at RT in Sodium dichromate 2 Sulfuric acid (96%) 7 Rinse thoroughly in deionized water; dry at 70°C for 30 min	Room temperature etch
	5. Form a paste using sulfuric acid – sodium dichromate solution and finely divided silica or fuller's earth. Apply; do not permit paste to dry. Time depends on degree of contamination (usually greater than 10 min at RT). Wash very thoroughly with deionized water, and air dry	Paste form of acid etch, useful when part cannot be immersed
	6. Phosphoric acid anodizing can be performed in accordance with ASTM D 3933	
	7. Sulfuric acid anodizing can be performed in accordance with MIL-A-8625	Found suitable for bare aluminum (nonclad), machined, or chemically milled parts which must be corrosion protected
Brass and bronze (see also copper and copper alloys)	Trichloroethylene	
	1. Etch for 5 min at 20°C in Zinc oxide 20 Sulfuric acid (96%) 460 Nitric acid (69%) 360 Rinse in water below 65°C and re-etch in the acid solution for 5 min at 49°C. Rinse in distilled water after washing and dry in air	Temperature must not exceed 65C when washing and drying

827

Substrates	Cleaning solvents	Substrate treatments	Comments
Chromium	Trichloroethylene	1. Abrasion. Grit or vapor blast, or 100 grit emery cloth, followed by solvent degreasing	Suitable for general purpose bonding
		2. Etch for 1–5 min at 90–95°C in	For maximum bond strength
		Hydrochloric acid (37%) 17	
		Water 20	
		Rinse in distilled water after cold/hot water washing and dry in hot air	
Copper and copper alloys	Trichloroethylene	1. Abrasion. Sanding, wire brushing, or 100 grit emery cloth, followed by vapor or solvent degreasing	Solution for general purpose bonding. Use 320 grit emery cloth for foil
		2. Etch for 10 min at 66°C in	For maximum bond strength. Suitable for brass and bronze. ASTM D 2651 sulfuric acid-dichromate-ferric sulfate etch
		Ferric sulfate 1	
		Sulfuric acid (96%) 75	
		Water 8	
		Wash in water at 20°C, and etch in cold solution of	
		Sodium dichromate 1	
		Sulfuric acid (96%) 2	
		Water 17	
		Etch until a bright clean surface has been obtained. Rinse in water, dip in ammonium hydroxide (s.g. 0.88) and wash in tap water. Rinse in distilled water and dry in warm air	
		3. Etch for 1–2 min at 20°C in	Room temperature etch. ASTM D 2651 nitric acid/ferric chloride etch
		Ferric chloride (42% solution in water) 15	
		Nitric acid (69%) 30	
		Water 197	

Rinse in distilled water after cold water wash and dry in air at 20°C

4. Etch for 30 sec at 20°C in

Ammonium persulfate	1
Water	4

Alternative etching solution to above where fast processing is required

Rinse in distilled water after cold water wash and dry in air at 20°C

5. Solvent degrease. Immerse 30 sec at RT in

Nitric acid (69%)	30
Deionized water	90

For copper alloys containing over 95% copper. Stable surface for hot bonding

Rinse in running water and transfer immediately to next solution; immerse for 1–2 min at 98°C in Ebonol C (Enthone, Inc., New Haven, CT) 24 oz and equivalent water to make 1 gal of solution. Rinse in deionized water and air dry

6. Solvent degrease and immerse in nitric acid solution as in 5 above. Immerse immediately for 2–3 mins at 93–102°C in the following solution in 1 gal of water:

Sodium chlorite (technical)	4.01 oz
Trisodium phosphate	1.34 oz
Sodium hydroxide	0.67 oz

ASTM D 2651 nitric acid, sodium chlorite etch. Suitable for copper alloys containing over 95% copper. It is not suitable for adhesives containing chlorides or for hot bonding polyethylene

Substrates	Cleaning solvents	Substrate treatments	Comments
		Rinse thoroughly in running water until a neutral test is produced when touched with indicator paper	
		Dry. Bond as soon as possible, but within the same working day	
Gold	Trichloroethylene	Solvent or vapor degrease after light abrasion with a fine emery cloth	
Iron		See steel (mild)	
Lead and lead based solders	Trichloroethylene	Abrasion. Grit or vapor blast, or 100 grit emery cloth followed by solvent degreasing	Apply the adhesive immediately after abrasion
Magnesium and magnesium alloys	Trichloroethylene	1. Abrasion with 100 grit emery cloth followed by solvent degreasing	Medium to high bond strength. ASTM D 2651 alkaline detergent treatment
		2. Vapor degrease. Immerse for 10 min at 60–70°C in:	
		Deionized water 95	
		Sodium metasilicate 2.5	
		Trisodium pyrophosphate 1.1	
		Sodium hydroxide 1.1	
		Sodium dodecylbenzene sulfonate, such as Nacconol 90G 0.3	
		Rinse in water and dry below 60°C	
		3. Vapor degrease. Immerse for 10 min at 71–88°C in	High bond strength. ASTM D 2651 chromic acid etch
		Water 4	
		Chromic acid 1	
		Rinse in water and dry below 60°C	

	4. Vapor degrease. Immerse for 5–10 min at 63–80°C in Water 12 Sodium hydroxide 1 Rinse in water. Immerse for 5–15 min at RT in Water 123 Chromic acid 24 Calcium nitrate 1.8 Rinse in water and dry below 60°C	ASTM D 2651 sodium hydroxide—chromic acid etch	
	5. Light anodic treatment and various corrosion preventive treatments have been developed by magnesium producers	Details may be obtained from the magnesium alloy producers or from ASM Handbook, Vol. 5, and Military Specification MIL-M-4502, Type I, Classes 1, 2, and 3	
	6. Some dichromate conversion coatings and wash primers designed for corrosion prevention are suitable for adhesive bonding	Preliminary tests should be conducted to determine suitability of these processes before acceptance. Details of the processes are found in the above referenced ASME Handbook and also in MIL-M-3171	
Nickel	1. Abrasion with 100 grit emery cloth followed by solvent degreasing	For general purpose bonding	Trichloroethylene
	2. Etch for 5 sec at 20°C in nitric acid (69%). Wash in cold and hot water followed by a distilled water rinse and air dry at 40°C	For general purpose bonding	
Silver	Abrasion with 320 grit emery cloth followed by solvent degreasing		Trichloroethylene

Substrates	Cleaning solvents	Substrate treatments	Comments
Steel (stainless)	Trichloroethylene	1. Abrasion with 100 grit emery cloth, grit or vapor blast followed by solvent degreasing.	Dry grit or sand blasting tends to warp thin sheet materials; these methods are suited only for thick section parts
		2. Solvent degrease and abrade with grit paper. Degrease again. Immerse for 10 min at 65–71°C in the following solution:	ASTM D 2651 sulfuric, nitric, hydrofluoric etch
		Water 90	
		Sulfuric acid (s.g. 1.84) 37	
		Sodium dodecylbenzene sulfonate, such as Nacconol 90G (Stephan Co., Northfield, IL) 0.2	
		Rinse thoroughly and remove smut with a stiff brush if necessary	
		Immerse for 10 min at RT in the following bright dip solution:	
		Water 88	
		Nitric acid (s.g. 1.42) 15	
		Hydrofluoric acid (35.35%, s.g. 1.15) 2	
		3. Immerse for 2 min at approximately 93°C in the following solution heated by a boiling water bath:	For maximum resistance to heat and environment. ASTM D 2651 hydrochloric, Orthophosphoric, hydrofluoric acid etch
		Hydrochloric acid (s.g. 1.2) 200	
		Orthophosphoric acid (s.g. 1.8) 30	
		Hydrofluoric acid (s.gg., 1.15) 10	
		4. Etch for 15 min at 63°C in	ASTM D 2651 sulfuric, sodium dichromate etch

Saturated sodium dichromate solution	30	
Sulfuric acid	100	Room temperature etch. Treatment may be followed by passivation for 20 min in 5–10% w/v chromic acid solution

Remove carbon residue with nylon brush while rinsing. Rinse in distilled water and dry in warm air at 93°C

5. Vapor degrease for 10 min and pickle for 10 min at 20°C in

Nitric acid (69%)	10	
Hydrofluoric acid (48%)	2	ASTM D 2651 sodium metasilicate treatment
Water	88	

Dry in air under 70°C

6. Immerse for 15 min at 63 ± 3°C in the following solution:

Water	47.2	
Sodium metasilicate	1.0	
Anionic surfactant, such as Triton X2000 (Rohm and Haas, Philadelphia, PA)	1.8	ASTM D 2651 hydrochloric sulfuric–dichromate etch

7. Immerse for 10 min at 60–65°C in the following solution:

Water	45	
Hydrochloric acid (s.g. 1.2)	50	
Formalin solution (40%)	10	
Hydrogen Peroxide (30–35%)	2	

Rinse thoroughly, and then immerse for 5 min at 50–65°C in sulfuric acid dichromate solution used for aluminum

Substrates	Cleaning solvents	Substrate treatments	Comments
Steel (mild, iron, and ferrous metals other than stainless)	Trichloroethylene	Rinse thoroughly, and then dry at not over 93°C	
		1. Abrasion. Grit or vapor blast followed by solvent degreasing with water free solvents	Xylene or toluene is preferred to acetone and ketone, which may be moist enough to cause rusting
		2. Etch in the following solution for 5 mins at 23°C: Deionized water 64.99% by vol Surfactant 0.01% by vol Phosphoric acid (85%) 30 % by vol Nitric acid (40 Baume) 5 % by vol	ASTM D 2651 nitric-phosphoric acid etch. Bonding should follow immediately after etching treatments since ferrous metals are prone to rusting. Abrasion is more suitable for procedure where bonding is delayed
		3. Etch for 5–10 min at 20°C in Hydrochloric acid (37%) 1 Water 1 Rinse in distilled water after cold water wash and dry in warm air for 10 min at 93°C	
		4. Etch for 10 min at 60°C in Orthophosphoric acid (85%) 1 Ethyl alcohol (denatured) 2 Brush off carbon residue with nylon brush while washing in running water. Rinse with deionized water and heat for 1 hr at 120°C	For maximum strength
Tin	Trichloroethylene	Solvent or vapor degrease after light abrasion with a fine emery cloth (320 grit)	

Material	Process	Remarks
Titanium and titanium alloys	Trichloroethylene	For general purpose bonding
	1. Abrasion. Grit or vapor blast or 100 grit emery cloth; followed by solvent degrease; or scour with a nonchlorinated cleaner, rinse, and dry	
	2. Etch for 5–10 min at 20°C in Sodium fluoride 2 Chromium trioxide 1 Sulfuric acid (96%) 10 Water 50 Rinse in water and distilled water. Dry in air at 93°C	
	3. Etch for 2 min at RT in Hydrofluoric acid (60%) 63 ml Hydrochloric acid (37%) 841 ml Orthophosphoric acid (85%) 89 ml Rinse in water and distilled water. Dry in air at 93°C	Suitable for alloys to be bonded with high temperature adhesives (e.g., polybenzimidasole). Bond within 10 min of treatment. ASTM D 2651
	4. Etch for 10–15 min at 38–52°C in Nitric acid (69%) 6 Hydrofluoric acid (60%) 1 Water 20 Rinse with water and distilled water. Dry in oven at 71–82°C for 15 min	Alternative etch for alloys to be bonded with polyimide adhesives is nitric: hydrofluoric: water in a ratio of 5:1:27 by wt. Etch 30 sec at 20°C
	5. Commercial etching liquids and pastes (PlasaJell 107C, Semco)	
	6. Immerse for 15 min at 76°C in the following solution to make 1 gal (3.6 l):	ASTM D 2651 nitric hydrofluoric etch

Substrates	Cleaning solvents	Substrate treatments	Comments
		Caustic cleaner, such as Vitro-Klene (Turco Purex Industries, Carson, CA) 6–8 oz Water Remainder Rinse in cold tap water, and immerse for 5 min at RT in the following solution: Nitric acid (s.g., 1.5) 48 Ammonium bifluoride (technical) 3 Water 49 Rinse in cold tap water and air dry at room temperature Immerse for 2 min at RT in the following solution with sufficient water to make 1 gal (3.6 l): Trisodium phosphate (technical) 50.0 gm Sodium fluoride (technical) 8.9 gm Hydrofluoric acid (48%) 26.0 ml Air dry at room temperature 8. Stainless steel surface processes have been generally found satisfactory for titanium	Sulfuric, sodium dichromate etch; sodium metasilicate; and Hydrochloric sulfuric-dichromate etch processes for stainless steel
Zinc and zinc alloys	Trichloroethylene	1. Abrasion. Grit or vapor blast or 100 grit emery cloth followed by solvent degreasing 2. Etch for 2–4 min at 20°C in Hydrochloric acid (37%) 10–20 Water 90–80	For general purpose bonding

Rinse with warm water and distilled water. Dry in air at 66–71°C for 30 min

3. Etch for 3–6 min at 38°C in

Sulfuric acid (96%)	2
Sodium dichromate	1
Water	8

Glacial acetic acid is an alternative to hydrochloric acid

Suitable for freshly galvanized metal

D.2 Surface Preparation Methods for Plastic Substrates

Substrates	Cleaning solvents	Substrate treatments	Comments
Acetal (copolymer)	Acetone	1. Abrasion. Grit or vapor blast, or medium grit emery cloth followed by solvent degreasing	For general purpose bonding
		2. Etch in the following acid solution for 10 sec at 25°C:	For maximum bond strength. ASTM D 2093
		Potassium dichromate 75	
		Distilled water 120	
		Sulfuric acid (96%) 1500	
		Rinse in distilled water, dry in air at RT	
Acetal (homopolymer)	Acetone	1. Abrasion. Sand with 280 grit emery cloth followed by solvent degreasing	For general purpose bonding
		2. Satinizing technique. Immerse the part in the following for 5–30 sec at 80–120°C:	For maximum bond strength. Recommended by DuPont
		Perchloroethylene 96.85	
		1,4 Dioxane 3.0	
		p-Toluenesulfonic acid 0.05	
		Cab-o-Sil (Cabot Corp.) 0.10	
		Transfer the part immediately to an oven at 120°C for 1 min. Wash in hot water. Dry in air at 120°C	
Acrylonitrile butadiene styrene	Acetone	1. Abrasion. Grit or vapor blast, or 220 grit emery cloth, followed by solvent degreasing	
		2. Etch in chromic acid solution for 20 min at 60°C	Recipe 2 for methyl pentene

Cellulosics: cellulose, cellulose acetate, cellulose acetate butyrate, cellulose nitrate, cellulose propionate, ethyl cellulose	Methanol, isopropanol	1. Abrasion. Grit or vapor blast or 220 grit emery cloth followed by solvent degreasing 2. After procedure 1, dry the part at 100°C for 1 hr and apply adhesive before the plastic cools to room temperature	For general purpose bonding
Diallyl phthalate, diallyl isophthalate	Acetone, methyl ethyl ketone	Abrasion. Grit or vapor blast, or 100 grit emery cloth followed by solvent degreasing	Steel wool may be used for abrasion
Epoxy	Acetone, methyl ethyl ketone	Abrasion. Grit or vapor blast or 100 grit emery cloth followed by solvent degreasing	Sand or steel shot are suitable abrasives
Ethylene vinyl acetate	Methanol	Prime with epoxy adhesive and fuse into the surface by heating for 30 min at 100°C	
Furane, ionomer, melamine resins, SAN, polysulfone, and rigid vinyl	Acetone, methyl ethyl ketone	Abrasion. Grit or vapor blast or 100 grit emery cloth followed by solvent degreasing	

Methyl pentene	Acetone	1. Abrasion. Grit or vapor blast or 100 grit emery cloth followed by solvent degreasing
		2. Immerse for 1 hr at 60°C in
		    Potassium chromate 7.5
		    Water 27.5
		    Sulfuric acid (96%) 65.0
		Rinse in water and distilled water. Dry in warm air
			For general purpose bonding

Substrates	Cleaning solvents	Substrate treatments	Comments
		3. Immerse for 5-10 min at 90C in potassium permanganate (saturated solution), acidified with sulfuric acid (96%). Rinse in water and distilled water. Dry in warm air	
		4. Prime surface with a lacquer based on urea formaldehyde resin diluted with carbon tetrachloride	Coatings (dried) offer excellent bonding surfaces without further treatment
Phenolic and phenolic melamine resins	Acetone, methyl ethyl ketone, detergent	1. Abrasion. Grit or vapor blast, or abrade with 100 grit emery cloth followed by solvent degreasing	Steel wool may be used for abrasion. Sand or steel shot are suitable abrasives. Glass fabric decorative laminate may be degreased with detergent solution
		2. Removal of surface layer of one ply of fabric previously placed on surface before curing. Expose fresh bonding surface by tearing off the ply prior to bonding	
Polyamide (nylon)	Acetone, methyl ethyl ketone	1. Abrasion. Grit or vapor blast or abrade with 100 grit emery cloth followed by solvent degreasing	Sand or steel shot are suitable abrasives
		2. Prime with a spreading dough based on the type of rubber to be bonded in an admixture with isocyanate	Suitable for bonding Polyamide textiles to natural and synthetic rubbers
		3. Prime with resorcinol formaldehyde adhesives	Good adhesion to primer coat with epoxy adhesives in metal to plastic joints
Polycarbonate	Methanol, isopropanol, detergent	Abrasion. Grit or vapor blast or 100 grit emery cloth followed by solvent degreasing	Sand or steel shot are suitable abrasives

Adherend	Solvent	Treatment	Notes
Fluorocarbons: polychlorotrifluoroethylene, polytetrafluoroethylene, polyvinyl fluoride, polymonochlorotrifluoroethylene	Trichloroethylene	1. Wipe with solvent and treat with the following for 15 min at RT: Naphthalene (128 g) dissolved in tetrhydrofuran (1 liter) to which is added sodium (23 g) during a stirring period of 2 hr Wash in acetone to remove excess organic materials, and subsequently wash with distilled or deionized water. Before bonding dry the treated adherends in an air circulating oven at 37 ± 3°C for about 1 hr	Sodium treated surface must not be abraded before use. Hazardous etching solutions require skillful handling. Proprietary etching solutions are commercially available (see procedure 2). PTFE film available in etched condition from various suppliers

		2. Wipe with solvent and treat as recommended in one of the following commercial etchings:
		
		    Bond Aid (W. S. Shamban and Co.)
		    Fluorobond (Joclin Mfg. Co.)
		    Fluoroetch (Action Associates)
		    Tetraetch (W. L. Gore Associates)	

		3. Prime with epoxy adhesives and fuse into the surface by heating for 10 min at 370°C	

		4. Expose to one of the following gases activated by corona discharge:
		
		    Air (dry) for 5 min
		    Air (wet) for 5 min
		    Nitrous oxide for 10 min
		    Nitrogen for 5 min
			Bond within 15 min of treatment

		5. Expose to electric discharge (50kV ac) for 4 min)	Bond within 15 min of treatment

Substrates	Cleaning solvents	Substrate treatments	Comments
Polyesters, polyethylene terphthalate (Mylar)	Detergent, acetone, methyl ethyl ketone	1. Abrasion. Grit or vapor blast or 100 grit emery cloth followed by solvent degreasing	For general purpose bonding

		2. Immerse for 10 min at 70–95°C in
		   Sodium hydroxide 2
		   Water 8
		Rinse in hot water and dry in hot air
			For maximum bond strength. Suitable for Mylar films

Chlorinated polyether	Acetone, methyl ethyl ketone	Etch for 5 min at 71 ± 3°C in
		   Potassium dichromate 75
		   Water 120
		   Sulfuric acid (96%) 1500
		Rinse in water and distilled water. Dry in air
			Suitable for film materials such as Pentane. ASTM D 2093

Polyethylene, polypropylene, polyformaldehyde	Acetone, methyl ethyl ketone	1. Solvent degreasing	Low bond strength applications, generally only with pressure sensitive or contact adhesives
		2. Expose surface to gas burner flame (or oxyacetylene oxidizing flame) until substrate is glossy	Surface must be bonded immediately after treatment

		3. Etch in the following:
		   Potassium dichromate 75
		   Water 120
		   Sulfuric acid (96%) 1500
		Polyethylene and polypropylene: 60 min at 25°C or 1 min at 71°C
		Polyformaldehyde: 10 sec at 25°C
			For maximum bond strength. ASTM D 2093

Plastic	Solvent	Surface preparation	Notes
Polymethylmeth-acrylate (acrylic)	Acetone, methyl ethyl ketone, detergent, methanol, trichloroethylene, isopropanol	Abrasion. Grit or vapor blast or 100 grit emery cloth followed by solvent degreasing	For maximum strength relieve stresses by annealing plastic for 5 hr at 100°C
Polyphenylene	Trichloroethylene	Abrasion. Grit or vapor blast or 100 grit emery cloth followed by solvent degreasing	
Polyphenylene oxide	Methanol	Solvent degrease	Plastic is soluble in xylene and may be primed with adhesive in xylene solvent
Polystyrene	Methanol, isopropanol, detergent	Abrasion. Grit or vapor blast or 100 grit emery cloth followed by solvent degreasing	Suitable for rigid plastic
Polysulfone	Methanol	Vapor degrease	
Polyurethane	Acetone, methyl ethyl ketone	Abrade with 100 grit emery cloth and solvent degrease	
Polyvinylchloride, polyvinylidene chloride, polyvinyl fluoride	Trichloroethylene, methyl ethyl ketone	1. Abrasion. Grit or vapor blast or 100 grit emery cloth followed by solvent degreasing 2. Solvent wipe with ketone	Suitable for rigid plastic. For maximum strength, prime with nitrile phenolic adhesive
Styrene acrylonitrile	Trichloroethylene	Solvent degrease	Suitable for plasticized material
Urea formaldehyde	Acetone, methyl ethyl ketone	Abrasion. Grit or vapor blast or 100 grit emery cloth followed by solvent degreasing	

D.3 Surface Preparation Methods for Elastomeric Substrates

Substrates	Cleaning solvents	Substrate treatments	Comments
Natural rubber	Methanol, isopropanol	1. Abrasion followed by brushing. Grit or vapor blast or 280 grit emery cloth, followed by solvent wipe	For general purpose bonding
		2. Treat the surface for 2–10 min with sulfuric acid (96%) at RT. Rinse thoroughly with cold water/hot water. Dry after rinsing in distilled water. Residual acid may be neutralized by soaking for 10 min in 10% ammonium hydroxide after hot water washing	Adequate pretreatment is indicated by the appearance of hair line surface cracks on flexing the rubber. Suitable for many synthetic rubbers when given 10–15 min etch at RT. Unsuitable for use on butyl, polysulfide, silicone, chlorinated polyethylene, and polyurethane rubbers
		3. Treat surface for 2–10 min with paste made from sulfuric acid and barium sulfate. Apply paste with stainless steel spatula and follow procedure 2 above	
		4. Treat surface for 2–10 min in Sodium hypochlorite 6 Hydrochloric acid (37%) 1 Water 200 Rinse with cold water and dry	Suitable for those rubbers amenable to treatments 2 and 3
Butadiene styrene	Toluene	1. Abrasion followed by brushing. Grit or vapor blast or 280 grit emery cloth followed by solvent wipe	Excess toluene results in swollen rubber. A 20 min drying time will restore the part to its original dimensions

Material	Solvent	Procedure	Notes
Butadiene nitrile	Methanol	2. Prime with butadiene styrene adhesive in an aliphatic solvent 3. Etch surface for 1–5 min at RT following method 2 for natural rubber	

Butyl and chlorobutyl rubber	Toluene	1. Abrasion followed by brushing. Grit or vapor blast or 280 grit emery cloth followed by solvent wipe
		2. Etch surface for 10–45 sec at RT following procedure 2 for natural rubber
		
		1. Solvent wipe
		2. Immerse in the following solution at 21–32°C for 90–150 secs:
		   Hydrochloric acid (37%) 0.3
		   Sodium hypochlorite (5.25%) 3.0
		   Distilled water 97.0
		Rinse in tap water followed by distilled water. Dry at 65°C maximum
		3. Prime with butyl rubber adhesive in an aliphatic solvent
			Solution life is 4 hr maximum

Chlorosulfonated polyethylene	Acetone, methyl ethyl ketone	Abrasion followed by brushing. Grit or vapor blast or 280 grit emery cloth followed by solvent wipe	General purpose bonding

Substrates	Cleaning solvents	Substrate treatments	Comments
Fluorosilicones	Methanol	Application of fluorosilicone primer to metal where intention is to bond unvulcanized rubber	Primer available from Dow Corning
Polyacrylic	Methanol	Abrasion followed by brushing. Grit or vapor blast or 100 grit emery cloth followed by solvent wipe	For general purpose bonding
Polybutadiene	Methanol	Solvent wipe	General purpose bonding
Polychloroprene (neoprene)	Toluene, methanol, isopropanol	1. Abrasion followed by brushing. Grit or vapor blast or 100 grit emery cloth followed by solvent wipe 2. Etch surface for 5–10 min at RT following procedure 2 for natural rubber	Adhesion improved by abrasion with 280 grit emery cloth followed by solvent wipe
Polysulfide	Methanol	Immerse overnight in strong chlorine water, wash and dry	
Polyurethane	Methanol	1. Abrasion followed by brushing. Grit or vapor blast or 280 grit emery cloth followed by solvent wipe 2. Incorporation of a chlorosilane into the adhesive elastomer system. 1% by weight is usually sufficient	Chlorosilane is available commercially. Addition to adhesive eliminates need for priming and improves adhesion to glass and metals. Silane may be used as a surface primer
Silicone	Acetone, methanol	1. Application of primer (e.g, Chemlok 607, Lord Chemical Co.) 2. Exposure to oxygen gas activated by corona discharge for 10 min	

D.4 Surface Preparation Methods for Miscellaneous Substrates

Substrates	Cleaning solvents	Substrate treatments	Comments
Brick and fired non glazed building materials	Methyl ethyl ketone	Abrade surface with a wire brush; remove all dust and contaminants	
Carbon graphite	Acetone	Abrasion. Abrade with 220 grit emery cloth and solvent degrease after dust removal	For general purpose bonding
Glass and quartz (non-optical)	Acetone, detergent	1. Abrasion Grit blast with carborundum wand water slurry, and solvent degrease. Dry for 30 min at 100C. Apply the adhesive before the glass cools to RT	For general purposes bonding. Drying process improves bond strength

		2. Immerse for 10–15 min at 20°C in
		   Sodium dichromate 7
		   Water 7
		   Sulfuric acid (96%) 400
		Rinse in water and distilled water, Dry thoroughly
			For maximum strength

Glass (optical)	Acetone, detergent	Clean in an ultrasonically agitated detergent bath. Rinse: dry below 38°C	
Ceramics and porcelain	Acetone	1. Abrasion Grit blast with carborundum and water slurry and solvent degrease 2. Solvent degrease or wash in warm aqueous detergent, rinse and dry	

Substrates	Cleaning solvents	Substrate treatments	Comments

		3. Immerse for 15 min at 20°C in
		
		    Sodium dichromate 7
		    Water 7
		    Sulfuric acid (96%) 400
		
		Rinse in water and distilled water. Oven dry at 66°C	

Concrete, granite, stone	Perchloroethylene, detergent	1. Abrasion. Abrade with a wire brush, degrease with detergent, and rinse with hot water before drying	For general purpose bonding
		2. Etch with 15% hydrochloric acid until effervescence ceases. Wash with water until surface is litmus neutral. Rinse with 1% ammonia and water. Dry thoroughly before bonding	Applied by stiff bristle brush. Acid should be prepared in a plastic pail. 10-12% hydrochloric or sulfuric acids are alternative etchants. 10% w/w sodium bicarbonate may be used instead of ammonia for acid neutralization
Wood, plywood		Abrasion. Dry wood is smoothed with a suitable emery paper. Sand plywood along the direction of the grain	For general purpose bonding
Painted surface	Detergent	1. Clean with detergent solution, abrade with a medium emery cloth, final wash with detergent	Bond generally as strong as the paint
		2. Remove paint by solvent or abrasion, and pretreat exposed base	For maximum adhesion

Appendix E

Suppliers to the Adhesive and Sealant Industries

E.1 Adhesive and Sealant Suppliers

Adhesive or sealant manufacturer	Address	Phone/web site	Recent Acquisitions	Product Types
3M	3M Center, Adhesives Div. Bldg. 2208E-05 St. Paul, MN 55144	612-737-601 www.mmm.com		Contact, cyanoacrylate, film, hot melt, pressure sensitive, rubber based, thermoplastic thermosetting, sealants, tapes
Aabbitt Adhesives, Inc.	2403 N. Oakley Ave. Chicago, IL 60647	773-227-2700		Cyanoacrylate, hot melt, pressure sensitive, protein based, rubber based, thermoplastic and vegetable adhesives and elastomeric sealants
Ablestik Electronic Materials and Adhesives	20021 Susana Rd. Rancho Dominguez, CA 90221	310-764-4600		Film, miscellaneous specialty adhesives (e.g., frozen single part adhesives)
Acheson Colloids Co.	1496 E. Francis Ontario, CA 91761	800-270-2228		Thermoplastic, thermosetting, and miscellaneous adhesives
Adchem Corp.	625 Main St. Westbury, NY 11550	516-333-3843		Film, adhesive tapes, miscellaneous adhesives
Adhesives Research, Inc.	400 Seaks Run Rd. Glen Rock, PA 17327	717-235-7979 www.adhesivesresearch.com		Pressure sensitive, rubber based-tapes
ADM Tronics Unlimited, Inc.	224-S Pegasus Ave. Northvale, NJ 07647	201-767-6040 www.admtroniocs.com		Pressure sensitive, and miscellaneous adhesives
Aremco Products, Inc.	PO Box 517 707-B Executive Blvd. Valley Cottage, NY 10989	914-762-0685 www.aremco.com		Hot melt, miscellaneous adhesives, elastomeric sealants, adhesive tape, labels and sheet

Company	Address	Phone/Web	Products
AtoFindley, Inc.	11320 Watertown Plank Rd. Wauwatosa, WI 53226	414-774-8071 www.atofindley.com	Hot melt, rubber-based, thermoplastic, vegetable adhesives
Avery Dennison Specialty Tape Div.	250 Chester St. N. Plainesville, OH 44077	800-262-2400	Film, pressure sensitive, adhesive tapes
Bacon Industries, Inc.	192 Pleasant St. Watertown, MA 02172	617-926-2550	Rubber-based, thermosetting, elastomeric sealants
Basic Adhesives, Inc.	25 Knickerbocker Ave. Brooklyn, NY 11237	201-438-8181	Contact, film, pressure sensitive, rubber-based, thermoplastic and vegetable adhesives, elastomeric sealants
B.F. Goodrich Adhesive Systems Div.	123 West Bartges St. Akron, OH 44311	216-374-2900 www.estane.bfgoodrich.com/	Hot melt, reactive hot melt, pressure sensitive, rubber-based, elastomeric sealants, miscellaneous adhesives
Bordon Chemical Co.	520-112 Ave. NE Bellevue, WA 98004	206-455-4400	Protein based, thermoplastic, thermosetting
Bostik, Inc., a unit of Total	211 Boston St. Middleton, MA 01949	978-750-7351 www.bostik.com	Contact, cyanoacrylate, film, hot melt, pressure sensitive, rubber-based, thermosetting, thermoplastic
Century International Adhesives and Coatings Corp.	802 Harmon Ave. Columbus, OH 43223	614-461-8415	Anaerobic, contact, film, hot melt, reactive hot melt, protein-based, rubber-based, vegetable, miscellaneous adhesives, elastomeric sealants
Eleckromek, Inc. R&A Specialty Chemical Co.			
Mydrin Adhesives			

Adhesive or sealant manufacturer	Address	Phone/web site	Recent Acquisitions	Product Types
ChemRex, Inc.	889 Valley Park Dr. Sharkopee, MN 55379	800-433-9517 www.chemrex.com		Thermoplastic, thermosetting, sealants
Ciba Specialty Chemicals	4917 Dawn Ave. East Lansing, MI 48823	517-351-5900		Thermosetting
Conap, Inc.	1405 Buffalo St. Olean, NY 14760	716-372-9650		Rubber-based, thermosetting, elastomeric sealants
Courtaulds Aerospace, Inc.	5454 San Fernando Rd. Glendale, CA 91209	818-240-2060		Thermosetting, rubber-based, elastomeric sealants
Craig Adhesives and Coatings Co.	80 Wheeler Point Rd. Newark, NJ 07105	201-344-1483		Film, pressure sensitive, protein based radiation curable, rubber-based, thermoplastic, thermosetting, vegetable
DAP Products, Inc.	PO Box 967 Conyers, GA 30012	770-483-9717		Rubber based, vegetable, thermosetting, thermoplastic, and miscellaneous adhesives
Devcon	30 Endicott St. Danvers, MA 01923	508-777-1100 www.devcon.com		Anaerobic, contact, cyanoacrylate, thermosetting, elastomeric sealants
Dexter Corp.	One Dexter Drive Seabrook, NH 03874	800-767-8786 www.Hysol.com		Cyanoacrylate, hot melt, thermoplastic, thermosetting
Dow Corning Corp.	P. O. Box 0994 Midland, MI 48686	517-496-6000		Pressure sensitive, rubber-based, thermosetting, elastomeric sealants

Company	Address	Phone/Web	Products
DYMAX Corp.	51 Greenwoods Rd. Torrington, CT 06790	860-482-1010 www.dymax.com	Radiation curable, thermosetting, miscellaneous adhesives, elastomeric sealants
Elf Atochem North America, Inc.	2000 Market St. Philadelphia, PA	215-419-7000 www.elf-atochem.com	Film, hot melt, pressure sensitive, rubber-based, thermoplastic, thermosetting
EMS-CHEMIE (North America), Inc.	2062 Corporate Way PO 1717 Sumter, SC 29151	803-481-6190	Hot melt, fusible fibers
Epic Resins	600 Industrial Blvd. Palmyra, WI 53156	800-242-6649	Thermosetting, miscellaneous, elastomeric sealants
Epoxies Etc.	PO Box 508 Greenville, RI 02828	401-232-7847 www.epoxies.com	Anaerobic, cyanoacrylate, radiation curable, thermoplastic and thermosetting
Epoxy Technology, Inc.	14 Fortune Dr. Billerica, MA 01821	978-667-3805	Radiation curable, and thermosetting
Essex Specialty Products, Inc.	1250 Harmon Rd. Auburn Hills, MI 48326	248-391-6300	Hot melt, rubber-based, thermoplastic, thermosetting, elastomeric sealants
Evans Adhesive Corp.	925 Old Henderson Rd. Columbus, OH 43220	800-888-0925	Hot melt, pressure sensitive, thermoplastic, vegetable
Fielco Industries, Inc.	1957 Pioneer Rd. Huntington Valley, PA 19006	215-674-8700 www.fielco.com	Protein based, thermoplastic, thermosetting
Franklin International	2020 Bruck St. Columbus, OH 43207	614-445-1555 www.franklin.com	Contact, pressure sensitive, protein based, rubber-based, thermoplastic, thermosetting, elastomeric sealants
Healthtec Adhesives			

Adhesive or sealant manufacturer	Address	Phone/web site	Recent Acquisitions	Product Types
GE Silicones	260 Hudson River Rd. Waterford, NY 12188	800-255-8886		Pressure sensitive, rubber-based, elastomeric sealants
Geocel Corp.	PO Box 398 53280 Marina Dr. Elkart, IN 46515	219-264-0645		Rubber-based, sealants
Gibson-Homans Co.	1755 Pkwy. Twinsburg, OH 44087	330-425-3255		Pressure sensitive, rubber-based, thermoplastic, thermosetting
H. B. Fuller Co.	1200 Willow Lake Blvd. St. Paul, MN 55164	612-236-5900 www.hbfuller.com	Datac Adhesives Limited Industrial Adhesives Limited	Contact film, reactive hot melt, pressure sensitive, protein-based, radiation curable, rubber-based, thermoplastic, thermosetting, sealants
Hardman Div. of Harcros Chemicals, Inc.	600 Cortlandt St. Belleville, NJ 07109	201-751-3000		Radiation curable, rubber-based, thermosetting, elastomeric sealants
Henkel Adhesives Corp.	1345 Gasket Drive Elgin, IL 60120	847-289-2427 www.henkelcorp.com	United Resin Products Loctite Corp. Manco	Anaerobic, contact, cyanoacrylate, hot melt, reactive hot melt, pressure sensitive, rubber-based, sealants, thermoplastic, thermosetting
Imperial Adhesives, Inc.	6315 Wiehe Cincinnati, OH 45237	513-351-1300		Contact, hot melt, reactive hot melt, pressure sensitive, thermoplastic

Company	Address	Phone / Web	Brands	Products
ITW Plexus	30 Endicott St. Danvers, MA 09123	800-851-6692 www.itwplexus.com		Thermoplastic
Le Page's, Inc.	147 Essex Ave. Gloucester, MA 01930	978-283-1000		Cyanoacrylate, protein-based, vegetable, tapes
Loctite Corp (a subsidiary of Henkel)	Hartford Square 10 Columbus Blvd. Hartford, CT 06106	860-520-5000 www.loctite.com/		Anaerobic, contact, cyanoacrylate, reactive hot melt, radiation curable, thermoplastic, thermosetting, elastomeric sealants
Lord Corp.	111 Lord Dr. Cary, NC 27512	919-468-5980 www.lordcorp.com	Castall Mavidon	Cyanoacrylate, radiation curable, rubber-based, thermosetting
Macco Adhesives	925 Euclid Ave. Cleveland, OH 44115	216-344-7360 www.liquidnails.com		Contact, rubber-based, elastomeric sealants
Macklanburg-Duncan Co.	10950 S. Pipeline Rd. Euless, TX 76404	817-267-1771		Rubber-based, sealants
MACtac (Morgan Adhesives Co.)	4560 Darrow Rd. Stow, OH 44224	330-688-1111 www.mactac.com		Film, hot melt, pressure sensitive, rubber-based, tapes
Morton International, Inc.	100 N. Riverside Plaza Chicago, IL 60606	312-807-2000 www.morton.com	Boston Advanced Technologies Surfacetek and Salins Du Midi & Pulverlac	Contact, hot melt, pressure sensitive, radiation curable, rubber-based, thermoplastic, thermosetting
National Casein Co., Inc.	601 W. 80th St. Chicago, IL 60620	773-846-7300		Contact, hot melt, pressure sensitive, protein-based, rubber-based, thermoplastic, thermosetting, vegetable

Adhesive or sealant manufacturer	Address	Phone/web site	Recent Acquisitions	Product Types
National Starch and Chemical Co.	10 Finderne Are. Bridgewater, NJ 08807	908-685-5000 www.nationalstarch.com	Sycan Acheson Industries Emerson & Cuming	Anaerobic, contact, cyanoacrylate, contact, film, hot melt, pressure sensitive, protein-based, radiation curable, rubber-based, thermoplastic, thermosetting
Neste Resins of North America	1600 Valley River Dr. Eugene, OR 97401	541-687-8840 www.nesteresins.com/usa.html		Thermosetting
Nusil Technology	1040 Cindy Ln. Carpinteria, Ca 93013	805-684-8780		Contact, film, pressure sensitive, rubber-based, thermosetting, sealants
Pacer Technology	9420 Santa Anita Ave. Rancho Cucamonga, CA 91730	800-538-3091 www.pacertechnology.com		Cyanoacrylate, pressure sensitive, radiation curable, thermoplastic, thermosetting, miscellaneous adhesives, elastomeric sealants
Para-Chem	Hwy. 14 PO Box 127 Simpsonville, SC 29681	800-763-7272 www.parachem.com/		Contact, rubber-based, thermoplastic, elastomeric sealants
Plymouth Rubber Co., Inc.	104 Revere St. Canton, MA 02021	781-828-0220 www.plymouthrubber.com	Brite-Line Technologies, Inc.	Adhesive tapes, labels, sheet and miscellaneous adhesives
Reichhold Chemicals, Inc.	PO Box 13582 Research Triangle Park, NC 27709	800-431-1920 www.reichhold.com	Chimica Cislago (Italy) Costenaro SpA (Italy)	Hot melt, reactive hot melt, pressure sensitive, radiation curable, thermosetting

Company	Address	Phone/Web	Products
Ricon Resins, Inc.	569 421/4 Rd. Grand Junction, CO 81505	970-245-8148	Film, pressure sensitive, rubber-based, thermosetting, elastomeric sealants, adhesive tapes
RPM, Inc.	2628 Pearl Rd. PO Box 777 Medina, OH 44258	330-273-5090 www.rpminc.com	Miscellaneous adhesives and sealants
Sika Corp.	201 Polito Ave. PO Box 297 Lyndhurst, NJ 07071	201-933-8800 www.sikaUSA.com	Hot melt, rubber-based, thermoplastic, thermosetting, elastomeric sealants
Smooth-On, Inc.	2000 St. John St. Easton, PA 18042	800-762-0744 www.somooth-on.com	Rubber-based, thermosetting
Thermoset Plastics, Inc.	5101 E. 65th St. Indianapolis, IN 46220	317-259-4161 www.thermoset.com	Thermosetting, elastomeric sealants
Tra-Con, Inc.	45 Wiggins Ave. Bedford, MA 01730	617-275-6363	Rubber-based, thermosetting, thermoplastic
Uniroyal Adhesives and Sealants	2001 W. Washington St. South Bend, IN 46628	800-999-GLUE	Gunther Mirror Mastics Contact, pressure sensitive, rubber-based, thermoplastic, thermosetting, elastomeric sealants
Upaco Adhesive Div. of Worthen Industries	3 E. Spit Brook Rd. Nashua, NH 03060	603-888-5443	Hot melt, protein based, rubber-based, thermoplastic, vegetable
Wacker Silicones Corp.	3301 Sutton Rd. Adrian, MI 49221	517-264-8500	Rubber-based, elastomeric sealants

E.2 Production Equipment

Autoclaves
   Advanced Process Technology, Inc.
   Gannon Equipment Corp.
   McGill Airpressure Corp.
   Tecknit
   WSF Industrials

Blenders
   Bematek Systems, Inc.
   Hockmeyer Equipment Corp.
   Jaygo, Inc.
   Charles Ross & Son Co.
   Tecknit

Conveyors
   Dymax Corp.
   Hudson Machinery Corp.
   ProSys-Innovative Packaging Equipment
   Reynolds Industries, Inc.
   Vibra-Screw, Inc.

Dispersing Equipment
   Ashby-Cross Co., Inc.
   Bematek Systems, Inc.
   Conn & Company
   Graco, Inc.
   Ingersoll Dresser Pumps

Extruders
   I & J Fisnar, Inc.
   Fluid Kinetics, Div. of Valco Cincinnati, Inc.
   Gannon Equipment Corp.
   Jayco, Inc.
   Polymer Systems, Inc.

Filling Equipment
   Ashby-Cross Co., Inc.
   I & J Fisnar, Inc.
   Kenmar Div. of Tiger Enterprises
   Schwerdtel Corp.
   Sealant Equipment & Engineering, Inc.

Mills
   Bematek Systems, Inc.
   CB Mills, Inc.
   Hockmeyer Equipment Corp.
   Charles Ross & Son Co.
   Tecknit

Mixers
Ashby-Cross Co., Inc.
Bematek Systems, Inc.
Conn & Company
ConProTec, Inc.
Jayco, Inc.
Proportioning Equipment
Fluid Kinetics, Div. of Valco Cincinnati, Inc.
Graco, Inc.
Liquid Control Corp.
Meter Mix Div. Of Accumetric, Inc.
Sealant Equipment and Engineering, Inc.
Pumps
Fluid Kinetics, Div. of Valco Cincinnati, Inc.
GS Manufacturing
Ingersoll Dresser Pumps
ProSys-Innovative Packaging Equipment
Schwerdtel Corp.
Solvent/Vapor Recovery Units
CB Mills, Inc.
Fenwal Safety Systems
Hockmeyer Equipment Corp.
Rho-Chem Corp.
Solvent-Kleene, Inc.
Tanks
Ashby-Cross Co., Inc.
CB Mills, Inc.
I & J Fisnar, Inc.
Hockmeyer Equipment Corp.
Myers Engineering, Inc.
Weighing Equipment
C. M. Ambrose Co.
BLH Electronics
Mettler-Toledo, Inc.
Pneumatic Scale Corp.
Tecknit

E.3 Application Equipment

Applicators
Advanced Coating & Converting Systems, Inc.
John W. Blair Div. of Accumetric, Inc.
ConProTec, Inc.
Graco, Inc.
Valco Cincinnati, Inc.

Brushes
  Black Bros. Co.
  Centrix, Inc.
  Diagraph Corporation
  Tecknit
  Henry Westphal & Co., Inc.
Caulking Guns
  Adhesive Packaging Specialties, Inc.
  Albion Engineering Co.
  Ashby-Cross Co., Inc.
  Graco, Inc.
  Kenmar Div. of Tiger Enterprises
Clamps
  Aeroquip Corp.
  Bel-Art Products
  Black Bros. Co.
  De-Sta-Co Industries
  Tecknit
Cleaners
  Ashby-Cross Co., Inc.
  Bel-Art Products
  The D. C. Cooper Corp.
  Impco, Inc.
  Loctite Corporation
Coating Equipment
  Advanced Coating and Converting Systems, Inc.
  Black Bros. Co.
  Chemsultants International Network
  Independent Machine Co.
  Koenert Corp.
Heating Equipment
  Blue M Electric/General Signal Co.
  Chromalox, Wiegand Ind. Div., Emerson Electric Co.
  Fluid Kinetics, Div. of Valco Cincinnati, Inc.
  Paul Mueller Co.
  Universal Process Equip., Inc.
Hot Melt Applicators
  John W. Blair, Div. of Accumetric, Inc.
  Graco, Inc.
  May Coating Technologies
  Nordson Corporation
  Slauterback Corp.

Hot Melt Extruders
Global Process Equipment, Inc.
Graco, Inc.
ITW Dynatec
Pam Fastening Technology, Inc.
Slautterback Corp.
Impregnators
Binks Manufacturing Co.
Hull Corp.
Loctite Corp.
Tecknit
United McGill Corp.
Laminators
Advanced Coating & Converting Systems, Inc.
Chemsultants International Network
Hudson Machinery Corp.
May Coating Technologies
Union Tool Corp.
Metering Equipment
Adhesives Packaging Specialties, Inc.
Dymax Corp.
Meter Mix Div. of Accumetric, Inc.
Liquid Control Corp.
Sealant Equipment & Engineering Co.
Mixing Equipment
Ashby-Cross Co., Inc.
EMI Corp.
Liquid Control Corp.
Myers Engineering, Inc.
TAH Industries, Inc.
Presses
Hockmeyer Equipment Corp.
Hull Corp.
Schwerdtel Corp.
United McGill Corp.
Universal Process Equipment, Inc.
Spray Guns
Albion Engineering Co.
Binks Manufacturing Co.
DeVilbiss Industrial Spray Equipment
Glue-Fast Equipment Co., Inc.
Nordson Corporation

E.4 Curing and Drying Equipment

Dryers
  The Bethlehem Corp.
  Energy Sciences, Inc.
  Gala Industries, Inc.
  Mohr Corp.
  Charles Ross & Son Co.

Ovens
  Binks Manufacturing Co.,
  Blue M Electric/General Signal Co.
  The Lanly Co.
  Trent, Inc.
  Webber Manufacturing Co., Inc.

Platens
  Drake
  Electric Specialty Heater Co.
  Hudson Machinery Crop.
  I. H. Co. Industrial Heater Div.
  Tecknit

UV
  Dymax Corp.
  Electro-Lite Corp.
  Fusion UV Systems, Inc.
  Spectronics Corp.
  Uvexs, Inc.

Visible Light
  Dymax
  EFOS, Inc.
  Electro-Lite Corp.

E.5 Plastic Welding Equipment

Film Sealing and Bag Making
  Amplas, Inc.
  Battenfield Gloucester Engineering Co., Inc.
  Packaging Aids Corp.
  Zed Industries, Inc.

Ultrasonic Welding
  Branson Ultrasonics Corp.
  Dukane Ultrasonics
  Mastersonics, Inc.
  Sonobond Ultrasonics, Inc.
  Ultra Sonics Seal

Thermal Assembly
  Bryant Assembly Technologies
  Forward Technology Industries, Inc.
  Laramy Products Co., Inc.
  Packaging Aids Corp.
  Zed Industries, Inc.
Dielectric/RF Welding
  Callanan Co.
  High Frequency Technology Co., Inc.
  Kabar Manufacturing Corp.
  Thermex Thermatron, Inc.
  USF Machine Systems, Inc.
Gas Welding
  Drader Injectiweld, Inc.
  Laramy Products Co., Inc.
Spin Welding
  Dukane Ultrasonics
  Kamweld Products Co., Inc.
  Mastersonics, Inc.
  Meccasonic Div., FTI, Inc.
  Sonics & Materials, Inc.
Electromagnetic Welding
  Emabond Systems, Ashland Chemical Co.
  Hellerbond Div., Alfred F. Leatherman Co.
Vibration Welding
  Branson Ultrasonics Corp.
  Forward Technology Industries, Inc.
  Meccasonic Div., FTI, Inc.
  Sonics & Materials, Inc.

E.6 Testing and Laboratory Equipment

Adhesion Testers
  Custom Scientific Instruments, Inc.
  Dillon Force Measurement Products
  Imass, Inc.
  Instron, Inc.
  Satec Systems, Inc.
Environmental Test Chambers
  Applied Test Systems, Inc.
  Engelhard Corp.
  Hemco Corp.
  Tenney Environmental
  Webber Manufacturing Co., Inc.

Flash Point Equipment
  Fisher Scientific Co.
  Paul N. Gardner Company, Inc.
  Humboldt Mfg. Co.
  Rascher & Betzold, Inc.
  Tarlin Scientific, Inc.
Flex Testers
  Applied Test Systems, Inc.
  Custom Scientific Instruments, Inc.
  Instron Corp.
  Satec Systems, Inc.
  Tinius Olsen Testing Machine Co.
Flow Meters
  American Sigma, Inc.
  Cole Parmer Instrument Co.
  EMCO Flowmeters a Div. of Engineering Measurements Co.
  Ingersoll Dresser Pumps
  Liquid Controls Corp.
Gelation Timers
  Automation Products, Inc.
  Paul N. Gardner Company, Inc.
  Micromet Instruments, Inc.
  Techne, Inc.
  Tecknit
Hardness Testers
  Paul N. Gardner Company, Inc.
  Instron Corp.
  Rex Gauge Co., Inc.
  SDS Co.
  Satec Systems, Inc.
Melt Testers
  Fisher Scientific Co.
  Humboldt Mfg. Co.
  Rheometric Scientific
  Satec Systems, Inc.
  Tarlin Scientific, Inc.
Ovens
  Blue M Electric/General Signal Co.
  Fisher Scientific Co.
  Paul N. Gardner Company, Inc.
  Hull Corp.
  Research, Inc.

Peel Strength Testers
Applied Test Systems
Chemsultants International Network
Custom Scientific Instruments, Inc.
Imass, Inc.
Testing Machines, Inc.
Presses (Curing)
Carver, Inc.
Dake
Mohr Corp.
Reliable Rubber & Plastic Machinery Co.
Tecknit
Salt Spray Cabinets
Atlas Electric Devices
Paul N. Gardner Company, Inc.
Q-Panel Lab Products
STR
Tecknit
Shear Testers
Applied Test Systems, Inc.
Chatillon
Instron Corp.
Rheometric Scientific
Testing Machines, Inc.
Strength Testers
Dillon Force Measurement Products
Instron Corp.
Rheometirc Scientific
Testing Machines, Inc.
Tinius Olsen Testing Machine Co.
Tack Testers
Chatillon
Chemsultants International Network
Satec Systems, Inc.
Tecknit
Testing Machines, Inc.
Tensile Testers
Chatillon
Dillon, Force Measurement Products
Instron Corp.
Testing Machines, Inc.
Tinius Olsen Testing Machine Co.

Thickness Gauges
  Cole-Palmer Instrument Co.
  Custom Scientific Instruments, Inc.
  Paul N. Gardner Company, Inc.
  Testing Machines, Inc.
  Werner Mathis, USA, Inc.
Vibration Testers
  GE Company
  Pyrometer Instrument Co., Inc.
  Robertshaw Controls Co., Tennessee Div.
  Tenney Environmental
  Testing Machines, Inc.
Viscometers
  C. W. Brabender Instruments, Inc.
  Brookfield Engineering Labs
  Cambridge Applied Systems, Inc.
  Paul N. Gardner Company, Inc.
  Rheometric Scientific
Weathering Equipment
  Atlas Electric Devices Co.
  Q-Panel Lab Products
  STR
  Tecknit
  Tenney Environmental
Weighing Equipment
  BLH Electronics
  Cole-Parmer Instrument Co.
  Dillon Force Measurement Products
  Fisher Scientific Co.
  Mettler-Toledo, Inc., Balances and Instruments

E.7 Consultants and Testing Laboratories

Chemical Analysis
  Adhesive Consultant Labs
  Battelle
  Herron Testing Laboratories, Inc.
  Arthur D. Little, Inc.
  David Sarnoff Research Center, Sub. of SRI International
Equipment Design
  Chemsultants International Network
  Hockmeyer Equipment Corp.
  Ingersoll Dresser Pumps
  Kenmar Div. of Tiger Enterprises
  Werner Mathis, USA

Formulae Development
Adhesive Consultants Labs
Chemsultants International Network
Epoxy Consulting, Inc.
Polymeric Systems, Inc.
Satas & Associates
Marketing Research and Analysis
Adhesive Information Services, Inc.
Chemsultants International Network
Arthur D. Little, Inc.
Skeist, Inc.
Strategic Analysis, Inc.
Patent Investigation
Business Communications Co., Inc.
Captan Associates, Inc.
EB Associates Laboratories
Satas & Associates
SelTech Consulting
Physical Testing
Adhesives Consultants Labs
Applied Research Laboratories
Chemsultants International Network
Detroit Testing Laboratory, Inc.
Q-Lab Weathering Research Service
Product Development
Adhesives Technologies, Inc.
Battelle
The ChemQuest Group, Inc.
Epoxy Technology, Inc.
Polysciences, Inc.
Research
Battelle
Chemsultants International Network
Innovative Formulations Corp.
David Sarnoff Research Center, Sub. of SRI International
Skeist, Inc.
Weather Testing
Applied Research Laboratories
Atlas Weathering Services Group
Delsen Testing Laboratories
Detroit Testing Laboratory, Inc.
Q-Lab Weathering Research Service

Conversion Factors

To convert....	To....	Multiply by....
Area		
in^2	mm^2	$6.451600 \times 10^{+2}$
in^2	cm^2	6.451600
ft^2	m^2	9.290304×10^{-2}
Force		
dyne	N	1.000000×10^{-5}
lbf	N	4.44822
Force per unit area		
lb/in^2 (psi)	MPa	6.894757×10^{-3}
MN/m^2	MPa	1.000000
kg/mm^2	MPa	9.806650
$dyne/cm^2$	Pa	1.000000
Force per unit length		
lb/in (piw)	N/m	$1.751268 \times 10^{+2}$
dyne/cm	mJ/m^2	1.000000
lb/ft	kg/m	1.488
Impact energy		
ft-lbf	J	1.355818
N-m	J	1.000000
Length		
A (Angstroms)	m	1.000000×10^{-10}
in	mm	25.4
mil	in	1.000000×10^{-3}
Mass		
lb	kg	4.5359237×10^{-1}
oz	gm	28.35
Pressure		
atm	Pa	$1.013250 \times 10^{+5}$
atm	torr	760
psi	Pa	$6.894757 \times 10^{+3}$
torr (mm Hg @ 0°C)	Pa	$1.333220 \times 10^{+2}$
Temperature		
°F	°C	$5/9 \times (°F - 32)$
°C	°F	$((°C \times 9)/5) + 32$
Viscosity		
poise	Pa-s (pascal second)	1.000000×10^{-1}
poise	g/cm-sec	1.000000
m^2/sec	stokes	$1.000000 \times 10^{+4}$

Index

Abrasion, mechanical, 223–226
Accelerated Aging, 713–717
Accelerator, 321
Acetal, bonding of, 629–631
Acrylic, modified, adhesive, 391–392
Acrylic, thermoplastic, adhesive, 410
Acrylic, thermoplastic, bonding of, 631–634
Acrylonitrile butadiene, 399–400
Acrylonitrile butadiene styrene, bonding of, 628–629
Adherend, properties
  important to adhesive selection, 425–437
  important to joint design, 102–104
Adhesion
  forces, 41, 49–51
  practical, 56
  science of, 8–10
  work of, 54
Adhesion promoters, 266–277
  chrome complex, 267–273
  silane, 273–277
  titanate, 273–277
  zirconate, 277
Adhesive, 280–412
  chemical composition, 281–286, 319–326
  classifications, 279–280, 345–355
  commercial formulation, 341
  cost, 316
  form, 307–315
  formulation, 326–338
  function, 280–281
  methods of reaction, 286–307
Adhesive, types, 280–412
  contact, 301
  elastomeric, 285, 352–353
  hot melt, 306–307, 315
  hybrid, 285, 330–333, 354
  moisture curing, 295–296
  multiple part systems, 291–292
  paste and liquids, 310
  pressure sensitive, 301–305
  reactivatable, 305
  resinous solvent, 306
  single part systems, 292–300
  solid, 299–300
  solvent based, 312–313
  solventless, 310–312
  thermoplastic, 284–285, 350–351
  thermoset, 281–284, 348–349
  UV/light curable, 296–297
  water based, 313–314
Adhesive bonding
  advantages, design, 26
  advantages, mechanical, 23
  advantages, other, 28
  advantages, production, 27
  competitive joining methods, 18–25, 419–420
  disadvantages, 18, 23
  disadvantages, design, 29
  disadvantages, mechanical, 28
  disadvantages, other, 30
  disadvantages, production, 30
  planning, 420–424
  processes, 417–419
  stages, 66–79
Adhesive, environmental resistant
  chemicals and solvents, 716–722
  combined effect, 713–717
  high temperatures, 693–702
  humidity and water, 707–717
  low temperatures, 703–707
  outdoor weathering, 717–719
  radiation, 723–724
  vacuum and outgassing, 722–723
Adhesive joints, 93–125
  adherend thickness, 102–104
  angle, 112–113
  bar, 112

Index

Adhesive joints (*Cont.*):
  beveled double strap, 109
  beveled lap, 107
  corner, 113
  cylindrical, 110–112
  depth of lap, 101–104
  double butt lap, 106
  double lap, 107
  double strap, 109
  end lap, 113
  flat adherends, for, 105–109
  joggle lap, 107
  mitered, 113
  mortise, 113
  plastic and elastomeric, 114
  recessed double strap, 109
  reinforced plastic, 115
  scarf butt, 106
  scarf, 115
  stiffening joints, 109
  strap, 109
  tongue and groove, 106
  tube, 112
  wood, 115
Adhesive thickness, 100
Adhesive, family
  acrylic, modified, 391–392, 706
  acrylic, thermoplastic, 410
  acrylonitrile butadiene, 399–400
  anaerobic resins, 386–390
  bismaleimide, 380–382, 700–701
  butyl rubber, 397–398, 706
  cellulosic resin, 408
  chloroprene (neoprene), 398–399, 706
  cyanoacylate, 390–391, 701
  epoxy, 355–366, 693–698, 704–705
  epoxy hybrid, 366–373, 706
  epoxy nylon, 372, 705
  epoxy phenolic, 371–372, 705
  epoxy polysulfide, 372–375, 705
  epoxy vinyl, 373
  ethylene vinyl acetate, 407
  glues of agricultural origin, 411
  glues of animal origin, 411–412
  litharge cement, 413
  melamine formaldehyde, 374
  natural rubber, 396–397
  neoprene, 706
  neoprene phenolic, 377, 698
  nitrile phenolic, 375–377, 698
  phenol resorcinol formaldehyde, 373–374
  phenolic, 374–375, 698, 705
  phenoxy, 409–410
  phosphate cement, 413
  polyamide, 408
  polyaromatic resins, 377–381, 699, 706
  polybenzimidazole, 382, 700
  polyester, thermoplastic, 408–409
  polyesters, 382–383
  polyimide, 378–381, 699–700, 706
  polyisobutylene, 400
  polyolefins, 409
  polysulfide, 400–401, 706
  polysulfone, 409, 701
  polyurethane, 383–386, 705
  polyvinyl acetal, 404
  polyvinyl acetate, 404–406
  polyvinyl alcohol, 407
  polyvinyl methyl ether, 400
  reclaimed rubber, 397
  resorcinol formaldehyde, 373–374
  silicone, 401–403, 698, 705–706
  sodium silicate, 412–413
  styrene butadiene rubber, 398
  sulfur cement, 413
  thermoplastic elastomers, 407
  toughened epoxy, 367–371
  urea formaldehyde, 374
  vinyl phenolic, 377
Aging
  accelerated, 713–717
  environmental, 38, 75–79, 445–449, 685–724
Air blade coater, 743–745
Air spray, 742
Airless spray, 742
Alkyd, bonding, 621–622
Aluminum, bonding, 595–602
  surface preparation, 257, 259, 261
Anaerobic resins, 386–390
Angle joints, 112
Antifungal agent, 325
Antihydrolysis agent, 325
Antioxidant, 325
Apparent contact angle, 68
Application
  equipment, 739–741
  film, 747–750
  hot melt, 750–751
  liquid, 741–745
  pastes and mastics, 746
  powders, 746–747
Applications, 17, 24
Auger electron spectroscopy, 194
Automation, 736–741
Backup materials, 508
Base resin, 320

Basic materials, 42–46
Bead coater, 743–745
Beryllium, bonding of, 602–606
Binder resin, 320
Bismaleimide, 381–382
Bitumen, sealant, 483
Bond failure
  energy, 55–57
  failure mode, 40
  mechanism, 39–41
Bond inspection, 183–190
Bonding equipment, 751–754
  heating, 752–754
  pressure, 752
Boundary failure region, 58
Boundary layer, 5, 64–66, 199
  theory, 88–89
Brazing, 25
Brookfield viscometer, 139
Butt joints
  adhesive, 105
  sealant, 122
Butyl rubber, adhesive, 397–398, 706
C-Scan, 188–189
Cadmium plate, bonding of, 606–607
Carrier, 323
Catalyst, 321
Caulk, dry rope, 485
Cellulosics, adhesive, 408
Cellulosics, bonding of, 634
Ceramics, bonding of, 677–679
  substrate, 205
Chemical cleaning, 221–222
Chemical resistance, 36, 719–722
Chemisorption, 50
Chlorinated polyether, bonding of, 634–635
Chlorinated polyolefin primer, 258–364
Chlorosulfonated polyethylene, sealant, 488
Chrome complex primers, 277
Cleavage, 156
Coating equipment, 743–745
Cohesion
  forces, 49–51
  work of, 54
Cohesive failure region, 58
Coinda scope, 189
Combined adhesive and mechanical fastening, 31
Combined environments, 713–717
Composites, thermoplastic, bonding of, 662
Composite, substrate, 205, 247–249

Composites, thermosetting, bonding of, 657–662
Computer selection programs, 759–764
  ADHESIVES, 763–764
  Adhesives Selector, 761–762
  AdhesivesMart, 761
  Sealant System, 762–763
Concrete, sealing of, 516
  substrate, 205
Contact angle, 53
Copper, bonding of, 607–608
Corner joints, 112
Corona treatment, 237–239
Corrosion resistance, 257
Cost
  adhesive, 316
  assembly, 27
  hidden, 30–31
Crack propagation, 99
Creep, 77, 160, 167–168
Critical surface tension, 54
  common solids, 56
  measurement, 54–55
Crosslinking, 83
Cryogenic applications, 702–707
CTBN adhesive formulation, 331–333
Cure rate, 141
Curing agent, 320
Cyanoacrylate, adhesive, 68, 390–391
  primers, 261
Cyanosilicone, sealant
Cylindrical joints, 110–112
Definitions, 2–6
  adherend, 5
  adhesion, 4
  adhesive, 3
  boundary layer, 5
  cohesion, 4
  interface, 5
  interphase region, 5
  joint, 5–6
  non-structural adhesive, 3
  primer, 5
  sealant, 3, 4
  structural adhesive, 3
Errors in design, 179
Desorption, 77–79, 710–713
Dialyl phthalate, bonding of, 622
Dielectric heating, 564–565
Diffusion constant, moisture, 707–710
Diluent, 322
Dip roll coaters, 743–745
DuNouy ring, 52
Dupre equation, 54

Elastomer, bonding of, 666–669
  Surface, 205, 249–250
Elastomeric adhesives, 285
Electrical conductivity, 336–338
Electrolytic cleaning, 223
Engraved roll coaters, 743–745
Environment
  factors, 754–757
  resistance to, 685–689
  service conditions, 446–449
  service stress, 447
  tests, 161–163, 163–169
Epoxy, adhesive, 355–366
  sealant, 484
Epoxy, bonding of, 622–623
Epoxy hybrid, adhesive, 366–373
Epoxy nylon, adhesive, 331, 372
Epoxy phenolic, adhesive, 371–372
Epoxy polysulfide, adhesive, 372–373
Epoxy vinyl, adhesive, 373
Equilibrium spreading pressure, 53–54
ESCA, 195
Ethylene vinyl acetate, adhesive, 407
  LaPlace-Kelvin equation, 67
  laser treatment, 242
Extenders, 322
Failure
  adhesive, 4, 39–40
  cohesive, 4, 40
  sealant, 79–81
Fatigue, 26
  test, 157–158
Fillers, 322
Film former, 325
Flame treatment, 239–240
Flow control, 327
Flow or curtain coater, 743–745
Fluoroalkyarylenesiloxanylene, sealant, 702, 706
Fluorocarbon, bonding of, 635–637
Fluoropolymer, sealant, 489–490
Foam, bonding of, 662–666
Fokker bond tester, 187
Forces
  adhesion, 49–51
  cohesion, 49–51
Form, physical, 307–315
  film, 314
  hot melt, 315
  liquid, 310
  paste, 310
  powder, 314
  solvent based, 312–313
  tape, 314

  water based, 313–314
Friction welding, 565–566
Gasketing, 509–510
  formed in place, 510–512
Glass, bonding of, 677–679
  sealing of, 517
Glass transition temperature, 83–86
  natural glues
    agricultural origin, 411
    animal origin, 411–412
Grout, 500
Hardener, 320
Hardness, 142
Heat sealing, 552
Heated tool welding, 543–553
Highway joints, 483
Historical development, 10–11
Honeycomb bonding, 679–680
Hot gas welding, 553–557
Hot spray, 742
Humidity resistance, 707–717
Hybrid adhesive, 285
Hydrocarbon rubber based, sealant, 485–487
Hydrogen bonding, 50, 84
Hydrolytic stability, 707–710
Impact, 158–160
Induction welding, 561–564
Information technology, 759–764
Infrared welding, 560
Inhibitors, 321
Interface, 5, 83–87
Internet web sites, 759–760
Interphase, 5, 89–90
Ion scattering spectroscopy, 194
Joint design
  adhesive selection, 438–439
  bond thickness, 439–441
  effect of bond area, 101–102
  efficiency, 98–104
  general rules, 104–105
  planning, 422–423
Kiss coater, 743–745
Knife spreader, 743–745
Lap joints
  adhesive, 105
  sealant, 122
  test, 149–150
Laplace-Kelvin equation, 67
Laser welding, 558–560
Lead, bonding of, 608–609
Litharge cement, 413
Magnesium, bonding of, 609–611
Manufacturing processes, 44–47

Index   875

Markets, 12
Masonry, sealing of, 516
Material properties
  importance of adhesion, 62–83
  importance to joint design, 99–102
  importance of cohesion, 83–87
Mechanical anchoring, 63
Mechanical fastening, 19, 22, 25, 31, 580–589
Melamine formaldehyde, 374
Metals, bonding of, 594–617
  sealing of, 517–518
  surface, 200–203, 246
Meter-mix-dispense equipment, 736–738
Metering, 733–735
Mixing, 735–738
Modulus, 103
Molecular weight, 83
Movement capability, 164–165
MS sealants, 491–493
Natural rubber, adhesive, 396–397
Neoprene phenolic, adhesive, 377
Nickel, bonding of, 611
Nitrile, sealant, 499–500
Nitrile phenolic, adhesive, 331, 375–377
Non-destructive testing, 184–191
  C-scan, 188–189
  Fokker bond tester, 187
  proof test, 187
  radiography, 187
  tap test, 186
  thermal transmission methods, 190
  ultrasonic inspection, 187–189
  visual inspection, 184–186
Nylon, bonding of, 637–638
Oil and resin based, sealants, 482
Organosilane, 267–273
Outdoor weathering, 717–719
  non-sea coast, 717–718
  sea coast, 718–719
Outgassing resistance, 722–723
Oxidation, 690–692
Oxidative cure, 482
Oxides, 202
Painted adherends, sealing of, 517
  Surface, 205
Part requirements, 441–442
Peel, 98
  test, 151–156, 165
Permeability coefficient, moisture, 707–710
Phenol resorcinol formaldehyde, 373–374
Phenolic, adhesive, 374–375
Phenoxy, adhesive, 409–410

Phosphate cement, 413
Plasma treatment, 240–242
Plastic
  joints, 114
  machine screws and bolts, 581
  mechanical fasteners, 581
  press fit design, 586–588
  rivets, 585
  self fastening, 586–589
  self threading screws, 583–585
  snap fit design, 588–589
  spring steel fasteners, 585–586
Plasticizer, 323
Plastic assembly, 537–538
Plastics, general assembly
  direct heat welding, 523–560
  indirect heating, 562–577
  joining methods
  mechanical fastening, 581–589
  solvent welding, 577–580
Plastics, joining methods
  dielectric heating, 564–565
  friction welding, 565–566
  heated tool welding, 543–553
  hot gas welding, 553–557
  induction welding, 561–564
  infrared welding, 560
  laser welding, 558–560
  mechanical fastening, 580–589
  resistance wire welding, 557–558
  solvent cementing, 577–580
  sources of information, 589–590
  spin welding, 566–569
  ultrasonic welding, 569–574
  vibration welding, 574–577
Plated adherends, bonding of, 611–612
  Surface, 205
Polyamide, adhesive, 408
Polyamide-imide, bonding of, 649–650
Polyaromatic resins, adhesives, 377–381
Polyaryletherketone, bonding of, 650
Polybenzimidazole, adhesive, 382
Polybutene, sealant, 485
Polycarbonate, bonding of, 640–643
Polychloroprene (neoprene), adhesive, 398–399
  sealant, 488
Polyether, sealant, 491–492
Polyethersulfone, bonding of, 653–654
Polyester, bonding of, 624–625
Polyester, thermoplastic, adhesive, 408–409
Polyester, thermoplastic, bonding of, 647–649

Polyesters, adhesive, 382–383
Polyetheretherketone, bonding of, 650
Polyetherimide, bonding of, 649
Polyetherketone, bonding of, 650
Polyimide, adhesive, 378–381
Polyimide, bonding of, 623–624, 649
Polyisobutylene, adhesive, 400
    sealant, 485
Polymeric surfaces, 203–205, 234–346
Polymeric materials, 81–82
Polyolefin, bonding of, 643–645
Polyolefin foam, sealant, 500
Polyolefins, adhesive, 409
Polyphenylene oxide, bonding of, 646–647
Polyphenylene sulfide, bonding of, 654
Polystyrene, bonding of, 650–652
Polysulfide, adhesive, 400–401
    sealant, 490–491, 706
Polysulfone, adhesive, 409
Polysulfone, bonding of, 652–653
Polyurethane, thermoset, bonding of, 626–627
Polyurethane, adhesive, 383–386
    sealant, 493–497, 706
Polyvinyl acetal, adhesive, 404
Polyvinyl acetate, adhesive, 404–406
    sealant, 483–484
Polyvinyl alcohol, adhesive, 407
Polyvinyl chloride, bonding of, 654–655
Polyvinyl chloride foam, sealant, 500
Polyvinyl chloride plastisol, sealant, 484
Polyvinyl methyl ether, adhesive, 400
Porosity sealing, 512–513
Precious metals, bonding of, 608
Pressure or squeeze coater, 743–745
Primary bonds, 62
Primers, 255–266
    advantages, 255–256
    aluminum, 257, 259, 261
    application, 256–259
    chlorinated polyolefin, 258, 264
    corrosion resistance, 257
    elastomers, 264–266
    for sealants, 517
    inserts, 264–266
    metals, 259–263
    plasma applied, 261
    polymers, 263–264
    surface exposure time, 257
    surface treatment, with, 242–246
Processing conditions, 421–422
Production requirements, 442–446
Proof tests, 187
Pyrolysis, 690–693

Quality control, 177–183
    bonding process, 182
    incoming materials, 180–181
    pre-manufacturing, 179
    surface treatment, 181–182
Radiation resistance, 723–724
Radiography, 197
Reclaimed rubber, adhesive, 397
Refrigerated storage, 731
Requirements, 36–39
Resistance wire welding, 557–558
Resorcinol formaldehyde, adhesive, 373–374
Retarder, 321
Reverse roll coater, 743–745
Reversion, 707–710
Riveting, 19–21, 27
RTV silicone, adhesive, 497–498
Safety factors, 754–757
Secondary bonds, 62
Sealant
    additives, 468–470
    applications, 33, 470–441
    application temperature, 525
    classifications, 453–462, 460–462
    characteristics, 477
    checklist of considerations, 34
    commercial formulations, 474–475
    composition, 460–470
    design considerations, 35
    dry rope caulk, 485
    families, 477
    fillers, 467–468
    formulation, 470–474
    functions, 31–36
    grouting material, 500
    hardening and non-hardening types, 453–457
    highway joints, 483
    oxidative cure, 482
    plasticizers, 468
    performance classification, 476
    production considerations, 36
    resin, 465
    single part system, 459–460
    solvent and water release, 460
    solvents, 467
    sources of information, 534–535
    tooling, 520
    transfer molding, 524
    two part systems, 457–459
    types, 522
Sealant, families
    bitumen, 483

Index 877

chloroprene (neoprene), 499
chlorosulfonated polyethylene, 488
cyanosilicone, 702, 706
epoxy, 484
fluoroalkyarylenesiloxanylene, 702, 706
fluoropolymer, 489–490
hydrocarbon rubber based, 485–487
nitrile, 499–500
oil and resin based, 482
polybutene, 485
polyether, 491–492
polyisobutylene, 486
polyolefin foam, 500
polysulfide, 490–491, 706
polyurethane, 493–497, 706
polyvinyl acetate, 483–484
polyvinyl chloride foam, 500
polyvinyl chloride plastisol, 484
silicone, 497–498, 698, 706
styrene butadiene copolymer, 498–499
thioether, 499
Sealant, form
 bulk materials, 518–522
 foam, 523–524
 preformed gaskets, 523–524
 tape, 522–523
Sealant, properties
 abrasion resistance, 530
 adhesion, 531–532
 adhesive properties, 34
 appearance, 534
 chemical effects, 36
 compressive strength, 531
 cure, 457–460
 cure rate, 525–526
 durability, 532–534
 hardness, 530
 high temperature resistance, 693–702
 low temperature resistance, 703–707
 mechanical properties, 530–531
 modulus, 530
 movement capability, 527–530
 performance classification, 476
 performance properties, 473–474
 properties, 478–481
 rheology, 524–525
 stress relaxation, 531
 tensile strength, 530
Sealant failure mechanisms, 79–81
Sealant joint design
 angle, 123
 backup materials, 508
 butt, 122
 compound butt, 124

corner, 124
efficiency, 115–122
 lap, 122
 lip, 125
 shear, 121
 stress distribution, 115–121
 threaded joints, 508–509
Secondary ion mass spectroscopy, 194
Selection, 424–425
 analysis form, 340
 factors, 6–8
 by computer, 760–761
Setting, 69–75
Setting methods, 44
Shear joints
 adhesive, 105
 sealant, 121
Shelf life, 139–140, 729–730
Silicone, adhesive, 401–403
 sealant, 497–498, 698, 706
Silicone, bonding of, 625–626
Soap, 325
Sodium silicate, adhesive, 412–413
Solidification, 36, 37, 69–75
Solids content, 143
Solvent, 321
Solvent cementing, 577–580
Solvent cleaning, 217–221
Solvent resistance, 719–722
Sources of information, 47, 589–590
Specifications, 173–177
Spin welding, 566–569
Spraying equipment, 741–743
Stabilizer, 325
Standards, 173–175
Static mixer, 733–735
Steel and iron, bonding of, 612–615
Stiffening joints, 109
Stone, sealing of, 516
Storage, 728–731
Storage life, 30
Strap joints, 107
Stress
 cleavage, 97–98
 combined with moisture and
  temperature, 713–717
 compressive, 94–96
 due to service environment, 75–79
 due to setting, 71, 74–75, 334–335
 due to thermal expansion, 71–76
 localized, 70
 peel, 97–98
 shear, 96–97
 tensile, 94–96

Index

types, 94–98
Stress distribution
  adhesive, 97–103, 108
  sealant, 115–121
Stressed aging, 713–717
Structural sandwich panels, 679–680
Stubmeter, 189
Styrene butadiene rubber, adhesive, 398
  sealant, 498–499
Substrates
  acetal, 629–631
  acrylic, 631–634
  acrylonitrile-butadiene-styrene, 628–629
  alkyd, 621–622
  aluminum, 595–602
  beryllium, 602–606
  cadmium plating, 606–607
  cellulosics, 634
  ceramic, 677–679
  chlorinated polyether, 634–635
  composite, 655–662
  concrete and masonry, 517
  copper, 607–608
  dialyl phthalate, 622
  elastomer, 666–669
  epoxy, 622–623
  fluorocarbon, 635–637
  foam, 662–666
  glass or porcelain438, 518, 677–679
  honeycomb, 679–680
  lead, 608–609
  magnesium, 609–611
  masonry, 516
  metal, 413–433, 518, 594–617
  nickel, 611
  nylon, 637–638
  painted surfaces, 518
  plastic, 433, 538–542, 617–620
  plated parts, 611–612
  polyamide-imide, 649–650
  polyaryletherketone, 650
  polycarbonate, 640–643
  polyethersulfone, 653–654
  polyester, thermoset, 624–625
  polyester, thermoplastic, 647–649
  polyetheretherketone, 650
  polyetherimide, 649
  polyetherketone, 650
  polyimide, 623–624, 649
  polyolefin, 643–645
  polyphenylene oxide, 646–647
  polyphenylene sulfide, 654
  polystyrene, 650–652
  polysulfone, 652–653
  polyurethane, thermoset, 626–627
  polyvinyl chloride, 654–655
  precious metal, 608
  silicone, 625–626
  steel and iron, 612–615
  stone, 517–518
  themoplastic, 627–628
  thermoplastic composite, 662
  thermoset, 620–621
  tin, 615
  titanium, 615–617
  unvulcanized elastomer, 435–436, 669
  vulcanized elastomer, 433–435, 669–673
  wood, 436–437, 673–677
  zinc, 617
Sulfur cement, 413
Surface attachment, 57–58
Surface cleanliness, 36–37
Surface energy, 51–54
Surface equilibrium, 215
Surface exposure time, 257
Surface interactions, 87–90
Surface preparation for sealants, 515–517
Surface roughness, 63, 68–69, 224
Surface tension
  common liquids, 56
  critical, 56
  methods of measurement, 51–55
Surface treatment
  active chemical treatment, 232–236
  active methods, 226–246
  adhesion abrasion, 236
  chemical cleaning, 221–222
  corona, 237–239
  elastomeric adherends, 249–250
  electrolytic cleaning, 223
  laser, 242
  flame, 239–240
  fluorine treatment, 235–236
  grafting, 236
  importance, 207–209
  mechanical treatment, 223–226
  metal surfaces, chemical treatment, 232–234
  metals, 246
  passive methods, 215–226
  plasma, 240–242
  plastics, 246–247
  polymeric composites, 247–249
  polymeric surfaces, chemical treatment, 234–236

Index 879

polymeric surfaces, physical treatment, 26–246
primers, 242–246
quality control, 164–165
reinforcing fibers, 270–272
sealants
selection, 209–213
solvent cleaning, 217–221
sources of information, 246
surface equilibrium, 215
ultrasonic cleaning, 222
UV radiation, 242
vapor honing, 222
Surfactant, 325
Tack, 140–141, 353–336
Tap test, 186
Tear strength, 166
Temperature
formulation factors, 329–330
high temperature resistance, 689–693
low temperature resistance, 702–703
Tensile test, 146
Testing
adherends, 143–145
basic principles, 128–135
material properties, 135–144
Testing, adhesives
180 deg peel, 155
button tensile, 147
cleavage, 156
climbing drum peel, 152
creep, 160
cross lap, 148
environmental, 161–163
fatigue, 157–158
impact, 158–160
lap shear, 149–150
peel, 151–156
standard exposure conditions, 162
standard test methods, 145–146
t-peel, 151–154
tensile, 146
tensile-shear, 149
Testing, adhesives and sealants
cure rate, 142
destructive, 183–184
government, 172–173
hardness, 142
hierarchy, 171
prototype, 170–173
non-destructive, 184–191
shelf-life, 139–140
solids content, 143
tack, 140

viscosity, 137–139
working life, 140
Testing, sealants
adhesion, 166
cohesion, 166
compressive set resistance, 167–168
creep, 167–168
environmental, 168–169
movement capability, 164–165
peel adhesion, 165
standard test methods, 163–164
tear strength, 166
Testing, surfaces
auger electron spectroscopy, 194
before and after surface preparation, 213–215
ESCA, 195
ion scattering spectroscopy, 194
secondary ion mass spectroscopy, 194
x-ray photoelectron spectroscopy, 195
Theories of adhesion
adsorption, 59–62
diffusion, 64
electrostatic, 64
mechanical, 62–64
weak boundary layer, 64–66
Thermal conductivity, 336–338
Thermal expansion coefficient
common materials, 72
lowering of, 333–334
adhesives, 72
common building materials
Thermal transmission testing, 190
Thermoplastic adhesives, 284–285
Thermoplastic bonding, 627–628
Thermoplastic composite bonding, 662
Thermoplastic elastomers, 407
Thermoplastic tradenames, 541–542
Thermoset plastic bonding, 620–621
Thermoset plastic tradenames, 540
Thermosetting adhesive, 281–284
Thickeners, 324
Thickness, adhesive or sealant, 100–101
Thioether, sealant, 499
Thixotrope, 324
Threaded joints, 508–509
Tin, bonding of, 615
Titanate primers, 273–277
Titanium, bonding of, 615–617
Toughened epoxy, adhesive, 367–371
Toughness, 330–333
Transfer of adhesive or sealant, 732, 738–739
Transition failure region, 58

Tropical climates, 713–717
Two roll reverse coater, 743–745
Ultrasonic cleaning, 222
Ultrasonic inspection, 187–189
Ultrasonic welding, 569–574
Unvulcanized elastomer bonding, 669
Urea formaldehyde, adhesive, 374
Urethane, adhesive, 332
UV radiation surface treatment, 242
Vacuum impregnation, 513
Vacuum resistance, 722–723
van der Waals forces, 50, 62
Vapor honing, 222
Vibration welding, 574–577
Vinyl phenolic, 377
Viscosity, 137–139, 327
Visual inspection, 184–186
Vulcanized elastomer bonding, 669–673
Water resistance, 707–717
Welding, 19–22
Wetting, 36–37, 59–62, 66–69
Wetting agent, 325
Wilhelmy plate, 52
Wood, bonding of, 673–677
Working life, 140
x-ray photoelectron spectroscopy, 195
Young equation, 53
Young-Dupre equation, 55
Zinc, bonding of, 617
Zisman plot, 54–55

ABOUT THE AUTHOR

Edward M. Petrie has worked for over 33 years with Westinghouse Corporation and Asea Brown Boveri (ABB) as a corporate expert on adhesive and sealant technology. He has consulted with many industrial operating divisions within these companies, and others, and gives courses in adhesive bonding at various colleges and institutions. He is a member of the Society of Plastics Engineers, IEEE, and ASM and has held various offices in these organizations. Currently, Mr. Petrie is Manager of Strategic Projects at the ABB Electric Systems Technology Institute in Raleigh, NC.